HANDBOOK OF HOLOCENE PALAEOECOLOGY AND PALAEOHYDROLOGY

INTERNATIONAL GEOLOGICAL CORRELATION PROGRAMME

**IUGS
UNESCO**

PROJECT 158B

HANDBOOK OF HOLOCENE PALAEOECOLOGY AND PALAEOHYDROLOGY

Edited by
B. E. BERGLUND
Department of Quaternary Geology, Lund University

with the assistance of
M. RALSKA-JASIEWICZOWA
Botanical Institute, Polish Academy of Sciences, Krakow

A Wiley–Interscience Publication

JOHN WILEY & SONS
Chichester – New York – Brisbane – Toronto – Singapore

Copyright © 1986 by John Wiley & Sons Ltd.

Library of Congress Cataloging in Publication Data:

Main entry under title:

Handbook of Holocene palaeoecology and palaeohydrology.

  'A Wiley–Interscience publication.'
  Includes index.
  1. Palaeontology—Recent—Handbooks, manuals, etc.
  2. Geology, Stratigraphic—Recent—Handbooks, manuals, etc.
  3. Palaeoecology—Handbooks, manuals, etc.
  4. Palaeolimnology—Handbooks, manuals, etc. 5. Glacial epoch—
Handbooks, manuals, etc. I. Berglund, Björn E.
II. Ralska-Jasiewiczowa, Magdalena.
QE741.H25 1985   560'.45   84-29094
ISBN 0 471 90691 3

British Library Cataloguing in Publication Data:

Handbook of Holocene palaeoecology and palaeohydrology.—(International
  Geological Correlation Programme; project 158B)
  1. Palaeontology—Quaternary   2. Palaeoecology
  I. Berglund, B. E.   II. Ralska-Jasiewiczowa, M.
  III. Series
560'.178   QE741

  ISBN 0 471 90691 3

Typeset by Activity Limited, Salisbury, Wiltshire.
Printed and bound in Great Britain.

To all friends of Holocene research

Contents

Contents   ix

PHYSICAL AND CHEMICAL METHODS

BIOLOGICAL METHODS

NUMERICAL TREATMENT OF BIOSTRATIGRAPHICAL DATA

List of Contributors

B. Aaby	Geological Survey of Denmark, Thoravej 31, DK-2400 Copenhagen NV, Denmark
S. T. Andersen	Geological Survey of Denmark, Thoravej 31, DK-2400 Copenhagen NV, Denmark
T. C. Atkinson	Climatic Research Unit and School of Environmental Sciences, University of East Anglia, Norwich NR4 7TJ, England
R. Arigo	Department of Geological Sciences, Brown University, Providence, R.I. 02912, U.S.A.
R. Battarbee	Palaeoecology Research Unit, Department of Geography, University College London, 26 Bedford Way, London WC1H 0AP, England
L. Bengtsson	Department of Limnology, Box 3060, S-220 03 Lund, Sweden
B. E. Berglund	Department of Quaternary Geology, Tornav. 13, S-223 63 Lund, Sweden
W. Bircher	Gerlisbrunnenstr. 48, CH-8121 Benglen, Switzerland
H. J. B. Birks	Botanical Institute, University of Bergen, P.O. Box 12, N-5014, Bergen, Norway
K. R. Briffa	Climatic Research Unit and School of Environmental Sciences, University of East Anglia, Norwich NR4 7TJ, England
·G. R. Coope	Department of Geological Sciences, University of Birmingham, Edgbaston, P.O. Box 363, Birmingham 15, England
G. Cronberg	Department of Limnology, Box 3060, S-220 03 Lund, Sweden
E. Daniel	Geological Survey of Sweden, Kiliansgatan 10, S-223 50 Lund, Sweden
J. A. Dearing	Department of Geography, Coventry (Lanchester) Polytechnic, Priory Street, Coventry CV1 5FB, England
J. Dickson	Department of Botany, University of Glasgow, Glasgow G12 8QQ, Scotland

G. Digerfeldt	Department of Quaternary Geology, Tornav. 13, S-223 63 Lund, Sweden
U. Eicher	Institute of Physics, University of Bern, CH-3012 Bern, Sidlerstrasse 5, Switzerland
T. Einarsson	Institute of Geology, University of Iceland, Hringbraut, IS-101 Reykjavik, Iceland
M. Enell	Department of Limnology, Box 3060, S-220 03 Lund, Sweden
I. Foster	Department of Geography, Coventry (Lanchester) Polytechnic, Priory Street, Coventry CV1 5FB, England
D. Frey	Department of Biology, Jordan Hall 138, Bloomington, Indiana 47405, U.S.A.
B. van Geel	Hugo-de-Vries Laboratory, Sarphatistraat 221, 1018 BX Amsterdam, the Netherlands
G. Grosse-Brauckmann	Department of Botany, Institute of Technology, Schnittspahnstrasse 3–5, D-6100 Darmstadt, F.R.G.
W. Hofmann	Max-Planck Institute for Limnology, Postfach 165, D-2320 Plön, F.R.G.
S. E. Howe	Department of Geological Sciences, Brown University, Providence, R.I. 02912, U.S.A.
M. J. Joachim	Department of Geology, University of Birmingham, Edgbaston, P.O. Box 363, Birmingham 15, England
L. Larsson	Institute of Archaeology, Krafts torg 1, S-223 50 Lund, Sweden
V. Ložek	Kořenského 1/1055, 150 00 Praha 5-Smíchov, Czechoslovakia
H. Löffler	Department of Limnology, Althanstrasse 14, A-1090 Wien, Austria
N. Malmer	Department of Plant Ecology, Östra Vallgatan 14, S-223 61 Lund, Sweden
P. Moore	Department of Biology, University of London, King's College, 68 Half Moon Lane, London SE24, England
A. V. Munaut	Laboratory of Palynology and Dendrochronology, 4 place Croix du Sud, 1348 Louvain-la-Neuve, Belgium
I. U. Olsson	Department of Physics, Box 530, S-751 21 Uppsala, Sweden
D. W. Perzy	Department of Geology, University of Birmingham, Edgbaston, P.O. Box 363, Birmingham 15, England
I. C. Prentice	Institute of Ecological Botany, Box 559, S-751 22 Uppsala, Sweden
M. Ralska-Jasiewiczowa	Institute of Botany, PAN, ul. Lubicz 46, 31-512 Krakow, Poland

G. Regnéll   Department of Plant Ecology, Östra Vallgatan 14, S-223 61 Lund, Sweden

M. Saarnisto   Department of Geology, University of Oulu, Oulu, Finland

W. Schoch   Swiss Federal Institute of Forest Research, CH-8903 Birmensdorf, Switzerland

U. Siegenthaler   Institute of Physics, University of Bern, CH-3012 Bern, Sidlerstrasse 5, Switzerland

L. Starkel   Institute of Geography, PAN, ul. Jana, 31-512 Krakow, Krakow, Poland

R. Thompson   Department of Geophysics, James Clerk Maxwell Bldg, Mayfield Road, Edinburgh EH9 3JZ, Scotland

K. Tolonen   Department of Botany, University of Helsinki, Unioninkatu 44, SF-00170 Helsinki 17, Finland

K. Wasylikowa   Institute of Botany, PAN, ul. Lubicz 46, 31-512 Krakow, Poland

T. Webb, III   Department of Geological Sciences, Brown University, Providence, R.I. 02912, U.S.A.

Preface

This handbook is devoted to palaeoecological methods applied mainly to organic lake and mire deposits. We refer mainly to Holocene environmental changes in the temperate zone, but most of the methods can be applied to older Quaternary deposits, and also to areas outside the temperate zone. An earlier version of this book was produced within the framework of project 158B of the International Geological Correlation Programme as described in the Foreword. This project emphasized the need for uniform methods of correlating stratigraphical data on a continental scale. Three volumes were published at the Department of Quaternary Geology in Lund during the years 1979–82. These were greatly appreciated by researchers and students, and copies have not been available for the last few years. Based on the experiences of the different coworkers within this project we found it necessary to revise and complete the preliminary handbook. Thirteen chapters have now been added. We were also convinced that, independent of project IGCP 158, there is an overall need of a methodological textbook in this field for both research students and professional specialists who are trying to acquaint themselves with methods outside their own speciality.

The handbook has eight main sections. Firstly, the *theoretical background* to environmental changes is described in three review papers. These attempt to give some glimpses into the relationships between physical and biological environments and how these are affected by changes in climate, hydrology and human impact. It is followed by the *research strategy* which can be applied in palaeoecological studies of lakes and mires. The concept of 'reference sites' for stratigraphical correlations is introduced, and the methods recommended for such sites are described in the following sections: *sampling and mapping techniques, stratigraphical methods, dating methods, physical and chemical methods, and biological methods*. Finally, the *numerical treatment of biostratigraphical data* is described and discussed in the final section.

It has been hard work to produce this volume. First of all I would like to thank all the contributors for their interest and efforts and for their great enthusiasm in this task. (I even received voluntary chapters!) I am very grateful to John Birks, Bergen, for many stimulating discussions concerning the contents of this book. I would like to thank especially my dear colleague,

Magdalena Ralska-Jasiewiczowa, Krakow, for her interest and advice. She acted as assistant editor for this book, and she has been the Secretary for the IGCP Project 158B. I am very grateful to Alix and John Dearing, Coventry, for their skilful help with the linguistic corrections. I also received many thoughtful editorial comments from John. The editorial work was done at the Department of Quaternary Geology in Lund, and many colleagues have been helpful. I would like to thank especially Gunnar Digerfeldt for many stimulating discussions. The original idea for this project appeared from such discussions ten years ago! My secretary, Karin Price, has been a great support in the editorial work — typing, and draft and proof checking — and I thank her for her generous help. Last but not least, I owe a great debt of gratitude to my family for their patience and support.

I am very grateful to the publishers, John Wiley & Sons Ltd., and especially to Dr Susan Hemmings and Mrs Helen Bailey, for their great interest in the production of this handbook.

BJÖRN E. BERGLUND

Foreword

LESZEK STARKEL

It is a challenge to reconstruct past environments since the last Ice Age from pieces of evidence found in the sediments of lakes, mires and seas. By applying a great variety of methods it is possible to establish environmental changes which have their causes in physical and biological factors, especially climatic and human. We are aware today of drastic changes in global environmental conditions around us, and many questions concerning future conditions arise. Knowledge about past environment may help us to understand more about the effects of modern climatic changes and human impact.

With this background, Unesco and the International Union of Geological Sciences (IUGS) initiated several research projects within the framework of the International Geological Correlation Program (IGCP) devoted to geological changes of importance to man. The IGCP Project 158, 'Palaeohydrology of the temperate zone', is one of these projects, dealing with environmental changes during the last 15,000 years. The temperate zone is characterized by drastic changes of the climate–vegetation belts. The former belts of ice sheets and tundra-steppe later became the belt of temperate forests, which even later were to become extensively explored by man. The aim of this project is to discover and explain these changes by studies in fluvial environments (subproject A) and lake and mire environments (subproject B). Studies on changes of great river valleys reveal material for constructing long-distance correlations and more general stratigraphies, especially in the perimarine and glaciated areas. But only lake and mire sediments can comprise complete and undisturbed sequences of change, which can be studied in detail. Palaeoecological methods, combined with various sedimentological and dating techniques, give the most solid background for investigating past environments.

The IGCP Project 158 was born among palaeoecologists already involved in the international collaboration of the International Union for Quaternary Research (INQUA). The general project proposal was formulated during a meeting in Czechoslovakia in 1976, and the first project meeting was organized at Szymbark field station in South Poland in 1977. The fluvial subproject is lead by myself, and the lake-mire subproject is lead by Björn Berglund.

This handbook will be a cornerstone for all research with aims common to those of Project 158B. It is based on a project manual already widely used in numerous palaeoecological laboratories; but this version is extended and completely revised in order to meet the demands of modern Holcene palaeoecology. A very high standard may be expected of this book from a glance at the list of contributors — some 43 experts from many countries. It also represents a piece of hard and good work by my friends, Björn Berglund and Magdalena Ralska-Jasiewiczowa.

Our thanks should be extended to the publisher, John Wiley and Sons, and its earth science editor, Dr Susan Hemmings. Their continuous assistance in editing volumes that reflect the activity of the IGCP Project 158 team is very valuable.

Cracow, March 1984

Professor LESZEK STARKEL.
Main leader of IGCP Project 158

Background to Palaeoenvironmental Changes During Holocene

Handbook of Holocene Palaeoecology and Palaeohydrology
Edited by B. E. Berglund
© 1986 John Wiley & Sons Ltd.

1

Late-Quaternary biotic changes in terrestrial and lacustrine environments, with particular reference to north-west Europe

H. J. B. BIRKS

Botanical Institute, University of Bergen, Norway

INTRODUCTION — PATTERN, PROCESS AND SCALE

Late-Quaternary biostratigraphy, when studied in taxonomic, stratigraphic and geographic detail, abounds in observable *patterns* of biotic change in time and space at various scales — changes in flora and fauna, population sizes and community composition; changes in rates, timings, durations and magnitudes of population and community change; extension and contraction of range-limits; etc. Observable patterns result from causative *processes*, but palaeoecological data only show patterns, not processes. The challenge to Quaternary palaeoecologists is to detect patterns *and* to interpret them in terms of underlying processes operative at the spatial and temporal scales reflected by palaeoecological data, and thus to link cause (process) with effect (pattern). In order to establish these links, it is often necessary to study contemporary patterns and processes as modern analogues for observed patterns and inferred processes in the past.

In Quaternary pollen analysis, interest is primarily in population and community changes over long time intervals (10^3–10^4 years) and large areas (c 10^8–10^{12} m^2) in response to fluctuations in physical environment (e.g. climate, soils, disturbances) and biota, including *Homo sapiens* (e.g. immigration, expansion, extinction). Patterns alone suggest little about processes unless the ecology of the taxa concerned, considered individually and collectively, is known and the past physical environment is reconstructed from independent evidence, such as stable-isotope ratios (Chapter 20), sediment lithology and chemistry (Chapters 12, 21), lake-level data (Chapter 5), and changes in organisms unrelated ecologically to the taxa of primary interest (Colinvaux, 1983; Howe and Webb, 1983). In the absence of independent environmental reconstructions, it is difficult, if not impossible, to distinguish unambiguously between biotic responses to changes in physical environment and responses to biological processes under conditions of constant environment. Environmental reconstructions and ecological interpretations must be considered simultaneously if the diverse patterns observable in late-Quaternary biostratigraphy are to be understood in terms of causal processes (Webb, 1981; Prentice, 1983).

Different patterns of ecological and palaeoecological complexity are discernible at different scales (Figures 1.1(b) and (c)). (See the review by Delcourt *et al.* (1983) and discussion of related scale problems in ecology and biogeography by Udvardy (1981), Wiens (1981) and Hengeveld (1982)). Different hypotheses about causal processes are possible and appropriate solely as a result of the scale of study (e.g. Stoddart, 1981; Allen and Starr, 1982; Hengeveld, 1982). For example, geographic-range patterns detectable at one scale may be controlled by processes very different from those important in influencing patterns of communities containing the same taxa but discernible at another scale; in turn, community patterns may be controlled by processes different from those determining the abundance and genetic constitution of populations and individuals of taxa within the community (Hengeveld, 1982). Some processes may only be relevant or specific to certain scales, and processes important at one scale cannot necessarily be generalized to other scales (Hengeveld, 1982).

Processes such as regional climatic change, soil retrogression, human interference, species immigration, competition, natural disturbances (e.g. fire, disease, windthrow), and local topographic and edaphic differentiation can all determine important patterns of population dynamics and vegetational change at particular scales (Figure 1.1(d)); Delcourt *et al.*, 1983; Prentice, 1983). In Quaternary palaeoecology, as in ecology and biogeography, it is important to identify and define spatial and temporal scales (Berglund, 1979a) or 'levels of variation' (*sensu* Hengeveld, 1982) represented by observed biostratigraphic patterns before attempting to infer causal processes. It is also essential to distinguish between what is observed and what is inferred, so that patterns detected at one scale are not used to infer processes for another scale.

Spatial and temporary scales should be considered together, if possible, because spatial patterns are dynamic and continually changing over a variety of time spans. Udvardy (1981), Wiens (1981) and Delcourt *et al.* (1983) suggest that there may, in certain instances, be a relationship between spatial and temporal scales (Figure 1.1(a)). Some patterns of biotic change discernible at one spatial scale may only be detectable at particular temporal scales and vice versa (Udvardy, 1981; Stoddart, 1981; Simberloff *et al.*, 1981; Delcourt *et al.*,

1983). For example, patterns of change in vegetation formations (10^{10}–10^{12} m^2) may only be detectable over time scales of 5–10 × 10^3 years (e.g. Huntley and Birks, 1983). In contrast, fine-scale patterns of change within a single vegetational stand (10^2–10^3 m^2) are discernible over 10^3 years or less (e.g. Andersen, 1979; Aaby, 1983).

Quaternary pollen analysis primarily operates within the macro-scale (*sensu* Delcourt *et al.*, 1983) with domains of 5 × 10^3–10^6 years ('millennial scale' — Udvardy, 1981) and 10^6–10^{12} m^2 (Figure 1.1(a)). Contemporary ecologists largely work within the micro-scale (*sensu* Delcourt *et al.*, 1983) of 10^0–5 × 10^3 years ('secular scale' — Udvardy, 1981; 'ecological time' — Simberloff *et al.*, 1981) and 10^0–10^6 m^2. Spatial scales of primary concern in Quaternary palaeoecology (Figure 1.1(b) are (1) *pollen-source areas* of bogs or lakes (> 500 m diameter) of about 10^9 m^2 (Tauber, 1965; Webb *et al.*, 1978; Berglund, 1979a); (2) *reference areas* (*sensu* Berglund, 1979a), *vegetation-landform units* recognized as 'discrete essentially geographical areas, on the basis of the predominant vegetation type ... and the prevalent landform' (Lichti-Feder-ovich and Ritchie, 1965), or *ecological districts* with a 'recurring pattern of landforms, soils, vegetation chronosequences and water bodies' (Ritchie, 1980), occupying 10^9–10^{11} m^2 and with more-or-less uniform regional pollen assemblages (Lichti-Federovich and Ritchie, 1968; Janssen, 1970); and (3) *vegetation formations, type regions* (*sensu* Berglund, 1979a), and *ecological regions* characterized by 'a distinctive regional climate expressed by the major (physiognomic) vegetation type, e.g. low-arctic shrub tundra' (Ritchie, 1980), with areas of 10^{10}–10^{13} m^2 and palynological patterns only discernible from networks of sites.

With the exception of pollen-stratigraphic records from small basins (10–30 m diameter) and soil profiles (Chapter 7) with pollen-source areas of 2–3 × 10^3 m^2 (Andersen, 1973) and recent formulations of a 'landscape ecology' (Forman and Godron, 1981; Romme and Knight, 1982) with a spatial domain of 10^5–10^9 m^2, there is little overlap in the operational spatial scales of Quaternary pollen analysis and contemporary ecology (Ritchie, 1981; Webb, 1981). Palynological data also cover time spans (10^3–10^5 years) very different from those of ecologists (10^0–10^2 years). They rarely have the temporal precision (10–50 years per sample) or sample density (1 sample per 100–500 years) of ecological data (1–5 years per sample, 1 sample per 1–10 years) (Webb, 1981). Exceptions are fine-resolution data from annually laminated sediments (Chapter 17), and very narrow, contiguous samples (e.g. Garbett, 1981; Green, 1982). Unfortunately the spatial precision of these data is coarse (c 10^9 m^2).

There are complex and intricate but, as yet, poorly understood interactions between scale, sampling, spatial, temporal and taxonomic precision, pattern detection, and process inference. Webb *et al.* (1978) show that decisions concerning, for example, taxonomic detail, geographic scale, sampling density

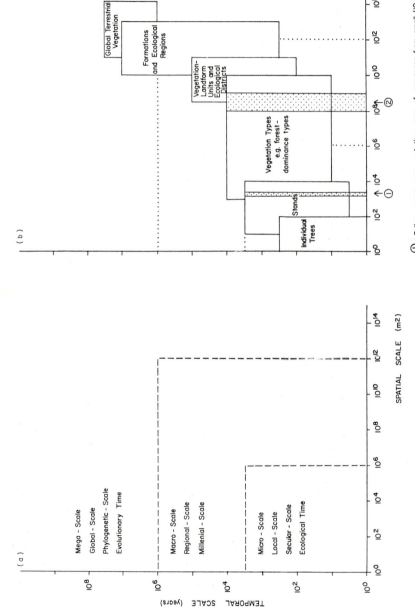

(a)

TEMPORAL SCALE (years)

Mega - Scale
Global - Scale
Phylogenetic - Scale
Evolutionary Time

Macro - Scale
Regional - Scale
Millenial - Scale

Micro - Scale
Local - Scale
Secular - Scale
Ecological Time

SPATIAL SCALE (m²)

(b)

Global Terrestrial Vegetation

Formations and Ecological Regions

Vegetation-Landform Units and Ecological Districts

Vegetation Types e.g. forest-dominance types

Stands

Individual Trees

① Pollen - source area and time span of sequences from small (10-20m diameter) hollows

② Pollen - source area and time span of sequences from lakes or bogs (>500m diameter)

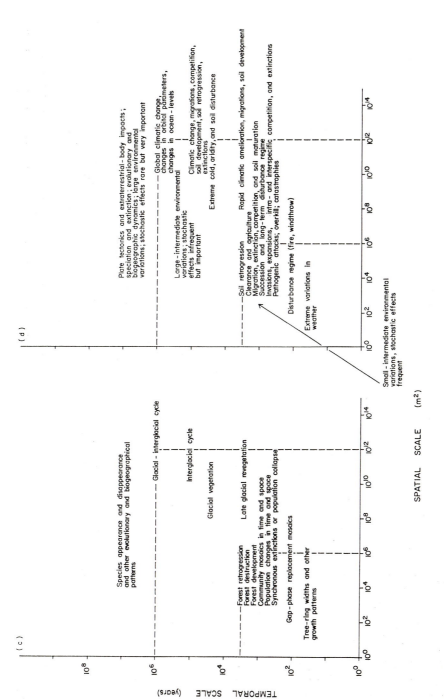

FIGURE 1.1. (a)–(d). See page 8 for caption.

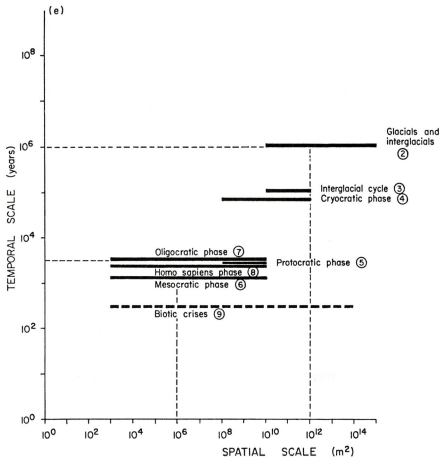

FIGURE 1.1.   Observed biotic patterns and inferred processes in relation to temporal and spatial scales (based on diagrams in Udvardy, 1981; Wiens, 1981; Delcourt *et al.*, 1983):

(a) Operational scales in ecology and palaeoecology.
(b) Types of contemporary vegetation patterns discernible at different scales. The shaded areas are the approximate spatial scales (pollen-source areas) and time spans of pollen-stratigraphic records from (1) small basins and (2) lakes and bogs.
(c) Major patterns of biotic change observed in the late-Quaternary. Positions within the time and space scales are approximate.
(d) Inferred processes for the observable patterns of biotic change. Positions within the time and space scales are approximate.
(e) Temporal and spatial scales discussed in the text. The numbers refer to the sections within the text.

and site type influence types and scales of patterns discernible from pollen data. Certain patterns are only detectable at particular scales with a specific sampling strategy.

This chapter reviews the main changes in late-Quaternary biotic patterns within north-west Europe, as recorded pollen-stratigraphically over time scales of 10^3–10^5 years, and discusses likely causative processes operative at relevant spatial and temporal scales (Figures 1.1(c)–(e)). It considers (1) glacial–interglacial patterns; (2) interglacial patterns; (3) glacial vegetation patterns; (4) late-glacial revegetation; (5) postglacial forest development; (6) forest retrogression; (7) forest destruction; and (8) short-term but widespread biotic crises.

GLOBAL GLACIAL–INTERGLACIAL PATTERNS
(1–2 × 10^6 years, 10^{10}–5 × 10^{14} m²)

Carbonate, oxygen-isotope and biostratigraphic patterns in ocean cores provide a continuous stratigraphy and independent global climatic record for the entire 1–2.5 million years of the Quaternary, dated, in part, by radiometric dating and palaeomagnetism (Shackleton and Opdyke, 1976). This record indicates multiple — seventeen or more — climatic cycles from cold to temperate conditions, eleven of which occurred since the first major continental glaciation and nine of which occurred in the last 730,000 years (Brunhes) (Kukla, 1977). Independent estimates suggest that each cold, often glacial, stage lasted 5–10 × 10^4 years, whereas temperate 'interglacials' occupy 1–2 × 10^4 years (Kukla *et al.*, 1972; Emiliani, 1972). At the onset of a cold stage there are rather gradual changes in O^{16}/O^{18} ratios of planktonic foraminifera, followed by fluctuations in these ratios within the stage, indicating changes in ice-sheet volumes and global temperature. Abrupt cooling and maximal ice-sheet extent occur near the end of the stage (Broecker and van Donk, 1970). For example, North Atlantic cores covering the last 10^5 years at 54°N suggest that glacial conditions occurred briefly from about 84,000 to 73,000 B.P., followed by a long but variable interstadial (73,000–30,000 B.P.). Climatic deterioration began at 30,000 B.P., resulting in maximum ice extent at about 20,000–18,000 B.P. Interglacial conditions developed rapidly at about 14,000 B.P., but were interrupted in the eastern North Atlantic by a brief, southward readvance of polar water between 11,000 and 10,000 B.P. (Sancetta *et al.*, 1973; Ruddiman *et al.*, 1977).

Oxygen-isotope stratigraphy of long marine cores indicates that for most of the Quaternary, ice-sheet volumes were greater and global temperatures lower than they are today. The Quaternary is thus primarily a cold period, often with extensive glaciation, disrupted by brief interglacial stages as warm as today (Emiliani, 1972). The underlying processes or 'forcing functions' for these global climatic changes are systematic variations with periodicities of 100,000, 41,000 and 23,000 years in the Earth's orbital geometry (eccentricity, obliquity, precession), leading to long-term fluctuations in insolation and cycles of glacial cooling and interglacial warming (Hays *et al.*, 1976; Imbrie and

Imbrie, 1980). This Milankovitch model predicts maximum summer insolation occurring at different times at different latitudes, for example as early as 10,000 B.P. at high latitudes in the Northern Hemisphere, a hypothesis supported by pollen-stratigraphic data from north-west Canada (Ritchie *et al.*, 1983). Despite its great potential, this powerful hypothetico-deductive approach is very rarely adopted in Quaternary palaeoecology (Edwards, 1983), even though it is an essential part of any mature (*sensu* Ball, 1975) science.

INTERGLACIAL PATTERNS — THE INTERGLACIAL CYCLE
(10–12×10^4 years, 10^{10}–10^{12} m^2)

Iversen (1958) proposed a simple model of ecological processes for observed patterns of interglacial vegetational history in north-west Europe, the so-called interglacial cycle (Figure 1.2). The *cryocratic* phase represents the cold, often glacial, stage with sparse assemblages of pioneer, arctic-alpine, steppe and ruderal herbs growing on base-rich, skeletal mineral soils, frequently disturbed by cryoturbation. Climate is cold, dry and continental (van der Hammen *et al.*, 1971).

At the onset of an interglacial, temperature rises and the *protocratic* phase begins. Basiphilous shade-intolerant herbs, shrubs and trees immigrate and expand rapidly to form widespread species-rich grasslands, scrub, and open woodlands growing on unleached, fertile soils of low humus content. This late-glacial phase supports a diverse flora of contrasting ecological and phytogeographic affinities (Iversen, 1954; Berglund, 1966). Immigration, population expansion, and even speciation (Iversen, 1958) are widespread and competition is generally low as light, nutrients, and space are ample. Lakes support a pioneer aquatic biota.

The *mesocratic* phase is characterized by development of temperate deciduous forests and fertile, brown-earth soils. Shade-intolerant species are rare or absent as a result of competition and habitat loss. The climate in the mesocratic may be the same as or warmer than in the protocratic. The various stages in the development of the deciduous forest may be direct, contemporaneous responses to climatic change, or may reflect delayed immigration of different trees following rapid climatic change at the beginning of the interglacial. These contrasting hypotheses are discussed later in this chapter. Lakes are fertile and frequently support a diverse biota.

Iversen (1958) termed the last, retrogressive phase the *telocratic* phase with open conifer-dominated woods, ericaceous heaths, and bogs growing on infertile, humus-rich podsols and peats. Local extinction of nutrient-demanding plants of the mesocratic occurs. Some protocratic plants expand as a result of decreasing shade. Lake productivity declines due to dystrophy. Iversen (1958) suggested that temperature falls at the onset of this phase, but Andersen (1964, 1966) has shown that climatic deterioration does not occur until well into

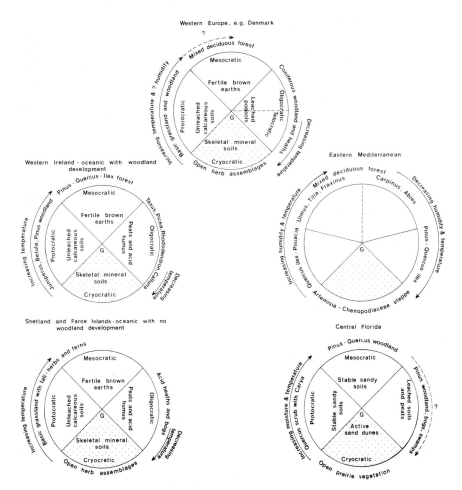

FIGURE 1.2. The interglacial cycle in Denmark (modified from Iversen, 1958), western Ireland, Shetland and Faroe Islands (modified from Birks and Peglar, 1979), eastern Mediterranean (modified from van der Hammen *et al.*, 1971) and central Florida. The glacial or cryocratic phase (G) in each area is shaded. The major observed changes in regional vegetation and soils are shown, along with inferred changes in regional climate (probable climatic changes are denoted by a solid line; unproven climatic changes by a broken line with a ?)

the phase of soil retrogression, suggesting that soils deteriorate independently of climatic change. Andersen calls the phase of impoverished soils prior to climatic deterioration the *oligocratic* phase. Major climatic changes occur at the onset of the cryocratic phase as forests disappear, frost action and solifluction destroy the leached acid soils, and herbs expand on to the newly exposed mineral soils. The glacial–interglacial cycle begins again (Figure 1.2).

This model of glacial–interglacial pattern and process suggests that broadly similar ecological–physiognomic assemblages recur in interglacials at the broad scales of 10^6–10^{12} m^2 and 10–12 $\times$ 10^4 years in response to regional climatic and edaphic changes and competition. The trees characteristic of the three interglacial phases differ in reproductive and population biology and ecological and competitive tolerances (see Table 1.1). Within these broad assemblages ('protocratic plants', e.g. herbs, *Betula*; 'mesocratic plants', e.g. *Quercus, Corylus*; 'oligocratic plants', e.g. *Picea, Calluna vulgaris*), the actual floristic and vegetational composition varies from interglacial to interglacial. Factors such as location of refugia in the cryocratic phase, migration rates, distances over which migration occurred, competition, succession, predation, ecotypes, and chance as it affects survival, dispersal and establishment, may all contribute to observed differences in interglacial patterns (Andersen, 1966, 1969; West, 1980). There is apparent order within interglacial vegetational patterns when viewed at the broad scale of an entire interglacial cycle (10–12 $\times$ 10^4 years), whereas within each phase of an interglacial (*c* 5 $\times$ 10^3 years) there is great variation between interglacials; hence the ability of pollen analysis to differentiate between interglacials.

Andersen (1964, 1966, 1969) provides convincing evidence for the interglacial cycle in Denmark. He presents palaeolimnological evidence for increased soil and lake acidification and associated decreases in lake productivity with the onset of the oligocratic phase independent of climate. This suggests natural soil succession from skeletal frost-disturbed soils *via* fresh unleached soils and fertile brown-earths with mull, to acid podsols with mor. Regional vegetation developed in parallel, from pioneer herbaceous assemblages via grasslands and open woodland, closed deciduous forest, to open coniferous forests, heaths and bogs.

The primary process influencing the contrast in observed pollen-stratigraphic patterns between glacial and interglacial stages at the scale of 10–12 $\times$ 10^4 years is global climatic change, particularly temperature. At the scale of 5 $\times$ 10^3 years, regional soil changes, immigration and competition influence observed patterns. However, once certain taxa (e.g. *Picea*) become established, vegetation itself influences soils, resulting in complex but poorly understood interactions between soil and vegetation (Andersen, 1966, 1969).

The interglacial-cycle model is applicable to other ecological and phytogeographic settings (Figure 1.2). Birks and Peglar (1979) propose an analogous cycle for northern oceanic regions such as the Shetland and Faroe Islands where tree growth is virtually absent during an interglacial because of wind exposure and low summer warmth. After a cryocratic phase of sparse herb assemblages with arctic-alpines growing on skeletal mineral soils, the protocratic and mesocratic phases are characterized by grassland with abundant tall-herbs and ferns growing on fertile, brown-earth soils. Soil deterioration is inevitable in oceanic environments. Extensive podsolization

TABLE 1.1. General ecological characteristics of the trees characteristic of the protocratic, mesocratic and oligocratic phases of an interglacial cycle

	Protocratic	Mesocratic	Oligocratic
Examples	*Betula* *Populus* *Salix*	*Quercus* *Ulmus* *Tilia*	*Picea* *Abies* *Fagus*
Reproductive rate	High	Low	Medium
Age of first seed setting	Young	Mature	Mature
Frequency of seed setting	High	Low	Low
Propagule-dispersal efficiency	Good	Poor	?Poor
Migration rate (m/year)	>1000	500–1000	<500
Competitive tolerances	Low	High	High
Longevity	Short	High	High
Seedling tolerances	Light-demanding	Shade-tolerant	Shade-tolerant
Seed production	High	High or low	High or low
Seedling mortality	High	?Low	Low
Growth rate	Fast	Slow	?Slow
Ability to regenerate under own canopy	Rare	Rare	Common
Shade production	Light	Dense	Dense
Crown geometry (Horn, 1971)	Multilayered	Monolayered	Multilayered
Rate of population increase (r)	High	Medium or low	Medium or low
Soil preferences	Fertile unleached	Brown earths with mull humus	Podsols with mor humus
Invasion behaviour	Large gaps essential	Small gaps essential	?Large gaps essential
Life-history traits	r-selected	K-selected	?
Demographic traits (Whittaker and Goodman, 1979)	Exploitation	Saturation	Adversity
Ecological traits (Grime, 1977)	Ruderal	Competitive	Stress-tolerant

and accumulation of mor humus and peat occur, resulting in widespread acid heath and bog communities dominated by, for example, *Bruckenthalia spiculifolia, Daboecia cantabrica, Erica mackaiana*, and, in the Holocene, *Calluna vulgaris* in the oligocratic phase. Here the primary process influencing observed interglacial vegetational patterns is soil change, especially retrogression. Watts (1967) outlines a related cycle for western Ireland (see Figure 1.2) with *Pinus, Betula* and *Populus* characteristic of the protocratic, *Pinus, Quercus* and *Ilex* forming a mesocratic phase, and *Abies, Picea, Rhododendron ponticum* and other Ericaceae dominating the oligocratic. See also Chapter 22, Fig. 22.6.

In the Mediterranean, van der Hammen *et al.* (1971) suggest, on the basis of long pollen-stratigraphic sequences, an interglacial cycle in which major processes determining the observed patterns are humidity and temperature changes. After a cryocratic phase of *Artemisia*–Chenopodiaceae steppe, *Quercus ilex–Pistacia* open woodland develops as temperature and humidity increase. Deciduous forest with *Quercus, Ulmus, Tilia*, and later *Carpinus*, dominates when humidity is maximal. As humidity and temperature fall, open *Pinus–Abies* forests with *Q. ilex* and *Pistacia* occur, prior to the onset of the next cold stage. Similarly in Florida, Watts (1971) proposes that moisture and, to a lesser extent, temperature changes are major determinants of vegetational change, perhaps due to rising water-tables associated with eustatic sea-level changes during interglacials. Dry prairie-like or dune vegetation with *Ceratiola* is widespread in the glacial stage (Watts, 1980). With increasing moisture and temperature, open oak scrub with *Carya* develops. This is replaced by pine–oak woodland with mesic associates such as *Fagus, Ulmus* and *Liquidambar*. Towards the end of an interglacial, southern pines and *Serenoa* expand, possibly in response to soil leaching, and evergreen-shrub bogs and cypress swamps of *Taxodium, Ilex* and Ericaceae develop on peat in low-lying areas.

The glacial–interglacial cycle provides the setting for the Quaternary 'palaeoeccological play' (cf. Hutchinson, 1965) in which the various glacials and interglacials are enacted, each with different characters, namely plant and animal combinations and abundances. I will discuss in more detail the various acts (*phases*), main characters and *patterns* they form, and plot (*processes*) of the most recent glacial–interglacial play. Of course, the plot is not directly analogous to that of a play, because interactions of the characters are not predetermined by an author, but occur as a result of complex and often unpredictable interactions between organisms and environment.

GLACIAL PATTERNS — THE CRYOCRATIC PHASE
(5–10×10^4 years, 10^8–10^{12} m^2)

Most of the Quaternary consists of cold, often glacial stages, as evidenced by oxygen-isotope stratigraphy of marine cores (Emiliani, 1972). There are, unfortunately, very few terrestrial sites with a continuous biostratigraphic record for the last (Weichselian) cold stage or any previous cryocratic phase in western Europe. Our knowledge of Weichselian biotas and their temporal and spatial patterns is fragmentary and based on sites beyond the extent of glaciation, occasionally within the periglacial zone, and usually with short stratigraphic sequences. Correlation and temporal ordering are dependent on radiocarbon dating. Problems of contamination become very critical in dating sediments older than about 40,000 B.P. (Chapter 14). Problems of vegetational and environmental reconstruction also arise because the fossil plant assemblages have no convincing widespread modern analogues.

The Weichselian began at about 115,000 B.P. (Kukla, 1980) and is characterized by primarily herbaceous pollen assemblages (Figure 1.3). At the close of the last interglacial, herb-dominated vegetation developed on mineral soils exposed by frost action and solifluction. Erosion, reworking and redeposition of interglacial sediments also occurred (Andersen, 1961). Interstadials, probably of short duration, occur at about 105,000 (Amersfoort), 100,000 (Brørup), and 82,000 B.P. (Odderade) (cf. Forsström, 1984; and Figure 1.3). They are characterized at latitudes 52–56°N by open, boreal-type woodland of *Pinus*, *Picea* (including *P. omorika*), *Larix*, *Betula* and *Alnus* and with *Bruckenthalia spiculifolia*. Both *P. omorika* and *B. spiculifolia* are confined today to south-east Europe. The herb and aquatic floras indicate a considerably warmer climate than is suggested by the trees (Andersen, 1961). There may not have been sufficient time for temperate trees (e.g. *Quercus, Ulmus*) to immigrate from southern refugia, suggesting that tree distributions were not in equilibrium with climate during these interstadials. Pollen of temperate trees are present in correlative sediments further south (Woillard, 1978).

After about 80,000 B.P. the vegetation of north-west Europe was sparse and herb-dominated, or even disappeared, to be replaced by polar-desert, particularly near the ice-sheet (Vorren, 1978). Brief interstadials with shrubs, including *Betula nana* and *Salix*, occur locally at 50,000 (Moershoofd), 40,000 (Hengelo), and 30,000 B.P. (Denekamp) (Kolstrup and Wijmstra, 1977). Ice reached its maximal extent at about 20,000–18,000 B.P. The pollen assemblages that occur widely from 60,000 to 14,000 B.P. were produced by a vegetation unknown today. It had structural and floristic affinities with steppe and tundra, and contained a mixture of ecological and geographic elements (Figure 1.3; — arctic, arctic-alpine, southern, steppe, ruderal, aquatic, marsh and halophyte taxa (Bell, 1969; Iversen, 1973)). This widespread 'no-analogue' vegetation type suggests a 'no-analogue' environment. Summers were possibly as warm as 10–15°C (Kolstrup, 1979, 1980), allowing thermophilous southern plants and northern plants tolerant of warm summers but intolerant of competition to coexist. Winters were probably cold, possibly as low as −25°C, causing extensive cryoturbation and unstable mineral soils (Williams, 1975;

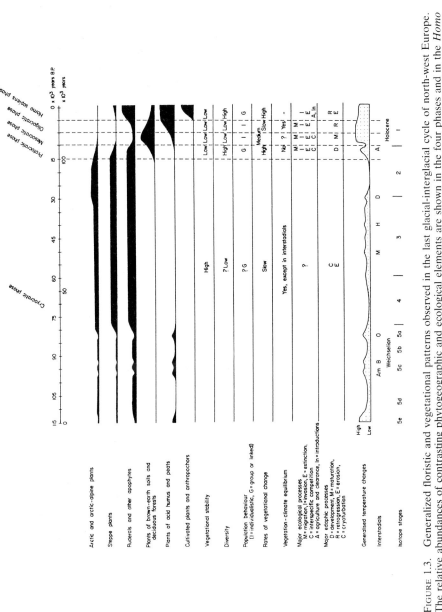

FIGURE 1.3. Generalized floristic and vegetational patterns observed in the last glacial-interglacial cycle of north-west Europe. The relative abundances of contrasting phytogeographic and ecological elements are shown in the four phases and in the *Homo sapiens* phase of the last 5000 years. The predominant ecological features and processes of these phases are summarized. The oligocratic and *H. sapiens* phase are shown for convenience as two separate phases, each of 5000 years duration, even though they occur in parallel in the Holocene. A very generalized temperature curve for the last 115,000 years is also shown. (Abbreviations: Am = Amersfoort; B = Brørup; O = Odderade; M = Moershoofd; H = Hengelo; D = Denekamp; A = Allerød)

Kolstrup, 1980). Precipitation was probably low, winds strong, and evaporation exceeded precipitation, resulting in an arid, open landscape, loess deposition, skeletal base-rich soils devoid of humus, and local saline areas that may have supported halophytes (Bell, 1969; cf. Adam, 1977). The growing season may have been short because of high albedos associated with sparse vegetation and extensive ice and snow (West, 1977). Although this vegetation contained plants with arctic or arctic-alpine distributions today (e.g. *Dryas octopetala, Saxifraga oppositifolia, Salix herbacea, S. polaris*) and may have had a tundra physiognomy, it lacked several taxa such as ericoid dwarf-shrubs that are abundant in modern tundra. It also contained many taxa absent from tundra today (e.g. *Helianthemum canum, Onobrychis viciifolia*).

Further south (46–48°N) vegetational cover was more extensive but treeless (Iversen, 1973; Kolstrup, 1980). It supported a diverse fauna with tundra and steppe affinities, including arctic fox, reindeer, steppe bison, mammoth, woolly rhinoceros, musk-ox, horse, wolf, bear, lion, and saiga antelope. *Artemisia* and Cheonopodiaceae pollen are particularly frequent in central and southern Europe, especially during the later part of the Weichselian, suggesting steppe-like soils, an extremely continental climate, and an arid landscape at the height of glaciation. This continentality continues into the early protocratic (van Geel and Kolstrup, 1978; Kolstrup, 1980) and contrasts with oceanic conditions prevalent in the closing telocratic phase of an interglacial and the early part of the cryocratic (Andersen, 1961; van der Hammen *et al.*, 1971).

Steppe-like vegetation with abundant *Artemisia* and Chenopodiaceae (including central Asian steppe genera *Kochia* and *Eurotia*; Smit and Wijmstra, 1970) occurred south of the Alps. It has long been assumed that temperate trees were widespread south and east of the Alps during the Weichselian. Temperate trees did survive in these regions, but as small, isolated populations in localities with favourable humidity, temperature, soils and topography — for example as a narrow zone at moderate elevations in the Balkan mountains between lowland arid steppe and alpine snow and ice (van der Hammen *et al.*, 1971). Comparable zonations occur today in the central Asian Highlands, in the south-west USA, and in Peru. It is possible that such local montane forest belts were the principal refugia for European deciduous trees. Survival of particular taxa in certain areas (see Huntley and Birks, 1983) may largely have resulted from chance, as the observed patterns of occurrence are indistinguishable from expected patterns based on random allocation of taxa to refugia (Birks, unpublished).

Many widespread vegetation types characteristic of the cryocratic in southern Europe and adjacent areas show little change, as recorded pollen stratigraphically, over long periods, often $2–3 \times 10^4$ years (e.g. Wijmstra, 1969; van Zeist and Bottema, 1977), just as pollen records from unglaciated regions of North America show little change for periods in excess of the duration of interglacials (e.g. Watts and Bradbury, 1982; Delcourt and Delcourt, 1983). The major climatic change that rapidly ended the last glacial (Broecker and van Donk, 1970; Imbrie and Imbrie, 1980) thus terminated long periods of apparent vegetational stability, as reflected by relatively constant pollen assemblages in time and space, and initiated rapid and very marked vegetational changes in the ensuing interglacial. Vegetational stability, at least at broad spatial scales, is a feature of the cryocratic (Figure 1.3), in contrast to rapid vegetational change in the protocratic and mesocratic. As cold stages, presumably with cryocratic biotic patterns similar to the Weichselian, dominate the Quaternary (Emiliani, 1972), long-term vegetational stability may well be the major biotic pattern in the Quaternary as a whole.

CLIMAP (1976) reconstructed global ice and snow cover, sea-surface temperatures, and vegetation cover at 18,000 B.P. Their reconstruction shows that ice and snow cover was twice that of today, ocean surface-temperatures were globally 2–3°C cooler, the North Atlantic Gulf Stream was displaced southwards to Iberia, western Europe was surrounded by polar water, and there was ocean cooling in equatorial regions resulting from increased upwelling of polar waters. Simulations of terrestrial climate at 18,000 B.P. (Gates, 1976) suggest a substantially cooler and drier climate in unglaciated regions than today, a prediction supported by available biostratigraphic and geologic data (Peterson *et al.*, 1979). Widespread aridity, especially in the tropics, is also predicted by Manabe and Hahn's (1978) simulations, as a result of strong surface-air flow from tropical continental regions to the oceans owing to larger reductions in air temperature over continents than over oceans. Lake-level data (Street and Grove, 1979) from tropical regions also indicate widespread aridity at 18,000 B.P., the time of glacial maximum. In other areas patterns are more complex, with high lake-levels at 18,000 B.P. in the south-west USA and parts of northern Africa. Lake productivity was low, and sediments are predominantly minerogenic as a result of low lake-levels, erosion, inwashing and aeolian activity.

LATE-GLACIAL PATTERNS — THE PROTOCRATIC PHASE
(5×10^3 years, 10^8–10^{10} m^2)

The last protocratic phase in north-west Europe began about 14,000 B.P. and includes the Weichselian late-glacial and earliest Holocene (10,000–*c* 9000 B.P.). Abrupt climatic amelioration occurred at about 14,000 B.P., possibly as a result of the 100,000-year eccentricity cycle being enhanced by the 23,000-year precession cycle (Imbrie and Imbrie, 1980). It led to rapid but variable retreat of Scandinavian ice, thereby creating newly deglaciated areas open to biotic colonization. It also initiated northward immigration of plants and animals from refugia in southern Europe.

The earliest vegetation on raw, base-rich mineral soils in deglaciated areas was sparse assemblages of opportunistic plants with contrasting geographic and ecological affinities today (e.g. *Oxyria digyna, Dryas octopetala, Rumex acetosella, Sagina procumbens, Plantago major, Armeria maritima, Helianthemum*) (see, for example, Iversen, 1954; Birks, 1973; Birks and Mathewes, 1978). Sites in former periglacial areas show an expansion of *Artemisia* pollen at about 13,500 B.P., possibly in response to climatic amelioration (van der Hammen *et al.*, 1971). Gramineae often dominates assemblages from these areas, whereas Cyperaceae dominates contemporaneous spectra from deglaciated areas (e.g. Vorren, 1978; Watts, 1979a). Vegetation in periglacial areas was probably similar to but more diverse than that further south during the late-cryocratic. It was herb dominated with a mixture of shade-tolerant, calcicolous, arctic, arctic-alpine, steppe, ruderal, oceanic, continental, northern and southern plants of low competitive ability growing on open, immature base-rich or neutral unleached mineral soils devoid of humus and in conditions of low competition and abundant light, space and lime. Several taxa with nitrogen-fixing nodules were present (e.g. *Dryas, Astragalus alpinus, Anthyllis vulneraria, Lotus corniculatus, Hippophaë rhamnoides*). Although this vegetation was common throughout north-west Europe, it has no convincing widespread modern analogue (Iversen, 1954, 1973). It presumably developed in response to a unique combination of factors that do not occur today, at least on a broad regional scale. For example, climate, in particular seasonality and radiation balance, may have been very different (Bryson and Wendland, 1966; Walker, 1978; Kutzbach, 1981; Kutzbach and Otto-Bliesner, 1982).

As soils stabilized, humus and nitrogen accumulated and *Empetrum nigrum, Salix, Betula nana* and *Juniperus* became increasingly important, just as in successions on deglaciated areas today after a period (50–100 years) of dominance by pioneer herbs and nitrogen-fixing plants (e.g. Faegri, 1933; Persson, 1964). By 13,000 B.P. a range of vegetation types was present; grass heaths and herb meadows, *Artemisia* communities, snow beds, willow scrub, stands of *B. nana, Empetrum* heaths, *Juniperus* stands, *Hippophaë* copses, tall-herb communities, wind-blasted 'fell-fields', sedge-dominated mires and gravel flushes, springs and flushes, scree and open gravel communities, and moss- and lichen-dominated communities (e.g. Firbas, 1949; Berglund, 1966; Birks, 1973; H. H. Birks, 1984). Bare ground (e.g. unstable screes, 'fell-field', snow-beds) was probably also extensive, especially in the early phases of revegetation (Ritchie, 1982). The distribution and extent of these vegetation types appear to be a function, in part, of geographic location and regional climate and, in part, of local topography, exposure, altitude and geology (e.g. Birks, 1973; Watts, 1977). Geographic location was particularly important as macro-climatic gradients, especially of oceanicity and continentality, may have been considerably steeper in the Weichselian late-glacial than today (Lamb, 1977; Huntley and Birks, 1983). The complex interactions between late-glacial vegetational patterns, regional location and climate, and local landform, altitude, topography and geology, remain to be elucidated in detail.

Lakes were abundant in deglaciated areas as a result of melting of buried-ice blocks. Sediments were initially inorganic. Lakes were rapidly colonized by pioneer, thermophilous aquatic animals and plants, many of which grow in streams and rivers today and tolerate silting. Their presence suggests that climate was warmer than is indicated by the contemporaneous regional treeless vegetation (Iversen, 1954, 1964a; Kolstrup, 1979; Berglund *et al.* 1984), presumably because of disequilibrium between tree distributions and climate.

By 12,500 B.P. *Betula pubescens* immigrated to southern Scandinavia and Britain to form birch copses and open 'park-tundra' (*sensu* Iversen, 1954, 1973). Some reduction in the extent of *Betula*, an expansion of herbs of disturbed soils, and increased erosion and inwashing of mineral material occurred locally in southern Scandinavia, the Netherlands and north Germany between 11,800 and 12,000 B.P. (e.g. Usinger, 1978; Berglund, 1979b). These changes may have resulted from temporary climatic deterioration (Older Dryas stadial), involving decreased summer temperatures (Iversen, 1954) or increased continentality with low winter temperatures, reduced snow cover, and increased drought (van Geel and Kolstrup, 1978; Kolstrup, 1982).

The widespread Allerød interstadial followed between 11,800 and 11,000 B.P. In some areas, particularly western Norway and much of the British Isles, there is no unambiguous evidence for a phase correlative with Older Dryas (cf. Watts, 1979b). Instead there is a long uninterrupted but variable interstadial from about 13,000 to 11,000 B.P., the so-called Windermere interstadial (Pennington, 1977), during which climate appears to have fluctuated markedly (see also Berglund, 1979b; Kolstrup, 1982). Several short climatic fluctuations, including oscillations contemporaneous with the Older Dryas stadial, can be detected at some sites (Pennington, 1977; Watts, 1977). Between 11,800 and 11,000 B.P., birch woodland with *Populus tremula*, *Sorbus aucuparia* and *Prunus padus* developed over much of north-west Europe, except in extreme northern and western areas, where shrubs, dwarf-shrubs, grasses or sedges remained abundant (e.g. Birks, 1973; Pennington, 1977; Watts, 1977), depending on regional climate and local conditions of geology, landform, exposure, snow cover (Birks, 1973), and possibly grazing pressure by large herbivores (Watts, 1977; Craig, 1978), as well as chance factors affecting plant dispersal and establishment. Several shade-intolerant herbs became rare or locally extinct, as a result of competition from taller, long-lived woody plants. Many heliophytes probably persisted, however, in treeless habitats such as inland cliffs and coastal situations. By 11,500 B.P. *Pinus sylvestris* reached north-west Europe for the first time since the Odderade interstadial. In central Europe it was the major tree from about 12,000 to 12,500 B.P. (Huntley and Birks, 1983).

The interstadial landscape was stabilized by an almost complete vegetation cover. Frost action, solifluction, aeolian activity and erosion were minimal and organic-rich lake-sediments began to accumulate. The presence of thermophilous beetles and marsh and aquatic plants (e.g. *Typha latifolia, Schoenoplectus lacustris*), which disperse rapidly, indicates that climate was warmer than is suggested by the terrestrial vegetation, and thus that terrestrial assemblages, particularly trees, were not in climatic equilibrium (Iversen, 1954, 1964a; Kolstrup, 1979). Soil maturation continued and humus accumulated. Some acidification and leaching occurred towards the end of the interstadial, especially in areas of acidic bedrock (Iversen, 1954; Berglund and Malmer, 1971), possibly favouring expansions of *Empetrum* and calcifuge herbs and ferns.

Vegetational revertence occurred widely between about 11,000 and 10,000 B.P. during the Younger Dryas stadial as a result of a major abrupt climatic deterioration associated with southward movement of polar water in the eastern North Atlantic (Ruddiman *et al.*, 1977). Temperatures decreased considerably in the North Atlantic (Sancetta *et al.*, 1973) and on the adjacent continental margin, as evidenced by, for example, oxygen-isotope ratios (Kolstrup and Buchardt, 1982). Scandinavian ice retreat was halted and some readvance occurred. In parts of the British Isles corrie glaciation occurred, and in the Scottish Highlands an ice-sheet redeveloped (Sissons, 1979). Woodland became more open, or even disappeared, particularly in the north and west (Watts, 1979b), where it was replaced by dwarf-shrub heaths, open grass heaths, sedge-dominated communities, or snow beds. The landscape was unstable again, with aeolian activity, solifluction, freeze–thaw activity, redevelopment of ground-ice formation, and extensive inwash of mineral material into lakes (e.g. Pennington, 1977; Watts, 1977; Hunt and Birks, 1982; Bennett, 1983a; H. H. Birks, 1984). Allerød humus was destroyed by frost action and washed into lakes (Berglund and Malmer, 1971). Soil disturbance and cryoturbation were prevalent and opportunistic plants of contrasting geographic distributions today but intolerant of competition and with a predilection for open, mineral soils flourished (e.g. *Artemisia*, Chenopodiaceae, *Plantago major, Helianthemum, Centaurea cyanus, Armeria maritima, Saxifraga oppositifolia, Koenigia islandica*).

Between 10,300 and 10,000 B.P. the Younger Dryas was terminated by rapid and permanent warming; new results indicate a climate amelioration already around 10,500 B.P. (Björck and Digerfeldt 1984, Berglund *et al.* 1984). Thermophilous beetles and water plants quickly colonized and upland plants already present expanded rapidly (e.g. Gramineae, *Rumex acetosa, Filipendula, Empetrum nigrum*) over a period of 100–250 years. *Salix* and *Juniperus* also persisted locally through the Younger Dryas, possibly as prostrate, poorly flowering shrubs protected by winter snow. With climatic amelioration, they responded rapidly by increased vegetative growth, improved flowering, and local population expansion for a period of 100–300 years. There is thus a series of short-lived (50–150 years) population peaks of Gramineae, *Rumex acetosa, Empetrum nigrum, Filipendula, Betula nana, Salix*, and *Juniperus* (e.g. Watts, 1977; Craig, 1978; Birks and Mathewes, 1978) before trees expanded. Soils ceased to be disturbed and were initially base-rich, shallow and generally low in organic content. With expansion of dwarf-shrubs, humus accumulated, soil maturation proceeded again, and new, longer-lived species became dominant. *Juniperus* humus is of mull type and may have provided favourable soil conditions for establishment of other taxa. This phase of rapid population changes and associated soil development is similar in temporal scale and ecological process to the 'nucleation' phenomena in primary succession described by Yarranton and Morrison (1974) (see also Ritchie, 1977), even though the spatial scales (10^8–10^9 m^2 (protocratic) and 10^5 m^2 (primary succession)) are different.

Betula, with its high reproductive rate, abundant wind-dispersed fruits, rapid growth rate, and young reproductive-maturity age, quickly immigrated and expanded (Table 1.1). It probably shaded out much of the *Empetrum* and *Juniperus* to form, within 300–400 years, birch-dominated woodlands from about 9700 to 9300 B.P. *Populus tremula, Sorbus aucuparia, Prunus padus, Salix* and *Viburnum opulus* were also present. The rate of population increase (*r*) of *Betula* is very high in southern England ($r = 1.2 \times 10^{-2}$), and population doubling times are low (*c* 50 years), suggesting that populations expanded as rapidly as possible (Bennett, 1983b). The expansion over 400 years must have involved several generations. The duration of the expansion phase increases northwards, suggesting some environmental constraints on expansion rates (Bennett, 1983a). *Pinus sylvestris*, also with wind-dispersed propagules, immigrated rapidly from areas further south and east and invaded and partly replaced birch as the dominant tree within a period of 300–500 years. Its population-increase rate is also high ($r = 1.1 \times 10^{-2}$) and doubling times low (65 years) in southern England (Bennett, 1983b). Its expansion appears to have been as rapid as possible and involved several generations (Bennett, 1983b). The duration of population expansion also increases northwards, suggesting some, as yet, unknown control on expansion rates (Bennett, 1983a). Birch, pine, aspen and willow favour open sites, have shade-intolerant seedlings, grow rapidly, and are relatively short-lived (Table 1.1). Between them, they dominated for nearly 10^3 years. In some areas the longer-lived pine assumed dominance, particularly on infertile soils such as sands, gravels, peats and limestones. However, they were all replaced as forest dominants as soon as shade-tolerant, competitive temperate deciduous trees of the mesocratic (Table 1.1) immigrated. Thermophilous aquatic and marsh taxa (e.g. *Typha latifolia, Ceratophyllum demersum, Cladium mariscus*) were present soon after 10,000 B.P. (Iversen, 1960), suggesting that summer temperatures were as high as they are today. The absence of temperate trees in north-west Europe at this time is a further example of tree distributions not being in equilibrium with climate for at least 10^3 years (Iversen, 1960, 1973). Their absence almost certainly results from their immigration from southern and south-east Europe not being instantaneous but taking hundreds or even thousands of years, depending on factors such as seed production, dispersal and establishment, population-expansion rates, age of reproductive maturity, competition and chance (Iversen, 1960; Faegri, 1963; Watts, 1973, 1982).

Several aquatic taxa (e.g. *Myriophyllum alterniflorum, Ceratophyllum demersum, Chara*) show enormous population expansions and maximum productivity between 13,000 and 12,500 B.P. or between 10,300 and 10,000 B.P., presumably in response to favourable lake-water chemistry, transparency, temperature and substrate, and low competition and predation (H. H. Birks, 1980). These population explosions are recorded by very high pollen values and abundant macrofossils (e.g. Iversen, 1954; Birks and Mathewes, 1978; Craig, 1978; Fredskild, 1983). They are then followed by a phase of reduced pollen and seed frequencies but in which vegetative remains (e.g. leaf spines) may be as common as before, suggesting that population sizes stayed constant but flowering and sexual reproduction declined (Watts, 1978).

These expansions may have resulted from sudden changes in availability of critical nutrients. As the landscape stabilized, inwashing of mineral material ceased, and organic sedimentation began. At Linsley Pond, Connecticut, Livingstone and Boykin (1962) proposed that phosphate which had previously been adsorbed on to mineral material was released and made available to limnic biota, thus allowing the logarithmic increase in lake productivity observed at the beginning of the Holocene (Livingstone, 1957). A similar explanation may apply to these European protocratic expansions. The subsequent decline in sexual reproduction may result from, for example, depletion of critical nutrients, changes in lake chemistry and trophic status, or inter- and intraspecific competition (Watts, 1978; H. H. Birks, 1980; Fredskild, 1983). Unfortunately little is known about the population and reproductive biology and ecological tolerances of aquatic biota, so these hypotheses must await testing.

The protocratic vertebrate fauna lacks many large mammals of the cryocratic (e.g. mammoth, woolly rhinoceros, steppe bison, musk-ox; Degerbøl, 1964; Stuart, 1977). It contained both northern (e.g. reindeer, wolf, lynx, tundra vole, elk, beaver, lemming, arctic fox, brown bear, wolverine) and steppe (e.g. suslik, bison, horse, pika) animals. Giant deer ('Irish elk') was particularly prominent in the interstadial of Ireland. It may have been important in determining the treelessness of the Irish interstadial (Watts, 1977; Craig, 1978). Many taxa became extinct in north-west Europe as woodland developed in the Allerød or early Holocene — for example bison (*Bison bonasus arbusto-tundrarum*) was replaced by aurochs, and reindeer by red deer. Other animals (e.g. elk) were favoured by woodland development. Giant deer, on the other hand, appear to have become extinct at the onset of the Younger Dryas, possibly in response to climatic change or, more likely, to vegetation change and habitat loss.

The climate of the European protocratic is characterized by generally rising but fluctuating temperatures and marked continentality. The main biological processes are rapid population and community changes in response to regional environmental changes, particularly climate (Figure 1.3). Populations of many taxa change together (e.g. Birks and Gordon, 1985), which suggests that they were limited by strong environmental controls. This contrasts with the predominantly individualistic behaviour of populations in the mesocratic phase. The protocratic is thus a time of linked population behaviour, vegetational instability, high rates of floristic turnover and population and community change, immigration and expansion of shade-intolerant plants, and disequilibrium between climate and tree flora (Figure 1.3). It is also a phase of maximal species diversity at the landscape scale, and possibly also of speciation due to rapid immigration and mixing of taxa previously separated (Iversen, 1958). Its inherent vegetational instability contrasts with the long-term stability of the cryocratic, even though there are many floristic and vegetational similarities between them. The same contrast exists in other phytogeographical settings. For example, Ritchie (1982), Watts (1979a, 1980) and Delcourt and Delcourt (1983) describe 'late-glacial' protocratic floras and vegetation that show great variations in their spatial and temporal patterns with rapid and short-lived population and vegetational changes and maximal diversity, in contrast to long-term stability and low diversity of the preceding cryocratic.

The last protocratic phase in north-west Europe can provide important insights, as yet poorly explored, about how communities respond to major, almost calamitous but contrasting environmental changes, namely rapid and widespread climatic amelioration and deterioration within only 10^3 years. It is an unparalleled example of 'coaxing history to conduct experiments' (Deevey, 1969). Processes of population expansion, contraction and collapse, local and regional immigration and extinction, and community compositional changes following invasion, species replacement and adjustment, and population decline may all have contributed to the observed patterns of biotic change at the landscape scale through addition, expansion, decline or loss of taxa. It may be possible, with biostratigraphic data of sufficient quality and temporal precision and appropriate data-analytical techniques, to reconstruct timings and extents of population and community changes and short-term population and community performances and responses to conditions, on the one hand, of severe, widespread environmental deterioration leading to population decline, persistence of small populations in favourable localities, chance survival or extinction, and community collapse and disintegration and, on the other hand, of rapid, widespread environmental amelioration and extreme ecological opportunism resulting in immigration, establishment, invasion, population expansion, competition, and community assembly and complexity. The

European lateglacial protocratic phase is truly 'a complex and eventful episode of a unique kind' (Watts, 1979b).

POSTGLACIAL FOREST DEVELOPMENT — THE MESOCRATIC PHASE (4×10^3 years, 10^3–10^{10} m^2)

The last mesocratic phase extends from about 9000 to 5000 B.P. It is characterized in north-west Europe by immigration, expansion and dominance of temperate deciduous trees and by development of fertile brown-earth forest soils with mull humus and dense mixed-deciduous forests dominated by *Corylus, Quercus, Ulmus* and *Tilia*. This widespread and distinctive vegetational formation did not develop as a single integrated unit but by progressive immigration and expansion of individual tree taxa, as demonstrated by the first continuous appearance and subsequent increase in pollen values of different taxa at different times. Competition for space and resources may have led to the development, over 2–4 $\times$ 10^3 years, of broad-scale competitive balances between taxa.

Corylus avellana is the first mesocratic shade-tolerant tree (Table 1.1) to immigrate and attain dominance. At about 9500–9000 B.P. it expanded extremely rapidly, often within 200–500 years, because there was no effective competition from other trees. In southern England its rate of population increase is high ($r = 1.5$–2.0×10^{-2}, doubling time = 40 years; Bennett, 1983b). Comparably rapid rates occur throughout much of north-west Europe, suggesting an absence of environmental constraints on its expansion (Bennett, 1983a). Comparisons between its low doubling time, age of first reproduction and longevity, suggest that hazel expanded as rapidly as possible and its expansion involved several generations (Watts, 1973; Bennett 1983b). It presumably flourished under the light shade of existing pine–birch–aspen woodlands and on fertile soils of the late protocratic. It rapidly became dominant, probably forming a low but dense canopy below which the shade-intolerant aspen, birch and pine were unable to regenerate (Iversen, 1960). The short-lived *Populus* and *Betula* rapidly declined, whereas *Pinus* persisted longer. In addition *Pinus* attained dominance on infertile sands and gravels unsuitable for *Corylus*. *Betula, Pinus* and *Salix* may have remained locally abundant on peats and other wet soils.

Ulmus soon followed and expanded rapidly within about 300–500 years ($r = 1.0 \times 10^{-2}$, doubling time = 70 years; Bennett, 1983b). Elm presumably invaded areas of fertile soils occupied by hazel. Elm's expansion, like hazel, appears to have occurred as rapidly as possible and involved several generations (Bennett, 1983b). *Quercus*, which is less shade-tolerant than elm or hazel, arrived at about 8500 B.P. in southern England and expanded more slowly over a 1000-year period ($r = 0.5 \times 10^{-2}$, doubling time = 140 years; Bennett, 1983b), possibly because of dense shade and interference from existing hazel and elm stands. Oak may have replaced pine on acid, sandy soils (*Q. petraea*) and, to a lesser extent, hazel on more fertile sites (*Q. robur*). *Q. robur* may also have been important in damp peaty areas (Iversen, 1960). It is likely, however, that the forests were largely dominated by *Corylus* with lesser amounts of *Ulmus* and *Quercus* until *Tilia* immigrated and expanded between 8000 and 6000 B.P. (Iversen, 1973; Andersen, 1978; Bennett, 1983a).

The expansion of *Tilia cordata* occurred within 300–800 years. Its rate of population increase was, like *Quercus*, low ($r = 0.7 \times 10^{-2}$, doubling time = 100 years; Bennett, 1983b), probably because of interference from existing vegetation and possibly also because of inherently low reproductive and population-expansion rates. This may explain *Tilia*'s slow migration rate compared with other trees (Huntley and Birks, 1983). *Tilia*, with its dense shade, great longevity and wide edaphic tolerances, probably replaced *Corylus* on a range of fertile soils and grew with *Ulmus* on richer sites and *Quercus* on poorer areas. Oak possibly dominated the poorest (*Q. petraea*) and wettest (*Q. robur*) soils, whereas elm probably favoured the most fertile, damp sites. Oak, elm and hazel may have occupied a wider range of edaphic conditions before *Tilia*'s expansion because of their broader ecological amplitudes in the absence of interference from *Tilia*. *Corylus* probably occupied a largely subordinate role as a shrub in the lime–elm–oak forests and in windthrow gaps and along forest edges with light-demanding shrubs (e.g. *Acer campestre*, *Cornus sanguinea*, *Viburnum opulus*, *V. lantana*, *Rhamnus cathartica*, *Sorbus aria*, *Crataegus* spp.).

Alnus glutinosa expanded between 8500 and 7000 B.P., presumably into damp or waterlogged sites previously occupied by *Salix*, *Betula* and *Pinus*. At some sites its expansion is extremely rapid, occurring within 200–400 years; at other sites it is slow, occurring over 500–800 years ($r = 0.4 \times 10^{-2}$, doubling time = 175 years; Bennett, 1983b), after a long period (800–1000 years) of consistently low values. Its main expansion may have been initiated by waterlogging and rising groundwater tables, possibly resulting from climatic change.

Fraxinus excelsior was the last mesocratic tree to expand, usually between 6000 and 5000 B.P. It probably favoured damp fertile sites along streams and by lakes, forming a transitional zone with *Q. robur* and *Ulmus* between alder stands in waterlogged areas and *Tilia* forests on well-drained soils. *Fraxinus* may also have replaced hazel in certain sites.

By 5000 B.P. much of north-west Europe was covered by dense, mixed-deciduous forest or 'wildwood' (*sensu* Rackham, 1980) growing on fertile, brown-earth soils. Regional-scale variations existed (Huntley and Birks, 1983), resulting from differences in tree composition (e.g. *Tilia*'s absence from Scotland and Ireland) and abundance (e.g. declining abundances towards range limits), competitive balances, regional environment (e.g. geology, soils, climate, altitude), and chance factors influencing dispersal, establishment and invasion (Palmgren, 1929). Local patterns of variation and forest mosaics also occurred at finer scales in response to altitude (e.g. Turner and Hodgson, 1979), soil (e.g. Watts, 1961), particularly moisture and nutrients, and natural disturbances such as storm damage, disease and possibly fire (e.g. Jones, 1945; White, 1979). Shade was maximal, and only shade-tolerant taxa grew within the forest. Many light-demanding shrubs may have persisted as 'fugitives' in transient open habitats such as gaps and forest edges. Shade-into-lerant herbs and animals of the protocratic became locally extinct as a result of habitat loss, but many persisted as small disjunct populations in treeless situations such as coastal cliffs, shingle and dunes, inland cliffs and screes, river gorges, steep limestone slopes and shallow soils, lake shores, calcareous fens and flushes, and mountains (Pigott and Walters, 1954). The fauna was very different from the protocratic, with elk, aurochs, red deer, roe deer, wild boar, bear, wolf, fox, wild cat, pine marten, beaver, polecat and lynx.

Many of the observed first appearances and expansions of major tree pollen types have been used to delimit pollen zones. Immigration times over large areas exceed the inherent errors in radiocarbon dating. Tree arrivals are demonstrably time transgressive over large areas, indicating that tree migration and expansion are not instantaneous processes. Pollen-zone boundaries thus only have chronostratigraphic validity over comparatively small areas.

The underlying causes for the observed temporal and spatial patterns of tree arrival and expansion are currently a topic of controversy within Quaternary palaeoecology. One hypothesis, favoured by von Post (1946) and Godwin (1966, 1975) and developed by Webb (1980) is his use of mathematical transfer-functions for reconstructing past climate from pollen-stratigraphic data (Chapter 40), proposes that regional climatic change is the primary determinant of many of the main pollen-stratigraphic changes in the early- and mid-Holocene, and thus that plant distributions, abundances and vegetation are in equilibrium with climate over the spatial and temporal scales of interest (Howe and Webb, 1983; Prentice, 1983). To provide support for this hypothesis, von Post (1926) and, more recently, Webb (1980) and Howe and Webb (1983), propose that spatial patterns of pollen-stratigraphic changes should be mapped. As broad-scale geographic patterns of modern pollen are similar to contemporary vegetation patterns, for example at the scale of vegetation formations (Chapter 39), and as broad-scale vegetation patterns often resemble broad-scale climatic patterns, geographically coherent and consistent patterns of pollen-stratigraphic change in one or more taxa over large areas are, according to Webb (1980), likely to result from regional climatic change — 'broad-scale consistency render improbable all but a climatic interpretation' (Howe and Webb, 1983). Many isopollen or isochrone maps display broad-scale geographically coherent patterns that appear to fulfil the criteria of Webb's (1980) test, but geographic consistency is not necessarily proof of direct climatic control (see, for example, maps by Moe, 1970; Huntley and Birks, 1983; Ralska-Jasiewiczowa, 1983; Birks, 1985).

An alternative hypothesis, proposed by Faegri (1940) and developed by Iversen (1960, 1973), suggests that observed pollen-stratigraphic patterns in the mesocratic reflect progressive migration of trees from glacial refugia under conditions of favourable climate and subsequent expansion (Huntley and Birks, 1983). These migrations were initiated by large and rapid, possibly step-like, climatic changes (Bryson and Wendland, 1967) at the onset of the Holocene. Each tree taxon may have then spread in different directions over thousands of kilometres from different glacial refugia at rates of 100–1000 m/year (Firbas, 1949; Huntley and Birks, 1983), until it reached its climatic limits and attained equilibrium with climate. Some taxa reached these limits early in the mesocratic, whereas others may not have reached them even after 10^4 years since deglaciation. Some distributions in climatic equilibrium today may not have been in equilibrium in the past. Differences in, for example, location of refugia, migration routes, inherent rates of spread, threshold responses, and ecological and competitive tolerances, and the presence of barriers to migration such as mountain ranges, existing vegetation, unsuitable soils and unfavourable climate, may account for the diverse patterns observed in time and space.

Patterns that record the unidirectional arrival and expansion into an area of a taxon that was limited in its spread by inherently slow migration rates or by non-climatic barriers cannot be assumed to be responses to regional climatic changes that were synchronous, or nearly so, with the observed palynological changes (Faegri, 1963, 1974; Watts, 1973, 1982). However, once a taxon has reached equilibrium with climate, distributional changes, either advance or retreat, or abundance changes may then be a direct but slightly lagged response to climatic change (Faegri, 1950, 1963; Hyvärinen, 1975; Birks, 1981; Eronen and Hyvärinen, 1982).

In north-west Europe mesocratic trees did not arrive in an order predictable from their climatic tolerances, as judged by modern distributions, or in the order in which they appeared in previous interglacials (Firbas, 1949; Iversen, 1960). For example, *Hedera* and *Viscum* immigrated into Denmark before *Alnus* and *Tilia*, even though the latter two grow further north today and are less sensitive to low temperatures, as shown by detailed autecological studies (Iversen, 1944; McVean, 1955; Pigott and Huntley, 1981; Pigott, 1981). In terms of any *simple* model of climatic control influencing arrival order, one would expect, *a priori*, *Alnus* and *Tilia* to arrive before *Hedera* and *Viscum*. Iversen (1960, 1973) argues that the sequence of arrivals cannot be a direct reflection of climatic change and proposes that they result from differential migration rates and lags over at least 10^3 years. Howe and Webb (1983) propose, however, that there is 'no unequivocal evidence for a thousand year or longer lags within Holocene vegetational change'. Walker (1978) discusses the likely importance of climatic 'syndromes', namely complexes of interacting variables that may be compensatory under different boundary conditions, in influencing and controlling plant distribution, growth, abundance, and reproduction. (See Hintikka (1963) and Neilson and Wullstein (1983) for examples of partial plant-climate syndromes.) Unfortunately we know little about present-day climatic syndromes and their relevance to environmental reconstructions.

Early-Holocene climate was different from today's — for example in seasonality, particularly insolation (c 7% greater than today), and in unique combinations and extremes of temperature, precipitation and insolation, as well as circulation patterns and temperature and precipitation means (Kutzbach, 1981; Kutzbach and Otto-Bliesner, 1982). Such differences may have resulted from variations in the solar constant or, more likely, from changes in solar radiation caused by changes in orbital parameters (Kukla, 1969). The problem is how to identify biotic responses to past seasonality differences and how to distinguish them from responses due to differences in means and from no-analogue assemblages resulting from non-climatic migrational lags using available palaeoecological data. Walker (1978) discusses this and concludes that it is 'impossible to distinguish between them' using current interpretative approaches. He emphasizes that 'we badly need to identify, in commonly available fossil materials, features which are the products of particular seasons and which have characteristics measuring aspects of the physical environment of those seasons, however complex they may be'. Although it is not clear how the hypothesis of differences in climatic seasonality can be tested or the associated biotic responses distinguished and quantified, Howe and Webb (1983) conclude that 'such changes can easily account for most of the changes in the composition of vegetation during the Holocene'.

Our knowledge of modern plant–pollen–climate relationships and of Holocene past climates is imperfect. It is thus important to distinguish clearly between observed patterns, inferred processes and assumed environmental changes, so as to avoid confusion between what is known, what is assumed, and what is testable. Walker (1978), in discussing environmental reconstructions, concludes: 'One of the great difficulties in comparing the reconstructions of different authors lies in discovering where analysis ends and intuition begins; both have their uses but it is important to know which is which'. The developing paradigm of seasonality differences to account for observed patterns of change in Holocene vegetational history highlights the importance of Walker's conclusion.

At present, falsification of these hypotheses concerning underlying processes for observed patterns in tree migration is an important problem in terrestrial Quaternary palaeoecology. It is important to avoid interpretative errors that can arise 'by confusing migration limits with climatic ones' (Faegri,

1950), and at the same time to avoid any circularity of using palynological data as the main basis for climatic reconstruction and then using this reconstruction to interpret observed pollen-stratigraphic patterns (Colinvaux, 1983). It is necessary to consider the ecology and climatic tolerances of each taxon separately to try to disentangle the likely importance and interactions of different climatic parameters on present and past distributions. It is also important to exploit and synthesize as much independent evidence for climatic change as possible; for example stable-isotope ratios (see Chapter 20; e.g. Lang, 1970; Lerman, 1974; Brenninkmeijer *et al.*, 1982), peat stratigraphy (Chapter 6; e.g. Aaby, 1976), lake-level data (Chapter 5; e.g. Digerfeldt, 1972; Street and Grove, 1979; Berglund, 1983), altitudinal and latitudinal fluctuations in tree-line (e.g. Markgraf, 1973; Karlén, 1976; Beug, 1982; Eronen and Hyvärinen, 1982), sedimentary and biotic changes in lakes (e.g. Wright, 1966; Karlén, 1976), montane-glacier changes (e.g. Denton and Karlén, 1973; Grove, 1979), and changes in other biota such as mollusca (Chapter 36; e.g. Kerney, 1968; Burleigh and Kerney, 1982), beetles (Chapter 34; e.g. Osborne, 1976), and fish (e.g. Casteel *et al.*, 1977). When this is attempted, it may then be possible to evaluate critically these competing hypotheses, and to test hypotheses about the role of seasonality changes and of combinations of climatic variables in influencing observed biostratigraphic patterns.

Mesocratic vegetation is, in all probability, a result, in part, of differential migration rates and lags of taxa behaving individualistically in their spread from different glacial refugia and, in part, of regional climatic change and differential plant responses and lags to these changes. Both vegetation and climate are dynamic and continuously change and interact, to differing degrees, over various spatial and temporal scales. The underlying processes for the observed pollen-stratigraphic patterns are thus likely to be extremely complex and almost certainly scale-dependent. In addition, soil-developmental processes, which themselves may operate over long time-spans, may be important at certain scales for particular taxa (Watts, 1973; Faegri, 1974). Many forest trees have mycorrhiza that may require particular soil conditions. Unfortunately little is known about the possible importance of soil maturation on mesocratic forest development (Faegri, 1963, 1974). Clearly climatic change, migrational lags and soil development may interact and influence observed patterns of change. We need to test these hypotheses as rigorously as possible in order to understand the underlying causal processes for observed patterns of vegetational change in the mesocratic.

We are similarly ignorant of the mechanisms that enable mesocratic trees to migrate at rates of 500 m/year or more (Pigott and Huntley, 1980; Davis, 1981a; Walker, 1982). Factors such as morphology and numbers of reproductive disseminules produced, frequency of reproduction, mechanisms of propagule dispersal and establishment, interference from existing vegetation, and age of first reproduction may all be important. However, many of these factors are poorly understood and are difficult, if not impossible, to study in the presently highly disturbed landscape of north-west Europe. Observed migration rates (500–1000 m/year) require propagule-dispersal distances of up to 10 km for taxa maturing in 20 years to be maintained regularly for long periods. Such dispersal ranges are possible for light, wind-dispersed (e.g. *Ulmus*) or water-dispersed (e.g. *Alnus*) propagules, but are difficult to envisage for heavy propagules such as *Quercus* acorns. Animal dispersal of propagules is probably of great importance in the spread of mesocratic forest trees (Skellam, 1951). Availability of gaps in existing vegetation and ability of immigrating taxa to colonize, grow and reproduce under the environmental conditions within gaps may also be important in influencing migration rates. Interference from existing vegetation and gap production may, in some instances, be more important in influencing observed migration rates than, for example, inherent dispersal distances of propagules and tree fecundity, reproductive age, and longevity (Green, 1982; Walker, 1982).

The ecological mechanisms by which new tree taxa invade and expand into existing forest vegetation are also largely unknown (Watts, 1973; Walker, 1982). Isochrone maps (see Figure 1.4; Moe, 1970; Davis, 1981a; Birks, 1985) give the impression that trees spread and advance as a continuously migrating front across the landscape and invade all possible sites. This hypothesis seems unlikely for many, if not all, mesocratic trees (Watts, 1973). An alternative hypothesis (Watts, 1973, and Figure 1.5; see also Faegri, 1949; Godwin, 1966; Walker, 1982) proposes that taxa migrate by long-distance chance dispersal of propagules into locally favourable openings caused by windthrow, death, fire or disease beyond the range of the taxon to form small outliers. These mature and in turn act as seed parents for local expansion of these outliers and further establishment in new gaps. In contrast to the initial, slow phase of seedling establishment with low population densities, the expansion phase may occur rapidly because local propagule deposition within the forest increases, probability of establishment rises, and population densities grow. Intraspecific competition may remain low at this stage. Over several generations, these populations expand and coalesce with the main population. Population growth then flattens off or even decreases owing to interspecific competition, density-dependent self-thinning and intraspecific competition, and the inability of many forest trees to regenerate under themselves. Harper (1967) proposes that 'the essential criterion for a balanced "co-occurrence" of two species is that the minority component should always be favoured'. Watts (1973) thus hypothesizes that during invasion, populations stabilize after a period when the less common invader has expanded vigorously at the expense of the commoner resident. Any invading seedlings would initially have a greater probability of establishment and achieving dominance in a gap than seedlings of the residents. As invasion progresses and the invader's population increases, intraspecific competition becomes more important, and eventually a relative

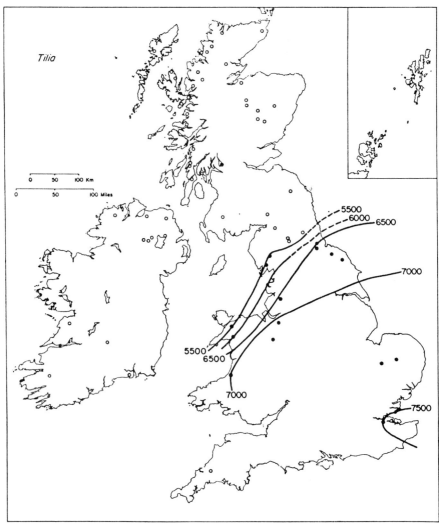

FIGURE 1.4.   Isochrone map for *Tilia* in the British Isles. The contours or 'isochrones' show the range limits of the genus at 500-year intervals from 7500 to 5500 B.P. The isochrones are based on pollen-stratigraphic data from sites indicated by solid circles. Sites where there is no pollen-analytical evidence for local presence of *Tilia* are shown as open circles (from Birks, 1985)

balance between seedling establishment of resident and invading taxa may occur, thereby enabling coexistence in a quasi-equilibrium until the next invasion (Forcier, 1975; Fox, 1977). By these processes, forest communities are assembled over long periods and many generations. Watts (1973) suggests that

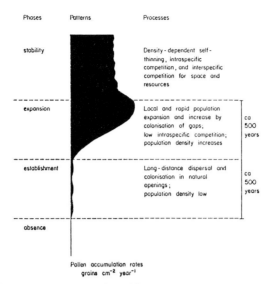

Phases   Patterns   Processes

stability   Density - dependent self -
                   thinning, introspecific
                   competition, and interspecific
                   competition for space and
                   resources

expansion   Local and rapid population
                   expansion and increase by   ca
                   colonisation of gaps;   500
                   low introspecific competition;   years
                   population density increases

establishment   Long - distance dispersal and
                   colonisation in natural   ca
                   openings;   500
                   population density low   years

absence

Pollen accumulation rates
grains cm^{-2} year^{-1}

FIGURE 1.5.   Observed pollen-stratigraphic patterns and inferred processes suggested by Watts (1973) in the establishment, invasion and expansion of a tree taxon in the mesocratic phase. The approximate durations of the establishment and expansion phases are also shown

major determinants of rates of population expansion are the time required for propagule dispersal and colonization of gaps in the existing forest and for seedlings to mature into seed parents, and interference in seedling establishment, survival and maturation. Green (1982) shows that invasion and expansion into existing forests in Nova Scotia coincide with rare, extensive natural disturbances, suggesting that availability of large openings may be an important determinant of invasion and expansion rates. Gaps are probably essential for invasion, as existing vegetation provides 'inertia' to vegetational change (Pearsall, 1959; Smith, 1965). Almost all mesocratic trees require openings for establishment and regeneration, possibly because of nutrient release, increased irradiance, relaxed root competition, or reduced host-specific insect predation. For example, in Białowieza Forest, Pigott (1975) shows that seedlings and saplings in gaps are often monospecific and are rarely the same as the trees around the gap. Mono-specific dominance can arise by chance combination of circumstances including, for example, fruit production and hence propagule availability in a particular year coinciding with gap formation, uneven dispersal of propagules, soil disturbance by animals, selective predation of propagules and seedlings, and interference between seedlings and field layer (Jones, 1945; Pigott, 1975).

Invasion and expansion can occur without climatic change (Watts, 1973). They may, however, be favoured by climatic change. For example, extensive gaps providing conditions of low competition for invaders may be commoner in vegetation following climatic change across a critical threshold that restricts, for example, seed production and regeneration of the residents (Walker, 1982). Storms, pathogen outbreaks, fire or other catastrophes may create more frequent and more extensive gaps in forests affected by, for example, prolonged drought, senescence, nutrient shortage, or parasites than in vegetation growing under optimal conditions. Rates of taxon replacement will clearly vary from taxon to taxon according to longevity, gregariousness, competition and environmental conditions, and can readily extend over many generations (Walker, 1982; cf. Woillard, 1979).

Interspecific competition was possibly less intense and intraspecific competition more intense in the early mesocratic than later. Trees that had recently immigrated may initially have had a uniform age-structure and fine-scale patchy distributions. As populations increased, these taxa would progressively colonize more gaps and their distributions would, over thousands of years, coalesce owing to local growth, clonal development and grouped regeneration, and develop a varied age-structure. Observed patterns of change may become more gradual as a result of tree replacement being largely by fine-scale gap-phase replacement (Jones, 1945; Williamson, 1975) rather than by invasion and broad-scale replacement within extensive openings (Green, 1982). Possibly by 5000 B.P. the mesocratic forests of north-west Europe were saturated, had a complex and varied age distribution, structure and dispersion (Jones, 1945; Pigott, 1975), and after 4×10^3 years were in broad-scale equilibrium with climate. Little is known, however, about fine-scale mainten-ance dynamics within these forests (Iversen, 1973) and the mechanisms by which several tree taxa could coexist. Local gap-phase replacement associated with senescence, wind-throw, and regeneration by monospecific groups of seedlings, often not the same species as the nearest canopy trees (Pigott, 1975), may have predominated, resulting in complex mosaics of monospecific groups of similar age, such as occur in surviving primeval forests of south-east Europe (Jones, 1945). Interestingly, Andersen (1975) demonstrates local gap-phase replacement every 350–400 years during 3×10^3 years of the mesocratic phase in the last interglacial with short-lived episodes of *Corylus*, *Acer* or *Fraxinus* alternating with long periods of *Tilia* dominance. These patterns presumably result from wind-throw, gap production, and regeneration followed by eventual replacement by *Tilia*. It is likely that similar processes were important within the mesocratic forests of the Holocene.

The climate in north-west Europe was probably warmer by about 2–3°C in summer and 1–2°C in winter during the mesocratic phase, at about 8000–5000 B.P., the so-called 'climatic optimum' (Iversen, 1944). This is evidenced by, for example, finds of *Hedera*, *Ilex* and *Viscum* pollen, by the occurrence of *Corylus*, *Cladium mariscus*, *Trapa natans*, *Emys orbicularis* and other thermophilous taxa beyond their present northern limits, and by the extension of plants such as *Lycopus europaeus* and *Phragmites communis* above their present altitudinal limits. It is not known if processes of forest development in the mesocratic were affected by this warming. Moreover, the magnitude, timing, climatic nature, geographic extent and spatial patterns of the 'climatic optimum' (e.g. means, combinations of variables, seasonality) are poorly known (Wright, 1976a; Tallantire, 1981).

Lakes during the mesocratic were generally base-rich and productive. Productivity is a function of nutrient input through precipitation and drainage from the catchment, interacting with buffering capacity and morphometry of the lake. Factors such as geology, climate, rates of weathering, run-off, flushing, and leaching, composition, productivity, and age-structure of vegetation, humus type and thickness, rainfall interception by trees, and nitrification can all interact and influence nutrient supply from a lake's catchment. For example, rates of phosphate release from soils will be influenced by soil pH which is, in turn, related to humus type, vegetation composition and geology. Although, in theory, nutrient input into lakes within a forested landscape is largely controlled by forest composition and dynamics (Gorham *et al.*, 1979) there are comparatively few examples of changes in lake productivity related to changes in forest composition in the mesocratic (Whitehead *et al.*, 1973; Binford *et al.*, 1983). Typically, mesocratic soils suffered progressive unidirectional and permanent loss of nutrients by leaching. Lakes similarly show correspondingly progressive decreases in nutrient status and productivity from productive eutrophic conditions in the protocratic to a less productive, mesotrophic state in the mesocratic as a result of meiotrophication (e.g. Digerfeldt, 1972). There are, however, many problems in reconstructing lake productivity from geochemical and biostratigraphical data (Livingstone, 1957; Likens and Davis, 1975; Binford *et al.*, 1983). Annual 'influx' measurements are essential but reliable estimates are difficult to obtain because of sediment redeposition and focussing within lakes (Lehman, 1975). During the mesocratic, many lakes accumulated organic gyttja ($Ld^{0-1}4$) that is particularly susceptible to resuspension and focussing (Davis and Ford, 1982). Aquatic plants were mainly floating-leaved nymphaeids in contrast to the submerged macrophytes of the protocratic (H. H. Birks, 1980). Fringing fen and swamp vegetation commonly developed, and hydroseral succession proceeded at different rates and along different routes at different sites (Walker, 1970), depending on local conditions.

Mesocratic phases can be distinguished in all previous interglacials in temperate latitudes by their broadly similar deciduous thermophilous tree flora. However, each mesocratic phase has a different biotic assemblage. Some taxa are absent from certain interglacials (e.g. *Tilia*, *Pterocarya*), whereas others are abundant in some and rare in others. Taxa immigrate in different orders each time, and most interglacials have assemblages which are markedly different qualitatively and quantitatively from any known today (e.g. the southern *Acer monspesulanum* grew with north-west European trees during the last interglacial in England). The reasons for these differences can be environmental, chiefly climatic, and biotic, chiefly due to factors affecting migration. In each cryocratic phase, each tree taxon would have been confined to one or a few suitable refuges, perhaps different each time. Migration would have subsequently occurred over different distances and by different routes and across different barriers. Rates of spread would have been influenced by pre-existing vegetation, presenting different amounts of interference to establishment. Genotypic evolution may also have altered competitive relationships and rates of invasion between interglacials. Chance would have affected survival, dispersal and establishment. It is now clear that the deciduous-forest formation of north-west Europe has no long Quaternary history. It separated into its components in each cryocratic phase, and redeveloped slightly differently in each mesocratic phase, within broad climatic and edaphic limits (Watts, 1973).

Mesocratic forests of different interglacials may be long-term, spatially widespread examples of 'multiple stable points' (*sensu* Sutherland, 1974), namely assemblages formed from similar biota but whose contrasting community structure is explicable in terms of specific historical or chance events that determine the presence or absence, order of arrival, expansion, and abundance of major taxa. Interest centres on why particular combinations and abundances of taxa ('stable points') developed, and on the generation of *testable* hypotheses concerning different mesocratic patterns and processes. Chance may play a larger role at these spatial and temporal scales than is often thought (Palmgren, 1929; Simberloff, 1978; cf. West, 1980).

Figure 1.3 illustrates the ecological characteristics of the mesocratic. Populations show strongly individualistic behaviour, expanding and contracting independently of populations of other taxa (Walker and Pittelkow, 1981). Diversity is lower, and rates of floristic and vegetational change slower than in the protocratic. Soils are fertile brown-earths under increasing and maximal shade. There is overall vegetational instability resulting from phases of short-lived ($5-10 \times 10^2$ years) vegetational change resulting from successive immigrations, invasions and expansions of new taxa and from continuously changing biotic composition and population sizes. These alternate with longer ($10-20 \times 10^2$ years) phases of little palynological change and hence of apparent vegetational stability (*sensu* Watts, 1973), at least at broad spatial and temporal scales, in which pollen values generally show small fluctuations around a mean with no consistent trend. However, Walker and Pittelkow (1981) and Green (1982) show that when individual pollen curves are analysed statistically, changes in population abundances occur continuously and not only at times of immigration and expansion. Changes in one taxon are largely independent temporally of changes in other taxa, in contrast to the linked behaviour of populations in the protocratic (Birks and Gordon, 1985).

The main patterns within the mesocratic primarily result from interactions between processes of immigration, invasion and expansion and of competition within and between resident and invading taxa within the limits of the prevailing climate. Rates and timings of these processes are probably related to contrasting shade tolerances of different taxa, their reproductive behaviour and dispersal mechanisms, the potential of taxa to colonize and exploit available gaps, and competitive abilities and ecological requirements for space, water and nutrients. Vegetational change, as detectable at the broad scale (c 10^9 m^2) of pollen-stratigraphic data, tends to become rarer and more gradual after the major trees have immigrated and expanded. Competitive balances among species may have developed, possibly involving complex plant–animal –environment interactions and even biotic feedback. Coevolution among taxa and 'community evolutionary history' seem unlikely to be important, however, over time scales of 2–3 $\times$ 10^3 years (McIntosh, 1981; cf. Claridge and Wilson, 1978; Boucot, 1983).

The testing of competing hypotheses concerning underlying processes that determine observed biotic patterns in the mesocratic remains an outstanding problem in late-Quaternary palaeoecology. As Prentice (1983) suggests: 'Climatology and ecology would both benefit from some way of disentangling the effects of climate and migrational lag on Holocene tree species distributions.'

POSTGLACIAL FOREST RETROGRESSION — THE OLIGOCRATIC PHASE (5 $\times$ 10^3 years, 10^3–10^{10} m^2)

At about 5000 B.P. vegetational patterns within north-west Europe were disrupted, leading to the initiation of retrogressive vegetational change within the deciduous-forest formation and development of an incompletely forested landscape and many non-forested vegetation types such as heaths, blanket mires and grasslands (Huntley and Birks, 1983). The onset of these changes coincides with the arrival of Neolithic people. The role of *Homo sapiens* in influencing vegetational patterns over the last 5000 years is discussed in the next section. In this section I consider vegetational patterns that probably result directly or indirectly from natural edaphic changes.

Progressive leaching of nutrients and associated soil deterioration are, under the cool, moist climate of north-west Europe, inevitable pedogenic processes independent of climatic change and man (Iversen, 1958; Andersen, 1964, 1966, 1969). Forest-soil deterioration occurs through several stages (Stockmarr, 1975; Andersen, 1979; Aaby, 1983). In the mesocratic, brown-earth soils with mull humus predominate. Litter is rapidly broken down by abundant soil flora and fauna, particularly earthworms. This releases nutrients and produces lumbricid-humus that mixes and combines with clay to give the characteristic crumb structure, aeration and drainage, and undifferentiated soil-profile of brown earths. Soil fertility is maintained by mixing activities of the fauna and by the ability of many deciduous trees to extract nutrients from deep soil layers, thereby replacing nutrients continually being lost by leaching (Andersen, 1966). As organic decomposition proceeds over long time periods, the weak acids formed from CO_2 in aqueous solution, together with humic acids, dissolve surface carbonate. This is readily leached and the upper layers become less base-saturated. Earthworms are eventually replaced by arthropods. With further acidification, other soluble minerals are leached from upper layers, the transitional arthropod-humus becomes more oligotrophic as its base-status declines, and the subhumus zone is mineralized and eventually leached, forming a podsoloid soil-profile. As these processes continue, the soil fauna declines, there is incomplete breakdown of litter, nutrients cease to be recycled, and acid mor humus (A_0) develops from litter rich in polyphenols and resistant to decomposition. Mor development and acid, percolating water cause podsolization. Iron and aluminium are dissolved and reprecipitated as an impermeable iron pan (B_2) overlain by humus pan (B_1), bleached sands (A_2), and humic sands (A_1), characteristic of a typical podsol profile.

This soil-retrogression model proposes that deciduous forests and associated soils in the mesocratic cannot maintain themselves permanently, because nutrient loss and soil degradation and acidification are inevitable in the moist climate of north-west Europe. Evidence from previous interglacials where human interference is absent demonstrates that 'forest vegetation in a humid climate may build up to maximum denseness and then retrogress into poorer vegetation types mainly or solely due to its own influence on the substrate' (Andersen, 1966). Such processes presumably also occurred in the Holocene, especially on well-drained soils where soil retrogression could proceed more rapidly and extensively than in low-lying areas with gley mull soils where seepage and flushing by groundwater would counteract acidification and mor formation.

The effect of these pedogenic processes in influencing vegetational patterns since 5000 B.P. can be discerned at two spatial scales. At the local scale of within-forest vegetational change, the progressive decline of dense, closed forests of *Tilia*, *Ulmus* and *Corylus* and expansion of more open forests of *Quercus*, *Ilex* and *Betula* on well-drained soils from about 6000 B.P. (e.g. Iversen, 1964b, 1969; Andersen, 1978) appear to result from natural soil retrogression independent of any human disturbance. Andersen (1979) and Aaby (1983) demonstrate complex but poorly understood interactions between rates and stages of soil change, litter and humus type, forest composition and structure, grazing, and human activity. Rates, directions and extent of soil degradation can be influenced at this scale by human interference through clearance, management or grazing, depending on site conditions (e.g. moisture, topography), critical thresholds, and tree and litter composition. Other changes within deciduous forests about 5000 B.P. that occurred as a result of natural soil retrogression include local extinction of plants of fertile well-drained and moist soils, replacement of *Alnus* and *Fraxinus* by *Betula* and *Calluna* on peaty soils, spread of shade-intolerant calcifuges, and accentuation of local vegetational differentiation by enhanced edaphic contrasts between low-lying, fertile moist sites and well-drained infertile areas (Iversen, 1964b, 1969).

At a broader, regional scale, accumulation of mor humus and formation of humus-iron pans in upland areas inhibited vertical drainage of soil water, decreased soil aeration, and resulted in increased surface run-off, waterlogged depressions, high groundwater levels, and widespread paludification (Pearsall, 1964; Iversen, 1964b; Smith and Taylor, 1969). In the British Isles, the end result was the extensive blanket mires characteristic of upland areas today. They occur widely on flat or gently sloping areas above 300 m, descending to sea-level in extreme western Scotland and Ireland. Some blanket mires developed as a result of natural soil degradation in the absence of human disturbance or climatic change (e.g. Tallis, 1964; H. H. Birks, 1975). In other areas climatic change at, for example, 4000 B.P. or 2500 B.P. may have caused hydrologic changes that initiated blanket-mire formation (e.g. Pennington, 1965) or expansion (e.g. Pennington *et al.*, 1972). Prehistoric people, through clearance of upland forests or forest thinning by browsing and grazing of domestic animals, may have altered local hydrological balances and accelerated rates of edaphic and vegetational change (Moore, 1975; Kaland in Berglund, 1983; Chapter 2). In some areas it is likely that 'the soil changes associated with peat initiation are most satisfactorily explained as a consequence rather than as a cause of the demise of forest and the development of peat' (Moore, 1975). Whatever the cause of blanket-mire initiation at a particular site, it led to the altitudinal descent of tree lines independently of any regional climatic change. Topography, altitude, climate, soil type, geology, intensity and type of human disturbance, and hydrologic and edaphic thresholds, all of which vary temporarily and spatially, may have influenced timings, rates and the extent of blanket-mire development. For example, within the British Isles blanket-mire initiation generally began earlier in Scotland than in England or Wales. Within an area, peat formation may have begun at any time between about 7500 and 2000 B.P., depending on local topography and altitude (Tallis, 1964; Chambers, 1981). Peat generally developed earlier on high-altitude plateaux than in valley bottoms and on gentler slopes at lower altitudes. Peat accumulation on exposed sites at a range of altitudes started even later (Tallis, 1964). Although Moore (1975) suggests that local and regional differences in ages of blanket-peat initiation may reflect differences in intensity of human disturbance, equally plausible hypotheses relating to climatic patterns and rates of edaphic change at different spatial and temporal scales can account for observed patterns in peat inception. There is a need for detailed investigations designed to test competing hypotheses about blanket-mire initiation at various spatial scales (Edwards and Hirons, 1982). Consideration of scale is particularly important because of contrasting patterns within and between areas.

Lakes with catchments containing blanket-mires frequently show a change from gyttja to dy (Ld^{4}4), reflecting a change from mesotrophic to oligotrophic or dystrophic conditions, and a reduction in lake biota and productivity. Hydroseral succession around many lowland lakes similarly shows acidification, with frequent convergence to *Sphagnum*-dominated mires in the late-Holocene (Walker, 1970). These lakes often show a change from mesotrophy to oligotrophy as a result of meiotrophication. Aquatic macrophytes change to those tolerant of low nutrients and intolerant of competition.

The importance of natural soil and climatic changes in influencing other vegetational patterns is unclear because of difficulties of isolating unambiguosly the effects of these changes from patterns influenced by human activity. Moreover, the end of the so-called 'climatic optimum' in north-west Europe is poorly dated and understood. It appears to have ended at different times (5000–2500 B.P.) in different parts of Fennoscandia (Berglund, 1983). The northward expansion of thermophilous deciduous trees beyond their present-day limits in Fennoscandia and their subsequent apparently asynchronous retreat southwards may result from regional climatic change, local or regional soil changes such as leaching, accumulation of acid humus from conifer litter, and podsolization, or forest clearance (Huttunen, 1980).

The westward, northward and southward migrations of *Picea abies* through Finland, Sweden and Norway from 6000 B.P. to the present (Moe, 1970; Huntley and Birks, 1983) may be a contemporaneous response to stepwise climatic change (Tallantire, 1972), or a delayed migration unrelated to any contemporaneous climatic change (Faegri, 1974; Birks, 1981). *Picea* is still migrating (Faegri, 1949). Its migration rate is higher (*c* 900 m/year) in areas of boreal pine–birch forest than in southern deciduous forests (*c* 500–600 m/year; Moe, 1970). At a local scale, spruce invasion and expansion may have been facilitated in some sites by forest disturbance, including burning and animal grazing and trampling, thereby providing gaps for colonization (e.g. Göransson, 1977; Tolonen, 1978; Huttunen, 1980). Different causal processes are probably important at different scales. Whatever its causes, the invasion of spruce into northern and central Fennoscandia over the last 6000 years resulted in important changes in forest composition and soil conditions, with widespread accumulation of mor humus, soil leaching and podsolization. These changes restricted regeneration of mesocratic deciduous trees and caused local or even regional extinction of some basiphilous taxa. The spruce invasion of Fennoscandia is an event that warrants further study and synthesis.

The general characteristics of the oligocratic phase (Figure 1.3) are low regional diversity, decreasing shade, infertile soils such as podsols and acid peats, infrequent and gradual vegetational change resulting from retrogression and late immigrations and invasions, predominantly individualistic behaviour of populations, and vegetational instability. Registration of these features varies from one region to another and from one spatial or temporal scale to another. Local and regional patterns are complementary, and both are important in elucidating underlying causative processes, operative at certain scales, that have influenced observed patterns of biotic change in the oligocratic. As Andersen (1964) concludes, 'soil development was indeed an important factor in the Quaternary'. The challenge is to evaluate its importance at particular scales in relation to other environmental changes occurring in the late-Holocene.

POSTGLACIAL FOREST DESTRUCTION — THE *HOMO SAPIENS* PHASE (5×10^3 years, 10^3–10^{10} m^2)

Since Iversen (1941) first showed the influence of prehistoric people on vegetation, the importance of *Homo sapiens* in facilitating and accelerating particular ecological processes, and hence in influencing observed patterns of biotic change within north-west Europe over the last 5000 years, has been increasingly recognized. Many data relating to human influence on vegetation are of a local nature and often linked to archaeological studies. This section does not attempt to review this wealth of information; it attempts instead to discuss the probable role of *H. sapiens* in influencing rates and directions of important causative processes and observed vegetational patterns.

Local forest-clearances or 'landnam' phases in the early Neolithic (*c* 4500 B.P.) were first detected and characterized by Iversen (1941). A 'landnam' consists of three episodes: (1) declining *Ulmus, Quercus, Tilia* and *Fraxinus* pollen values, increasing Gramineae and *Pteridium* percentages, and abundant charcoal; (2) increased *Corylus, Betula, Populus* and *Salix,* and *Plantago lanceolata,* grass, and other herb pollen values, and the presence of cereal pollen; and (3) high *Corylus* and increasing *Fraxinus* and, later, *Ulmus* and *Quercus* pollen percentages (Iversen, 1973). These distinctive patterns are interpreted as reflecting:

(1) Clearance of elm–oak–lime–ash forest by nomadic Neolithic people using flint axes and fire to create openings, so-called 'slash-and-burn' clearance. Large trees were probably killed by ring-barking.
(2) Short-term cultivation and harvesting of cereals in recently burnt, cleared areas, and free cattle, pig and goat browsing and grazing on rough pasture within the forest. Cultivation, trampling and grazing created disturbed habitats and encouraged spread of weeds and ruderals. Burnt, uncultivated areas were colonized by 'pioneer' trees such as birch, willow and aspen. Hay and elm and ash leaves were collected for winter fodder. Hazel flowering was enhanced by the marked reduction in shade, and hazel nuts were probably harvested. Hazel may have been coppiced, and areas of scrub-pasture created at the expense of high forest.

(3)   Abandonment of clearings as fertility declined, thereby allowing rapid regeneration by *Fraxinus*, a shade-tolerant tree with high fecundity, fast growth but short longevity, and *Corylus*, probably by regrowth from coppice stools. Both were later replaced by longer-lived, more shade-to-lerant *Tilia, Quercus* and, to a much lesser extent, *Ulmus*.

   The ecological validity of this model of vegetation pattern and causative process and its archaeological feasibilty in relation to available early Neolithic technology have been elegantly demonstrated by the Draved forest-clearance and crop-cultivation experiments (Iversen, 1956, 1973).

   Although the causative processes within a landnam are not in doubt, questions arise concerning the spatial and temporal scales of such phases (Turner, 1964a; Edwards, 1979). Close-interval radiocarbon dating (Pilcher *et al.*, 1971; Pilcher and Smith, 1979) indicate that Irish landnams may last 250–600 years, and thus that observed patterns result from the cumulative effects of many small, short-term slash-and-burn and cultivation episodes rather than a single episode, possibly resulting from continuous immigration of many, small nomadic groups into the pollen-source area of a site over several hundred years. Clearance episodes are very rapid (*c* 20 years) compared with cultivation episodes (150–500 years). Some of these appear to reflect predominantly arable land-use, whereas others are largely pastoral in character. It is possible that formerly cultivated areas were kept open by grazing, thereby inhibiting tree colonization and forest regeneration. Once initiated, regeneration was rapid (50–150 years). There were probably different types of land-use in different areas in the early Neolithic. Clearly the hypothesis that Danish landnams are 50–100 years duration (Iversen, 1956) requires testing. We are also ignorant of the spatial extent of landnams. Existing pollen-dispersal models for forested environments (Tauber, 1965) suggest large source areas (*c* 10^9 m^2) for lakes and bogs of 500–1000 m diameter (Figure 1.1(b)). Comparatively large clearances within the source area are required to produce the magnitude of change observed pollen stratigraphically. There is a need for simple quantitative modelling of the effects on pollen deposition of clearances of different duration, frequency, spatial extent and distribution within representative source areas (Edwards, 1979, 1982).

   Besides the 'landnam culture(s)', other early Neolithic cultures (e.g. Ertebølle) may have influenced vegetation locally before the first landnam by removing elm bark and feeding enclosed domestic animals on young elm and ash shoots produced by pollarding or ring-barking mature trees (Troels-Smith, 1960; Iversen, 1973). Elm crowns would be killed, and the more open canopy would favour oak, which would produce more acorns for pig winter-food. The flowering of elm would be largely prevented, and the observed decline of *Ulmus* pollen and parallel increase in *Quercus* pollen at about 5000 B.P. would result (Iversen, 1973). Again problems of spatial and temporal scale and ecological and archaeological feasibility arise in evaluating these processess as likely causes for the observed patterns. Alternative hypotheses for the widespread and often spectacular primary elm decline are discussed in the next section.

Continued forest clearance and agriculture, interspaced by phases of abandonment and secondary forest regeneration, occurred as the result of development and expansion of more permanent land-use practices (e.g. animal husbandry, ploughing and cultivation, woodland management) since late Neolithic, Bronze Age, Iron Age, Roman, Viking, Medieval and recent times. Forests initially became more open, and wood- and scrub-pasture and hazel coppice expanded. However, increased human interference led ultimately to widespread deforestation of much of north-west Europe and development of extensive pastures or 'commons', fields and settlements. Almost all extensive, naturally forested areas surviving today have been intensively managed by selective sylviculture over many centuries (Rackham, 1980).

Turner (1965, 1970) considers three important questions in studying human influence on vegetational patterns in time and space:

(1) What vegetational changes are represented by observed pollen-stratig- raphic patterns? Answers to this require detailed fine-resolution, close-interval samples and a basis for interpreting observed pollen assemblages in terms of past land-use practices.
(2) Where and what-sized areas were cleared? This is a question of spatial resolution and scale that can only be answered by analysing geographic arrays of contemporaneous pollen data (e.g. along transects).
(3) When did particular clearances occur? This question requires reliable and independent chronology, such as radiocarbon dating.

Three contrasting forest-clearance patterns are commonly observed (Tur- ner, 1964b, 1965): (1) small temporary clearances ('landnams' — *sensu* Iversen, 1941) of short duration (100–200 years) in which a small but unknown area (?10% of the pollen-source area) is cleared, agriculture is rudimentary, and economy is based on shifting agriculture; (2) extensive clearances with Gramineae pollen exceeding 50% total pollen, suggesting over half the source-area is cleared and the economy is based on settled agriculture; and (3) complete clearance, with very high grass and other NAP values, reflecting widespread deforestation and extensive agriculture.

Small temporary and extensive clearances appear to have been localized in their geographic distribution and extent, as indicated by observed spatial patterns in contemporaneous pollen assemblages within a single site (Turner, 1965, 1970). Particular assemblages (e.g. *Artemisia*, Cereals, Chenopo- diaceae) are interpreted on the basis of surface-samples as reflecting arable activity, whereas high Gramineae, *Plantago lanceolata* and *Pteridium* values are suggestive of pastoral economy (Turner, 1964b). Behre (1981) discusses the problems of using individual pollen types, assemblages, and arable:pastoral indices as indicators of past land-use practices and associated weed communities, particularly of practices that are extinct or no longer widely used, such as litter and leaf-gathering ('Laubheu'), mixed rotational ley and three-field farming of winter cereals, summer cereals and root crops, and fallow, plaggen soils and infield–outfield systems.

Radiocarbon correlations of observed patterns of forest clearance and inferred land-use practices at sites within and between different areas provide a basis for local and regional comparisons in time and space of land-use patterns, and for correlations with archaeological and agricultural patterns (e.g. Berglund, 1969; Turner, 1965; Godwin, 1975; Núnèz and Vuorela, 1978). There is considerable potential in using appropriate multivariate data-analytical methods (Chapters 37, 38) to quantify pollen stratigraphic patterns in terms of human influence and to reconstruct land-use patterns in time and space.

There are many, complex, but poorly understood ecological affects of *H. sapiens* on the biota of north-west Europe. Forests initially became more open by local clearances, fire, and free browsing and grazing by domestic animals. Forest gaps were larger than in the mesocratic, thereby altering the regeneration balance between trees and allowing light-demanding trees and shrubs (e.g. birch, hazel, ash) to increase locally. The range of *Fraxinus excelsior* expanded regionally following forest disturbance (Birks, 1985). Forest clearance became more widespread, possibly with selective destruction of particular trees (e.g. *Tilia* for bast fibre and leaf fodder — Andersen, 1978; Aaby, 1983; *Ulmus* for leaf fodder — Iversen, 1973) or even regional extinction (e.g. *Pinus* in Ireland — Watts, 1984). Forest habitats diminished and many woodland animals became regionally extinct (e.g. bison, wild boar, lynx, beaver, bear, wolf).

Rates of natural local soil degradation were accelerated by forest clearance or changes in forest composition following disturbance through, for example, changes in groundwater level or litter and humus type (Andersen, 1979; Aaby, 1983). Mor-humus accumulation and podsolization became locally widespread (e.g. Berglund, 1962; Pennington, 1965; Stockmarr, 1975; Andersen, 1979; Aaby, 1983). Forest clearance resulted in soil deterioration, nutrient loss and erosion, especially in areas of high relief, as indicated by changes in lake sediments and biota (e.g. Pennington, 1965, 1981; Tolonen, 1978; Binford *et al.*, 1983). Nutrient input from soil inwashing and land-use practices (e.g. hemp and flax retting, ploughing, manuring) often increased lake productivity, resulting in 'cultural eutrophication'. Many lakes that became oligotrophic by meio-trophication were eutrophic again, or even meromictic, as a result of changing land-use (e.g. Huttunen and Tolonen, 1975; Tolonen, 1978).

In some areas, forest clearance and subsequent dereliction of clearings facilitated local colonization and expansion of new immigrants such as *Fagus sylvatica*, *Picea abies* and possibly *Carpinus betulus*. It seems likely that the rapid migration of *Fagus* across north-west Europe since 4000 B.P. (Huntley and Birks, 1983) may have only been possible by the creation of abundant, large clearings within *Tilia*- or *Quercus*-dominated forests on well-drained soils. In some areas mixed beech–oak–holly forests developed (e.g. Iversen, 1969; Stockmarr, 1975; Aaby, 1983); in other areas there was a rapid change from *Tilia*- or *Quercus*-dominance to *Fagus*-dominance. This remarkable change commonly occurred after an extensive phase of human activity involving clearance and grazing followed by abandonment of cleared and cultivated areas (e.g. Andersen, 1973, 1978). The abandonment phase may have occurred as a result of local population collapse following, for example, cultural change, emigration, or over-exploitation of environmental resources (Iversen, 1973). Alternatively, the introduction, in some areas, of improved ploughing may have allowed cultivation of heavier, deeper and more fertile soils and dereliction of cleared areas on well-drained sites that were rapidly colonized by beech (Godwin, 1966). It is not known what ecological processes permitted beech to attain total dominance in some areas, whereas in other areas oak and holly regenerated to a sufficient extent to form mixed beech–oak–holly forests (Andersen, 1978, 1979).

Other types of secondary woodland developed in areas beyond the then natural geographic range of beech; for example, woods of pure ash, oak, yew, birch or holly became established on particular soil types following abandonment of cleared or cultivated areas, relaxation in grazing pressure, or reduction in fire frequency. Many of these were subsequently cleared or modified through burning, grazing, or woodland management by later cultures. Only a few stands survive today, often in areas protected in some way from fire, intense grazing or other disturbances. Scrub, often resistant to fire, browsing and grazing, also developed under particular edaphic conditions and management regimes (e.g. *Crataegus*, *Corylus*, *Cornus sanguinea-Rhamnus catharticus-Acer campestre*, *Prunus spinosa*, *Ulex europeaus*, and *Juniperus* scrub).

Heathlands developed locally as rough pasture became derelict. They expanded as human pressure increased and open forests on acid soils were cleared (e.g. Iversen, 1964b, 1969). Further soil degradation and podsolization favoured heathland development. Burning, winter grazing and cutting kept *Calluna* heaths treeless. Grasslands also developed on a range of fertile soils following extensive forest clearance and intensive grazing. On poor soils they replaced heathland in response to repeated burning, trampling and overgrazing. Many, if not all, heathlands and grasslands of lowland north-west Europe owe their origins, directly or indirectly, to anthropogenic activities.

Several vegetation types, distinguishable at a range of scales, were destroyed as a result of human activities, the composition of others was severely altered, and many new communities appear to have developed in direct or indirect response to the almost infinite permutations in time and space of clearance patterns and land-use regimes, intensity and frequency of burning, trampling, browsing and grazing, soil fertility and moisture, climate, available flora, predation, manuring, and chance factors (Berglund, 1962; Iversen, 1964b, 1969; Janssen, 1970; Göransson, 1977 for example).

Other cleared areas, particularly on fertile soils, were ploughed and intensively cultivated. Crops were intentionally introduced (e.g. cereals, *Cannabis sativa*, *Fagopyrum sagittatum*, *Linum usitatissimum*) along with accidental introduction of alien weeds that are now widely naturalized in north-west Europe (anthropochors). Several ruderals and other heliophilous, competition-intolerant plants that were common in the protocratic and rare and restricted to naturally treeless situations in the mesocratic (Figure 1.3) expanded into the many open habitats created by forest clearance and grazing of domestic animals (e.g. *Plantago lanceolata*, *Rumex acetosella*) or on to mineral soils disturbed by agriculture and human settlement (e.g. *Artemisia*, Chenopodiaceae, *Plantago major*, *Polygonum aviculare*), so-called archaeophytes (*sensu* Faegri, 1963). Several plants, although originally part of the indigenous forest-vegetation, also became commoner in response to anthropogenic influence, so-called apophytes (e.g. *Calluna*, *Juniperus*, *Pteridium*, *Urtica dioica*). Plants such as *Centaurea cyanus* and *Spergula arvensis* were present in the protocratic, possibly absent in the mesocratic, but accidentally reintroduced by man. In the absence of evidence for local persistence during the mesocratic, they are usually regarded as anthropochors. Plants formerly native in earlier interglacials in north-west Europe (e.g. *Azolla filiculoides*, *Rhododendron ponticum*, and, in the case of Britain, Denmark, etc., *Picea abies*) were also reintroduced. Much of the present flora of north-west Europe (*c* 40%) is hemerophilous and dependent, in one way or another, on man.

The major patterns of biotic change within north-west Europe in the last 5000 years result from natural soil changes (see the previous section), delayed migration, and direct or indirect effects of *H. sapiens*. As summarized in Figure 1.3, local and regional forest destruction and creation of clearings, cultivated areas, heaths and pastures, and other open habitats resulted in numerous, minor, but often rapid floristic and vegetational changes, frequent linked population changes, presumably in response to strong overriding environmental and biotic controls, increased biotic diversity at a regional scale, at least until recent times, high vegetational instability associated with invasion and expansion of taxa, many of which were prominent in the protocratic, introduction of aliens, decline or local extinction of taxa characteristic of the mesocratic, and soil degradation and increased expansion of taxa common in the oligocratic. Many of these patterns reflect irreversible changes in the natural landscape. The natural rates and directions of vegetational change that occurred in response to migration and natural climatic, edaphic and other biotic changes in the first 5000 years of the Holocene were disrupted and changed by the appearance and ever-increasing importance of *Homo sapiens* and agriculture in north-west Europe. It is, however, extremely difficult to identify and distinguish between natural and human-induced patterns. Climatic change almost certainly occurred; the problem is to isolate unambiguously its effects on biotic patterns from other processes. Many plants are reliable indicators of human influence because of their specific habitat requirements and wide climatic tolerances (Iversen, 1941). On the other hand, there are few reliable indicators of climatic change alone that are not influenced, to some extent, by cultural activity (cf. Trocls-Smith, 1960; Iversen, 1964a). The timing, magnitude and ecological effects of the end of the 'climatic optimum', and of the 'Little Ice Age' of the 15th–18th centuries in north-west Europe, require critical elucidation (e.g. Walker, 1984). It is not clear if the telocratic phase (*sensu* Iversen, 1958), with its declining temperatures, has been reached in the present interglacial (cf. Kukla, 1969, 1980; Kukla *et al.*, 1972).

Similarly, the complex interactions between regional climatic change, cultural spread and change, expansion, stability, and contraction of human populations, extent and type of land-use, local and regional hydrology, and local soils and topography over the last 5000 years, are largely unresolved (Berglund, 1969, 1983; Iversen, 1973; Welinder, 1973). Questions concerning palaeoecology of *H. sapiens* in relation to the natural environment represent an important task for future research (Faegri, 1974). Careful and critical consideration of the many patterns of biotic change observable over a range of spatial and temporal scales and the generation of testable hypotheses concerning causative processes are required to evaluate the influence and interaction between *H. sapiens*, climate, soil, and plant migration, all of which have varied in space and time, on particular scale-dependent processes. The understanding of these interactions remains a major challenge to the present methodology and interpretive paradigms of Quaternary palaeoecology. As Iversen (1973) noted: 'Plant communities and animal life can only be understood in a historical perspective'. The same applies to the role of *H. sapiens* in north-west Europe over the last 5000 years.

BIOTIC CRISES AND CATASTROPHES
($2–10 \times 10^2$ years, 10^3–10^{14} m^2)

A biotic crisis is an event whose duration is short relative to the total time under study. It is rare, unpredictable and often unique, and results in major changes in the biota of interest (Raup, 1981). There are very few documented examples of widespread synchronous biotic crises in the late-Quaternary. Recognition of biotic change as a synchronous event is dependent on time scales and precision of study. Within a span of 60×10^6 years and dating precision of 10^3 years, changes occurring over 5×10^3 years appear 'synchronous', whereas in the context of 10^5 years of the late-Quaternary and with a dating precision of 10^2

years, changes have to occur within 5–10 $\times$ 10^1 years to be regarded as synchronous.

The widespread regional 'mass extinction' of many genera of large mammals and birds at about 11,000 B.P. was a broadly synchronous biotic crisis of great importance, perhaps comparable ecologically to the extinction of dinosaurs at the end of the Cretaceous. The 11,000 B.P. protocratic crisis appears unique in geologic records of mass extinctions in being confined to terrestrial vertebrates. In North America, for example, 67 mammal genera became extinct during the Quaternary; of these 32 (48%) disappeared between 13,000 and 10,000 B.P., including large herbivores such as mammoths, mastodons, ground sloths, elephants, horses, camels and glyptodonts. Twenty-two bird genera went extinct in the Quaternary; 10 (45%) of these disappeared between 13,000 and 10,000 B.P. (Grayson, 1977). No extinct vertebrate genera are known after 10,000 B.P. (Meltzer and Mead, 1983). In Eurasia the scale of extinction was less than in the Americas or Australasia, but several large mammals disappeared by 11,000 B.P., including mammoth, woolly rhinoceros and giant deer.

There are two competing hypotheses to explain the observed extinction patterns (Martin and Wright, 1967). The first proposes climatic change at the end of the last cryocratic phase, resulting in direct environmental stress, vegetational change, habitat loss, food shortage and extinction. This hypothesis is falsified by the absence of comparable extinctions at the close of previous cryocratic phases when climatic and habitat changes of comparable extent and rapidity presumably occurred. The second hypothesis proposes that the observed extinctions result from 'prehistoric overkill' following immigration of specialized Upper Palaeolithic 'big-game' hunters. A human-population explosion occurred, resulting in intense and excessive hunting and predation of a fauna that provided easy prey, as it had not previously been exposed to human-type predators armed with flint, bone and antler-pointed weapons (Martin, 1973). Janzen (1983) suggests that human hunters were 'helped' by large predatory carnivores, because as large herbivores were decimated by human overkill, the prey available to carnivores was drastically reduced. This encouraged intense predator pressure on surviving large herbivores and their offspring. Janzen (1983) notes that of the surviving large herbivores, many have 'group defences' (e.g. caribou, muskox) or are 'very shy, wary and fleet' (e.g. deer, tapir).

The major drawbacks in the 'prehistoric-overkill' hypothesis are, like the fodder-gathering hypothesis for the elm-decimation 5000 years ago, archaeological and ecological feasability, absence of any direct evidence, and archaeological detection at relevant time and space scales (Martin, 1973). Other problems are raised by, for example, Grayson (1977), Gillespie *et al.* (1978) and McLean (1981). Mosimann and Martin (1975) have shown by mathematical simulation that the model of 'Pleistocene overkill' is theoretically possible.

These simulations do not, of course, prove that overkill occurred, only that it is possible.

The ecological effects of rapid extinction of over 75% of the New World's large herbivores such as gomphotheres and glyptodonts must have been profound, for example on seed dispersal, browsing, grazing, trampling, and tree regeneration (Janzen and Martin, 1982). Large grazers and browsers such as bison, mammoth, and woolly rhinoceros in north-west Europe may have been important in delaying or even inhibiting tree growth during interstadials in the last cryocratic and protocratic (Iversen, 1964a). For example, giant deer in Ireland may have maintained the late-glacial treeless vegetation (Watts, 1977; Craig, 1978). Perhaps giant-deer populations were particularly large in a near-island such as Ireland where natural predators may have been rare or absent. One can only speculate on whether forests would have developed in the Irish Holocene if giant deer had survived the rapid climatic deterioration and habitat change at the onset of the Younger Dryas stadial. This extinction, like other extinctions of large vertebrates about 11,000 B.P., represents a crisis in late-Quaternary biotic history. Much remains to be discovered about the detailed patterns in time and space of vertebrate extinctions before we can fully understand the underlying causes of these biotic crises. The true explanation probably involves both climatic and habitat change *and* human-predation, and may be different for different taxa.

Local and regional extinctions of plants and invertebrates in the late-Quaternary are extremely poorly understood and warrant detailed study. Notable botanical extinctions include *Naias marina* in the mid-Holocene of Ireland (Watts, 1978), *Pinus* in the late-Holocene of Ireland (Watts, 1984) and the Hebrides (Wilkins, 1984), and *Corylus* and *Tilia* in the late-Holocene of central Scandinavia (Tallantire, 1981; Huntley and Birks, 1983).

A widespread, catastrophic decline in *Ulmus* populations occurred at about 5000 B.P. in north-west Europe (Huntley and Birks, 1983). Important features of this biotic crisis are: (1) its synchroneity and universal occurrence at all sites throughout the geographic range of *Ulmus*, even in regions where there is little or no archaeological evidence for Neolithic peoples (e.g. northern Scotland); (2) its rapidly declining pollen values from 20–25% to 1–5% in 50–100 years (e.g. O'Connell, 1980; Garbett, 1981), which is considerably shorter than the natural longevity of elm, suggesting that the decline occurred rapidly and without recruitment, rather than gradually over several generations (Bennett, 1983a); (3) its consistency from site to site and lack of correspondingly consistent pollen-stratigraphic patterns in other trees; and (4) its long-term ecological effects, as elm populations rarely re-established their former size.

Several hypotheses have been presented to explain the *Ulmus* decline (Smith, 1961; Ten Hove, 1968; Iversen, 1973), including climatic change, competition, soil deterioration, selective collection of leaf-fodder following pollarding or ring-barking (see the previous section), and disease. The only hypothesis that satisfactorily explains the scale and observed temporal and spatial patterns of the elm demise proposes coincidental spread of early Neolithic agriculture *and* a widespread fatal pathogen specific to elm (Watts, 1961; Rackham, 1980). The disease hypothesis was originally proposed by Iversen in 1955 (Troels-Smith, 1960) and discussed in detail by Watts (1961), Smith (1961), Iversen (1973), Rackham (1980), Groenman-van Waateringe (1983) and Göransson (1984).

Specific pathogenic attack is the only absolutely selective process that accounts for the rapid and widespread death of 50% or more of elm populations throughout north-west Europe in less than 250 years without affecting other trees. Areas decimated by elm disease with abundant dead or dying elms would provide large natural openings in otherwise continuous mid-Holocene forest cover. *Quercus, Betula, Fraxinus* and *Corylus* could have colonized these openings, resulting in the frequently observed increases in their pollen. Creation of large gaps and associated disturbed soils would have encouraged apophytes. These areas may have been preferentially exploited by early Neolithic agriculturalists because they could be readily cleared. Elms damaged by *H. sapiens* as a result of burning, pollarding or ring-barking would, as today, be more susceptible to infection than undamaged trees. The spread and effects of the epidemic may thus have been accelerated by migrating Neolithic peoples and their animals exploiting the forests around openings. The appearance of cereal and other anthropochorous pollen types just after the elm decline cannot be explained as a result of natural gap production following disease: agriculture is implicated. A close but possibly varying interaction between disease and human interference is thus proposed to explain the observed patterns in time and space (Watts, 1961; Rackham, 1980; Huntley and Birks, 1983; Göransson, 1984), perhaps 'with civilization helping the spread of the disease and the disease helping the spread of civilization' (Rackham, 1980). As Iversen (1973) concluded, disease 'is an explanation which can solve the problem in a most elegant way'.

What pathogen could have been involved is unfortunately not known. Today elm disease is caused by *Ceratocystis ulmi* which is transmitted by bark-boring *Scolytis* beetles. Rates of pathogen spread of 4 km/year are required to account for observed patterns of elm decimation 5000 years ago (Huntley and Birks, 1983). These are possible judging by present rates (Sarre, 1978; Strobel and Lanier, 1981). Although Heybroek (1963) concluded that the elm decline could not have been caused by any pathogen known today, it is possible that the pathogen responsible for the decline 5000 years ago is now extinct. Not all pathogens are specific through time or space — for example, various virus die-backs (Smith, 1961). Heybroek (1963) rejected the disease hypothesis because there were, at that time, no historic records of elm disease before this century and because modern elms do not show any disease resistance, such as would be expected if the disease had been endemic for 5000 years. However, elm disease is much older than

Heybroek suggested and elms show resistance to it (Rackham, 1980). Fossil *Scolytus* beetles have recently been found just below the *Ulmus* decline (Moore, 1984).

There is considerable geographic variation in patterns of elm recovery after the *Ulmus* decline (Huntley and Birks, 1983). In Ireland, for example, large populations were present again on fertile soils by 4500 B.P. (e.g. O'Connell, 1980). Watts (1982) suggests that this lag may reflect the time required for individuals that survived the epidemic because of isolation, natural immunity, or reduced pathogenic virulence in a cool, oceanic climate to build up new populations, perhaps by ecological processes not dissimilar to those in tree invasion in the early-Holocene (see Figure 1.5) or for injured trees to recover by suckering. Population resistance to the pathogen at that time would be favoured and evolution of resistant strains may also have occurred over longer time spans. In other areas, for example Sweden and England, elm populations never recovered after 5000 B.P., probably as a result of forest clearance, early agriculture, soil deterioration, competition, and decimation of populations to a level too low for any regional population recovery.

The outbreak of chestnut blight (*Endothecia parasitica*) in eastern North America between 1904 and 1950 is an impressive, well-documented case of a rapid, catastrophic demise of a major forest tree, *Castanea dentata*. Its decline is clearly recorded pollen-stratigraphically (Anderson, 1974), and led to an expansion of *Betula* and, subsequently, *Quercus* (Brugam, 1978). Rates of *Castanea* population decline are extraordinarily high ($r = -2$ to -5×10^{-2}; Tsukada, 1983).

A similar pathogenic-attack hypothesis is proposed by Davis (1981b) to explain the widespread, synchronous decline of *Tsuga canadensis* at about 4800 B.P. throughout eastern North America. Hemlock populations declined very rapidly ($r = -2 \times 10^{-2}$; Tsukada, 1983) to 10% of their size within about 50 years, possibly as a result of a widespread fungal pathogen or massive outbreaks of hopper moths (Davis, 1981b). Ecological effects of this catastrophe were considerable, leading to regional, long-term changes in forest dominance (Davis, 1981b) and local changes in lake biota and productivity (Whitehead *et al.*, 1973).

Falsifying the hypothesis of widespread pathogenic attack on elm in north-west Europe or hemlock in eastern North America is extremely difficult (Iversen, 1973; Davis, 1981b). One possible approach involves spatial diffusion and epidemological models (e.g. Bailey, 1957; Gregory, 1968) to simulate geographic and temporal patterns of widespread pathogenic attack at scales appropriate for pollen-analytical data. Such models could be used in an analogous way to Mosimann and Martin's (1975) models to simulate 'prehistoric overkill', namely to model 'prehistoric epidemics' and to establish whether observed patterns of widespread synchronous decline in tree populations could result from a particular hypothetical causative process. 'Simulations can prove nothing about prehistory' (Mosimann and Martin,

1975); they can, however, indicate whether a specific process provides a plausible hypothesis for observed patterns. Such striking catastrophic population decimation and even local or regional extinction deserve further detailed study, using, for example, annually laminated lake-sediments (Chapter 17) to provide a precise chronology for the observed patterns (Watts, 1982).

Another type of synchronous biotic change results from widespread short-term geological catastrophes such as volcanic eruptions. These are recorded as tephra layers in lake or bog sediments (Chapter 16), and are events of short duration that can, in cases of extensive ash deposition, destroy vegetation over a wide area. Patterns of vegetation development following ash falls can be reconstructed by detailed close-interval pollen analyses (Mehringer *et al.*, 1977). There is, at present, little evidence for ash falls influencing long-term temporal patterns of biotic change in the western USA, an area of extensive Holocene volcanic activity. The main effects are local vegetation destruction and short-term, rather minor changes in population abundance and vegetation composition (Mehringer *et al.*, 1977; Mack *et al.*, 1979). The Laacher See volcanic eruption in the Eifel Mountains (*c* 10980 B.P.) had little apparent effect on late-glacial vegetation. Ash has been found at least 900 km from Laacher See (Wegmüller and Welten, 1973; Usinger, 1978). It would be of interest to study this event in more detail and to evaluate what, if any, ecological effects it had.

Quaternary palaeocologists, in common with many palaeobiologists, largely ignore crises and catastrophes as important determinants of observed patterns of biotic change at particular temporal and spatial scales. As Raup (1981) suggests: 'Natural systems cannot be understood nor fully interpreted without taking disturbance into account ... . To consider crises as a legitimate and important part of the formative processes in ecology and evolution is new to many of us and anathema to some.' Late-Quaternary biotic crises may be important but unpredictable components in the plot of the 'palaeoecological play'.

CODA

In this chapter I have reviewed the main patterns of late-Quaternary biotic change in north-west Europe and presented hypotheses about underlying causal processes operative at relevant temporal and spatial scales. Problems abound, however, for as Iversen (1964a) concluded: 'The glacial–interglacial and postglacial macrosuccessions are the result of a complicated pattern of environmental changes. To assess and separate all the factors is indeed a fascinating undertaking. So far it may, however, be sound to face the difficulties.'

The importance of scale, of generating testable hypotheses, and of detailed biostratigraphic investigations cannot be overemphasized as 'no amount of ecological theorizing about succession, competition and migration can replace the solid evidence represented by biostratigraphy' (Wright, 1976b). This chapter attempts to highlight some of the more outstanding problems in elucidating and understanding late-Quaternary biotic change in north-west Europe that require detailed palaeoecological studies as a means of testing hypotheses about patterns and processes. We are a long way from satisfactorily understanding the mechanisms influencing the diverse patterns of biotic change and vegetation dynamics observable at the local, regional and continental scales over the last 100,000 years. The wide range of techniques and approaches discussed in this book provide the Quaternary palaeoecologist with the means of studying in increased detail and precision many fascinating aspects of the most recent glacial–interglacial 'palaeoecological play'.

Acknowledgements

This chapter reviews and integrates ideas and data from a diverse published literature. In addition some of the ideas discussed have developed, directly or indirectly, as a result of stimulating discussions with many people, including Bent Aaby, Svend Andersen, Cathy and Tony Barnosky, Keith Bennett, Björn Berglund, Hilary Birks, Hazel and Paul Delcourt, Knut Faegri, Francis Gilbert, Rob Hengeveld, Brian Huntley, Roel Janssen, Henry Lamb, Jan Mangerud, Jim Ritchie, Bill Watts, Tom Webb and Herb Wright. The chapter could not have been completed without the support, assistance and patience of Björn Berglund, Hilary Birks, Francis Gilbert and Sylvia Peglar. To all these friends I am extremely grateful.

I dedicate this chapter to G. Clifford Evans in recognition of his many contributions to scholarship, his concern for people and truth, and his courage of mind, and in appreciation of his support and friendship.

BIBLIOGRAPHY

Aaby, B. (1976). Cyclic climatic variations in climate over the past 5,500 yr reflected in raised bogs. *Nature*, **263**, 281–284.
Aaby, B. (1983). Forest development, soil genesis and human activity illustrated by pollen and hypha analysis of two neighbouring podzols in Draved Forest, Denmark. *Danmarks Geologiske Undersøgelse*, II, **114**, 114 pp.
Adam, P. (1977). The ecological significance of 'halophytes' in the Devensian flora. *New Phytologist*, **78**, 237–244.
Allen, T. F. H., and Starr, T. B. (1982). *Hierarchy: Perspectives for Ecological Complexity*, University of Chicago Press, Chicago.
Andersen, S. Th. (1961). Vegetation and its environment in the Early Weichselian Glacial (Last Glacial). *Danmarks Geologiske Undersøgelse*, II, **75**, 175 pp.
Andersen, S. Th. (1964). Interglacial plant successions in the light of environmental changes. *Report VI International Congress on Quaternary, Volume II: Palaeobotanical Section*, pp. 359–367.

Andersen, S. Th. (1966). Interglacial vegetation succession and lake development in Denmark. *Palaeobotanist*, **15**, 117–127.

Andersen, S. Th. (1969) Interglacial vegetation and soil development. *Meddelelser fra Dansk Geologisk Forening*, **19**, 90–102.

Andersen, S. Th. (1973). The differential pollen productivity of trees and its significance for the interpretation of a pollen diagram from a forested region. In: *Quaternary Plant Ecology* (Eds. H. J. B. Birks and R. G. West), Blackwell, Oxford, pp. 109–115.

Andersen, S. Th. (1975). The Eemian freshwater deposit at Egernsund, South Jylland, and the Eemian landscape development in Denmark. *Danmarks Geologiske Undersøgelse Årbog* 1974, 49–70.

Andersen, S. Th. (1978). Local and regional vegetational development in eastern Denmark in the Holocene. *Danmarks Geologiske Undersøgelse Årbog* 1976, 5–27.

Andersen, S. Th. (1979). Brown earth and podzol: soil genesis illuminated by microfossil analysis. *Boreas*, **8**, 59–73.

Anderson, T. W. (1974). The chestnut pollen decline as a time horizon in lake sediments in eastern North America. *Canadian Journal of Earth Sciences*, **11**, 678–685.

Bailey, N. T. J. (1957). *The Mathematical Theory of Epidemics*, Griffin, London.

Ball, I. R. (1975). Nature and formulation of biogeographical hypotheses. *Systematic Zoology*, **24**, 407–430.

Behre, K-E. (1981). The interpretation of anthropogenic indicators in pollen diagrams. *Pollen et Spores*, **23**, 225–245.

Bell, F. G. (1969). The occurrence of southern steppe and halophyte elements in Weichselian (last-glacial) floras from southern Britain. *New Phytologist*, **68**, 913–922.

Bennett, K. D. (1983a). Devensian late-glacial and Flandrian vegetational history at Hockham Mere, Norfolk, England. I. Pollen percentages and concentrations. *New Phytologist*, **95**, 457–487.

Bennett, K. D. (1983b). Postglacial population expansion of forest trees in Norfolk, UK. *Nature*, **303**, 164–167.

Berglund, B. E. (1962). Vegetation på ön Senoren. I. Vegetationshistoria. *Botaniska Notiser*, **115**, 387–419.

Berglund, B. E. (1966). Late-Quaternary vegetation in eastern Blekinge, south-eastern Sweden — a pollen analytical study. I. Late-glacial time. *Opera Botanica*, **12**(1), 180pp.

Berglund, B. E. (1969). Vegetation and human influence in south Scandinavia during prehistoric time. *Oikos Supplement*, **12**, 9–28.

Berglund, B. E. (1979a). Definition of investigation areas. In *Palaeohydrological Changes in the Temperate Zone in the last 15000 years. Subproject B: Lake and Mire Environments* (Ed. B. E. Berglund), Department of Quaternary Geology, University of Lund, vol. 1, pp. 11–22.

Berglund, B. E. (1979b). The deglaciation of southern Sweden 13,500–10,000 B.P. *Boreas*, **8**, 89–118.

Berglund, B. E. (1983). Palaeoclimatic changes in Scandinavia and on Greenland — a tentative correlation based on lake and bog stratigraphical studies. *Quaternary Studies in Poland*, **4**, 27–44.

Berglund, B. E., Lemdahl, G., Liedberg-Jönsson, B., and Persson, T. (1984). Biotic response to climatic changes during the time span 13,000–10,000 B.P. — A case study from SW Sweden. In *Climatic Changes on a Yearly to Millenial Basis* (Eds. N. A. Mörner and W. Karlén). D. Reidel Publ. Comp., pp. 25–36.

Berglund, B. E., and Malmer, N. (1971). Soil conditions and late-glacial stratigraphy. *Geologiska Föreningens i Stockholm Förhandlingar*, **93**, 575–586.

Beug, H-J. (1982). Vegetation history and climatic changes in central and southern Europe. In *Climatic Changes in Later Prehistory* (Ed. A. F. Harding), Edinburgh University Press, Edinburgh, pp. 85–102.

Binford, M. W., Deevey, E. S., and Crisman, T. L. (1983). Paleolimnology: an historical perspective on lacustrine ecosystems. *Annual Review of Ecology and Systematics*, **14**, 255–286.

Birks, H. H. (1975). Studies in the vegetational history of Scotland. IV. Pine stumps in Scottish blanket peats. *Phil. Trans. Royal Soc. London* B, **270**, 181–226.

Birks, H. H. (1980). Plant macrofossils in Quaternary lake sediments. *Ergebnisse der Limnologie*, **15**, 60pp.

Birks, H. H. (1984). Late-Quaternary pollen and plant macrofossil stratigraphy at Lochan an Druim, north-west Scotland. In *Lake Sediments and Environmental History* (Eds. E. Y. Haworth and J. W. G. Lund), University of Leicester Press, Leicester, pp. 377–405.

Birks, H. H., and Mathewes, R. W. (1978). Studies in the vegetational history of Scotland. V. Late Devensian and early Flandrian pollen and macrofossil stratigraphy at Abernethy Forest, Inverness-shire. *New Phytologist*, **80**, 455–484.

Birks, H. J. B. (1973). *Past and Present Vegetation of the Isle of Skye: a Palaeoecological Study*, Cambridge University Press, London.

Birks, H. J. B. (1981). The use of pollen analysis in the reconstruction of past climates: a review. In *Climate and History* (Eds. T. M. L. Wigley, M. J. Ingram and G. Farmer), Cambridge University Press, Cambridge, pp. 111–138.

Birks, H. J. B. (1985). Flandrian (post-glacial) isochrone maps and tree migration patterns in the British Isles (manuscript in preparation).

Birks, H. J. B., and Gordon, A. D. (1985). *Numerical Methods in Quaternary Pollen Analysis*, Academic Press, London (in press).

Birks, H. J. B., and Peglar, S. M. (1979). Interglacial pollen spectra from Sel Ayre, Shetland. *New Phytologist*, **83**, 559–575.

Björck, S., and Digerfeldt, G. (1984). Climatic changes at Pleistocene/Holocene boundary in the Middle Swedish endmoraine zone, mainly inferred from stratigraphic indications. In *Climatic Changes on a Yearly to Millenial Basis* (Eds. N. A. Mörner and W. Karlén). D. Reidel Publ. Comp., pp. 37–56.

Boucot, A. J. (1983). Area-dependent-richness hypotheses and rates of parasite/pest evolution. *American Naturalist*, **121**, 294–300.

Brenninkmeijer, C. A. M., van Geel, B., and Mook, W. G. (1982). Variations in the D/H and $^{18}O/^{16}O$ ratios in cellulose extracted from a peat bog core. *Earth and Planetary Science Letters*, **61**, 283–290.

Broeker, W. S., and van Donk, J. (1970). Insolation changes, ice volumes, and the O^{18} record in deep-sea cores. *Reviews of Geophysics and Space Physics*, **8**, 169–198.

Brugam, R. B. (1978). Pollen indicators of land-use change in southern Connecticut. *Quaternary Research*, **9**, 349–363.

Bryson, R. A., and Wendland, W. M. (1967). Tentative climate patterns for some late glacial and post-glacial episodes in central North America. In *Life, Land and Water* (Ed. W. J. Mayer-Oakes), University of Manitoba Press, Winnipeg, pp. 271–298.

Burleigh, R., and Kerney, M. P. (1982). Some chronological implications of a fossil molluscan assemblage from a Neolithic site at Brook, Kent, England. *Journal of Archaeological Science*, **9**, 29–38.

Casteel, R. W., Adam, D. P., and Sims, J. D. (1977). Late-Pleistocene and Holocene remains of *Hysterocarpus traski* (Tule perch) from Clear Lake, California, and inferred Holocene temperature fluctuations. *Quaternary Research*, **7**, 133–143.

Chambers, F. M. (1981). Date of blanket peat initiation in upland South Wales. *Quaternary Newsletter*, **35**, 24–29.

Claridge, M. F., and Wilson, M. R. (1978). British insects and trees: a study in island biogeography or insect/plant coevolution?' *American Naturalist*, **112**, 451–456.

CLIMAP (1976). The surface of the Ice-Age Earth. *Science*, **191**, 1131–1137.

Colinvaux, P. A. (1983). The meaning of palaeoecology. *Quarterly Review of Archaeology*, **4**, 8–9.

Craig, A. J. (1978). Pollen percentage and influx analyses in south-east Ireland: a contribution to the ecological history of the late-glacial period. *Journal of Ecology*, **66**, 297–324.

Davis, M. B. (1981a). Quaternary history and stability of forest communities. In *Forest Succession: Concepts and Applications* (Eds. D. C. West, H. H. Shugart and D. B. Botkin), Springer Verlag, New York, pp. 132–153.

Davis, M. B. (1981b). Outbreaks of forest pathogens in Quaternary history. *Proceedings of IV International Palynological Conference Lucknow*, **3**, 216–227.

Davis, M. B., and Ford, M. S. (1982). Sediment focusing in Mirror Lake, New Hampshire. *Limnology and Oceanography*, **27**, 137–150.

Deevey, E. S. (1969). Coaxing history to conduct experiments. *BioScience*, **19**, 40–43.

Degerbøl, M. (1964). Some remarks on late- and post-glacial vertebrate fauna and its ecological relations in northern Europe. *Journal of Ecology*, **52**(Supplement), 71–85.

Delcourt, H. R., Delcourt, P. A., and Webb, T. (1983). Dynamic plant ecology: the spectrum of vegetational change in space and time. *Quaternary Science Reviews*, **1**, 153–175.

Delcourt, P. A., and Delcourt, H. R. (1983). Late-Quaternary vegetational dynamics and community stability reconsidered. *Quaternary Research*, **19**, 265–271.

Denton, G. H., and Karlén, W. (1973). Holocene climatic variations — their pattern and possible cause. *Quaternary Research*, **3**, 155–205.

Digerfeldt, G. (1972). The post-glacial development of Lake Trummen. Regional vegetation history, water level changes and palaeolimnology. *Folia Limnologica Scandinavica*, **16**, 104pp.

Edwards, K. J. (1979). Palynological and temporal inference in the context of prehistory, with special reference to the evidence from lake and peat deposits. *Journal of Archaeological Science*, **6**, 255–270.

Edwards, K. J. (1982). Man, space and the woodland edge — speculations on the detection and interpretation of human impact in pollen profiles. In *Archaeological Aspects of Woodland Ecology* (Eds. M. Bell and S. Limbrey), British Archaeological Reports International Series **146**, Oxford, pp. 5–22.

Edwards, K. J. (1983). Quaternary palynology: consideration of a discipline. *Progress in Physical Geography*, **7**, 113–125.

Edwards, K. J., and Hirons, R. K. (1982). Date of blanket peat initiation and rates of spread — a problem in research design. *Quaternary Newsletter*, **36**, 32–37.

Emiliani, C. (1972). Quaternary hypsithermals. *Quaternary Research*, **2**, 270–273.

Eronen, M., and Hyvärinen, H. (1982). Subfossil pine dates and pollen diagrams from northern Fennoscandia. *Geologiske Föreningens i Stockholm Förhandlingar*, **103**, 437–445.

Faegri, K. (1933). Über die Längenvariationen einiger Gletscher des Jostedalsbre und die dadurch bedingten Pflanzensukzessionen. *Bergens Museums Årbok* 1933, **7**, 255pp.

Faegri, K. (1940). Quatärgeologische Untersuchungen im westlichen Norwegen. II: Zur spätquartären Geschichte Jaerens. *Bergens Museums Årbok* 1939–40, **7**, 201pp.

Faegri, K. (1949). Studies on the Pleistocene of western Norway. IV. On the immigration of *Picea abies* (L.)Karst. *Bergen Universitetet Årbok* 1949, **1**, 52pp.

Faegri, K. (1950). On the value of palaeoclimatological evidence. *Centenary Proceedings of the Royal Meteorological Society* 1950, 188–195.

Faegri, K. (1963). Problems of immigration and dispersal of the Scandinavian flora. In *North Atlantic Biota and their History* (Eds. A. Löve and D. Löve), Pergamon, Oxford, pp. 221–232.

Faegri, K. (1974). Quaternary pollen analysis — past, present and future. *Advances in Pollen Spore Research*, **1**, 62–69.

Firbas, F. (1949). *Spät- und nacheiszeitliche Waldgeschichte Mitteleuropas nördlich der Alpen*, Fischer, Jena.

Forcier, L. K. (1975). Reproductive strategies and the co-occurrence of climax tree species. *Science*, **189**, 808–810.

Forman, R. T. T., and Godron, M. (1981). Patches and structural components for a landscape ecology. *Bioscience*, **31**, 733–740.

Forsström, L. (1984). Eemian and Weichselian correlation problems in Finland. *Boreas*, **13**, 301–318.

Fox, J. F. (1977). Alternation and coexistence of tree species. *American Naturalist*, **111**, 69–89.

Fredskild, B. (1983). The Holocene development of some low and high arctic Greenland lakes. *Hydrobiologia*, **103**, 217-224.

Garbett, G. G. (1981). The elm decline: the depletion of a resource. *New Phytologist*, **88**, 573–585.

Gates, W. L. (1976). Modelling the Ice-Age climate. *Science*, **191**, 1138–1144.

Gillespie, R., Horton, D. R., Ladd, P., Macumber, P. G., Rich, T. H., Thorne, R., and Wright, R. V. S. (1978). Lancefield Swamp and the extinction of the Australian megafauna. *Science*, **200**, 1044–1048.

Godwin, H. (1966). Introductory address. In *World Climate from 8000 to 0 B.C.* (Ed. J. A. Sawyer), Royal Meteorological Society, London, pp. 3–14.

Godwin, H. (1975). *The History of the British Flora* (2nd edn), Cambridge University Press, Cambridge.

Göransson, H. (1977). *The Flandrian Vegetational History of Southern Östergötland*. Thesis **3**, Dept. of Quat. Geol., University of Lund.

Göransson, H. (1984). Pollen analytical investigations in the Sligo area. In *The Archaeology of Carrowmore* (Ed. G. Burenhult), Theses and Papers in North-European Archaeology **14**, Stockholm, pp. 154–193.

Gorham, E., Vitousek, P. M., and Reiners, W. A. (1979). The regulation of chemical budgets over the course of terrestrial ecosystem succession. *Annual Review of Ecology and Systematics*, **10**, 53–84.

Grayson, D. K. (1977). Pleistocene avifaunas and the overkill hypothesis. *Science*, **195**, 691–693.

Green, D. G. (1982). Fire and stability in the postglacial forests of southwest Nova Scotia. *Journal of Biogeography*, **9**, 29–40.

Gregory, P. H. (1968). Interpreting plant disease dispersal gradients. *Annual Review of Phytopathology*, **6**, 189–212.

Grime, J. P. (1977). Evidence for the existence of three primary strategies in plants and its relevance to ecological and evolutionary theory. *American Naturalist*, **111**, 1169–1194.

Groenman-van Waateringe, W. (1983). The early agricultural utilization of the Irish landscape: the last word on the elm decline? *British Archaeological Report British Series*, **116**, 217–232.

Grove, J. M. (1979). The glacial history of the Holocene. *Progress in Physical Geography*, **3**, 1–54.

Harper, J. L. (1967). A Darwinian approach to plant ecology. *Journal of Ecology*, **55**, 247–270.

Hays, J. D., Imbrie, J., and Shackleton, N. J. (1976). Variations in the Earth's orbit: pacemaker of the Ice Ages. *Science*, **194**, 1121–1132.

Hengeveld, R. (1982). *Problems of Scale in Ecological Research*. Thesis, University of Leiden.

Heybroek, H. M. (1963). Diseases and lopping for fodder as possible causes of a prehistoric decline of *Ulmus*. *Acta Botanica Neerlandica*, **12**, 1–11.

Hintikka, V. (1963). Über das Grossklima einiger Pflanzenareale in zwei Klimakoordinatensystemen Dargestellt. *Annales Botanici Societatis Zoologicae Botanicae Fennicae 'Vanamo'*, **34**(5), 64pp.

Horn, H. S. (1971). *The Adaptive Geometry of Trees*, Princeton University Press, Princeton.

Howe, S., and Webb, T. (1983). Calibrating pollen data in climatic terms: improving the methods. *Quaternary Science Reviews*, **2**, 17–51.

Hunt, T. G., and Birks, H. J. B. (1982). Devensian late-glacial vegetational history at Sea Mere, Norfolk. *Journal of Biogeography*, **9**, 517–538.

Huntley, B., and Birks, H. J. B. (1983). *An Atlas of Past and Present Pollen Maps for Europe: 0–13000 years ago*, Cambridge University Press, Cambridge.

Hutchinson, G. E. (1965). *The Ecological Theater and the Evolutionary Play*, Yale University Press, New Haven and London.

Huttunen, P. (1980). Early land use, especially the slash-and-burn cultivation in the commune of Lammi, southern Finland, interpreted mainly using pollen and charcoal analyses. *Acta Botanica Fennica*, **113**, 1–45.

Huttunen, P., and Tolonen, K. (1975), Human influence in the history of Lake Lovojärvi, S. Finland. *Finskt Museum* 1975, 68–105.

Hyvärinen, H. (1975). Absolute and relative pollen diagrams from northernmost Fennoscandia. *Fennia*, **142**, 23pp.

Imbrie, J., and Imbrie, J. Z. (1980). Modelling the climatic response to orbital variations. *Science*, **207**, 943–953.

Iversen, J. (1941). Land occupation in Denmark's Stone Age. *Danmarks Geologiske Undersøgelse*, II, **66**, 68 pp.

Iversen, J. (1944). *Viscum, Hedera* and *Ilex* as climatic indicators: a contribution to the study of the post-glacial temperature climate. *Geologiske Föreningens i Stockholm Förhandlingar*, **66**, 463–483.

Iversen, J. (1954). The late-glacial flora of Denmark and its relation to climate and soil. *Danmarks Geologiske Undersøgelse*, II, **80**, 87–119.

Iversen, J. (1956). Forest clearance in the Stone Age. *Scientific American*, **194**, 36–41.

Iversen, J. (1958). The bearing of glacial and interglacial epochs on the formation and extinction of plant taxa. *Uppsala Universiteit Årsskr*, **6**, 210–215.

Iversen, J. (1960). Problems of the early post-glacial forest development in Denmark. *Danmarks Geologiske Undersøgelse*, IV, **4**(3), 32pp.

Iversen, J. (1964a). Plant indicators of climate, soil, and other factors during the Quaternary. *Report VI International Congress on Quaternary. Volume II: Palaeobotanical Section*, 421–428.

Iversen, J. (1964b). Retrogressive vegetational succession in the post-glacial. *Journal of Ecology*, **52**(Supplement), 59–70.

Iversen, J. (1969). Retrogressive development of a forest ecosystem demonstrated by pollen diagrams from fossil mor. *Oikos Supplement*, **12**, 35–49.

Iversen, J. (1973). The development of Denmark's nature since the last glacial. *Danmarks Geologiske Undersøgelse* V, **7-C**, 126pp.

Janssen, C. R. (1970). Problems in the recognition of plant communities in pollen diagrams. *Vegetatio*, **20**, 187–198.

Janzen, D. H. (1983). The Pleistocene hunters had help. *American Naturalist*, **121**, 598–599.

Janzen, D. H., and Martin, P. S. (1982). Neotropical anachronisms: the fruits the gomphotheres ate. *Science*, **215**, 19–27.

Jones, E. W. (1945). The structure and reproduction of the virgin forest of the north temperate zone. *New Phytologist*, **44**, 130–148.

Karlén, W. (1976). Lacustrine sediments and tree-limit variations as indicators of Holocene climatic fluctuations in Lappland, northern Sweden. *Geografiska Annaler*, **58A**, 1–34.

Kerney, M. P. (1968). Britain's fauna of land mollusca and its relation to the post-glacial thermal optimum. *Symposium of the Zoological Society of London*, **22**, 273–291.

Kolstrup, E. (1979). Herbs as July temperature indicators for parts of the pleniglacial and late-glacial in the Netherlands. *Geologie en Mijnbouw*, **58**, 377–380.

Kolstrup, E. (1980). Climate and stratigraphy in northwestern Europe between 30,000 B.P. and 13,000 B.P., with special reference to the Netherlands. *Mededelingen Rijks Geologische Dienst*, **32**, 181–253.

Kolstrup, E. (1982). Late-glacial pollen diagrams from Hjelm and Draved Mose (Denmark) with a suggestion of the possibility of drought during the Earlier Dryas. *Review of Palaeobotany and Palynology*, **36**, 35–63.

Kolstrup, E., and Buchardt, B. (1982). A pollen analytical investigation supported by an ^{18}O-record of a late glacial lake deposit at Groenge (Denmark). *Review of Palaeobotany and Palynology*, **36**, 205–230.

Kolstrup, E., and Wijmstra, T. A. (1977). A palynological investigation of the Moershoofd, Hengelo, and Denekamp interstadials in the Netherlands. *Geologie en Mijnbouw*, **56**, 85–102.

Kukla, J. (1969). The cause of the Holocene climate change. *Geologie en Mijnbouw*, **48**, 307–334.

Kukla, J. (1977). Pleistocene land-sea correlations. I. Europe. *Earth-Science Reviews*, **13**, 307–374.

Kukla, J. (1980). End of the last interglacial: a predictive model of the future? *Palaeoecology of Africa*, **12**, 395–408.

Kukla, G. J., Matthews, R. K., and Mitchell, J. M. (1972). The end of the present interglacial. *Quaternary Research*, **2**, 261–269.

Kutzbach, J. E. (1981). Monsoon climate of the early Holocene: climatic experiment with the Earth's orbital parameters for 9000 years ago. *Science*, **214**, 59–61.

Kutzbach, J. E., and Otto-Bliesner, B. L. (1982). The sensitivity of the African–Asian monsoonal climate to orbital parameter changes for 9000 years B.P. in a low-resolution general circulation model. *Journal of the Atmospheric Sciences*, **19**, 1177–1188.

Lamb, H. H. (1977). Climatic analysis. *Phil. Trans. Royal Soc. London* B, **280**, 341–350.

Lang, G. (1970). Florengeschichte und mediterran-mitteleuropäische Florenbeziehungen. *Feddes Repertorium*, **81**, 315–335.

Lehman, J. T. (1975). Reconstructing the rate of accumulation of lake-sediment: the effect of sediment focusing. *Quaternary Research*, **5**, 541–550.

Lerman, J. C. (1974). Isotope 'palaeothermometers' on continental matter: assessment. *Colloques Internationaux du C.N.R.S.*, **219**, 163–181.

Lichti-Federovich, S., and Ritchie, J. C. (1965). Contemporary pollen spectra in central Canada. II. The forest–grassland transition in Manitoba. *Pollen et Spores*, **7**, 63–87.

Lichti-Federovich, S., and Ritchie, J. C. (1968). Recent pollen assemblages from the Western Interior of Canada. *Review of Palaeobotany and Palynology*, **7**, 297–344.

Likens, G. E., and Davis, M. B. (1975). Post-glacial history of Mirror Lake and its watershed in New Hampshire, U.S.A.: an initial report. *Verhandlungen der International Vereinigung für Theoretische und Angewandte Limnologie*, **19**, 982–993.

Livingstone, D. A. (1957). On the sigmoid growth phase in the history of Linsley Pond. *American Journal of Science*, **255**, 364–273.

Livingstone, D. A., and Boykin, J. C. (1962). Vertical distribution of phosphorus in Linsley Pond mud. *Limnology and Oceanography*, **7**, 57–62.

McIntosh, R. P. (1981). Succession and ecological theory. In *Forest Succession: Concepts and Applications* (Eds. D. C. West, H. H. Shugart and D. B. Botkin), Springer-Verlag, New York, pp. 10–23.

McLean, D. M. (1981). Size factor in the late Pleistocene mammalian extinctions. *American Journal of Science*, **281**, 1144–1152.

McVean, D. N. (1955). Ecology of *Alnus glutinosa* (L.)Gaertn. I. fruit formation. *Journal of Ecology*, **43**, 46–60.

Mack, R. N., Rutter, N. W., and Valastro, S. (1979). Holocene vegetation history of the Okanogan Valley, Washington. *Quaternary Research*, **12**, 212–225.

Manabe, S., and Hahn, D. G. (1977). Simulation of the tropical climate of an ice age. *Journal of Geophysical Research*, **82**, 3889–3911.

Markgraf, V. (1974). Palaeoclimatic evidence derived from timberline fluctuations. *Colloques Internationaux du C.R.N.S.*, **219**, 67–77.

Martin, P. S. (1973). The discovery of America. *Science*, **179**, 969–974.

Martin, P. S., and Wright, H. E. (1967). *Pleistocene Extinctions — the Search for a Cause*, Yale University Press, New Haven and London.

Mehringer, P. J., Blinman, E., and Petersen, K. L. (1977). Pollen influx and volcanic ash. *Science*, **198** 257–261.

Meltzer, D. J., and Mead, J. I. (1983). The timing of late Pleistocene mammalian extinctions in North America. *Quaternary Research*, **19**, 130–135.

Moe, D. (1970). The post-glacial immigration of *Picea abies* into Fennoscandia. *Botaniska Notiser*, **123**, 61–66.

Moore, P. D. (1975). Origin of blanket mires. *Nature*, **256**, 267–269.

Moore, P. D. (1984). Hampstead Heath clue to historical decline of elms. *Nature*, **312**, 103.

Mosimann, J. E., and Martin, P. S. (1975). Simulating overkill by paleoindians. *American Scientist*, **63**, 304–313.

Neilson, R. P., and Wullstein, L. H. (1983). Biogeography of two south-west American oaks in relation to atmospheric dynamics. *Journal of Biogeography*, **10**, 275–297.

Núñez, M. G., and Vuorela, I. (1978). A tentative evaluation of cultural pollen data in early agrarian development research. *Suomen Museo* 1978, 5–36.

O'Connell, M. (1980). The developmental history of Scragh Bog, Co. Westmeath and the vegetational history of its hinterland. *New Phytologist*, **85**, 301–319.

Osborne, P. J. (1976). Evidence from the insects of climatic variation during the Flandrian period: a preliminary note. *World Archaeology*, **8**, 150–158.

Palmgren, A. (1929). Chance as an element in plant geography. *Proceedings of the International Congress of Plant Sciences*, **1**, 591–602.

Pearsall, W. H. (1959). The ecology of invasion: ecological stability and instability. *New Biology*, **29**, 95–101.

Pearsall, W. H (1964). After the ice retreated. *New Scientist*, (383), 757–759.

Pennington, W. (1965). The interpretation of some post-glacial vegetation diversities at different Lake District sites. *Proc. Royal Soc. of London* B, **161**, 310–323.

Pennington, W. (1977). The late Devensian flora and vegetation of Britain. *Phil. Trans. Royal Soc. of London* B, **280**, 247–271.

Pennington, W. (1981). Records of a lake's life in time: the sediments. *Hydrobiologia*, **79**, 197–219.

Pennington, W., Haworth, E. Y., Bonny, A. P., and Lishman, J. P. (1972). Lake sediments in northern Scotland. *Phil. Trans. Royal Soc. of London* B, **264**, 191–294.

Persson, Å. (1964). The vegetation at the margin of the receding glacier Skaftafells-jökull, southeastern Iceland. *Botaniska Notiser*, **117**, 323–354.

Peterson, G. M., Webb, T., Kutzbach, J. E., van der Hammen, T., Wijmstra, T. A., and Street, F. A. (1979). The continental record of environmental conditions at 18,000 B.P.: an initial evaluation. *Quaternary Research*, **12**, 47–82.

Pigott, C. D. (1975). Natural regeneration of *Tilia cordata* in relation to forest-structure in the forest of Białowieza, Poland. *Phil. Trans. Royal Soc. of London* B, **270**, 151–179.

Pigott, C. D. (1981). Nature of seed sterility and natural regeneration of *Tilia cordata* near its northern limit in Finland. *Annals Botanici Fennici*, **18**, 255–263.

Pigott, C. D., and Huntley, J. P. (1980). Factors controlling the distribution of *Tilia cordata* at the northern limits of its geographical range. II. History in northwest England. *New Phytologist*, **84**, 145–164.

Pigott, C. D., and Huntley, J. P. (1981). Factors controlling the distribution of *Tilia cordata* at the northern limits of its geographical range. III. nature and causes of seed sterility. *New Phytologist*, **87**, 817–839.

Pigott, C. D., and Walters, S. M. (1954). On the interpretation of the discontinuous distributions shown by certain British species of open habitats. *Journal of Ecology*, **42**, 95–116.

Pilcher, J. R., and Smith, A. G. (1979). Palaeoecological investigations at Ballynagilly, a Neolithic and Bronze Age settlement in County Tyrone, northern Ireland. *Phil. Trans. Royal Soc. of London* B, **286**, 345–369.

Pilcher, J. R., Smith, A. G., Pearson, G. W., and Crowder, A. (1971). Land clearance in the Irish Neolithic: new evidence and interpretation. *Science*, **172**, 560–562.

Prentice, I. C. (1983). Postglacial climatic change: vegetation dynamics and the pollen record. *Progress in Physical Geography*, **7**, 273–286.

Rackham, O. (1980). *Ancient Woodland: its History, Vegetation and Uses in England*, Edward Arnold, London.

Ralska-Jasiewiczowa, M. (1983). Isopollen maps for Poland: 0–11,000 years B.P. *New Phytologist*, **94**, 133–175.

Raup, D. M. (1981). Introduction: what is a crisis? In *Biotic Crises in Ecological and Evolutionary Time* (Ed. M. H. Nitecki), Academic Press, New York, pp. 1–12.

Ritchie, J. C. (1977). The modern and late Quaternary vegetation of the Campbell--Dolomite uplands, near Inuvik, N.W.T. Canada. *Ecological Monographs*, **47**, 401–423.

Ritchie, J. C. (1980). Towards a late-Quaternary palaeoecology of the Ice-Free Corridor. *Canadian Journal of Anthropology*, **1**, 15–28.

Ritchie, J. C. (1981). Problems of interpretation of the pollen stratigraphy of northwest North America. In *Quaternary Paleoclimates* (Ed. W. C. Mahaney), Geo Abstracts, Norwich, pp. 377–391.

Ritchie, J. C. (1982). The modern and late-Quaternary vegetation of the Doll Creek area, north Yukon, Canada. *New Phytologist*, **90**, 563–603.

Ritchie, J.C., Cwynar, L.C., and Spear, R.W. (1983). Evidence from north-west Canada for an early Holocene Milankovitch thermal maximum. *Nature*, 305, 126–129.

Romme, W. H., and Knight, D. H. (1982). Landscape diversity: the concept applied to Yellowstone Park. *BioScience*, **32**, 664–670.

Ruddiman, W. F., Sancetta, C. D., and McIntyre, A. (1977). Glacial/interglacial response rate of subpolar North Atlantic waters to climatic change: the record in oceanic sediments. *Phil. Trans. Royal Soc. of London* B, **280**, 119–142.

Sancetta, C., Imbrie, J., and Kipp, N. G. (1973). Climatic record of the past 130,000 years in North Atlantic deep-sea core V23–82: correlation with the terrestrial record. *Quaternary Research*, **3**, 110–116.

Sarre, P. (1978). The diffusion of Dutch elm disease. *Area*, **2**, 81–85.

Shackleton, N. J., and Opdyke, N. D. (1976). Oxygen-isotope and palaeomagnetic stratigraphy of Pacific core V28–239 Late Pliocene to Latest Pleistocene. *Geological Society of American Memoir*, **145**, 449–464.

Simberloff, D. (1978). Using island biogeographic distributions to determine if colonization is stochastic. *American Naturalist*, **112**, 713–726.

Simberloff, D., Heck, K. L., McCoy, E. D., and Connor, E. F. (1981). There have been no statistical tests of cladistic biogeographical hypotheses. In *Vicariance Biogeography: a Critique* (Eds. G. Nelson and D. E. Rosen), Columbia University Press, New York, pp. 40–63.

Sissons, J. B. (1979). The Loch Lomond stadial in the British Isles. *Nature*, **280**, 199–203.

Skellam, J. G. (1951). Random dispersal in theoretical populations. *Biometrika*, **38**, 196–218.

Smit, A., and Wijmstra, T. A. (1970). Application of transmission electron microscope analysis to the reconstruction of former vegetation. *Acta Botanica Neerlandica*, **19**, 867–876.

Smith, A. G. (1961). The Atlantic–Sub-Boreal transition. *Proceedings of the Linnean Society of London*, **172**, 38–49.

Smith, A. G. (1965). Problems of inertia and threshold related to post-glacial habitat changes. *Proc. Royal Soc. of London* B, **161**, 331–342.

Smith, R. T., and Taylor, J. A. (1969). The post-glacial development of soils in northern Cardiganshire. *Transactions of the Institute of British Geographers*, **48**, 75–96.

Stockmarr, J. (1975). Retrogressive forest development, as reflected in a mor pollen diagram from Mantingerbos, Drenthe, the Netherlands. *Palaeohistoria*, **17**, 37–51.

Stoddart, D. R. (1981). Biogeography: dispersal and drift. *Progress in Physical Geography*, **5**, 575–590.

Street, F. A., and Grove, A. T. (1979). Global maps of lake-level fluctuations since 30,000 yr B.P. *Quaternary Research*, **12**, 83–118.

Strobel, G. A., and Lanier, G. N. (1981). Dutch elm disease. *Scientific American*, **245**, 40–50.

Stuart, A. J. (1977). The vertebrates of the last cold stage in Britain and Ireland. *Phil. Trans. Royal Soc. of London* B, **280**, 295–312.

Sutherland, J. P. (1974). Multiple stable points in natural communities. *American Naturalist*, **108**, 859–873.

Tallantire, P. A. (1972). The regional spread of spruce (*Picea abies* (L.)Karst.) within Fennoscandia: a reassessment. *Norwegian Journal of Botany*, **19**, 1–16.

Tallantire, P. A. (1981). Some reflections on hazel (*Corylus avellana* L.) and on its boundary in Fennoscandia during the post-glacial. *Acta Palaeobotanica*, **21**, 161–171.

Tallis, J. H. (1964). The pre-peat vegetation of the Southern Pennines. *New Phytologist*, **63**, 363–373.

Tauber, H. (1965). Differential pollen dispersion and the interpretation of pollen diagrams. *Danmarks geologiske Undersøgelse* II, **89**, 69pp.

Ten Hove, H. A. (1968). The *Ulmus* fall at the transition Atlanticum–Subboreal in pollen diagrams. *Palaeogeography, Palaeoclimatology, Palaeoecology*, **5**, 359–369.

Tolonen, M. (1978). *Palaeoecological Studies on a Small Lake, S. Finland, with Special Emphasis on the History of Land Use.* Thesis, University of Helsinki.

Troels-Smith, J. (1960). Ivy, mistletoe and elm climatic indicators — fodder plants. *Danmarks Geologiske Undersøgelse,* IV, **4**(4), 32pp.

Tsukada, M. (1983). Late-Quaternary spruce decline and rise in Japan and Sakhalin. *Botanical Magazine,* **96**, 127–133.

Turner, J. (1964a). Surface pollen samples from Ayrshire. *Pollen et Spores,* **6**, 533–592.

Turner, J. (1964b). The anthropogenic factor in vegetational history. I. Tregaron and Whixall Mosses. *New Phytologist,* **63**, 73–90.

Turner, J. (1965). A contribution to the history of forest clearance. *Proc. Royal Soc. of London* B, **161**, 343–354.

Turner, J. (1970). Post-Neolithic disturbance of British vegetation. In *Studies in the Vegetational History of the British Isles* (Eds. D. Walker and R. G. West), Cambridge University Press, London, pp. 97–116.

Turner, J., and Hodgson, J. (1979). Studies in the vegetational history of the Northern Pennines. I. Variations in the composition of the early Flandrian forests. *Journal of Ecology,* **67**, 629–646.

Udvardy, M. D. F. (1981). The riddle of dispersal: dispersal theories and how they affect vicariance biogeography. In *Vicariance Biogeography: a Critique* (Eds. G. Nelson and D. E. Rosen), Columbia University Press, New York, pp. 6–29.

Usinger, H. (1978). Bölling–interstadial und Laacher Bimstuff in einem neuen Spätglazial-Profil aus dem Vallensgård Mose/Bornholm. Mit pollengrossenstatiti-scher Trennung der Birken. *Danmarks Geologiske Undersøgelse Årbog* 1977, 5–29.

Van der Hammen, T., Wijmstra, T. A., and Zagwijn W. H. (1971). The floral record of the late Cenozoic of Europe. In *The Late Cenozoic Glacial Ages* (Ed. K. K. Turekian), Yale University Press, New Haven and London, pp. 391–424.

Van Geel, B., and Kolstrup, E. (1978). Tentative explanation of the late glacial and early Holocene climatic change in north-western Europe. *Geologie en Mijnbouw,* **57**, 87–89.

Van Zeist, W., and Bottema, S. (1977). Palynological investigations in western Iran. *Palaeohistoria,* 29–19–85.

Von Post, L. (1926). Einige Aufgaben der regionalen Moorforschung. *Sverige Geologiske Undersøgelse* C, 337.

Von Post, L. (1946). The prospect for pollen analysis in the study of the Earth's climatic history. *New Phytologist,* **45**, 193–217.

Vorren, K-D. (1978). Late and Middle Weichselian stratigraphy of Andøya, north Norway. *Boreas,* **7**, 16–38.

Walker D. (1970). Direction and rate in some British post-glacial hydroseres. In *Studies in the Vegetational History of the British Isles* (Eds. D. Walker and R. G. West), Cambridge University Press, London, pp. 117–139.

Walker, D. (1978). Envoi. In *Biology and Quaternary Environments* (Eds. D. Walker and J. C. Guppy), Australian Academy of Sciences, Canberra, pp. 259–264.

Walker, D. (1982). Vegetation's fourth dimension. *New Phytologist,* **90**, 419–429.

Walker, D., and Pittelkow, Y. (1981). Some applications of the independent treatment of taxa in pollen analysis. *Journal of Biogeography,* **8**, 37–51.

Walker, M. J. C. (1984). A pollen diagram from St. Kilda, Outer Hebrides, Scotland. *New Phytologist,* **97**, 99–113.

Watts, W. A. (1961). Post-Atlantic forests in Ireland. *Proceedings of the Linnean Society of London,* **172**, 33–38.

Watts, W. A. (1967). Interglacial deposits in Kildromin Townland, near Herbertstown, Co. Limerrick. *Proceedings of the Royal Irish Academy,* **65B**(15), 339–348.

Watts, W. A. (1971). Postglacial and interglacial vegetation history of southern Georgia and central Florida. *Ecology*, **52**, 676–690.

Watts, W. A. (1973). Rates of change and stability in vegetation in the perspective of long periods of time. In *Quaternary Plant Ecology* (Eds. H. J. B. Birks and R. G. West), Blackwell, Oxford, pp. 195–206.

Watts, W. A. (1977). The Late Devensian vegetation of Ireland. *Phil. Trans. Royal Soc. of London* B, **280**, 273–293.

Watts, W. A. (1978). Plant macrofossils and Quaternary palaeoecology. In *Biology and Quaternary Environments* (Eds. D. Walker and J. C. Guppy), Australian Academy of Sciences, Canberra, pp. 53–67.

Watts, W. A. (1979a). Late Quaternary vegetation of central Appalachia and the New Jersey coastal plain. *Ecological Monographs*, **49**, 427–469.

Watts, W. A. (1979b). Regional variation in the response of vegetation to lateglacial climatic events in Europe. In *Studies in the Late-glacial of north-west Europe* (Eds. J. J. Lowe, J. M. Gray and J. E. Robinson), Pergamon, Oxford, pp. 1–21.

Watts, W. A. (1980). The Late Quaternary vegetation history of the southeastern United States. *Annual Review of Ecology and Systematics*, **11**, 387–409.

Watts, W. A. (1982). Response of biotic populations to rapid environmental and climatic changes. *Abstracts Seventh AMQUA Conference*, 19–21.

Watts, W. A. (1984). The Holocene vegetation of the Burren, western Ireland. In *Lake Sediments and Environmental History* (Eds. E. Y. Haworth and J. W. G. Lund), University of Leicester Press, Leicester, pp. 359–376.

Watts, W. A., and Bradbury, J. P. (1982). Palaeoecological studies at Lake Patzcuaro on the west-central Mexican Plateau and at Chalco in the Basin of Mexico. *Quaternary Research*, **17**, 56–70.

Webb, T. (1980). The reconstruction of climatic sequences from botanical data. *Journal of Interdisciplinary History*, **10**, 749–772.

Webb, T. (1981). The past 11,000 years of vegetational change in eastern North America. *BioScience*, **31**, 501–506.

Webb, T., Laseski, R. A., and Bernabo, J. C. (1978). Sensing vegetational patterns with pollen data: choosing the data. *Ecology*, **59**, 1151–1163.

Wegmüller, S., and Welten, M. (1973). Spätglaziale Bimstufflagen des Laacher Vulkanismus im Gebiet der westlichen Schweiz und der Dauphiné (F). *Eclogae Geologicae Helveticae*, **66**, 533–541.

Welinder, S. (1975). Prehistoric agriculture in eastern middle Sweden. *Acta Achaeologica Lundensia*, **4**, 102pp.

West, R. G. (1977). Early and middle Devensian flora and vegetation. *Phil. Trans. Royal Soc. of London* B, **280**, 229–246.

West, R. G. (1980). Pleistocene forest history in East Anglia. *New Phytologist*, **85**, 571–622.

White, P. S. (1979). Pattern, process, and natural disturbance in vegetation. *Botanical Review*, **45**, 229–299.

Whitehead, D. R., Rochester, H., Rissing, S. W., Douglass, C. B., and Sheehan, M. C. (1973). Late glacial and postglacial productivity changes in a New England pond. *Science*, **181**, 744–747.

Whittaker, R. H., and Goodman, D. (1979). Classifying species according to their demographic strategy. I. Population fluctuations and environmental heterogeneity. *American Naturalist*, **113**, 185–200.

Wiens, J. A. (1981). Scale problems in avian censusing. *Studies in Avian Biology*, **6**, 513–521.

Wijmstra, T. A. (1969). Palynology of the first 30 metres of a 120 m deep section in northern Greece. *Acta Botanica Neerlandica*, **18**, 511–527.

Wilkins, D. A. (1984). The Flandrian woods of Lewis (Scotland). *Journal of Ecology*, **72**, 251–258.

Williams, R. B. G. (1975). The British climate during the last glaciation: an interpretation based on periglacial phenomena. *Geological Journal Special Issue*, **6**, 95–120.

Williamson, G. B. (1975). Pattern and seral composition in an old-growth beech–maple forest. *Ecology*, **56**, 727–731.

Woillard, G. M. (1973). Grande Pile Bog: a continuous pollen record for the last 140,000 years. *Quaternary Research*, **9**, 1–21.

Woillard, G. M. (1979). Abrupt end of the last interglacial s.s. in north-east France. *Nature*, **281**, 558–562.

Wright, H. E. (1966). Stratigraphy of lake sediments and the precision of the paleoclimatic record. In *World Climate from 8000 to 0 B.C.* (Ed. J. A. Sawyer), Royal Meteorological Society, London, pp. 157–173.

Wright, H. E. (1976a). The dynamic nature of Holocene vegetation: a problem in paleoclimatology, biogeography, and stratigraphic nomenclature. *Quaternary Research*, **6**, 581–596.

Wright, H. E. (1976b). Pleistocene ecology — some current problems. *Geoscience and Man*, **13**, 1–12.

Yarranton, G. A., and Morrison, R. G. (1974). Spatial dynamics of a primary succession: nucleation. *Journal of Ecology*, **62**, 417–428.

Note added at proof stage

Since completion of this chapter, several very important contributions have been published that are of direct relevance to the topics discussed above. These include:

Andersen, S. Th. (1984). Forests at Løvenholm, Djursland, Denmark at present and in the past. *Det Kongelige Danske Videnskarbenes Selskab Biologiske Skrifter*, **24**, 208pp.

de Beaulieu, J-L., and Reille, M. (1984). A long Upper Pleistocene pollen sequence from Les Echets, near Lyon, France. *Boreas*, **13**, 111–1132.

Cole, K. (1985). Past rates of change, species richness, and a model of vegetational inertia in the Grand Canyon, Arizona. *American Naturalist*, **125**, 289–303.

Finegan, B. (1984). Forest succession. *Nature*, **312**, 109–114.

Maloney, B. K. (1984). Disease and the elm decline: a method of testing the hypothesis. *Circaea*, **2**, 91–96.

Martin, P. S., and Klein, R. G. (eds.) (1984). *Quaternary Extinctions, a prehistoric revolution*. University of Arizona Press, Tucson.

Ritchie, J. C. (1984). *Past and Present Vegetation of the Far Northwest of Canada*. University of Toronto Press, Toronto.

Welinder, S. (1983). The ecology of long-term change. *Acta Archaeologica Lundensia*, **9**, 115pp.

Wright, H. E. (1984). Sensitivity and response time of natural systems to climatic change in the Late Quaternary. *Quaternary Science Reviews*, **3**, 91–132.

2

Lake sediments and palaeohydrological studies

J. A. Dearing and I. D. L. Foster

*Department of Geography,
Coventry (Lanchester) Polytechnic, U. K.*

INTRODUCTION

Palaeohydrological studies have traditionally focused on relict fluvial land-forms, such as buried channels and terrestrial stratigraphic sequences, in order to reconstruct palaeodischarge and to infer climatic controls (cf. Collinson and Lewin, 1983; Gregory, 1983). The main purpose of this chapter is to review evidence available from an analysis of lake sediments which may be used, directly or indirectly, to infer palaeohydrological conditions; it does not aim to provide an exhaustive literature review of methodologies and interpretations but examines the underlying assumptions of such studies and explores the limitations and scope of different approaches to the determination of palaeohydrological environments. The ensuing discussion is centred on small lakes (area of water surface 100 ha or smaller) and their catchments within which hydrological and limnological processes, although complex in their nature and interaction, are perhaps most easily assessed in terms of sedimentary response.

LAKE-LEVEL FLUCTUATIONS

Many studies have attempted to reconstruct former lake levels on the basis of sedimentary evidence from which palaeohydrological conditions have been inferred. Studies undertaken on closed lakes, mainly in semi-arid and tropical zones, have made a great impact on palaeoclimatic research (Street and Grove, 1979), but studies based on sediments from open or outlet lakes, with their intrinsically more complex systems and covering timescales with less dramatic climatic response, have not provided detailed or unequivocal information about palaeoclimates. Within the context of open lake systems in humid temperate zones, two sets of questions may be posed by the palaeohydrologist. First, does the sedimentary analysis provide convincing evidence that the lake levels have fluctuated, or are there alternative explanations? Secondly, if the evidence is convincing, to what extent or degree of precision can a palaeohydrological interpretation be made, given the present, and frequently poor, level of understanding of the direct relationships between catchment hydrology and lake level? The methodology for sedimentary studies in this context is dealt with by Digerfeldt (Chapter 15), and it is our intention here to consider some of the ways in which the above questions may be tackled so that lake-level data may usefully contribute to our understanding of palaeohydrological processes. As a first step, we attempt to place lakes and their water levels in the framework of hydrological systems, and to identify the major factors and pathways of water movement by which lake levels are controlled.

Factors of lake-level change

Figure 2.1 is a simplified diagram showing the major factors and water movements that may operate in determining a lake level. Clearly, the effects of changes in basin permeability, crustal movements (tectonic, neotectonic or isostatic) and threshold levels have to be identified and evaluated before palaeohydrological and palaeoclimatological inferences can be made. Basin permeability will depend on lake sediment properties and the structure of the underlying bedrock or strata. Once a lake is established, deep seepage is unlikely to represent a major pathway for lake water (see below) except when crustal movements initiate a seepage zone through faulting. Those lakes situated in areas where the crust is unstable, or whose basin comprises weakly jointed strata, will be particularly susceptible to deep seepage; the lake level of Lake Azigza in the Middle Atlas range, set in Jurassic limestone, has diminished by about 7 m since 1981 by the process of deep seepage (authors' observations). Crustal movements, especially those of isostatic uplift, can cause the lake and its catchment to tilt. A dramatic example of lake-level change caused by tilting is provided by Kukkonen (1973), but very small angular changes from the horizontal can be predicted to cause significant changes in relative lake level. Simple calculations indicate that a ¼° shift from the horizontal will produce a relative lake level some 4 m above or below the previous shoreline over a horizontal distance of only 1000 m. Identifying such effects by levelling past and present shorelines at different points around the lake is a logical step, but may be frustrated in small lakes where the morphology of shorelines is not well developed and liable to degradation by erosional and biological processes over reasonably short time periods. Threshold adjustment

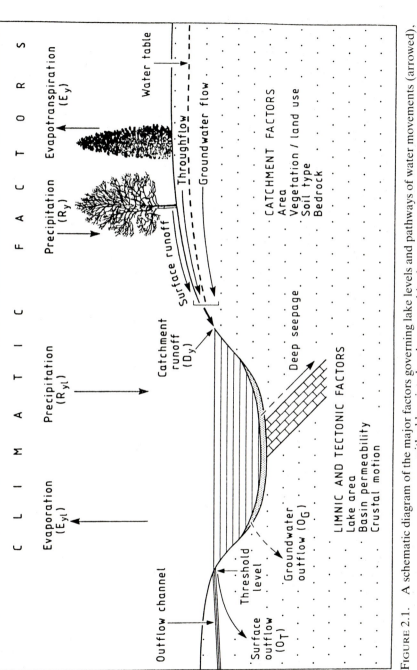

FIGURE 2.1. A schematic diagram of the major factors governing lake levels and pathways of water movements (arrowed), with abbreviated codes as used in the text

is a major control on lake level, and only where the outlet stream has drained continuously over a solid bedrock can it be assumed to have little effect on past lake levels. In drift covered areas it can be supposed that threshold adjustment was originally rapid, and perhaps pulsed, during the time taken for the erosion of unconsolidated deposits overlying the bedrock. Measurements of mean deposit thickness in the vicinity of the channel may be the only basis for estimating the likely extent of downcutting. Once these factors have been assessed or excluded, palaeohydrological interpretations of lake-level change may be advanced according to the sedimentary evidence.

Factors governing the sedimentary evidence

Lake-level changes can be constructed from several lines of sedimentary evidence including hiatuses, macrofossil distributions, coarse minerogenic or organic material distributions and changes in gross sediment stratigraphy (Chapter 15). In order that such sedimentary data may be used to infer lake-level changes, it is necessary to understand the relationships between lake levels and sedimentation patterns for all past periods. It cannot be assumed that the modern relationships have persisted throughout the lake's history. As with all palaeohydrological and palaeoecological enquiry, contemporaneous processes can only be inferred. However, consideration of the following factors may help the investigator in evaluating the quality of the sedimentary evidence:

(1) Wind-driven currents are normally the mechanism by which a sedimentation limit is defined. Historical fluctuations in wind direction and strength may therefore have altered previous relationships between water depth and sedimentation for any point on the lake bed. Both Simola (1981) and Dearing (1983) provide evidence to demonstrate the effects of wind on sedimentation in small lakes (see also Håkansson and Jansson, 1983). Such effects may not represent climatic change, but rather increased or decreased levels of exposure caused by changes in the catchment vegetation or in the type and form of littoral aquatic vegetation.

(2) Thermal stratification has been shown to effect a movement of sediment from littoral to profundal zones as a consequence of overturn following the breakdown of the thermocline (Davis, 1976). Lakes which do not exhibit thermoclines today may have done so in earlier stages of development when water depths were greater. A combination of increased wind exposure, the former existence of a thermocline and steeper underwater slopes during perhaps the pre-Boreal period could have caused sediment to focus into the deepest zones and to set sedimentation limits in relatively deep water.

(3) An absolute sediment chronology of the highest resolution is a prerequisite to an accurate assessment of the timing and longevity of lake level, and hence palaeohydrological changes. With the exception of annually

laminated sediments, the resolution of normal dating methods is at best in terms of decades, and at worst in centuries. Sedimentary features, such as sediment hiatuses and minerogenic layers, can therefore usually be dated only within a fairly coarsely resolved timescale, thus biasing interpretations towards apparently long-lived palaeohydrological change.

(4) Sediment composition will reflect the combined effect of hydrological and limnological processes. It will therefore be difficult to infer a singular explanation in terms of water-level fluctuation. The presence of marl, for instance, in lower sediment sequences may be evidence for shallow water, higher concentrations of dissolved calcium and high rates of biogenic preciptation. But alternatively, marl could signify the greater availability of calcium in the water during an early stage of ecosystem development (cf. Crocker and Dickson, 1957). Minerogenic layers (or loss-on-ignition residues) may simply demonstrate changes in the flux of allochthonous material, and even where such layers can be traced in littoral sequences the possibility of mass movements or slope wash as the result of, for instance, the aftermath of forest fire has to be considered (cf. Bloemendal *et al.*, 1979; Rummery *et al.*, 1979). Vertical stratigraphic changes which indicate hydroseral development are found in lakes whose catchments are small enough to have had water levels controlled by groundwater. In such circumstances, care should be taken to evaluate the consequences of both vegetation/land use and climatic change on runoff (both surface and subsurface). Deforestation that increases runoff (see below) could cause a substantial increase in the water depth of a lake with no permanent outlet threshold. But similarly, groundwater and hence water levels could rise as a consequence of iron pan formation or clay translocation in catchment soils, thus making the lake level (or bog level) part of a perched water table isolated from the regional groundwater level.

Water balance considerations in palaeohydrological reconstructions

Calculated changes in lake level provide, in theory, empirical estimates of changes in lake volume, which in turn might be assumed to represent the balance between catchment runoff (D_y), precipitation to the lake area (R_{yl}), evaporation from the lake area (E_{yl}) and outflow losses (O_T, O_G). Thus:

$$V_l = D_y + R_{yl} - E_{yl} - O_T - O_G \qquad (2.1)$$

where V_l is the change in lake volume (more or less equivalent to the change in water depth in some lakes). Unfortunately, empirical estimates for parameters on the right-hand side of equation 2.1 will invariably be unknown, and hence a water balance will be virtually impossible to reconstruct. The role of lake-level change studies in palaeohydrological and palaeoclimatic investigations is therefore one of providing supporting evidence to other lines of enquiry. In this sense, the data obtained have to be evaluated for their potential to provide substantive evidence of hydrological response to climatic and ecological change. The range of magnitude of such responses is not easy to predict for even one environment, but some guidance as to what to expect can be achieved by consideration of reconstructed palaeoclimatic data and contemporary catchment studies. Factors such as climate, vegetation and soil constitute extrinsic controls on lake levels, and identification of their relative importance is usually the aim of the study. There may also be, however, intrinsic controls that reduce the possibility of making direct comparisons between lake-level data obtained for different periods in the lake's history, and so reduce the chance of realizing initial aims. Intrinsic controls are largely bound up with the hydrological and limnological mechanisms that translate catchment water balances into lake volumes, and a lake level may respond differently to similar climatic conditions according to the antecedent conditions in the lake. In humid temperate zones $R_{yl} \geqslant E_{yl}$ (Figure 2.1) over an annual period, except in severe drought years. Therefore all lakes in this climatic region will normally have water levels controlled by outflow losses (O_T or O_G), and both may have occurred in the same lake during the last 15,000 years. It might be justified to assume that lakes with well-defined outflow streams (O_T) today, have a minimal O_G component; but in the past the reversed state could have prevailed. For example, a kettle-hole lake in sandy deposits could originally have had a low water level controlled by O_G into the surrounding permeable material. As fine sediment with a low hydraulic conductivity accumulated at the lake bed, the total area of the lake basin with high percolation rates would decrease and water levels could be expected to rise independently of surface water inflow and the level of the surrounding water table. Lake levels would rise until the basin morphology dictated an O_T component, and adjustment of the threshold level by outflowing water could be expected to result in a continuous lowering of water level until water inflow, lake volume and water outflow reached equilibrium. And as Reynolds (1979) points out, a true groundwater lake may well become isolated from the water table of the catchment if the lake level falls below the lip of impermeable sediment. With the impermeable layer reducing O_G, the lake level might remain relatively high even in dry climatic phases.

In order to identify the likely effect of extrinsic controls on lake levels in different lakes, it may be useful to reconstruct palaeohydrological conditions from palaeoclimatic and water balance data that is presently available. Using Lamb's (1977) and Lockwood's (1979) palaeoclimate estimates, changes in lake levels have been estimated for central England during selected periods over the last 9000 years. Table 2.1 shows calculations of outflow water (O) from lakes with small, medium and large catchments (lake area:catchment area ratios 1:1, 1:10 and 1:100 respectively), using the simple formula:

$$O = D_y k + R_{yl} - E_{yl} \tag{2.2}$$

TABLE 2.1. Calculated changes in water levels (± millimetres) for selected periods throughout the last 9000 years, showing variability with lake-area:catchment-area ratio, based on palaeoclimatic data presented by Lamb[1] (1977) and Lockwood[2] (1979)

Years B.P.	Lake-area:catchment-area		
	1:1	1:10	1:100
9000[1]	+684	+4167	+38997
6500	+868	+5251	+49081
4500	+759	+4674	+43824
2450–2900	+861	+5172	+48282
450–800	+784	+4780	+44740
250–400	+707	+4307	+40307
0–50	+771	+4686	+43836
6500[2]	+46	+838	+8758
0–50	+237	+1866	+18156

where k is the ratio of catchment area to lake area and all other parameters (Figure 2.1) are expressed as millimetre equivalents. E_{yl} has been calculated using Lamb's and Lockwood's figures for evaporation or potential evapotranspiration adjusted for 20% higher rates from a water surface (Prior, 1980). All the outflow values in Table 2.1 are positive, indicating that broad climatic change, as presented by Lamb and Lockwood for Flandrian climates, would not be expected to produce permanent (i.e. lasting more than a year) reductions in lake level below the outlet threshold. The lower losses at 6500 B.P. calculated from Lockwood's data (1979) take into account the reduced runoff from deciduous forest and the associated higher rates of winter evaporation from a forest canopy. Table 2.2 shows the effect of a drought year on outflow losses. The same palaeoclimatic data has been used again, but adjusted for 43% less R_y, 26% more E_y and 40% less D_y in order to duplicate the effects of the 1975/76 drought years in central England (data from Elmdon airport, Birmingham, calibrated to the 1940–70 climatic averages). Lake-level lowerings (as shown by negative values in Table 2.2) are predicted for Lockwood's hydrological regime, with size of catchment greatly influencing the magnitude of lowering. At around 6500 B.P. the mean annual lake level during a drought is calculated to be about 0.5 m below the outlet threshold in lakes with small catchments, and at or just below the outlet threshold in lakes with larger catchments. The effect on lake level of a sequence of '1975/76 drought years' can be calculated by simply multiplying the positive or negative water depth increments in Table 2.2 by the number of years in the sequence. So for 6500 B.P. a decade of droughts would reduce water levels by about 5 m and 0.25 m in 1:1 and 1:10 lake-catchments respectively. Four general points emerge from a consideration of these calculations:

TABLE 2.2. Calculated impact of drought on water levels (± millimetres) for selected periods throughout the last 9000 years, showing variability with lake-area:catchment-area ratio, based on palaeoclimatic data presented by Lamb[1] (1977) and Lockwood[2] (1979)

Years B.P.	Lake-area:catchment-area		
	1:1	1:10	1:100
0–50[1]	+51	+2400	+25890
0–50[2]	−276	+705	+10515
6500[2]	−502	−25	+4745

(1) For climates similar to those in central England the water balance models predict that broad climatic change during the last 9000 years would not have induced prolonged periods of low water levels. Lamb's (1977) discussion of climatic change in the rest of the temperate zone would indicate that rainfall and evaporation rates are not thought to have been significantly different from those in central England; at least, not to such an extent that outflow losses shown in Tables 2.1 and 2.2 would radically alter. Apparent increases in outflow losses relative to present day (0–50 B.P.) losses are of the order of 5–10%, and are therefore likely to be accommodated by the configuration of present thresholds, and not to result in prolonged periods of raised water levels.

(2) At any time, drought years can produce low lake levels below the outlet threshold, but probably only in lakes with relatively small catchments (i.e. 1:1–1:10) (cf. Thom and Ledger, 1976). Prolonged periods of drought, lasting several years, would lower lake levels owing to diminished groundwater storage, and some shallow lakes would undoubtedly dry up. Though droughts may have quite short return periods (cf. Thom and Ledger, 1976) they are characterized as being relatively short-lived (a few years at most). Even allowing for the time taken for groundwater stores to recover from moisture depletion, it is difficult to imagine the circumstances under which a lake level remains low for a period exceeding a decade, without there being a modification to the form and structure of the natural vegetation and its response in the pollen record.

(3) More important than either broad climatic change or extreme climatic events is probably the seasonal variability of past climates, for which data are sparse. Annual figures of rainfall, evaporation etc. as used to construct Tables 2.1 and 2.2 mask changes in their seasonal distribution which can clearly alter the magnitude of annual runoff (cf. Dury, 1965). All the calculations presented here have been expressed as annual depth-equivalents to outflow losses. During climatic phases with hot summers, lakes could have low water levels for several months, but still show O_T losses over the year. Such seasonal

fluctuations in water level could be expected to reach their greatest amplitude when the frequency or duration of anticyclonic conditions was high.

(4) Comparison of the two sets of outflow losses for 6500 B.P. (Table 2.2) shows the importance of vegetation in modifying runoff. Lockwood (1979) used a runoff value of 88 mm in his palaeohydrological reconstruction for Atlantic forest, a 49% reduction of the present day value of 181 mm runoff from grassland (cf. Aario's (1969) 20–30% reduction in runoff during the Atlantic in S. Finland). Data obtained from contemporary catchment studies and simulation models suggest that maximum reductions in runoff caused by afforestation of grassland are about 40–50% (e.g. Swift *et al.*, 1975), and similar percentage increases in runoff could be expected for the initial period following deforestation (e.g. Bormann *et al.*, 1974). In contrast to the short-lived effects of droughts on lake levels, a reduction of runoff by 50% caused by afforestation might be expected to produce a prolonged period of low water. However, climatically induced changes in vegetation communities are likely to be too gradual to alter runoff to this extent, with the exceptions of vegetation change caused by fire or during colonization of deglaciated landscapes. Certainly clearance phases could be expected to increase runoff, but this is likely to be accommodated by the threshold or to trigger adjustment of the threshold level. In the latter case, it is possible that rapid adjustment might lead to a real water level lowering. Generally speaking, decreased runoff will tend to increase the amplitude of seasonal water level fluctuations, and increased runoff will tend to diminish seasonal fluctuations.

Implications

(1) Lake levels are controlled through a complex interaction of extrinsic and intrinsic factors, and although the existence of lake-level change through the Holocene is not in doubt, the magnitude of the role played by climatic and hydrologic factors at any site may be unclear; threshold adjustment, crustal movements, climatic change, vegetation change and land-use change can all induce lake-level fluctuations.

(2) Identification of synchronous changes in lake levels within a region may be the best means for isolating the effects of climate (Chapter 15), but true synchroneity will frequently be difficult to achieve given typical dating errors. Regional lake-level fluctuations may be the result of macro-crustal movements.

(3) Sediment deposition in lakes at an early stage of their history may well have been subjected to a number of intrinsic controls (e.g. efficient sediment focusing) which later on either do not operate, or operate with less effect. Consequently, even greater scrutiny of the sedimentary evidence is needed at these times. Water levels in 'young' lakes may also have been subjected to intrinsic controls, with a greater role played by groundwater and threshold adjustment.

(4) Modelling lake-level response from palaeoclimatic data is a useful means for predicting the magnitude of water level lowerings in different sizes of lake–catchment systems. Data for central England suggest that long-term climatic change in the Flandrian has been too subtle to induce long-term responses in the levels of lakes with permanent outflows. More likely, lake levels are affected by changes in the seasonal variability of climate, or short-lived extreme climatic events such as droughts. It is apparent that some types of sedimentary evidence for lake-level change, such as hiatuses and coarse minerogenic layers, can be the consequence of different scales of climatic change, and very precise dating will be needed in order to identify the true climatic cause. The models also suggest that increased runoff will not normally cause permanent higher water levels, except in those basins controlled by groundwater; what may be recorded is the sedimentation limit associated with a rise in minimum summer levels or recovery from drought levels.

(5) Reconstructed vegetation/land-use change can be usefully compared with present-day analogues in monitored catchments, in order to predict the likely effects on runoff and hence lake levels. From a brief survey of analogue data it would appear that these effects may be potentially greater than those induced by long-term climatic change. In order to detect the rapid vegetation changes that are predicted to cause dramatic fluctuations in runoff, constructed pollen records through a period of lake-level change should be of the highest resolution possible.

(6) The opportunity to calibrate sedimentary evidence against known water level fluctuations should be taken whenever possible. There is a lack of testing of methodologies, and yet lakes with documented artificial lowerings are common, and there are some examples of lakes for which geomorphological evidence indicates an outlet threshold control on water level caused by adjustments to isostatic uplift (e.g. Kukkonen, 1973; Renberg, 1978).

LAKE-SEDIMENT-DERIVED SEDIMENT YIELDS

The minerogenic content of lake sediments will often be the result of hydrological processes transporting suspended sediment into the limnic system. An historical record of lake-sediment-derived sediment yields is potentially useful for identifying the form of and response to palaeohydrological or man-induced change. The four examples presented in Figure 2.2 are all based on multiple coring exercises relating synchronous sedimentation levels to dated master cores, and should provide accurate information on absolute changes in sediment yield comparable with stream monitored estimates (cf. Figure 2.2(d)). Although it may be possible to derive a temporal picture of sediment yield from sedimentation rates of single 'master' cores, problems of variable sedimentation patterns make interpretation of the data difficult and only qualitative; that is to say, providing a trend of sediment yield. Preferably,

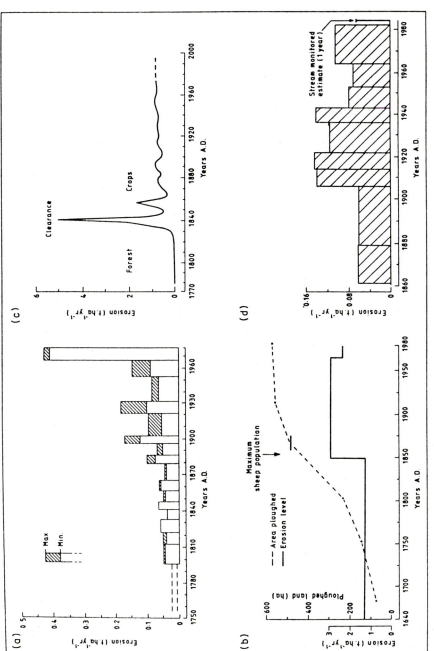

FIGURE 2.2. Lake sediment based estimates of sediment yield: (a) Llyn Peris, Gwynedd, U.K. (Dearing *et al.*, 1981) (b) Havgårdssjön, Scania, Sweden (Oldfield *et al.*, 1983) (c) Frains Lake, Michigan, U.S.A. (Davis, 1976) (d) Merevale Lake, Warwickshire, U.K. (Foster *et al.*, 1985). Reproduced from Petts, G. E. and Foster, I. D. L. (1985), copyright by Edward Arnold

sediment yields should be calculated from total volumes of deposited sediment by using core correlation methods in a large number of cores (Chapter 13). The following sections examine the problem of interpreting such data in terms of contemporary drainage basin studies, and how contemporary analogues might be used to infer palaeohydrological change.

Evaluation

Calculations of total dry minerogenic mass should be evaluated for components derived from non-catchment sources, i.e. eroded lake bank material, atmospheric inputs and reworked subaqueous deposits. (Diagenic and biogenic components are considered in Chapter 13). Additionally, consideration should be given to the magnitude of sediment loss through the lake outflow.

Many lakes exhibit signs of bank erosion caused by wave, ice or trampling actions. Very few quantitative data exist for rates of bank retreat; but a guide to the extent of contribution by bank material to sediment yield calculations is given here in the form of a nomograph (Figure 2.3). Use of the nomograph requires the following statistics: an estimate of probable sediment yield from the catchment (kg ha^{-1}yr^{-1}) obtained from lake-based studies or published data, the lake area (ha), the catchment area (ha) and the length (m) of erodible bank (or lake width:length ratio). An example of using the nomograph is shown in Figure 2.3 and is explained in the caption. Also shown is the maximum or tolerable mean annual bank retreat that is required to limit bank contributions to less than 5% of the total sediment yield for recent historical periods at seven sites. Values for maximum bank retreat range from less than 0.01 cm yr^{-1} (Mirror Lake) to just over 0.3 cm yr^{-1} (Seeswood Pool), i.e. about 10–300 cm per 1000 years. These figures may need substantial modification for some of the sites according to field conditions (especially height and length of erodible bank), but they serve to show that imperceptible erosional processes can produce significant allochthonous inputs. Catchments with low sediment yields and small lakes are clearly more sensitive to these inputs, but periods of fluctuating water levels might lead to excessive bank erosion in many lakes.

Seasonal or continuous reworking of subaqueous littoral sediments by wave action and overturn processes may lead to erroneous calculations of sediment yield. In theory, high-density sampling strategies which include both littoral and profundal zones should overcome problems of sediment redistribution; losses in some cores being compensated by gains in others. However, correlation of synchronous depths between littoral and profundal cores may prove to be difficult (Chapter 13). More serious is the problem of mass slumping of deposited sediment on slopes greater than 4–5% (Håkansson and Jansson, 1983), which may result locally in unconformable or inverted

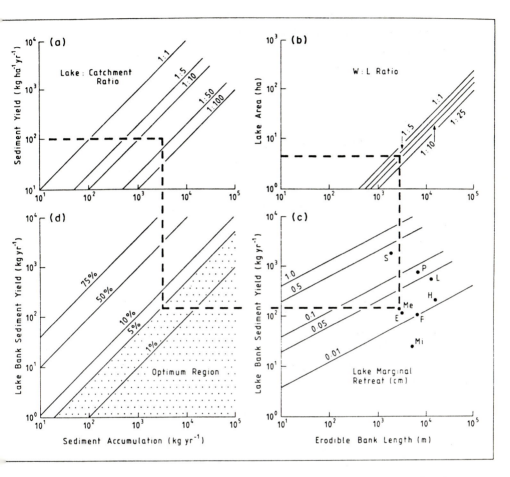

FIGURE 2.3. Nomograph of bank contributions to lake sediment dry mass, showing: (a) Relationships between lake-area:catchment-area ratio and sediment yield to provide dry sediment accumulation (kg per yr) (assuming that the lake has 100% trap efficiency); (b) relationships between lake area (ha) and the lake width:length ratio to give estimate of erodible bank (or obtainable by field measurements); (c) relationships between percentage contribution and bank length to give values of lake bank erosion or marginal retreat (cm per yr) (lake bank sediment yields based on a bank height of 40 cm with a dry bulk density of 2.65 g cm^{-3} which can be adjusted for site conditions); (d) percentage contributions to dry sediment accumulation by bank material. Also shown are the extrapolations (dotted lines) used to assess bank material in Merevale lake (Me; Foster *et al.*, 1985), and 'tolerable' marginal retreat values in Seeswood Pool, Central England (S; unpublished data), Frain's Lake, Michigan (F; Davis, 1976), Loe Pool, Cornwall, U.K. (L; O'Sullivan *et al.*, 1982), Llyn Peris, Gwynedd, Wales (P; Dearing *et al.*, 1981), Havgårdssjön, S. Sweden (H; Oldfield *et al.*, 1983), Mirror Lake, New Hampshire (Mi; Likens and Davis, 1975), and Lake Egari, Papua New Guinea (E; Oldfield unpublished data)

sediment sequences and problematic dating. The occurrence of reworked pollen grains or stratigraphic changes may give an indication of the timing and extent of reworking but are unlikely to provide accurate assessments of the quantitites of sediment involved. Lakes with steep underwater contours at their margins and flat bottoms will minimize these effects.

Atmospheric fallout of dust particles to the lake surface may make a significant contribution to the annual increment of deposited minerogenic material. In Great Britain, the range of figures for fallout are 2000–8000 kg ha^{-1} yr^{-1} in urban areas, and 40–400 kg ha^{-1} yr^{-1} in rural areas (Simmons, 1974). At the Merevale Lake site in central England (cf. Figure 2.2) an atmospheric input of about 330 kg ha^{-1} yr^{-1} required adjustment of estimated sediment yields by about 9% (Foster *et al.*, 1985). Site-specific data should be sought wherever possible, both for the present environment and analogues of past environments.

Outflow losses of suspended sediment depend on a complexity of factors, including sediment concentrations, flow conditions in the lake, lake morphology and river or stream discharge. Normally it is possible to identify those lakes with poor trap efficiencies by utilizing standard empirically derived graphs of trap efficiency versus water residence time (Figure 2.4). Care should be taken in using such graphs, however, as seasonal variability in water residence times may distort the apparent trap efficiency percentage: high stream discharges are usually associated with high sediment yields but correspondingly low water residence times and hence low trap efficiencies. It would seem sensible to choose lakes that have minimum water residence times of 9–12 months.

Interpretation

An interpretation of lake-sediment-derived sediment yields requires consideration of the following points, each of which is discussed below.

(1) the relationship between stream transported sediment yields and the erosion process;
(2) the magnitude–frequency characteristics of transported sediment;
(3) the relationship between flow velocity or discharge and sediment transport processes;
(4) the environmental controls on discharge and sediment yield.

The sediment yield of a stream contributing to lake inputs will be a function of hillslope erosional processes and of the balance between erosion and deposition in stream channels. In lakes receiving no channelled inputs, direct soil erosion processes will to a large extent control the allochthonous component of the sedimentary record. For the former system, the relationship between erosion and sediment yield is complex since not all material detached from hillslopes and channels will reach lake basins. It is estimated that the

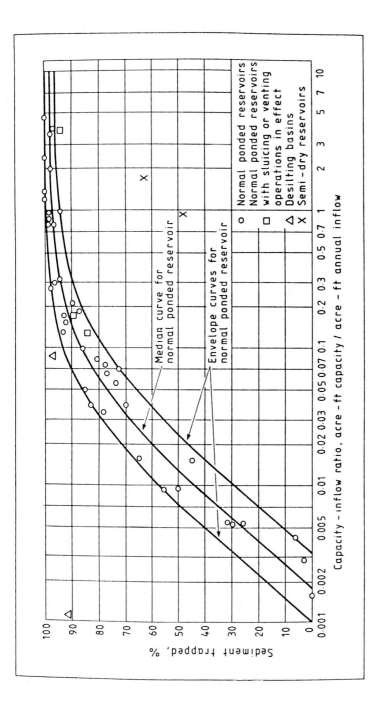

FIGURE 2.4. The trap efficiency of reservoirs (after Brune, 1953). Trap efficiency (%) is plotted against the capacity inflow ratio (equivalent to annual residence time) and may be used to approximate trap efficiency in natural lakes. *Reproduced from Brune (1953), Trans. Amer. Geophys. Union, **34**, 407–418, copyright by the American Geophysical Union*

amount of particulate material that arrives at a downstream point will often be less than 25% of the amount eroded (Vanoni, 1975). Recent theoretical studies by Trimble (1981) have indicated that this figure may fall as low as 6%. This suggests that the majority of sediment is deposited at some intermediate location within the drainage basin, such as in flood plains, especially where entrainment and transport velocities fall below a critical threshold. The difference between sediment yield (Y) and erosion (E) is usually expressed as the sediment delivery ratio (SD):

$$SD = \frac{Y}{E}$$

The most useful model for predicting SD has been empirically verified by the US Soil Conservation Service (1971). This model expresses SD as a simple function of basin area which, though derived from data exhibiting considerable scatter, may be used where direct estimates are not available. Thus:

$$SD = 0.36\, A^{-0.2}$$

where A is the drainage basin area in km^2. Attempts to use this equation should be made with some caution, since it has not been verified in Europe and has been criticized in many North American studies (cf. Trimble, 1981).

Sediment-yield data provide no information on the relative contribution of hillslope and channel erosion processes, and to date no generally applicable predictive models are available. Recent work utilizing the phenomenon of enhanced magnetic susceptibility in soil profiles (Oldfield *et al.*, 1979; Walling *et al.*, 1979) shows promise as an indirect method for assessing the relative proportions of topsoil and stream bank erosion.

The transportation of particulate material will be a function of the magnitude and frequency of storm events. In their examination of the significance of magnitude and frequency in relation to geomorphic events in fluvial systems, Wolman and Miller (1960) suggest that although the high magnitude event can move large quantities of sediment in a short period of time, this amount is relatively small in comparison with total transport in a given area. Their data showed that 78–98% of sediment movement occurs at discharges with recurrence intervals of between 1:5 and 1:10 years. As drainage basin size decreases, however, the high-magnitude/low-frequency event becomes more important in sediment transport. Other studies have confirmed Wolman and Miller's observations (e.g. Piest, 1965; Webb and Walling, 1982). Any change in the magnitude and frequency of storm events in a region, as exemplified by Dury (1965), may increase or decrease sediment yields even where mean annual rainfalls and temperatures are constant. A further problem is that similar magnitude-frequency events may induce different catchment responses (cf. Newson, 1980) because of variable antecedent moisture conditions or the existence of presently unknown intrinsic thresholds. Geomorphic thresholds have been discussed in detail (Schumm, 1977, 1979; Brunsden and Thornes, 1979), but of more concern here are the thresholds relating to sediment transport. It is unlikely that analysis of lake-sediment-based erosion rates will permit magnitude-frequency interpretations unless stratigraphical horizons (e.g. laminations of variable thickness) can be correlated with documented hydrological data (cf. Simola, 1981).

The suspended-load component of rivers is often not a capacity load and is controlled by sediment availability on hillslopes and in channels. Broad relationships between grain diameter and water velocity have been established (e.g. Hjulström, 1935; Sundborg, 1956), but other controls such as grain density and the cohesive nature of fine sediments impose an important secondary control (Figure 2.5). Additionally, the presence of considerable quantities of organic matter, especially in forested ecosystems, may complicate these relationships because of density differences. Likens *et al.* (1977) have shown that in watershed 6 of the Hubbard Brook Experimental Forest over 33% of the total particulate transport is organic. Furthermore, Foster *et al.* (1985) have shown that although the organic content of suspended sediment derived from a forested catchment in central England is less than 5% of the total output, the relative proportion of organic matter in suspended sediment increases to more than 90% of the total at low sediment concentrations and hence low velocities. The relative contribution of bedload to total sediment transport is usually small in humid temperate regions of low relief. The ratio between bed sediment and suspended sediment transport for the mountain rivers of Great Britain is approximately 4:1 (Lewin *et al.*, 1974), whereas in piedmont zone rivers the bed:suspended proportion ranges from 0.02 to 0.11 (McManus and Al Ansari, 1975; Grimshaw and Lewin, 1980).

For individual basins, a direct relationship may be derived between stream discharge and sediment yield, although this usually results in a large scatter in the regression relationship (Gregory and Walling, 1973). Furthermore, no past environment can be assumed to possess the same statistical relationship as today, and the use of contemporary analogue data to model the relationship in past environments will lead to circular arguments. What may be useful is to relate the porportion of sand (>63 μm), in lake sediment, semi-quantitatively, to the stream power, which as Figure 2.5(c) shows can exhibit a linear relationship. Care should be taken to sample a sediment core that represents mean accumulation rates through time and which has not been affected by slumping or inputs from the shore zone. Yamamoto (1977) used this approach to identify broad changes in palaeoprecipitation.

Many attempts have been made to examine the relationships between sediment yield and climate (Langbein and Schumm, 1958; Fournier, 1960; Douglas, 1967, 1976; Fleming, 1969, 1982; Meybeck, 1977; Edwards, 1979; Walling and Webb, 1983). Such analyses have made use of broad climatic

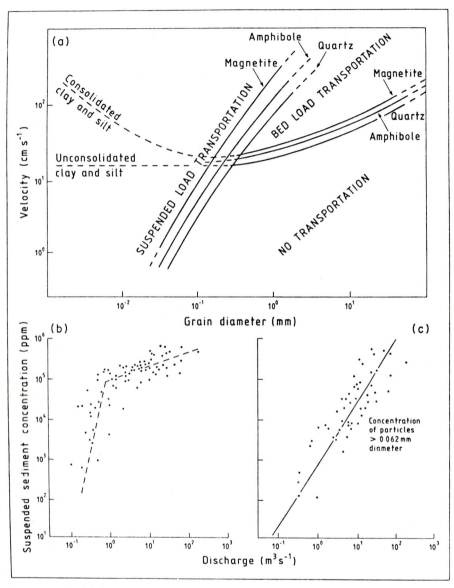

FIGURE 2.5. (a) (after Sundborg, 1956, and Ljunggren and Sundborg, 1968, redrawn in Collinson and Thompson, 1983) Erosion, transport and deposition in fluvial systems as a function of grain diameter and flow velocity, for minerals of varying specific gravity (applicable to velocities 1 m above the bed). (b) and (c) (after Gregory and Walling, 1973) Suspended sediment concentration as a function of stream discharge for all particle sizes (b) and for particles >0.062 mm diameter (c). *(a) Reproduced by permission of George Allen & Unwin and Almqvist & Wiksell Publ. (b) and (c) Reproduced by permission of the authors and American Geophysical Union*

parameters such as annual precipitation, annual runoff and effective precipitation, often modified by consideration of vegetation, temperature and catchment size (Figure 2.6 (a)–(d)). Curves based on global data are unlikely to be of value owing to the variety of critical precipitation and runoff thresholds identified as inducing sediment yield increases under different climatic regimes (cf. Walling and Webb, 1983). For long records of sediment yield regional curves modified by land-use, vegetation and geology may be usefully employed as a means of gauging the broad climatic and stream discharge variability that causes sediment yield to fluctuate in natural systems.

Estimates of contemporary sediment yields on the basis of stream monitoring frequently obscure the large variation in annual statistics. Data from Poland (Jarocki, 1957) and south-west England (Webb and Walling, 1982) give calculated coefficients of variation of 131.25% and 57.9% respectively, over time periods of 4 and 7 years respectively. Such high variations over short timescales will be smoothed by lake sediment analyses, which may complicate the reconstruction of palaeodischarge from the sedimentary record, and restrict interpretation of lake-based data to broad climatic change. Separation of the impact of human activities from broad climatic change in the lake sediment record can only be substantiated on the basis of documentary evidence or detailed palaeoecological reconstructions. Under constant climatic conditions, clearance and cultivation may increase sediment yields by an order of magnitude over natural rates (Golubev, 1982). The exact increase in sediment yield, however, will be a function of soil erodibility, crop management practice, slope angle and slope length, as exemplified by the Universal Soil Loss Equation, in addition to the erosivity of raindrops and overland flow (cf. Kirkby and Morgan, 1980; De Boodt and Gabriels, 1980; Walling, 1982).

CONCLUSIONS

(1) Evaluation of the non-catchment-derived component in lake sediments is essential prior to an interpretation of trends in sediment yield or sediment accumulation.

(2) Some attempt could be made to estimate the sediment delivery ratio of the catchment from published ratios for similar environments and to adjust the sediment yields accordingly.

(3) Consideration of the magnitude and frequency of sediment transport in contemporary rivers suggests that changes may occur in either the channel or the catchment which will be independent of climatic change, and will not usually be recorded in sediment stratigraphy.

(4) Velocity–sediment-concentration relationships on a particle-size basis in contemporary streams could be used to reconstruct palaeodischarges more effectively than total-concentration–discharge models.

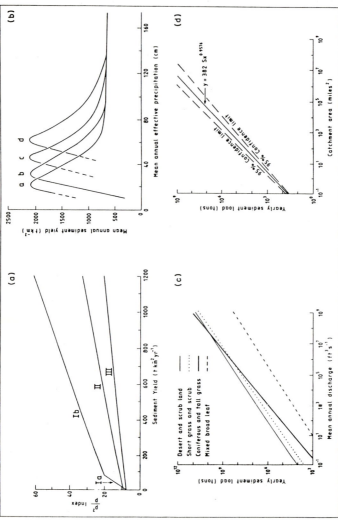

FIGURE 2.6. Precipitation runoff, climate and vegetation controls on sediment yields. (a) (after Fournier, 1960) p^2/P index (p = rainfall in wettest month in mm, P = mean annual rainfall in mm), subdivided on the basis of relief and climate: Ia low relief/temperate; Ib low relief/tropical, subtropical, semi-arid; II high relief/humid; III high relief/semi-arid. (b) Mean annual sediment yield plotted against mean annual effective precipitation. These curves are modified by mean annual temperature $a = 4.5°C$, $b = 10°C$, $c = 15°C$, $d = 20°C$. The reduction in sediment yield on each curve represents the control on erosion by increasing vegetation density. (c) and (d) (after Fleming, 1969) Mean annual sediment yields and discharge for several different vegetation zones (c) and modified for the impact of basin area on sediment yield (d). (b) *Reproduced from Langbein and Schumm (1958), Trans. Amer. Geophys. Union, 39, 1076–1084, copyright by the American Geophysical Union*

(5) Although sediment load is some function of stream discharge, and hence climate, it may equally relate to a modification of the landscape by clearance and cultivation. Comparison of observed data with the known range of climatically controlled sediment yields (e.g. Figure 2.6) may indicate the deviation from the climatic norm that might be associated with human impact.

Acknowledgements

We are grateful to B. Berglund, K. Gregory, R. Batterbee, A. Brown, A. Dawson, G. Digerfeldt and N. Roberts for helpful comments.

REFERENCES

Aario, R. (1969). The northern discharge channel of ancient Päijänne and the palaeohydrology of the Atlantic period. *Bull. Geol. Soc. Finland*, **41**, 3–20.

Bloemendal, J., Oldfield, F., and Thompson, R. (1979). Magnetic measurements used to assess sediment influx at Llyn Goddionduon. *Nature*, **200**, 50–53.

Bormann, F. H., Likens, G. E., Siccama, T. G., Pierce, R. S., and Eaton, J. S. (1974). The export of nutrients and recovery of stable conditions following deforestation at Hubbard Brook. *Ecol. Mono.*, **44**, 2–5–277.

Brune, G. M. (1953). Trap efficiency of reservoirs. *Trans. Amer. Geophys. Union*, **34**, 407–418.

Brunsden, D., and Thornes, J. B. (1979). Landscape sensitivity and charge. *I.B.G. Trans. New Series*, **4**, 463–484.

Collinson, J. D., and Lewin, J. (Eds.) (1983). *Modern and Ancient Fluvial Systems*, Blackwell, Oxford.

Collinson, J. D., and Thompson, D. B. (1982). *Sedimentary Structures*, George Allen and Unwin, London.

Crocker, R. L., and Dickson, B. A. (1957). Soil development on the recessional moraines of the Herbert and Mendenhall glaciers, south-eastern Alaska. *J. Ecol.*, **45**, 169–185.

Davis, M. B. (1973). Redeposition of pollen grains in lake sediment. *Limnology and Oceanography*, **18**, 44–52.

Davis, M. B. (1976). Erosion rates and land-use history in Southern Michigan. *Environmental Conservation*, **3**, 139–148.

De Boodt, M., and Gabriels, D. (Eds.) (1980). *Assessment of Erosion*, John Wiley, London.

Dearing, J. A. (1983). Changing patterns of sediment accumulation, in a small lake in Scania, southern Sweden. *Hydrobiologia*, **103**, 59–64.

Dearing, J. A., Elner, J. K., and Happey-Wood, C. M. (1981). Recent sediment flux and erosional processes in a Welsh upland lake-catchment based on magnetic susceptibility measurements. *Quat. Res.*, **16**, 356–372.

Douglas, I. (1967). Man, vegetation and the sediment yield of rivers. *Nature*, **215**, 925–928.

Douglas, I. (1976). Erosion rates and climate : geomorphological implications. In: *Geomorphology and Climate*, (Ed. E. Derbyshire), John Wiley, London, pp. 269–287.

Dury, G. H. (1965). General theory of meandering valleys. *U.S. Geological Survey Professional Paper* 452, 43 pp.

Edwards, K. A. (1979). Regional contrasts in rates of soil erosion and their significance with respect to agricultural development in Kenya. In: *Soil Physical Properties and Crop Production in the Tropics* (Eds. R. Lal and D. J. Greenland), John Wiley, New York.

Fleming, G. (1969). Design curves for suspended load estimation. *Proc. Inst. Civ. Engrs.*, **43**, 1–9.

Fleming, G., and Kadhimi, A. A. (1982). Sediment modelling and data sources: a compromise in assessment. In: *Recent Developments in the Explanation and Prediction of Erosion and Sediment Yield* (Ed. D. E. Walling), IAHS Pub. No. 137, pp. 251–260.

Foster, I. D. L., Dearing, J. A., Simpson, A. D., and Appleby, P. G. (1985). Estimates of contemporary and historical sediment yields in the Merevale catchment, Warks. U.K. *Earth Surface Processes and Landforms* **10**, 45–68.

Fournier, F. (1960). Debit solide des cours d'eau. Essai d'estimation de la perte en terre subie, par l'ensemble de globe terrestre. *Int. Assn. Sci. Hyd.*, Pub. 53, pp. 19–22.

Golubev, G. N. (1982). Soil erosion and agriculture in the world : an assessment and hydrological implications. In: *Recent Development in the Explanation and Prediction of Erosion and Sediment Yield* (Ed. D. E. Walling), IAHS Pub. No. 137, pp. 26 1–268.

Gregory, K. J. (Ed.) (1983). *Background to Palaeohydrology. A Perspective*, John Wiley, London.

Gregory K. J., and Walling, D. E. (1973). *Drainage Basin Form and Process*, Edward Arnold, London

Grimshaw, D. L., and Lewin, J. (1980). Reservoir effects on sediment yield. *J. Hyrol.*, **47**, 163–171.

Hjulström, F. (1935). Studies of the morphological activities of rivers as illustrated by the River Fyris. *Bull. Geol. Inst. Univ. Uppsala*, **25**, 221–527.

Håkansson, K., and Jansson, M. (1983). *Principles of Lake Sedimentology*, Springer-Verlag, Berlin.

Jarocki, W. (1957). *A Study of Sediment*, Wydawnictwo Morskie, Poland.

Kirkby, M. J., and Morgan, R. P. C. (Eds.) (1980). *Soil Erosion*, John Wiley, London.

Kukkonen, E. (1973). Sedimentation and typological development in the basin of the lake Lohjanjärvi, South Finland. *Geological Survey of Finland*, Bulletin 261, 67 pp.

Lamb, H. H. (1977). *Climate, Present, Past and Future. Vol. 2: Climatic History and the Future*, Methuen, London.

Langbein, W. B., and Schumm, S. A. (1958). Yield of sediment in relation to mean annual precipitation. *Trans. Amer. Geophys. Union*, **39**, 1076–1084.

Leopold, L. B., and Maddock, T. (1953). The hydraulic geometry of stream channels and some physiographic implications. *U.S. Geol. Survey Prof. Paper* 252, 56 pp.

Lewin, J., Cryer, N., and Harrison, D. I. (1974). Sources for sediments and solutes in mid-Wales. In: *Fluvial Processes in Instrumented Watersheds*, Gregory, K. J. and Walling, D. E. (Eds.), 73–85, I.B.G., London.

Likens, G. E., Bormann, F. H., Pierce, R. S., Easton, J. S., and Johnson, N. M. (1977). *Biogeochemistry of a Forested Ecosystem*, Springer-Verlag, New York.

Likens, G. E., and Davis, M. B. (1975). Post-glacial history of Mirror lake and its watershed in New Hampshire, USA : an initial report. *Verh. Internat. Verein. Limnol.*, **6**, 982–993.

Ljunggren, P., and Sundborg, A. (1968). Some aspects of fluvial sediments and fluvial morphology. II. A study of some heavy mineral deposits in the valley of the river Lule Älv. *Geogr. Ann.*, **50A**, 121–135.

Lockwood, J. G. (1979). Water balance of Britain, 50,000 B.P. to the present day. *Quat. Res.*, **12**, 297–310.

McManus, J., and Al-Ansari, N. A. (1975). Calculation of sediment discharge in the River Earn, Scotland. *Proc. 9th Int. Congr. Sediment*, Nice, pp. 113–118.

Meybeck, M. (1977). Dissolved and suspended matter carried by rivers : composition, time and space variation, and world balance. In: *Interaction between Sediments and Fresh Water* (Ed. H. L. Golterman), Dr. W. Junk, The Hague.

Newson, M. D. (1980). The geomorphological effectiveness of floods — a contribution stimulated by two recent events in mid-Wales. *Earth Surface Processes*, **5**, 1–16.

Oldfield, F., Battarbee, R. W., and Dearing, J. A. (1983). New approaches to recent environmental change. *The Geographical Journal*, **149**, 167–181.

Oldfield, F., Rummery, T. A., Thompson, R., and Walling, D. E. (1979). Identification of suspended sediment source by means of magnetic measurements : some preliminary results. *Water Resources Research*, **15**, 211–218.

O'Sullivan, P. E., Coard, M. A., and Pickering, D. A. (1982). The use of laminated lake sediments in the estimation and calibration of erosion rates. In: *Recent Developments in the Explanation and Prediction of Erosion and Sediment Yield*, IAHS Pub. No. 137, pp. 383–396.

Petts, G. E., and Foster, I. D. L. (1985). *River and Landscape*. Edward Arnold, London.

Piest, R. F. (1965). The role of the large storm as a sediment contribution. *Proc. Fed. Int. Agency, Sed. Conf. U.S.D.A. Misc. Pub* 970, pp. 97–108.

Prior, M. J. (1980). Evaporation and soil moisture deficit. In: *Atlas of Drought in Britain 1975–76* (Eds. J. C. Doornkamp and K. J. Gregory). I.B.G., London, pp. 29–30.

Renberg, I. (1978). Palaeolimnology and varve counts of the annually laminated sediment of Lake Rudetjärn, northern Sweden. *Early Norrland*, **11**, 63–92.

Reynolds, C. S. (1979). The limnology of the eutrophic meres of the Shropshire–Cheshire Plain. *Field Studies*, **5**, 93–173.

Rummery, T. A., Bloemendal, J., Dearing, J. A., Oldfield, F., and Thompson, R. (1979). The persistence of fire-induced magnetic oxides in soils and lake sediments. *Ann. Geophys.*, **35**, 103–107.

Schumm, S. A. (1977). *The Fluvial System*, J. Wiley, New York.

Schumm, S. A. (1979). Geomorphic thresholds: the concept and its applications. *I.B.G. Trans. New Series*, **4**, 485–515.

Simmons, I. G. (1974). *The ecology of natural resources*, Edward Arnold, London.

Simola, H. (1981). Sedimentation in a eutrophic stratified lake in S. Finland. *Ann. Bot. Fennici*, **18**, 23–36.

Street, F. A., and Grove, A. T. (1979). Global maps of lake level fluctuations since 30,000 B.P. *Quat. Res.*, **12**, 83–118.

Sundborg, A. (1956). The river Klaralven : a study of fluvial processes. *Geogr. Ann.*, **38**, 127–316.

Swift, L. W., Swank, W. T., Markin, J. B., Luxmoore, R. J. and Goldstein, R. A. (1975). Simulation of evapotranspiration and drainage from mature and clear-cut, deciduous forests and young pine plantation. *Water Resources Research*, **11**, 667–673.

Thom, A. S., and Ledger, D. C. (1976). Rainfall, runoff and climatic change. *Proc. Instn. Civ. Engrs.*, **61**, 633–652.

Trimble, S. W. (1981). Changes in sediment storage in the Corn Creek Basin, drifters area, Wisconsin. 1853–1975. *Science*, **214**, 181–183.

Vanoni, V. A. (Ed.) (1975). *Sedimentation Engineering* (American Society of Civil Engineering Manuals and Reports on Engineering Practice No. 54), ASCE, New York, NY.

Walling, D. E. (Ed.) (1982). Recent developments in the explanation and prediction of erosion and sediment yield. *Proc. 1st. Scient. Gen. Assmb.*, Exeter, UK, IAHS Pub. No. 137.

Walling, D. E., Peart, M. R., Oldfield, F., and Thompson, R. (1979). Suspended sediment sources identified by magnetic measurements. *Nature*, **281**, 110–113.

Walling, D. E., and Webb, B. W. (1983). Patterns of sediment yield. In: *Background to Palaeohydrology* (Ed. K. J. Gregory), John Wiley, Chichester, pp. 69–100.

Webb, B. W., and Walling, D. E. (1982). The magnitude and frequency characteristics of fluvial transport in a devon drainage basin and some geomorphological implications. *Catena*, **9**, 9–23.

Wolman, M. G., and Miller, J. C. (1960). Magnitude and frequency of forces in geomorphic processes. *Journal of Geology*, **68**, 54–74.

Yamamoto, A. (1977). The structure of density and grain size variations seen near the 130 m layer in the 200 m long core sample from Lake Biwa. In: *Palaeolimnology of Lake Biwa and the Japanese Pleistocene* (Ed. S. Horie), **5**, 125–126.

Handbook of Holocene Palaeoecology and Palaeohydrology
Edited by B. E. Berglund
© 1986 John Wiley & Sons Ltd.

3

Hydrological changes in mires

P. D. MOORE

*Department of Biology,
King's College,
London, U.K.*

INTRODUCTION

Mires are peat-producing ecosystems that develop in sites of abundant water supply. The term mire is used here in a very general sense (as it is by Gore, 1983) to include bogs, fens and swamps, which differ from one another in such features as the position of their water table relative to the peat surface and the pattern of water influx into the mire, which in turn influences nutrient supply. In the case of swamps, which have a water table permanently above the peat surface, the mire ecosystem may merge with the lake (Chapter 2), the two being distinguished merely by the depth of water at which emergent aquatic plants are able to survive. In general, swamps seldom develop if the growing season water table is more than 2 m above the soil surface (Spence, 1982), though some emergent, swamp species may persist as submerged aquatics at even greater depths. At the other end of the mire spectrum, bogs may have their soil surface several metres above the general groundwater table, though the impermeable nature of the peat mass below them (Ingram, 1982), together with the water supply coming direct from precipitation, ensures an abundance of water over their surface and the water table is rarely deeper than 50 cm — and that only in times of drought.

All mires are sensitive to changes in hydrology. Figure 3.1 shows how such a change can influence processes within the ecosystem which may have far-reaching and complex effects on its overall balance. The palaeoecologist faces two problems: first the detection of hydrological change in the stratigraphic record of the mire, and second the elucidation of the environmental causes of the change.

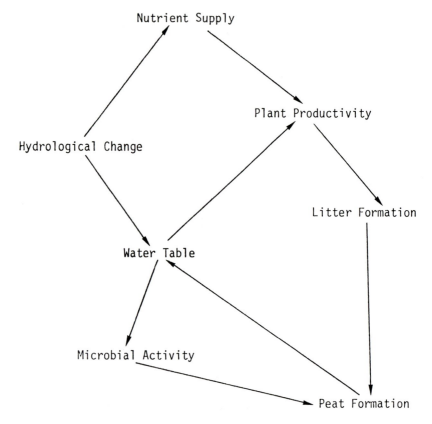

FIGURE 3.1.   Relationship between mire hydrology and other processes in the
mire ecosystem

MIRE CLASSIFICATION AND HYDROLOGY

Before these two problems can be analysed in detail, it is necessary to consider
some general aspects of mire hydrology. A simplified outline only will be given
here; for exhaustive reviews of this subject Ingram (1983) and Ivanov (1981)
should be consulted.

Mires can be subdivided into two major types, depending on the influence of
groundwater (Moore and Bellamy, 1974; Gore, 1983; Moore, 1984). Where
the surface of a mire receives water from outside that mire's own limits, it can
be referred to as *rheotrophic* (literally 'fed by flow') and can be given the broad
term *fen*. Where the surface of the mire receives water only directly from the
atmosphere in the form of rain or snow, it is *ombrotrophic* ('fed by the rain')
and can be termed a *bog*. A simplified model of the hydrological relationships
of such mires is given in Figure 3.2.

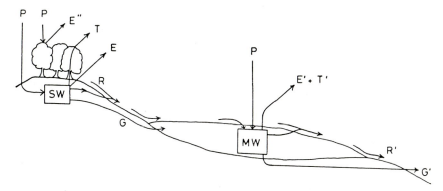

FIGURE 3.2. Simplified hydrological model of a mire. P = precipitation; R = runoff water from mire catchment; R' = runoff water from mire; G = groundwater seepage; G' = seepage from mire peat mass; E = evaporation from ground surface; E' = evaporation from mire surface; E'' = evaporation of intercepted water from vegetation canopy in catchment; T = transpiration in catchment; T' = transpiration of mire vegetation; SW = soil water reservoir; MW = mire water reservoir

The hydrological budget for this model mire is given by:

$$\Delta MW = \text{influx} - \text{efflux}$$
$$= (R + G + P) - (E' + T' + R' + G')$$

where ΔMW is the change in the mire water reservoir.

If the centre of the mire is elevated above the influence of runoff from surrounding slopes (R) and groundwater seepage (G), then water influx is entirely due to precipitation (P) and the mire can be termed a bog. If groundwater seepage and runoff contribute to the surface water of the mire, then it is a fen.

Changes in the mire-water reservoir (ΔMW) can be caused by changes in efflux or influx. Changes in *efflux* may be brought about in several ways:

(1) The accumulation of peat in the mire itself alters drainage patterns and can lead to reduced efflux and hence, very often, additional peat accretion. Moore and Bellamy (1974) use the term *secondary peat* to describe this material. This can be regarded as the consequence of autogenic successional processes which will be found in most developing mires.

(2) Erosion of the peat mass. Some mires, particularly those developed on sloping surfaces, may be subject to erosion in which streams cut back at their headwaters and permit a more rapid surface-water drainage following storms. Such erosion can lean to increased overall efflux and a lowered water table in the mire.

(3) Drainage on the part of man is usually carried out by the construction of ditches, thereby increasing efflux.

(4) A change in overall temperature at the site will have an effect on evaporation and transpiration from the mire surface.

Influx of water to the mire can similarly be modified as a result of changes in a number of variables:

(1) A change in precipitation is felt directly by the mire surface as an influx change. This is the case in both ombrotrophic and rheotrophic systems, though in the latter case its effect will be magnified since the response will also be felt in the groundwater draining from the surrounding catchment.

(2) A change in temperature will modify evaporation and transpiration rates throughout the catchment, thus affecting the amount of water available for influx into the mire. Interception losses and evapotranspiration efflux from the mire surface will also be affected. Other influx changes will only be experienced by rheotrophic mires.

(3) Changes in vegetation of the catchment, either as a result of climatic processes, successional developments or human activity, will change the overall transpirational losses in the catchment and also the interception and re-evaporation losses, E'' (Swank and Douglass, 1974; Hornbeck *et al.*, 1975). Both of these biotic changes will modify the groundwater of the catchment and hence the influx of water to the rheotrophic mire.

(4) Loss of litter from the surface of the catchment soils as a consequence, for example, of erosion or burning, reduces the capacity of those soils to retain incident water (Reynolds and Knight, 1973). A reduced soil reservoir of this type means that there will be less buffering capacity against the effects of storms, leading to sudden addition to the mire influx (Kochel and Baker, 1982; Finley and Gustavson, 1983).

HYDROLOGICAL CHANGE AND MIRE STRATIGRAPHY

The model that has been described here demonstrates the complexity of hydrological interactions in mire ecosystems. The palaeoecologist is faced with the task of detecting changes in past mire hydrology on the basis of stratigraphic evidence. There are several sources of such evidence to be found in the stratified peat, but there is one further complication which needs to be considered, that is the spatial scale of the hydrological changes which are being sought.

The surface of a mire is not normally uniform, but has a pattern of wetter and drier areas developed in response to a whole range of environmental factors such as climate, slope, vegetation interactions and drainage processes (see reviews by Sjörs, 1961; Ruuhijärvi, 1960; Barber, 1981; Moore, 1982; Foster *et al.*, 1983). When examining mire stratigraphy, scale must be kept firmly in mind, for quite different explanations may be necessary to account for small-scale local events relating to drainage patterns than those needed to account for gross morphological change in the entire system. The use by Aario (1932) of the terms 'Grossform' and 'Kleinform' for these two scales of mire features is to be recommended (Goode, 1974).

One important practical implication of this scale problem is that individual corings of peat deposits are of little value in elucidating past hydrological changes in mires, for it is impossible to determine from them whether any observed variations in stratigraphy belong to the Grossform or the Kleinform scale of development. The problem is best overcome either by using a series of closely spaced corings (e.g. Moore, 1977; Smart, 1982) or, preferably, by the excavation of sections of the mire (e.g. Walker and Walker, 1961; Stewart and Durno, 1969; Aaby, 1976; Barber, 1981).

The main sources of evidence on which hydrological changes can be detected are the following:

(1) changes in macrofossil stratigraphy;
(2) changes in peat humification;
(3) the initiation of peat deposition over ground surfaces;
(4) changes in microfossil stratigraphy;
(5) alterations in peat chemistry;
(6) changes in the ratio of inorganic to organic materials in the peat.

The occurrence and interpretation of these sources of evidence will now be discussed.

Macrofossil stratigraphy

Perhaps the most widely used source of evidence for hydrological change in peat stratigraphy is its macrofossil stratification. Changes in the relative abundance of *Sphagnum* mosses, *Eriophorum vaginatum, Narthecium ossifragum, Rhacomitrium lanuginosum* and *Calluna vulgaris* can be extremely useful in the reconstruction of local hydrological events. Some problems may arise, however, where woody plants actually replace a former peat (Grosse-Braukmann, 1979). Detailed studies involving the identification of *Sphagnum* species can be very informative (e.g. Smart, 1982). The interpretation of such data is based on a knowledge of the present ecology of the plant, such as the tendency for *Rhacomitrium lanuginosum* to grow on well-drained hummocks or erosion features (Moore, 1977), or the physiological demands of the species, such as the requirement of *Narthecium ossifragum* for well-oxygenated flowing water (Summerfield and Rieley, 1975).

The change in macrofossil content may be associated with a change in sediment type, such as the invasion of *Phragmites* peat over ombrogenous peat as at Elan Valley, shown in Figure 3.3 (Moore and Chater, 1969), or the development of algal detritus layers within ombrogenous peats (Walker and Walker, 1961), or even calcareous marl deposits developed over *Sphagnum*

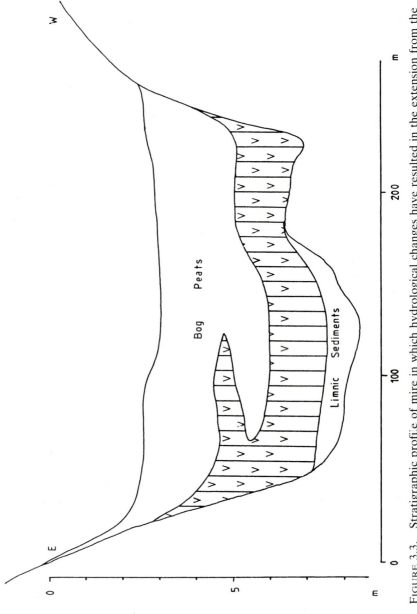

FIGURE 3.3. Stratigraphic profile of mire in which hydrological changes have resulted in the extension from the east of *Phragmites* and wood (rheotrophic) peat over *Sphagnum* (ombrotrophic) peat. Elan Valley, Wales

peat (Hohenstatter, 1972). The distinction between gross changes in the total stratigraphy and therefore hydrology of the mire, and local changes in pool patterns, can easily be made by studying the mire profile. In the case of the changes at Elan Valley there has clearly been a major hydrological change involving the groundwater component of influx. In the case of algal pool sediments found at Claish Moss (Figure 3.4), changes are evidently local and do not necessarily imply influx changes unless the pattern can be demonstrated to repeat itself across the mire surface (Barber, 1981).

The connection of macrofossil changes with causative factors is extremely difficult because, by their very nature, they tend to reflect only local alterations in hydrology. If the alteration occurs in an ombrotrophic mire and the developing peat remains in an ombrotrophic state, then either the event is local (Kleinform) or, if widespread, is likely to be a reaction of the mire to climatic change, either in the form of increased precipitation or decreased temperature or both. The quest for significant, regional changes of this sort is well-documented (e.g. Godwin, 1946; Barber, 1981). Conway (1948) elaborated on von Post's views of climatic thresholds in ecology and this point has been applied by Barber (1982) to the question of changing hydrology and vegetation of ombrotrophic mires. Depending on their hydrological balance, some bogs could be in a more sensitive condition than others to respond to changes in the influx/efflux balance resulting from climatic shifts. Different bogs, in other words, may have different response thresholds. This proposal needs further development both in terms of its theoretical basis and in its demonstration in the field, but it is an interesting possibility.

Where macrofossil evidence indicates a change from ombrotrophic to rheotrophic conditions, the likelihood is that influx has increased, at least partly as a result of runoff or groundwater input (components R and G in Figure 3.2). It is possible that precipitation has increased at the same time and is, indeed, responsible for the entire increase in influx. So, the extension of reed-swamp over bog peat in Figure 3.3 could be a response to an increase in climatic wetness resulting in surface flooding. It could, on the other hand, be associated with vegetation change in the catchment, perhaps as a result of woodland clearance. There may be archaeological sources of evidence to indicate such interaction (e.g. Oldfield, 1967), one of the best examples of this being the flooding of the Polish Iron Age site at Biskupin, where flooded peatland settlement almost certainly resulted from local changes in forest cover (Skarzynska, 1965).

But there are times when the development of the peat mass itself can lead to reduced drainage (efflux) and hence additional peat build up (secondary peat, *sensu* Moore and Bellamy, 1974), or coalescence of raised bogs can raise local water tables (Casparie, 1972). Alternatively changes may occur in the local drainage pattern of the land surface leading to local swamping of the mire, as appears to have happened at Hohenstatter's (1972) Bavarian mire, where *Sphagnum* peat is replaced by lake marl, muds and *Carex* peats.

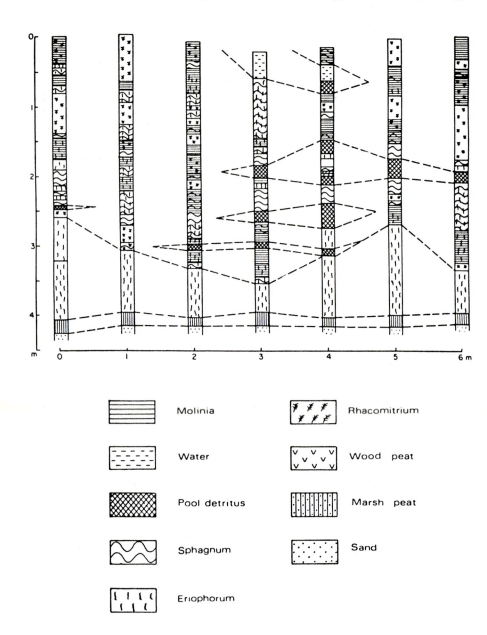

FIGURE 3.4.   Closely spaced peat borings through a peat deposit at Claish Moss, Scotland, showing Kleinform variations in peat stratigraphy. Pools extend and contract in time, as do hummocks bearing *Rhacomitrium lanuginosum*

Humification changes

Even where macrofossil changes are not strongly evident in the peat stratigraphy, alterations in the level of decomposition of the peat may be clearly marked in the form of colour bands in the section. Decomposition in peats and the factors which influence it have been reviewed in some detail by Martin and Holding (1978) and by Dickinson (1983). Some of these factors are summarized in Figure 3.5. As can be seen, alteration in the hydrological regime is likely to be a major contributor to any change in decomposition rates and hence humification.

Although colour (Aaby, 1976) and hand testing are the most frequently used guides to the state of humification, some attempts have been made to attain greater precision by using pollen density (e.g. Dickinson, 1975). There are considerable technical problems attached to this, however, and also questionable assumptions are necessary concerning the general pollen influx.

Humification changes, together with macrofossil changes, are the basis of most studies on pool patterns and on climatic variation using raised bog sites, the two sources of evidence normally being used to supplement one another.

Mire initiation

The commencement of peat accumulation at a site is in itself an indication of a changing hydrological budget such that more water is retained in the incipient mire water reservoir (MW in Figure 3.2). This may occur in lowland valley sites where drainage impedence, perhaps coupled with additional influx from the catchment, can result in the development of valley mires. Sometimes it is possible to link both of these processes to a common factor, such as the clearance of forest by man. Destabilized soils lead to erosion, and the decomposition of mineral material including clays in the valley may enhance its water retention capacity. At the same time, any land use which maintains a low vegetation cover in the catchment, such as grazing, will maintain a higher influx into the system. In combination, this can lead to valley mire initiation, as has been demonstrated in some southern English valley mires (Moore and Willmot, 1976).

On upland plateau sites, deforestation and consequently increased influx, sometimes coupled with podzolization and pan formation (Taylor and Smith, 1981), followed by intensive human land-use by grazing and burning, is now well established as the major stimulus to the replacement of woodland by blanket mires (Moore, 1975; Moore *et al.*, 1984). Stratigraphically this is demonstrated by the development of ombrogenous peat over mor humus or, especially in basins and valleys, over wood peats (Chapman, 1964a).

Human activity need not always be involved in the initial development of peat deposits, of course. In the lowlands of north-west Germany, Behre (1983) relates peat formation to rising groundwater tables caused by a rising sea level. And in Alaska, Ugolini and Mann (1979) consider soil maturation and

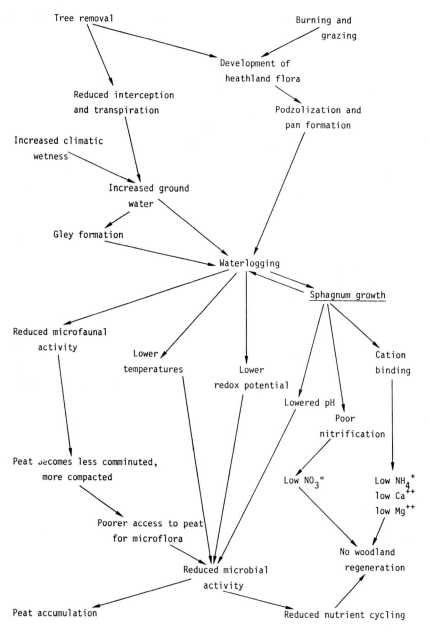

FIGURE 3.5.   Model showing the sequence of events leading to peat initiation in
an upland site

podzolization as the prime agent in the development of some low-lying tundra peatlands. They regard bog as the final stage in the vegetation succession of the area, replacing coniferous forest.

Microfossil stratigraphy

Microfossils in the peat may provide additional evidence of hydrological change, both in a local and in a regional context. The pollen of types which are more easily characterized to generic or specific level are on the whole of more value than those which cannot be determined so precisely. For example, the use of *Narthecium* pollen by Wiltshire and Moore (1983) proved more useful than Cyperaceae in denoting changes in water regime. Taxa such as *Hydrocotyle* may be used as indicators of groundwater input (Godwin, 1975). Often it is possible to correlate microfossil with macrofossil or humification changes, as in the case of the Peel raised bogs in the southern Netherlands, where Janssen and Ten Hove (1971) found a close correlation between peaks in *Sphagnum* spores and peat stratigraphy.

Regional changes in hydrology will be reflected in the pollen fallout, but may not be easy to interpret. Forest clearance and land-use changes, or burning within the catchment, will affect the overall pollen assemblage, but its hydrological implications often have to remain speculative. Where the peat surface receives drainage water from the catchment, some characteristically dry-land taxa may increase with additional runoff as the local soils are eroded and redeposited on the mire surface. Taxa such as *Ilex* may prove useful indicators in this respect as they have done in lake sediments (Pennington, 1979), their presence suggesting the erosion and redeposition of woodland mor humus layers.

Other microfossils may prove valuable indicators of local hydrological conditions, such as fungal remains, which van Geel (1972) found to be generally associated with drier phases in a bog's history. Rhizopod tests survive well in peat and have been used extensively by van Geel, as by Aaby (1976) and others. Corbet (1973) has shown how well these may be correlated with environmental conditions, and Heal (1962) has demonstrated them to respond to changes in wetness by encysting rather than migrating, hence retaining their place in the stratigraphic sequence.

Peat chemistry

This is a source of evidence that has not been adequately exploited by palaeoecologists involved in peat analysis. Much is known about the water chemistry of peatland types (e.g. Sjörs, 1950, Gorham, 1956), and it can be demonstrated that at least some of these chemical features are effectively fossilized in the peat stratigraphy (Bellamy and Rieley, 1967). But there are few peatland palaeoecological studies that have fully taken advantage of this most valuable potential.

One example of the possibilities of peat chemistry in detecting hydrological change is the work of Chapman (1964b) at Coom Rigg Moss in northern England. By means of stratigraphic changes in peat chemistry he was able to reconstruct the development of an ombrotrophic surface from a rheotrophic one and was even able to speculate, on the basis of Ca:Mg ratio, on the influence of precipitation and the degree of oceanicity in the prevailing climate. This is an area of research which palaeoecologists urgently need to follow up.

Inorganic materials in peat

Severe hydrological changes in a catchment can result in a considerable additional water influx to rheotrophic mires and sometimes even to ombrotrophic ones. Such high-energy influx may carry eroded inorganic materials from the catchment soils and redistribute them, often unevenly, over the mire surface.

An example of this is shown in Figure 3.6 which is derived from a study of a peaty hollow in oak woodland in southern England. Two phases of inorganic inwash are shown in this site, in early zone KG 2 and in zone KG 3. On both occasions the inwash is accompanied by changes in the balance of tree pollen, involving a decrease in *Quercus* and an increase in *Betula*. These changes, together with alterations in some herbaceous components not shown here (Moore, 1983), suggest that a forest succession has been interrupted by intense disturbance (probably human) on two occasions and that these perturbations have sufficiently upset the hydrological balance to cause the destabilized soils to be washed over the mire surface. Here the microfossil profile supplements the inorganic input to provide a fairly precise picture of the palaeoecology of the site. The use of pollen preservation and damage can also supply useful information concerning this kind of mineral inwash (Cushing, 1967).

On occasions, the macrofossil record can be of equal value in the interpretation of inorganic inwash. Figure 3.7 shows the stratigraphic profile of a valley mire at Acebron in southern Spain. The full profile of this mire enables one to ascertain the extent of the inwash episode responsible for the extensive clay and sand layers in the upper part of the peat. Clearly this was a major event affecting the entire valley. One can also discern the pattern of water movement during the event as a result of examining the textural characteristics of the sediment, for the high-energy water movements down the centre of the valley have been responsible for a sand-rich deposit, whereas the marginal areas have largely received silts and clays. All these sediments are rich in wood charcoal fragments indicating that the forested catchment has been disturbed by fire, leading to destabilized soils and also to additional water influx to the valley as a consequence of vegetation destruction.

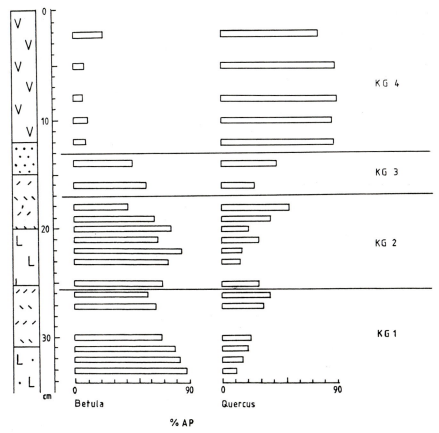

FIGURE 3.6. Stratigraphic column and the pollen content of *Betula* and *Quercus* in a woodland hollow deposit at Kingswood Glen, Bromley, Kent. Two phases of mineral inwash occurred and these are accompanied by reductions in the *Quercus:Betula* pollen ratio

Both of these examples involve human disturbance, but this is not always the cause of mineral inwash. Godwin (1975) shows an inorganic sediment layer at Borth Bog in west Wales which resulted from a marine transgression, and a whole series of such sediments interdigitating with peats is recorded from north Germany by Behre (1983). Marine sediments of this type may be recognized by characteristic microfossils, such as foraminifera or pollen grains of Chenopodiaceae.

In alpine areas, mires may receive inorganic matter carried by snow melt waters, as has been described in the stratigraphy of Rotmoos in southern Austria by Rybnicek and Rybnickova (1977). Macrofossil content and general topography and mire morphology should make it apparent when this type of hydrological influence is encountered.

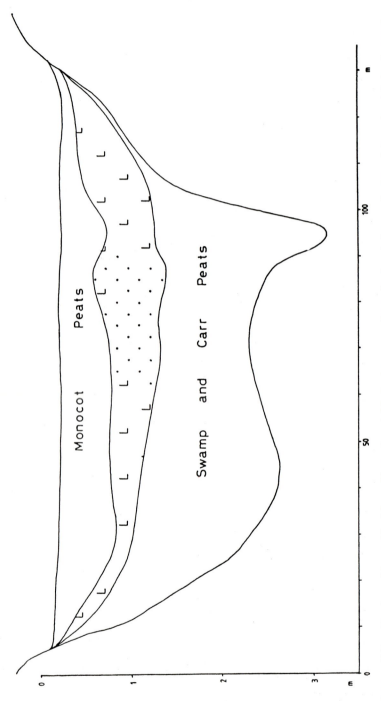

FIGURE 3.7. Stratigraphic profile of a valley mire at El Acebron, southern Spain. A layer of mineral inwash, rich in charcoal fragments, extends across the mire in the upper peat layers. The sand content of this inwash layer is greatest in the centre where water flowed with highest energy during the flood event

CONCLUSIONS

The hydrological relationships of mires form a complex system and the alteration of any component in the system can have extensive repercussions. Two areas in which such repercussions are often felt are the accretion of sediment or litter on the mire surface and the activity of micro-organisms on those materials. The outcome of these interactions is a change in the nature of the material which accumulates.

The palaeoecologist is faced with the task of elucidating the nature of a past hydrological change on the basis of the sediments which survive, and this can best be done by the combination of all possible approaches outlined here. By considering all available evidence it is often possible to pinpoint the prime source of the hydrological event in terms of which component of the influx and/or efflux has been modified (Figure 3.2). It may then be possible to speculate on the factors which underlie the hydrological change, such as climatic fluctuations, local drainage pattern alterations or some form of human impact. Finally, the hypotheses generated from such a study may receive confirmation from other studies of lakes, mires or soils in the region.

REFERENCES

Aaby, B. (1976). Cyclic climatic variations in climate over the past 5500 yr reflected in raised bogs. *Nature*, **263**, 281–284.

Aario, L. (1932). Pflanzentopographische und palaogeographische Moorunter-suchungen in N. Satakunta. *Fennia*, **55**, 1–179.

Barber, K. E. (1981). *Peat Stratigraphy and Climatic Change*. Balkema, Rotterdam.

Barber, K. E. (1982). Peat-bog stratigraphy as a proxy climatic record. In: *Climatic Change in Later Prehistory* (Ed. A. F. Harding), Edinburgh University Press, pp. 103–113.

Behre, K. E. (1983). An interdisciplinary research project on the development of landscape, prehistoric settlements and the history of vegetation in the N. W. German lowlands. *Quat. Studies in Poland*, **4**, 223–228.

Bellamy, D. J., and Rieley, J. (1967). Ecological statistics of a miniature bog. *Oikos*, **18**, 33–40.

Casparie, W. A. (1972). *Bog Development in Southeastern Drenth (The Netherlands)*, Dr. W. Junk, The Hague.

Chapman, S. B. (1964a). The ecology of Coom Rigg Moss, Northumberland. I. Stratigraphy and present vegetation *J. Ecol.* **52**, 299–313.

Chapman, S. B. (1964b). The ecology of Coom Rigg Moss, Northumberland. II. Chemistry of peat profiles. *J. Ecol.*, **52**, 315–321.

Conway, V. M. (1948). Von Post's work on climatic rhythms. *New Phytol.*, **47**, 220–237.

Corbet, S. A. (1973). An illustrated introduction to the testate rhizopods in *Sphagnum*, with special reference to the area around Malham Tarn, Yorkshire. *Field Studies*, **3**, 801–838.

Cushing, E. J., (1967). Evidence for differential pollen preservation in late Quaternary sediments in Minnesota. *Rev. Palaeobotan. Palynol.*, **4**, 87–101.

Dickinson, W. (1975). Recurrence surfaces in Rusland Moss, Cumbria (formerly North Lancashire). *J. Ecol.*, **63**, 913–936.

Dickinson, C. H. (1983). Micro-organisms in peat. In: *Ecosystems of the World. Vol. 4A Mires : Swamps, Bog Fen and Moor* (Ed. A. J. P. Gore), Elsevier Scientific, Amsterdam, pp. 225–245.

Finley, R. J., and Gustavsen, T. C. (1983). Geomorphic effects of a 10-year storm on a small drainage basin in the Texas Panhandle. *Earth Surface Processes and Landforms*, **8**, 63–77.

Foster, D. R., King, G. A., Glaser, P. H., and Wright, H. E. (1983). Origin of string patterns in boreal peatlands. *Nature*, **306**, 256–258.

Godwin, H., (1946). The relationship of bog stratigraphy to climatic change and archaeology. *Proc. Prehist. Soc.*, **12**, 1–11.

Godwin, H. (1975). *History of the British Flora.* Cambridge University Press, Cambridge.

Goode, D. A. (1974). The significance of physical hydrology in the morphological classification of mires. *Proc. Int. Peat. Soc. Symp., Glasgow (1973)*, I.P.S., Helsinki.

Gore, A. J. P. (1983). Introduction. In: *Ecosystems of the World. Vol. 4A Mires : Swamp, Bog, Fen and Moor* (Ed. A. J. P. Gore), Elsevier Scientific, Amsterdam, pp. 1–34.

Gorham, E. (1956). On the chemical composition of some waters from the Moor House Nature Reserve. *J. Ecol.*, **44**, 377–384.

Grosse-Brauckmann, G. (1979). Zur Deutung einiger Makrofossil — Vergesellschaftungen unter dem Gesichtspunkt der Torfbildung. In: *Werden und Vergehen von Pflanzengesellschaften* (Ed. O. Wilmanns and R. Tuxen), J. Cramer, Braunschweig, pp. 111–132.

Heal, O. W. (1962). The abundance and micro-distribution of testate amoebae (Rhizopoda : Testacea) in *Sphagnum. Oikos*, **13**, 35–47.

Hohenstatter, E. (1972). Ein ungewöhnliches Moorprofil. *Telma*, **2**, 65–72.

Hornbeck, J. W., Likens, G. E., Pierce, R. S., and Bormann, F. H. (1975). Strip cutting as a means of protecting site and streamflow quality when clearcutting northern hardwoods. In: *Forest Soils and Forest Land Management* (Ed. B. Bernier and C. H. Winget), Les Presses de l'Université Laval, Quebec.

Ingram, H. A. P. (1982). Size and shape in raised mire ecosystems : a geographical model. *Nature*, **297**, 300–303.

Ingram, H. A. P. (1983). Hydrology. In: *Ecosystems of the World. Vol. 4A Mires : Swamp, Bog, Fen and Moor* (Ed. A. J. P. Gore), Elsevier Scientific, Amsterdam, pp. 67–158.

Ivanov, K. E. (1981). *Water Movements in Mirelands* (Translated by A. Thompson and H. A. P. Ingram), Academic Press, London.

Janssen, C. R., and Ten Hove, H. A. (1971). Some late-Holocene pollen diagrams from the Peel raised bogs (Southern Netherlands). *Rev. Palaeobotan. Palynol.* **11**, 7–35.

Kochel, R. C., and Baker, V. R. (1982). Palaeoflood hydrology. *Science*, **215**, 353–360.

Martin, N. J., and Holding, A. J. (1978). Nutrient availability and other factors limiting microbial activity in the blanket peat. In: *Production Ecology of British Moors and Montane Grasslands* (Ed. O. W. Heal and D. F. Perkins), Springer Verlag, Berlin, pp. 113–135.

Moore, P. D. (1975). Origin of blanket mires. *Nature*, **256**, 267–269.

Moore, P. D. (1977). Stratigraphy and pollen analysis of Claish Moss, north-west Scotland : significance for the origin of surface pools and forest history. *J. Ecol.*, **65**, 375–398.

Moore, P. D. (1982). Pool and ridge patterns in peat mires. *Nature*, **300**, 110.

Moore, P. D. (1983). Palynological evidence of human involvement in certain palaeohydrological events. *Quat. Studies in Poland*, **4**, 97–105.

Moore, P. D. (1984). The classification of mires : an introduction. In: *European Mires* (Ed. P. D. Moore), Academic Press, London, pp. 1–10.

Moore, P. D. and Bellamy, D. J. (1974). *Peatlands*, Paul Elek, London.

Moore, P. D., and Chater, E. H. (1969). Studies in the vegetational history of mid-Wales. I. The post-glacial period in Cardiganshire. *New Phytol.*, **68**, 183–196.

Moore, P. D., Merryfield, D. L., and Price, M. D. R. (1984). The vegetation and development of blanket mires. In: *European Mires* (Ed. P. D. Moore), Academic Press, London. pp. 203–235.

Moore, P. D., and Willmot, A. (1976). Prehistoric forest clearance and the development of peatlands in the uplands and lowlands of Britain. *Proc. 5th Int. Peat Congr.*, Poznan, Poland.

Oldfield, F. (1967). The palaeoecology of an early Neolithic waterlogged site in northwestern England. *Rev. Palaeobotan Palynol.*, **4**, 67–70.

Pennington, W., (1979). The origin of pollen in lake sediments : an enclosed lake compared with one receiving inflow streams. *New Phytol.*, **83**, 189–213.

Reynolds, J. F., and Knight, D. H. (1973). The magnitude of snowmelt and rainfall interception by litter in lodgepole pine and spruce fir forests in Wyoming. *North-west Science*, **47**, 50–60.

Ruuhijärvi, (1960). Über die regionale Einteilung der nordfinnischen Moore. *Ann. Bot. Soc. "Vanamo"*, **31** (1), 1–360.

Rybnicek, K., and Rybnickova, E. (1977). Moor untersuchungen in oberen Gurgltal, Ötztaler Alpen. *Folia Geobot. Phylotax. Praha*, **12**, 245–291.

Sjörs, H. (1950). On the relation between vegetation and electrolytes in North Swedish mire waters. *Oikos*, **2**, 241–258.

Sjörs, H. (1961). Surface patterns in boreal peatlands. *Endeavour*, **20**, 217–224.

Skarzynska, K. (1965). Paleohydrological research in the territory of ancient Poland. *Bull. de l'Acad. Polonaise des Sciences Ser. Geol. et Geogr.*, **13**, 237–247.

Smart, P. J. (1982). Stratigraphy of a site in the Munsary Dubh Lochs, Caithness, Northern Scotland : development of the present pattern. *J. Ecol.*, **70**, 549–558.

Spence, D. H. N. (1982). The zonation of plants in freshwater lakes. *Adv. Ecol. Res.*, **12**, 37–125.

Stewart, J. M., and Durno, S. E. (1969). Structural variations in peat. *New Phytol.*, **68**, 167–182.

Summerfield, R. J., and Rieley, J. O. (1975). Growth of *Narthecium ossifragum* in relation to the dissolved oxygen concentration of the rooting substrate. *Plant and Soil*, **41**, 701–705.

Swank, W. T., and Douglass, J. E. (1974). Streamflow greatly reduced by converting deciduous hardwood stands to pine. *Science*, **185**, 857–859.

Taylor, J. A., and Smith, R. T. (1981). The role of pedogenic factors in the initiation of peat formation and in the classification of mires. *Proc. 6th Int. Peat. Congr.*, Duluth, Michigan, pp. 108–118.

Ugolini, F. C., and Mann. D. H. (1979). Biopedological origin of peatlands in south east Alaska. *Nature*, **281**, 366–368.

van Geel, B. (1972). Palynology of a section from the raised peat bog 'Wietmarscher Moor' with special reference to fungal remains. *Acta Bot. Neerl.*, **21**, 261–284.

Walker, D., and Walker, P. M. (1961). Stratigraphic evidence of regeneration in some Irish bogs. *J. Ecol.*, **49**, 169–185.

Wiltshire, P. E. J., and Moore, P. D. (1983). Palaeovegetation and palaeohydrology in upland Britain. In: *Background to Palaeohydrology : A Perspective* (Ed. K. J. Gregory), John Wiley, Chichester, pp. 433–451.

Research strategy

4

Palaeoecological reference areas and reference sites

Björn E. Berglund

*Department of Quaternary Geology,
Tornavägen 13,
Lund, Sweden*

INTRODUCTION

Past environmental changes in temperate and subarctic regions are often recorded in the deposits of lakes and mires. This applies to biotic as well as physical changes which are mainly governed by climate and human activities. The international project IGCP 158B has formulated, as its main goals, descriptions of such changes, correlations of these on a regional and continental scale, and the search for their causes. Parallel work is performed within project IGCP 158A dealing with the changes in fluvial environments. The specific aims can be summarized in the following way, which may be regarded as an overall definition of palaeoecological research with a holistic approach:

(1)   to provide palaeoecological and stratigraphical research with continental reference sites related to an absolute chronology covering the last 15,000 years;
(2)   to apply a variety of uniform palaeoecological methods to obtain fully accurate information on biotic and environmental conditions on the continents during the last 15,000 years;
(3)   to describe biotic changes in both time and space. Biotic changes refer to local ecosystems of lakes and mires as well as to regional, terrestrial ecosystems in selected reference areas;

111

(4)   to describe hydrological and limnological changes in both time and space.
        These changes include quantitative aspects of the hydrological cycle, such
        as water-level changes, as well as qualitative aspects of lake ecosystems,
        such as nutrient status and productivity;
(5)   to describe climatic changes in both time and space and to assess their
        impact on biological and hydrological systems;
(6)   to describe human activity in both time and space and to assess its impact
        on biological and hydrological systems;
(7)   to correlate hydrological changes in lake catchment areas with conditions
        in fluvial environments in order to describe the total hydrological regime
        and its relationship to climatic and human factors — a synthetic approach
        involving collaboration between the projects IGCP A and B.

The research strategy of this project has earlier been described, more fully,
by Berglund (1979a) and in a synthetic form by Berglund and Digerfeldt (1976)
and Berglund (1979b, 1983). As can be seen from these presentations, we have
emphasized the importance of palaeoecological research in lakes and their
catchments. The potential of lake sediments to reveal past environmental
changes has been underlined by many research workers (cf. Haworth and
Lund, 1984). The theoretical background to the recording of biological and
physical activities and processes occurring within a lake drainage basin has
been described by Oldfield (1977). His simplified sedimentation scheme is
reproduced here in Figure 4.1. Palaeohydrological studies of lakes are
described by Digerfeldt (Chapter 5). Mires and soil profiles give complementary information which is described in detail by Aaby (Chapter 6) and Andersen
(Chapter 7).

In order to provide the palaeoecological correlation work with comparable
material it is important to define uniform criteria for selection of field sites
which are representative for the surrounding region. We have therefore
introduced the concepts of *reference sites* which are representative for a
uniform *type region*. It is also important to apply uniform palaeoecological
methods to make proper correlations. Methods that are applicable at different
kinds of reference sites are proposed in this chapter.

When field areas are unaffected by human activity, the deposits in lakes and
mires may monitor environmental changes. If such sites are within nature
conservation areas, they may also provide the opportunity for the study of
future changes. Some of our proposed research areas should therefore be
protected as environmental reference areas with the potential of tracing past as
well as studying future biotic and physical changes (cf. the program for the
environmental control (PMK) in Sweden and related planning in other
countries, Bernes *et al.*, in press).

PALAEOECOLOGICAL TYPE REGIONS AND REFERENCE AREAS

Each country can be subdivided into *palaeoecological* and *ecological type*

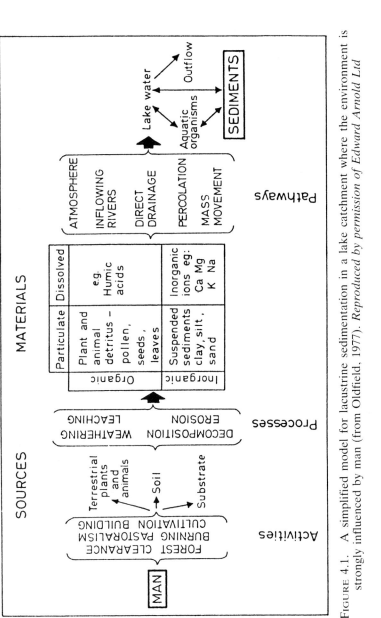

FIGURE 4.1.   A simplified model for lacustrine sedimentation in a lake catchment where the environment is strongly influenced by man (from Oldfield, 1977). *Reproduced by permission of Edward Arnold Ltd*

regions which are more or less uniform as regards geology, geomorphology, climatology and biotic conditions. An extract from the European project map illustrates this subdivision for North Europe (see Figure 4.3). Owing to the diversity of the landscape, such regions are larger in lowlands than in areas with high relief. Approximate size variations are indicated in Table 4.1.

TABLE 4.1.   Size of type regions, reference areas and reference sites; space and time resolutions of sites (approximate figures based on conditions in Scandinavia)

	Size	Spatial resolution	Accumulation rate (mm/yr)	Temporal resolution (yrs)
Type region	30,000–100,000 km^2	—		—
Reference area (with primary ref. site)				
(a) Lowlands	2,000–10,000 km^2	200–500 km	—	
(b) Mountains	Smaller	As above	—	
Reference site (primary or secondary site)				
(a) Forested lowland areas	25–50 ha, 600–800 m diam.	200 ± 50 km	0.5–1	1 cm = 10–20
(b) Open mountain areas	< 10 ha, 100–300 m diam.	As above, or denser	0.2–0.5	1 cm = 20–50

Within each region type at least one reference site should be selected. A *reference site* is defined as a lake or mire with a continuous sequence of deposits covering the time span after the last deglaciation, and situated in an area that is representative for the whole region, so that a palaeoecological study of the site will, it is hoped, reveal a representative record of environmental changes. When the palaeoecological studies are restricted to mainly pollen analysis and dating at *one* site, it will be *a reference site for palaeovegetation and chronology*, or *a secondary reference site*. Such sites will form a densely spaced network of well-dated, continuous sediment profiles with good pollen diagrams.

Some areas that are suitable with regard to representativity and variation of palaeoecological and ecological source material — lakes, mires, soils, archaeological material, natural or semi-natural vegetation, etc. — should be chosen as *palaeoecological reference areas*. Within these areas several sites may be selected for palaeoecological research. One main site, from a lake or a mire, should be chosen as *a palaeoecological reference site*, or *a primary reference site*, where a great variety of methods can be applied in order to obtain rich information on biotic as well as physical environmental changes. Water-level changes and bog humification changes are emphasized in these studies. When possible, the human impact should also be studied in cooperation with archaeologists and historians. Often a team of specialists has to collaborate in order to cover the multidisciplinary approach. Therefore the reference areas with primary reference sites will form a sparse network — they will not represent all region types but they will be enough to cover the different biotic and climatic regions. An idealized scheme for correlation between region types and reference sites is given in Figure 4.2. The geographic division for Scandinavia is shown in Figure 4.3, where type regions as well as reference sites are indicated.

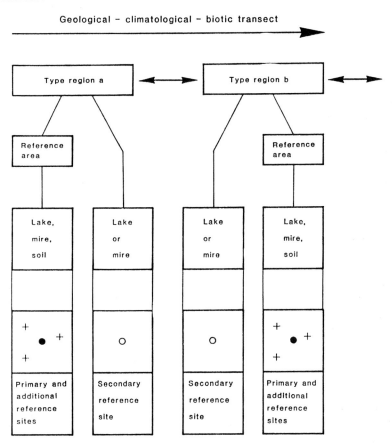

FIGURE 4.2. Schematic illustration of the proposed correlation between type regions based on palaeoecological studies in reference areas with primary (and additional) reference sites and at secondary reference sites

All reference sites will form the basis for computerized correlation and for the mapping of biotic zones and their shifts (cf. Chapter 37). By using transfer functions such data may give information on climate and human impact in time

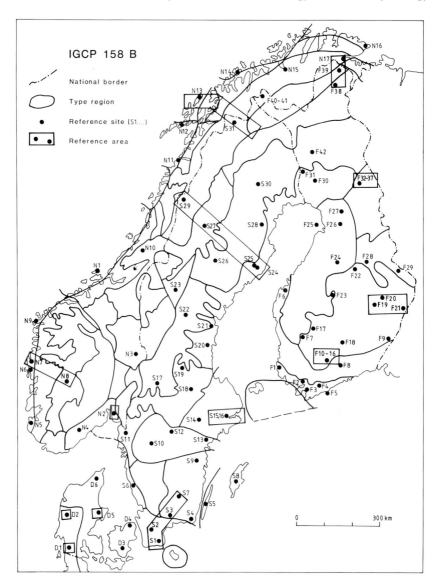

FIGURE 4.3. Research plan for the project IGCP 158B in Scandinavia. Type regions and reference sites are indicated — reference sites are numbered within each country, e.g. S1 – S31 for Sweden. Palaeoecological reference areas are shaded

and space. The primary reference sites will give additional information about terrestrial as well as lacustrine environments, hydrological changes, timberline changes etc. The different sets of data will be stored to provide the basis for explaining environmental changes in the past.

DEFINITION OF REFERENCE AREAS

Choice of palaeoecological reference areas

A reference area should be representative for the type region as regards geology/geomorphology, climate and vegetation (vegetation – landform unit). In a region with great landform diversity, more than one reference area may be chosen for further studies. The size will depend on the geological/ geomorphological situation — normally it will vary between 2000 and 10,000 km^2, with the smallest areas in mountain situations. In a lake district the area will cover one or several catchment areas. The distance between reference areas will vary depending on the size of type regions, i.e. the landscape homogeneity — and research ambitions!

An ideal reference area should include organic deposits of lakes, mires and soils. When choosing reference sites, careful reconnaissance studies are recommended. Several lake/mire sites must be studied before choosing the main site(s) for multidisciplinary stratigraphic studies. A set of surface pollen spectra from different sites may be of some help to test the site representativity in relation to modern vegetation. Lake sites are preferred owing to the continuous sedimentation in small-to-medium sized basins and to the fact that the sediments reflect physical/chemical changes of the catchment area as well as within the basin itself, e.g. hydrological changes (Chapter 5). The ideal situation is an area with one or several lakes in the upper part of a river system — groundwater lakes with small inlets/outlets are to be preferred. When lake basins are not available, a mire/bog is chosen as a reference site. In addition to palaeolimnological studies of lake sediments, studies of the palaeoecology and stratigraphy of peat bogs will give valuable information on humidity changes (Chapter 6). Therefore it is of interest to study lake sediments as well as bog stratigraphy in the same reference area. Humus layers in terrestrial soil profiles of forests will reveal information about soil formation, forest succession, local human impact etc. (Chapter 7). Examples of ideal reference areas in lowlands and in mountains are illustrated in Figures 4.4 and 4.5 respectively (cf. Jacobson and Bradshaw, 1981).

Quality of reference sites

The requirements of pollen analysis and radiocarbon dating will influence the choice of reference sites since these methods are the main methods for palaeoecological and chronological correlation. This has an impact on basin morphology and basin size, sediment geochemistry and sediment time resolution etc. Non-calcareous areas with ± isodiametric lakes of moderate size (25–50 ha) and depth (5–15 m), having gentle slopes and some shallow bays, will normally reveal source material of good quality for pollen analysis, palaeolimnological and palaeohydrological studies. In order to obtain annually laminated sediments the lakes should be deeper. Details concerning quality of

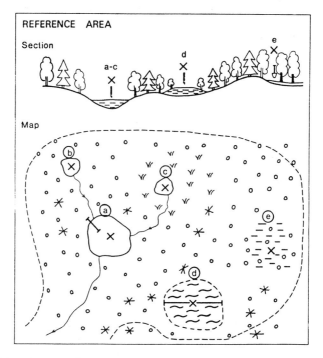

FIGURE 4.4. A generalized sketch of an ideal reference area in a lowland area. Catchment area indicated with a broken line. (a) indicates a primary reference site of a lake studied by means of a main core and at least one stratigraphical cross-section (b) and (c) indicate additional lake reference sites, where pollen-analytically studied cores may contribute to the palaeoecological knowledge of different environments — (b) an undisturbed forest area, (c) a landscape with early cultivation. (d) indicates a raised bog, an additional or alternative reference site, where cross-sections give information on humification changes etc. (e) indicates a forest humus site in a podsolized forest area which will give information on local forest and soil successions

lake and mire sites are discussed in Chapters 5 and 6 (cf. Jacobson and Bradshaw, 1981).

There is an important relation between basin size and source data for pollen spectra. This is discussed by Prentice in Chapter 39. Based on his studies, pollen diagrams representative for a larger area (10–30 km radius) may be based on cores from basins with sizes 25–100 ha in forest regions, from smaller basins in open regions such as subarctic areas and deforested culture landscapes. General size conditions for reference sites and reference areas are summarized in Table 4.1.

Radiocarbon dating requires non-calcareous, organic deposits (gyttja, peat, wood). Therefore, when possible, the reference sites should be chosen in oligotrophic areas. When lakes are calcareous, bogs may be used as reference sites or dates may be obtained from bog stratigraphy and transferred to the lake

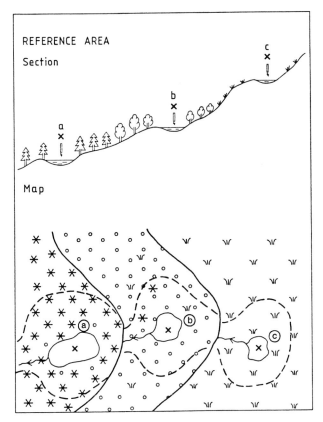

FIGURE 4.5. A generalized sketch of a reference area with a vegetation and climate zonation produced by altitudinal gradients, e.g. coniferous woodland, woodland tundra and alpine tundra in a transect from a valley bottom to the mountain tops. A reference area should preferably include sites which are representative for the different zones, here indicated by the alternative reference sites (a), (b) and (c). Site (a) would probably be chosen as the most suitable primary reference site

stratigraphy by applying pollen-analytical correlations. For methodological details and interpretational problems, see Chapter 14.

All palaeoecological analyses including datings demand continuous, rapid sediment accumulation or peat growth to obtain an adequate time resolution of each sample. It is recommended that sites should be chosen with a mean decomposition peat growth of 0.5–1 mm per year. In lakes 1 mm/year is optimal, with 0.5 mm/year the normal accumulation rate. However, sometimes, especially in arctic–subarctic conditions, lower accumulation rates are common, perhaps as low as 0.1–0.2 mm/year. In bogs, peat growth normally varies between 0.2 and 1 mm/year. Values for the time resolution of samples of different thicknesses are summarized in Tables 4.1 and 4.2.

TABLE 4.2. Palaeoecological methods — proposed sample size and time resolution for sediments with accumulation rate 0.5–1 mm/yr

Methods	Mimimum demands at all reference sites	Potential methods applied at primary reference sites	Sampling density (yrs)	Sample thickness (cm)	Temporal resolution of each sample (yrs)	Reference (Chapter) number
Description of site catchment						
Geological mapping	x	—	—	—	—	9
Vegetational mapping	—	x	—	—	—	10
Archaeological mapping	—	x	—	—	—	11
Stratigraphical methods						
Characterization acc. to TS-system	x	—	—	—	—	12
Microscopic sediment analysis	—	x	—	—	—	12
Photo documentation	—	x	—	—	—	12, 17
X-radiographing	—	x	—	—	—	12, 17
Cross-section correlations	x	—	—	—	—	5, 13
Dating methods						
^{14}C-dating	x	—	Min. 1/2000 Max. 1/500	2.5–5	50–100	14
^{210}Pb-dating	—	x	1/20	2.5	<5	14
^{137}Cs-dating	—	x	Continuous	2.5	<5	14

Palaeomagnetic dating	15	—	Continuous	—	×	—
Tephrachronology	16	—	—	—	×	—
Annually laminated sediments	17	1	Continuous	—	×	—
Dendrochronology	18, 19	1	Continuous	—	×	—
Physical analyses						
Particle size	12	40	1/100	—	×	—
Sediment influx	13	—	—	—	×	—
Bulk density	21	40	1/100 or continuous	2	—	×
Dry weight and water content	21	40	1/1000 or continuous	2	—	×
Organic content (incl. org. carbon)	21	40	1/100 or continuous	2	—	×
Carbonate content	21	40	1/100 or continuous	2	—	×
$^{18}O/^{16}O$ analyses	20	40	1/100	2	×	—
Humification (peat)	6	40	Continuous	2	—	×
Chemical analyses						
Phosphorus	21	40	1/200	2	×	—
Nitrogen	21	40	1/200	2	×	—
Plant pigments	21	40	1/200	2	×	—
K, Ca, Mg	21	40	1/200	2	×	—
Heavy metals	21	40	1/200	2	×	—
Pollen/spore analysis						
Present-day pollen deposition	4, 39	—	—	—	—	×
Percentage diagram	22	20–40	1/100	1–2	—	×
Influx diagram	22	20–40	1/100	1–2	×	—
Fungal spores	24	20–40	1/100	1–2	×	—
Charred particles	23	20–40	1/100	1–2	—	×

cont.

TABLE 4.2. *cont.*

Methods	Minimum demands at all reference sites	Potential methods applied at primary reference sites	Sampling density (yrs)	Sample thickness (cm)	Temporal resolution of each sample (yrs)	Reference (Chapter) number
Algae analysis						
Present-day diatom flora	—	x	—	—	—	26
Diatom diagram	—	x	1/200	1–2	20–40	26
Other algae	—	x	1/200	1–2	20–40	24, 25
Plant macrofossil analysis						
Fruits and seeds	—	x	Continuous	2.5–5	50–100	27
Vegetative remains	—	x	Continuous	2.5–5	50–100	28
Bryophytes	—	x	Continuous	2.5–5	50–100	30
Wood and charcoal	—	x	—	—	—	29
Palaeozoological analysis						
Rhizopods	—	x	1/100	1	20	31
Cladocera	—	x	1/200	1	20	32
Ostracods	—	x	Continuous	2.5–5	50–100	33
Coleoptera	—	x	Continuous	5–10	100–200	34
Chironomids	—	x	1/200	1	20	35
Molluscs	—	x	Continuous	5–10	100–200	36

RECOMMENDATIONS FOR WORKING METHODS

A palaeoecological correlation project with a holistic goal has to be supported by data obtained from a variety of methods. These are listed and grouped in Table 4.2 in the following way:

(a) methods applied at all lake/mire reference sites — so-called minimum demands for primary as well as secondary reference sites. Data will form the basis for palaeovegetation and chronology;

(b) potential methods applied at primary reference sites or at sites with special sediment conditions. Data will form the basis for palaeohydrology, palaeolimnology, palaeochemistry, palaeoclimatology, palaeovegetation, and settlement history etc.

Standard sample size and time resolution for each sample is given when annual accumulation rates are 0.5–1.0 mm, which is assumed to be the normal range for temperate forest regions. Accumulation conditions may change within a reference profile and therefore demand different sampling densities, e.g. denser sampling during periods of rapid climatic change and vegetation successions during late-glacial periods, during periods of changed bog growth, e.g. at recurrence surfaces, or during periods corresponding to the very last centuries through which important environmental changes may be registered. Sampling for ^{210}Pb- and ^{137}Cs-dating requires decomposition rates exceeding the normal ones — the values in the table are based on 5–10 mm/year.

RESEARCH PLANNING AT A PRIMARY REFERENCE AREA

Description of lake/mire catchment

When a primary reference site has been chosen, the site and its surroundings (the catchment area) should be documented to include at least the geology, modern vegetation, archaeology and land use history. Details are described in Chapters 9–11.

Description of lake/mire site

The sampling site, modern lake or mire should be described in detail, especially with regard to hydrology, vegetation and limnological features. The stratigraphy can be documented by reference to cross-sections; for palaeolimnological studies a grid of corings is needed which can form the basis for calculation of sediment influx. Details are described in Chapters 5–6 and 12–13.

A scheme for describing and subsampling sediment cores

Figure 4.6 illustrates different steps in the treatment of a sediment core from a reference site. Firstly, the sediment character is described directly by using the Troels–Smith system; secondly, the sediments are described on the basis of

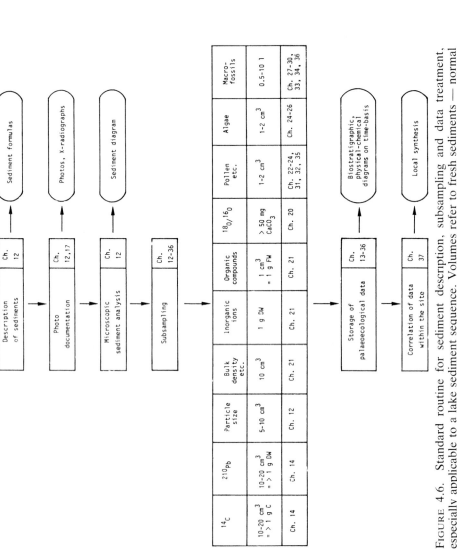

FIGURE 4.6. Standard routine for sediment description, subsampling and data treatment, especially applicable to a lake sediment sequence. Volumes refer to fresh sediments — normal values are given for Holocene sediments in a lowland region. For details, see the relevant chapters as indicated in the scheme by chapter numbers. For 'Pollen etc.' and 'Macrofossils'

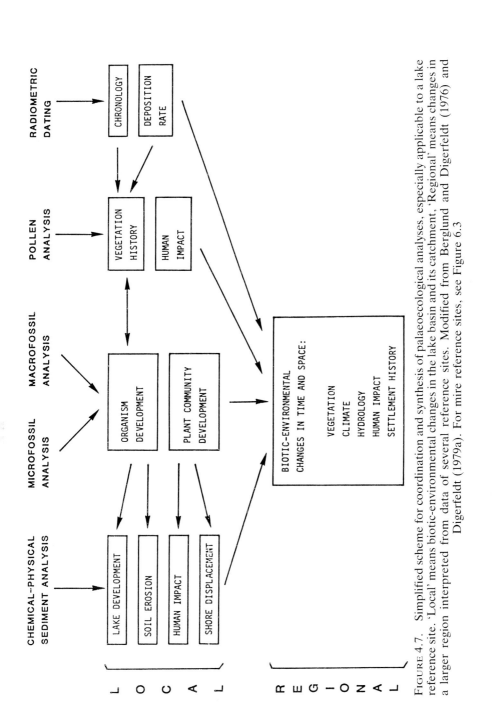

FIGURE 4.7. Simplified scheme for coordination and synthesis of palaeoecological analyses, especially applicable to a lake reference site. 'Local' means biotic-environmental changes in the lake basin and its catchment, 'Regional' means changes in a larger region interpreted from data of several reference sites. Modified from Berglund and Digerfeldt (1976) and Digerfeldt (1979a). For mire reference sites, see Figure 6.3

microscopic analysis ('Strukturanalys') (see Chapters 12 and 17). Then follows the subsampling for different parallel analyses, which are performed depending on the research aim and the character of the sediment. The scheme gives the standard routine for subsampling which is especially applicable to lake sediments. Values of sample quantities are taken from the methodological chapters mentioned in the figure.

Coordination of analysis and synthesis

A simplified scheme in Figure 4.7 illustrates the coordination of palaeoecological analyses at one reference site — normally obtained from one main sediment core as well as from additional cores stratigraphically correlated with each other. Palaeobiological and palaeophysical/chemical data reflect biotic and environmental changes in the lake/mire basin and its catchment. The interpretation will therefore deal initially with the local development at the site and its surroundings. In this simple scheme, lake development, shore displacement (water-level changes), soil erosion, vegetation history and human impact are mentioned. Correlation of several reference sites may make it possible to exclude local conditions and to make regional conclusions about biotic–environmental changes in the past, especially to trace their causes in terms of climate, hydrology and human impact.

REFERENCES

Berglund, B. E. (Ed.) (1979a). *Palaeohydrological Changes in the Temperate Zone in the Last 15,000 Years: Subproject B, Lake and Mire Environments.* Project guide vol. I, Dept. Quat. Geol., Lund University, 123pp.

Berglund, B. E. (1979b). Presentation of the IGCP Project 158B: Palaeohydrological changes in the temperate zone in the last 15,000 years — Lake and mire environments. *Acta Univ. Ouluensis*, **A82**, Geol. 3:39–48.

Berglund, B. E. (1983). Palaeohydrological studies in lakes and mires — a palaeoecological research strategy. In: *Background to Palaeohydrology* (Ed. K. J. Gregory), John Wiley, Chichester, pp. 237–254.

Berglund, B. E., and Digerfeldt, G. (1976). Environmental changes during Holocene — a geological correlation project on a Nordic basis. *Newsl. Stratigr.*, **5**, 1, 80–85.

Bernes, C., Giege, B., Johansson, K., and Larsson, J. E. (in press). Design of an integrated monitoring programme in Sweden. *Environmental Monitoring and Assessment.*

Haworth, E. Y., and Lund, J. W. G. (Eds.) (1984). *Lake Sediments and Environmental History.* Studies in palaeolimnology and palaeoecology in honour of Winifred Tutin, Leicester Univ. Press, Leicester, England.

Jacobson, G. L., and Bradshaw, R. H. W. (1981). The selection of sites for palaeovegetational studies. *Quat. Res.*, **16**, 80–96.

Oldfield, F. (1977). Lakes and their drainage basins as units of sediment-based ecological study. *Progress in Physical Geogr.*, **1**, 3, 460–504.

Handbook of Holocene Palaeoecology and Palaeohydrology
Edited by B. E. Berglund

5

Studies on past lake-level fluctuations

GUNNAR DIGERFELDT

*Department of Quaternary Geology,
Tornavägen 13,
Lund, Sweden*

INTRODUCTION

A palaeoecologist may be interested in studying Holocene lake-level fluctuations for various reasons. Information on the occurrence and the effects of past fluctuations may be necessary for a correct palaeolimnological understanding of lake development — for example, trophic changes and changes in productivity indicated by biological and chemical records in sediments. Past fluctuations in lake level, particularly in shallow and moderately deep lakes, may have resulted in extensive erosion and redeposition of sediment. Information on such disturbances in sedimentation is certainly necessary if stratigraphical records are not to be misinterpreted.

However, most of the interest in Holocene lake-level studies is explained by the possibility of using recorded past fluctuations for reconstructing and interpreting regional palaeohydrological changes, and — provided there is evidence for a climatic causation — palaeoclimatic changes. The reconstruction of regional palaeohydrology and palaeoclimate is a major aim in Holocene palaeoecological research, and a prerequisite for the correct interpretation and understanding of most environmental changes studied by palaeoecologists.

Some reviews of regionally recorded Holocene lake-level fluctuations, indicating their usefulness in palaeohydrological reconstruction, have been presented by Berglund *et al.* (1983), Ralska-Jasiewiczowa and Starkel (1985) and Gaillard (1985). The reviews are included in the IGCP project 158, in which reconstruction is intended to be based on a combination of evidence obtained from past lake-level fluctuations, mire stratigraphy and past changes in river discharge (Berglund 1983, Starkel 1983).

127

This chapter deals with the recognition of lake-level fluctuations in sedimentary records, and describes a method for their thorough and reliable reconstruction.

SEDIMENTARY EVIDENCE OF LAKE-LEVEL FLUCTUATIONS

Past fluctuations in lake level may be recorded from various lines of geomorphological evidence; for example, past higher levels may be indicated by raised shore ridges and wave-cut cliffs. However, much evidence is to be found in sediment stratigraphy. A review of sediment stratigraphical evidence has been given by Richardson (1969), in which a large number of studies based on various lines of evidence are discussed.

Principally, lake-level fluctuations may be stratigraphically recorded in as many ways as fluctuations may affect the variety of limnological and sedimentological processes in lakes. The commonly used sedimentological, biological, sediment-physical and sediment-chemical records are described by Richardson. Biological records are dealt with elsewhere in this handbook — e.g. diatoms (Chapter 26), cladocera (Chapter 32), ostracods (Chapter 33) and chironomids (Chapter 34).

Despite the fairly large amount of potential stratigraphical evidence, the palaeohydrologist is often confronted with the problem that most evidence is susceptible to alternative interpretations. The stratigraphically recorded changes often cannot be convincingly linked to past fluctuations in lake level, but can be explained by other environmental changes — for example, natural successional changes in a lake, or even external environmental changes in the lake catchment area. Moreover, with some lines of evidence there are alternative interpretations with respect to the character of palaeohydrological change — recorded evidence may indicate either a past rise or a past lowering in lake level. The problem of alternative interpretations of records has been thoroughly discussed by, among others, Ralska-Jasiewiczowa and Starkel (1985).

METHOD FOR RECONSTRUCTION OF PAST FLUCTUATIONS

Principles of recognition and interpretation

In studies of Holocene lake-level fluctuations in southern Sweden a method was used in which the interpretation of recorded changes was based on a combination of different types of stratigraphical evidence. The method was designed to increase the likelihood of constructing positive and convincing links between the sediment records and lake-level fluctuations.

The different forms of evidence comprised changes in (1) the distribution of lake vegetation, (2) the sediment composition, and (3) the level of the sediment limit. Changes were established by macrofossil and pollen analyses, and by analysis of the sediment composition in a transect of sediment cores outwards from the shore (which should provide a more reliable construction than is usually possible from the study of single sediment cores).

The method is partly inspired from the pioneer studies on lake development by Lundqvist (1925, 1927), which are still fundamental to the understanding of lake sedimentology and sedimentation processes. The method is intended for fairly small lakes (< 100 ha) in which past fluctuations in lake level can be assumed to be most distinctly recorded and easily reconstructed, owing to the generally more regular sedimentation and hydrodynamics.

Changes in distribution of lake and shore vegetation

The distribution of the macrophyte vegetation is largely determined by the water depth, resulting in the characteristic zonation of emergent, floating-leaved and submerged vegetation outwards from the shore. The distribution can sometimes be directly or indirectly affected by exposure. However, in most small and sheltered lakes the effects of variation in exposure can be assumed to be insignificant.

A decrease in water depth (a lowered lake level) will normally result in an outward spread of macrophytes, while an increase in water depth (a raised level) will result in their inward displacement.

A recorded inward displacement of the vegetation can usually be regarded as positive evidence for a past rise in lake level — at least, it is difficult to find any other explanation for such a change. However, in the case of a recorded outward spread there are alternative explanations. The spread may be due to a lowering in lake level, but it may also merely record the natural outward spread of vegetation associated with the progressive infilling and overgrowing of the lake. A combination of these two explanations may also be possible.

Changes in sediment composition

Littoral sediments are normally characterized by a larger quantity of reworked coarse minerogenic matter, and also coarse organic matter originating from the macrophyte vegetation along the shore. The quantity will largely depend on the exposure of the shore to wave action and wind-induced water currents, and on the character of shore deposits. From the shore outwards there is normally a gradual decrease in the quantity, and a particle-size change of the coarse matter contained in the sediment.

A past lowering in lake level may, because of the associated outward displacement of the shore, be recorded by an increase in the amount of coarse matter contained in the sediment. A rise in lake level and the associated inward shore displacement may be recorded by a decrease in this quantity.

Changes in level of sediment limit

The sediment limit is defined as the highest limit for deposition of predominantly organic sediment (Lundqvist, 1925). The definition may appear somewhat imprecise, but in modern lakes it is usually an easily observed limit. Like the quantity of coarse matter, the limit is determined by the exposure of the shore. Above the sediment limit fine detrital organic matter cannot permanently accumulate owing to the disturbing effect of waves and wind-induced water currents. In lakes characterized by predominantly minerogenic sedimentation the sediment limit may be compared with the limit between erosion and transportation bottoms as defined by Håkansson and Jansson (1983).

A past rise in lake level will usually be recorded by a rise in the sediment limit. A lowering in lake level will result in a lowered limit, but also in the erosion and redeposition of older sediment which had accumulated at a higher level before the lowering.

With respect to both the coarse matter in sediment and the sediment limit, there are certainly alternative responses to past lake-level fluctuations depending on local conditions — for example, the character and distribution of shore deposits, shore morphology, and the character of shore and lake vegetation. In some cases, and then largely depending on the shore morphology, past fluctuations may have significantly affected the exposure, which may lead to different records.

Records of coarse minerogenic matter may not always be unambiguously linked to past fluctuations in lake level, but may be due to external changes in lake catchment that affect the quantity of minerogenic matter supplied to marginal sediment. Frequently an increase in minerogenic matter in sediment may represent increased soil erosion due to human impact in the lake catchment. External environmental changes influencing the exposure of the lake — forest clearances or other changes in forest vegetation — may affect both the sediment limit and the coarse matter in the sediment.

Most records may therefore offer alternative explanations, and the interpretation and reconstruction of past lake-level fluctuations based on one type of record may be problematic and unreliable when no supporting information is available.

However, the important theme in the method described here is that convincing links to past lake-level fluctuations and their reliable interpretation can be obtained by using a combination of lines of evidence. A combination of records showing an outward spread of vegetation, an increase in coarse matter in the sediment, and a lowered sediment limit, must be positive evidence for a past lowering in the lake level. Alternative interpretations could be applied to the individual records, but only a lowering in lake level can explain such a combination of evidence. Likewise, a combination of records for an inward displacement of vegetation, a decrease in coarse matter in the sediment, and a raised sediment limit, must be regarded as convincing evidence for a past rise in the lake level.

The ideal of complete multiple records cannot be expected in all lakes studied. The records available largely depend on the character of the lake, and in most lakes one or other lines of evidence may be missing or less distinctly recorded. Moreover, owing to local variations in shore character, useful records are mostly found only in particular parts of the lake. Past vegetational changes are commonly more distinctly recorded and easily recognized in eutrophic lakes — because of the higher production and diversity in macrophyte vegetation — than in oligotrophic lakes. In eutrophic lakes past changes in the distribution can usually be fairly easily reconstructed from macrofossil diagrams. In oligotrophic lakes vegetational changes may be recorded primarily by changes in coarse organic matter originating from the sparse macrophytic vegetation. In lakes characterized by sandy and open shores past fluctuations in lake level are likely to be recorded primarily by changes in coarse minerogenic matter and in the sediment limit.

However, the aim of achieving the complete combination of evidence must be pursued until a recorded fluctuation in lake level has been demonstrated to have regional significance, which means that a fluctuation must be convincingly evidenced in a number of lakes in the region studied. After the regional significance has been demonstrated, however, it should be reasonable to base an interpretation and reconstruction on individual lines of stratigraphical evidence. For example, a separate increase in coarse minerogenic matter or a separate record of a lowered sediment limit should be reliable evidence for correlation with earlier evidence of regional lowerings in the lake level.

Some examples of the application of the method described, demonstrating both combined and individual records of past lake-level fluctuations, are given in Figures 5.2–5.6.

FIELDWORK AND LABORATORY ANALYSES

The various aspects of the work involved in the method are shown schematically in Figure 5.1. Selection of a suitable lake should always be based on reconnaissance stratigraphical studies to ensure the availability of useful records of past lake-level fluctuations. Lakes in which entirely local conditions can be suspected to have affected lake level must be excluded, when the objective is to reconstruct regional palaeohydrological changes. If hydrological information of a lake and surrounding catchment is available this, too, should be used. Information on hydrological regimes is important for the evaluation and understanding of lake-level response to regional palaeohydrological change, but such information is usually not available.

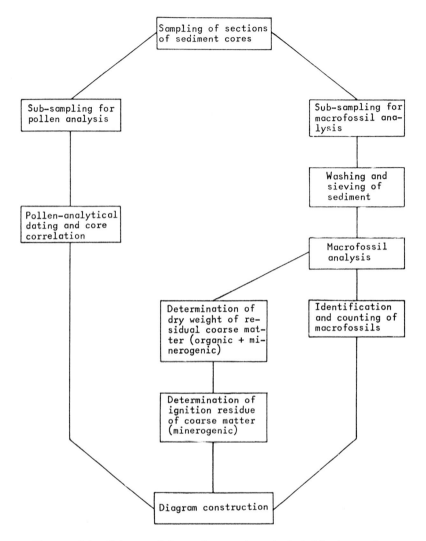

FIGURE 5.1.   Scheme of the various analyses included in the studies

In lakes with surface outlets, palaeohydrological changes in water inflow from the catchment will be partly adjusted by changes in discharge through the outlet; lakes without surface outlets are therefore to be preferred. However, from studies so far carried out in southern Sweden, the presence of a surface outlet will not usually seriously restrict the possibility for reconstructing past lake-level fluctuations.

The lake should have a moderate water depth and a regular bottom morphology. In shallow lakes with very gently sloping margins, past lowerings in lake level are often associated with extensive erosion, resulting in stratigraphical disturbance and complicated recording. Deep lakes with steeply sloping margins will be unsuitable, particularly for the recording of past changes in the distribution of shore and lake vegetation.

The number of sediment cores required in the section, as well as the distance between the cores, largely depends on the character and the slope of the lake margin. The section must be continued to such a distance from the shore and to such a water depth that any water-level fluctuations are no longer recorded in the outermost sediment core. When deciding the distance between the cores, the principle should be followed that the depth increase between adjoining cores (the depth to the bottom of the cores corresponding to the original slope of the lake margin) should be about 1 m. Sediment cores that are too widely spaced will affect the possibility of reconstructing recorded changes thoroughly enough, and particularly changes in the level of the sediment limit.

The sediment volume required depends on the character of the sediment and the concentration of macrofossils. Since the studies also include determination of the quantity of coarse matter, usually a somewhat larger volume is required than recommended for analysis of macrofossils alone (Chapter 27). For sampling sediment cores, a piston sampler with a tube diameter of about 10 cm is recommended. Though somewhat heavier than the conventional sediment sampler, such a sampler is still fairly easily handled and operated. If each macrofossil sample comprises 5 cm of the core, a sediment volume of about 400 cm^3 is obtained.

The technique for macrofossil analysis is dealt with in Chapter 27, where dispersion and washing of sediment, separation and identification of macrofossils are thoroughly described. In cases where the macrofossil analysis appears to be very time-consuming, it may be reasonable to restrict the washing to the larger sieve, and to exclude the recommended supplementary 0.2 mm sieve. In studies of lake level, the macrofossil analysis is used to reveal past changes in the distribution of vegetation, which does not always mean that a complete reconstruction is necessary. Most seeds and fruits of interest for the reconstruction and interpretation of past distributional changes in lake vegetation will usually be obtained when using only a 0.5 mm sieve. However, whenever enough time is available, the recommendation of a supplementary 0.2 mm sieve for washing should be followed.

The washing residue after macrofossil analysis consists of non-identifiable coarse organic matter and coarse minerogenic matter. The combined quantity of organic and minerogenic matter is determined by weighing after drying at 105°C, and the separate quantity of coarse minerogenic matter by weighing after ignition at 550°C. If two sieves have been used for washing, the quantities of both fractions can be determined.

In some earlier studies (e.g. Digerfeldt, 1972, 1974) the quantity of coarse matter was determined on volume (after compaction under some pressure) and dry-weight bases.

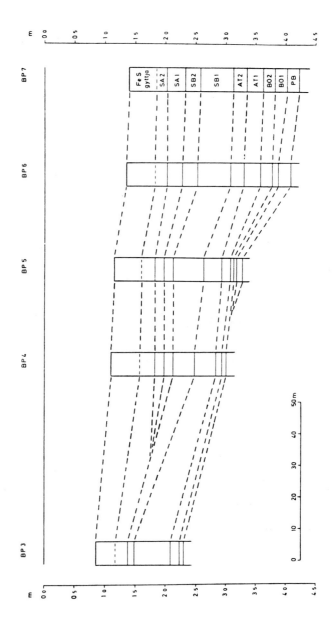

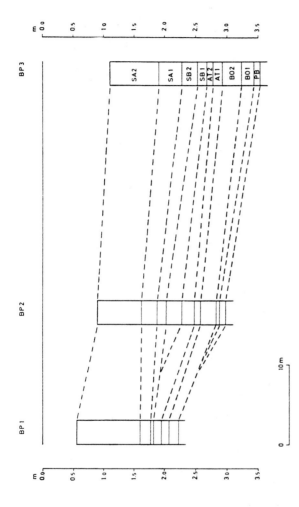

FIGURE 5.2.   Pollen-analytically correlated sediment section from Lake Trummen (above) and Lake Immeln (below), southern Sweden. The sections show two periods of lower sediment limit, indicating lowered lake level, viz. in SB 1–2 and in PB-BO 1. The zonation refers to the South-Swedish pollen-zone system described by Nilsson (1935, 1964), in which the boundaries have been ^{14}C dated as follows: PB/BO 1, 9600 B.P.; BO 1/BO 2, 8500 B.P.; BO 2/AT 1, 7900 B.P.; AT 1/AT 2, 6400 B.P.; AT 2/SB 1, 5000 B.P.; SB 1/SB 2, 3500 B.P.; SB 2/SA 1, 2100 B.P.; SA 1/SA 2, 1200 B.P. (from Digerfeldt, 1972, 1974)

BP 1

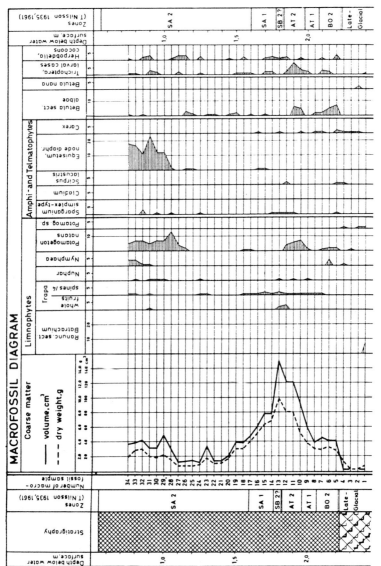

FIGURE 5.3. Macrofossil diagrams from BP 1 and BP 2 in the section from Lake Immeln (Figure 5.2). The lake is oligotrophic and fruits and seeds were fairly sparse. However, at BP 2 an outward spread of *Potamogeton natans* is indicated during the low-level period in PB-BO 1, and finds of *Trapa natans*, *Nuphar luteum* and *Nymphaea alba* are made somewhat more regularly in association with the low-level period in SB 1–2. At BP 1 both low-level periods are indicated by comprehensive hiatuses. An outward spread of *Potamogeton natans* is indicated in association with the low-level period in SB 1–2, and also *Trapa natans* is then found more regularly.

BP 2

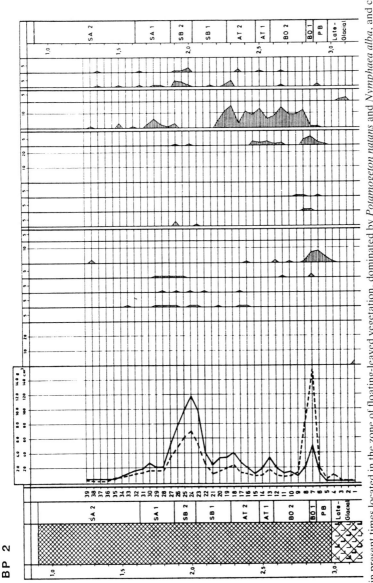

BP 1 is in present times located in the zone of floating-leaved vegetation, dominated by *Potamogeton natans* and *Nymphaea alba*, and close to a reed of *Equisetum fluviatile*. The recorded outward spread of this vegetation partly represents the natural succession associated with the progressive infilling. However, the spread may have been accelerated by recent regulations of lake level.

Both low-level periods are most distinctly recorded by an increase of the coarse matter in sediment. The distinct increase in dry-weight at BP 2 during the low-level period in BP-BO 1 is largely due to minerogenic matter. The same changes are indicated in a diagram from BP 3 (from Digerfeldt, 1974)

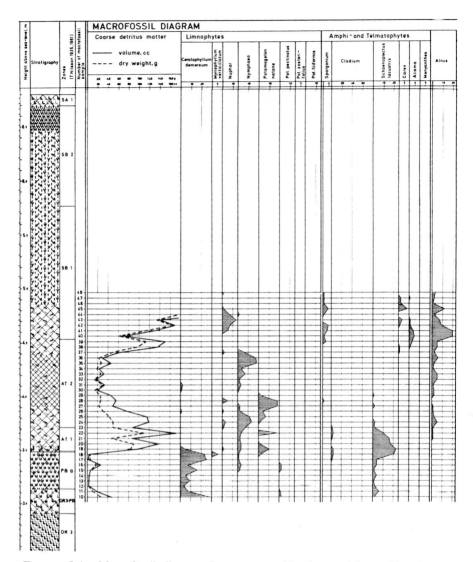

FIGURE 5.4. Macrofossil diagram from a eutrophic, former lake at Torreberga, southern Sweden. The stratigraphy shows a comprehensive hiatus, comprising the later part of PB, BO 1–2, and earlier part of AT 1. From other studies the hiatus may represent two low-level periods, and accordingly 'hide' an intervening rise in lake level.

In the macrofossil diagram the low lake level is indicated by an outward spread of *Scirpus (Schoenoplectus) lacustris*. Before this — in the earlier part of PB — the coring area was occupied by *Ceratophyllum demersum*, indicating deeper water. Owing to a rise in lake level the reed of *Scirpus lacustris* was later displaced towards the shore, and replaced by a floating-leaved vegetation of *Nymphaea alba* and *Potamogeton natans*. The overgrowing of the lake occurred in earlier part of SB 1 by an unusually rapid spread of *Alnus*. An *Alnus*-carr peat is almost immediately overlying the detritus gyttja, which may indicate that the overgrowing was accelerated by a lowering in lake level. The complete study of the former lake at Torreberga includes four macrofossil diagrams (from Digerfeldt, 1971)

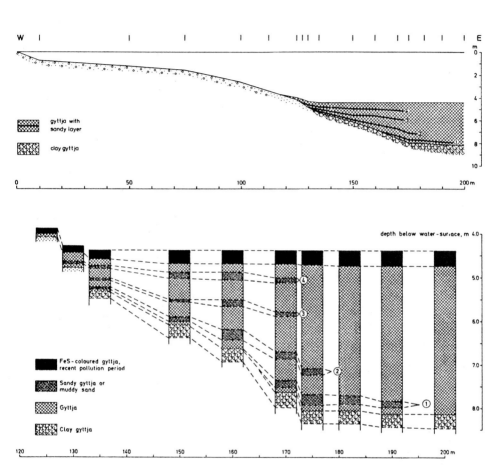

FIGURE 5.5.   Stratigraphic section from Lake Växjösjön, southern Sweden, showing sandy–silty layers indicating periods of lowered lake level. The layers are pollen-analytically dated and correlated at 160 m, 175 m and 182 m from the shore. The layers 1 and 3 correspond to the low-level periods in PB-BO 1 and SB 1–2 demonstrated in Figures 5.2 and 5.3. Macrofossil analysis was not included in the study. However, local vegetational changes and displacement of lake vegetation are pollen-analytically indicated — e.g. changes in the spread of *Isoëtes*.

The sandy–silty layers are distinctly developed only in part of the lake. The recording of past lake-level fluctuations by changes in coarse minerogenic matter in sediment depends on local factors. The studied shore section shows a fairly ideal situation, since even minor fluctuations in lake level are likely to have resulted in significant displacement of the shore and changes in the quantity of reworked minerogenic matter in littoral sediment (from Digerfeldt, 1975)

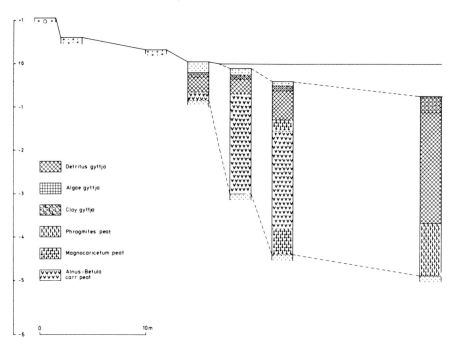

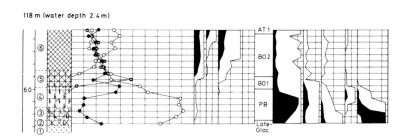

118 m (water depth 2.4 m)

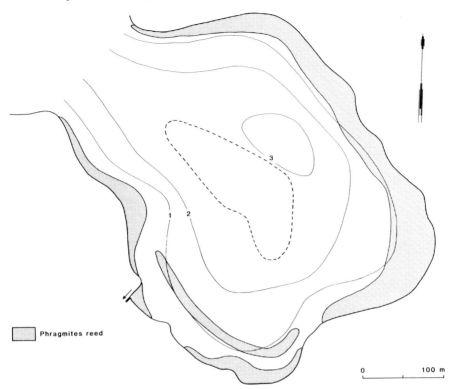

Phragmites reed

0   100 m

FIGURE 5.6. Part of a stratigraphic section from Lake Lyngsjö, south-western Sweden (left, above). The lake was artificially lowered by about 1 m at the end of the last century. The lowering is close to the shore recorded by a sandy and silty layer. The layer can be followed out to about 40 m from the shore, where it is found 10–15 cm below the sediment surface.

Close to the shore the detritus gyttja is underlain by *Alnus-Betula*-carr peat and magnocaricetum peat, indicating a past lower lake level and partial overgrowing. Owing to a strong local disturbance it was not possible to get a more precise pollen-analytical dating of the carr peat.

The central part of the lake has not been overgrown. However, a layer of *Phragmites* peat (layer 4) indicates a significantly lower lake level in PB-BO 1 (left, below). The map (above) shows the distribution of *Phragmites* reeds and the water depth (in m) in the eastern part of present Lake Lyngsjö. The dashed line indicates the outward spread of *Phragmites* reeds during the low-level period in PB-BO1. In the central part of the lake the *Phragmites* peat is found about 6 m below present lake level (from Digerfeldt, 1976)

The map (above) shows the distribution of *Phragmites* reeds and the water depth (in m) in the eastern part of present Lake Lyngsjö. The dashed line indicates the outward spread of *Pharagmites* reeds during the low-level period in PB-BO1. In the central part of the lake the *Phragmites* peat is found about 6 m below low present lake level (from Digerfeldt, 1976).

The reader is referred to Chapter 27 for diagram construction. For several reasons absolute diagrams are preferred, expressing macrofossils in number per unit volume. The use of macrofossils for reconstructing past lake vegetation has been thoroughly discussed by Birks (1980), in which a large number of analytical macrofossil studies of lake sediments are also reviewed. The sub-samples for pollen analysis are taken centrally in each macrofossil sample. The pollen analysis is primarily meant for dating and correlating the cores in the section. However, in most cases it will serve also as a useful supplement to the macrofossil analysis in the interpretation of local vegetational changes. The pollen diagrams from the innermost cores in the section may be strongly disturbed by local vegetation. Moreover, owing to the occurrence of hiatuses, the diagrams will mostly be incomplete. However, the pollen diagram from the outermost core should represent a mainly complete sequence, and provided it is not locally disturbed the diagram can also contribute to the interpretation of regional vegetation changes.

REFERENCES

Berglund, B. E. (1983). Palaeohydrological studies in lakes and mires — a palaeoecological research strategy. In: *Background to Palaeohydrology* (Ed. K. J. Gregory), John Wiley, Chichester, pp. 237–254.

Berglund, B. E., Aaby, B., Digerfeldt, G., Fredskild, B., Huttunen, P., Hyvärinen, H., Kaland, P. E., Moe, D., and Vasari, Y. (1983). Palaeoclimatic changes in Scandinavia and on Greenland — a tentative correlation based on lake and bog stratigraphical studies. *Quaternary studies in Poland*, **4**, 27–44.

Birks, H. H. (1980). Plant macrofossils in Quaternary lake sediments. *Ergebnisse der Limnologie*, **15**, 1–60.

Digerfeldt, G. (1971). The post-glacial development of the ancient lake at Torreberga, Scania, South Sweden. *Geologiska Föreningens i Stockholm Förhandlingar*, **93**, 601–624.

Digerfeldt, G. (1972). The post-glacial development of Lake Trummen: regional vegetation history, water-level changes and palaeolimnology. *Folia limnologica scandinavica*, **16**, 1–96.

Digerfeldt, G. (1974). The post-glacial development of the Ranviken bay in Lake Immeln. I. The history of the regional vegetation; II. The water-level changes. *Geologiska Föreningens i Stockholm Förhandlingar*, **96**, 3–32.

Digerfeldt, G. (1975). Post-glacial water-level changes in Lake Växjösjön, central southern Sweden. *Geologiska Föreningens i Stockholm Förhandlingar*, **97**, 167–173.

Digerfeldt, G. (1976). A Pre-boreal water-level change in Lake Lyngsjö, central Halland. *Geologiska Föreningens i Stockholm Förhandlingar*, **98**, 329–336.

Gaillard, M.-J. (1985). Postglacial palaeoclimatic changes in Scandinavia and central Europe — a tentative correlation based on the studies of lake level fluctuations. *Ecologia Mediterranea*, Marseilles.

Håkansson, L., and Jansson, M. (1983). *Principles of Lake Sedimentology*, Springer, Berlin, pp. 1–316.

Lundqvist, G. (1925). Utvecklingshistoriska insjöstudier i Sydsverige. *Sveriges Geologiska Undersökning*, C 330, 1–129.

Lundqvist, G. (1927). Bodenablagerungen und Entwicklungstypen der Seen. *Die Binnengewässer*, **2**, 1–122.

Nilsson, T. (1935). Die pollenanalytische Zonengliederung der spät — und postglazialen Bildungen Schonens. *Geologiska Föreningens i Stockholm Förhandlingar*, **57**, 385–562.

Nilsson, T. (1964). Standardpollendiagramme und C^{14}-Datierungen aus dem Ageröds mosse im mittleren Schonen. *Lunds universitets årsskrift*, N. F. 2, **59**:7, 1–52.

Ralska-Jasiewiczowa, M., and Starkel, L. (1985). Stratigraphical records of Holocene hydrological changes in lake, mire, and fluvial deposits in Poland (in press).

Richardson, J. L. (1969). Former lake-level fluctuations — their recognition and interpretation. *Communications International Association of Theoretical and Applied Limnology*, **17**, 78–93.

Starkel, L. (1983). The reflection of hydrologic changes in the fluvial environment of the temperate zone during the last 15,000 years. In: *Background to Palaeohydrology* (Ed. K. J. Gregory), John Wiley, Chichester, pp. 213–236.

Handbook of Holocene Palaeoecology and Palaeohydrology
Edited by B. E. Berglund
© 1986 John Wiley & Sons Ltd.

6

Palaeoecological studies of mires

BENT AABY

Geological Survey of Denmark, Thoravej 31,
Copenhagen, Denmark

INTRODUCTION

Mires are essential elements in north-west European landscapes today as they cover between 32% (Finland) and 3% (Denmark, England) of the land area. For centuries they have been regarded as valuable sources for geological and palaeoecological studies, and much of our knowledge about climate, vegetation and human activity in the past is derived from peat deposits.

Different types of mires have their own advantages depending on the aims of the project. Raised bogs, for examples, are unique for palaeoclimatological research, whereas variations in trophic conditions or changes in groundwater supply are studied in various fen types. Pollen analysis of swamp peat without wood remains may furnish information on the regional forest composition which is difficult to obtain from peat deposits once covered by local tree pollen sources, e.g. *Alnus*, *Betula* and *Quercus*. Thus it is important to determine the purpose of an actual project and the problems to be solved, before deciding on the investigation site.

FIELD WORK

Measuring of open sections

Peat exposures in artificially drained mires may convey excellent opportunities for stratigraphical studies. Structural features and variations in the degree of decomposition may often be easier to distinguish in the weathered section than in fresh material, and once destroyed they will not reappear for some weeks. It is therefore recommended that a survey sketch be made of the weathered section before more thorough analyses are initiated.

After inspection, the weathered section is cleaned up and divided by vertical lines at convenient intervals, e.g. 33.3 cm. Also, a horizontal 0-line is laid down. The exposed section is examined in detail until various layers of deposits are recognized, and matchsticks are pushed in on the vertical lines at levels that reveal variation in sediment composition or humification. The position of each matchstick is measured and transferred to graph paper, and observed boundaries between different layers are drawn in by hand. Solid and hatched lines may be used to indicate distinct and more diffuse boundaries.

A pencil drawing of a section can also be made by pegging a metre square frame with 10 cm wire deviation on to the section and transferring the stratigraphy to graph paper. The content of determinable plant remains should be noted for the various deposits in order to obtain an overall impression of the 'mother-formation' (the vegetation, which produced the plant remains), thereby permitting a description of the genesis of the mire. The identification of macrofossils is dealt with in Chapter 28.

The various plant layers should be described in a consistent way and the characterization system elaborated by Troels-Smith (see Chapter 12) is recommended. The colour of the exposed deposits is one of the parameters for description. Standard cards (Munsell cards) may be used in the field, but often they are of little use, as recently exposed peat needs some time to stabilize in its colour. The impression of the 'stable' colour depends on the water content of the peat, as dry material often appears lighter than similar wet material. The type of daylight may also influence the impression of peat colours, especially the position of the sun and cloud formations, and reddish-brown colours become more common late in the afternoon. Accordingly, colours are difficult to describe objectively in the field. 'Objective' colour information can be obtained by rubbing or squeezing peat on rough, white paper and examining the colours after the paper has dried. These stains may be compared with colour cards.

In drying out profiles it is difficult to determine the degree of humification by squeezing the peat and examining the colour of the extracted water (von Post, 1924, Troels-Smith, 1955). Instead, examination of the colour and the texture of the peat can be made directly (e.g. Casparie, 1972). After cleaning the layer in question, possible changes in colour are observed, and the quicker the colour changes, the more humified is the peat. The variations in relative humification obtained directly often comply with the accuracy demanded. A laboratory procedure for determining the degree of humification is described later.

For special purposes it may be advantageous to use a more detailed humification scale than the 5-class scale in the Troels-Smith system. A 10-class scale for determining the degree of humification in ombrotrophic peat is shown in Table 6.1.

TABLE 6.1. A 10-class scale of humification (after von Post, 1924)

Type of peat	Humification value
Yellow–light-brown peat, often with undamaged *Sphagnum* leaves	
Yellow, pale and whitish	1
Yellow	2
Light-brown	3
Brown peat, with *Sphagnum* leaves more or less damaged	
Milk chocolate-brown	4
Brown	5
Dark chocolate-brown	6
Dark-brown peat, the *Sphagnum* leaves are badly damaged	
Dark brown, plant structures can still be observed in matrix, and wood remains, seeds, etc. are determinable	7
Dark coffee-brown, with a distinct bluish-black glow and only few plant structures remaining in the matrix. Wood fragments, seeds etc. are distintegrated and difficult to determine	8
Plant structures are observed in the matrix occasionally. Macroscopic remains, hardly determinable	9
Blackish-brown peat	
Totally destroyed organic material and charred material	10

Levelling

Levelling is necessary for orographic studies and for correlating stratigraphy, sample depth etc. at a number of investigation sites. A relative depth scale may be given when measuring parameters of only local interest, but often it is preferable to establish a given level in absolute values, e.g. metres above sea level.

It is difficult to level on mires because the surface is sensitive to compaction. To obtain reasonably good results, it is important to ensure that the measuring instrument remains in a horizontal position. The wooden staff may be stabilized if placed on the end of a 1 m pole that has been rammed into the peat deposit.

The measuring accuracy should in general be about ±2 cm per 1000 m. The accuracy should always meet actual demands; and since boundaries between layers and even the mire surface may be weakly defined, a reduced measuring accuracy may be tolerated, but it should be better than ±5–7 cm per 1000 m.

LABORATORY WORK

Slicing technique

The collected material is subjected to various treatments, such as extrusion of cores, cutting of subsamples etc., before analyses can be made. A flow diagram for laboratory processing is shown in Figure 6.1.

When monoliths or large-diameter (10 cm) cores are available it is suggested that they be cut longitudinally, and one of the halves be saved for pollen analytical sampling and various physical and chemical treatments. The remaining half is stored and later used for reference purposes and sampling for radiocarbon dating at selected levels. A knife or thin steel wire may be useful cutting tools.

The stratigraphy is studied on the newly exposed and cleaned peat surface. The procedures of deposit characterization are discussed later.

Further cutting of the first subcolumn can be done with the same tools if the peat structure is not too heterogeneous or resistant. If so, freezing of the subcolumn may be advantageous. This procedure is only possible when the material is not water-saturated, otherwise the volume increases. Freezing facilitates a high accuracy of slicing as the ice stabilizes the material, e.g. weakly decomposed *Sphagnum* peat. Thus, persistent wood remains like *Calluna* twigs and highly tangled tussocks of *Eriophorum vaginatum* are easily cut without displacing the adjacent matter.

Larger peat-bricks (1–2 cm thickness) are cut from the frozen subcolumn with an ordinary saw, which may also be used for additional subsampling from the bricks. A knife is also useful when cutting smaller samples where the thickness of the brick does not exceed about 1.5 cm.

Thin bricks (3–5 mm) of frozen peat are better cut on a small circular saw. This type of machine may not produce sawdust, which is an advantage for close-interval sampling (Aaby *et al.*, 1979).

Storing

Samples for pollen analysis need no further protection from drying out if they are stored in corked glass tubes and used within a few months. For permanent storage, a small amount of glycerine is added to keep the sediment soft, or the cork may be sealed with a layer of wax. It is important to prevent drying out as shrinkage of the sample causes deformation of the pollen grains, which seriously reduces the quality of the record. Secondary disturbances are hindered by freeze-drying the samples, and the material will remain unaffected for years.

It is often necessary to store monoliths, cores or larger peat bricks for considerable periods of time, before they can be analysed. To prevent any secondary disturbances while in storage, the samples should be kept effectively sealed (e.g. in heavy polyethylene bags or tubes) to prevent water loss and the development of aerobic conditions, which encourage microbial activity and raise the rate of decomposition. Storage in darkness at 4–8°C can be

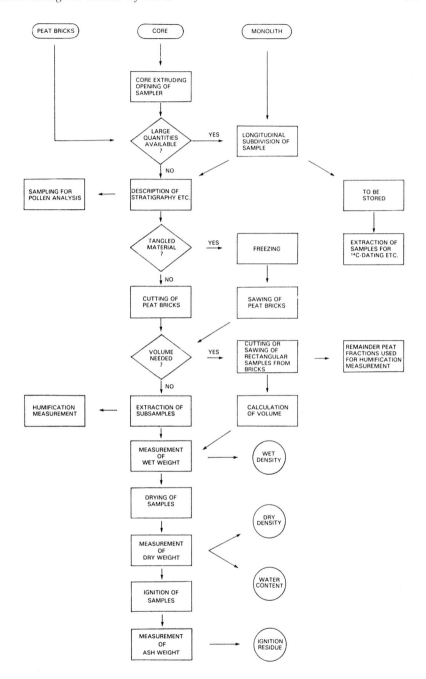

FIGURE 6.1.   Flow diagram, showing laboratory processing methods of peat samples

recommended, and at such a temperature samples may be kept undamaged for many years.

Freezing may distort the stratigraphy if the peat is water-saturated. Rigidly held cores are therefore not recommended for freezing, because the water content remains high after sampling, whereas monoliths and peat bricks can be frozen because the water content has decreased after removal from the exposure.

Physical properties

Ignition residue

The content of inorganic material is measured by drying 8–10 cm^3 of fresh sediment in a drying cabinet for at least 24 hours at 105°C; it is then weighed and ignited at 550°C for 4 hours. The weight of the ash is calculated as a percentage of the sediment dry-weight. Ignition residue calculations are generally performed from the same sample as is used for density measurements (see Figure 6.1).

Sediment density

A known volume, approximately 8–12 cm^3, is cut from the bricks. The density can be expressed as wet density and dry density.

The uncertainty of measuring volumes may be considerable and different methods can be applied to secure a high degree of accuracy:

(1) In almost homogeneous sediments, which are medium or strongly decomposed, rectangular samples can be cut with a sharpened knife (Figure 6.2). The uncertainty of volume measurement is about ±5–7%, although in unfavourable conditions the uncertainty may exceed this value.

(2) The volume may be measured by immersing the peat sample in water. This procedure is possible only if the peat is saturated or the ability to imbibe water is low. Much caution is needed as the sample may easily disintegrate. The measuring accuracy is possibly lower than for the first method.

(3) The slicing technique with the use of frozen samples has already been mentioned, and this method is recommended if loose, heterogeneous or tangled material has been collected. The measuring uncertainty may be low, about ±5%.

The degree of humification

This may be determined visually in a subjective way (see above). To obtain exact values for the decomposition, a chemical–optical determination method can be applied. A useful method has been developed by Overbeck (1947) and modified by Bahnson (1968). It consists of a colorimetric determination of an alkali extract of the peat. The alkali absorption is proportional to the amount of humic matter dissolved, and the degree of humification is calculated relative to 0.2 g of a standard (=100%).

Bahnson's modified method is given below (translated from the original Danish):

The peat sample is comminuted and dried in an open dish under an infra red lamp. *Eriophorum vaginatum*, if present, is cut with a pair of scissors as this material may otherwise form a felty mass. The peat is then ground in a mill and dried in a drying cabinet at 105°C.

0.200 g of the sample is placed in a 200 ml volumetric flask and 100 ml of a 0.5% NaOH solution is added with a pipette. The flask is heated to boiling point on an electrical plate that preferably should be large enough to accommodate a set of up to 24 flasks. When the liquid begins to boil, foam is usually formed and may overflow the neck of the flask. This can be avoided by rapidly removing the flask from the hot plate. After a few seconds, when the comminuted peat sample is thoroughly wetted, the flask can be safely placed on the hot plate again and boiled for an hour. After cooling, distilled water is added to the mark, and the suspension is filtered through a medium filter. 50 ml of the well-shaken filtrate is placed in a 100 ml volumetric flask and diluted with distilled water to the mark, after which the flask is thoroughly shaken. The absorption is then measured on an EEl colorimeter with a filter (540 nm). The zero point of the instrument is determined with distilled water. If the colorimetric reading exceeds six, the sample is diluted once again; in this way a uniform measuring accuracy is obtained. In order to compare results from different series it is necessary to adhere to a strict time-schedule as some fading of the intensities occurs with time. The largest source of error in this method arises from inadequate shaking of the flasks; this applies particularly to flasks with a high and thin neck. Repeated measurements, made on two or more portions of each sample, should be made in order to check the reproducibility of results. In each series a pulverized humic acid standard (from Fluka Ltd., Switzerland), treated in the same way as the samples, is included.

Dilution experiments have shown that the humic acid standard forms an almost true solution. The calibration curve for the humic acid standard may therefore be calculated. For the linear curve the equation $y=0.12x - 0.1$ was found; x=degree of humification, y=colorimetric reading.

FIGURE 6.2. Part of a peat core prepared for sampling after splitting the PVC-tube. Material for pollen analysis is first collected by pressing glass tubes into the peat. Afterwards, rectangular subsamples are cut from the bricks and used for determination of wet- and dry-density, and ignition residue. Humification measurements are performed on the remaining material

PALAEOECOLOGICAL IMPLICATIONS DERIVED FROM MIRE INVESTIGATIONS

Peat bogs furnish information on floristic and environmental conditions, and results concerning local as well as extralocal and regional parameters are obtained. The regional and extralocal data (terminology: Janssen, 1973) derive mainly from the analysis of pollen, and as the mire vegetation itself may contribute considerable amounts of pollen the location of the source area may become obscured. The extent and the importance of this problem will vary with the local vegetation. Grass species are not inhabitants of Scandinavian ombrotrophic mires, and pollen of this family will therefore be separated out as a regional or extralocal component. Reedswamps produce large quantities of pollen, particularly grass and sedge pollen, which cannot always be distinguished from the now-local pollen input to the swamp.

The influence of the local flora on pollen input is high only near the source and falls off rapidly to represent regional values. This was demonstrated, for example, by Turner (1964), who found that the influence of a local vegetation type surrounding a mire could hardly be traced on the mire by pollen analysis at a distance of about 200 m from the source. The location of an investigation site therefore has a considerable influence on the validity of the general conclusions drawn (see Chapter 4).

On the contrary, the strong representation of the local pollen spectra may provide a detailed insight into the vegetation on sites where pollen analysis is hindered, (e.g. on fertile mineral soils) when sampling close to the border of the mire.

A simplified scheme for planning a study of peat deposits is shown in Figure 6.3, with separate diagrams for local and regional (extralocal) interpretations to stress the importance of the source areas of the fossils.

Local floristic or stratigraphical relationships are investigated by detailed analysis of microfauna, micro- and macroflora, and chemical–physical properties of the peat. These results give information on plant community development, hydrology, ecological changes etc., which are used to infer climate, environment and possible human impact on the mire ecosystem. If radiometric dating results are available (^{14}C, ^{210}Pb) a chronology can be established and rates of peat accumulation determined (Aaby *et al.*, 1979, El-Daoushy *et al.*, 1982).

Descriptions of mire history often require a series of investigation sites to enable, for instance, the rates of mire transgression to be calculated.

Conclusions about the landscape outside the mire are based mainly on a sequence of pollen analyses, from which relative and absolute pollen frequencies are calculated. They provide a broad basis for interpretation of vegetation and settlement history, environmental changes, climatic conditions etc.

Man's activity may also be registered by the presence of inorganic material blown into the mire from tilled or otherwise exposed soils (see later). Natural processes such as volcanic activity may be detected in some areas. At some investigation sites chemical analysis may furnish important information about ancient or present-day metal deposition (e.g. Aaby and Jacobsen, 1979; Aaby *et al.*, 1979; Pakarinen *et al.*, 1980; Oldfield *et al.*, 1981; Jones, 1983).

PEAT FORMATION

Natural mire ecosystems have a positive energy budget, with the production of organic material exceeding the loss of matter. The annual production of organic

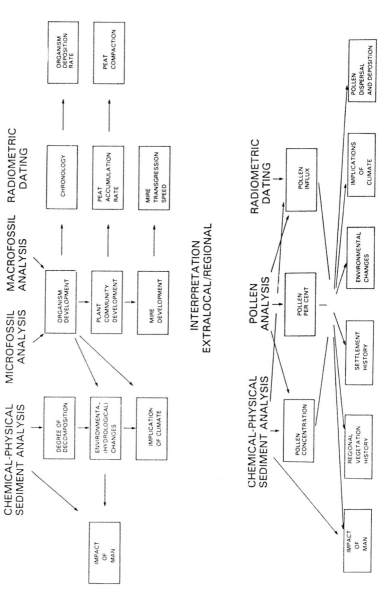

FIGURE 6.3. Flow diagram, showing the possible palaeoecological implications of mire investigations. The diagram is divided into two parts according to source area

matter is often similar to that of many mineral soils; some ombrotrophic mires yield 2–10 t ha^{-1}, or just as much as oak forests (Aaby and Jacobsen, 1979; Burmeister *et al.*, 1980). The loss of organic matter is considerable; and depending on the type of plant remains, and chemical and hydrological conditions, about 10–25% may become fossilized (Holling and Overbeck, 1961; Overbeck, 1975; Clymo, 1984). Humification takes place mainly in the upper part of the peat body (Clymo, 1965), and the degree of humification essentially depends on how long it takes before the peat becomes anaerobic as determined by the water level. The amount of peat accumulated is therefore influenced mainly by the net production, the rate of aerobic decay, and the length of period within the aerobic zone (Clymo, 1975).

Because of their high sensitivity to hydrological changes, mire ecosystems, and especially the ombrotrophic type, have furnished valuable knowledge on water-level changes, and hence on palaeoclimates (see Chapter 3).

PEAT ACCUMULATION RATES

Peat accumulation processes in the aerobic zone have been studied by measuring the annual increment of selected plant material or from established age–depth curves based on a number of ^{210}Pb or ^{14}C datings.

In 1902, Weber measured the annual vertical growth of *Trichophorum austriacum* as 20–30 mm, whereas Gross (1912) obtained values of only about 10 mm. The internodal linear growth is generally 10–60 mm per year for most peat-forming herbs (Bertsch, 1925). Linear growth of *Sphagna* has been studied mainly by Overbeck and Happach (1957) and Rudolph (1963a, b, 1964) and distinct variations within species were obtained as well as variations between *Sphagnum* species. The annual growth variability is possibly due to genetic differences and to hydrological and ecological conditions (Overbeck and Happach, 1957).

Shrinkage, or autocompaction of the peat deposit, is ignored by measuring the internodal growth. This process was taken into consideration by, for example, Heikurainen (1953), who calculated the annual increase in peat depth from young *Pinus* trees, by measuring the distance from the actual mire surface to the position of the seed from which the tree once originated and dividing this length by the number of tree rings counted.

Using the ^{210}Pb dating method (Figure 6.4), the annual increment of the uppermost peat layer on a Danish ombrotrophic mire was about 7.5 mm/yr, and this is considered to represent the annual gross deposition of organic matter from the *Sphagnum* vegetation (Aaby *et al.*, 1979). In accordance with this, Hansen (1966) estimated an increment of *Sphagnum cuspidatum* from the same mire, by measuring the growth of *Drosera intermedia* at 6–14 mm/year ($n=4$).

The ^{210}Pb-dating curve shows considerable autocompaction in the aerobic

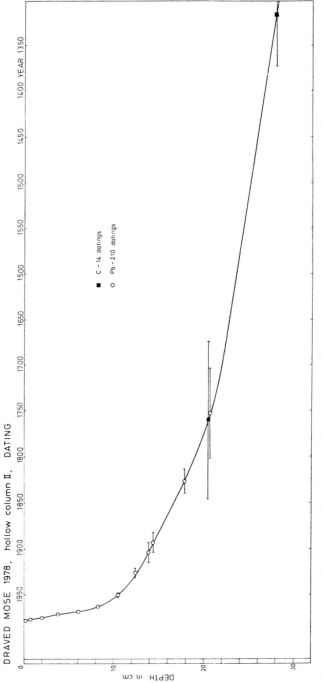

FIGURE 6.4.   Age–depth curve from weakly decomposed *Sphagnum cuspidatum* peat based on ¹⁴C and ²¹⁰Pb datings. Horizontal bars ±SD. Sample thickness is 3 mm (Aaby *et al.*, 1979)

zone; the increment values have decreased by 50% at 6–8 cm. Below 8 cm the increment values gradually become smaller, and from 20.5–28.0 cm the value is 0.2 mm/year, or only about 3% of the original annual deposition (Aaby *et al.*, 1979).

The ^{210}Pb-dating method has also been applied to increment investigations in Finland (El-Daoushy *et al.*, 1982).

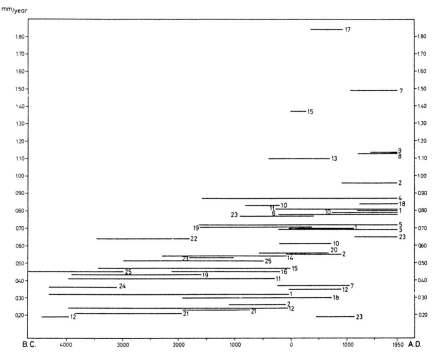

FIGURE 6.5.   Mean rates of peat formation for a number of north European raised bogs, calculated from calibrated ^{14}C dates (Damon *et al.*, 1973). The growth rates have been calculated for the longest time intervals possible in order to minimize the uncertainties caused by errors in the ^{14}C dates (Aaby and Tauber, 1975). *Reproduced by permission of Universitelsforlaget*

Knowledge about accumulation rates in deeper peat deposits are today based mainly on ^{14}C datings. Results obtained from a number of north European ombrotrophic mires are compiled in Figure 6.5. The measured accumulation rates were generally low in the Atlantic and in the early part of the Subboreal, gradually increasing and reaching a maximum in the Subatlantic. Although the accumulation rates obtained from some bogs may be considered minimum values because of artificial drainage and shrinkage of the peat body, it is assumed that the general tendency to higher accumulation rates for younger peat layers mostly is a function of autocompaction (Aaby and Tauber 1975).

THE INFLUENCE OF HUMIFICATION AND AUTOCOMPACTION ON PEAT ACCUMULATION

Granlund (1932) arrived at the opinion that strongly humified peat was formed in periods with a low peat accumulation rate, while slightly decomposed peat originated in periods with rapid peat growth. Low precipitation, which gave rise to a vegetation of less water-demanding species and to a greater oxidation of the upper peat layers, was considered the main cause of the strong humification.

Following the introduction of the ^{14}C dating method much more accurate measurements of the rate of peat formation in ombrotrophic mires, and its relation to the degree of humification, have been obtained. A number of mire investigations seem to corroborate the opinion of Granlund (Florschütz, 1957; Nilsson, 1964; Tolonen, 1973), but other investigations suggest no clear correlation between rate of peat accumulation and degree of humification (Schneekloth, 1965; Olausson, 1957; Aaby and Tauber, 1975). The complexity of peat formation is illustrated in a 2.4 m deep ombrotrophic peat section in Denmark (Aaby and Tauber, 1975). Here, accumulation rates during the last

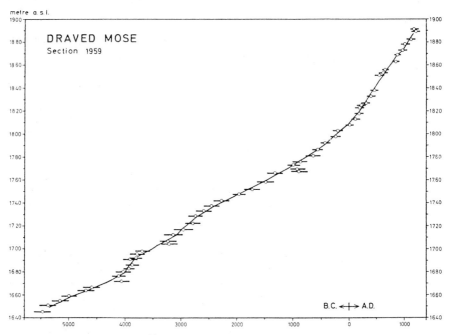

FIGURE 6.6. Calibrated ^{14}C dates from an ombrotrophic mire with statistical counting errors and the error of calibration indicated (Damon *et al.*, 1973). The curved line is based on sliding means of 5 successive dating results (see also Figure 6.7) (Aaby and Tauber, 1975). *Reproduced by permission of Universitetsforlaget*

DRAVED MOSE section 1959

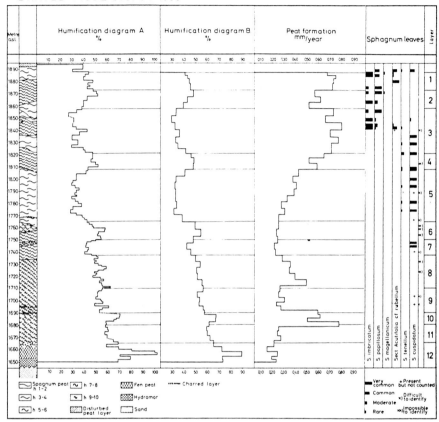

FIGURE 6.7. Degree of humification, rates of peat formation, and the content of *Sphagnum* leaves in a Danish ombrotrophic peat section. The time interval covered is 5400 B.C. to A.D. 1200 in calibrated ^{14}C years. Humification diagram A: Values of the individual peat samples. Humification diagram B: Mean values within the levels used for calculation of accumulation rates (Aaby and Tauber, 1975). *Reproduced by permission of Universitetsforlaget*

6500 years have been calculated from 59 ^{14}C dates (Figure 6.6). Comparison of variations in humification intensity and rate of peat formation show (Figure 6.7) a negative correlation between the mentioned parameters in the upper part of the peat section (layers 1–4), whereas the correlation is positive in the lowermost part (layers 7–11). No clear relationship between growth rate and degree of humification is registered in intermediate layers 5 and 6. Also in this section, autocompaction is presumed to have had a serious influence on the peat body. This is consistent with the fact that calculated growth rates for *Cuspidatum* peat formed in similar wet conditions (layers 3, 5, 7, 9 and 11) decrease with depth in the upper part of the peat section and then remain constant, corresponding to a maximum compaction of the layer. The decrease in accumulation rates with depth is also valid for peat formed in dry local conditions, but the decrease is only half of the decrease shown by the *Cuspidatum* peat. It thus appears that peat formed in moist conditions is more compressible than peat formed in dry conditions (Aaby and Tauber, 1975). This may be explained by the lack of stabilizing elements in *Cuspidatum* peat, while the vegetation on the dry parts of raised bogs is dominated by, for instance, *Calluna vulgaris*, *Empetrum nigrum* and *Eriophorum vaginatum* together with various bryophytes, and the wooden elements are rather resistant to compaction (Kaye and Barghoorn, 1964). A more intensive aerobic decomposition, taking place under dry conditions, may also, in itself, result in a greater primary consolidation. The relationships between the degree of humification and rate of peat formation in the mentioned peat section (Figure 6.7) are accordingly explained by little compaction in the upper part, where weakly decomposed peat shows higher accumulation rates than moderately decomposed material. In the lower part, further compaction has given the more stable peat, formed in dry conditions, the highest accumulation rates. Thus, detailed investigations have highlighted the importance of compaction processes going on in deep peat deposits; and because the compressibility varies in different peat types a clear correlation between measured growth rates and degrees of humification cannot always be expected.

ANTHROPOGENIC ACTIVITY TRACED BY ASH CONTENT IN PEAT DEPOSITS

In sandy soil landscapes, field erosion by wind is often reported, especially in spring, when tilled fields lie bare and dust particles can be transported over considerable distances. Similar incidents related to agriculture in the past are identified in mire deposits by an increase in the ash content of the peat.

Ombrotrophic mire-types offer the best possibilities for research of this type, as the mire does not interact with the minerogenic groundwater and all input of mineral matter is supplied by the atmosphere through precipitation or dry fallout.

The amount of dust particles incorporated may vary depending on the structure of the vegetation, as the filtering efficiency of, for instance, a tall herb vegetation with numerous erect thin leaves and stalks is much higher than the almost smooth surface of a *Sphagnum* vegetation. The concentration of airborne particles within the peat is a function not only of its initial deposition rate, but also of the rate of peat decay, and the degree of peat compaction. Furthermore, there is also a 'local' source of inorganic matter derived from the microflora — e.g. diatoms — which may cause anomalies when originating in considerable amounts, as demonstrated by Vuorela (1983) in minerotrophic peat. In ombrotrophic *Sphagnum* peat, she found that the content of silicoflagellates in the ignition residue was too small to affect the ash values.

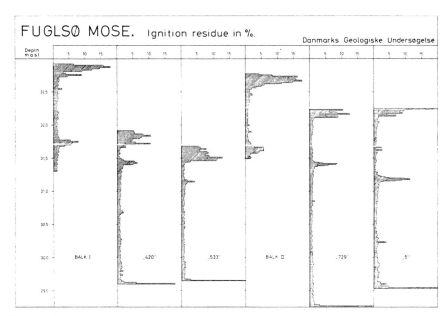

FIGURE 6.8. Ignition residue diagrams from six sections in a raised bog. The bottom maximum represents the mineral soil, the following peak at about 31.5 m is dated to the early Iron Age and the high values at the top originate from medieval and modern time, with the exception of sections 420 and 533, where the surface layer has been disturbed (from Bahnson, 1973)

Traces of anthropogenic activity have been demonstrated from six sections in an ombrotrophic mire (Bahnson, 1973). The peat consists mainly of *Sphagnum*, with some *Eriophorum vaginatum* and *Calluna vulgaris* remains. The ash content was expressed as weight % of dry matter, and values below 1% were obtained from pre- and inter-agrarian deposits (Figure 6.8). In prehistoric agricultural phases, the values rose to 2–5%, and maxima of 5–10% were calculated for the early Iron Age. The ash maxima could be shown to exist all over the bog, and additional ash peaks were found in the marginal sections.

Using the same procedure, very similar results have been demonstrated from Finnish ombrotrophic mires (Vuorela, 1983).

Kramm (1978) quantified the content of inorganic dust particles in peat from a north-west German ombrotrophic mire by counting the number of particles in a 0.01 mm^2 area of a slide preparation. The amount of dust particles registered for prehistoric cultivation phases are two to three times the amount in intervening periods, whereas much higher values originated from the Middle Ages and modern time.

Quantified dust contents may confer valuable knowledge about the extension of arable fields, or other areas with exposed soils, whereas, for instance, permanent pastures contribute no additional amount of dust. Comparisons between the trends in the curves of some pollen types and dust content may give

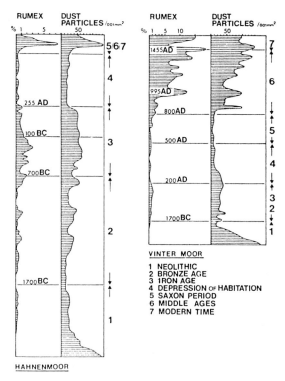

HAHNENMOOR

FIGURE 6.9.　The relationship between the *Rumex* pollen curve and the quantity of airborne inorganic dust particles in two north-west German ombrotrophic mires. The dust particle curve is based on the number of particles counted in a 0.01 mm² area of a slide preparation. The chronology is based on uncalibrated ¹⁴C dates (after Kramm, 1978, in Behre, 1981). *Reproduced by permission of Pollen et Spores*

an indication of the source area for the dust. Kramm (1978) showed a close relationship between the dust particle curve and that of *Rumex* pollen, which consisted mainly of *Rumex acetosella* (Figure 6.9). This supports an interpretation of *Rumex* as an indicator of disturbed biotopes, such as crop fields, and heaths which are being stripped as part of the plaggen process, as well as newly cleared woodlands (Behre, 1981).

Vuorela (1983), in her attempt to reconstruct the economic development and the pattern of human settlement, used Cerealia and weed pollen (*Artemisia*, Chenopodiaceae, *Plantago* and *Rumex*) as culture indicators. In all diagrams, the surface layers show a discrepancy between the increasing ash content and the decreasing Cerelia and weed frequencies, although the sites investigated are surrounded by intensively cultivated fields. This was probably due to the small amount of pollen liberated from *Hordeum* and *Avena* crops, as well as the small area of fallow fields. Vuorela therefore suggested that the ash

Palaeoecological studies of mires 163

content of peat deposits is an even more useful indicator of the existence of arable land than are the selected pollen types.

Measurements of the ash content may thus provide valuable information on an environment, and complement pollen analytical research.

REFERENCES

Aaby, B., and Jacobsen, J. (1979). Changes in biotic conditions and metal deposition in the last millennium as reflected in ombrotrophic peat in Draved Mose, Denmark. *Danm. Geol. Unders. Årbog* 1978, 5–44.
Aaby, B., Jacobsen, J., and Jacobsen O. S. (1979). Pb-210 dating and lead deposition in the ombrotrophic peat bog Draved Mose, Denmark. *Danm. Geol. Unders. Årbog* 1978, 45–68.
Aaby, B., and Tauber, H. (1975). Rates of peat formation in relation to degree of humification and local environment, as shown by studies of a raised bog in Denmark. *Boreas*, **4**, 1–17.
Bahnson, H. (1968). Kolorimetriske bestemmelser af humifeceringstal i højmosetørv fra Fuglsø mose på Djursland. *Medd. Dan. Geol. Foren.*, **18**, 55, 63.
Bahnson, H. (1973). Spor af muldflugt i keltisk jernalder påvist i højmoseprofiler. *Danm. Geol. Unders. Årbog* 1972, 7–12.
Behre, K.-E. (1981). The interpretation of anthropogenic indicators in pollen diagrams. *Pollen et Spores*, **23**, (2), 225–245.
Bertsch, K. (1925). Das Brunnenholzried. *Veroff. d. Staatl. Stelle G. Württ. Landesamt f. Denkmalspflege*, **2**, 67–172.
Burmeister, E. G., Gottlich, K., Grospictsch, T., and Kaule, G. (1980). Begriffsbestimmungen anhand der Moortypen Mitteleuropas. In: *Moor und Torfkunde* (2nd edn.) (Ed. K. Gottlich), Schweizerbartsche Verlagsbuchhandlung, Stuttgart, pp. 1–46.
Casparie, W. A. (1972). *Bog Development in South-Eastern Drenthe (the Netherlands)*, Dr. Junk, The Hague.
Clymo, R. S. (1965). Experiments on breakdown of *Sphagnum* in two bogs. *J. Ecol.*, **53**, 747–758.
Clymo, R. S. (1975). A model of peat bog growth. In: *Ecology of Some British Moors and Montane Grasslands* (Ed. O. W. Heal), Springer-Verlag, Berlin.
Clymo, R. S. (1984). The limits of bog growth. *Phil. Trans. R. Soc. Lond.*, **B303**, 605–654.
Damon, P. E., Long, A., and Wallik, E. I. (1973). Dendrochronologic calibration of the carbon-14 scale. *Proc. Int. Dating Conf.*, New Zealand, pp. A28–A43.
El-Daoushy, F., Tolonen, K., and Rosenberg, R. (1982). Lead-210 and moss increment dating of two Finnish *Sphagnum* hummocks. *Nature*, **296**, 429–431.
Florschütz, F. (1957). Over twee "geijkte" pollen- en sporendiagrammen uit de omgeing van Vriezenveen. *Boor en Spade*, **8**, 174–178.
Granlund, E. (1932). De svenska högmossarnas geologi. *Sver. Geol. Unders. Afh.*, **26**, 1–193.
Gross, H. (1912). Ostpreussens Moore mit besonderer Berücksichtigung ihrer Vegetation. *Schr. d. Phys.-ökonom. Ges., Königsberg i. Pr.*, **53**, 184–264.
Hansen, B. (1966). The raised bog Draved Kongsmose. *Bot. Tidskr.*, **62**, 146–185.
Heikurainen, L. (1953). Die Kiefernbewachsenen eutrophen Moore Nordfinlands. *Ann. Bot. Soc. 'Vanamo'*, **26**, 2.
Holling, R., and Overbeck, F. (1961). Über die Grösse der Stoffverluste bei der Genese von Sphagnumtorfen. *Flora*, **150** (2/3), 191–208.

Janssen, C. R. (1973). Local and regional pollen deposition. In: *Quaternary Plant Ecology* (Eds. H. J. B. Birks and R. G. West), Blackwell.

Jones, J. M. (1983). Heavy metals in ombrotrophic peat: a magnetic approach to monitoring historical deposition. *Ecol. Bull.*, **14**(4), 166(Abs.).

Kaye, C. A., and Barghoorn, E. S. (1964). Late Quaternary sea-level change and crustal rise at Boston, Massachusetts, with notes on the autocompaction of peat. *Bull. Geol. Soc. Am.*, **75**, 63–80.

Kramm, E. (1978). Pollenanalytische Hochmooruntersuchungen zur Floren- und Siedlungsgeschichte zwischen Ems und Hase. *Abhandl. Landesmuseum f. Naturkunde, Münster*, **40**(4), 1–44.

Nilsson, T. (1964). Standardpollendiagramme und C14-Datierungen aus dem Ageröds Mosse im Mittelen Schonen. *Acta Univers. Lund.*, N. F. 2, **59**(7), 1–52.

Olausson, E. (1957). Das Moor Roshultsmyren. *Acta Univers. Lund.*, N.F. 2, **53**(12), 1–72.

Oldfield, F., Tolonen, K., and Thompson, R. (1981). History of particulate atmospheric pollution from magnetic measurements in dated Finnish peat profiles. *Ambio*, **10**, 185–188.

Overbeck, F. (1947). Studien zur Hochmoorentwicklung in Niedersachsen und die Bestimmung der Humifizierung bei stratigraphish-pollenanalytischen Moor untersuchungen. *Planta Archiv für wissenschaftl. Botanik.*, **35**, Hft. 1/2.

Overbeck, F. (1975). *Botanish-Geologishe Moorkunde*. Karl Wachholtz Verlag, Neumünster.

Overbeck, F., and Happach, H. (1957). Über das Wachstum und den Wasserhausholt einiger Hochmoorsphagnen. *Flora*, **144**, 335–402.

Pakarinen, P., Tolonen, K., and Soveri, J. (1980) Distribution of trace metals and sulfur in the surface peat of Finnish raised bogs. *Proc. 6th Int. Peat Congr. Aug. 17–23, 1980*, Duluth.

Rudolph, H.-J. (1963a). Die Kultur von Hochmoor-Sphagnen unter definierten Bedingungen. *Beitr. z. Biologie der Pflanzen*, **39**(2), 153–177.

Rudolph, H.-J. (1963b). Die Kultur der Sphagnen unter definierten Bedingungen als Weg zur Klärung physiologisher und ökologischer Probleme im Hochmoor. *Ber. Dtsch. Bot. Ges.*, **76**, 16–20.

Rudolph, H.-J. (1964). Zur Frage der Membranochromie bei Sphagnen. I. Welche Faktoren bestimmen den Farbwechsel?, *Flora*, **155**.

Schneekloth, H. (1965). Die Rekurrenzfläche in Grossen Moor bei Gifhorn—eine zeitgleiche Bildung?. *Geol. Jahrb.*, **83**, 477–496.

Tolonen, K. (1973). On the rate and pattern of peat formation during the postglacial time. *Suo*, **24**, 83–88.

Troels-Smith, J. (1955). Karakterisering af løse jordarter. *Danm. Geol. Unders.*, IV Series, **3**(10), 1–73.

Turner, J. (1964). Surface sample analyses from Ayrshire, Scotland. *Pollen et Spores*, **6**, 583–592.

von Post, L. (1924). Das genetische System der organogenen Bildungen Schwedens. *Comite Internat. de Pedologie.*, IV Commission No. 22.

Vuorela, I. (1983). Field erosion by wind as indicated by fluctuations in the ash content of *Sphagnum* peat. *Bull. Geol. Soc. Finland*, **55**, (1), 25–33.

Weber, C. A. (1902). *Über die Vegetation und Entstehung des Hochmoores von Augstumal im Memeltal, mit vergleichen den Ausblicken auf andere Hochmoore der Erde*, Berlin.

Handbook of Holocene Palaeoecology and Palaeohydrology
Edited by B. E. Berglund
© 1986 John Wiley & Sons Ltd.

7

Palaeoecological studies of terrestrial soils

S. T. ANDERSEN

*Geological Survey of Denmark,
Thoravej 31,
Copenhagen, Denmark*

INTRODUCTION

Soils form an integral part of the terrestrial environment. Lakes and mires may offer indirect information on the development of the soils in their catchment areas; more direct information is obtained by the study of the soils themselves. It is therefore recommended that the study of terrestrial soils be included in the reference areas. In the Holocene, man has widely disturbed the terrestrial soils by tilling, but original soil profiles have been preserved in areas that have not been ploughed — such as pastures, heaths and woodlands, or sites buried beneath archaeological earthworks or dunes.

A characterization of soils according to general pedological principles may be included in the description of a reference area (see Chapter 4). Palaeoecological studies of selected soil sections may give information on vegetational development, stages in soil development, and the impact of land-use by man. Pollen analyses of terrestrial soils have been performed notably in Britain, the Netherlands, Belgium, Austria and Denmark.

CLASSIFICATION OF TERRESTRIAL SOILS

Brown earths consist of parent mineral material mixed with more or less completely mineralized organic matter. Organic litter has been digested and mixed into the soil by large earthworms (Lumbricida) and other soil fauna ('Muld' in the terminology of Müller, 1878, 1884). Iron and aluminium compounds are preserved in the topmost soil layer (Figure 7.1).

165

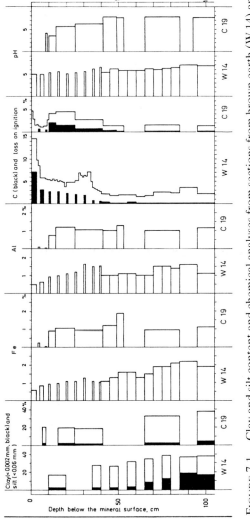

FIGURE 7.1. Clay and silt content and chemical analyses from sections from brown earth (W 14) and podzol (C 19). Percentages of dry weight. Fe = acid-soluble iron, Al = acid-soluble aluminium, C = organic carbon. pH was measured in aqueous extractions (from Andersen, 1979)

Podzoloids consist of parent mineral material, where iron and aluminium compounds have been partly leached from the uppermost layers. Large earthworms are absent, and more or less mineralized organic matter is mixed into the uppermost layers mainly by arthropods, and minor earthworms.

Podzols consist of parent mineral material, where iron and aluminium compounds have been removed from the uppermost layer (the eluvial or A_{1+2}-horizon) and deposited at a lower depth (the illuvial or B-horizon). Soluble humus may also precipitate in the illuvial horizon (Figure 7.1). Decomposed litter is deposited on top of the mineral soil as a humus layer (the A_0-horizon). Müller (1878, 1884) and Hartmann (1965) distinguished zoogenous (coprogenous) and mycogenous (vegetable) humus. Zoogenous humus has been digested by soil fauna, whereas mycogenous humus has been decomposed mainly by fungi. The latter deposit was called 'Mor' by Müller (1878, 1884), but he also used this term to describe zoogenous deposits.

PLANT FOSSILS IN TERRESTRIAL SOILS

Plant debris may be buried in terrestrial soils by downward transport by soil fauna or percolation, or may become incorporated in accumulating humus. Macrofossils are rarely preserved. Pollen grains are destroyed in soils with pH values above 6 (Dimbleby, 1957, Havinga 1971). Dark-coloured fungal hyphae are very resistant to decomposition and are present in nearly all soils. Pollen and hypha assemblages buried in the mineral soil during former soil stages may be preserved in podzols.

Pollen and pteridophyte spores in soils

Pollen assemblages deposited on the land surface mainly reflect the vegetation at or near the sampling site; pollen diagrams from the terrestrial soils therefore reflect strictly local changes in vegetation due to natural succession or influence by man. In this way pollen diagrams from soils highlight or pinpoint single vegetation processes which become masked in diagrams of regional significance. The assemblages buried in soils are more or less mixed vertically owing to the former activity of soil fauna (see later).

Pollen deterioration

The pollen grains and spores preserved in terrestrial soils may be modified by deterioration. Among the several deterioration categories distinguished by Cushing (1967) and Delcourt and Delcourt (1980), corrosion may completely destroy pollen grains. Havinga (1964, 1967) distinguished two types of corrosion, the perforation and the thinning types; transitions may, however, occur, and the two types may be difficult to distinguish in severe cases.

Andersen (1984) found that the average frequency of corroded pollen grains is low at pH values less than 5.5; the average corrosion was higher in pollen assemblages buried in mineral soils, indicating higher pH values at the time when the pollen grains were buried. Havinga (1967) noticed that perforation corrosion prevailed in an unmodified podzol and that thinning predominated in old podzols. Aaby (1983) distinguished perforated ('corroded') and thinned grains in two podzols.

Havinga (1967) found selective removal of *Quercus* pollen in podzols where thinning corrosion was strong, whereas Aaby (1983) found only slight differences between less modified pollen spectra. The original pollen assemblages may thus have changed in strongly modified (oxidized?) podzols.

Dark-coloured fungal hyphae in soils

Litter decaying on the soil surface is infected by fungal hyphae. Living hyphae occur mainly in the litter layer and decompose rapidly, except for pigmented hyphae, which develop when decomposition is retarded (Andersen, 1984). The dark-coloured hyphae become fragmented by soil fauna, which feed on this litter and become incorporated in the soil. Macroarthropods such as Isopoda (woodlice) and Diplopoda (millipedes) living at the soil surface initially cut the hyphae into fragments predominantly 20–50 μm long (Figure 7.2), whereas micro-arthropods such as Cryptostigmata (oribatids) cut the hyphae into still smaller fragments predominantly 10–30 μm long, when feeding on litter or faecal pellets from the larger animals (Figure 7.3). The length of the hypha fragments therefore decreases with increasing treatment from oribatids; hypha fragments shorter than 21 μm thus increase from about 10% in samples affected only by macro-arthropods to more than 45% in samples that have been completely comminuted by oribatids.

Hypha fragments produced at the surface of a brown earth by macro-arthropods become increasingly comminuted by oribatids living in the soil when transported downwards by earthworms. The frequency of short fragments thus characteristically increases with depth; a former brown-earth stage can therefore be recognized in the mineral soil of a present podzol by increased frequencies of short hypha fragments with depth (Andersen, 1979, 1984). When earthworm activity ceases and organic matter is transported downwards to only a few centimetres depth (podzoloid stage), the activity of oribatids becomes concentrated in the topmost soil levels, and short hypha fragments are very frequent. The hypha fragments found in terrestrial humus layers usually have low frequencies of short fragments at the surface, and increased frequencies below the topmost centimetres owing to the activity of oribatids living in the topmost part of the humus.

Former changes in soil conditions can thus be traced by measurements of the fragments of dark-coloured hyphae found in the soil samples (see later).

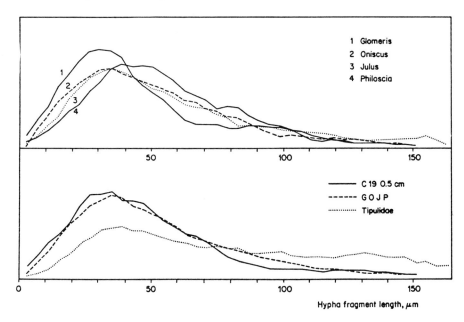

1 Glomeris
2 Oniscus
3 Julus
4 Philoscia

——— C 19 0.5 cm
------- G O J P
·········· Tipulidae

Hypha fragment length, μm

FIGURE 7.2.   Frequency distribution of the length of hypha fragments in faecal pellets from macroarthropods and Tipulidae larvae, and in a soil sample (C 19). GOJP = average curve for pellets from the four macroarthropods (from Andersen, 1984). *Reproduced by permission of Leicester University Press*

FIELD METHODS

Soil sections are most conveniently studied in excavations, where stratification, structure and colour of the various horizons may be noted. Samples may be secured by cutting thin slices (0.5–2.0 cm) in a vertical column. Accurate sampling may, however, be impeded by the presence of roots, mouseholes or rocks. Aaby (1983) was able to cut out a monolith from coherent humus layers.

LABORATORY PROCEDURES

Samples from soil sections may be examined for dry density, degree of humification, organic content, chemical elements and grain size distribution.

Dry density and *degree of humification* may be determined by methods similar to those used for peat samples (Aaby, 1983; Chapter 6).

Organic content is determined as the loss on ignition at 555°C. If clay is present in considerable quantity, changes in organic content may be detected by determination of organic carbon. Andersen (1979, 1984) illustrated volumetric relations by using a logarithmic scale for ignition residue. Aaby (1983) calculated the volume percentage of the ignition residue using a fixed dry

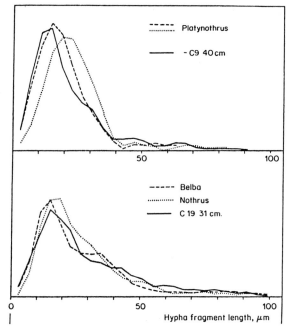

FIGURE 7.3.   Frequency distribution of the length of
hypha fragments in faecal pellets from oribatids and
in two samples (−C 9 and C 19, from Andersen,
1984). *Reproduced by permission of Leicester Uni-*
versity Press

density for mineral matter (1.60). The volume percentage of the organic matter
cannot be determined exactly, as pore space enters the volume of the ignitable
matter (Aaby, 1983).

    Chemical elements such as iron and aluminium and *Grain size distribution* in
mineral deposits are determined by standard procedures.

MICROFOSSIL ANALYSIS OF SOILS

Pollen analysis

Qualitative and quantitative pollen analysis are performed by the procedures
described in Chapter 22. Aaby (1983) related pollen concentrations to the
sample volume and the weight of organic matter, and calculated pollen
deposition rates for dated sections.

Fungal hypha analysis

Andersen (1979, 1984) recorded the length distribution of hypha fragments in slides prepared for pollen analysis (the samples should not be sieved during preparation) by drawing the individual fragments with a drawing apparatus attached to the microscope and measuring the fragments with a ruler. Iversen (1964) and Aaby (1983) measured hypha coverage in relation to pollen coverage (total pollen or tree pollen) by the micrometer scale method. This procedure should be replaced by measurement of total hypha length in relation to numbers of pollen grains. Aaby (1983) calculated total hypha length per weight and volume unit and calculated rates of hypha production (Aaby 1983).

Charcoal

The numbers of microscopic charcoal particles may be recorded in relation to the pollen sum during pollen analysis (Iversen, 1964; Aaby, 1983; cf. Chapter 23). The method thus includes only those charcoal particles that are so small that they do not hamper the preparation of thin slides. Larger charcoal particles may have become crushed during the sample preparation or at the production of the microscope slides, and thus enter the amount of charcoal dust recorded.

DATING OF SOIL PROFILES

Iversen (1964) was able to date large charcoal particles isolated from soils by the radiocarbon method. The dating of organic matter in soil profiles by the radiocarbon method is otherwise unreliable owing to the mobility of the humus and the possible presence of modern rootlets (Aaby, 1983). Pollen diagrams from soils must therefore be dated by indirect methods, if charcoal is absent.

A minimum age may be obtained for soil profiles present under archaeologically datable monuments (Dimbleby, 1962). In other cases, events seen in pollen diagrams may be dated by historical records (Andersen, 1979; Aaby, 1983) or by comparison with pollen diagrams from small hollows in the vicinity (Andersen, 1984). Somewhat more uncertain dates may be obtained by a comparison with pollen diagrams of regional significance. Aaby calculated ages of soil pollen diagrams by extrapolation based on an assumed constant pollen sedimentation rate (Figure 7.4).

MODIFICATION OF POLLEN DIAGRAMS FROM SOILS

As mentioned above, pollen assemblages deposited on the surface of mineral soils become mixed vertically during burial by soil fauna. Andersen (1979) compared a pollen diagram from terrestrial humus with contemporaneous diagrams from two podzols and a brown earth in the vicinity. A *Calluna* maximum, which was very distinctive in the humus, could still be recognized in the topmost part of the mineral soil of the podzols but was nearly obliterated in the brown earth (Figure 7.5). Increased obliteration of details could also be observed for other curves from the pollen diagrams (Figure 7.6). It could also be

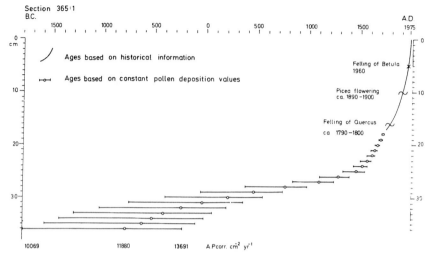

FIGURE 7.4.   Relation between sediment age and depth in a section from a podzol, calculated from historically dated changes in pollen composition A.D. 1750–1975 and by assumed constant pollen deposition rate prior to 1750 (from Aaby, 1983)

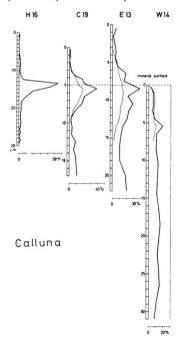

FIGURE 7.5.   Frequencies of *Calluna* pollen (in % of total pollen) in terrestrial humus (H 16), in two neighbouring podzols (C 19 and E 13), and a brown earth (W 14, heavy lines), and curves for H 16 smoothed by increasing running means (thin lines, from Andersen, 1979)

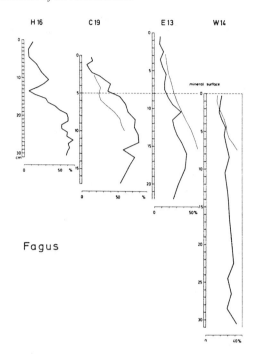

FIGURE 7.6.   Frequencies of *Fagus* pollen, otherwise as Figure 7.5

shown that *Picea* pollen, which was nearly lacking below the *Calluna* maximum in the humus deposit, occurred to considerable depths in the podzols (more than 20 cm), presumably owing to percolation, and down to 30 cm depth in the brown earth (Figure 7.7). Hence, original features of the vegetational development can be only faintly recognized in brown earths and in podzols, which contain pollen assemblages from former brown-earth stages, whereas the pollen curves are somewhat better differentiated in the topmost part of podzols (the former podzoloid stage).

FEATURES OF SOIL DEVELOPMENT RECOGNIZED IN SOILS

Pedologists have claimed that present-day podzols have developed from former brown earths by degradation (e.g. Müller, 1878, 1884; Mückenhausen, 1957; Duchaufour, 1965), a supposition that has been confirmed by pollen analysts in various ways. Andersen (1979, 1984) showed that former soil stages can be recognized in present podzols by means of variations in hypha fragment length, organic content and pollen corrosion (Figures 7.8 and 7.9). (Aaby used different names for identical stages.) The former soil stages can be linked with vegetational development and anthropogenic influence by means of pollen diagrams.

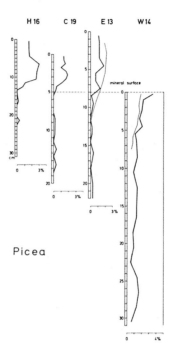

Picea

FIGURE 7.7.   Frequencies for *Picea* pollen, otherwise as Figure 7.5

Former brown-earth stages

Former neutral brown earths have left few traces in podzols, because the pollen grains have been eliminated and dark-coloured hyphae were scarce owing to rapid decomposition of the litter. Pteridophyte spores, which are less easily destroyed than pollen grains, may have accumulated in the soil (Andersen, 1979).

Traces of a former oligotrophic brown-earth stage may be recognized in the A_2- and B-horizons of present podzols by moderately to strongly corroded pollen assemblages, increased frequencies with depth of short hypha fragments produced by oribatids, and low organic content. The amounts of hypha fragments may vary according to the composition of the leaf litter at the time of deposition (Andersen, 1984). The pollen assemblages often reflect forest vegetation undisturbed by man.

Former podzoloid stage

A former podzoloid stage may be recognized in the A_1-horizon of present podzols by decreased pollen corrosion, high frequencies of short hypha fragments and increased organic content. The pH value was thus less than 5.0;

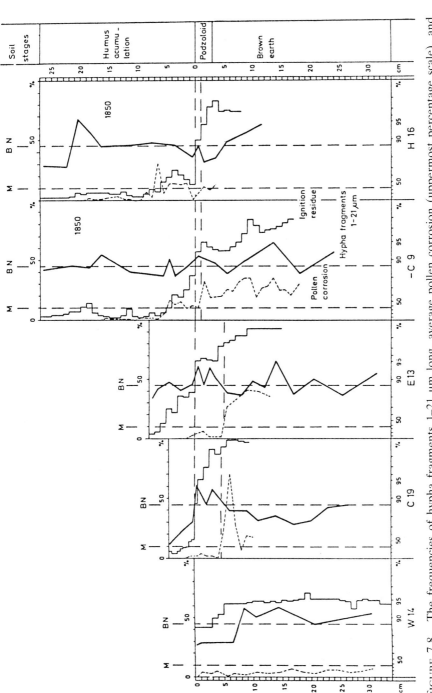

FIGURE 7.8. The frequencies of hypha fragments 1–21 μm long, average pollen corrosion (uppermost percentage scale), and ignition residue (lowermost percentage scale), in brown earth (W 14) and in humus layers above and mineral soil below the mineral surface (at 0 cm) in four podzols (C 19, E 13, –C 9 and H 16). M indicates the frequency of short hypha fragments in macroarthropod pellets (10%), and BN the lower limit for oribatid pellets (45%, from Andersen, 1984). *Reproduced by permission of Leicester University Press*

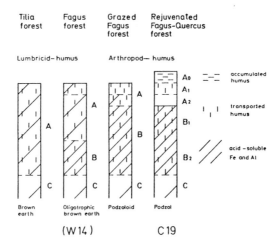

FIGURE 7.9.   Soil development in a Danish forest (from Andersen, 1979)

an arthropod community dominated; organic material was transported to only a shallow depth, and the pollen assemblages were only slightly modified by mixing. The change from the oligotrophic brown-earth stage to the podzoloid stage has proved in some cases to be due to a change to less decomposable leaf litter or tree felling or grazing. The podzoloid stage may have been rather short, or it may have lasted from a few centuries up to two millennia (Andersen, 1979, 1984; Aaby, 1983).

Humus accumulation (podzol) stage

Terrestrial humus layers are characterized by low pollen corrosion, varying frequencies of short hypha fragments, and low content of mineral matter. There was, practically speaking, no mixing of the pollen assemblages, and the production of dark-coloured hyphae was high (Aaby, 1983). Aaby distinguished this stage as a 'raw humus stage' from the preceding 'arthropod humus stage'. However, as pointed out by Andersen (1979), the plant debris in the accumulated humus, including the fungal hyphae, has been digested by arthropods. The accumulated humus is, accordingly, truly arthropod humus. Terrestrial humus layers up to 1 m deep have been recorded (Faegri, 1954) and may date back to Atlantic time (Iversen, 1969). Such layers are extremely well suited to studies of local vegetational development. The onset of humus accumulation seems often to have been caused by tree felling, or to have followed on cessation of human activity which caused reforestation (Iversen, 1969; Andersen, 1979; Aaby, 1983).

Podzolization of the mineral soils was completed during the humus accumulation stage. This process may have been completed in about 100 years (Andersen, 1979).

REFERENCES

Aaby, B. (1983). Forest development, soil genesis and human activity illustrated by pollen analysis and hypha analysis of two neighbouring podzols in Draved Forest, Denmark. *Geol. Surv. Denmark.* 2, **114**, 1–114.

Andersen, S. T. (1979). Brown earth and podzol: soil genesis illuminated by microfossil analysis. *Boreas*, **8**, 59–73.

Andersen, S. T. (1984). Stages in soil development reconstructed by evidence from hypha fragments, pollen and humus contents in soil profiles. In: *Lake Sediments and Environmental History* (Eds. E. Haworth and J. W. G. Lund), Leicester University Press, Leicester, England.

Cushing, E. J. (1967). Evidence for differential pollen preservation in late-Quaternary sediments in Minnesota. *Rev. Palaeobot. Palynol.*, **4**, 87–101.

Delcourt, P. A., and Delcourt, H. R. (1980). Pollen preservation and Quaternary environmental history in the south-eastern United States. *Palynology*, **4**, 215–231.

Dimbleby, G. W. (1957). Pollen analysis of terrestrial soils. *New Phytol.*, **56**, 2–28.

Dimbleby, G. W. (1962). The development of British heathlands and their soils. *Oxf. For. Mem.*, **23**, 1–120.

Duchaufour, P. (1965). *Précis de Pédologie*, Masson, Paris.

Faegri, K. (1954). On age and origin of the beech forest (*Fagus silvatica* L.) at Lygrefjorden, near Bergen (Norway). *Geol. Surv. Denm.* 2, **80**, 230–249.

Hartmann, F. (1965). *Waldhumusdiagnose auf Biomorphologischer Grundlage*, Springer-Verlag, Wien/New York.

Havinga, A. J. (1964). Investigation into the differential corrosion susceptibility of pollen and spores. *Pollen et spores*, **6**, 621–635.

Havinga, A. J. (1967). Palynology and pollen preservation. *Rev. Palaeobot. Palynol.*, **2**, 91–98.

Havinga, A. J. (1971). An experimental investigation into the decay of pollen and spores in various soil types. In: *Sporopollenin* (Eds. J. Brooks, P. R. Grant, M. Muir, P. van Gijzel and G. Shaw), Academic Press, London, pp. 446–479.

Iversen, J. (1964). Retrogressive vegetational succession in the postglacial. *J. Ecol.*, **52** (Suppl.), 59–70.

Iversen, J. (1969). Retrogressive development of a forest ecosystem demonstrated by pollen diagrams from fossil mor. *Oikos Suppl.* 12, 35–49.

Mückenhausen, E. (1957). Die wichtigsten Böden der Bundesrepublik Deutschland. *Wissens. Schriftenr. des AID*, **14**, 1–146.

Müller, P. E. (1878). Studier over skovjord, som bidrag til skovdyrkningens theori, I. *Tidsskr. Skovbrug*, **3**, 1–124.

Müller, P. E. (1884). Studier over skovjord, som bidrag til skovdyrkningens theori, II. *Tidsskr. Skovbrug*, **8**, 1–232.

Sampling and mapping techniques

Handbook of Holocene Palaeoecology and Palaeohydrology
Edited by B. E. Berglund
© 1986 John Wiley & Sons Ltd.

8

Sampling techniques for lakes and bogs

BENT AABY

*Geological Survey of Denmark,
Thoravej 31,
Copenhagen, Denmark*

and

GUNNAR DIGERFELDT

*Department of Quaternary Geology,
Tornavägen 13,
Lund, Sweden*

INTRODUCTION

After a representative and suitable sampling site has been located, successful coring is necessary in order to ensure a sediment or peat sequence as complete and undisturbed as possible. Taking into consideration the long time usually required for later analytical work, it is clear that the time spent making a sampling is worth while; no laboratory or analytical techniques can compensate for poor sampling techniques. Disturbances and incompleteness of the sediment or peat core obtained, which may not be apparent in the field, may lead to serious misinterpretations.

A description will be given of some available and commonly used chamber samplers and piston samplers, and types suitable for sampling in different conditions will be recommended. Sampling from open sections, which may sometimes be possible particularly in peat deposits, will also be described.

CHAMBER SAMPLERS

The chamber samplers are filled from the side, and two types will be described — the Hiller sampler and the Russian sampler (Figure 8.1).

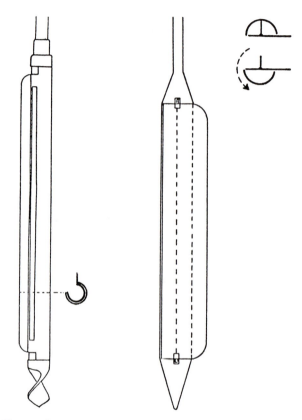

FIGURE 8.1.   The Hiller sampler (left) and the Russian sampler (right)

Hiller sampler

This is operated by twisting the sampler and filling the chamber by cutting out a segment of sediment, which is scraped into the chamber. A gentle twist in the opposite direction closes the chamber.

This commonly used sampler is named after the Swedish peat engineer K. Hiller, but it should in fact derive its name from the Norwegian forester P. Chr. Asbjørnsen, who first described the prototype in 1868 (Fries and Hafsten, 1965). Its main feature is its robustness. It is usually capable of penetrating very compact sediments and peats, and even rather deep deposits with a high minerogenic content. Therefore, the Hiller sampler is very useful for extensive or survey investigations in lakes and bogs. It should, however, only be used for these purposes as there is considerable risk of contamination.

Especially during penetration, the projecting flange of the rotating chamber is liable to catch resistant plant remains, which are transferred to a deeper position, causing contamination. Also the auger head disturbs the deposit as it penetrates, so that the actual stratigraphy may be disturbed. Another disadvantage of the Hiller sampler is that intact cores cannot be removed and samples must be taken in the field.

In order that the sampler remains rigid only a tiny slit is open to the chamber, which complicates the sampling procedure and makes for difficult cleaning.

Various modifications have been made. Thomas's (1964) modification allows the removal of intact cores by having a removable auger head. Although the Thomas modification is a great improvement, some of the serious disadvantages still remain.

The common length of the sampler is 0.5 m or 1 m, and the diameter is about 4 cm. The diameter should not be increased, since sampling by scraping presupposes instantaneous filling of the chamber, so as not to cause serious contamination.

Russian sampler

This sampler was originally described by Belekopytov and Beresnevich (1955) and later by Jowsey (1966). It is commonly used, and while the Hiller sampler scrapes the sediment into the chamber, the Russian sampler does not disturb the deposit. Penetration to the required depth allows the sampler to slide past the material which will be sampled. The chamber is rotated 180° and the edge of the semicylinder cuts around the material to be sampled, thus maintaining it in its original position. Sediment disturbances are thus minimized.

The Russian sampler has several advantages. The construction of the auger head and the fin flange prevent trapping fibrous remains. Sample collection in the field or removal of intact cores is easily undertaken as the semicylindrical core can be fully exposed on the fin. This exposure of the sample is also a considerable advantage because stratigraphy, humification, colour etc. can be described in detail, after the sediment surface has been cleaned.

The sampler is suitable for most types of homogeneous sediment and peat. However, in peat with larger wood remains or in compact minerogenic sediments, the auger head may easily be lodged and on rotating only the upper part of the semicylindric chamber turns, causing damage. Wood and small stones may also be caught between the fin and the cylinder edge, so causing the chamber to remain unlocked or to become damaged.

The common length of the sampler is 0.5 m or 1 m, and the diameter is about 5 cm. However, both can be modified if a larger sample quantity is wanted. Smith *et al.* (1968) have designed a useful modification — length 1.5 m and diameter 12 cm — for collection of an adequate quantity of material for close-interval radiocarbon dates.

Provided the sediment is sufficiently consolidated the Russian sampler can sometimes be used for studies on annual laminations and palaeomagnetism. In such studies continuous cores as long as possible are usually required. In homogeneous sediments the length of the sampler can be increased to at least 2 m, and due to its construction the orientation of the core required in palaeomagnetic studies can be obtained.

PISTON SAMPLERS

Principles of construction and operation

Piston samplers have been found to be the best for use in most types of lake sediments. Successful coring with a piston sampler requires (1) a knowledge of the major principles of sediment coring and sampler construction, and (2) a knowledge of the techniques of handling and operating a sampler. The former can be learned from available papers on coring and sampler construction. The latter can partly be learned from studying descriptions; however, the best knowledge about how to handle and operate samplers can be obtained only by practice and experience in fieldwork.

Thorough descriptions of the principles of sediment coring and sampler construction may be found in papers by, for example, Hvorslev (1949), Kjellman *et al.* (1950), Kullenberg (1947, 1955), Piggot (1941) and Emery and Dietz (1941). Some very good and detailed manuals of sediment coring by different types of piston samplers have been given by Wright *et al.* (1965) and Wright (1980). The present description of piston samplers will concentrate on summarizing some principles of sediment coring, and on surveying the different types of samplers available for coring in different conditions.

In Figure 8.2 a piston sampler is compared with an open drive sampler. The open sampler is simply a tube driven into the sediment. The hydrostatic pressure on the sediment core inside the sampling tube is the same as on the outside sediment. During coring the sliding resistance between the core and the inside tube wall will cause an increasing pressure on the sediment at the lower tube mouth. As the tube continues to be driven into the sediment, the progressively increasing pressure will lead to a downward deflection, stretching and thinning, and finally lateral displacement of the sediment. Gradually smaller and thinner increments of sediment will be added to the core, and when finally the inside sliding resistance exceeds the strength of the sediment

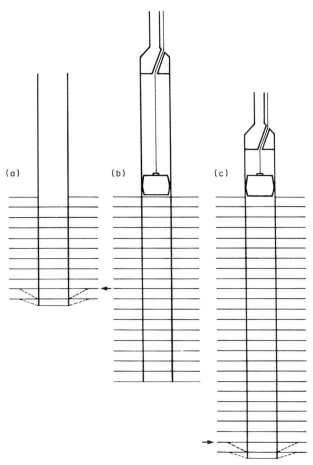

FIGURE 8.2. Simplified drawings showing coring with (a) an open drive sampler, and (b) and (c) a piston sampler. The arrow indicates beginning of downward deflection and thinning of sediment

no more sediment will enter the sampling tube. Further penetration by the tube will result in a cone of sediment formed in front of the tube mouth, and the sampler with the core will simply act as a solid pole, displacing all the sediment.

The 'safe length of sample' is, according to Hvorslev (1949), defined as the length of core that can be obtained before downward deflection of sediment begins. The 'limiting length of sample' is the length that can be obtained before a permanent cone is formed, when no sediment is added to the core. The 'total recovery ratio' can be calculated by dividing the length of core by the length of sampler penetration. It must be considered that below the 'safe length of sample' any core obtained will be shortened, and will not represent a complete and correct sequence of sediment.

The important principle in the construction of the piston sampler is elimination of the hydrostatic pressure over the core inside the tube. The tight-fitting piston is kept immobilized immediately above the sediment, and the hydrostatic pressure on the sediment outside the sampling tube will oppose the inside sliding resistance and the creation of a vacuum between the sediment core and the piston. Not until the inside sliding resistance exceeds the hydrostatic pressure will no more sediment enter the sampling tube.

This means that piston samplers permit longer sediment cores to be obtained, and that the 'safe and limiting lengths' are related to the water depth and the hydrostatic pressure. If no vacuum has been created below the piston, the recovery and the representation of sediment must be complete. However, if a vacuum has been created, the penetration has exceeded the 'safe length of sample' and the lower part of the core will be shortened and disturbed.

In shallow and moderately deep lakes the positive (core length increasing) effect of the piston is mostly less than usually thought. Most of the light-weighted piston samplers, which are suitable for such lakes, have a sampling tube length of about 1 m. This length is determined partly by the convenience of handling and operation, but also by the possible 'safe length of sample' that can be obtained. The tube length can fairly easily be modified and increased; however, then it must always be checked that the safe length is not exceeded.

To obtain a complete sediment sequence, sampling of successive cores will usually be necessary. Such sampling should be made in alternate holes. To be sure that a complete sediment sequence is obtained, and to enable checking of possible disturbances of the uppermost and lowermost parts of the cores, the successive sampling should include 10–20 cm overlaps.

Besides the water depth and hydrostatic pressure, the 'safe length of sample' also depends on the diameter of the sampling tube, the properties of the sediment, and on the specific construction and operation of the sampler.

· The length of core will increase by increasing the tube diameter; however, it must then be kept in mind that the ratio between the thickness of the tube wall and the diameter should always remain as low as possible. A thicker tube wall means that an increased quantity of sediment at the tube mouth has to be displaced inwards or outwards — in the former case increasing the inside sliding resistance. An increase of the tube diameter will also increase the risk of loss of core — or part of it — during core withdrawal. Most piston samplers are designed for a tube diameter of 5–6 cm, but in most sediments this can, if required, be increased to 10 cm without there being a great risk of core loss.

With regard to the character of sediment, it is evident that several properties can fairly significantly affect the length of core — for example, the organic and minerogenic content, the water content and consolidation — influencing the cohesion and strength of the sediment and the inside sliding resistance. The effects of different properties are not always precisely known. However, even if they were, this knowledge would be difficult to apply in each separate coring. After some corings in sediments of varying character the effects of different properties will be empirically learned.

The effects of various methods for driving the sampling tube into the sediment have been thoroughly discussed by Hvorslev (1949). It is recommended that, whenever possible, the driving is performed in one fairly fast and uninterrupted pushing. This to prevent the development and building-up of inertial adhesion and friction, increasing the inside sliding resistance. To reduce the risk of loss of the core — or part of it — a pause should be made before the withdrawal and lifting of the sampler. In compact sediments where the sampler has to be operated by a chain-hoist or by hammering, it should be attempted to make the driving as steady and uninterrupted as possible.

For various specific details of sampler construction the reader is referred to the descriptions in Hvorslev (1949) and Kjellman *et al.* (1950) (e.g. shape of cutting edge of tube, inside and outside clearances of the tube mouth). The effects of these details on light-weight and hand-operated piston samplers will be fairly insignificant.

As described, piston samplers will, particularly in deep water, permit the retrieval of long, continuous and undisturbed cores. However, by increasing the tube length and penetration, the inside sliding resistance will sooner or later be a limiting factor. A piston sampler with metal foils constructed by Kjellman *et al.* (1950) is at present the only available sampler in which the sliding resistance is completely eliminated. The core is surrounded by a number of foils fixed to the piston, and accordingly does not slide against the inner tube wall. Unfortunately, the foil sampler is expensive and requires heavy and complicated equipment for operation, which means that its use can usually only be considered in special coring work. Owing to its construction the sampler can be used only on land or in fairly shallow water.

Livingstone (1967) has described another use of foils for reducing the sliding resistance, which is apparently effective provided that strong enough foils can be obtained.

Many drawings of different types of piston samplers may give the impression that constructing a sampler is difficult. However, all that is required for a useful light-weight and hand-operated sampler is a sampling tube, a piston and some driving rods. Knowing the principles of operation, it should be possible for most people to design and construct their own sampler, which on the evidence of the large number of modifications would seem to be the case.

Samplers for shallow lakes

The choice of sampler type is largely determined by the water depth of the lake.

In shallow lakes, samplers operated by rods can be used. Such samplers are recommended, whenever possible, since they mostly allow a thorough and continuous control of the coring; depth of sampler penetration and core recovery can easily be determined and checked. A light-weight rod-operated piston sampler has been described by Livingstone (1955). Modifications and improvements have been presented by among others Vallentyne (1955), Walker (1964), Cushing and Wright (1965), Wright (1967) and Merkt and Streif (1970). A fairly light-weight rod-operated sampler has also been described by Wieckowski (1970). All these samplers are designed for taking approximately 1 m cores, but this length can be modified. The sampler described by Wright is to be recommended, because it is both simple in operation and strong in construction, which permits coring even in very compact sediments. A further modification for taking longer cores (3.5 m and 7.5 m) has also been described (Wright, 1980).

The water depth in which rod-operated piston samplers can be used varies somewhat depending on the character of the sediment. When coring in compact sediments rods may bend even in fairly shallow water. In such cases — and always in deeper water — casing tubes should be used to stabilize the rods, so preventing them from bending when the sampler is driven into the sediment.

Since rod-operated samplers are to be preferred, the use of casing is also generally recommended for water depths down to 20 m. The use of casing is possible in deeper water, but then additional equipment and other means of assistance are required, which are not usually available in general coring exercises.

Most rod-operated piston samplers can also be used for peat coring in bogs, provided that the peat is below the water table and completely saturated. The samplers work best in highly to moderately humified peat, but it is important to use tubes as thin-walled as possible with sharp cutting edges. Coring in slightly humified and fibrous peat, and in peat rich in large wood pieces, is usually problematic or impossible.

When sampling in drained peat, not completely water-saturated, the core will mostly become significantly compacted. Smith *et al.* (1967) constructed an open-drive sampler in which the tube had a longitudinal slit to prevent compaction. Another sampler that may sometimes be useful in peat coring was described by Couteaux (1962). The sampler is made of two half-tubes, and in construction it is something between a side-cutting and an open-drive sampler. In cases when peat coring by use of a piston sampler is impossible or problematic, a large-capacity Russian sampler may also be recommended as a useful alternative.

Samplers for deep lakes

In very deep lakes certain types of piston sampler have to be chosen which are lowered by cables, and in which a force other than that of driven rods is used for pushing the sampling tube into the sediment. In the Kullenberg piston sampler (Kullenberg, 1947, 1955) the driving force is obtained by a combination of loading and free-fall. The exact water depth need not be known, since the piston is automatically locked and immobilized when the sampler is some distance above the sediment surface. The original Kullenberg sampler was constructed for deep-sea coring, and the equipment required is both heavy and complicated. However, light-weight and more easily operated modified designs can be constructed, which are suitable for coring in deep lakes (Wright *et al.*, 1965).

In the deep-water sampler constructed by Mackereth (1958; Smith, 1959) compressed air provides the driving force. The compressed air is used both for fixing an anchoring chamber and for driving the sampling tube into the sediment. The equipment is fairly complicated, but the sampler can be operated from a small boat.

In difficult terrain, where it may be impossible to transport any heavy equipment, some simple constructions of light-weight deep-water samplers can be recommended. These samplers, described by Huttunen and Meriläinen (1975) and Digerfeldt (1978), are lowered by cables and then smoothly hammered down into the sediment by a moveable load of moderate weight (in soft sediments a few kilograms may be enough).

The higher hydrostatic pressure in deep lakes means that longer continuous cores can be obtained. However, whenever possible the recovery must be checked, and the user should be aware that the lower part of the core may be shortened and incomplete. In using the Mackereth sampler or a modified Kullenberg sampler the length of the sampling tube can be varied, but the driving has to begin at the sediment surface and the operation completed if possible in one attempt. An advantage with the Digerfeldt sampler is that sampling of successive sediment sequences is possible, since the piston, provided that the sediment is not too compact, can be locked at the tube mouth by a break-pin. This is important when the hydrostatic pressure is not high enough to permit a continuous sampling of the whole sediment sequence in one drive. Another advantage is that the sampler can be withdrawn and lifted by a ball-clamp, which is lowered and clamped to a short rod attached to the sampler head. A problem in deep-water coring is that usually the sediment thickness cannot be determined in advance of sampling, so that the correct length of tube has to be guessed. If the whole length of tube is not driven into the sediment, and the piston is accordingly not displaced and locked at the sampler head, the piston cable on the Digerfeldt sampler need not be used for withdrawal and lifting. This means that possible disturbance of the upper part of the core by suction can be avoided.

Surface sediment sampling

To obtain undisturbed cores of the frequently loose and unconsolidated surface sediment (the upper 1 m approximately) special samplers are sometimes required, and several different types have been described. In cases where fine structural sediment features have to be recorded — e.g. annual laminations — and a large quantity of sediment is not required, the use of frozen-core samplers is recommended. The different types available are described in Chapter 17.

Other surface sediment samplers can be divided into two major types — open samplers (i.e. without pistons) and samplers with stationary pistons. The former type can be used when a core of only the uppermost 20–30 cm is needed. Most of these samplers are lowered and operated by rope, and are often loaded to ensure sufficient penetration. The samplers are therefore easily handled, and can be used in both shallow and deep water. To prevent the sediment core from slipping out during withdrawal and lifting, the open upper end of the sampling tube must be closed by a stopper or similar efficient closing mechanism. The 'safe length of sample' will vary depending on the character of sediment, but heavy loading to obtain a longer core should generally be avoided. Owing to the slight consolidation of the surface sediment, the sliding resistance inside the tube will, sooner than is usually thought, cause a downward deflection and lateral displacement of the sediment, resulting in incomplete and disturbed samples. In cases where it is intended that the surface sediment core is to be used, for example, for thorough dating and reconstruction of changes in accumulation rate and different influx calculations, such problems may lead to serious misinterpretations.

Håkansson and Jansson (1983) have discussed various aspects of surface sediment sampling and described some available types of samplers. The well-known Jenkin sampler (Mortimer, 1942) and Kajak sampler (Kajak, 1966) belong to this category of open samplers, but the former seems to be unnecessarily complicated.

Wright (1980) recommends a very simple and efficient surface sediment sampler constructed by Hongve (1972). Probably even simpler, but equally efficient, is the sampler designed by Benoni and Enell (Figure 8.3).

For sediment X-ray work and special studies of structural features, Axelsson .and Håkansson (1972) have designed a surface sediment sampler with rectangular sides.

Most of the rod-operated piston samplers can also be used for surface sediment sampling. However, some special modifications have been constructed (Rowley and Dahl, 1956; Brown, 1956; Davis and Doyle, 1969). The problems most likely to be met are connected with high water contents and slight consolidation. In order to avoid disturbances and lateral displacement of the sediment, the driving of the sampling tube should be made as smoothly as possible. Subsampling of the core has to be made with the tube in the vertical position and usually by upwards extrusion of the sediment. Good techniques for extrusion are described by Wright *et al.* (1965) and Håkansson and Jansson (1983).

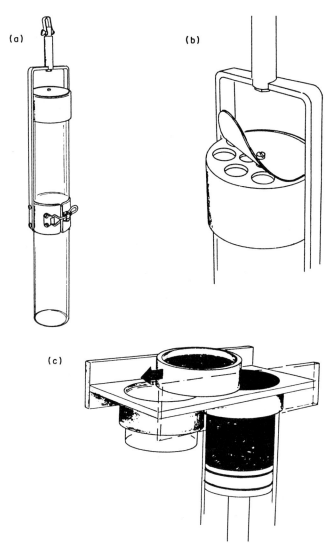

FIGURE 8.3. The surface sediment sampler designed by S. Benoni and M. Enell (Institute of Limnology, Lund, Sweden). (a) The sampling tube is made of transparent plastic, 50 cm in length and 7.5 cm in diameter, and it is easily exchangeable. (b) Detail of valve which is put on top of sampling tube. The valve, which is made of 1–2 mm rubber sheet, prevents the sediment core from slipping out as the sampler is withdrawn. (c) Detail of sectioning device, which can be put on top of sampling tube. Subsampling is made by upwards extrusion of sediment core

Cable-operated deep-water samplers, as described above, which allow the tube to be smoothly driven, can also be used for reliable surface sediment sampling. Mackereth (1969) has designed a modification of the compressed-air sampler for undisturbed sampling of the upper 1 m of the surface sediment. When a particularly large quantity of surface sediment is required, Digerfeldt and Lettevall (1969) have described a square-formed (10 × 10 cm) piston sampler. The core is prevented from slipping out by automatically released closing plates at the sampler mouth. Undisturbed subsampling is made possible by inserting partition plates, after which the sampler can be placed horizontally and one side opened. This sampler can be operated by either rods or cable (Digerfeldt, 1978), and accordingly can be used in both shallow and deep water.

SAMPLING FROM OPEN SECTIONS

In many bogs, old peat exposures, drainage ditches or other features may represent valuable sites for palaeoecological studies, since the stratigraphy can be inspected in detail. What may look homogeneous in a core may sometimes be divided into two or more distinct layers in an open section. Similarly, former irregular sediment accumulation features or artificial disturbances may be detected in open sections, whereas they may remain undiscovered in cores. In addition, the risk of contamination is reduced and the amount of sediment obtainable at each level is unlimited when sampling from open sections. Because of these advantages, sampling from open sections is recommended instead of coring, whenever possible.

Samples are collected directly in the field, if conditions are convenient, or exposures can be examined and complete monoliths removed to the laboratory before extraction of samples takes place. Especially if the content of resistant material (e.g. wood remains) is considerable, or if the peat is loose and only weakly decomposed, it is desirable to dig monoliths and freeze them before further treatment (see Chapter 6). Sediments with a high content of minerogenic matter are best collected directly from the section, as they are fragile.

Open sections may be excavated even at places with a high water table, but no general recommendation can be given as local conditions may vary considerably in the same mire type. Practical experience shows that monoliths can be extracted from undisturbed ombrotrophic mires to a depth of approximately 2 m if the stratum is moderately or strongly humified, whereas digging is almost impossible in weakly humified peat. Investigation holes may also be difficult to excavate in artificially drained peat deposits, because waterlogged cracks are common and, by perforation, water quickly pours out and impedes any water control.

Therefore automatic pumping is generally needed when carrying out successful investigations in newly excavated pits. It should be stressed that buttress constructions are required in deeper holes to prevent collapsing of the peat walls.

When collecting an entire monolith it is important to be aware of compaction after the column has been released from the section and is still in a vertical position. The total length should therefore be measured and matchsticks inserted at intervals (e.g. 10 cm) before the column is released. Compaction is insignificant in partly drained deposits where shrinkage of the peat has already progressed and in deeper deposits which have already been exposed to autocompaction. On the other hand, shrinkage may be considerable in the upper part of weakly decomposed peat where there is a high water content. After removing the monolith it is easy to compensate for an actual compaction.

The size of the peat column may vary, but for most investigations a 10 × 10 × 150 (max.) cm column will be ideal. In longer sections, alternating columns are sampled and the overlap should be at least 10 cm. Metal boxes with a few perforations in the back are useful for sampling.

REFERENCES

Axelsson, V., and Håkansson, L. (1972). A core sampler with rectangular coring tubes for soft sediments. *Geografiska Annaler*, **54**A, 32–33.

Belokopytov, I. E., and Beresnevich, V. V. (1955). Giktorf's peat borers. *Torf. Prom*, **8**, 9–10.

Brown, S. R. (1956). A piston sampler for surface sediments of lake deposits. *Ecology*, **37**, 611–613.

Couteaux, M. (1962). Notes sur le prélèvement et la préparation de certain sédiments. *Pollen et spores*, **IV** (2), 317–322.

Cushing, E. J., and Wright, H. E. (1965). Hand-operated piston corers for lake sediments. *Ecology*, **46**, 380–384.

Davis, R. B., and Doyle, R. W. (1969). A piston corer for upper sediment in lakes. *Limnology and Oceanography*, **14**, 643–648.

Digerfeldt, G. (1978). A Simple Corer for Sediment Sampling in Deep Water, *Dep. Quaternary Geology, Lund, Report* **14**, pp. 1–10.

Digerfeldt, G., and Lettevall, U. (1969). A new type of sediment sampler. *Geologiska Föreningens i Stockholm Förhandlingar*, **91**, 399–406.

Emery, K. O., and Dietz, R. S. (1941). Gravity coring instrument and mechanics of sediment coring. *Bull. Geological Soc. of America*, **52**, 1685–1714.

Fries, M., and Hafsten, U. (1965). Asbjørnsen's peat sampler the prototype of the Hiller sampler. *Geologiska Föreningens i Stockholm Förhandlingar*, **87**, 307–313.

Håkansson, L., and Jansson, M. (1983). *Principles of Lake Sedimentology*, Springer, Berlin, pp. 1–316.

Hongve, D. (1972). En bunnhenter som er lett a lage. *Fauna*, **25**, 281–283.

Huttunen, P., and Meriläinen, J. (1975). Modifications of a rodless core sampler for investigating lake sediments. *Publications University of Joensuu*, **B II:4**, 1–4.

Hvorslev, M. J. (1949). *Subsurface Exploration and Sampling of Soils for Civil Engineering Purposes*, Committee on sampling and testing, American Society of Civil Engineers, Vicksburg, pp. 1–521.

Jowsey, P. C. (1966). An improved peat sampler. *New Phytologist*, **65**, 245–248.

Kajak, Z. (1966). Field experiment in studies on benthos density of some Mazurian lakes. *Gewässer und Abwässer*, **41/42**, 150–158.

Kjellman, W., Kallstenius, T., and Wager, O. (1950). Soil sampler with metal foils. *Proc. Royal Swedish Geotechnical Inst.*, **1**, 1–76.

Kullenberg, B. (1947). The piston core sampler. *Svenska Hydrografisk-Biologiska Kommissionens Skrifter*. III, *Hydrografi* **1:2**, 1–46.

Kullenberg, B. (1955). Deep-sea coring. *Reports of the Swedish Deep-Sea Expedition*. IV, *Bottom investigations*, **2**, 35–96.

Livingstone, D. A. (1955). A lightweight piston sampler for lake deposits. *Ecology*, **36**, 137–139.

Livingstone, D. A. (1967). The use of filament tape in raising long cores from soft sediment. *Limnology and Oceanography*, **12**, 346–348.

Mackereth, F. J. H. (1958). A portable core sampler for lake deposits. *Limnology and Oceanography*, **3**, 181–191.

Mackereth, F. J. H. (1969). A short core sampler for subaqueous deposits. *Limnology and Oceanography*, **14**, 145–151.

Merkt, J., and Streif, H. (1970). Stechrohr-Bohrgeräte für limnische und marine Lockersedimente. *Geologisches Jahrbuch*, **88**, 137–148.

Mortimer, C. H. (1942). The exchange of dissolved substances between mud and water in lakes, II. *Journal of Ecology*, **30**, 147–201.

Piggot, C. S. (1941). Factors involved in submarine core sampling. *Bull. Geological Soc. of America*, **52**, 1513–1524.

Rowley, J. R., and Dahl, A. O. (1956). Modifications in design and use of the Livingstone piston sampler. *Ecology*, **37**, 849–851.

Smith, A. J. (1959). Description of the Mackereth portable core sampler. *Journal of Sedimentary Petrology*, **29**, 246–250.

Smith, A. G., Pilcher, J. R., and Singh, G. (1968). A large capacity hand-operated peat sampler. *New Phytologist*, **67**, 119–124.

Thomas, K. W. (1964). A new design for a peat sampler. *New Phytologist*, **63**, 422–425.

Vallentyne, J. R. (1955). A modification of the Livingstone piston sampler for lake deposits. *Ecology*, **36**, 139–141.

Walker, D. (1964). A modified Vallentyne mud sampler. *Ecology*, **45**, 642–644.

Wieckowski, K. (1970). New type of lightweight piston core sampler. *Bulletin de l'Academie Polonaise des Sciences* **XVI:I**, 1, 57–62.

Wright, H. E. (1967). A square-rod piston sampler for lake sediments. *Journal of Sedimentary Petrology*, **37**, 975–976.

Wright, H. E. (1980). Cores of soft lake sediments. *Boreas*, **9**, 107–114.

Wright, H. E., Livingstone, D. A., and Cushing, E. J. (1965). Coring devices for lake sediments. In: *Handbook of Palaeontological Techniques* (Eds. B. Kummel and D. M. Raup), Freeman, San Francisco, pp. 494–520.

9

Geological survey mapping

ESKO DANIEL
Geological Survey of Sweden,
Kiliansgatan 10,
Lund, Sweden

INTRODUCTION

Palaeoecological interpretations based on stratigraphic studies of lake and mire deposits require information on the geology of the area. Modern geological standard maps, like those produced in Sweden and many other countries, also have the information needed for reconstructions of potential vegetation (see Chapter 10).

The Geological Survey of Sweden publishes three main series of large-scale geological maps of the land area; maps of Quaternary deposits, maps of the bedrock, and hydrogeological maps, all with comparatively comprehensive descriptions. Only the first mentioned are discussed here.

The Quaternary maps are published at a scale of 1:50,000 with the different Quaternary deposits, as well as other geological information, printed in colours on topographical standard maps. The field-maps, on which the primary geological information is found, are drawn at a scale of 1:10,000.

The Quaternary maps are representative of the deposits at a depth of about 0.5 m below the ground surface, and the topsoil is normally not taken into account. The deposits are classified according to a standard system applied by the Geological Survey since 1963. The basic colours represent genetically different deposits, and thus the primary information given to the users of the maps is how the deposits were formed — i.e. whether it is, for example, a till, glaciofluvial deposit or a beach deposit. The grain size composition of the deposits is shown by using different shades of the colours as well as symbols printed on the colours. Similar scales, methods and classification systems are used in all the Scandinavian countries as well as in Finland. In some other countries topsoil maps are produced (e.g. in England), and these are also very useful for vegetation reconstructions. The mapping used to be based mainly on work in the field, especially in cultivated areas with a complex geology. However, it has become evident that aerial photograph interpretation is a good complement in making Quaternary maps in areas with a comparatively uniform and well-known geology. In such areas Quaternary maps based on aerial photograph interpretation are almost as reliable as those made more or less exclusively by field mapping. As aerial photograph interpretation is not a suitable method for mapping in the major part of Skåne, the southern-most part of Sweden, because of the very complex geology and the intense agricultural influence on the topsoil, this method is used very little here.

An auger is the most important tool used for mapping the Quaternary deposits in Skåne, and the geological information collected with the auger is drawn on a topographical–economical field-map at a scale of 1:10,000. Lately there has been a great interest in these field-maps because of their larger scale, and — in some cases — the more detailed information available from the field-maps compared with the final Quaternary maps at a scale of 1:50,000. As yet no such field-maps have been published, but they can be redrawn and made available for those interested in them.

CLASSIFICATION OF THE QUATERNARY DEPOSITS

The genetic classification presented below is the one used by the Geological Survey of Sweden, and the following types of deposits are taken into consideration on the modern maps.

Glacial deposits

Till is subdivided into gravelly, sandy, silty to fine sandy, clayey till and clay till with varying clay content. The boulder frequency of the till surface is subdivided into four different types, from low boulder frequency to a high frequency of large boulders. A special symbol is used when the till has a wave-washed surface layer.

Glaciofluvial deposits are subdivided into eskers, glaciofluvial gravel, glaciofluvial sand and unspecified glaciofluvial deposits.

Glaciolacustrine sediments are normally of a fine sandy composition.

Glacial fine-grained sediments are subdivided into glacial coarse silt, glacial fine silt and glacial clay (which also can be divided into different types).

Postglacial deposits

Lake and sea sediments are subdivided into beach sediments, such as cobbles, gravel and sand–fine sand; and fine-grained lake and sea sediments, such as coarse silt, fine silt, clay and gyttja clay.

Fluvial sediments are subdivided into fluvial gravel, fluvial sand and fluvial clay–silt.

Aeolian sediments were formerly subdivided into sand with or without dune-forms. Nowadays aeolian sand without dune-forms are not represented on the maps, but are marked simply as postglacial sand.

Peat is subdivided into bogs, fens and gyttja, and furthermore there is a special symbol for the thin peat cover (less than 0.5 m) on other types of Quaternary deposits.

The occurrence of bedrock outcrops, that is to say, where Quaternary deposits are lacking or thinner than 0.5 m, is naturally an important part of the information given on the maps. These also contain information on the thicknesses of the deposits, glacial striae, springs, marshes, artificial fill and the local stratigraphy where possible. The information also includes geomorphological elements such as esker-ridges and moraine-ridges, end-moraines and dunes. Other geomorphological data are normally compiled in the description.

The original field-maps include information for potential maps at different scales, e.g. between 1:10,000 and 1:100,000, and for many users the classification listed above is too precise and comprehensive. It is, however, easy to modify the classification and to redraw the maps, thus making them more suitable for the specific purposes of the user — for example, for palaeoecological studies.

FIELD-MAPPING

As mentioned above there are various ways of mapping in the field depending, among other factors, on which type of geology is expected, whether the area is forested or not, and other specific problems of the area in question. A part of a Quaternary map of an intensively cultivated part of Skåne will be used as an example — an area where the distribution of different types of deposits is mostly highly complex (cf. Daniel, 1977; Ringberg, 1980). Those planning a palaeoecological project within this area have shown great interest in special maps for selected reference areas around ancient lakes.

When mapping in the field, the geologist walks over the area and uses an auger, or occasionally a spade, to obtain a sample from an adequate depth (0.4–0.7 m), in order to classify the soil and to check its thickness. At the same time he must observe the frequency of boulders and stones, the dominant types of rocks among the boulders, and so on. The frequency of observations made with the auger depends on the complexity of the geology and the experience of the geologist. On average, one must examine the composition of the soil at least every 100 m when moving along transects lying 125–200 m apart. Thus the net of observations will be rather dense (Figure 9.1), and the more varied the geology is, the denser the net of observations must be. It has, however,

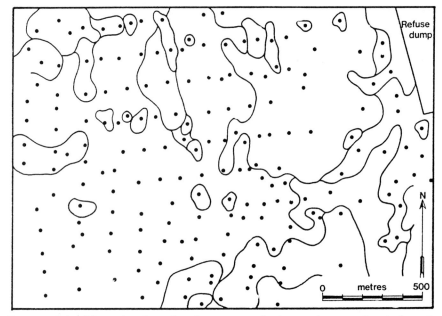

FIGURE 9.1. Map showing the frequency of observations made during the field-mapping of an area with complex geology in southern Skåne. Every dot represents a boring made with the auger. The geological contour-lines are interpolated between the observation points or, if possible, on a morphological basis. The area is a part of the mapped area in Figure 9.2

been proved that the complexity of the geology can be so extreme that the resulting picture of the distribution of different deposits will be greatly simplified and generalized. This is especially the case where the geological borders and the morphology do not coincide, and where biological activity, cultivation and other human activities, as well as postglacial erosion and deposition, have made the original geological borders invisible on the ground surface.

MAP COMPILATION

The field geologist has to reconstruct the geological borders in the field by direct observations or by interpolating the contour lines between observations of different types of deposits. The result of his work will then be a field-map with contour lines and symbols for the geological units of such accuracy that it can be redrawn and reduced to a scale of 1:50,000 without any significant modifications. The mode of field-mapping and the printing techniques are the limiting parameters. Normally the smallest reproducible area is 50 m in diameter in the field, that is 1 mm on the map. Some important geological units, such as very small bedrock outcrops and small deep depressions filled with gyttja, can be enlarged to the minimum area on the map.

Before the final map is printed the field-map is checked against and completed with data compiled from well-drillings, seismic investigations and other geotechnical investigations. The field-mapping is also supplemented by a

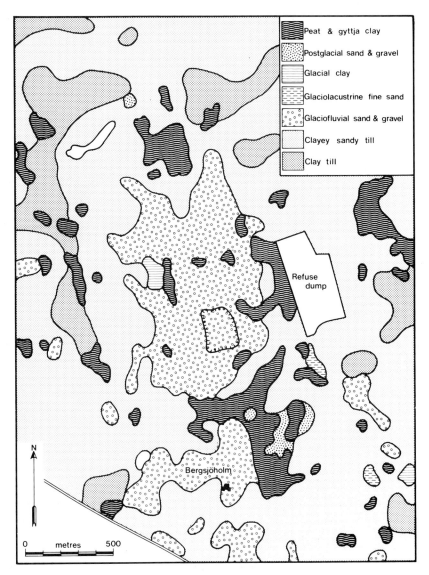

FIGURE 9.2. Principal cross-section through the area in Figure 9.2

limited number of laboratory analyses. The most important of these is the analysis of the grain-size composition of different types of deposits. Besides the grain-size composition, the lime content and the rock types of the gravel fraction and — in some cases — the clay minerals are analysed. As far as possible within the set time and economical limits the stratigraphy of the area in question is investigated and presented in the description published together with each map sheet.

Figure 9.2 is an example of a field-map from an area some kilometres north-west of the town of Ystad on southern-most Skåne. This map area also includes a palaeoecological reference site, the ancient lake basin Bjär-sjöholmssjön (cf. Nilsson, 1961). As the geology is very complex in the area a stratigraphical cross-section is an important complement to the map. A schematic section is presented in Figure 9.3. The same area is discussed in Chapter 11 from the point of view of archaeological survey mapping, and in Chapter 10 from the point of view of potential vegetation reconstruction.

The geology in this area can very briefly be described as follows. It is part of a typical hummocky moraine landscape covering large areas in south-western Skåne. The terrain is characterized by hummocks with a relative height of 5–15 m and depressions of varying size filled with peat. The hummocks consist of till, glaciofluvial sediments and glaciolacustrine sediments, the distributions of which are impossible to predict.

The till in the area is clayey sandy or a clay till (5–15% and 15–25% clay respectively). The boulder frequency of the till surface is very low, as is the

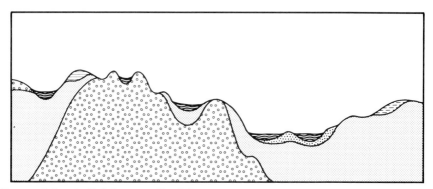

FIGURE 9.3.   A slightly generalized Quaternary map of an area mentioned in the text. The map is based exclusively on field-mapping, cf. Figure 9.1

frequency of stones. The lime content of the till is 10–20% (analysed on fractions smaller than 0.06 mm) in the unleached till. The leaching has normally reached a depth of at least 1–1.5 m depending on which topographical position the samples are taken from.

In the central part of the area there is a glaciofluvial stony gravel whose stratigraphical position seems to be below the till. In the area, quite a few observations of glaciofluvial sediments covered by till were made. The morphology of the gravel is even more broken than it is in the till-covered areas, and dead-ice depressions seem to be more frequent in the gravel than in the till. However, the greater permeability in the glaciofluvial deposit prevents the depressions from being filled with water and peat as is the case in the till-covered parts of the area. The sub-till glaciofluvial sediments are also found in a superficial position in small areas spread out in the till-covered hummocks. In other places there are small patches of glacial clay (in Sweden called plateau-clay) and silt as well as somewhat coarser glaciolacustrine sediments on the hummocks.

The youngest sediments are the peat and the gyttja clays found in the depressions and dead-ice hollows where former lakes were located. The largest peat area is situated in the ancient lake Bjärsjöholmsjön, where even during historical time a shallow and irregular lake existed according to older topographical maps.

REFERENCES

Nilsson, T. (1961). Ein neues Standardpollendiagramm aus Bjärsjöholmsjön in Schonen. *Lunds Universitet Årsskrift.*, **N.F. 56.**
Daniel, E. (1977). Beskrivning till jordartskartan Trelleborg NO (Description to the Quaternary map Trelleborg NO, with English summary). *Sveriges Geol. Unders*, **Ae 33.**
Ringberg, B. (1980). Beskrivning till jordartskartan Malmö SO (Description to the Quaternary map Malmö SO, with English summary). *Sveriges Geol. Unders.*, **Ae 38.**

10

Mapping present and past vegetation

NILS MALMER

AND

GÖSTA REGNÉLL

Department of Plant Ecology,
Lund University,
Östra Vallgatan 14,
Lund, Sweden

INTRODUCTION

The interpretation of fossil plant records derived from microfossils (e.g. pollen) or macrofossils (e.g. wood remains and seeds) provides initially a list of the species which were likely to have been growing in the area surrounding the sampling point. A more thorough palaeoecological understanding of an area calls for information on the ecological conditions prevailing and their variation between different parts of the landscape. This is especially important when tracing the effects of man's activity in the landscape. The intention here is to present a method for obtaining the necessary basic information for a more thorough palaeoecological interpretation.

THEORETICAL BACKGROUND

The plant species identified through different kinds of fossils represent a sample of the flora in the region. 'Flora' here means all the plant species occurring in a certain biotic region or province of any kind. All the plants growing together within an area collectively form the vegetation of that area. For a long time it has been common in palaeoecology to try to reconstruct the vegetation from an interpretation of the plant fossils preserved.

Vegetation is never homogenous. It consists of patches with different vegetational attributes, such as species composition, species abundance, age structure of the species populations, vertical layering resulting from the growth forms of the species, and the spatial distribution of individual plants. This pattern is never an irregular one but is determined by biotic as well as abiotic conditions.

The conditions influencing the vegetation and its characteristics could be summarized in the following way:

$$V = f(fl, fz, e, c, s, t)$$

where V designates the vegetation as a function, f, of fl, the flora of the region, fz the fauna, mainly the herbivores, in the landscape, e the ecological demands of the plants related mainly to their ecophysiology, c the climate, s the soil conditions (or more general edaphic conditions), and t time.

Climate and soil enter the function as an integration of all environmental factors influencing plant growth. As both the microclimate and the edaphic conditions may change from place to place in the landscape, they result in a corresponding patchiness of the vegetation. Time has to be included in the function, as biological processes on both the individual and the population levels continuously bring about changes in the vegetation, resulting in perpetual small-scale vegetational successions — a shifting mosaic steady state. Fires, windfalls, pests or other extensive catastrophies immediately result in secondary successions starting at earlier successional stages. Such secondary successions are usually different from a primary succession starting from juvenile soils, e.g. after deglaciations or on newly formed sand dunes.

In general, plant successions are considered predictable, i.e. it is possible to describe the major vegetational types which may develop in a region or a landscape. For instance, the identification of different biomes and biotic regions is one expression of this idea. In discussing successions, long-term vegetational changes should also be mentioned; they may be caused by climatic changes, by changing soil conditions due to, for example, weathering or paludification (peat initiation), or by the colonization and spread of new species — a new tree species may, for example, change the structure and influence the vegetation considerably.

As a consequence of all these conditions the pattern in the vegetation of the landscape is far from irregular. A very common way to describe this pattern is to use the concept of plant community, defined as regularly recurring combinations of plant species. In a plant community, the plant species are growing together, sharing the resources of the site and having a strong mutual influence on the environmental conditions. Within a certain phytogeographical region, characterized by a given flora, the main species composition of a plant community as well as most of its other vegetational attributes will be the same, provided the successional stage and site conditions are the same. The most common variables used for characterizing plant communities are the structure and the species composition of the vegetation, sometimes supplied with quantitative measures of abundance of some or all species present.

Man's influence on the vegetation and the landscape, e.g. through clearings and small fields, can be looked on as a succession of catastrophies bringing plant successions back to earlier stages (Figure 10.1). When such catastrophies are not too extensive, they will considerably increase the variation and patchiness of the vegetation. The vegetational types created in this way by man are often designated as semi-natural. They differ from the mature natural vegetational types mainly through the layering of the vegetation and the species abundances. New species are occasionally introduced, but are rather few in number and unimportant, in terms of plant biomass. Only when the influence from man is very strong, destroying most of the natural and semi-natural vegetation over vast areas, does it create a completely new vegetation and landscape with the introduction of many new species.

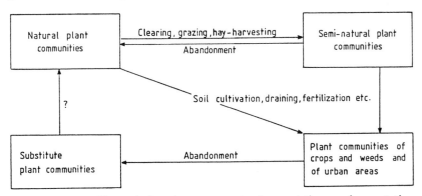

FIGURE 10.1. The relation between natural vegetation and vegetation influenced by man. The question-mark emphasizes that heavy impact of cultivation may change the conditions (especially the hydrologic and edaphic ones) so much that natural plant communities will not reappear in the foreseeable future

The relation between the spontaneously occurring natural vegetation, in an area, and the edaphic conditions can be described by relating the different plant communities to the soil water conditions and the soil reaction combined with other related soil conditions. As an example, this relationship is shown for southern Sweden in Figure 10.2. The semi-natural plant communities can be treated in the same way (Figure 10.3). These two figures, when compared, give an idea of the recurrent successions in the vegetation of the man-made landscape. The comparison can even be extended to a heavily exploited agricultural landscape. However, in that case the soil has often been drastically changed and it is not clear whether the successions result in a plant community of natural vegetation or not. But even in such 'substitute communities' (Figure

NATURAL VEGETATION

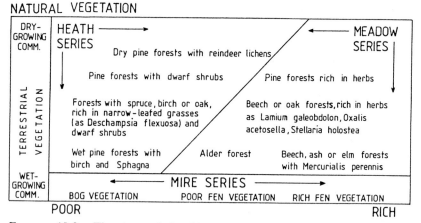

FIGURE 10.2. The interrelationship between the natural vegetation of southern Sweden and the soil nutrient status and soil water regime (schematized, with lakes omitted)

SEMINATURAL VEGETATION

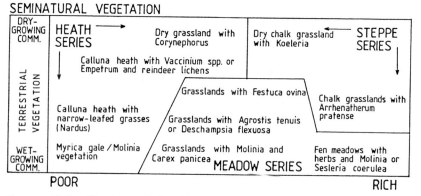

FIGURE 10.3. The interrelationship between the semi-natural vegetation of southern Sweden and the soil nutrient status and soil water regime (schematized)

10.1) the structure and layering of the vegetation is still thought to develop, albeit slowly, towards the natural vegetation formed on similar sites.

MAPPING AND RECONSTRUCTION

The general purpose of a palaeoecological reconstruction is to present a description of the general structure and the variation of the vegetation in the landscape. It must refer to a certain period of time and should consider not only the variation due to the edaphic conditions and spontaneously created successional stages in the natural vegetation, but also the variation due to

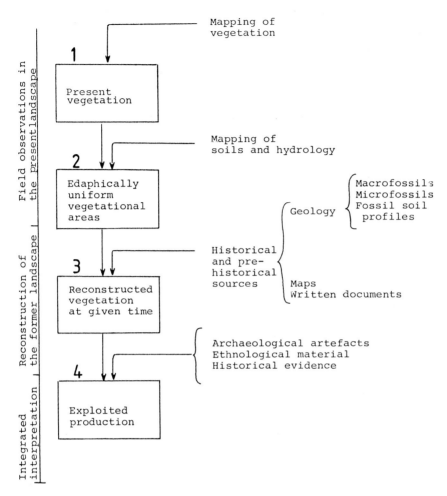

FIGURE 10.4. Schedule for reconstructing vegetation and production

man's land-use and management of the vegetation. The general outline for a procedure for such a reconstruction is given in Figure 10.4. It has to be modified according to local conditions and sources available, but the procedure is independent of scale, whether the area considered is small or large.

The first step in the procedure is to map the present vegetation. In this mapping, not only the layering of the vegetation (forest, shrubs etc.), but also its species composition and some quantitative estimate of cover or abundance, must be considered for the delimitation of the vegetational units/plant communities. Then any system of classification of vegetation can be applied, provided the recognized units have known edaphic demands and successional interrelationships. All kinds of vegetation may be mapped, even weed communities of cultivated soils; but forests and semi-natural vegetation, be it only along hedges or ditches, give more useful information.

Except for some distinct but generally small vegetation units (e.g. springs), vegetation units recognized for mapping should always cover at least 400 m² in the terrain. If smaller, they may be noted as elements, but on the map they must be included in another unit or, if several units regularly occur together, they may be considered a mosaic, i.e. a vegetational complex unit.

For practical reasons, it is rarely possible to indicate something on a map smaller than 0.5 cm² or narrower than 2–3 mm, i.e. areas of 0.08 ha, 0.5 ha and 12.5 ha on maps with scales of 1:4000, 1:10,000 and 1:50,000 respectively. Therefore, in the selection of the vegetational units to be mapped, the scale of the map has to be considered. It has been proved most useful to carry out the whole mapping procedure at scales from 1:4000 to 1:10,000 and then from these maps to reduce the scale and simplify the designations. It is recommended that, initially, those types of vegetation which clearly have successional relationships to each other can be combined, e.g. forest clearances with corresponding mature forest types.

The second step is to delimit areas, here called *vegetational areas*, which are assumed to be uniform with regard to the result of continued succession; different stages are supposed to develop towards a common plant community with a uniform species composition. Left undisturbed, such an area would be supposed to be uniformly covered by the same mature stages in the vegetational successions. Within such an area the edaphic conditions should be so uniform that remaining differences are not reflected in the vegetation.

In a landscape with a natural or semi-natural vegetation, a vegetation map will closely reflect the edaphic conditions — the soil conditions and soil water regime. It is more difficult to work in an intensively cultivated landscape where the natural or semi-natural vegetation only occurs in isolated patches surrounded by large fields. These patches are often so small and so impoverished in species that their vegetation is difficult to characterize in an informative way. Of course, even the information obtained from weed communities can be interpreted, but those communities are much more variable and strongly depend on the management of the fields.

The best way to delimit the vegetational areas on the map is to combine the information obtained from the vegetational map with that from the soil map (see Chapter 9). Standard subsoil maps are always useful but, if available, topsoil maps will also supply important information. In a landscape dominated by natural vegetation there is a rather close resemblance between the two maps, as long as the topography and hydrology are not too variable. In such a case the main stress is laid on the vegetational map for the delimitation of the vegetational areas, as this map will clearly reflect not only the soil type but also the water regime and any effects of topography. Only if the soil conditions in an area with uniform vegetation are so varying that the soils might be expected to react differently to climatic changes affecting the water regime are there reasons to separate an area with uniform vegetation into two or more vegetational areas.

Where the present vegetation mainly consists of semi-natural vegetation or vegetation types even more modified by management, it is necessary to rely more on the soil map when delimiting vegetational areas. However, even in such a case, the small patches of vegetation found will indicate the relation between vegetation and soil conditions in the area. Conclusions can then be drawn about the potential vegetation existing on the different soil types. The borders between the soil types can be used to delimit the areas with uniform vegetation, at least in an approximate way.

The third step in the procedure is to combine the information given in the maps of vegetational areas with the available information about plant species and vegetation from historical sources: geological (e.g. fossils, fossil soil profiles) or written documents (e.g. maps, tax records). Such a combination of data must refer to a specific period of time. The distribution in the landscape of the different tree species represented in the pollen and wood assemblages can be derived and the kind of forest community can be determined from the edaphic conditions indicated by the vegetational areas. Furthermore, it is possible to recognize areas lacking tree growth because of water flooding (organic soils) or thin soil. Pollen assemblages, maps and written records can all, when combined and referred to the vegetational areas, provide information about land-use and vegetation types. For prehistoric periods the pollen assemblages can be used to indicate what kind of plant community (e.g. wet meadow, dry meadow, heathland, fields) developed after the forest clearance. Conclusions can then be drawn both about the land-use and the soils that were preferred for certain types of management.

These possibilities for the reconstruction of the vegetation and landscape around a prehistoric settlement have been used by Bartholin *et al.* (1981). The whole procedure, with its three steps, is now illustrated in a case study from southern Sweden.

CASE STUDY

This case study refers to an area around Bergsjöholm, a castle situated 3 km from the Swedish south coast, north-west of the town of Ystad. The study follows the general outline of Figure 10.4 but is restricted to the period 2000–1500 B.C. (non-calibrated radiocarbon chronology) with special interest paid to the influence of man. The period corresponds archaeologically to late Middle Neolithic and Late Neolithic.

Information available from the area is mainly:

(1) a map of the present vegetation, also showing the topography (Figure 10.5; original in 1:10,000);

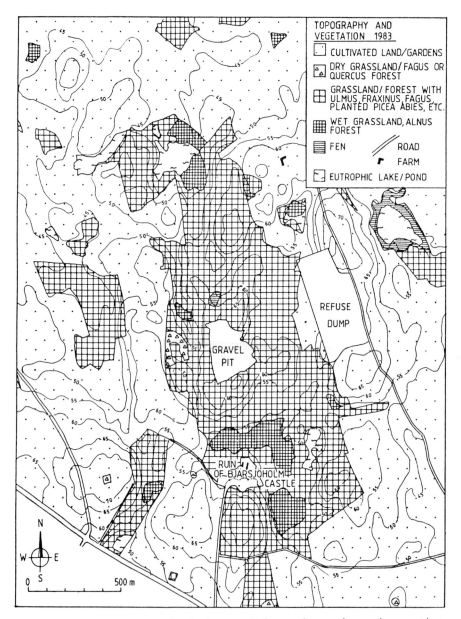

FIGURE 10.5.   Present vegetation in the case-study area (mapped as a minor part in a work, described by Bengtsson *et al.*, 1984). When studying the maps, note that the castle denoted is the medieval one, now ruined. The map in Chapter 9 shows the modern one, situated about 300 m further to the south

(2) a geological survey map, showing the subsoil at 0.5 m depth (Daniel, Chapter 9; original in 1:10,000);

(3) 'Skånska rekognosceringskartan' (1812–20) — a military reconnaissance map, giving valuable information about the hydrological conditions before the period of extensive draining in the nineteenth century (Figure 10.6; original in 1:20,000);

(4) a complete pollen diagram (Figure 10.8) from the lake Bergsjöholmssjön, near the castle, which is now drained (Berglund, 1969; Nilsson, 1961).

Present-day vegetation

Vegetation was mapped according to a slightly modified standard method (Påhlsson, 1972). Mainly judged on the composition of the field layer, the vegetation is recognized as belonging to different 'series' and moisture types, as in Figures 10.2 and 10.3. Where trees and shrubs are present, the dominant types are denoted on the map (not shown here). Waste-land vegetation, gardens, cultivated fields etc. are generally not studied.

Cultivated fields are found to dominate the area and its neighbourhood. Pastures and leys are few and usually heavily fertilized, showing a uniform and simple flora. During the nineteenth and twentieth centuries many natural depressions, small ponds and swamps were drained or filled. Continuous erosion still influences the landscape, e.g. gradually covering small deposits of peat in depressions with clay topsoil.

The forested areas are dominated by *Ulmus glabra*, *Fagus silvatica*, *Fraxinus excelsior*, *Quercus* species, planted *Picea abies* and, in the wettest parts, *Alnus glutinosa*. The distribution of the tree species is much the result of direct or indirect action by man. The composition of the field layer reflects the variation in soil conditions better than does the tree layer. It reveals an overall high nutrient status of the soils; e.g. *Mercurialis perennis* occurred in most of the forest sections, except where spruce had been planted.

The wettest parts of the forests have been drained, so real alder carrs (with, for example, *Iris pseudacorus*) do not appear. Dry forest types, however, were found in small areas on hilltops or on slopes. There, *Poa nemoralis* is a characteristic species. We can be sure that the present distribution of the dry forest types has significance for reconstructing past vegetational distributions. The main sites of alder stands can also be deduced, but the hydrological changes in the landscape have been so extensive that the precise extent of wet forest types cannot be reconstructed from present vegetation only.

Soil conditions and hydrology

This part of the case study is based on sources (2) and (3), mentioned above.

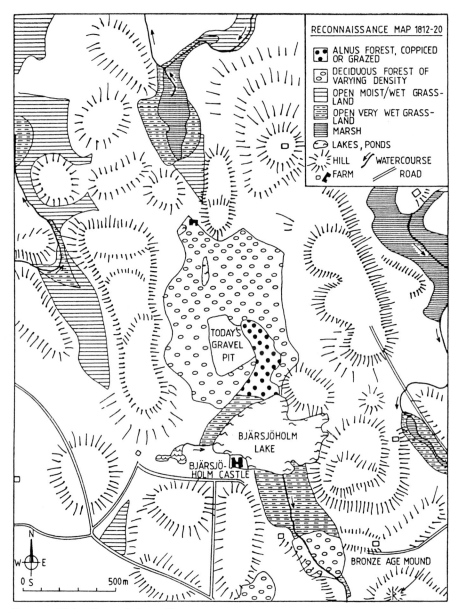

FIGURE 10.6. Part of the military reconnaissance map ('Skånska rekognoscerings-kartan') 1812–20, redrawn for the case-study area. Legend according to the interpretations by Emanuelsson and Bergendorff (1983). When comparing this map to the others, consider the imperfections of the surveying equipment of the time

Water and wetlands

At the beginning of the nineteenth century, the shallow lake Bergsjöholms-sjön covered a rather large area (Figure 10.6). It was later completely drained, though digging has given new surface water in some 5% of the original lake. Three other very small lakes, or rather ponds, were also indicated in 1812–20. They are now associated with peat and gyttja clay deposits on the subsoil map (Figure 9.2) and with still open water or *Betula/Salix* carr in the present landscape. At the beginning of the nineteenth century the lake was to a great extent surrounded by open wetlands, which are now partly covered with forest (*Alnus glutinosa, Betula* species, *Fraxinus excelsior* and *Fagus silvatica*).

In a former wetland area with marsh in the north-east the ground water table has been raised recently, producing a pond containing such species as *Iris pseudacorus*. In the western part wetlands have mainly been drained for grazing or as arable land. Some clay pits have resulted in very small bodies of open water in the landscape (not mapped).

In summary, the hydrological conditions have changed drastically since the period 1812–20. The peat and gyttja clay deposits and the 1812–20 situation are, however, giving consistent information and are likely to give the best idea about the prehistoric distribution of open water and wetlands. The interpretation in Figure 10.7 assumes that most present-day peat and gyttja clay deposits were still corresponding to open water in the period 2000–1500 B.C.

Dry areas

Nowadays, areas distinguished by dry-growing vegetation are few. Most of them are situated on deposits of glaciofluvial sand/gravel. However, other areas with similar vegetation may have existed. For example, some hills of gravel, which were formerly likely to have been characterized by dry-growing vegetation, have been exploited. Also, the castle itself was situated on an area of these deposits which protruded slightly into the former lake. More difficult is to judge where dry-growing vegetation may have existed in the present-day cultivated hilly till areas. Erosion has levelled the hilltops, and fertilizers and irrigation have camouflaged the effects of the original differentiation in soil conditions and moisture. The present mainly forested area on glaciofluvials north of the lake is bordered to the west and east by cultivated hills, the slopes of which must have been rather dry in parts.

In summary, the number of former sites with dry conditions is likely to be underestimated from a study of the present landscape. Some are of special interest as probable sites of prehistoric settlements. For example, the location of the original castle on the lake is strikingly favourable and might also have attracted earlier inhabitants of the landscape.

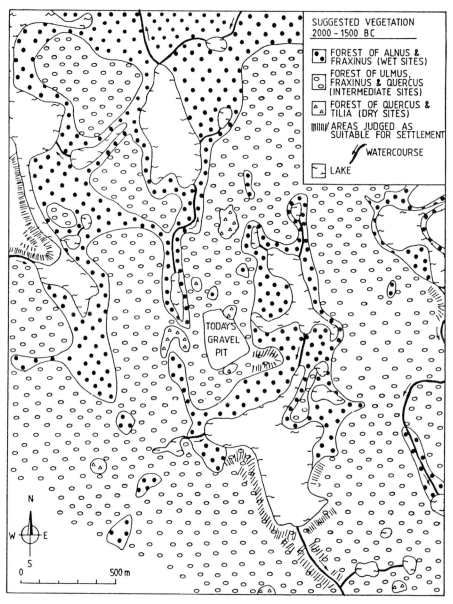

FIGURE 10.7.   Vegetation and suitable settlement areas suggested for the period
2000–1500 B.C.

The pollen diagram (Figure 10.8)

The whole period discussed, until 1500 B.C., is prior to the time when *Fagus* and *Carpinus* found their way northwards to Sweden, and also prior to the great deforestations during the Bronze Age and the Iron Age.

The landnam phase, when cattle and/or cultivation start playing an important part in the economy, is generally considered to occur at or after the *Ulmus* decline, here at about 3300 B.C. The simultaneous declines in *Tilia* and, though less pronounced, *Fraxinus*, confirm this interpretation.

Grasses become more abundant, and around 3000 B.C. *Plantago lanceolata* also appears, indicating established grazing swards, mainly on rather dry soils.

Quercus shows moderate fluctuations, perhaps only the result of the variation in total pollen, in which *Corylus* plays an important role. Since oak regenerates poorly after felling and takes a long time to grow to a fertile age, this may indicate that oak did not decline very much.

Alnus, on the other hand, regenerates quickly when felled or coppiced, even in grazed areas. The almost constant level of *Alnus* pollen, therefore, does not immediately imply that the alder was an unexploited species. However, as the indications of grasslands in wet areas are rather weak, there is no reason to assume that grassland occupied areas other than those where *Ulmus*, *Fraxinus* and *Tilia* lost their position. The decline in *Fraxinus*, a species which originally held a strong position in a mixed *Alnus–Fraxinus* forest, may have been due to coppicing or felling of this comparatively good fodder producer.

An interesting role is played by *Betula* and *Corylus*. Sharp rises occur after the *Ulmus* decline in the Early Neolithic period. Both species, especially *Betula*, gain in a colonizing situation; and their decline around 2500 B.C. (followed by a rise in *Ulmus*, *Tilia* and *Fraxinus*) strongly suggests a weaker human impact and a denser forest, though perhaps with an 'over-exploited' short period in between, suggested by the sharp peak in *Plantago* around 2500 B.C.

The period until 1500 B.C. is characterized by an increasing activity of man with his animals. More extensive swards are indicated by the rise in *Plantago* and the appearance of *Rumex acetosella*. *Calluna* stays low until there is a slight rise at the end of the period; the dominant soils in the area are not very sensitive to leaching. Cultivation increases (*Cerealia*), and presumably coppiced *Corylus* woods gain while the oak goes down again.

The reconstruction map (Figure 10.7)

Alnus forest, mixed with *Fraxinus*, must have occupied the wettest areas around the lakes, along water courses and in other places where peat/gyttja clay deposits still exist. (There are no signs of significant ombrotrophic mires or fens too wet for trees.)

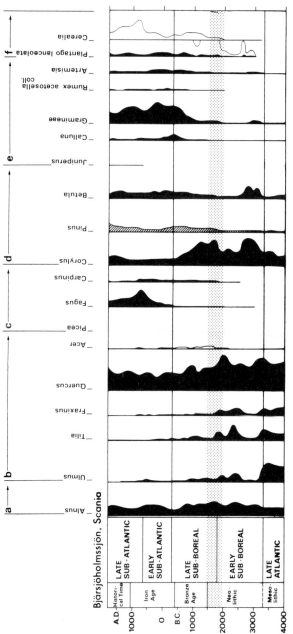

FIGURE 10.8. Pollen diagram (Nilsson, 1961; revised by Berglund, 1969). The shaded area denotes the period of main discussion. Time-scale refers to non-calibrated radiocarbon years. *Reproduced by permission of Oikos*

The distribution of the other trees is more difficult to settle in this case, as Figure 10.5 only indicates small variations in the soil reaction and the amount of plant mineral nutrients. The reason for this is that most of the sandy deposits here are situated in depressions in the till deposits and consequently have not been greatly impoverished by leaching. However, it is possible to consider two originally dominating forest types, *Quercus(–Tilia)* in the drier sites and *Ulmus(–Fraxinus–Quercus)* in the intermediate ones. This interpretation assumes hydrology as the main differentiating factor in this landscape of generally rich soils. The difference between glaciofluvial deposits and till is consequently considered to be less important, though the more sandy soils give a more distinct vegetational differentiation where dry parts occur. Species indicating leaching *(Calluna, Rumex acetosella)* do not play an important part here until late in the pollen diagram. However, the small amounts of *Calluna* appearing must be associated with cultivation (cf. the increasing amounts of cereals; Figure 10.8) or with sustained, heavy grazing and the start of leaching in some of the driest areas near settlements.

Originally, *Corylus* would be associated mainly with the *Quercus* forest; but in places where human impact opened the canopy, both the *Quercus* and the *Ulmus* forests must have been suitable for both *Betula* and *Corylus*. Suitable dry areas for settlement are indicated in Figure 10.7, but nothing can be said about their exact locations.

Concluding remarks

This case study is only one example of applying the procedure outlined here. Further similar studies are presented in Klötzli 1967 and Bartholin *et al.* 1981. A more detailed result could be obtained if further sources were used (e.g. historical maps, showing the hydrological conditions; charcoal remains, giving information of the firewood species; palaeoecological investigations dating the bottom of the numerous peat deposits or giving details of the pollen sedimentation in a smaller, local basin at some distance from the lake). However, the important idea has been to demonstrate how these maps may open up discussion and criticism, and create a basis for hypothesis formulation which other workers, such as archaeologists, can attempt to confirm or falsify.

REFERENCES

Bartholin, T. S., Berglund, B. E., and Malmer, N. (1981). Vegetation and environment in the Gårdlösa area during the Iron Age. *Acta Regiae Soc. Hum. Litt. Lund.*, **75**, 45–53.

Berglund, B. E. (1969). Vegetation and human influence in south Scandinavia during prehistoric time. In: Impact of Man on the Scandinavian Landscape during Late Post-Glacial (Ed. B. E. Berglund), *Oikos Suppl.*, **12**, 9–28.

Bengtsson, S., Regnéll, G., and Risinger, B. (1984). *Översiktliga vegetationsbeskrivningar av områden kring Ystad*, Växtekologiska institutionen, Lunds universitet.

Emanuelsson, U., and Bergendorff, C. (1983). Skånes natur vid 1800–talets början — en växtekologisk utvärdering av den skånska rekognosceringskartan. *Ale*, **4**/1983, 18–40.

Klötzli. F. (1967). Die heutigen und neolithischen Waldgesellschaften der Umgebung des Burgäschisees mit einer Übersicht über nordschweizerische Bruchwälder. In: *Seeberg-Burgäschisee-Süd*, Teil 4. *Acta Bern*, **2**(4), 105–123.

Nilsson, T. (1961). Ein neues Standardpollendiagramm aus Bjärsjöholmssjön in Schonen. *Lunds Univ. Årsskr.*, N.F. 2, **56**:18.

Påhlsson, L. (1972). *Översiktlig vegetationskartering*, Naturvårdsbyrån, Statens naturvårdsverk, Stockholm.

11

Archaeological survey mapping

Lars Larsson

Institute of Archaeology,
University of Lund,
Krafts torg 1,
Lund, Sweden

INTRODUCTION

The documentation and charting of prehistoric human activity has a long history in the context of southern Scandinavian archaeological research. The foundations for this were laid as early as the seventeenth century, when a considerable number of ancient monuments came to be officially registered in Denmark and Sweden (Nielsen, 1981; Ebbesen, 1983). Later, in Denmark, documentation became more systematic during the late 1800s, whereby entire provinces were minutely surveyed with the result that all visible or locally remembered monuments came to be recorded.

In the early part of this century an organization was set up within the framework of the antiquarian authorities in both Sweden and Denmark, the purpose of which was and still is the documentation and preservation of ancient remains (Hyenstrand, 1979; Nielsen, 1981).

The inventory being carried out in Scandinavia has been primarily directed towards the documentation of ancient monuments visible above ground. This is highly justified for tracts of land where impediments and pastureland dominate the scene. The kind of land exploitation conducted in areas like these is not detrimental to the survival of ancient constructions because, conversely, the latter are no hindrance to the pursuit of the former. A radical contrast to these conditions exists in those areas where the soil is particularly well suited to agriculture. Here, extremely few ancient monuments have survived unscathed, where, indeed, they have not been utterly obliterated. It is therefore only to be expected that the number of ancient remains visible above ground is small in such areas; for which reason the results of an inventory based on these will hardly reflect the actual intensity of early settlement, but will instead reflect the development and form of agriculture during the later centuries.

PLANNING OF SURVEY MAPPING

It is clear from the above that an inventory, however thoroughly conducted, cannot always fully satisfy the demands imposed on it and expectations of a total documentation of ancient and prehistoric activities are almost never realized. A number of factors, other than the destructive effects of modern exploitation, contribute to this situation, and these must be borne in mind even in the case of a large-scale inventory. In certain periods of prehistoric times it was not customary to erect visible memorabilia, and, of course, traces of wooden house constructions are only to be found below the surface.

It is therefore vitally important to establish, at the planning stage, the specific aims. Where the objective is to obtain information concerning the relationship between man and environment — and this must be regarded as being an overriding consideration in connection with palaeoecological research — the study will come to be directed not only towards an appreciation of how great an effect this has had on the surroundings. In such cases it must also be established from the outset as to what measure an archaeological inventory of palaeoecological conditions should embrace the entire span of both the prehistoric and historic periods, or only parts of these.

Some form of model of the effects of human activity is required in order to study the relationship between settlement and environment. This is of particular significance if the analysis stems from a sampling point which has been the object of a thorough study of palaeoecological conditions. Models concerning resource exploitation have been proposed in recent years for the purpose of archaeological research. These models are based on what is called site catchment analysis, where their content and form are dependent on the societies' varying economic systems (Vita-Finzi, 1978). For a society with an exclusively catchment economy a resource area is defined by the distance that a person on foot can achieve between dawn and dusk, starting and finishing at base camp. The resource area in this case is usually expressed as a circle having its centre at base camp and with an optimum radius of about 10 km. For agricultural societies, on the other hand, the limits of the resource area are considered to be markedly less — a circle with a radius of 1–2 km is regarded as sufficient to express the area exploited (Jarman *et al.*, 1982).

If an inventory of archaeological remains is to be conducted around a sample point which illustrates the local vegetation, it is necessary first to have a conception of the period's economic base in order to decide the extent of the inventory area and the *modus operandi* to be employed. In most cases, however, the basis of an inventory is concerned with more general aims. Archaeological inventories are quite often conducted on the basis of limitations imposed by geographical conditions — for example, areas with particular soil types, or areas restricted by natural impediments such as waterways, marshland, rock formations etc. In these cases palaeoecological conditions will come to be included in the analysis, but at a later stage of the work.

Even though site catchment analysis is very often applied by archaeological researchers in their study of the relationship between human activity and the physical environment, a close correspondence between archaeology and certain natural sciences relevant to this type of analysis appears to be lacking. Certain features ought to be considered vital to a study of the physical environment — such as mapping of the present floral make-up of the research area, and quaternary-geological soil type mapping. A correlation of the basic facts obtained by means of palaeoecological analysis of core-samples in the context of geologic/plant-ecologic mapping would provide an extremely good foundation on which to base a retrospective study of the floral community's composition and distribution.

It is this aspect that has quite often been overlooked in site catchment analyses, with the result that it does not figure in the final evaluation, in spite of the obvious fact that the composition of the faunal population is conditioned by the nature and distribution of the flora, even — and possibly especially — where the latter has been altered as a result of human activity.

VARIOUS ELEMENTS IN AN INVENTORY

An inventory of prehistoric and historic activities is most successfully compiled by thoroughly investigating all possible approaches. The various elements can be subdivided as follows:

(1) collection of earlier inventory material;
(2) map studies;
(3) examination of museum collections and the inventory of private collections;
(4) inventory in the field;
(5) other factors.

The initial stages of an inventory should be concentrated on the collection and study of source material already available. By this is meant earlier inventories, and public and private artifact collections. The circumstances of the find are important — as a general rule, large, intact objects most often indicate depots or graves, while fragmentary objects and waste products are usually evidence of occupation.

Museum collections generally constitute the basis for a broad appreciation of an area's prehistory, but they often lack detailed descriptions of the find circumstances. Private collections, primarily farmhouse collections, usually comprise a more modest range; but in many instances the find conditions and location are well known. This situation may, however, change for the worse in the future, in so far as the majority of finds were made at a time prior to the mechanization of agriculture, i.e. when ploughing, sowing and reaping were done in intimate contact with the surface. This type of contact has all but disappeared since the introduction of modern farming methods, while at the same time the work tempo has greatly increased. Sitting on a tractor high above ground and with one's back to the ploughing rig, one cannot observe objects brought up to the surface with the same ease as one could when the ploughshare was foremost and the tiller's attention was undividedly directed towards the cut of the furrow. The members of the generation that belonged to the pre-mechanized period are for the most part of such an age today that they no longer take an active part in the working of the land, with the result that vital information regarding find circumstances is on the point of becoming lost to us.

The study of all existing map material can often simplify the survey work. Older as well as newer survey maps can provide useful information, and a comparison between older and newer maps usually gives a good insight into two factors of vital interest to the inventory.

One factor concerns the degree of exploitation. If it can be definitely established that an area has been intensively cultivated over a considerable time, then it is hardly reasonable to expect visible traces of, for example, primitive agriculture. Instead, this search ought to be concentrated on impediments or pastureland. In this way, a certain appreciation of culture--geographical conditions may be obtained with the aid of maps.

The second factor concerns the relationship between firm ground and marshland. The map material can, in certain cases, be used in a retrospective analysis of drainage conditions. Areas of marshland can more often than not be excluded from the field reconnaissance, if the intention is to establish the extent of settlement.

When the examination of maps, earlier inventories and artifact collections has been completed, the results may be correlated and a plan of action drawn up. Only then can work in the field be initiated. This can take one of two forms — either a general reconnaissance of all the accessible terrain, or a more selective search where, for example, the objective is to document particular settlement areas from a chosen period of prehistory.

Regardless of the method employed, it is important that the field work be assigned to a season of the year when conditions are at their most favourable. By this is meant a period in which the land is lying fallow, and preferably some time after raking activities following the last harvest. Late winter and early spring are the periods in which conditions are usually most propitious; considerable tracts of land lie waste and the surface is washed free of superfluous dust, so that surface traces of ancient activity in the form of artifacts are easily observed. In addition, the high moisture content present at these times increases the possibility of observing colour deviations in the soil, which may well indicate traces of occupation layers, or other forms of buried constructions.

The reconnaissance area should be inspected by means of parallel lines set at close intervals. If possible the procedure should be repeated on separate occasions, as the degree of moisture content and light conditions etc. can vary. The field reconnaissance should be complemented by interviews with the landowners on the subject of any remembered soil colour inconsistencies or stone concentrations.

Settlement remains from different prehistoric periods cannot always be traced by one inventory method. Certain raw materials are sufficiently resistant to survive even where the area of ancient activity has been subjected to intensive cultivation for a great length of time. It is, for example, relatively easy to establish settlement remains from those periods when flint was the raw material used in artifact production. The raw materials for objects characteristic of other periods, primarily ceramics, are, however, by no means equally as resistant, which means that settlement remains from periods when minerogenic material was not used to any great extent may come to be under-represented in the inventory. For the documentation of these periods, one must rely partly on observing dark miscolourings in the soil, which may indicate hearths or sections of an earlier intact occupation layer which the plough has brought to the surface.

Inconsistencies in the vegetation (growth rate, colour variations etc.), indicative of a heightened humus content in, for example, buried constructions, are sometimes directly observable at ground level. An important complement to the observation of these, as well as insignificant rises and depressions, not apparent at ground level, is the employment of aerial photography (Wilson 1982).

Another method which has found wide application in Sweden and Norway (Bakkevig, 1980) involves the extraction of phosphate samples from immediately below the topsoil layer. A localized high phosphate content may indicate the presence of a prehistoric settlement. Other methods include drilling with a screw borer and/or test excavation in the form of small pits at regular intervals. The former method is suitable for pastureland. The latter method, apart from being time-consuming, oversteps the border between reconnaissance and actual archaeological excavation, with all the special preconditions and restrictions — not least of which is the legal aspect — implicit in this method.

It will be apparent from earlier comments that several critical factors require careful consideration when one is conducting an archaeological inventory. It is of the greatest importance that one is conscious from the outset of the many sources of potential error inherent in such an inventory. This is, of course, not intended to discourage an undertaking of this kind, but rather to emphasize the limits of what an archaeological survey can achieve, and the restrictions by which it is bound.

Unfortunately, no comprehensive correlation yet exists between palaeoecological investigations on the one hand and archaeological inventories, conducted according to the methods described above, on the other. Such a correlation is, therefore, only possible in the context of specific requirements.

An example of palaeoecological/archaeological correlation within a limited period of prehistoric time is an inventory planned in conjunction with the South Scanian project entitled 'The Cultural Landscape during 6000 years' (Berglund and Stjernquist, 1982).

The study of Neolithic settlement patterns has been based on the results of archaeologically orientated surveying methods of the type described above. Palaeoecological studies have been conducted concurrent with, and parallel to, the archaeologically surveyed area. The field area comprises considerable variations in terms of topography, hydrology and soil types. Several sediment cores were taken at specific basins, for the purpose of Quaternary biological analysis.

An area near the coast in the eastern section of the area in question has been the object of an intensive inventory (Figure 11.1). This included the registration of local farmhouse collections. Some 5000 acres of open fields were subjected to close scrutiny. Results relating to the surface finds and dated prehistoric settlements have been correlated with those obtained in the Quaternary biological investigation (at site 1, Hjelmroos 1985). Man-induced interference can be seen here in the form of both settlement remains and pollen patterns, each of which is traceable to an early part of the Neolithic period. By virtue of a correlation of the two sciences, it has been possible to reconstruct the landscape in this region of the research area as it was during the Early Neolithic (see Chapter 10).

An archaeological survey was made within a radius of 2 km from two palaeoecological sample points, i.e. the boundaries of that area judged to have been exploited by a Neolithic settlement. The sample points are situated at 7 and 12 km, respectively, from the present-day coastline (Figure 11.1, sites 2 and 3). Very few traces of Neolithic settlement were found within the 7 km radius of the sample point located in the southern part of the Krageholmsjö lake (site 2). This result is further confirmed by the palaeoecological analysis, which shows that plants, indicative of an induced interference in the make-up of the landscape, first make their appearance during the Late Neolithic (Gaillard 1984).

On the other hand, abundant traces of Early Neolithic settlement were found within the 12 km radius of the other sample point, located in the region of the

Vasasjö lake (site 3). Here, again, the results of the archaeological survey and palaeoecological analyses are complementary. The latter reflects the slight impact of a Neolithic economy on the environment of an area, best described as a belt located somewhat between the coast itself and the hinterland.

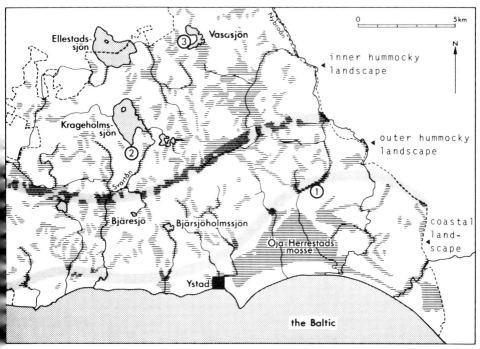

FIGURE 11.1.   Area of the project 'The Cultural Landscape during 6000 Years' in southernmost Sweden. Three areas of intensive study are specially marked (Müller-Wille 1984 with modifications)

A concrete example of cooperation between archaeological and palaeoecological expertise is supplied by the results of a thorough-going analysis of Hullsjön lake in Western Sweden (Digerfeldt and Welinder, 1979, 1985). The lake's environs were the subject of an archaeological inventory of the extent and development of settlement in the area. This inventory was, however, exclusively based on the inspection of several large museum collections and of the existing inventory of ancient monuments. Distribution of the latter in relation to that of ancient artifacts varies greatly in the analysed area, i.e. occurrences of one category are often countered by a comparative lack of occurrences of the other, in any given area (Figure 11.2). In addition, urn graves without visible markings at surface level are characteristic of this part of the country. These have only been encountered in connection with gravel mining or other forms of exploitation, for which reason they reflect more the

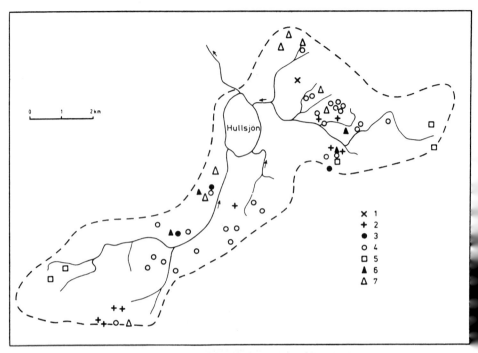

FIGURE 11.2.   Map showing the archaeological source material in the catchment area of Lake Hullsjön in western Sweden (Digerfeldt and Welinder, 1979). 1 = Middle Neolithic axes, 2 = Late Neolithic and Early Bronze Age shaft-hole axes, 3 = cairns, 4 = cup mark sites, 5 = stone cists, 6 = urn fields, 7 = grave fields

extent of modern interference than that of prehistoric settlement. In so far as the material for analysis was based mainly on surface artifact finds and ancient monuments in the form of grave mounds, a direct chronologic and chorologic relationship was assumed to exist between these and the intensity of the settlement, a postulation which cannot be accepted without certain reservations.

In this project a difficulty is the choice of correlation factors, where grave constructions from different periods are required to relate to each other. In order to handle this, special means of calculation were applied. A basis for an estimation of settlement intensity in a wide prehistoric perspective was obtained by means of this calculation, which was expressed graphically. This proposes that a marked increase in settlement took place during the Late Neolithic period and Iron Age.

A zonal subdivision of the area, expressed as concentric circles 2 km apart and with Hullsjön lake as the central point, formed the basis for a chorological analysis of the intensity of prehistoric settlement. The distribution of each type of artifact and monument within the respective zones are shown in Figure 11.3.

This proposes that a redisposition of settlement from peripheral areas, resulting in an increased intensity of settlement in the near vicinity of the lake, should have taken place during a late part of prehistoric time. Both of these conclusions are based on the total number of artifacts and ancient monuments in the neighbourhood of Hullsjön lake. If, on the other hand, these witnesses

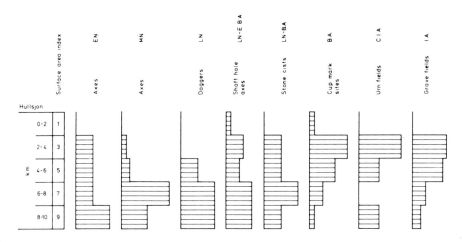

FIGURE 11.3. The distribution of the archaeological source material, expressed by index numbers, in rings with a 2 km width and centre in Hullsjön. The diagram shows the amount of source material per ring. The amount per unit area can be calculated by use of the indicated surface area index (Digerfeldt and Welinder, 1979)

to human activity are limited to that area which comprises the lake's catchment area *per se*, there exists no distinct archaeological evidence for any great increase in settlement intensity here since the Late Neolithic period.

An obvious increase in human activity during the Late Neolithic manifests itself in the results of the palaeoecological analysis in the form of clay deposits on the lake bottom, and this may be directly correlated to evidence of intensified settlement witnessed by the archaeological material from the same period. A decrease in soil erosion is noticeable in the period embracing the beginning of our present epoch, which may be interpreted as being due to a restructuring of settlement patterns.

These palaeoecological indications of a decrease in human influence on the environment find no support in the archaeological material from the same period, which instead suggests an uninterrupted level of activity. In order to make fuller sense of these contradictions, however, a more thorough archaeological inventory than that presented in this particular instance would be required.

REFERENCES

Bakkevig, S. (1980). Phosphate analysis in archaeology — problems and recent projects. *Norwegian Archaeological Review*, **13**, 73–100.

Berglund, B., and Stjernquist, B. (1981). Ystadprojektet — det sydsvenska kulturlandskapets förändringar under 6000 år. *Skrifter från Luleälvsprojektet*, **1**, Universitet i Umeå, 161–185.

Digerfeldt, G., and Welinder, S. (1979). Settlement development and human impact in the Hullsjön area, Västergötland, W. Sweden. University of Lund, *Department of Quaternary Geology, Report* **15**.

Digerfeldt, G., and Welinder, S. (1985). An example of the establishment of the Bronze Age cultural landscape in SW Scandinavia. *Norwegion Archaeological Review*, **18**.

Ebbesen, K. (1983). Fortidsmindreregistreringen. *Fortid og nutid*, **XXX**,(3), 173–190.

Gaillard, M.-J. (1984). A Palaeohydrological Study of Krageholmssjön (Scania, South Sweden). *LUNDQUA Report*, **25**, 1–40.

Hjelmroos, M. (1985). Vegetation history of Fårarps mosse, southeast Scania, in the early subboreal. In Larsson, L, Hanging triangles and multiple arcs. A settlement of the Early and Late Funnel Beaker Culture at Karlsfält, southern Scania, Sweden. *Acta Archaeologica*, 54.

Hyenstrand, A. (1979). *Arkeologisk regionindelning av Sverige*, Stockholm.

Jarman, M. R., Bailey, G. N., and Jarman, H. N. (1982). *Early European Agriculture: its Foundation and Development*, Cambridge.

Malmer, N. and Regnéll, G. (1985). Mapping present and past vegetation. *Handbook of Holocene Palaeoecology and Palaeohydrology*, 203–217.

Müller-Wille, M. (1984). Siedlungsarchäologische Forschungsprojekte in Schweden. *Praehistorische Zeitschrift*, **59**. 1984, 2.

Nielsen, P. O. (1981). Hundredtusind fortidsminder. Om den arkaeologisk kortlaegning i Danmark siden 1807. *Nationalmuseets Arbejdsmark* 1981, 61–69.

Vita-Finzi, C. (1978). *Archaeological Sites in their Setting*, London.

Wilson, D. R. (1981). *Air Photo Interpretation for archaeologists*, London.

Stratigraphical methods

Handbook of Holocene Palaeoecology and Palaeohydrology
Edited by B. E. Berglund
© 1986 John Wiley & Sons Ltd.

12

Characterization of peat and lake deposits

BENT AABY

*Geological Survey of Denmark,
Thoravej 31,
Copenhagen, Denmark*

and

BJÖRN E. BERGLUND

*Department of Quaternary Geology,
Lund University,
Tornavägen 13,
Lund, Sweden*

INTRODUCTION

Unconsolidated sediments often have common names which refer to certain qualities such as colour, appearance and usefulness; but the names are not used consistently. The first scientific attempt to classify organic deposits was carried out by the Swedish geologist Hampus von Post (1862), who used the genesis of the different deposits as a basis for classification. The idea of using the formation processes as a basis for separating sediments was quickly accepted by most European countries.

During the following 50 years this system of classification became more elaborate, and in 1924 a milestone was reached with publication of Lennart von Post's *Das genetische System der organogenen Bildungen Schwedens*. A more complete version appeared two years later (von Post and Granlund, 1926). It was stated that theoretically the genetic point of view was the most rational basis for a standard sediment nomenclature, but von Post admitted that in practice it was almost impossible to use. A practical system of classifying organic deposits should be based on (1) the genesis; (2) the content of determinable plant remains; and (3) some physical properties. Thus, he proposed an operative classification model based on the determination of the 'mother-formation' (the vegetation, which produced the plant remains) combined by genetic principles. The von Post system was widely accepted and is still among the most favoured today (Schneekloth and Schneider, 1972; Overbeck, 1975).

A limitation in using the combined genetic-floristic system is the lack of consistency within extensive areas, because environmental and climatological conditions vary, as does the flora in each plant community. In addition, investigators inexperienced in the determination of macrofossils have difficulty in using the system.

In order to create a universal classification system a purely descriptive approach was described by J. Troels-Smith (1955). The approach is based on the assumption that almost all deposits are mixtures of various components (deposit elements). The number of deposit elements is limited and the objective is to provide a simple means of characterizing any one unconsolidated deposit in much the same way as a chemical compound is characterized by the number of its elements.

Various advantages are obtained over the genetic deposit system: the deposits are described through formulae, making several combinations possible, hence the investigation is not restricted to dealing only with a limited number of fixed groups. Besides, the system allows a deposit to be characterized without considering its genesis; conversely, this characterization may assist in determining the position of the deposit within the genetic system.

The Troels-Smith system is briefly described below together with some techniques for microscopic analysis of different deposit elements.

CHARACTERIZATION OF UNCONSOLIDATED SEDIMENTS

Terminology

In order to avoid terminological confusion of this system with other systems that have been proposed, the terms are given Latin names. The terminology for deposit elements corresponds in principle to that used for names of plants and animals; in fact the Linnaean system of two-term symbols. Thus the first term denotes the genus, and is written with an initial capital letter, and the second term denotes the species, usually written with a small initial letter.

Characterization of deposits

In the description of deposits, three factors are important: (1) physical properties; (2) humicity, i.e. the degree of decomposition of the organic substance; and (3) the component parts, i.e. the nature, as well as the proportion of the elements of which the deposit is composed. For all three classes of parameters a 5-class scale (0–4) is used for characterization, zero implying the absence of, and four the maximum presence of, or the sole occurrence of, the element concerned. The value is given as an estimate based on simple test methods and experience gained mainly in the field.

Physical properties

Four categories of physical qualities are described: (1) nigror: the degree of darkness; (2) stratificatio: the degree of stratification; (3) elasticitas: the degree of elasticity; and (4) siccitas: the degree of dryness. The four categories are estimated according to the 5-class scale. In addition to this, the colour may be described by using the standard Munsell charts.

Humicity

The term humicity (or degree of humification) was defined by von Post as 'the degree of disintegration of the organic substance, regardless of the way this disintegration has taken place, and of what substances resulted from it' (von Post and Granlund, 1926). The humicity is indicated by an index on the 5-class scale and given to certain deposit elements (see Table 12.1).

Deposit elements

Most deposits are mixtures of various deposit elements. In practice fifteen deposit elements have proved useful to include in the characterization system. They are divided into five main groups: *Turfa, Detritus, Limus, Argilla,* and *Grana* (see Table 12.1).

A special element, *Substantia humosa* (Sh), has been included because in certain cases it may be difficult to determine whether a homogeneous, organic black substance is the result of complete disintegration of *Turfa bryophytica, Turfa lignosa, Turfa herbaceae,* or whether it consists of *Limus humosus,* or is the remains of partly oxidized organic matter, as in arable soil. In these cases the neutral term *Substantia humosa* may be used. It is possible through this element to express the degree of humification in a deposit as the proportions of both *Substantia humosa* and the remaining non-disintegrated part of the deposit. Using Sh, the humicity index of the deposit elements should be zero.

The proportion of elements in a given deposit is estimated on the 5-class scale (0–4), which implies that the total of deposit elements *must* always be four. In practice, however, it has proved profitable to be able to indicate the presence of traces, i.e. very slight quantities of a given element, without having to indicate one or more of the main components in fractional numbers. This is

TABLE 12.1. Deposit elements partly according to Troels-Smith (1955)

	Substantia humosa		Humous substance, homogeneous microscopic structure
	Sh		
I Turfa	*Tb*$^{0-4}$	*T. bryophytica*	Mosses +/− humous substance
	Tl$^{0-4}$	*T. lignosa*	Stumps, roots, intertwined rootlets, of ligneous plants +/− trunks, stems, branches, etc. connected with these +/− humous substance
	Th$^{0-4}$	*T. herbacea*	Roots, intertwined rootlets, rhizomes, of herbaceous plants +/− stems, leaves, etc. connected with these +/− humous substance
II Detritus	*Dl*	*D. lignosus*	Fragments of ligneous plants > 2 mm
	Dh	*D. herbosus*	Fragments of herbaceous plants > 2 mm
	Dg	*D. granosus*	Fragments of ligneous and herbaceous plants, and, sometimes, of animal fossils (except molluscs) < 2 mm > c. 0.1 mm
III Limus	*Ld*$^{0-1}$	*L. detrituosus*	Plants and animals (except diatoms, needles of spongi, siliceous skeletons, etc. of organic origin), or fragments of these. Particles < c. 0.1 mm. +/− humous substance
	Lso	*L. siliceus organogenes*	Diatoms, needles of spongi, siliceous skeletons, etc. of organic origin, or parts of these. Particles < c. 0.1 mm
	Lc	*L. calcareus*	Marl, not hardened like calcareous Tufa; lime and the like. Particles < c. 0.1 mm
	Lf	*L. ferrugineus*	Rust, non-hardened. Particles < c. 0.1 mm
IV Argilla	*As*	*A. steatodes*	Particles of clay < 0.002 mm
	Ag	*A. granosa*	Particles of silt 0.06 to 0.002 mm
V Grana	*Gmin*	*G. minora***	Particles of sand 2–0.06 mm
	Gmaj	*G. majora***	Particles of gravel 60–2 mm

*New terms

done by using the plus sign (+). The number of traces must not exceed about one-eighth of the given deposit volume.

The deposit may also contain elements which are regarded as foreign components. Thus, the bones of a drowned animal, embedded in gyttja, can be considered a foreign element within the gyttja. Foreign elements are estimated separately, in proportion to the total of all components, including the foreign ones.

The use of formulae

The following are examples of deposits characterized by means of formulae:

Non-humified *Sphagnum* peat	$Tb^0 4$, or $Tb^0(Spha.)4$
Moderately humified *Sphagnum* peat	$Tb^2(Spha.)4$ or $Sh2, Tb^0 (Spha.)2$.
Calcareous gyttja with small seeds and mollusc shells (shells not included in the sum)	$Lc3, Ld^0 1, Dg+$, [Test. et Part. test.(mol.)1]
The same sediment with shells included in the sum	$Lc2, Ld^0 1,$ Test. et Part. test.(mol.)1, Dg+

Further examples are shown in Figure 12.1, where sediments deposited under different hydrological conditions are genetically classified and given a signature.

DEPOSIT SYMBOLS

A system of symbolic notation was proposed by Troels-Smith to express the following features:

(1) composition of deposits;
(2) degree of humification;
(3) boundaries between strata;
(4) presence and thickness of laminations;
(5) physical nature of deposits.

In forming symbols it is desirable to express as many features as possible at the same time, which is difficult in practice without loss of clarity. The symbols are prepared in such a way that it should be possible to express the features mentioned under points (1)–(4) in the same section drawing. The symbols used for the fifth property may be regarded as supplementary information, to be drawn separately.

The choice of symbols has been made with respect to previous symbols

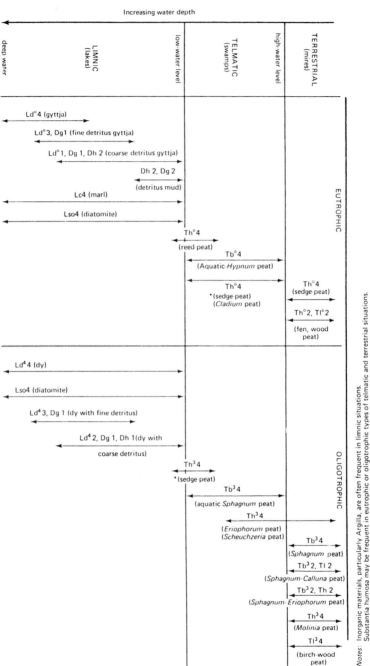

FIGURE 12.1.   Scheme illustrating a series of sediment types deposited under different hydrological conditions (from Birks and Birks, 1980). *Reproduced by permission of the authors and Edward Arnold*

introduced by von Post (e.g. von Post and Granlund, 1926), and with principles proposed by Faegri and Gams (1937).

In order to express the proportion of the components of the individual deposits, all symbols have been divided into four degrees of density corresponding to values (1)–(4) in the 5-class scale for characterization of deposits. The degree of humification is indicated by the linear thickness of the symbol; since the index zero expresses absence of humification there will be five degrees of thickness (cf. Table 12.1). Selected examples of element symbols are illustrated in Figure 12.2. A stratigraphical application is shown in Figure 12.3.

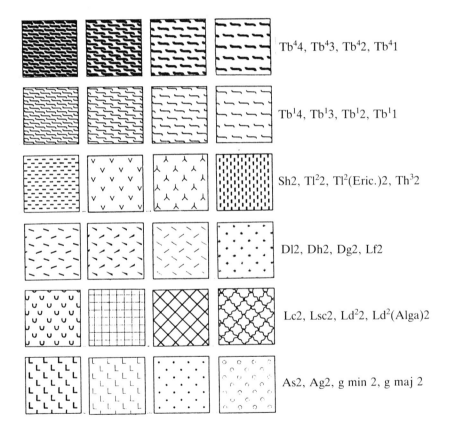

Tb^44, Tb^43, Tb^42, Tb^41

Tb^14, Tb^13, Tb^12, Tb^11

Sh2, Tl^22, Tl^2(Eric.)2, Th^32

Dl2, Dh2, Dg2, Lf2

Lc2, Lsc2, Ld^22, Ld^2(Alga)2

As2, Ag2, g min 2, g maj 2

FIGURE 12.2.   Examples of deposit element symbols according to the Troels-Smith system with the proposed modifications

MODIFICATIONS OF THE TROELS-SMITH SYSTEM

One of the advantages of the Troels-Smith system is its flexibility. Depending on the actual aim, the sediment diagnosis can be given very specifically or with only general information — for example, when characterizing simplified longer sections. Thus, instead of using the whole range of the 5-class scale, the alternatives are a 3-class scale (e.g. class values 0, 2 and 4) or a 2-class scale. In accordance with the diagnosis, the number of density symbols may be restricted (Aaby, 1979). Similarly, abbreviation of descriptions by omitting some of the physical properties in the diagnosis is often seen.

Since its appearance in 1955, the Troels-Smith system has proved to be a useful instrument for scientists on all continents. The system is universally used for describing organic deposits, whereas experience has shown that authors often use alternative methods for the characterization of mineral deposits. The reason for this is mainly that most international lithological standards have changed since 1955, when the Troels-Smith system appeared. Today each of the principal grain size fractions (silt, sand and gravel) is often divided into three sub-fractions (see Table 12.2). The Troels-Smith system is designed to be monopartite or bipartite for the fractions mentioned. In addition, only the fine and medium sized gravel is included in *Grana glareosa minora* (2.0–6.0 mm) and *Grana glareosa majora* (6.0–20.0 mm) whereas coarse gravel is undefined.

In order to standardize the different definitions of mineral particles, we intend to change the definitions, nomenclature and symbols for some size fractions (see Table 12.1), as follows:

Grana minora. Mineral particles 0.06 mm to 2.0 mm. The symbol is the same as that used for *Grana saburralia* (Troels Smith, 1955). *Grana minora* may be further subdivided.
Grana majora. Mineral particles 2.0 mm to 60.0 mm. The symbol is the same as that used for *Grana glareosa* minora (Troels-Smith, 1955). *Grana majora* may be further subdivided.

The symbol system for deposit elements was already very elaborate when the characterization system appeared, and only a few changes have been proposed since. The system allows additional symbols to be included, and a notable omission was a symbol for dwarf shrub roots. By using the common symbol for *Turfa lignosa*, it has been uncertain whether a stratum contained remains of trees and/or dwarf shrubs. For ecological reasons the Tl-symbol was accordingly not desirable for use on strata from open atlantic mires. An inverted-Y sign is proposed for *Turfa lignosa* originating from *Ericales* (Aaby, 1979); it is visually related to the Tl-symbol and it is part of the symbol often used for *Ericales* in pollen diagrams (Jonassen, 1950).

A further elaboration of the symbol system is proposed by including a sign for gyttja composed mainly of algae material. Algae gyttja is easily recognized by its characteristic appearance, jelly-like and with wine-red or green colours. The high content of hyaline matter produces unique properties, one of which is the high propensity to shrinkage when dried. This type is named *L. detrituosus (algarum)*. The sign is cross-hatching with regular bent lines (see Figure 12.2). The proposed symbol is similar to that of Berglund (1966) and Aaby (1979).

DRAWING PROCEDURE FOR TROELS-SMITH SYSTEM

Drawing of the larger sections of lithological columns in pollen diagrams is time-consuming and requires high-quality draughtsmanship. Attempts have been made (Aaby, 1979) to simplify the drawing procedure and the guidelines mentioned may still be recommended, if drawing is made by hand.

Recently, pre-designed deposit element signatures have been elaborated by Troels-Smith and Nilausen (granted by the National Science Research Council of Denmark) and are now available for non-commercial use. Master sheets of most symbols are distributed by the Department of Botany of the University of Bergen.* Copy-proofs are made from the masters and up to four copy-proof sheets may be superimposed, without seriously reducing the drawing quality (see Figure 12.3), as the sheets are transparent with glue on the back.

MICROSCOPIC SEDIMENT ANALYSIS

The Troels-Smith method as described here is a field method for a semi-quantification of the volumetric proportions of the main elements in lake and mire deposits. However, there is a need for a complementary analysis to quantify deposit elements more accurately. In the field it is already possible to apply a simple technique for differentiation of minerogenic particles and organic components of a sediment suspension in a graduated cylinder. In combination with macrofossil analysis it is also possible to get an estimate of the amount of coarse organic and minerogenic matter helpful for the characterization of the deposit (cf. Chapter 5). For a classification of the organic component, a complete macrofossil analysis is still the best method (cf. Chapters 27 and 28).

However, it is recommended that the Troels-Smith method be complemented by a simple microscopic sediment analysis for calculation of the relationship between different deposit elements. This has already been

*Postal address: P.O. Box 12, N-5014 Bergen, Norway.

Sermermiut B

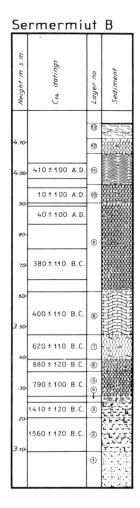

13. Recent moss peat with *Sphagnum*.
    Comp.: $Tb^0 3$, Tb^0sphagni 1, $Th^0 +$.

8. Light-brown moss peat.
   Nig. 2, strfc. 1, elas. 3–4, sicc. (1)–2.
   Colour: Light yellow-brown.
   Struc.: Felted.
   Comp.: Tb^1 sphagni 3, $Th^1 1$, Gmin. (+).

5. Humified peat.
   Nig. 3, strfc. 3, elas. 1, sicc. 2.
   Colour: Dark grey-brown, dries up brighter, more
   red–brown.
   Struc.: Laminar.
   Comp.: $Th^3 4$. Gmin. +, Ag+, Dl+, Dh+.

1. Stony sand.
   Nig. 2, strfc. 0, elas. 0, sicc. 2.
   Colour: Greyish-brown when wet, greyish-yellow
   when dry.
   Struc.:Heterogeneous, in places sandy and stony,
   in places clayey.
   Comp.: Gmin. 3, Ag1.

FIGURE 12.3. Diagnosis and lithological column produced from predesigned signatures, using copy-proof sheets. Description and diagnosis by S. Jørgensen and J. Troels-Smith (Fredskild, 1967). The original diagnoses have been changed in accordance with the modifications proposed

proposed by the Swedish geologist and palaeolimnologist Gösta Lundqvist, and he called the method 'Strukturanalyse' (Lundqvist, 1926, 1927). In a large number of papers published between 1925 and 1940 Lundqvist described the sediment characteristics of Swedish lakes by applying the Strukturanalyse method systematically in order to correlate lake sediments both within and between lakes. His method was more objective than previous ones, in which the frequencies of different elements were only estimated. He

TABLE 12.2. Definition of mineral particle fractions according to various characterization systems. Lithological standard refers to international classification systems, cf. the geotechnical system proposed in Sweden (Karlsson *et al.*, 1981)

Lithological standard				Troels-Smith's system	
				Modified	Original
Clay		<0.002	mm	*Argilla steatodes*	*Argilla steatodes*
Silt	fine,	0.002–0.006	mm	*Argilla granosa*	*Argilla granosa*
	medium,	0.006–0.02	mm		
	coarse,	0.02–0.06	mm		
Sand	fine,	0.06–0.2	mm	*Grana minora*	*Grana arenosa*
	medium,	0.2–0.6	mm		*Grana saburralia*
	coarse,	0.6–2.0	mm		
Gravel	fine,	2.0–6.0	mm	*Grana majora*	*Grana glareosa minora*
	medium,	6.0–20.0	mm		*Grana glareosa majora*
	coarse,	20.0–60.0	mm		Not defined

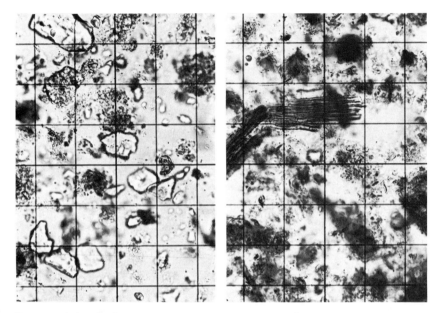

FIGURE 12.4.   Sediment suspensions, clay gyttja (Ld[0]1, As2, Ag1) of Alleröd age to the left, clayey fine detritus gyttja (Ld[0]3, As1, Dg+) of Preboreal age to the right. The net micrometer makes it possible to calculate the area of each deposit element. This is repeated for several eye-fields in order to complete the microscopic sediment analysis. *Photo by Thomas Persson*

applied the method to lake sediments only, but it is also useful for peat deposits. He also constructed a 'Struktur diagramm' (a cumulative histogram) to illustrate the lithostratigraphy of a sediment column. To date, only a few scientists have applied this method, and in a more or less modified form (e.g. Iversen, 1954; Fjerdingstad, 1954; Berglund, 1966).

The main principle of the method is to calculate the frequencies of different element areas. Lundqvist made slides from unit volumes (2 mm^3) of the sediment, in order that a statistical area analysis might give approximate volume-based relationships between different elements, such as mineral particles, detritus, different organism groupings etc. It is often difficult to sample small sediment and peat volumes, so the following method is proposed:

(1)   A water–glycerol suspension of a small quantity of sediment is made on an object glass. For peat deposits, heating in KOH (the so-called alkali preparation) is needed to obtain a suspension, but this means that humus colloids are dissolved. To facilitate differentiation between fine detritus and mineral particles, one may add KI$_3$ (iodine-potassium iodide) to the suspen-

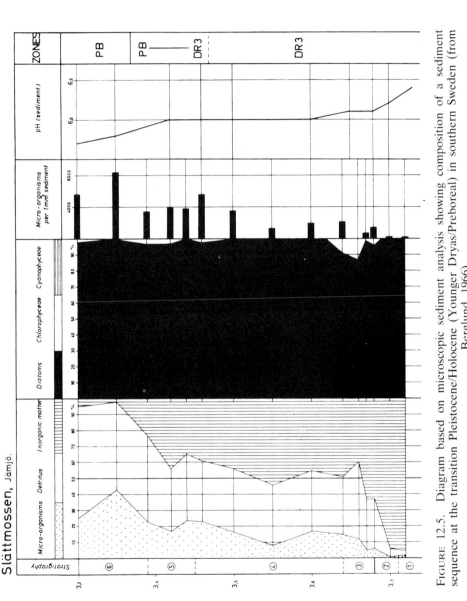

FIGURE 12.5. Diagram based on microscopic sediment analysis showing composition of a sediment sequence at the transition Pleistocene/Holocene (Younger Dryas/Preboreal) in southern Sweden (from Berglund, 1966)

KYLEN LAKE   Sediment components

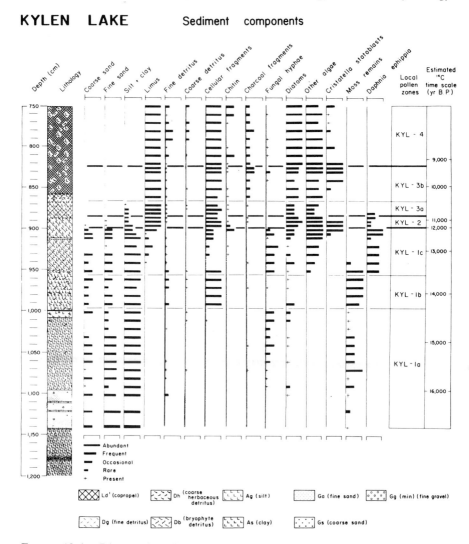

FIGURE 12.6.   Diagram based on a simplified microscopic sediment analysis showing the estimated deposit elements in a sediment column covering the Late Wisconsin and the Early Holocene in Minnesota. Troels-Smith's symbols are applied in the lithological column (from Birks, 1981). *Reproduced by permission of Academic Press Inc*

sion. The quantity of calcareous mineral particles may be estimated if an analysis is made before as well as after hydrochloric acid treatment. Normal cover glasses (e.g. 24 × 32 mm) are used.

(2)   Microscopic analysis is carried out at ×250 magnification (Figure 12.4). By using a net-micrometer (100 μm mesh) placed in one of the oculars

the degree of cover for each element is calculated. The number of squares covered gives an approximate area. This calculation is made for 20–30 eye fields for each sample. The results are added and expressed as a percentage of the total sum.

(3) The results obtained from the analysed sediment column can be summarized in a total diagram with the main deposit elements differentiated, complemented by histograms for individual elements. Iversen (1954) simplified the presentation of analyses from the classical Bølling section by a diagram illustrating the relationships between micro-organisms, detritus and minerogenic matter. A similar diagram is reproduced in Figure 12.5, showing the sediment changes at the Pleistocene/Holocene boundary in South Sweden (Berglund, 1966).

An alternative method has been applied by Birks (e.g. 1981). A water suspension of each sample was examined in the microscope and the relative abundances of different elements estimated on a 5-point scale. The results were illustrated in a special sediment diagram (Figure 12.6). This is a useful method, perhaps less time-consuming than the one previously described.

These methods may be applied for lake sediments as well as for peat deposits. However, and particularly for more compact peat layers, it is essential to use the peat microtome technique for preparation of slides with peat sections (cf. Cohen and Spackman, 1972). This method implies paraffin embedding of 1 cm^3 peat cubes, which are then cut into sections about 5 μm thick.

For minerogenic sediments — clay to clay gyttja — it is often important to analyse the particle size distribution (granulometry). Conventional methods include sieving and sedimentation analysis. Less time-consuming, and suitable for long series of samples, is the use of a sedigraph analyser, where minerogenic particles are measured by an X-ray technique.

PHOTO DOCUMENTATION

Photographic documentation of fresh peat walls or sediment cores is important for a detailed stratigraphic description. Layer boundaries and colour differences are most objectively documented in this way. In some cases it is important to obtain photographs from fresh as well as from oxidized sediment cores, e.g. FeS-coloured lake sediments. X-radiographing may also be helpful. This applies, for example, to laminated sediments (see Chapter 17).

REFERENCES

Aaby, B. (1979). Characterization of peat and lake deposits. In: *Palaeoecological Changes in the Temperate Zone in the Last 15,000 Years: Subproject B, Lake and Mire Environments*, Project guide vol. I (Ed. B. E. Berglund), Dept. Quat. Geol., Lund University, 1–140.

Berglund, B. E. (1966). Late Quaternary vegetation in eastern Blekinge, south-eastern Sweden: a pollen-analytical study. I. Late-glacial time. *Opera Bot.*, **12**, 1–190.

Birks, H. J. B. (1981). Late Wisconsin vegetational and climatic history at Kylen Lake, north-eastern Minnesota. *Quaternary Research*, **16**, 322–355.

Birks, H. J. B., and Birks, H. H. (1980). *Quaternary Palaeoecology*, Edward Arnold, London, pp. 1–289.

Cohen, A. D., and Spackman, W. (1972). Methods in peat petrology and applications to reconstruction of palaeoenvironments. *Geol. Soc. Ann. Bull.*, **83**, 129–142.

Faegri, K., and Gams, H.-(1937). Entwicklung und Vereinheitlichung der Signaturen für Sediment- und Torfarten. *Geol. Fören. Förhandl. Stockholm.*

Fjerdingstad, E. (1954). The subfossil algal flora of the Lake Bølling Sø and its limnological interpretation. *Kgl. Danske Vidensk. Selsk. Biol. Skr.*, **7**, 56pp.

Fredskild, B. (1967). Palaeobotanical investigations at Sermermiut, Jakobshavn, West Greenland. *Medd. Grönland*, **178**, 1–54.

Iversen, J. (1954). The late-glacial flora of Denmark and its relation to climate and soil. *Danm. Geol. Unders.*, **2**, 87–119.

Jonassen, H. (1950). Recent pollen sedimentation and Jutland heath diagrams. *Danske Bot. Arkiv.*, **13**, 1–168.

Karlsson, R., Hansbo, S., and the Swedish Geotechnical Society (1981). *Soil classification and identification*. Swedish Council for Building Research, S:t Göransg. 66, S-112 33 Stockholm, Sweden, 49pp.

Lundqvist, G. (1926). En metod för mikroskopiska sedimentanalyser. *Geol. Fören, Förhandl.*, Stockholm, **48**.

Lundqvist, G. (1927). *Bodenablagerungen und Entwicklungstypen der Seen.* Die Binnengewässer II (Ed. A. Thienemann), Stuttgart, pp. 1–124.

Munsell Soil Color Charts. Munsell Color, Macbeth Division of Kollmorgen Corp. 241 North Calvert Street, Baltimore, Maryland 21218, USA.

Overbeck, F. (1975). *Botanisch-Geologische Moorkunde*, Karl Wachholtz Verlag, Neumünster.

von Post, H. (1862). Studier öfver Nutidens koprogena Jordbildningar, Gyttja, Dy, Torf och Mylla. *Kgl. Svenska Vetenskapsakad. Handl.*, **4**, 1–59.

von Post, L. (1924). *Das Genetische System der Organogenen Bildungen Schwedens*, Comité internat. d. Pedologie IV, communication 22.

von Post, L., and Granlund, E. (1926). Södra Sveriges torvtillgångar I, *Sver. Geol. Unders.*, C **335**, 1–127.

Schneekloth, H., and Schneider, S. (1972). Vorschlag zur Klassifizierung der Torfe und Moore in der Bundesrepublik Deutschland. *Telma*, **2**, 57–63.

Troels-Smith, J. (1955). Karakterisering af løse jordarter (Characterization of unconsolidated sediments). *Danm. Geol. Unders.* IV, **3**, 1–73.

13

Core correlation and total sediment influx

J. A. Dearing

*Department of Geography,
Coventry Polytechnic, U.K.*

INTRODUCTION

Broadly speaking, palaeoecology remains a descriptive science where attempts to identify and quantify past ecological processes are faced with innumerable problems. The representation of palaeoenvironments by fossil organisms is virtually always partial, and reconstruction is characterized by its imprecision. Many lake basins, however, efficiently trap inflowing particulates and certain dissolved elements which can theoretically be linked to mean rates and forms of 'palaeoprocesses' in the catchment. But while lake sediment *sequences* have been widely used to identify qualitative changes in the biotic and abiotic components of palaeoenvironments, few studies have considered quantitative changes based on *total* sediment analyses (Oldfield, 1977).

The problems of reconstructing and delimiting local plant communities derived from lake sediment pollen records (Oldfield, 1970) make total sediment analyses particularly relevant to quantitative studies of *abiotic* processes acting within a lake catchment. Studies of nutrient flow in modern terrestrial ecosystems which have been subjected to change indicate that inputs from atmospheric and weathering sources vary little in comparison to outputs in the form of particulate erosion and dissolved elements in runoff (Bormann *et al.*, 1974). Data for particulate and dissolved losses based on total sediment analyses can therefore provide useful insight into the dynamics and stability of palaeoenvironments. Evaluation of erosion rates and nutrient flows in palaeoenvironments, although unavoidably imprecise, is necessary if palaeoecologists are to replace conventional palaeoecological explanation based largely on inference from contemporary ecosystems and to approach the discovery of causative linkages between organisms and environment (Birks

247

and Birks, 1980). Such linkages are perhaps most necessary in studies aimed at assessing the impact of human activity on past biological and hydrological systems.

The practicalities of total sediment analysis are daunting in their complexity, but recent work (Bloemendal *et al.*, 1979; Davis, 1976; Dearing *et al.*, 1981; Foster *et al.*, 1985; O'Sullivan *et al.*, 1982) suggests that total influx values are obtainable which, though crude in comparison with data from contemporary ecosystem studies, provide a sound basis for examining the internal dynamics of past environments. What follows is a summary of methods showing how sediment volumes and their qualities can be estimated in given periods of time, and includes as an illustration a case study in which the combined use of total sediment analysis, pollen and diatom sequences and documentary evidence has allowed nutrient flow and soil erosion to be partially quantified for the past 350 years.

CONSIDERATIONS

Total sediment volumes have been estimated in reservoirs by comparing periodic measurements of water depth with original basin form (eg. Borland and Miller, 1958; Pais-Cuddon and Rawel, 1969; Rapp *et al.*, 1972). Such studies provide mean influx data over the period in which the reservoir has trapped sediment, usually a period in the order of decades. The effects of road construction and forest fire on sediment yield were inferred from reservoir sediment studies in California by Anderson (1974). Sediment thicknesses can also be ascertained from seismic-reflection profiles which identify pronounced changes in sediment composition — for example, the transition between minerogenic late-glacial sediments and postglacial gyttja. But where studies require sediment volume calculations over relatively short periods (10–100 years) and over long timespans, these techniques are usually inappropriate. Calculations are then more likely to be based on sediment thickness found between stratigraphic or dated horizons in sediment cores. The non-uniform nature of sedimentation in many, if not all, lakes negates the extrapolation of sedimentation rates in single or few cores to the whole lake (Dearing 1983), and hence sediment volumes are estimated from mean sediment thicknesses found in many cores. *Core correlation* is the commonly used method for identifying synchronous sediment levels in different cores from which sediment thicknesses can be measured, and utilizes sediment properties which should ideally fulfil the following conditions:

(1) *areal continuity*, i.e. the 'sediment property' should be deposited in similar proportions to other sediment constituents over the whole lake bed surface so that suites of downcore sediment records are in parallel;

(2) *areal synchroneity*, i.e. the 'sediment property', once deposited, should

either be persistent and immobile, or (excepting properties also used in absolute dating, e.g. radioisotopes) change in respect of its form or position at a relatively constant rate over the lake bed surface, thus making fluctuations in parallel records synchronous and correlatable.

While internal mixing processes in small (<100 ha) lakes can be assumed to disperse more or less inflowing suspended particulates (including mineral matter, organic matter, planktonic forms, pollen) and dissolved elements throughout the water body, the final concentration in sedimented material will, as explained in Chapter 2, also depend on factors such as resuspension/deposition cycles, thermal stratification and the deposition of coarse material from high-energy flow (Brush and Brush, 1972; Davis, 1968, 1973; Davis and Brubaker, 1973; Davis *et al.*, 1971; Flower, 1980; Holmes, 1968). In addition, varying rates of bioturbation, sediment slumping and diagenesis may alter the parallelism in sediment records. Many sediment properties may therefore not fulfil the above conditions and core correlations based on different lines of evidence should be sought whenever possible. Environmental gradients of limnological phenomena such as redox potential in bottom water, post-depositional disturbance and sediment mixing, are generally steepest at shelving lake margins. Consequently correlations between cores from littoral and profundal zones are likely to be the most problematic. It follows that the most suitable lakes for obtaining accurate assessments of total sediment volumes will be steep-sided and circular where the area of shallow margins is small in comparison with the rest of the lake bed area. Very small, shallow (<2–3 m) or highly exposed lakes may possess sediments that have all been subjected to some degree of disturbance as a result of intense and widespread water turbulence, and should be avoided (see Chapter 2).

SAMPLING

A sampling design for correlation studies should be based on preliminary analyses of cores taken from potentially different depositional environments (e.g. deep holes, extreme ends of long lakes). It may then be necessary to stratify the sampling according to the area of each depositional environment. In 'simple' basins it may be sufficient to sample on a systematic grid basis (see below). The difficulties in locating sampling points on open water effectively excludes the use of random sampling strategies, except where automated techniques for sample site positioning (Battarbee *et al.*, 1983) are available or where ice-cover remains long enough for multiple coring to be undertaken. Floating rope stretched between shores and/or securely anchored buoys have been successfully used to mark out transects up to 500 m long in calm conditions, sampling points being marked by anchored buoys, or by poles pushed into the sediment. The density of sampling depends on the lake area,

TABLE 13.i.   Coring densities used in published total sediment analyses

Site	Lake area (ha)	Coring density	Reference
Llyn Goddionduon, N. Wales	6.2	1 per 0.0035 ha/0.04 ha	Bloemendal *et al.*, 1979
Frains Lake, Michigan	6.7	1 per 0.3 ha	Davis, 1976
Havgårdssjön, Sweden	55	1 per 1.0 ha	This chapter
Llyn Peris, N. Wales	20[1]	1 per 1.3 ha	Dearing *et al.*, 1981
Merevale lake, Warwickshire, England	6.5	1 per 0.08 ha	Dearing *et al.*, 1984
Loe pool, Cornwall, England	44	1 per 1.0 ha	O'Sullivan *et al.*, 1983

[1]Refers to lake bed area over which sediment was deposited.

the irregularity of sedimentation, the cost and time available — hence no firm advice can be given. Published studies have used various sampling densities (Table 13.1).

CORE CORRELATION METHODS

Visible stratigraphy

Visible stratigraphic markers can be classified as regular or irregular, according to their temporal variability.

Regular

Annually laminated sediments perhaps offer the most accurate method for correlating cores. Unfortunately their occurrence does not appear to be widespread, and with few exceptions (e.g. Simola, *et al.*, 1981) they are mainly restricted to deep meromictic or holomictic lakes with small catchments in areas of continental climate. The various types of laminations and sampling methods are described in Chapter 17. Bioturbation and turbulence in shallow areas of lakes may destroy the continuity of laminations; hence flat-bottomed and steep-sided lakes offer the greatest potential for correlation studies. The annual nature of laminations can be confirmed by microstratigraphical analysis (Simola, 1979).

Irregular

Changes in sediment colour or texture frequently occur and can provide a means for correlating synchronous levels. Distinct colour changes may be caused by a shift in the balance of allochthonous or autochthonous particulate material reaching the lake bed, such as the effect of short-lived influx events caused by environmental disturbance (e.g. Dearing, 1979; Digerfeldt *et al.*, 1975); but they may also be the result of changes in water chemistry, such as the black colouration caused by iron sulphide precipitation in anoxic bottom waters (e.g. Battarbee and Digerfeldt, 1976; Pennington *et al.*, 1976). Textural changes can be identified by classifying sediment according to the schemes presented in Chapter 12, and have been used by Digerfeldt (1972a, 1976) in core correlation. Two kinds of problem can hinder the use of visible marker horizons in correlation. Firstly a colour change boundary may be smeared from coring or extrusion, or may be gradual and thus difficult to correlate. The former difficulty can be overcome by using a non-piston type corer, or by noting the colour change in the central portion of sediment slices. Secondly, spatial variations in the deposition of allochthonous matter (e.g. Dearing, 1983) or in chemical conditions may affect the continuity of a marker horizon. Thus coarse textured bands in littoral areas may be of local distribution and not synchronous with similarly positioned bands of silt in less marginal cores (see also below), and onset of iron sulphide production may be non-synchronous according to the spatial extent and intensity of anoxic conditions at the mud–water interface (Figure 13.1). Tephra layers are good sediment properties for core correlation, especially in large lakes (Watkins *et al.*, 1978); their largely atmospheric origin ensures widespread deposition, and their appearance and distinct magnetic properties makes them easy to detect (Oldfield *et al.*, 1980; Figure 13.2). X-ray radiography and densitometry have been used to identify stratigraphic changes in unextruded cores (Koivisto and Saarnisto, 1978).

Other stratigraphies

This section deals with stratigraphies obtained through chemical, physical or magnetic analyses, or by microscopic examination. Features in sediment records on which correlations can be based are of three types: peaks and troughs, appearance and disappearance, and 'assemblage' boundaries incorporating the fluctuations of more than one sediment record. Peaks and troughs which are broad may give correlations with large errors, and those which are of relatively small amplitude may be artefacts of spatially variable processes acting at or after the time of deposition. Dramatic peaks and troughs, unambiguous first and last records, and distinct assemblage zone boundaries provide the most useful features for correlation.

Micro- and macrofossil records

Microfossil analyses from more than one core have usually been undertaken to identify water level fluctuations or spatial variations in sedimentation rates.

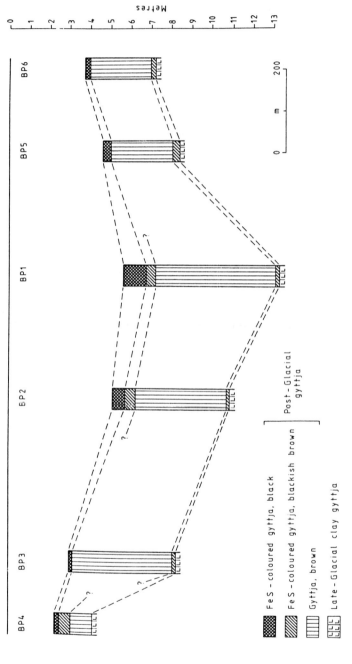

FIGURE 13.1. Stratigraphic section from Växjösjön, Sweden. Note the poor correlation based on FeS colourations (Battarbee and Digerfeldt, 1976). *Reproduced by permission of E. Schweizerbark'sche Verlagsbuchhandlung*

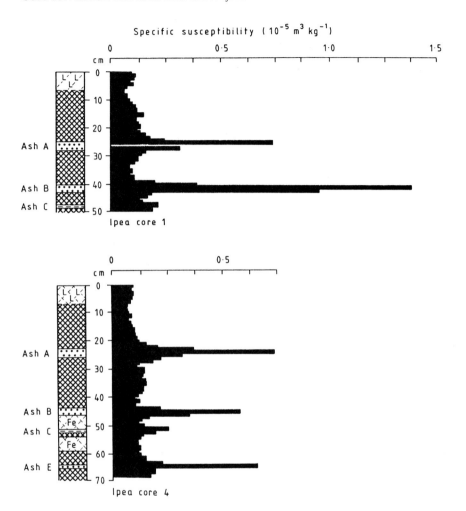

FIGURE 13.2.   Stratigraphy and specific magnetic susceptibility of single samples from
Lake Ipea, Papua New Guinea, cores 1 and 4 (Oldfield *et al.*, 1980). *Reproduced by
permission of Blackwell Scientific Publications Ltd*

Core correlations are based on either microfossil zone boundaries in pollen or
diatom records (e.g. O'Sullivan *et al.*, 1973; Figure 13.3(*a*)), or in fluctuations
of the records of one taxon; the appearance and peak frequency values of
Ambrosia (ragweed) pollen have been widely used as marker horizons in
recent sediments from lakes in eastern U.S.A. (e.g. Bortleson and Lee, 1976;
Bruland *et al.*, 1975; Davis, 1976; Maher, 1977). Water level studies make use
of correlations based on pollen zone boundaries to identify hiatuses in
sediment deposition at the lake margin (e.g. Digerfeldt, 1971, 1972b, 1974,

(a)

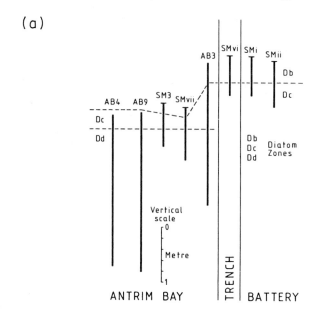

(b)

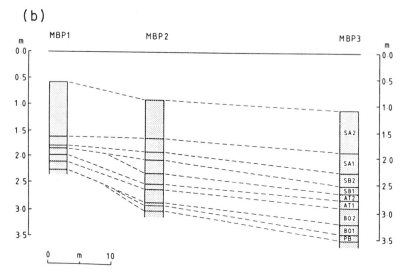

FIGURE 13.3.   Core correlations based on microfossil zone boundaries. (a) Diatom assemblage zones in Lough Neagh, N. Ireland, cores sampled over distances of up to about 9 km (O'Sullivan *et al.*, 1973). *Reproduced by permission of Blackwell Scientific Publications Ltd.*
(b) Pollen assemblage zones in a section of profiles from Lake Immeln, Sweden (Digerfeldt, 1974). *Reproduced by permission of the Publishing House of the Swedish Research Councils*

1977; Figure 13.3(*b*)) and illustrate how cores on transects through marginal zones can be correlated, albeit by rather time-consuming analytical methods. An additional number of studies have used pollen assemblage zones to correlate cores for purposes of extending sequences (e.g. O'Sullivan, 1975). Numerical methods for zoning microfossil assemblages (see Chapter 37) seem particularly attractive where a zone boundary would otherwise be dictated by the fluctuations in the frequences of a single taxon (e.g. the decline in elm pollen frequencies). Macrofossils, because of their larger size, are less likely to be distributed over the whole lake bed and might have to be considered as being of lesser importance in correlation studies, except perhaps in terms of their contribution to total organic matter (see below). Fossils of many benthic faunas are likely to provide erroneous correlations owing to the frequently discrete nature of the living populations.

Chemical, organic-matter and particle-size records

Spatial variations in redox potential and pH conditions which may affect the parallelism of chemical profiles present the greatest problem when using such data for correlation. Additional problems may arise through irregular relationships between particle size and chemical composition (Thompson and Morton, 1979), and the local variability of sediment chemistry caused by inputs of sediment from different lithologies. Unfortunately most studies of chemical profiles are based on the results from a single core, and assessment for core correlation is reliant on the few published results from multiple core analysis (e.g. Bortleson and Lee, 1975) and theoretical considerations. Stable elements possibly include the alkali metals existing as oxides in mineral particles, and heavy metals which tightly adsorb to fine particulates. In a small shallow lake in north Wales core correlations have been successfully constructed using records of total Fe and Mn (Bloemendal, personal communication). Elements reaching the lake primarily in rain-out are likely to have a wide distribution, and in the case of certain elements or chemical species (e.g. Pb, ^{137}Cs) their presence in recent sediments may register occurrence in the atmosphere as pollutants and thus provide both correlatable and dateable marker horizons (Appleby and Oldfield, 1979; Pennington *et al.*, 1976; Figure 13.4).

Sediment dating based on the radioisotope decay of ^{210}Pb or ^{14}C is not likely to be utilized in the correlation of large numbers of cores owing to the prohibitive cost, and still unresolved problems of interpretation (Appleby and Oldfield, 1978; Oldfield *et al.*, 1978).

There is evidence to suggest that in many lakes organic matter determinations can provide a sound basis for core correlation. Analyses of total carbon (Pennington *et al.*, 1976), total masses of organic matter derived from macrofossil analyses (Digerfeldt 1972b; Figure 13.5(a)), and loss-on-ignition (Davis, 1976; Tolonen *et al.*, 1975; Figures 13.5(b) and (c)) all indicate a

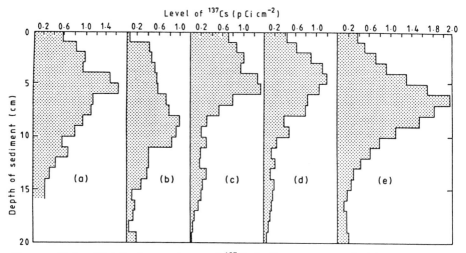

FIGURE 13.4. Distribution patterns of ^{137}Cs in five cores sampled from Blelham Tarn, English Lake District: peak levels attributable to a date of 1963 A.D. (Pennington *et al.*, 1976). *Reproduced by permission of Blackwell Scientific Publications Ltd*

potential for core correlation. In this respect loss-on-ignition determinations which are rapid, cheap and easy to perform appear particularly useful and worthy of further evaluation.

Particle-size records may be used for core correlation in lakes or over certain lake zones where allochtonous sediments are well mixed before sedimenting. Problems arise with correlations between profundal and littoral cores unless an assumption can be made that the relationship between particle size and distance from inflow or shore is inverse and more or less linear. Hence percentage peaks of fine silt in profundal cores could be linked to coarse silt in littoral cores. The synchroneity of coarse layers in littoral areas may be affected by sediment resuspension following water level fluctuations. Wet-density values, though obtained by means of routine sediment analysis, are frequently unvarying below the uppermost few centimetres of labile sediment and will normally be of little use in core correlation (Foster *et al.*, 1985).

Magnetic records

Core correlation may be made on the basis of both non-naturally remanent properties (e.g. magnetic susceptibility κ and χ, IRM, SIRM, ARM) and naturally remanent properties (i.e. declination, inclination, NRM). Definitions, details and units of measurements and further explanation are given in Chapter 15, All the measurements are rapid, cheap and non-destructive, and in the case of κ, NRM and relative declination can be measured 'continuously' on

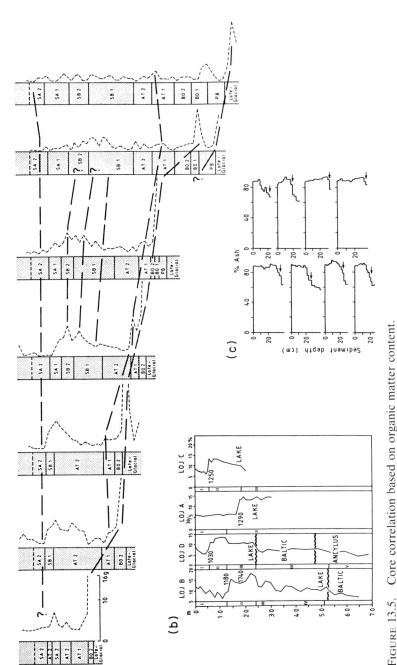

FIGURE 13.5. Core correlation based on organic matter content.
(a) Total organic mass (macrofossil analyses) and pollen zone boundaries in cores from Lake Trummen, Sweden (after Digerfeldt, 1972b). *Reproduced by permission of Gleerup Publ.*
(b) Loss-on-ignition profiles, radiocarbon dates and stratigraphy for cores from Lojärvi, Finland (Tolonen *et al.*, 1975). *Reproduced by permission of the Finnish Botanical Publ. Board.*
(c) Loss-on-ignition profiles (% ash) and pollen marker horizons (←) for cores from Frains Lake, Michigan (Davis, 1976). *Reproduced by permission of the Foundation for Environmental Conservation*

unextruded cores at intervals down to 1 cm. One drawback with continuous measurements of κ is poor definition of the extreme 1–5 cm of a core where the measuring bridge scans both sediment and empty tube or air, or in the case of very recent sediments scans labile sediment with a characteristically high water content. Successful core correlations based on continuous κ meaᶜurements were made by Bloemendal *et al.* (1979), Thompson *et al.* (1975) (see Figure 15.3) and are used in the case study outlined below. Palaeomagnetic directional data (declination and inclination) are not well preserved in coarse or disturbed (e.g. bioturbated) sediments, and correlations using relative declination have been most successful in deep oligotrophic lakes such as those of the English Lake District (see Chapter 15 for further discussion). Specific non-naturally remanent properties (especially χ) have been used in correlation both singly (Dearing, 1979; Dearing *et al.*, 1981, 1984; Thompson, 1975; Figure 13.6) and in combination as ratios (especially χ/SIRM) (Bloemendal *et al.*, 1979; Figure 13.7). As described above, Oldfield *et al.* (1980) have illustrated the use of χ in correlating tephra layers (see Figure 13.2).

The problem of identifying synchroneity between χ profiles increases where postulated correlation levels have different particle size distributions. Björck *et al.* (1982) have shown that for minerogenic particles derived from igneous and metamorphic rocks there is a non-linear relationship between χ and particle size. Only for fine sediments (generally <150 μm or 63 μm) can a positive and linear relationship be assumed. Correlations between cores of fine sediment from deep water zones are likely to be more easily identified than between cores in which some of the sediments to be correlated are dominated by coarser particles (>150 μm). Consequently correlation with cores from littoral areas or between cores from small shallow lakes may be difficult.

The need for independent sets of analyses to corroborate core correlations of any kind is obvious, and focuses attention on correlation methods such as loss-on ignition and magnetic analyses which are rapid and economically feasible for a large number of cores. Thompson and Edwards (1982) have convincingly demonstrated the use of multiple sediment analysis in providing accurate core correlation with fine resolution (Figure 13.8).

TOTAL INFLUX CALCULATIONS

Sediment influx

Correlation levels are dated using radioisotope methods, palaeomagnetism, microfossil assemblage chronozones or dated laminations. Total wet-sediment volumes per unit time are calculated using planimetric methods (Dearing *et al.*, 1984) or by multiplying *mean* wet sediment volume by the number of years represented by the sediment thickness. Calculation of mean wet-density values by oven drying (110°C) constant volumes of wet sediment enables total

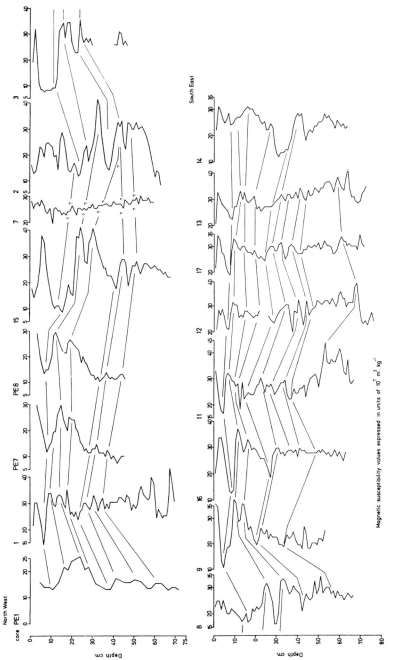

FIGURE 13.6.   Core correlations between sixteen 1 m cores from Llyn Peris, N. Wales, based on single magnetic susceptibility values (Dearing *et al.*, 1981). *Reproduced by permission of Academic Press Inc*

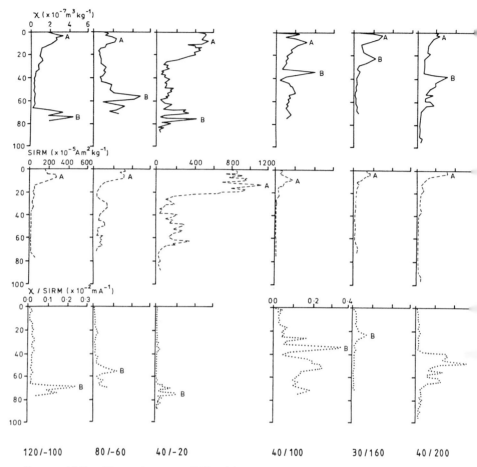

FIGURE 13.7.   Magnetic susceptibility (χ), saturation isothermal remanence (SIRM) and χ/SIRM profiles in six cores from Llyn Goddionduon, N. Wales. Postulated correlations between χ peaks A and between χ peaks B are confirmed from SIRM values and χ/SIRM ratios (Bloemendal *et al.*, 1979). *Reprinted by permission from Nature, 280, 50–53. Copyright (c) 1979 Macmillan Journals Ltd*

wet-sediment volumes per unit time to be converted to total dry-sediment mass per unit time. This value represents material of allochthonous sources, inorganic and organic, and may include products of authigenesis. In calculating total sediment influx, consideration should be given to the extent of deposition in deltas and reedswamps, to the magnitude of particulate losses through the outflow, and to the possible occurrence of redeposited sediment introduced to central zones during episodes of water level fluctuations (see Chapter 2 for further discussion).

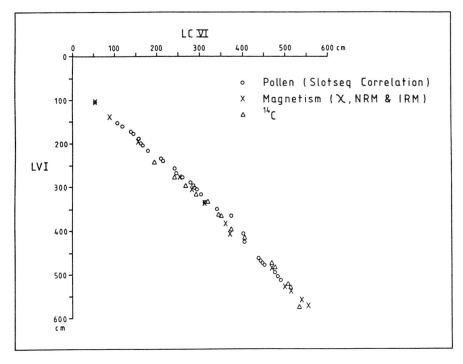

FIGURE 13.8. Multiple core correlations between two cores (LVI and LCVI) from Lough Catherine, N. Ireland, based on pollen, magnetic and radiocarbon analyses (Thompson and Edwards, 1982). *Reproduced by permission of Universitetsforlaget, Oslo*

Catchment-derived sediment yields

Conversion of annual sediment mass values to sediment yields normally involves the making of gross assumptions about the proportion of allochthonous material in the sediments. Loss-on-ignition at 500°C or 630°C for 30 minutes may be used to estimate the inorganic content; but weight loss may arise from the dehydration of clay minerals and the destruction of carbonates as well as from the ignition of organic material. Ash weight was used by Davis (1976) to estimate erosion rates of inorganic material. Sediment yield estimates based on loss-on-ignition values may also be erroneous due to the inclusion of precipitated minerals at the mud–water interface or in organisms; diatom silica made up about 30% of the sedimenting material by weight in Lough Neagh, Northern Ireland (Flower, 1980). Total volume influx of diatoms can be assessed using the methods outlined in Chapter 26, and Davis (1976) used values of concentrations of frustules and sediment accumulation rates to gain an impression of the relative down-core changes in diatom silica content. Alternatively, diatom silica may be estimated by analysis of the alkali-soluble component (10% $NaCO_2$) after pretreatment with hydrogen peroxide and conc. hydrochloric acid (Flower, 1980).

Estimates of allochthonous inputs will be difficult in lakes where levels of chemical precipitation are high — for example, lakes in calcareous basins. The estimated value of allochthonous inorganic sediment divided by the drainage basin area gives an estimated rate of sediment yield. Loss-on-ignition data can be used to estimate allochthonous carbon influx in oligotrophic lakes where it can be assumed that levels of autochthonous carbon remaining in buried sediment are low (Mackereth, 1966). Present-day monitoring of erosional rates and processes in the drainage basin provides information about the spatiality of erosion, and palaeohydrological interpretation of sediment yield should consider the points discussed in Chapter 2.

Chemical influx

Values for total allochthonous sediment influx through time allows (in theory) the calculation of total chemical influx through time. The choice of chemical elements which can be accurately studied in terms of total influx depends on the history of redox and pH conditions in the lake, both at the mud–water interface and in the sediment column, and on the levels of biological uptake (cf. diatom silica) and losses through the outflow.

Suitable chemical elements for influx calculations are those which tightly adsorb to clay particulates, have low solubility, and which can be assumed to remain stable in the sediments over long periods. Phosphorus seems to possess these properties and is of great ecological interest. It seems that even where reducing conditions at the mud–water interface effect a dissolution of P in bottom water, co-precipitation with Fe^{2+} or complete uptake by plankton in all but the most P-loaded lakes will ensure subsequent precipitation in the sediments. Other suitable elements are the alkali metals K, Mg and Na existing in mineral particles; however, calculation of influx may be of limited ecological value as the dissolved nutrient phases of those elements are not likely to be completely precipitated in sediments but pass through the outflow. Heavy metals are apparently stable in all but the most intensely reducing conditions, but influx calculations should take into account atmospheric fallout of pollutants as an additional source (Livett *et al.*, 1979). Estimates of total chemical influx have been made for P, Cu and organic matter at a Welsh upland site (Dearing *et al.*, 1981), and for P at a southern Swedish site (see below).

CASE STUDY OF SEDIMENT INFLUX

The following is an outline of preliminary work (Oldfield *et al.*, 1983) aimed at quantifying rates and forms of influx over time from total analyses of recent lake sediments. The site (Havgårdssjön, Scania) comprises a subcircular

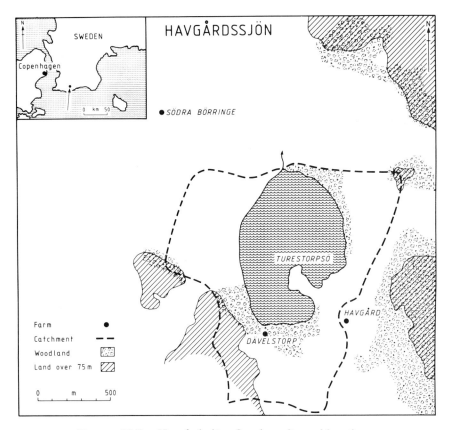

FIGURE 13.9. Havgårdssjön, Sweden: site and location

kettle-hole lake (max. depth 5 m, area 55 ha) set within the low undulating landscape of Scania in Sweden (Figure 13.9). Arable farmland with some improved pasture dominates the land-use of the drainage basin (area 141 ha), a situation largely unchanged over the last 350 years. One metre cores of sediment were sampled on a 100 m × 100 m grid using a modified Livingstone type corer from a boat (Figure 13.10 inset). Clear plexiglas tubes were used to check that mud–water interfaces had been sampled undisturbed, and perforated or notched rubber bungs were slipped down on to the sediment surface to ensure no vertical mixing of the sediment during transport and measurement.

Continuously measured magnetic susceptibility (κ) at 2 cm intervals was undertaken on about 55 cores, the time taken being less than two working days. The similarity in major susceptibility fluctuations (Figure 13.10) indicated their potential use for correlation, especially as one major trough coincided with a stratigraphic change. Three levels were initially correlated for sediment influx

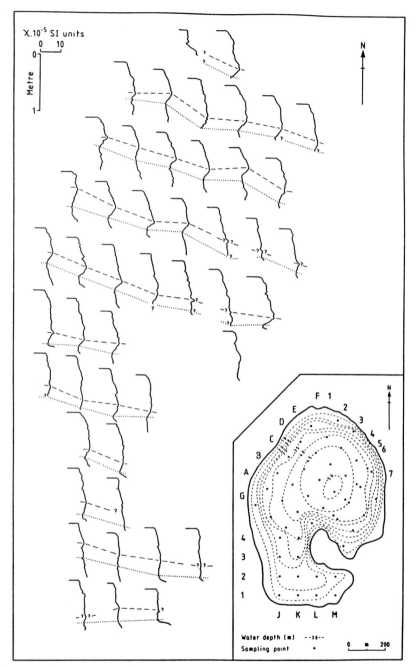

FIGURE 13.10. Magnetic susceptibility profiles of forty-seven 1 m cores from Havgårdssjön, showing postulated correlations between the main peak (--) and between the main trough (....). A stratigraphic change occurs in many cores at the level of the main trough. Inset: bathymetry and sampling grid (Dearing, 1983). *Reproduced by permission of W. Junk Publ*

calculations. A timescale for the susceptibility fluctuations was constructed using palaeomagnetic relative declination data (Dearing, 1983); and by using wet-density and loss-on-ignition measurements, sediment volumes were converted to estimates of erosion during three periods. These erosion levels are shown in Figure 13.11 together with the calculated area of ploughed land in the farms of the drainage basin since 1603. The three-fold increase in erosion apparently reflects both the five- to six-fold increase in the area ploughed annually and the changes in animal husbandry which took place during the nineteenth century.

Analyses of total P in sediments deposited in the seventeenth and twentieth centuries showed little variation. But when transformed to influx values of total P the rise from the earlier to the later period is of the order of 3.5 times. Figure 13.12 shows an attempt to recreate the P cycles in the drainage basin for 1682 (the date farm records were begun) and 1980, using the influx data for erosional losses, together with other published data of P movements within similar agricultural ecosystems. Between these dates changes in inputs via rainfall and weathering can be assumed to be small in comparison with changes in fertilizer inputs and erosional losses. Monitoring of tile drain water in the drainage basin has shown that dissolved P accounts for only about 0.5% of the total P entering the lake. Problems of evaluating P losses through the outflow, and P movements in the 1682 ecosystem, are as yet unresolved but not insoluble, and it is intended that the study will provide a detailed P cycle for the 1682 system on which to assess the impact of more recent human activity. Pollen and diatom records from a central core support the picture of intensifying agricultural practices hitherto obtained; cereal pollen increase from 5% to over 10% of the AP sum, and the diatom *Asterionella formosa*, often indicative of increased P loadings to a water body, appears only in the uppermost sediment which was deposited during the last few decades.

FINAL COMMENTS

The case study outlined here relies on documentary evidence for the reconstruction of past environments. For older periods or at sites where documentation of land-use is unavailable, microfossil, macrofossil and archaeological records would be used instead. In such cases imprecise environmental reconstruction combined with the difficulties of total sediment analyses will normally limit the useful comparison of sediment influx or chemical flow to that which existed in relatively well-defined and contrasting environments. But even at this coarsely resolved level of study total sediment analyses provide a basis for examining both soil stability, through the calculation of particulate losses, and soil deterioration, through the calculation of chemical losses, during periods of vegetation change such as forest clearance or the well-documented change from deciduous forest to blanket

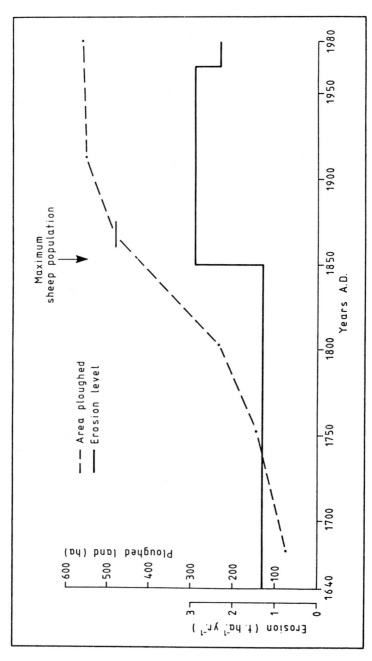

FIGURE 13.11. Erosion levels and area annually ploughed in the Havgårdssjön catchment, 1640–1980 A.D. Also shown is the timing of the maximum sheep population in the neighbourhood (Oldfield *et al.*, 1983). *Reproduced by permission of the Royal Geographical Society*

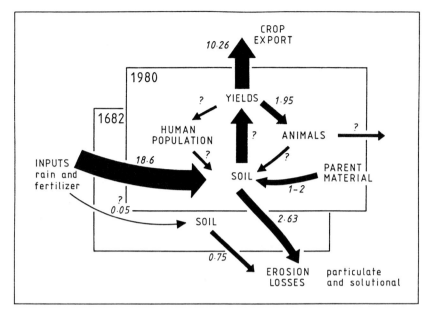

FIGURE 13.12. Partial reconstruction of total P movements in the 1682 and 1980 A.D. catchments based on total sediment analyses and published data for P in agricultural ecosystems (Oldfield *et al.*, 1983). *Reproduced by permission of the Royal Geographical Society*

peat in highland Britain. Results of total sediment analyses could provide a new input of information to traditional palaeoecological dilemmas, such as the cause of the Elm Decline, and in some environments may well represent the only means of providing base-line values of denudation and chemical loss for contemporary observations of geomorphological and pedological processes. The potential of any set of sediments to provide quantitative information about abiotic processes in the catchment will depend largely on the identification and the resolution of suitable correlation and dating methods; but on present evidence the techniques now available could be successfully used at many lake-catchment sites world-wide.

Acknowledgements

My thanks go to Kevin Edwards, Ian Foster, Bob Jones, Frank Oldfield and Paddy O'Sullivan for making comments on an earlier draft.

REFERENCES

Anderson, H. W. (1974). Sediment deposition in reservoirs associated with rural roads, forest fires and catchment attributes. In *Effects of Man on the Interface of the*

Hydrological Cycle with the Physical Environment, Symposium of the International Association of Hydrological Sciences, IAHS-AISH Pub. 113, Paris.

Appleby, P. G., and Oldfield, F. (1978). The calculation of lead-210 dates assuming a constant rate of supply of unsupported ^{210}Pb to the sediment. *Catena*, **5**, 1–8.

Appleby, P. G., and Oldfield, F. (1979). Letter to *Environmental Science and Technology*, **13**, 478–480.

Battarbee, R. W., and Digerfeldt, G. (1976). Palaeoecological studies of the recent development of Lake Växjösjön: I. Introduction and chronology. *Arch. Hydrobiol.*, **77**, 330–346.

Battarbee, R. W., Titcombe, C., Donnelly, K., and Anderson, J. (1983). An automated technique for the accurate positioning of sediment cores sites and the bathymetric mapping of lake basins. *Hydrobiologia*, **103**, 71–74.

Birks, H. J. B., and Birks, H. H. (1980). *Quaternary Palaeoecology*, Edward Arnold, London.

Björck, S., Dearing, J. A., and Johnsson, A. (1982). Magnetic Susceptibility of late Weichselian deposits in S. E. Sweden. *Boreas*, **11**, 99–111.

Bloemendal, J., Oldfield, F., and Thompson, R. (1979). Magnetic measurements used to assess sediment influx at Llyn Goddionduon. *Nature*, **280**, 50–53.

Borland, W. M., and Miller, C. R. (1958). Distribution of sediment in large reservoirs. *Journal of the Hydraulic Division, Proc. of the Am. Soc. of Civil Engineers*, HY2, paper 15877-1-1587-9.

Bormann, F. H., Likens, G. E., Siccama, T. G., Pierce, R. S., and Eaton, J. S. (1974). The export of nutrients and recovery of stable conditions following deforestation at Hubbard Brook. *Ecological Monographs*, **44**, 255–277.

Bortleson, G. C., and Lee, G. F. (1975). Recent sedimentary history of Lake Monona, Wisconsin, *Water, Air and Soil Pollution*, **4**, 89–98.

Bruland, K. W., Koide, M., Bowser, C., Maher, L. J., and Goldberg, E. D. (1975). ^{210}Lead and pollen geochronologies on Lake Superior sediments. *Quaternary Research*, **5**, 89–98.

Brush, G. S., and Brush, L. M. (1972). Transport of pollen in a sediment laden channel: a laboratory study. *Am. J. of Science*, **272**, 359–381.

Davis, M. B. (1968). Pollen grains in lake sediments: redeposition caused by seasonal water circulation. *Science*, **162**, 796–9.

Davis, M. B. (1973). Redeposition of pollen grains in lake sediment. *Limnology and Oceanography*, **18**, 44–52.

Davis, M. B. (1976). Erosion rates and land use history in southern Michigan. *Environmental Conservation*, **3**, 139–148.

Davis, M. B., and Brubaker, L. B. (1973). Differential sedimentation of pollen grains in lakes. *Limnol. Oceanogr.*, **18**, 635–646.

Davis, M. B., Brubaker, L. B., and Beiswenger, J. M. (1971). Pollen grains in lake sediments: pollen percentages in surface sediments from southern Michigan. *Quaternary Research*, **1**, 450–467.

Dearing, J. A. (1983). Changing patterns of sedimentation in a small lake in Scania, S. Sweden. *Hydrobiologia*, **103**, 59–64.

Dearing, J. A., Elner, J. K., and Happey-Wood, C. M. (1981). Recent sediment flux and erosional processes in a Welsh upland lake-catchment based on magnetic susceptibility measurements. *Quaternary Research*, **16**, 356–372.

Digerfeldt, G. (1971). The post-glacial development of the ancient lake at Torreberga, Scania, S. Sweden. *Geologiska Föreningen i Stockholm Förhandlingar*, **93**, 601–624.

Digerfeldt, G. (1972a). A preliminary report of an investigation of Littorina transgressions in the Barsebäck area, western Skåne. *Geologiska Föreningen i Stockholm Förhandlingar*, **94**, 537–548.

Digerfeldt, G. (1972b). The post-glacial development of Lake Trummen. *Folio Limnologica Scandinavica*, No. 16. Gleerup, Lund.

Digerfeldt, G. (1974). The post-glacial development of the Ranviken bay in Lake Immeln. I: The history of the regional vegetation, and II: The water-level changes. *Geologiska Föreningen i Stockholm Förhandlingar*, **96**, 3–32.

Digerfeldt, G. (1976). A pre-boreal water-level change in Lake Lyngsjö, central Halland. *Geologiska Föreningens i Stockholm Förhandlingar*, **98**, 329–336.

Digerfeldt, G., Battarbee, R. W., and Bengtsson, L. (1975). Report on annually laminated sediment in Lake Järlasjön, Nacka, Stockholm. *Geologiska Föreningens i Stockholm Förhandlingar*, **97**, 29–40.

Flower, R. J. (1980). *A Study of Sediment Formation, Transport and Deposition in Lough Neagh, Northern Ireland, with Special Reference to Diatoms*, unpublished Ph.D. thesis, New University of Ulster.

Foster, I. D. L., Dearing, J. A., Simpson, A., Carter, A. D., and Appleby, P. G. (1985) Estimates of contemporary and historical sediment yields in the Merevale catchment, Warks. U.K. *Earth Surface Processes and Landforms*, **10**, 45–68.

Holmes, P. W. (1968). Sedimentary studies of late Quaternary material in Lake Windermere (Great Britain). *Sediment Geol.*, **2**, 201–224.

Koivisto, E., and Saarnisto, M. (1978). Conventional radiography, xeroradiography, tomography and contrast enhancement in the study of laminated sediments: premliminary report. *Geografiska Annaler*, **60**, Series A, 55–61.

Livett, E. A., Lee, J. A., and Tallis, J. H. (1979). Lead, zinc and copper analyses of British blanket peats. *J. Ecol.*, **67**, 865–892.

Mackereth, F. J. H. (1966). Some chemical observations on post-glacial lake sediments. *Phil. Trans. R. Soc. B.*, **250**, 165–213.

Maher, L. J. (1977). Palynological studies in the western arm of Lake Superior. *Quaternary Research*, **7**, 14–44.

Oldfield, F. (1970). Some aspects of scale and complexity in pollen-analytically based palaeoecology. *Pollen et spores*, **12**, 163–171.

Oldfield, F. (1977). Lakes and their drainage basins as units of sediment based ecological study. *Prog. Phys. Geog.*, **1**, 460–504.

Oldfield, F., Appleby, P. G., and Battarbee, R. W. (1978). Alternative ^{210}Pb dating: results from the New Guinea Highlands and from Lough Erne, N. Ireland. *Nature*, **271**, 339–342.

Oldfield, F., Appleby, P. G., and Thompson, R. (1980). Palaeoecological studies of lakes in the Highlands of Papua new Guinea. I: The chronology of sedimentation. *J. Ecol.*, **68**, 457–478.

Oldfield, F., Battarbee, R. W., and Dearing, J. A. (1983). New approaches to environmental change. *Geogr. J.*, **149**, 167–181.

O'Sullivan, P. E. (1975). Early and Middle-Flandrian pollen zonation in the Eastern Highlands of Scotland. *Boreas*, **4**, 197–207.

O'Sullivan, P. E., Coard, M. A., and Pickering, D. A. (1982). The use of laminated lake sediments in the estimation and calibration of erosion rates. In: *Recent Developments in the Explanation and Prediction of Erosion and Sediment Yield*, IAHS Publ. 137.

O'Sullivan, P. E., Oldfield, F., and Battarbee, R. W. (1973). Preliminary studies of Lough Neagh sediments. I: Stratigraphy, chronology and pollen analysis. In: *Quaternary Plant Ecology* (Eds. Birks and West), Blackwell Scientific Publications, Oxford.

Pais-Cuddon, I. C. dos M., and Rawal, S. N. C. (1969). Sedimentation of reservoirs. *Journal of the Irrigation and Drainage Division, Proc. of the Am. Soc. of Civil Engineers*, IR3, paper 6789, 415–429.

Pennington, W., Cambray, R. S., Eakins, J. D. and Harkness D. D. (1976). Radionuclide dating of the recent sediments of Blelham Tarn. *Freshwater Biology*, **6**, 317–331.

Rapp, A., Murray-Rust, D. H. Christiansson, C., and Berry, L. (1972). Soil erosion and sedimentation in four catchments near Dodoma, Tanzania. *Geografiska Annaler*, **54**, Series A, 255–318.

Simola, H. (1979). Microstratigraphy of sediment laminations deposited in a chemically stratifying eutrophic lake during the years 1913–1976. *Holarctic Ecology*, **2**, 160–168.

Simola, H., Coard, M. A., and O'Sullivan, P. E. (1981). Annual laminations in the sediments of Loe Pool, Cornwall. *Nature*, **290**, 238–241.

Thompson, R. (1975). Long period European geomagnetic secular variation confirmed. *Geophys. J. R. Astr. Soc.*, **43**, 847–859.

Thompson, R., Battarbee, R. W., O'Sullivan, P. E., and Oldfield, F. (1975). Magnetic susceptibility of lake sediments. *Limnol. Oceanogr.*, **20**, 687–698.

Thompson, R., and Edwards, K. J. (1982). A Holocene palaeomagnetic record and a geomagnetic master curve from Ireland. *Boreas*, **11**, 335–349.

Thompson, R., and Morton, D. J. (1979). Magnetic susceptibility and particle size distribution in recent sediments of the Loch Lomond drainage basins, Scotland. *J. Sedimentary Petrology*, **49**, 801–818.

Tolonen, K., Siiriäinen, A., and Thompson, R. (1975). Prehistoric field erosion sediment in Lake Lojarvi, S. Finland and its palaeomagnetic dating. *Ann. Bot. Fennici*, **12**, 161–164.

Watkins, N. D., Sparks, R. S. J., Sigurdsson, H., Huang, T. C., Federman, A., Carey, S., and Ninkovitch, D. (1978). Volume and extent of the Minoan tephra from Santorini Volcano: new evidence from deep-sea sediment cores. *Nature*, **271**, 122–126.

Dating Methods

Handbook of Holocene Palaeoecology and Palaeohydrology
Edited by B. E. Berglund
© 1986 John Wiley & Sons Ltd.

14

Radiometric dating

Ingrid U. Olsson

Department of Physics,
Uppsala, Sweden

Radiometric dating is based on the disappearance or development of an isotope because of radioactive decay. Cosmogenic isotopes (e.g. ^{14}C, ^{3}H, ^{10}Be, ^{39}Ar, ^{32}Si, ^{36}Cl, ^{26}Al), artificially produced isotopes (e.g. ^{137}Cs, ^{239,240}Pu, ^{241}Am, and members of the natural radioactive decay series (e.g. ^{210}Pb, ^{230}Th, ^{226}Ra, ^{228}Th) may be trapped in some medium to constitute a closed system (there is no gain or loss of the isotope except for radioactive decay). The content of an isotope will decrease by a certain fraction per unit time (Figure 14.1). By measuring the remaining activity one can determine the time that has elapsed since the initial isolation, provided the initial activity is known.

If the daughter is somehow removed from the mother in a series decay, the daughter will grow to reach secular equilibrium with the mother. The degree of this increase may be measured and used as a dating method. It has been used, for example, for ^{230}Th and ^{231}Pa. The concentrations are usually assessed by low-activity measurements using various detectors, pulse-height analysers and, often, sophisticated shields to reduce the influence of environmental radioactivity which would otherwise make the measurements impossible.

It is now possible to measure single atoms by means of accelerators. The main advantage of this is that very small samples can be measured, but many isotopes can be used which are difficult to measure with conventional methods.

An isotope's half-life is the principal factor deciding the range for which it can be used in age determinations. Geochemistry is another important factor governing the usefulness of the isotope.

273

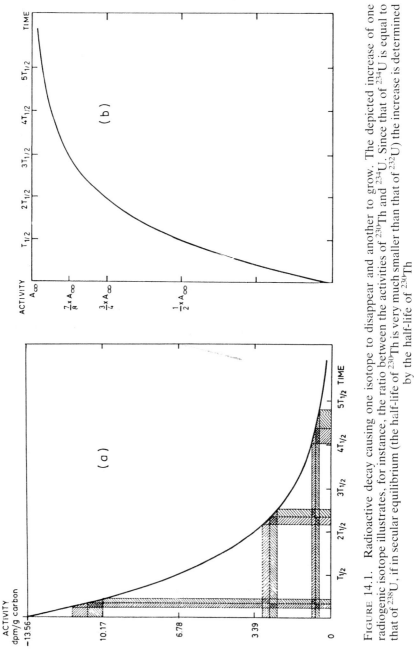

FIGURE 14.1. Radioactive decay causing one isotope to disappear and another to grow. The depicted increase of one radiogenic isotope illustrates, for instance, the ratio between the activities of ^{230}Th and ^{234}U. Since that of ^{234}U is equal to that of ^{238}U, if in secular equilibrium (the half-life of ^{230}Th is very much smaller than that of ^{232}U) the increase is determined by the half-life of ^{230}Th

I.   RADIOCARBON DATING

INTRODUCTION

A radiocarbon dating, far from being an isolated laboratory result, is based on the sample's deposition in nature, its collection, the relation of the sample to an event or a process, measurement of the isotope activity, and an interpretation of the result. The activity measurement is an objective measurement, but the quality of the final result is to a great extent dependent on how the sample was collected and processed. This guide has been compiled to facilitate interpretation, and to improve the necessary collaboration between ^{14}C laboratories, field workers and various other institutions.

An absolute chronology such as that based on the ^{14}C method allows us to determine whether or not processes were contemporaneous. The total uncertainty is dependent not only on the uncertainty in the activity measurement, but also on the precision of other measurements, such as pollen analyses. There is, for example, no point in determining the radiocarbon age with an uncertainty of ±50 years if the relevant pollen-analytical level has an uncertainty of ±200 years. Contamination is another source of error. Different conditions which cause the radioactivity to vary with time and site should be considered and their influence on the final date evaluated.

Figure 14.2 illustrates the radiocarbon dating method and some of the problems to be discussed here. Radiocarbon is produced in the atmosphere by the interaction of cosmic rays. Since these are modulated by the Earth's

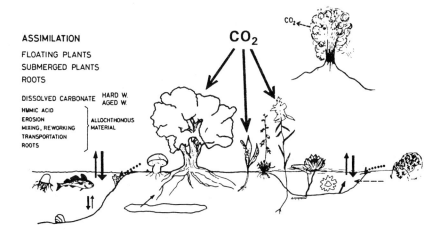

FIGURE 14.2.   A schematic illustration of the uptake of ^{14}C by plants and animals, and the interchange between different reservoirs and sources of contamination. The broader arrow, when two are parallel, indicates that a larger amount of radiocarbon passes in that direction

magnetic field one can expect long-term variations. The charged particles emitted by the sun also modulate the cosmic-ray intensity. This may cause short-term variations. This needs further investigation for a full understanding. When a plant or animal dies, or the cellulose and lignin is incorporated in a tree-ring, the radiocarbon content starts to decrease since no new carbon is absorbed. The decreasing ^{14}C content is a measure of age.

RADIOCARBON YEARS AND CONVENTIONAL YEARS

Variations in the $^{14}C/^{12}C$ ratio of carbon dioxide in the atmosphere are now well established. These variations were discovered by dating samples of known ages. In the main, dendrochronologically dated samples were used, but historically known samples were included too — for example, Egyptian samples from as far back as 2800 B.C., which proved to be consistent with the tree-ring samples.

Many efforts have been made since the late 1950s to construct a curve giving the variations. Until recently most of this work has been done in three laboratories in the U.S.A. Olsson compared three interpretations of the calibration points (1974a). These interpretations were submitted by Damon *et al.* (1973, 1974), Michael and Ralph (1973) and Suess (1970). The curve from the Pennsylvanian laboratory was revised by Ralph *et al.* (1973). A statistician (Clark, 1975) has since produced a very smooth curve, and several radiocarbon workers have produced their own curves (e.g. Switsur (1973), who, however, reversed the axes of his diagram, and Olsson (1974b)). McKerrel (1975a, b) has compiled various tables. Suess' curve contained many oscillations and was later compared with a new revised curve (Suess, 1979). Stuiver (1978) published a curve clearly indicating pronounced short-term variations of the $^{14}C/^{12}C$ ratio from A.D. 1500 to A.D. 1950. These measurements are extended back to 2000 B.P. by Stuiver (1982) and Pearson and Baillie (1983), and still farther back to almost 6000 B.P. by Pearson *et al.* (1983). Shorter intervals are covered by de Jong *et al.* (1979), de Jong and Mook (1980), and Bruns *et al.* (1980a).

The uncertainty in these high-precision measurements is close to ±20 years. In comparisons of results from different laboratories, any bias must, however, be considered, as illustrated by Stuiver (1982).

It might be argued that the curves from the Belfast laboratory are based on a dendrochronology with the weakest link at about 2550 B.C. (Baillie *et al.*, 1983). Here only one tree constitutes the bridge between the older and younger sections. The correlation is, however, strong.

One group of trees was recently correlated with the long chronology from 5300 to 940 B.C. and the shorter one from 940 B.C. to 220 B.C. This correlation confirms the wiggle-match against the bristlecone-pine calibration curves. Moreover the earlier gap at 1900 B.C. is now well bridged. Thus the

Belfast chronology is continuous from 5300 B.C. to about 200 B.C. with an error of less than 20 years. The uncertainty ±10 years is obtained by wiggle-matching, and comparisons with the results from Groningen (de Jong, 1981). Comparisons with results from some other laboratories and earlier conversion curves are given by Baillie *et al.* (1983) and Pearson and Baillie (1983). The agreement between Belfast, Groningen and Seattle is so good that older calibration curves now have a mainly historical interest. The German oak chronology (Becker, 1983) is extended back to 2804 B.C. by the continuous Hohenheim master chronology. Beyond that age there are floating chronologies back to 7600 B.C. Samples from as far back as 7200 B.C. have been used for calibration (Bruns *et al.*, 1983). The results from 800 to 2800 B.C. are summarized by Becker (1983), who discusses some disagreement between wiggle-matched ages and dendroages. The figures are 73 and 67 years. No conclusions regarding this offset should be drawn until the ^{14}C content is also measured by Stuiver in Seattle. The Hohenheim oak chronology is recently extended back to 4066 B.C. (Linick *et al.*, 1985).

As an illustration the author's calibration curve and band from 1974 is shown here, together with a smoothed curve based on results published by Pearson and Baillie (1983) and Pearson *et al.* (1983) (Figures 14.3(*a*) and (*b*)). The author's band is drawn because of the difficulty of depicting a calibration curve. The trend, seen here, that oak samples yield less deviation than do bristle-cone pine samples was observed by Freundlich and Schmidt (1983). This, and a possibly varying response to solar cycle variations and global variations, must be studied further.

The ages determined from radiocarbon measurements using the half-life 5568 (or 5570) years, normalized to δ^{13}C = −25 per mil in the PDB-scale and related to the international dating standard, are called radiocarbon years. If a calibration is made using a diagram or table for radiocarbon years as a function of calendar years, the ages are said to be 'calibrated'. The curve used for the calibration should always be stated.

As a first estimate, calendar years are read off corresponding to the radiocarbon year given (the age), the age plus σ and the age minus σ. Because of the shape of the curve the values obtained may correspond to a broad or a narrow range, or even to more than one range (Figure 14.4). The sigma value used should be based on all uncertainties of the measurements in the laboratory. The uncertainty in the calibration curve must be taken into consideration.

Other principles of working have been given by Clark (1975), Damon *et al.* (1974), Klein *et al.* (1982), Ralph *et al.* (1973), and Olsson (1972b).

In view of the existence of high-precision curves already published or expected to be released in the near future, a recommendation was made at the 1982 Seattle radiocarbon conference that researchers should try to use the most appropriate curve available. The efforts made by a committee (Klein *et al.*, 1982) to combine all old measurements in order to construct a calibration

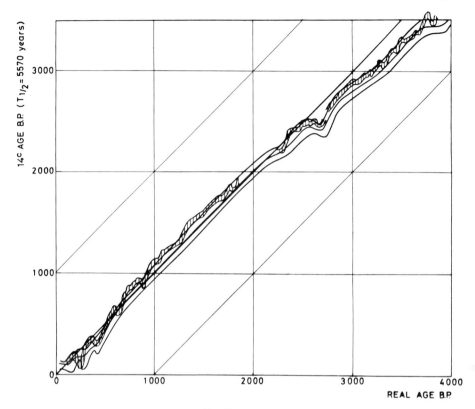

FIGURE 14.3.   The variations in the $^{14}C/^{12}C$ ratio plotted as a calibration diagram. On top of Olsson's band (1974b), covering about 87% of the dendrochronological calibration points available then, are new results by Pearson and Baillie (1983) and Pearson *et al.* (1983), introduced as a narrow hatched band

tool was acknowledged and appreciated by the conference delegates, although it was recognized that certain periods had already been better investigated by the high-precision laboratories. In this connection it must be recalled that detailed calibration curves should be used for short-lived samples such as grains, straws and single tree-rings, but smoothed curves for peat, most charcoal samples etc. deriving from a longer period (Olsson, 1974b) The influence of the choice of different curves has been further studied by de Jong (1981) and also by Mook (1983) and discussed by Pearson *et al.* (1983). This may add another uncertainty to the calibrated result (Pearson *et al.*, 1983). An application has been described by Olsson *et al.* (1984), using lichen for climatic studies. Some species may have a biological age of 100 years or more. The calibration may be still further complicated if the ^{14}C measured samples have a reservoir age (see below) or are contaminated by material with a

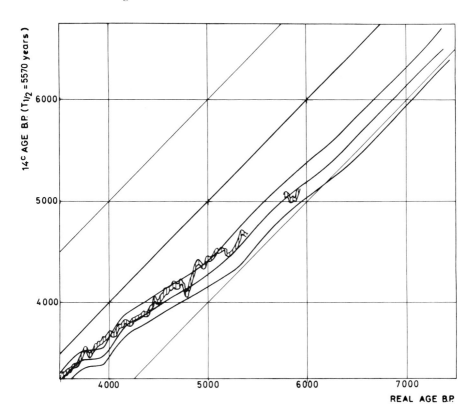

different age. If, for instance, the measured activity of a sample derives totally from autochthonous material, but 10% of the material emanates from graphite which seems to be infinitely old, the date will (see below) appear to be about 800 years older than for an uncontaminated sample. The calibration should be made on the ^{14}C corrected for the contamination, although it may be very difficult to determine the corrected ^{14}C age.

A standard has been common since 1959, being 95% of the activity in 1950 of the 1957 oxalic-acid standard from the National Bureau of Standards when this has a $\delta^{13}C$ value of -19 per mil (Godwin 1959, 1962). This is called SRM-4990 by Mann (1983). Because the supply of reference material was practically exhausted a few years ago, a new batch was prepared. The activity was carefully measured in several laboratories, using the standard SRM-4990 as reference material (Mann, 1983), and this is now used as a standard (Stuiver, 1983).

Since different standards were in use before 1959, and even later in some laboratories, the earlier results should now be adjusted to the oxalic-acid standard.

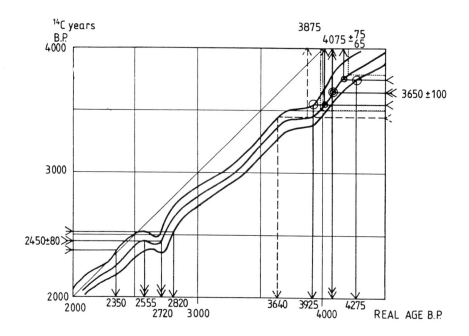

FIGURE 14.4.   Different principles for calibrating ^{14}C results. Objections can be raised against all. Given is the ^{14}C age 3650 ± 100 ^{14}C years B.P.

(a) Use the curve and read the values corresponding to 3550, 3650 and 3750. The result is 4075^{+75}_{-65}. The uncertainty is underestimated since the curve is regarded as exact. The value corresponding to 3650 − 2σ (3450) is 3875 as indicated.

(b) Read the values as in (a) but increase the final uncertainty by the uncertainty of the curve, here assumed to be ±100. $\sqrt{(75^2 + 100^2)} = 125$ and $\sqrt{(65^2 + 100^2)} = 120$. Objection: statistical uncertainties have been added statistically to systematic uncertainties.

(c) Increase the given uncertainty of the result to include some uncertainty in drawing the curve. Read as in (a) (dotted lines). Here an example with ±100 years is illustrated, yielding $\pm\sqrt{(100^2 + 100^2)} = \pm 140$ years. Although the curve is based on values with statistical uncertainties the curve itself has systematic uncertainties. Thus the final uncertainty cannot be the true statistical uncertainty.

(d) Use the curve for the midpoint (3650) and the boundaries of the band for (the age − σ) and (the age + σ). The result is 4075, 3925 and 4275, thus 4075^{+200}_{-150}. Similarly the value for (age − 2σ) will be 3640. The uncertainty seems overestimated. The author has used this (e.g. Olsson, 1972b), but considering the difference seen in Figure 14.3 the estimate seems realistic.

(e) Use different bands, drawn by Klein *et al.* (1982), dependent on the given uncertainty, to obtain a range −2σ to +2σ. The range −σ to +σ is not given. Any asymmetry is lost. Another value such as 2450 ± 80 ^{14}C years B.P. will yield two values for the mid-point seen in the figure. A range of 220 years is produced if the date is 2440 ± 15 ^{14}C years, and the curve is exact. The curve and band in the figure are drawn as an illustration but based on Olsson's curve (1974b)

ISOTOPIC FRACTIONATION

Although their chemical behaviour is identical, different isotopes of an element may react slightly differently because of their different masses. The translation, rotation and vibration energies will diverge slightly. The vibration energy in particular determines the isotopic fractionation. This is more easily seen for lighter elements such as hydrogen, carbon and oxygen than for the heavier elements. The dissolution of carbon dioxide in water is one instance where the heavy carbon isotopes are enriched, while the assimilation of carbon dioxide enriches the lighter carbon isotopes. The thermodynamic laws stipulate that the heavier isotope ^{14}C will be twice as enriched as ^{13}C (Craig, 1954). Since such fractionation is dependent on temperature, pH and physiological processes, the ^{14}C values should be adjusted to a certain $^{13}C/^{12}C$ ratio with the help of the ^{13}C values. The standard used in radiocarbon work is a carbonate sample, and when stated in relation to this the values are said to be given in the PDB scale. The deviation is given as the millesimal deviation, $\delta^{13}C$. Since no attention was paid to this value at the beginning of the radiocarbon era no adjustment was made from the start. The standard used in the different laboratories for modern activity was usually wood, a normal $\delta^{13}C$ value for this being -25 per mil. For this reason all radiocarbon activities, when measured for dating purposes, should be adjusted so as to correspond to a $\delta^{13}C$ value of -25 per mil (Figure 14.5). For a marine carbonate this usually means a correction of the ^{14}C activity by about 50 per mil, corresponding to 410 years. If the $\delta^{13}C$ value is reduced by 25 per mil and the $\delta^{14}C$ value by 50 per mil, the age will appear 410 years older than when no normalization is made. For samples from the North Atlantic Sea this correction happens to equal approximately the apparent age or reservoir age of the water (see below). The two corrections should, however, be kept separate even when they are roughly the same but with different signs. For a marine mammal the $\delta^{13}C$ value is about -16 per mil so that the $\delta^{13}C$ correction corresponds to -18 per mil in the $\delta^{14}C$ value, or about 150 years to be added to the non-normalized age. If the reservoir age is 400 years the net correction will be 250 years. Further examples are given in Figure 14.5.

A survey of the range of $\delta^{13}C$ values for different materials is presented in Figure 14.6. This shows that the normalization for sediment samples, with $\delta^{13}C$ values from -16 to -35 per mil, dated in the Uppsala Laboratory, as given in the lists published from 1959 to 1972, varies from about -160 to $+145$ years. Different fractions, obtained by chemical separation, of a sample usually have different $\delta^{13}C$ values. Thus the holocellulose differs significantly from the extractives. Some detailed examples are given by Olsson and Osadebe (1974), who also discuss the normalization to -25 per mil instead of the value of carbon dioxide in the air (approximately -7 per mil). Similar considerations were presented by Stuiver and Robinson (1974). Typical corrections due to

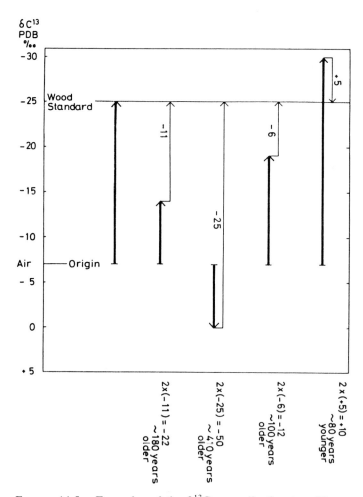

FIGURE 14.5. Examples of the $\delta^{13}C$ normalization to -25 per mil and the corresponding ^{14}C and age corrections. The $\delta^{13}C$ normalization is made to eliminate the effect of fractionation at the uptake of ^{14}C, and can be regarded as a bias, a positive or negative apparent age, when the carbon was incorporated. This bias will always have the same value in years

normalization are illustrated in Table 14.1. Similar tables appear in Lerman (1973), Stuiver and Polach (1977) and Gulliksen (1980).

Selected sample types are included in Table 14.1 to illustrate typical values, differences and similarities. The table is a guide to selection of adequate corrections when the results have not been normalized for isotopic fractionation, and of the uncertainties if the $\delta^{13}C$ values are not measured. The table was compiled from the first eleven lists issued by the Uppsala Laboratory, and includes recent results from bones and concretions. The age correction is to be added to a result not yet normalized for the $\delta^{13}C$ deviation from -25 per mil.

The spread of the $\delta^{13}C$ values for a particular series of sediments of one type, from one core, is much smaller than that given in the table. If single measurements are missing from such series the uncertainties thereof can usually be estimated as 1–3 per mil. Similar considerations are appropriate for peat and other samples. When single samples of unidentified species are dated without $\delta^{13}C$ determination, an uncertainty of ± 3 per mil is postulated for marine shells, whales, seals, wood and charcoal. A slightly larger statistical uncertainty should be selected for brackish water shells, freshwater carbonates and resin. A single peat or gyttja sample should be ascribed an uncertainty of ± 4 and ± 6 per mil respectively, in the $\delta^{13}C$ value. An uncertainty of ± 5 per mil in $\delta^{13}C$ is equivalent to ± 80 radiocarbon years.

Whenever samples with the C_4 pathway for assimilation (such as *Desmostachya bipinnata* and corn) are dated, -10 to -19 per mil is a typical $\delta^{13}C$ range. Succulents often have a $\delta^{13}C$ value around -17 per mil.

With three exceptions the range determined in Pennsylvania for 117 samples of *Sequoia gigantea* is from -20 to -26 per mil. The mean value is -22.6 per mil. The range, also determined in Pennsylvania, for 202 samples of *Pinus aristata* is from -17 to -28 per mil, with five exceptions. The mean value is -22.9 per mil. These two species have a significantly smaller negative $\delta^{13}C$ value than the Scandinavian species.

A warning must be given against applying $\delta^{13}C$ values from one laboratory to samples from another if different fractions are used. This is especially obvious for what is called collagen and must be remembered also when stable isotope values are used for revealing the origin of the food eaten by prehistoric man (Tauber, 1983; Chisholm *et al.*, 1983).

If a sample is mixed, containing components of different ages, the correction is more complicated. An example may be given from Uppsala. Carbon dioxide from the air is greatly depleted when statically absorbed in NaOH for determination of the atmospheric ^{14}C activity. Instead of being -7 per mil the value approaches -25.5 per mil. In 1973 the gas collected close to the new volcano on Heimaey, Iceland, had a measured $\delta^{13}C$ value of $+7.9$ per mil and the ^{14}C activity was a few per cent of the modern (Olsson, 1979). Thus the inactive carbon dioxide from the volcano had a $\delta^{13}C$ value $+8.9$ per mil or about 34 per mil heavier than the atmospheric carbon dioxide. This value is significantly different from the values reported by others (Saupé *et al.*, 1980; Bruns *et al.*, 1980b).

INFLUENCE OF CONTAMINANTS

It is easily understood that contamination to a degree of 50% with very old

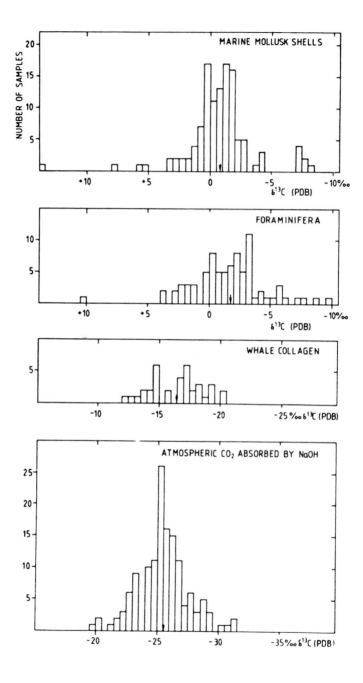

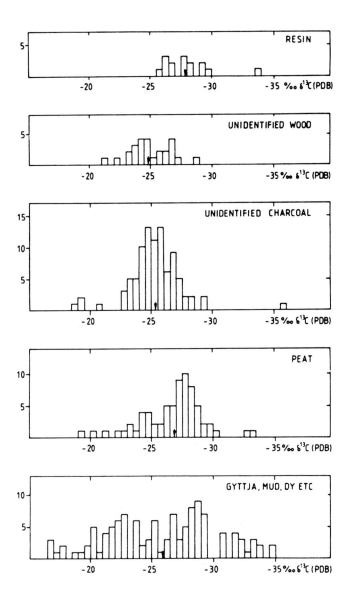

FIGURE 14.6. $\delta^{13}C$ variations in nature as determined on groups of samples dated at the Uppsala Laboratory and published in the first 11 lists, and in papers on bone dating. Some recent results on concretions are also included

TABLE 14.1.   Some $\delta^{13}C$ values of samples dated in Uppsala (rounded figures when appropriate, some outliers excluded)

Sample	Number of samples	Mean (‰)	Age correction range (years)	Mean age correction
Marine mollusc shells	116	-0.8	515-270	400
Foraminifera	79	-1.7	480-280	380
Fresh-water shells	2	-10	270-210	245
Fresh-water concretions	4	-13.9	190-170	180
Brackish-water shells	14	-9	330-210	260
Brackish-water concretions	6	-2.3	385-360	375
Whales (collagen)[1]	36	-16.5	200-80	140
Seals (collagen)[1]	4	-15.5	200-115	155
Terrestrial animals[1]	4	-23	50-10	30
Wood, not identified	26	-25	65- -65	0
Leaves (*Betula, Salix*)	7	-28	-30- -70	-50
Leaves (*Solanum, Typha*)	7	-25	50- -40	0
Leaves (*Desmostachya bipinnata*)	8	-11	235-210	95
Charcoal, not identified	86	-25.5	40- -70	-10
Resin (archaeological)	15	-28	0- -145	-50
Peat	69	-27	95- -145	-30
Gyttja, mud and dy	117	-26	145- -160	-15
Atmospheric CO_2 when absorbed by NaOH	136	-25.5	60- -70	-10

Pretreatment accepted by the author (Olsson *et al.*, 1974, El-Daoushy *et al.*, 1978)

[1]Pretreatment accepted by the author (Olsson *et al.*, 1974, El-Daoushy *et al.*, 1978)

carbon will cause a sample to appear one half-life too old. Because of the exponential decay, 1% corresponds to 80 years and 10% to 846 years if 5568 years denotes $T_{1/2}$. For our discussions very old carbon is older than 50,000 or 100,000 years. No detectable radiocarbon then remains in this contaminant. Graphite and old carbonate deposits are examples of very old carbon detected in samples for radiocarbon dating. Naturally occurring contaminants may be slightly (or much) older than the sample, although not 'infinitely older'. The influence on the age is illustrated in Figure 14.7.

When the contaminant is younger than the sample itself the error may be large, as seen from the ascending curves in Figure 14.7. One per cent of modern material will reduce the apparent age of an infinitely old sample to 37,000 radiocarbon years. If the real sample is 10,000 years old and 1% of modern material is present, the sample will be dated at 9800 years, but 10% will cause the sample to appear 8200 years old.

PRETREATMENT

Mechanical pretreatment

The outer parts are usually removed to eliminate surface contaminants. Whenever a large piece of wood or charcoal is submitted and the size of the sample allows removal of the surface layer, this is done. Samples collected with corers should be trimmed before despatch. All corers should be free from oil and carefully cleaned before use to eliminate contaminants.

Samples should first be checked for roots and rootlets. Sometimes, but by no means always, the roots are almost dissolved by a chemical pretreatment. Since roots may penetrate deep into the sediment it is essential that measures be taken to remove them, for instance by wet sieving.

Handpicking roots is often advisable. Whenever pieces of wood, cones etc. are detected in sediments it is advisable to remove them, and sometimes they are worth dating separately. It should be remembered that wood may sink into loose sediment and be recovered at a level much older than the sample itself. For example, a plough was dated at 970±80 radiocarbon years B.P. although the pollen analysis indicated that it derived from the Bronze Age.

Pretreatment of shells and carbonate concretions

Since especially the outer parts of shells are easily contaminated by chemical exchange with carbon in groundwater, or by penetration of carbon dioxide, the carbonate samples are usually leached by acid. As a rule more than 20%, but preferably 50%, of the sample is removed. The rest of the sample is divided into two or more fractions to allow comparison of the ages of the innermost parts with those of the outer layers. A significant difference naturally indicates

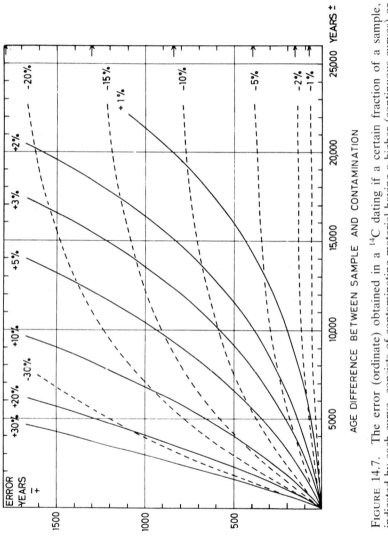

FIGURE 14.7. The error (ordinate) obtained in a ^{14}C dating if a certain fraction of a sample, indicated by each curve, consists of contaminating material having a higher (continuous curves) or lower activity (dashed curves) than the original sample; expressed as an age difference (abscissa) between this sample and the contaminant

that the sample is contaminated, although the inner fraction may yield an age which is correct within the margin of error. A severely contaminated sample may occasionally yield the same ages within the margin of error. Secondary calcite may, however, be more resistant to acid than the shell, making the contamination worse (Vita-Finzi and Roberts, 1984). If recrystallization of aragonite to calcite has occurred there is a significant risk of contamination by material of different age. Most species deposit the carbonate as aragonite. The same applies to the greater part of the shells of the remaining species. Thus it is often advisable to check the crystallization for information to be used in the discussion. It has, however, been demonstrated (Olsson *et al.*, 1968) that carbon dioxide can penetrate a shell, and when adsorbed on the inner surfaces this type of contamination cannot be detected by X-ray analysis. Deterioration of shells may be studied by scanning electron microscopy. Poor shell samples, with a pitted surface seeming weathered to the naked eye, are usually too seriously contaminated to yield a reliable age.

Foraminifera samples and mollusc shells should be cleaned in boiled distilled water, or water acidified to a pH value of 3, to avoid contamination. This is very important when the sample is small, so preventing proper pretreatment.

In this connection it may be mentioned that the organic fraction of shells (e.g. Håkansson, 1969) can be used for dating. In some cases, however, the ages do not agree when determined on the inorganic and organic fractions (Taylor and Slota, 1979).

The risk of contamination by carbon dioxide in the air means that the samples should be stored dry and preferably sealed from the atmosphere from the time of collection until the dating process. Samples stored in museums for decades yielded, for instance, a radiocarbon age of about 34,000 years, corresponding to 1.5% of modern material, although 'infinitely old' (Olsson *et al.*, 1968).

The separation of sea sediments into fractions according to grain sizes is to be recommended when material can be transferred from land into the sea. Old carbonate grains may be carried by the wind and deposited in the sea, not too far from the coast, in certain size fractions according to the distance (Olsson and Eriksson, 1965). For the organic fraction of the sediment, regarded by Geyh as the more reliable, Geyh *et al.* (1974) reported a significant disagreement between ages as a function of depth. They explained this as being due to bacterial influence during storage.

Chemical pretreatment of wood, charcoal, peat, gyttja, etc.

The chemical processing performed in radiocarbon laboratories usually consists of leaching with dilute hot acid to remove carbonates and calcium; rinsing in water; treatment with dilute sodium hydroxide overnight at $+80°C$ to extract humic acid; and subsequent washing with water. Finally the insoluble fraction from the alkali treatment is acidified to remove absorbed carbon dioxide and the soluble fraction acidified to a pH value of approximately 3 to

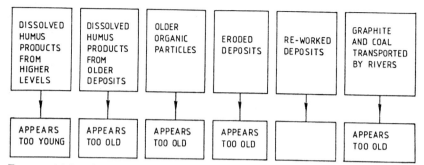

FIGURE 14.8. A general survey of the influence of possible allochthonous material of different origin on age determinations of deposits

precipitate the humic acid. The humic acid may be older or younger than the real sample, as shown in Figure 14.8.

The above treatment is recommended even if it is assumed that no contamination with humic acid has occurred. The alkali treatment is sometimes repeated to ensure satisfactory pretreatment (Olsson, 1979a). As a rule, however, one extraction is sufficient.

Chemical pretreatment of bones

The inorganic fractionation of bones should be avoided since the dates obtained are unreliable. Usually the collagen is dated, since all evidence indicates that no exchange occurs although the collagen gradually disappears by hydrolysis due to the enzyme collagenase. The collagen obtained at different pretreatments of bones is more or less pure. Different treatments with HCl and EDTA are used (e.g. Longin, 1971; Berglund *et al.*, 1976; Olsson *et al.*, 1974; El-Daoushy *et al.*, 1978).

Separation into amino acids has also been tested (Ho *et al.*, 1969). Another amino-acid method in current use should also be mentioned — it is based on the epimerization phenomenon (Hassan and Hare, 1978), and the basic principles were given by Bada and Schroeder (1975). A survey of different treatments has been given by Taylor (1980).

APPARENT AGES OF LAKE SEDIMENTS — RESERVOIR EFFECT AND CONTAMINATION

Lake sediments such as dy and gyttja may consist of decayed organic matter from leaves etc., submerged plants, and material washed out from land by, for example, rivulets or snowslides. The pollen flora at a certain depth reflects contemporary flora on land as well as in the lake. When the sediment is dated a false age may be ascribed it because of contamination or the reservoir effect.

Both effects were considered by Ingmar and Olsson (Olsson, 1972a) when discussing the apparent age of sediments from Uppland, Sweden.

If the sediment is to some extent inorganic there is always a risk that some old organic material has been added to the contemporary sediment. In some cases it may be graphite or coal, but occasionally the contaminating old material is simply older. Comparisons between ages determined for peat and those from gyttja have revealed that gyttja tends to be dated at an older age than the peat. The difference may be very small, but can amount to thousands of years. Although the alkali-soluble fraction used to be rejected when dating gyttja, it now seems evident that this fraction often, or rather mostly, yields an age which is more reliable than that from the insoluble fraction. This was also confirmed by radiocarbon dating of material deposited in annual varves. In some cases it was demonstrated that the insoluble fraction could be divided into two fractions on combustion. The first fraction obtained on degassing seems to be more reliable than the last fraction from the combustion. The modern combustion bombs do not allow such fractionation. Examples are given by Olsson (1972a, b, 1973).

Moreover, because of the risk of a lower $^{14}C/^{12}C$ ratio in lake water than in the atmosphere (the reservoir effect), it is often preferable to date peat rather than lake sediments. If submerged plants in the lake contribute to the sediment, the ^{14}C determination will indicate an earlier event than the real one. The low $^{14}C/^{12}C$ ratio may derive from dissolved old carbonate and the lake may be a hard-water lake. The effect is then called the hard-water effect; but the $^{14}C/^{12}C$ ratio may also be low because of a supply of old groundwater with a content of dissolved carbonate so low that it hardly can be called hard, but rather soft water. Even in such lakes some submerged plants may contribute to the sediment. Säynäjälampi is a good example of a hard-water lake (Donner *et al.*, 1971; Olsson *et al.*, 1983). The calcium content was given as 8.2 mg/l, the top sediment dated at about 2000 radiocarbon years, and submerged plants yielded results indicating a much lower $^{14}C/^{12}C$ ratio than in the atmosphere.

The exchange rate over the lake surface is probably slow enough to cause the water to have a somewhat lower activity than the atmosphere. If peat is chosen the age may be expected to be more realistic, but some uncertainty remains because there may be roots penetrating deep into the peat. Naturally these are younger and should be removed. A pure *Sphagnum* peat seems ideal. In the main, however, the material is mixed. A fen peat and a telmatic peat will yield ages slightly too old because of the inclusion of some submerged plants. Olsson measured the ^{14}C content of different constituents of water from a lake, a fen, a lagg and a raised peat bog and found that the activity was usually lower than in the contemporary atmosphere. Submerged plants may be used for studies of the ^{14}C concentration of the water. Some plants assimilate from the bicarbonate, others from the dissolved CO_2, and still others, such as *Lobelia dortmanna*, obtain their CO_2 from the sediment through the roots. The ^{14}C content thus may vary according to the species and sometimes according to local conditions (the sediment accumulation rate etc.). Nowadays, it seems that the pH of a non-hard-water lake barely affects its ^{14}C content. The exchange over the surface needs further study. The mechanism and the speed must be included in the discussions. The reservoir age of Swedish lakes must be assumed to be 300–400 years as a first approximation, unless specific reasons can be given for other assumptions.

Dates in error because of contamination by erosion and redeposition etc. have been discussed by, for example, Hörnsten and Olsson (1964), Olsson (1972a, 1973, 1978, 1979b, 1983b), Huttunen and Tolonen (1977), Donner and Gardemeister (1971), Donner and Jungner (1973, 1974) and Renberg (1978). Erosion may often be caused by ditching for agriculture.

Part of the ignition loss may be due to removal of crystal water. This may be a serious source of error when the organic carbon content is determined by ignition for samples with low carbon content.

Extraction of lipids is sometimes recommended (Grant-Taylor, 1973) but the greatest care must be taken to remove the organic solvents.

APPARENT AGE OF SEA WATER, MARINE PLANTS, MARINE MAMMALS AND FISH-EATING ANIMALS

When dealing with marine samples a somewhat lower $^{14}C/^{12}C$ ratio should be expected in the samples, while still alive, than in the contemporaneous atmosphere. The main reasons are the slow mixing between the reservoirs, and the fact that the deep-water reservoir contains between 10 and 100 times as much carbon as does the surface reservoir. Usually the surface water down to a depth of about 100 m is regarded as a well-mixed reservoir and the deep water as another. The exchange between the atmosphere and the surface water was formerly regarded as molecular, and that between the surface water and deep water as turbulent; but the biological activity is now believed to play a major role (Bolin, 1976). Notwithstanding these considerations, the carbon in deep water is exchanged so slowly with that of the surface water that the radiocarbon will decay to an appreciable extent before exchange. Thus, the ascending carbon will contain less radiocarbon than that descending from the surface layer to the deep water. Consequently the water, and the samples taking their carbon directly or indirectly (such as fishermen, polar bears and some birds) from the water, will have a lower $^{14}C/^{12}C$ ratio than contemporaneous terrestrial samples (Tauber, 1979). We then say that a reservoir effect is present. It should be stressed that this discussion tacitly assumes that the correction for isotopic fractionation is performed.

Since many papers on the reservoir effect are marred by erroneous conclusions only a few are mentioned here, but it is emphasized that the list of papers with correct discussions or results is not complete. A few results on the reservoir effect have been given by, for example, Håkansson who measured shells and bones (1969, 1970, 1974). Five samples from the Baltic, submitted by Berglund, were measured at the Stockholm Laboratory (Engstrand, 1965). Various samples were published by Mangerud (1972) and recalculated by Mangerud and Gulliksen (1975). Several samples from the coastal environment of the United Kingdom were given by Harkness (1983). Olsson (1980) critically analysed some earlier results together with new measurements taking into account the variations of $^{14}C/^{12}C$ in the atmosphere during the last 300 years. The reservoir effect was then estimated to be about 25–85 years less than previously. The large spread seen earlier for Iceland was changed into more consistent values by six new results yielding 365 ± 20 years (Håkansson, 1983).

It must be remembered that we still do not know the apparent age of the waters of long ago, since the samples used are all less than a few hundred years old.

LOW-ACTIVITY REGIONS

^{14}C-free carbon dioxide is released from volcanoes and in areas with gaseous emanations. Although the impact on the global $^{14}C/^{12}C$ ratio seems negligible, the consequences for certain regions are serious. The errors introduced may amount to over 1000 years. Chatters *et al.* (1969) reported ^{14}C deficiency from Hawaii; Šilar (1976) from Bohemia; Olsson (1979c and unpublished) from Iceland; Saupé *et al.* (1980) from the Monte Amiata district, Italy; Bruns *et al.* (1980b) from the Eifel area, western Germany and from Thera, Greece; and Sulerzhitzky (1970) from Kamchatka and the Kurile Islands.

The ^{14}C activity seems to be lower in the southern than in the northern hemisphere, although the effect is so small that it is difficult to prove this deficiency (Lerman *et al.*, 1970). If this is due to the exchange between the atmosphere and the top layer of the ocean, a deficiency is also to be expected over isolated islands. Olsson (1979c, 1983a) is studying this phenomenon to explain apparently too old radiocarbon ages for the *landnám* of Iceland.

OTHER UNCERTAINTIES

Sample choice when composition of sediment is changing

Because of the fact that a radiocarbon sample often corresponds to an accumulation period of 150 years or longer, and the accumulation rate may change rather suddenly with a factor of 2 or more, it is inadvisable to sample over such a boundary. If the composition of a sample changes such that part is gyttja and part is peat, the reservoir effect in the gyttja may seriously affect the mean value. The same risk is present if the sample is taken over a lake isolation layer. Thus it is advisable to select samples below and above such boundaries. This will also facilitate discussion. Wide variations in the organic content cause uncertainties in the level to be used when a sample is cored, and the section dated may have accumulated during a period longer than that covered by the statistical uncertainty.

Statistical uncertainties

Since radioactive decay follows statistical laws it follows that the statistical uncertainty must be given. The normal distribution is shown in Figure 14.9. Since any age measurement includes factors other than the decay, the mode of calculation may also be discussed.

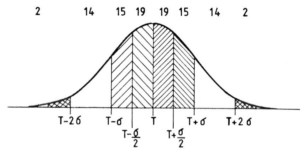

FIGURE 14.9. The statistical uncertainty, $\pm\sigma$, indicates the range, (age $-\sigma$) to (age $+\sigma$). There is a 68% probability that a measurement will fall within this range and a 95% probability of it falling within the range (age -2σ) to (age $+2\sigma$). If 100 laboratories were to measure the age of one sample and all were to ascribe the uncertainty of 100 years, and the mean age is 3000 years, we would expect about 38 laboratories to obtain results between 2950 and 3050, 15 laboratories results between 2900 and 2950, another 15 laboratories between 3050 and 3100. It would also be expected that one laboratory will obtain a value of over 3200, and another a value of less than 2800, while two groups, each consisting of 15 laboratories, would obtain values lying in the ranges 3100 to 3200 and 2800 to 2900 respectively

Unfortunately there is no agreement between radiocarbon workers concerning the errors which should be incorporated in the stated uncertainty. Some laboratories give only that deriving from the decay. Others also include uncertainties in the δ^{13}C value, the size of the sample, the stability of the equipment etc.; i.e. the uncertainty in the *physical determination*. In any case it must be remembered that the major error in most laboratories derives from the determination of the decay rates when dealing with samples of moderate age. For very old samples with activities close to zero the stability of the background is of great importance. This is sometimes overlooked when the age limit is calculated.

As an example, results from three samples (two of them dated as three fractions) from Finnsjön consisting of carbonate concretions are given as ages greater than certain values. Two of the fractions were rather small and had to be diluted. They may be older or younger than the others, even though they were given as >26,900 and >35,900 respectively and the others as greater than values between 39,400 and 42,200. The limit given is calculated on any detectable activity plus 2σ. When all the measurements were compared with the background it was found that these samples had lower activity than the background, although still within the limits of error. Indeed the difference was $\sigma/4$.

A radiocarbon laboratory, when reporting results, should always give the physical measurement, including a correction for isotopic fractionation, but errors caused by the reservoir effect, contamination etc. should be stated separately in the conclusions. In the date-list, laboratories usually state which errors are included. There was a time when some laboratories rounded the values to include uncertainties for previous variations. This obviously caused difficulties with comparative work since the physical measurements could not then be correctly weighted and discussed. For all data, the uncertainty in the determination of the standard should be included; but when a series of dates measured at a single laboratory are compared, this uncertainty should be regarded as a systematic error, excluded at first, and then added statistically to the mean value if such a value is given. When calculating rates, and in internal comparisons, the uncertainties which are systematic and common should be excluded. This is the reason why the uncertainty of the half-life should never be included in the physical result. The value of the half-life is irrelevant to calibration of the results, provided that the radiocarbon age and the calibration curves are calculated with known fixed half-life values.

When comparing results, especially between different laboratories, one must remember that there might be a bias between the sets of results. The expected statistical spread of the measurements of the international common standard will undoubtedly cause some of the values used to be too high and others too low. An evaluation of some results was given by Stuiver (1982), and a comparison between 20 laboratories based on eight samples was presented in 1982 in *Nature* (**298**, 619–623.) It is apparent that some laboratories estimate their statistical uncertainties satisfactorily while others underestimate the errors. It is difficult for the consumer to judge whether the given uncertainties should be taken as real, or increased. Studies of publications from the laboratories giving long-term stability and statistical estimates may help.

The basic rule for statistical addition of errors (σ_1, σ_2, ... σ_n) is to take the square root of the sum of the squares of the different errors:

$$\sigma = (\sigma_1^2 + \sigma_2^2 + ... + \sigma_n^2)^{1/2}$$

Independent values are weighted according to the inverted squares of the errors when a mean value is calculated:

$x_1 \pm \sigma_1$ is given the weight $1/\sigma_1^2$

$x_2 \pm \sigma_2$ is given the weight $1/\sigma_2^2$

$x_n \pm \sigma_n$ is given the weight $1/\sigma_n^2$

The mean value is:

$$\bar{x} = \frac{(1/\sigma_1)^2 x_1 + (1/\sigma_2)^2 x_2 + \ldots + (1/\sigma_n)^2 x_n}{(1/\sigma_1)^2 + (1/\sigma_2)^2 + \ldots + (1/\sigma_n)^2}$$

and the error of the mean value is:

$$\sigma_m = \left[\frac{1}{(1/\sigma_1)^2 + (1/\sigma_2)^2 + \ldots + (1/\sigma_n)^2} \right]^{1/2}$$

Final error in date ascribed to pollen-analytical level

It must be remembered that radiocarbon dating of a certain level of sediment or peat accumulation is subject to error, because of the uncertainty not only of the date but also in the level. If, for instance, pollen analyses are performed every 5 cm, and the accumulation rate is 1 cm in 30 years, one should ascribe to the date an uncertainty corresponding to 2.5 cm, i.e. 75 years. This value is of the same order as an ordinary statistical uncertainty of a radiocarbon date. The final uncertainty of the age of the level to be used for further discussion is thus 105 radiocarbon years. If the accumulation rate is imprecisely determined the uncertainty is even greater. If it is given as 1 cm in 30±6 years, 2.5 cm indeed corresponds to 75±15 years. If the value ±15 years is added statistically the uncertainty becomes 110 years. Thus it is recommended that pollen analyses should be tightened for the levels where radiocarbon samples are taken.

SIZE OF THE SAMPLE

Most laboratories have different detectors suitable for samples equivalent to 1 g or more of carbon. If the extremely small counter described by Stoenner *et al.* (1979) is employed, 10–20 mg of carbon can be used, but then the statistical uncertainty will also be higher. If, however, measuring times of 70 days are used, as reported by Harbottle at the Archeometric Conference in London in March 1979, reasonable uncertainties can be achieved. Much smaller samples can be used when dating with accelerators instead of ion detectors, such as proportional or scintillator counters. This technique, developed well enough to allow 'routine' measurements in a few laboratories with results comparable with conventional datings, was described during the Third International Symposium on Accelerator Mass Spectrometry, Zurich, 10–13 April 1984.

Loss during treatment

Peat

When wet material is collected and taken directly from the corer, 50–150 g or 150 cm^3 usually gives a sample of 1 g of carbon as the insoluble fraction and a smaller amount as a soluble fraction.

Gyttja, etc.

The amount of organic material varies so widely as to exclude a general figure. If, however, the sample is regarded as organic, 10–20 cm^3 may suffice to yield 1 g of carbon. Thus a sample of around 20 g is a minimum when the sample is organic. Ten times this amount is often needed. The ignition loss may serve as a guide.

The yield of gas from the insoluble fraction is frequently around 40%. The remainder sometimes proves to be diatoms. Since the degree for the decay also varies, the figure given here can only be regarded as a guide.

Samples containing very little carbon should be dated on the soluble fraction.

Wood

When the wood is dry and in good condition, half is usually removed at the chemical pretreatment stage, almost 3 g being needed for the combustion in order to obtain a sample equivalent to 1 g of carbon. Thus 6 g is a normal starting amount after the mechanical treatment. Sometimes, however, the condition of the wood is so poor that 60 g or even more is suitable.

Charcoal

As a rule, slightly more carbon is present in charcoal than in wood, but essentially the same figure can be used.

Resin

No pretreatment with sodium hydroxide is applied and about 3 g or slightly less is equivalent to 1 g of carbon.

Shells

Since 12 g of carbonate is equivalent to 1 g of carbon, a sample of 50 g is recommended to allow the removal of 20–25 g and a preparation of two fractions.

Bones

The condition varies greatly. In many areas it is possible to date the samples when 200 g is supplied. Laboratories usually prefer much more material because of the varying state of degradation seen in bone samples.

II. METHODS FOR LATE HOLOCENE DEPOSITS

The ^{210}Pb dating method was devised by Goldberg (1963) but did not come into common use until a few years ago. It is based on the escape of radon from the Earth, and the subsequent decay of this radioactive gas with a half-life of 3.8 days. The daughters have fairly short half-lives until the isotope ^{210}Pb is reached (Figure 14.10). Since its half-life is 22.26±0.22 years, ^{210}Pb should be suitable for dating within the range of 1–150 years. The decay

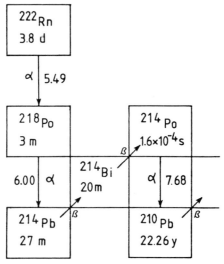

FIGURE 14.10. The uranium series from radon to lead-210

products of radon are solid and will descend to be embedded in sediments. Because of the decay the lead thus supplied will disappear, so that little remains after 150 years. By studying the remaining lead as a function of the depth the period since the lead was deposited can be determined and, thereby, the sediment accumulation rate.

As in all dating methods there are complications. One is that the sediment in a lake usually contains minerals incorporating an amount of uranium, or its daughters, which is not negligible. This causes a continuous supply of ^{210}Pb to the sediment at all depths. This lead must be subtracted from the total to find the lead relevant to the dating — the unsupported lead. This extra continuous supply is counteracted by the decay. In the ideal case the lead from this source, called the supported lead, is in radioactive equilibrium, the parents being ^{226}Ra and uranium. But since the method is based on the escape of radon, a daughter of ^{226}Ra, from the Earth, it is clear that the supported lead is not always in equilibrium with the radium in the sediment. If the radium is out of equilibrium with Th and U, the supported lead may vary with depth. There are two methods now in use to determine the supported lead. One measures the lead at levels older than 150 years and assumes that the supported ^{210}Pb is the same higher up. But since this is not always the case, it is advisable to measure the radium content at several levels and discuss the numerical treatment of the data when the ^{226}Ra content and the ^{210}Pb content on some levels are known. The number of parallel determinations should also be fixed when results are available and the desired accuracy is determined.

It must be emphasized that the unsupported lead deposited per unit area and time varies from place to place. Thus a few determinations from every core must be performed from the top down to a level corresponding to 100–150 years, besides a few for levels older than 150 years to determine the accumulation rate of the sediment (Figure 14.11).

If the sediment is purely inorganic the method may fail since the lead should be scavenged from the water within a reasonable time; this applies especially if organic matter is present to which the lead is attached. Since the ^{210}Pb deposition may vary from year to year, it is advisable to use samples from a few years, to eliminate the short-term variations.

If the sediment is inhomogeneous the normalization may present problems. Usually the ^{210}Pb content is determined per gram of dry sediment. It appears that an extra irregular supply of, for example, clay from agricultural activities such as ditching will give a smaller unsupported ^{210}Pb concentration per gram. Different compositions of minerogenic material may also involve variation in the supply of supported lead.

THE C.I.C. AND C.R.S. MODELS FOR ^{210}Pb DATING

Two models are in current use. The simpler is called the 'constant initial

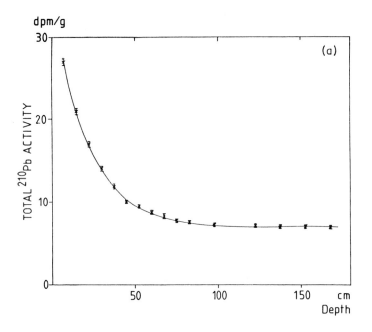

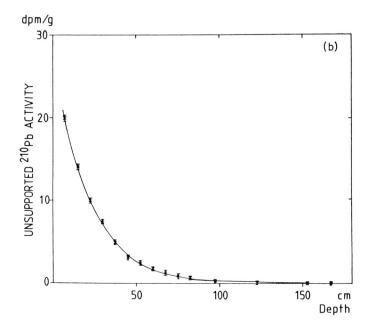

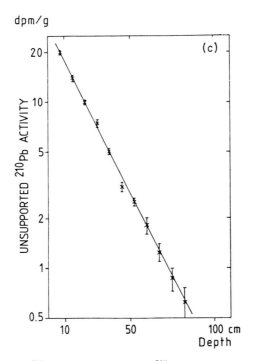

FIGURE 14.11.   Total ^{210}Pb and unsupported ^{210}Pb in a sediment. An ideal case is depicted where the supported lead can be determined from sediment which is about 150 years old or more, and subtracted from the total ^{210}Pb to yield the unsupported lead. The compaction is neglected here but it may modify results unless a correction is made. Bioturbation frequently disturbs the uppermost sediment. Whenever the sediment is inhomogeneous, normalization may present problems

concentration' (c.i.c.) model. The results given in Figure 14.11 can be calculated with this model. It is obvious that a homogeneous sediment with a constant accumulation rate can be discussed using this model, since the amount of unsupported ^{210}Pb precipitated each year in one specific area should be constant. There may be some extra lead close to areas used for testing nuclear weapons.

The 'constant rate of supply' (c.r.s.) model was devised by Goldberg (1963) and Crozaz et al. (1964). It means that the same amount of unsupported lead (A_0) is supplied per chronological unit. After time t this has decayed to:

$$A_x = A_0 e^{-\lambda t_x}$$

and is then buried at the depth x. The total activity of unsupported lead down to a depth corresponding to about 200 years, where the remaining activity is small, is then:

$$A_{\text{total}} = \int_0^{200} A_0 e^{-\lambda t_x} \, dt = \frac{1}{\lambda} A_0(1 - e^{-200\lambda}) \approx \frac{1}{\lambda} A_0$$

At depth x:

$$A_{x\,\text{total}} = \frac{1}{\lambda} A_0(1 - e^{-\lambda t_x})$$

$$t_x = -\frac{1}{\lambda} \ln\left(1 - \frac{A_{x\,\text{total}}}{A_{\text{total}}}\right) = \frac{1}{\lambda} \ln\left(\frac{A_{x\,\text{total}}}{A_{\text{total}} - A_{x\,\text{total}}}\right)$$

Since A corresponds to the deposition during one year and the measurements are usually normalized to unit dry weight, the measured activity N_x is given as:

$$N_x = \frac{A_x}{r} = \frac{A_0 e^{-\lambda t_x}}{r}$$

where r is the rate expressed as unit dry weight per chronological unit. Hence:

$$N_x = \frac{A_0 - \lambda A_{x\,\text{total}}}{r}$$

$$\approx \frac{\lambda}{r}(A_{\text{total}} - A_{x\,\text{total}})$$

$$r = \frac{\lambda}{N_x}(A_{\text{total}} - A_{x\,\text{total}})$$

It is clear that many measurements are needed to integrate the activity over the whole core.

Oldfield *et al.* (1978) applied the c.r.s. method on some sediments and obtained much better results than with the c.i.c. method. For two sites this was seen from the dating of an ash layer. The calculation and the c.r.s. model are also given by Appleby and Oldfield (1978, 1983). The c.i.c. model falls short of the c.r.s. model when the results are compared with absolute dates of laminated lake sediments (Appleby *et al.*, 1979) and *Sphagnum* dated with the moss increment method (El-Daoushy *et al.*, 1982).

Figures for density, dry weight, organic content, pH and grain size distributions are necessary for accurate interpretation of the data. These are used to translate the depth in centimetres per year into grams per year. Information concerning geochemical conditions should be provided if known. Information on drainage etc. will also help to fix years for certain levels and for data processing. The compaction must be taken into consideration.

The sample required for a ^{210}Pb dating consists of a few grams of dry material. It should preferably be delivered freeze-dried, or dried at a temperature lower than 100°C to remove most of the water; or as a wet sample to avoid troublesome grinding. As with radiocarbon dating, the information should be given on special sheets, and communication with the laboratory is recommended before the samples are submitted. The practical laboratory work for one determination of lead or radium at present takes almost one day.

The technique of measurement for ^{210}Pb determinations has been described by El-Daoushy (1978). The problems involved in dating are discussed by El-Daoushy (1978), Oldfield *et al.* (1978) and others (see below, under Bioturbation). Early datings were published by Pennington (1973) and Pennington *et al.* (1973, 1976).

SOME COMPLICATIONS WITH ^{210}Pb DATING

If the top of the core is missing the c.i.c. model seems to be useful, but the total time for which unsupported lead can be detected is significantly less than 150 years. Sometimes the curves showing the unsupported ^{210}Pb concentration as a function of depth show a flat section at the top. This may be due to physical turbation or bioturbation. In laminated varves from Baja California a decrease in the ^{210}Pb concentration is seen. The varves are thick with a high water content. Koide *et al.* (1973) suggested some mobilization of lead in the interstitial water and also suggested that disturbance further down the core was due to slumping.

RELATED ISOTOPES

The ^{228}Th/^{232}Th ratios can be used for periods up to about 10 years, determined by the half-life of 1.9 years for ^{228}Th (Koide *et al.*, 1973), which is separated from the water and the predecessor ^{228}Ra. It is the excess ^{228}Th which is used, but a normalization to ^{232}Th is made.

Similarly the ^{234}U/^{230}Th and ^{235}U/^{231}Pa ratios can be used, but because of the half-lives these methods are especially effective for samples somewhat older than 10,000 years (Broecker, 1963; Kaufman *et al.*, 1971; Ivanovich and Harman, 1982).

^{137}Cs, ^{239,240}Pu AND ^{241}Am DATING

Isotopes produced artificially as a consequence of nuclear weapon tests can be used as an injection which can be traced in sediments. ^{14}C has been used as a tracer for studies of the exchange between different reservoirs and also for dating on a short-term scale. ^{137}Cs, with a half-life of 30 years, has been detectable since 1945. In 1954 the first pronounced increase occurred in the Northern Hemisphere, in 1960 there was a minimum, and in 1963 the maximum, but by 1965 the activity was about one-third of that in 1963. This isotope was used by Pennington *et al.* (1973) for studies of lakes in the English Lake District and was used in combination with palaeomagnetic measurements, ^{210}Pb and ^{14}C measurements (Pennington, 1973; Pennington *et al.*, 1976). As with ^{14}C, the peaks are less pronounced in the Southern Hemisphere and exhibit another pattern (Longmore *et al.*, 1983). In Hidden Lake, the ^{137}Cs increases steadily towards the surface. Physical and biological mixing seems negligible. It is possible that the sediments reach the lake with some delay since they were precipitated from the atmosphere on to the catchment area. The accumulation rate in the lake is low. Furthermore the total ^{137}Cs in a core is less than expected from the fallout. Transport deeper into the sediment is apparent. But since some zoo plankton contain excess ^{137}Cs in comparison with the vegetation etc., it has also been suggested (Longmore *et al.*, 1983) that ^{137}Cs is released from the sediments and taken up from the water by the biota. Oldfield *et al.* (1979) concluded that they could detect diffusion of ^{137}Cs and its uptake by living plants in ombrotrophic peat. Similarly Jaakkola *et al.* (1983) using a time scale fixed by annual lamina, state that the ^{137}Cs migrates in a sediment. ^{137}Cs has sometimes been detected using the β^--decay after chemical separation, but it is usually measured on the bulk sediment, the 661 keV γ-peak and either a NaI (Tl) or a Ge(Li) detector.

^{239,240}Pu, with half-lives of 2.4×10^4 and 6.5×10^3 years, should exhibit about the same distribution in time as ^{137}Cs, and have been tested in some laboratories. Jaakkola *et al.* (1983) found a migration, although less than for ^{137}Cs. Both isotopes can be used for dating, but the uncertainty must be given in years. ^{241}Pu and ^{241}Am, from fallout, have also been detected in recent sediments (Koide *et al.*, 1980). The concentration follows that of ^{239,240}Pu.

^{32}Si AND ^{39}Ar DATING

Cosmic-ray-produced ^{32}Si and ^{39}Ar may bridge the gap between ^{210}Pb and ^{14}C. The half-life of ^{32}Si was long given as a few hundred years, but was recently revised to be about 300 years (Clausen, 1973; DeMaster, 1980) based on counted annual layers in ice or a combination of varved sediment and ^{210}Pb. Very recently, however, two new measurements have appeared 108 ± 18 and 101 ± 18 years respectively, using accelerators (Elmore *et al.*, 1980; Kutschera *et al.*, 1980). The half-life of ^{39}Ar is 269 years.

^{32}Si has been determined by measurements of the daughter ^{32}P. ^{32}Si yields a potential dating method for sediments and ice. It is scavenged by rain and snow. Thus a similar technique to that for ^{210}Pb will be applied by determining the activity as a function of depth for a depth profile. Some ^{32}Si is produced by nuclear-bomb testing.

^{39}Ar is apparently produced at a constant rate, and artificially produced ^{39}Ar is negligible (Loosli, 1983). Being a noble gas the geochemical complications are reduced. Production of ^{39}Ar in granitic rocks is one complication. The isotope seems suitable for glaciers, oceans and groundwater. The normalizations are made against the total argon content. The determinations are difficult and made in underground laboratories because of the very low background radiation levels needed.

GENERAL REMARKS ON SOME DATING ERRORS

Bioturbation

When dating sediments the risk of bioturbation should be considered. Preserved varves are an indication that no bioturbation has occurred. Studies of the geochemical conditions may yield other signs. Recent studies using data from ^{14}C, ^{210}Pb and ^{137}Cs measurements, as well as earlier papers, reveal that bioturbation may occur and distort the dates. In the theoretical case of complete mixing down to a certain depth, and no bioturbation below this sharp boundary, it is obvious that the top layer will be dated too old, because of old material transported upwards, whereas the deepest layer of the mixed zone will be dated too young. In consequence any layer farther down in the sediment, once disturbed by bioturbation, is dated somewhat too young. The bioturbation is discussed by, *inter alia*, Robbins and Edgington (1975), Robbins *et al.* (1977), Peng *et al.* (1976), Nozaki *et al.* (1977), Håkansson and Källström (1978), and Berger and Johnson (1978).

A simple box model is usually described to provide a qualitative explanation. Since the mixing process is particle-selective a diffusion model has been applied in many discussions. Certain disturbances may also derive from evolved gases, and wind and wave action if the water is shallow.

Thus, if the radiocarbon ages of different layers of the top sediment are the same, within the margin of error, the reason is not always apparent age — a combination of contamination by old eroded material and a reservoir age of the water — bioturbation may also have played a part. In such cases a combination of datings with different methods offers the best solution.

Another consequence of bioturbation is that some species — old material such as coccoliths — can be detected high up in a sequence, as pointed out by McIntyre *et al.* (1967), and by Bilal Ul Haq in Olausson *et al.* (1971). This material may be avoided (by means of grain size selection) by radiocarbon dating of the less mobile foraminifera (Olsson, 1972a).

Frost and other mechanical action

Errors may arise from disturbances by frost of the stratigraphy in soil. Material may fill up ice wedges. In sorted circles shells may move upwards by continued freezing and thawing so that rather old material can be detected on the surface. This was seen on the raised beaches at Kapp Linné, Spitsbergen (Olsson *et al.*, 1976).

Furthermore, charcoal may be displaced downwards along cavities left by roots (Florin, 1975).

Movements on another scale occur when deposits are transported by glaciers; shells of widely different ages may be collected from such assemblages. Difficulties in dating raised beaches with the help of shells are discussed by Donner and Jungner (1980) and Donner *et al.* (1977). These investigations suggest that the shells should be separated, whenever possible, according to species. Storm ridges indicate that samples may be deposited at much higher levels than the mean sea level implies (Birkenmajer and Olsson, 1971). This is of importance in shore displacement studies.

Correlation between cores

Whenever there is a need to use material from two or more cores, collected close to each other, to provide enough material to date, the cores must be carefully correlated with each other. This can sometimes be done visually, especially when distinct varves are seen, or by more elaborate investigations such as pollen and diatom analysis, or sometimes even by X-ray analysis (cf. Chapter 13).

Communication

The requirements for obtaining reliable dates are that the sample is carefully collected, that the system is closed, that pretreatment is undertaken to remove all possible contaminations, and that the activity measurement is made with care so as to eliminate possible spurious errors. One fundamental requirement for collaboration between a researcher submitting material and a radiocarbon analyst is that the researcher gives information about the sample. This is normally done on specially prepared sheets provided by the different laboratories. These then provide a basis not only for further discussion but also for assessing correct treatments in the laboratory, for publication of results in lists to be regarded as abstracts and for comprehensive surveys of the work. This chapter was written to show the scope of dating methods, to provide the basis for a better understanding of the problems encountered, and to elucidate the possible uncertainties that accompany the dates.

REFERENCES

Appleby, P. G., and Oldfield, F. (1978). The calculation of lead-210 dates assuming a constant rate of supply of unsupported ^{210}Pb to the sediment. *Catena*, **5**, 1–8.
Appleby, P. G., and Oldfield, F. (1983). The assessment of ^{210}Pb data from sites with varying sediment rates. *Hydrobiologia*, **103**, 29–35.
Appleby, P. G., Oldfield, F., Thompson, R., Huttunen, P., and Tolonen, K. (1979). ^{210}Pb dating of annually laminated lake sediments from Finland. *Nature*, **280**, 53–55.

Bada, J. L., and Schroeder, R. A. (1975). Amino acid racemization reactions and their geochemical implications. *Naturwissenschaften*, **62**, 71–79.

Baillie, M. G. L., Pilcher, J. R., and Pearson, G. W. (1983). Dendrochronology at Belfast as a background to high-precision calibration. *Radiocarbon*, **25**, 171–178.

Becker, B. (1983). The long-term radiocarbon trend of the absolute German oak tree-ring chronology, 2800 to 800 BC. *Radiocarbon*, **25**, 197–203.

Berger, W. H., and Johnson, R. F. (1978). On the thickness and the ^{14}C age of the mixed layer in deep-sea carbonates. *Earth and Planetary Science Letters*, **41**, 223–227.

Berglund, B. E., Håkansson, S., and Lagerlund, E. (1976). Radiocarbon-dated mammoth (*Mammuthus primigenius* Blumenbach) finds in South Sweden. *Boreas*, **5**, 177–191.

Birkenmajer, K., and Olsson, I. U. (1971). Radiocarbon dating of raised marine terraces at Hornsund, Spitsbergen, and the problem of land uplift. *Norsk Polarinstitutt — Årbok 1969*, 17–43.

Bolin, B. (1976). Modeling the oceans and ocean sediments and their response to fossil fuel carbon dioxide emissions. In: *The Fate of Fossil Fuel CO_2 in the Oceans*. Marine Sciences 6 (Eds. N. R. Andersen and A. Malahoff), Plenum Press, New York, pp. 81–95.

Broecker, W. S. (1963). A preliminary evaluation of uranium series in equilibrium as a tool for absolute measurement on marine carbonates. *J. Geophys. Res.*, **68**, 2817–2834.

Bruns, M., Levin, I., Münnich, K. O., Hubberten, H. W., and Fillipakis, S. (1980b). Regional sources of volcanic carbon dioxide and their influence on ^{14}C content of present-day plant material. *Radiocarbon*, **22**, 532–536.

Bruns, M., Münnich, K. O., and Becker, B. (1980a). Natural radiocarbon variations from AD200 to 800. *Radiocarbon*, **22**, 273–277.

Bruns, M., Rhein, M., Linick, R. W., and Suess, H. E. (1983). The atmospheric ^{14}C level in the 7th Millenium BC. *PACT*, **8**, 511–516.

Chatters, R. M., Crosby, J. W. III, and Engstrand, L. G. (1969). *Fumarole Gaseous Emanations: their Influence on Carbon-14 Dates*, Washington State Univ., Technical extension service, circ, 32.

Chisholm, B. S., Nelson, D. E., and Schwarz, H. P. (1983). Dietary information from δ^{13}C and δ^{15}N measurement on bone collagen. *PACT*, **8**, 391–400.

Clark, R. M. (1975). A calibration curve for radiocarbon dates. *Antiquity*, **49**, 251–266.

Clausen, H. B. (1973). Dating of polar ice by ^{32}Si. *J. Glac.*, **12**, 411–416.

Craig, H. (1954). Carbon 13 in plants and the relationships between carbon 13 and carbon 14 variations in nature. *J. Geol.*, **62**, 115–149.

Crozaz, G., Picciotto, E., and de Breuck, W. (1964). Antarctic snow chronology with ^{210}Pb. *J. Geophys. Res.*, **69**, 2597–2604.

Damon, P. E., Ferguson, C. W., Long, A., and Wallick, E. I. (1974). Dendrochronologic calibration of the radiocarbon time scale. *Am. Antiquity*, **39**, 350–366.

Damon, P. E., Long, A., and Wallick, E. I. (1973). Dendrochronologic calibration of the carbon-14 time scale. In: *Proc. 8th Int. Conf. Radiocarbon Dating*, Lower Hutt, 18–25 Oct. 1972, Royal Society of New Zealand, Wellington, pp. A28–A43 (44–59).

DeMaster, D. J. (1980). The half life of ^{32}Si determined from a varved Gulf of California sediment core. *Earth and Planetary Science Letters*, **48**, 209–217.

Donner, J., Eronen, M., and Jungner, H. (1977). The dating of the Holocene relative sea-level changes in Finnmark, North Norway. *Nor. Geogr. Tidsskr.*, **31**, 103–128.

Donner, J. J., and Gardemeister, R. (1971). Redeposited Eemian marine clay in

Somero, south-western Finland. *Bull. Geol. Soc. Finland*, **43**, 73–88.

Donner, J. J., and Jungner, H. (1973). The effect of re-deposited organic material on radiocarbon measurements of clay samples from Somero, south-western Finland. *Geol. Fören. Stockh. Förh.*, **95**, 267–268.

Donner, J. J., and Jungner, H. (1974). Errors in the radiocarbon dating of deposits in Finland from the time of deglaciation. *Bull. Geol. Soc. Finland*, **46**, 139–144.

Donner, J. J., and Jungner, H. (1980). Radiocarbon ages of shells in Holocene marine deposits. *Radiocarbon*, **22**, 556–561.

Donner, J. J., Jungner, H., and Vasari, Y. (1971). The hard-water effect on radiocarbon measurements of samples from Säynäjälampi, north-east Finland. *Comm. Physico-Math.*, **41**, 307–310.

El-Daoushy, M. F. A. F. (1978). *The Determination of Pb-210 and Ra-226 in Lake Sediments and Dating Applications*, Uppsala University Institute of Physics Report 979.

El-Daoushy, M. F. A. F., Olsson, I. U., and Oro, F. H. (1978). The EDTA and HCl methods of pre-treating bones. *Geol. Fören. Stockh. Förh.*, **100**, 213–219.

El-Daoushy, F., Tolonen, K., and Rosenberg, R. (1982). Lead 210 and moss-increment dating of two Finnish *Sphagnum* hummocks. *Nature*, **296**, 429–431.

Elmore, D., Anantaraman, N., Fulbright, H. W., Gove, H. E., Hans, H. S., Nishiizumi, K., Murrell, M. T., and Honda, M. (1980). Half-life of ^{32}Si from tandem-accelerator mass spectrometry. *Phys. Rev. Letters*, **45**, 589–592.

Engstrand, L. G. (1965). Stockholm natural radiocarbon measurements, VI. *Radiocarbon*, **7**, 257–290.

Florin, M.-B. (1975). Microfossil contents of two soil profiles from western Kolmården, southern central Sweden. *Geol. Fören. Stockh. Förh.*, **97**, 135–141.

Freundlich, J. C., and Schmidt, B. (1983). Calibrated ^{14}C dates in Central Europe — same as elsewhere?. *Radiocarbon*, **25**, 279–286.

Geyh, M. A., Krumbein, W. E., and Kudrass, H.-R. (1974). Unreliable ^{14}C dating of long-stored deep-sea sediments due to bacterial activity. *Marine Geol.*, **17**, M45–M50.

Godwin, H. (1959). Carbon-dating conference at Groningen, September 14–19, 1959. *Nature*, **184**, 1365–1366.

Godwin, H. (1962). Half-life of radiocarbon. *Nature*, **195**, 984.

Goldberg, E. D. (1963). Geochronology with Pb^{210}. In: *Radioactive Dating*, Proc. Symp. Athens Nov. 19–23, 1962, IAEA, Vienna, pp. 121–131.

Grant-Taylor, T. L. (1973). The extraction and use of plant lipids as a material for radiocarbon dating. In: *Proc. 8th Int. Conf. Radiocarbon Dating*, Lower Hutt, 18–25 Oct. 1972, Royal Society of New Zealand, Wellington, pp. E58–E67 (439–448).

Gulliksen, S. (1980). Isotopic fractionation of Norwegian materials for radiocarbon dating. *Radiocarbon*, **22**, 980–986.

Håkansson, L., and Källström, A. (1978). An equation of state for biologically active lake sediments and its implications for interpretations of sediment data. *Sedimentology*, **25**, 205–226.

Håkansson, S. (1969). University of Lund radiocarbon dates, II. *Radiocarbon*, **11**, 430–450.

Håkansson, S. (1970). University of Lund radiocarbon dates, III. *Radiocarbon*, **12**, 534–552.

Håkansson, S. (1974). University of Lund radiocarbon dates, VII. *Radiocarbon*, **16**, 307–330.

Håkansson, S. (1983). A reservoir age for the coastal waters of Iceland. *Geol. Fören. Stockh. Förh.*, **105**, 64–67.

Harkness, D. D. (1983). The extent of natural ^{14}C deficiency in the coastal environment of the United Kingdom. *PACT*, **8**, 351–364.

Hassan, A. A., and Hare, P. E. (1978). Amino acid analysis in radiocarbon dating of bone collagen. In: *Proc. 6th Am. Chem. Soc. Symposium on Archeological Chemistry*, Chicago, 1977, pp. 109–116.

Ho, T. Y., Marcus, L. F., and Berger, R. (1969). Radiocarbon dating of petroleum-impregnated bone from tar pits at Rancho La Brea, California. *Science*, **164**, 1051–1052.

Hörnsten, Å., and Olsson, I. U. (1964). En C^{14}-datering av glaciallera från Lugnvik, Ångermanland. *Geol. Fören. Stockh. Förh.*, **86**, 206–210.

Huttunen, P., and Tolonen, K. (1977). Human influence in the history of Lake Lovojärvi, S. Finland. *Finskt Museum* **1975**, 68–105.

Ivanovich, M., and Harmon, R. S. (Eds.) (1982). *Uranium Series Disequilibrium: Applications to Environmental Problems*. Clarendon Press, Oxford.

Jaakkola, T., Tolonen, K., Huttunen, P., and Leskinen, S. (1983). The use of fallout ^{137}Cs and $^{239,240}Pu$ for dating of lake sediments. *Hydrobiologia*, **103**, 15–19.

de Jong, A. F. M. (1981). *Natural ^{14}C variations*. Thesis Groningen, 119pp.

de Jong, A. F. M., and Mook, W. G. (1980). Medium-term atmospheric ^{14}C variations. *Radiocarbon*, **22**, 267–272.

de Jong, A. F. M., Mook, W. G., and Becker, B. (1979). Confirmation of the Suess wiggles: 3200–3700 B.C. *Nature*, **280**, 48–49.

Kaufman, A., Broecker, W. S., Ku, T.-L., and Thurber, D. L. (1971). The status of U-series methods of mollusc dating. *Geochim. Cosmochim. Acta*, **35**, 1155–1183.

Klein, J., Lerman, J. C., Damon, P. E., and Ralph, E. K. (1982). Calibration of radiocarbon dates: Tables based on the consensus data of the Workshop on Calibrating the Radiocarbon Time Scale. *Radiocarbon*, **24**, 103–150.

Koide, M., Bruland, K. W., and Goldberg, E. D. (1973). Th-228/Th-232 and Pb-210 geochronologies in marine and lake sediments. *Geochim. Cosmochim. Acta*, **37**, 1171–1187.

Koide, M., Goldberg, E. D., and Hodge, V. F. (1980). ^{241}Pu and ^{241}Am in sediments from coastal basins off California and Mexico. *Earth and Planetary Science Letters*, **48**, 250–256.

Kutschera, W., Henning, W., Paul, M., Smither, R. K., Stephenson, E. J., Yntema, J. L., Alburger, D. E., Cumming, J. B., and Harbottle, G. (1980). Measurement of the ^{32}Si half-life via accelerator mass spectrometer. *Phys. Rev. Letters*, **45**, 592–596.

Lerman, J. C. (1973). Carbon 14 dating: origin and correction of isotope fractionation errors in terrestrial living matter. In: *Proc. 8th Int. Conf. on Radiocarbon Dating*, Lower Hutt, 18–25 Oct. 1972, Royal Society of New Zealand, Wellington, pp. H16–H28 (612–624).

Lerman, J. C., Mook, W. G., and Vogel, J. C. (1970). C^{14} in tree rings from different localities. In: *Radiocarbon Variations and Absolute Chronology*, Proc. Twelfth Nobel Symposium, Uppsala, 11–15 Aug. 1969 (Ed. I. U. Olsson), Almqvist & Wiksell, Stockholm, and John Wiley & Sons, New York, pp. 275–301.

Linick, T. W., Suess, H. E., and Becker, B. (1985). La Jolla measurements of radiocarbon in South German oak tree-ring chronologies. *Radiocarbon*, **27**, 20–32.

Longin, R. (1971). New method of collagen extraction for radiocarbon dating. *Nature*, **230**, 241–242.

Longmore, M. E., O'Leary, B. M., and Rose, C. W. (1983). Caesium-137 profiles in the sediments of a partial-meromictic lake on Great Sandy Island (Fraser Island), Queensland, Australia. *Hydrobiologia*, **103**, 21–27.

Loosli, H. H. (1983). A dating method with ^{39}Ar. *Earth and Planetary Science Letters*, **63**, 51–62.

Mangerud, J. (1972). Radiocarbon dating of marine shells, including a discussion of apparent age of recent shells from Norway. *Boreas*, **1**, 143–172.

Mangerud, J., and Gulliksen, S. (1975). Apparent radiocarbon ages of recent marine shells from Norway, Spitsbergen and Arctic Canada. *Quaternary Res.*, **5**, 263–273.

Mann, W. B. (1983). An international reference material for radiocarbon dating. *Radiocarbon*, **25**, 519–527.

McIntyre, A., Bé, A. W. H., and Preikstas, R. (1967). Coccoliths and the Pliocene-Pleistocene boundary. In: *Progress in Oceanography* (Ed. M. Sears), **4**, Pergamon Press, Oxford, pp. 3–25.

McKerrel, H. (1975a). Correction procedures for C-14 dates. In: *Radiocarbon: Calibration and Prehistory* (Ed. T. Watkins), Edinburgh Univ. Press, Edinburgh, pp. 47–100.

McKerrel, H. (1975b). Conversion Tables. In: *Radiocarbon: Calibration and Prehistory* (Ed. T. Watkins), Edinburgh Univ. Press, Edinburgh, pp. 110–127.

Michael, H. N., and Ralph, E. K. (1973). Discussion of radiocarbon dates obtained from precisely dated sequoia and bristlecone pine samples. In: *Proc. 8th Int. Conf. Radiocarbon Dating*, Lower Hutt, 18–25 Oct. 1972, Royal Society of New Zealand, Wellington, pp. A11–A27 (27–43).

Mook, W. G. (1983). [14]C calibration curves depending on sample time-width. *PACT*, **8**, 517–525.

Nozaki, Y., Cochran, J. K., Turekian, K. K., and Keller, G. (1977). Radiocarbon and ^{210}Pb distribution in submersible-taken deep-sea cores from project Famous. *Earth and Planetary Science Letters*, **34**, 167–173.

Olausson, E., Bilal Ul Haq, U. Z., Karlsson, G. B., and Olsson, I. U. (1971). Evidence in Indian Ocean cores of Late Pleistocene changes in oceanic and atmospheric circulation. *Geol. Fören. Stockh. Förh.*, **93**, 51–84.

Oldfield, F., Appleby, P. G., and Battarbee, R. W. (1978). Alternative ^{210}Pb dating: results from the New Guinea Highlands and Lough Erne. *Nature*, **271**, 339–342.

Oldfield, F., Appleby, P. G., Cambray, R. S., Eakins, J. D., Barber, K. E., Battarbee, R. W., Pearson, G. W., and Williams, J. M. (1979). [210]Pb, ^{137}Cs and [239]Pu profiles in ombrotrophic peat. *Oikos*, **33**, 40–45.

Olsson, I. U. (1972a). The pretreatment of samples and the interpretation of the results of ^{14}C determinations. In: *Symposium of Climatic Changes in Arctic Areas during the Last Ten-thousand Years*, Oulanka and Kevo, 4–10 Oct. 1971. *Acta Univ. Oulensis*, **A3**, 9–37.

Olsson, I. U. (1972b). The C[14] dating of samples for botanical studies of prehistoric agriculture in northern Ångermanland. In: *Early Norrland*, **1**, Kungl. Vitterhets Historie och Antikvitets Akademien, Stockholm, pp. 35–41.

Olsson, I. U. (1973). A critical analysis of ^{14}C datings of deposits containing little carbon. In: *Proc. 8th Int. Conf. Radiocarbon Dating*, Lower Hutt, 18–25 Oct. 1972, Royal Society of New Zealand, Wellington, pp. G11–G28 (547–564).

Olsson, I. U. (1974a). The Eighth International Conference on Radiocarbon Dating. *Geol. Fören. Stockh. Förh.*, **96**, 37–44.

Olsson, I. U. (1974b). Some problems in connection with the evaluation of C[14] dates. *Geol. Fören. Stockh. Förh.*, **96**, 311–320.

Olsson, I. U. (1978). A discussion of the C[14] ages of samples from Medelpad, Sweden. *Early Norrland*, **11**, 93–97.

Olsson, I. U. (1979a). The importance of the pretreatment of wood and charcoal samples. In: *Radiocarbon Dating*, Proc. Ninth Int. Radiocarbon Conference, Los Angeles and San Diego, June 1976 (Eds. R. Berger and H. E. Suess), University of California Press, pp. 135–146.

Olsson, I. U. (1979b). A warning against radiocarbon dating of samples containing little carbon. *Boreas*, **8**, 203–207.

Olsson, I. U. (1979c). The radiocarbon contents of various reservoirs. In: *Radiocarbon Dating*, Proc. Ninth Int. Radiocarbon Conference, Los Angeles and San Diego, June 1976 (Eds. R. Berger and H. E. Suess), University of California Press, pp. 613–618.

Olsson, I. U. (1980). Content of ^{14}C in marine mammals from northern Europe. *Radiocarbon*, **22**, 662–675.

Olsson, I. U. (1983a). Radiocarbon dating in the Arctic region. *Radiocarbon*, **25**, 293–294.

Olsson, I. U. (1983b). Dating non-terrestrial materials. *PACT*, **8**, 277–294.

Olsson, I. U., El-Daoushy, M. F. A. F., Abd-El-Mageed, A. I., and Klasson, M. (1974). A comparison of different methods for pretreatment of bones, I. *Geol. Fören. Stockh. Förh.*, **96**, 171–181.

Olsson, I. U., El-Daoushy, F., and Vasari, Y. (1983). Säynäjälampi and the difficulties inherent in the dating of sediments in a hard-water lake. *Hydrobiologia*, **103**, 5–14.

Olsson, I. U., and Eriksson, K. G. (1965). Remarks on C^{14} dating of shell material in sea sediments. In: *Progress in Oceanography* (Ed. M. Sears), **3**, Pergamon, Norwich, pp. 253–266.

Olsson, I. U., Göksu, Y., and Stenberg, A. (1968). Further investigations of storing and treatment of Foraminifera and molluscs for C^{14} dating. *Geol. Fören. Stockh. Förh.*, **90**, 417–426.

Olsson, I. U., Holmgren, B., and Skye, E. (1984). Questions arising when using lichen for ^{14}C measurements in climatic studies. In: *Proc. 2nd Nordic Symp. on Climatic Changes and Related Problems*, Stockholm, May 1983 (Eds. N.-A. Mörner and W. Karlén), D. Reidel, Dordrecht and Boston, pp. 303–308.

Olsson, I. U., and Osadebe, F. A. N. (1974). Carbon isotope variations and fractionation corrections in ^{14}C dating. *Boreas*, **3**, 139–146.

Olsson, I. U., Stenberg, A., and Göksu, Y. (1967). Uppsala natural radiocarbon measurements, VII. *Radiocarbon*, **9**, 454–470.

Pearson, G. W., and Baillie, M. G. L. (1983). High-precision ^{14}C measurement of Irish oaks to show natural atmospheric ^{14}C variations of the AD time period. *Radiocarbon*, **25**, 187–196.

Pearson, G. W., Pilcher, J. R., and Baillie, M. G. L. (1983). High-precision ^{14}C measurement of Irish oaks to show the natural ^{14}C variations from 200 BC to 4000 BC. *Radiocarbon*, **25**, 179–186.

Peng, T.-H., Broecker, W. S., Kipphut, G., and Shackleton, N. (1976). Benthic mixing in deep sea cores as determined by ^{14}C dating and its implications regarding climate stratigraphy and the fate of fossil fuel CO$_2$. In: *The Fate of Fossil Fuel CO$_2$ in the Oceans*. Marine Sciences 6 (Eds. N. R. Andersen and A. Malahoff), Plenum Press, New York, pp. 355–373.

Pennington, W. (1973). The recent sediments of Windermere. *Freshwat. Biol.*, **3**, 363–382.

Pennington, W., Cambray, R. S., Eakins, J. D., and Harkness, D. D. (1976). Radionuclide dating of the recent sediments of Blelham Tarn. *Freshwat. Biol.*, **6**, 317–331.

Pennington, W., Cambray, R. S., and Fisher, E. M. (1973). Observations on lake sediments using fallout ^{137}Cs as a tracer. *Nature*, **242**, 324–326.

Ralph, E. K., Michael, H. N., and Han, M. C. (1973). Radiocarbon dates and reality. *Masca Newsletter*, **9**, 1–20.

Renberg, I. (1978). Palaeolimnology and varve counts of the annually laminated sediment of Lake Rudetjärn, northern Sweden. *Early Norrland*, **11**, 63–92.

Robbins, J. A., and Edgington, D. N. (1975). Determination of recent sedimentation rates in Lake Michigan using Pb-210 and Cs-137. *Geochim. Cosmochim. Acta*, **39**, 285–304.

Robbins, J. A., Krezoski, J. R., and Mozley, S. C. (1977). Radioactivity in sediments of the Great Lakes: post-depositional redistribution by deposit-feeding organisms. *Earth and Planetary Science Letters*, **36**, 325–333.

Saupé, F., Strappa, O., Coppens, R., Guillet, B., and Jaegy, R. (1980). A possible source of error in ^{14}C dates: volcanic emanations (examples from the Monte Amiata District, Provinces of Grosseto and Sienna, Italy). *Radiocarbon*, **22**, 525–531.

Šilar, J. (1976). Radiocarbon ground-water dating in Czechoslovakia — first results. *Věstnik Ústředního ústavu geologického*, **51**, 209–220.

Stoenner, R. W., Harbottle, G., Sayre, E. V. (1979). Carbon 14 dating of small samples by proportional counting. Lecture given at the 19th International Symposium on Archaeometry and Archaeological Prospection 1979, 28–31 March.

Stuiver, M. (1978). Radiocarbon timescale tested against magnetic and other dating methods. *Nature*, **273**, 271–274.

Stuiver, M. (1982). A high-precision calibration of the AD radiocarbon time scale. *Radiocarbon*, **24**, 1–26.

Stuiver, M. (1983). International agreements and the use of the New Oxalic Standard. *Radiocarbon*, **25**, 793–795.

Stuiver, M., and Polach, H. A. (1977). Reporting of ^{14}C Data. *Radiocarbon*, **19**, 355–363.

Stuiver, M., and Robinson, S. W. (1974). University of Washington Geosecs North Atlantic carbon-14 results. *Earth and Planetary Science Letters*, **23**, 87–90.

Suess, H. E. (1970). Bristlecone-pine calibration of the radiocarbon time-scale 5200 B.C. to the present. In: *Radiocarbon Variations and Absolute Chronology*, Twelfth Nobel Symposium, Uppsala, 11–15 Aug. 1969 (Ed. I. U. Olsson), Almqvist & Wiksell, Stockholm, and John Wiley & Sons, New York, pp. 303–311.

Suess, H. E. (1979). A calibration table for conventional radiocarbon dates. In: *Radiocarbon Dating*, Proc. Ninth Int. Radiocarbon Conf., Los Angeles and San Diego, June 1976 (Eds. R. Berger and H. E. Suess) University of California Press, pp. 777–784.

Sulerzhitzky, L. D. (1970). Radiocarbon dating of volcanoes. *Bull. Volcanologique*, **35**, 85–94.

Switsur, V. R. (1973). The radiocarbon calendar recalibrated. *Antiquity*, **47**, 131–137.

Tauber, H. (1979). ^{14}C activity of Arctic marine mammals. In: *Radiocarbon Dating*, Proc. Ninth Int. Radiocarbon Conf., Los Angeles and San Diego, June 1976 (Eds. R. Berger and H. E. Suess), University of California Press, pp. 447–452.

Tauber, H. (1983). ^{14}C dating of human beings in relation to dietary habits. *PACT*, **8**, 365–376.

Taylor, R. E. (1980). Radiocarbon dating of Pleistocene bone: toward criteria for the selection of samples. *Radiocarbon*, **22**, 969–979.

Taylor, R. E., and Slota, P. J. (1979). Fraction studies on marine shell and bone samples for radiocarbon analyses. In: *Radiocarbon Dating*, Proc. Ninth Int. Radiocarbon Conf., Los Angeles and San Diego, June 1976 (Eds. R. Berger and H. E. Suess), University of California Press, pp. 422–432.

Vita-Finzl, C., and Roberts, N. (1984). Selective leaching of shells for ^{14}C dating. *Radiocarbon*, **26**, 54–58.

Handbook of Holocene Palaeoecology and Palaeohydrology
Edited by B. E. Berglund
© 1986 John Wiley & Sons Ltd.

15

Palaeomagnetic dating

Roy Thompson

*Department of Geophysics,
University of Edinburgh, U.K.*

INTRODUCTION

Magnetic measurements have many uses in Quaternary studies and are attractive because they are rapid and non-destructive, preclude few other measurements and involve iron minerals which are sensitive to anthropogenic changes. They have been mostly applied to gyttja and clay but are potentially useful in all natural Quaternary deposits ranging through fluvioglacial, marine, limnic and pedogenic deposits to loess, ombrotrophic peat and ice cores. Applications have included:

(1) correlating between sediment cores using magnetic mineralogy variations (e.g. Thompson *et al.*, 1975; see also Chapter 13);
(2) dating and correlating between Holocene sites using palaeomagnetic directional data and a geomagnetic master curve (e.g. Mackereth, 1971; Thompson, 1977);
(3) investigating the behaviour of the geomagnetic field on a time scale of 10^2–10^6 years (e.g. Kent and Opdyke, 1977);
(4) tracing the movement of iron minerals between sites, for example, during erosion, transport and deposition cycles (e.g. Oldfield *et al.*, 1979).

This chapter will concentrate on applications (1) and (2), namely the magnetic mineral correlation and palaeomagnetic dating aspects. Other applications have been reviewed by Oldfield (1978).

FIELD METHODS AND EQUIPMENT

Choosing sediment types

Correlation studies

Material in any form can be used for magnetic correlations which are based on magnetic mineral type and concentration. Both fresh or dried material can be analysed. As in any stratigraphic investigation, longer and more complete and continuous sequences produce more reliable correlations.

Dating studies

Gyttja rather than peat or silt has produced the most useful data. For palaeomagnetic direction measurements and dating, material should be (*a*) orientated, (*b*) fine grained (< 63 µm), and (*c*) fresh. The more orientation information available, the more reliable the magnetic dating. In cliff or pit sections complete orientation should be possible. In present lake deposits a long, continuous, straight, vertical core, even without detailed orientation information, is generally sufficient for palaeomagnetic studies.

Sampling and coring equipment

Correlation studies

Any sampling method can be used.

Dating studies

In order to obtain palaeomagnetic direction information, undeformed samples must be collected. In cliff or pit sections box corers can be pushed into the deposit, and the strike and dip of the box noted. Alternatively, plastic sample holders (volume ~ 10 ml) may be pushed directly into the face. An orientation accuracy of about 2° can be obtained using an ordinary magnetic compass and dip circle. In present lake deposits, piston corers (e.g. Mackereth or Livingstone) and straight push–pull corers (e.g. Jowsey or Russian) take suitable cores (see Chapter 8). Coring methods which involve vibration or rotation of the core bit are liable to disturb the sediment too much. Long cores (> 3 m) are most suitable. If only short core sections (~ 1 m) are available, two complete, but overlapping, series should be taken. Palaeomagnetic land coring is more difficult, the most suitable equipment being the Swedish foil corer. With all coring equipment great care must be taken to ensure that the cores are not twisted.

Sampling strategy

Palaeomagnetic directional data in sediments are not necessarily a true reflection of ancient field directions (Watkins, 1971). Palaeomagnetic results from a single core can thus be ambiguous. However, if an adjacent core is available, interpretation is greatly simplified and dating is much more reliable. Two adjacent cores or sections should always be collected for palaeomagnetic studies.

LABORATORY WORK

Subsampling

Fresh material for directional data

The usual method is to take subsamples using small, rectanglar sided, plastic boxes of volume roughly 10 ml. A small hole allows air to escape during subsampling and can be subsequently sealed with waterproof tape.

Dried material

Dried sediment allows magnetic concentration calculations on a specific dry-weight basis to be made. Drying at low temperature, e.g. 40°C, is preferable, as possible alteration of magnetic minerals is avoided. The dried sediment can be packed firmly into small plastic boxes, using clean foam rubber, for magnetic measurements. Distortion during drying may affect the direction of natural remanence and so fresh samples are to be preferred for natural directional data. However, all other magnetic measurements can be made without difficulty on dried material.

Fine suspensions

Suspended particles may be collected on *clean* filter papers, dried, reweighed, slightly moistened and packed into plastic boxes for magnetic measurements (Oldfield *et al.*, 1979).

Peat

Peat samples are generally low in magnetic mineral content (Oldfield *et al.*, 1978b). Compression of the peat, using a normal vice, will increase the sensitivity of magnetic analyses.

Measurement

The main magnetic measurements of value to the palaeolimnologist are as follows:

(1) *whole-core susceptibility*, which permits correlation based on concentration of magnetic minerals (Radhakrishnamurty *et al.*, 1968; Molyneux and Thompson, 1973);
(2) *single-sample susceptibility* (χ), which permits correlation based on concentration of magnetic minerals;
(3) *single-sampling saturation isothermal remanent magnetization* (SIRM), which permits correlation based on concentration of magnetic minerals. SIRM also allows correlation based on mineral types using SIRM/χ ratios, calculated from measurements (2) and (3);
(4) *coercivity* of SIRM, which permits correlation based on type of magnetic minerals;
(5) *Whole-core remanence*, which permits dating and correlation based on palaeomagnetic vectors (Molyneux *et al.*, 1972; Dodson *et al.*, 1974);
(6) *single-sample remanence*, which permits dating and correlation based on palaeomagnetic vectors;
(7) *single-sample cleaned remanence*, which permits dating and correlation based on stable palaeomagnetic vectors.

Instruments

A great range of instruments is available for magnetic analyses (Collinson, 1983). Suitable palaeomagnetic equipment is widely available in geophysics institutes and universities throughout the world.

Suseptibility bridges

Most instruments permit subsample susceptibility measurements at a rate of 50–100 samples per hour. For measurements of recent sediments the noise level should be less than about 10^{-5} S.I. units. Inexpensive portable battery-operated equipment suitable for field and laboratory use has been described by Scollar (1968) and is available commercially.

Magnetometers

The natural remanence of recent sediments can be measured on many types of instruments, provided the sample is not rotated at too high a rate (e.g. >50 Hz), and that the noise level is about 10^{-3} Am^{-1} or lower using a sample measuring time of about 5 minutes. SIRM can be measured on magnetometers with a noise level two orders of magnitude higher than this. If large numbers of samples are to be processed, computerized instruments (e.g. Molyneux, 1971) are invaluable. Portable fluxgate and thin-film magneto-meters became commercially available in 1981.

Field generation

Uniform fields of strength up to 1 T can be produced for SIRM and coercivity studies by conventional electromagnets with flat 50 mm diameter pole pieces. Pulse dischargers can also produce equivalent fields for short periods and have proved most useful in mineral magnetic studies.

Zero field storage

Low magnetic fields for removal of viscous magnetic components are a useful extra facility. They can be conveniently provided by mumetal shielding or Helmholtz coil systems.

Field prospecting equipment

Field instruments capable of detecting concentrations of magnetic minerals, for example, in soil profiles, in exposures of bedrock, or in stream bedload tracing experiments, have been described by Howell (1966) and Colani (1966), and are available commercially as metal/mineral detectors.

CORRELATION

Magnetic susceptibility has proved to be a very useful parameter on which core correlations can be based. Peaks in susceptibility correspond to horizons with high heavy-mineral concentrations.

Figure 15.1 shows an example of between-core correlation based on magnetic susceptibility variations. Core 1 has been matched to core 2 by stretching the depth scale by a factor of 1.003 and then offsetting the depth scale by 25.4 cm. These optimal parameters were determined by minimizing the residual sum of square deviations from a cubic spline fitted to the combined susceptiblity data following the method of Clark and Thompson (1979). The smooth curves in Figure 15.1 show cubic splines fitted to the individual cores. These curves are used to test the significance of the correlation using a standard F test and produce 95% confidence limits. Clark and Thompson's method can be used to test any correlation pattern and is not restricted to linear relationships between rates of deposition.

Other examples of correlating sets of cores using magnetic parameters are to be found in Figures 13.2, 13.6, 13.7 and 13.8.

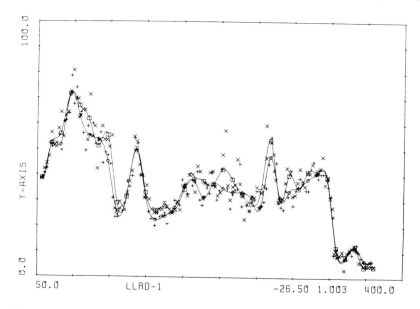

FIGURE 15.1. Loch Lomond susceptibility (*y* axis) versus depth (cm) plots. Core 1 stretched according to method described in Clark and Thompson (1979). Susceptibility data from Turner and Thompson (1979). Upright and diagonal crosses: susceptibility data from separate cores. Smooth curves are least-square cubic spline fits with knot intervals chosen by cross-validation

DATING

Figure 15.2 shows master geomagnetic curves for dating recent sediments. By matching palaeomagnetic direction variations to the master patterns a magnetic age may be assigned to the sediment. The age scales are based on conventional ^{14}C ages. Master curves are likely to be applicable to sediments found up to one to two thousand kilometres from the type sites. Matching of direction variations with the master curves is subjective and care must be taken not to mismatch major cycles. High-frequency variations (e.g. the inflection between turning points g and f of the western Europe declination curve) can help in distinguishing between the major cycles.

It is important to ensure that any palaeomagnetic directions measured in lake sediments are a true record of the ancient magnetic field. Reproduction of results in two or more cores and recognition of historically documented field changes (Thompson and Barraclough, 1982) in the uppermost sediments are the most reliable guides to checking the likelihood that the palaeomagnetic remanence in the sediment has been predominantly aligned by the geomagnetic field.

The difficulties of assigning ages to palaeolimnomagnetic master curves are

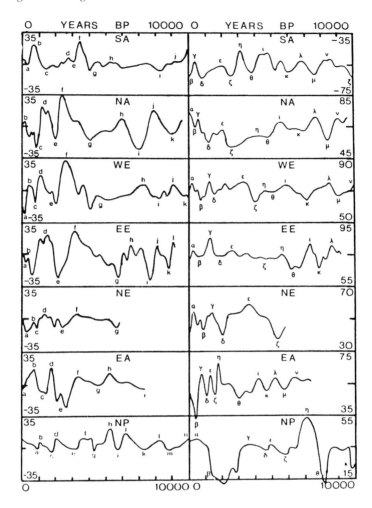

FIGURE 15.2. Regional declination and inclination master curves. Tree-ring calibrated time-scale in calendar years B.P. (0 B.P. = 1950 A.D.). South Australia (SA), data from Barton and McElhinny (1982). North America (NA), data from Banerjee *et al.* (1979). Western Europe (WE), data from Mackereth (1971) and Turner and Thompson (1979). Eastern Europe (EE) data from Huttonen and Stober (1980) and Tolonen *et al.* (1975). Near East (NE) data from Thompson *et al.* (1985). Eastern Asia (EA), data from Horie *et al.* (1980). North Pacific (NP) data from McWilliams *et al.* (1982)

illustrated in Figure 15.3, where ^{14}C ages from seven British lakes are plotted against palaeomagnetic features. The horizontal spread of age determinations shows the disagreement between sites. The dating spread increases in sediments deposited after the elm decline (i.e. after oscillation g–h), presumably on account of greater natural carbon contamination in the younger sediments, associated with increased soil erosion. All the palaeomagnetic master curves of Figure 15.2 are likely to have errors of several hundred years on account of ^{14}C dating difficulties. Palaeomagnetic work on laminated sediments (particularly carbonate laminated sequences) holds the greatest prospect for reducing such dating errors.

The likely geographic rate of variation of secular variation magnetostratigraphic features is illustrated in Figure 15.4. Changes in declination and inclination during the last 600 years are plotted as a set of curves at 10° longitude and 5° latitude intervals across Europe. The area covered ranges from the Faroes to the White Sea in the north, to Casablanca in the south-west and Mesopotamia in the south-east. The curves are built up from a series of spherical harmonic coefficients calculated from the master curves of Figure 15.2 and from the archaeomagnetic data of Kovacheva (1982). The most intense, longer-wavelength palaeomagnetic features can be seen to occur over relatively large geographical regions, whereas the lower-amplitude, more rapid fluctuations can be seen to have smaller geographic ranges. The patterns of Figure 15.4 indicate that over distances of a few hundred to a thousand kilometres from type sites palaeomagnetic master curves should be able to be used for dating without significant additional difficulties, although the ages of all secular variation magnetostratigraphic features necessarily change with location.

APPENDIX

Initial low-field apparent reversible susceptibility

Definition

Susceptibility is the ratio of induced magnetization to applied field, or, more simply, a measure of the 'magnetizability' of a sample, i.e. its degree of attraction to a magnet. Susceptibility is largely a function of the volume of ferrimagnetic minerals (e.g. magnetite and maghaemite) in a sample, but it is affected by the grain size and shape of the magnetic particles, their spontaneous magnetization, internal stress and other non-trinsic parameters (Thompson *et al.*, 1975). Only where ferrimagnetic minerals are very sparse or absent can the presence of haematite or paramagnetic minerals contribute significantly to susceptibility.

Units

Volume susceptibility (κ)   1 S.I. unit $= 7.96 \times 10^{-2}$ GOe^{-1}
Specific susceptibility (χ)   1 m^3 kg $= 79.6$ Gcm^{3}g^{-1}Oe^{-1}

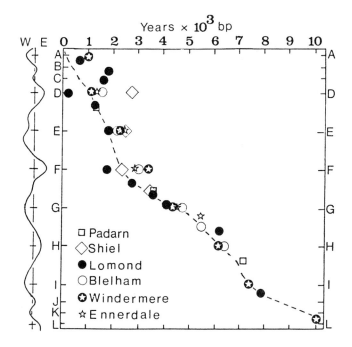

FIGURE 15.3. Geomagnetic declination variations (A–L) versus conventional ^{14}C age for 7 British lakes. Dashed line shows preferred ages of geomagnetic fluctuations. Discrepancies of ^{14}C ages from dashed line result partly from difficulties in resolving palaeomagnetic fluctuation, partly from errors in ^{14}C assessment, but are mainly attributed to post-*Ulmus* decline inwash of 'old carbon' with soil peat erosion. [14]C age determinations and palaeomagnetic data: Windermere (Mackereth, 1971), Lomond (Turner and Thompson, 1979), Shiel (Thompson and Wain-Hobson, 1979), Padarn (Elner and Happey-Wood, 1979), Blelham (Thompson, 1975), Ennerdale (Mackereth's palaeomagnetic data reported in Thompson, 1975)

Measurements

Volume susceptibility (κ) has been measured in sediments using either whole cores, or wet samples of approximately constant volume. Only dry-weight-based lake sediment samples provide susceptibility values which can be expressed as absolute numbers or can be compared quantitatively with soils, stream sediment and rock susceptibility measurements.

Continuous-core volume-susceptibility measurements have the virtue of speed, and since they can precede all other studies without constraining any, they are a valuable tool in the initial prospecting and core correlation, although inevitably they tend to smooth out any variations which have a narrow depth range.

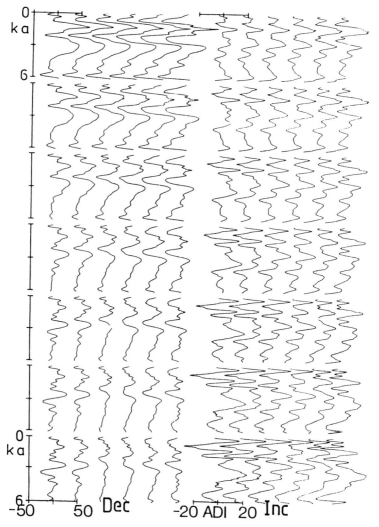

FIGURE 15.4. Palaeomagnetic declination and inclination variations since 6000 years B.P. at 10° longitude and 5° latitude intervals across Europe based on disjoint spherical harmonic analyses of lake sediment and archaeological samples before 1550 A.D. and historical observations since 1550 A.D. Upper left declination and inclination curves for locality 10°W, 65°N; upper right — 40°E, 65°N; lower left — 10°W, 35°N; lower left — 40°E, 35°N. Declination and inclination ranges 50° and 20° about axial dipole directions respectively. The quality and dating of the curves depends largely on the accuracy of the magnetic data from the three regions of (a) north-western Britain (Mackereth, 1971; Turner and Thompson, 1979), (b) eastern Finland (Tolonen *et al.*, 1975; Huttonen and Stober, 1980), and (c) Bulgaria (Kovacheva, 1982)

Wet-volume-based susceptibility measurements have the advantage of being easily combined with other palaeomagnetic studies. Provided relative orientation is retained in any suite of samples taken from the same core, a combination of susceptibility, declination and inclination measurements on lake sediments seems potentially very useful, and wet subsamples form the most suitable basis on which to achieve this.

Soil and stream sediment samples have been either air- or oven-dried and their susceptibility expressed as m^3kg^{-1}.

Rock samples have simply been weighed and their susceptibility expressed in the same way. Susceptibility may decrease when measured in higher frequency alternating fields. Such frequency dependence is characteristic of many topsoils with enhanced susceptibilities.

Influx

Magnetic susceptibility can be used in influx calculations (see Chapter 13; and Bloemendal *et al.*, 1979). Magnetic mineral influx or sediment influx is measured in kilograms or cubic meters (Thompson, 1980) and is calculated by totalling up the dry-mass or wet-volume of material that accumulated between two dated horizons. Note that the word influx is widely used in palynological studies in a sense which is likely to be misleading to the general reader or physical scientist, when 'influx' is inappropriately substituted for the straightforward phrase 'accumulation rate' (Thompson, 1980).

Isothermal remanent magnetization (IRM)

Definition

Isothermal remanent magnetization is the magnetic moment induced in and retained by a sample after it has been placed, at room temperature, in a magnetic field. Isothermal remanence increases non-linearly with the strength of the applied field until saturation isothermal remanent magnetization (SIRM) is reached; beyond this level an increase in the field will not lead to any increase in isothermal remanence. For most purposes saturation remanence is determined by using a saturation field of 1 tesla (10 kOe). The saturating field helps to differentiate between different sizes and types of magnetic minerals, as does determination of the coercivity of remanence $(B_0)_{CR}$. This is the reversed field strength required to reduce the saturation remanence to zero remanence. Low coercivities are characteristic of large-grained magnetite, high coercivities of fine-grained haematite. In practice, coercivity curves appear to differentiate assemblages of soil and sediment samples very sensitively without, however, permitting more than rather general, non-quantitative estimates of the basis of the variation.

In a suite of samples in which a single type of magnetic mineral is dominant or in which the magnetic minerals present occur throughout in constant proportion, there will be a direct linear relationship between susceptibility and saturation remanence (Thompson et al., 1975).

Units

Intensity of isothermal remanence (M)	$1\,Am^{-1} = 1\,mG$
Specific isothermal remanence (σ)	$1\,Am^2kg^{-1} = 1\,Gcm^2g^{-1}$

Measurement

Isothermal remanence has been measured in soils and sediments using dried and weighed samples or wet samples of approximately constant volume. Only in the former case have values been calculated on an absolute basis ($Am^2\,kg^{-1}$).

Anhysteretic remanent magnetizaton (ARM)

A related laboratory imparted remanence is an anhysteretic remanence which can be used for similar mineral magnetic purposes (Banerjee et al., 1981).

Natural remanent magnetization (NRM)

Definition

NRM is the fossil magnetism of rocks and sediments. The magnetic remanence of a sample is defined by a vector, i.e. a direction and a magnitude. The intensity of magnetization of a sediment is characteristically considerably less than one-millionth that of a toy magnet. The mechanism by which NRM is acquired in sediments depends on their mode of formation. A chemical remanent magnetization (CRM) is produced by chemical action and the growth of authigenic oxides at low temperatures. A detrital remanent magnetization (DRM) occurs through the alignment of previously magnetized allogenic particles. Magnetization acquired after consolidation is referred to as a secondary component.

Units

The magnetization of a material is its magnetic moment per unit volume given in units of ampere per metre (Am^{-1}). Directions of magnetization are specified in degrees by the declination (D) eastwards from true north and the inclination (I) reckoned positive downwards.

Geomagnetism from palaeomagnetism

If the fossil magnetism of a sediment was aligned by the ancient geomagnetic field, then palaeomagnetic measurements reveal past changes in direction of the

Earth's magnetic field. Furthermore, if past geomagnetic direction variations have been established and linked to an existing chronology (e.g. ^{14}C or K–Ar), then palaeomagnetic measurements can be used as a dating tool.

Directions

Geomagnetic direction changes which have been recognized and are potentially useful for dating are:

(1) *Polarity reversal:* both declination and inclination change through 180° and then remain stable. Polarity reversals are world-wide phenomena. A polarity subzone has an approximate duration of 10^4–10^5 years while a zone lasts approximately 10^5–10^6 years.
(2) *Polarity excursion:* a change in direction in which the virtual geomagnetic pole migrates through more than 45° for a short time (e.g. 10^2–10^4 years) and then returns to the original polarity. It has yet to be demonstrated that such behaviour is globally synchronous.
(3) *Secular variation:* smaller changes in direction than (2) but of a similar time scale. Some direction fluctuations are recognizable over regions of 'continental extent'. For dating purposes secular inclination fluctuations are likely to be largest and of most value at low latitudes, whereas secular declination changes are of greater importance at higher latitudes (see Figure 15.4).

Intensities

NRM intensities in sediments reflect many other parameters than simply ancient geomagnetic field intensity. Raw intensity data are most closely related to the magnetic mineralogy and degree of alignment of magnetic particles. With careful normalization and correction (e.g. Levi and Banerjee, 1976), sediments may yield useful palaeointensity data.

REFERENCES

Banerjee, S. K., King, J., and Marvin, J., (1981). A rapid method for magnetic granulometry with application to environmental studies. *Geophys. Res. Lett.*, **8**, 333–336.
Banerjee, S. K., Lund, S. P., and Levi, S., (1979). Geomagnetic record in Minnesota lake sediments — absence of the Gothenburg and Erieau excursions. *Geology*, **7**, 588–591.
Barton, C. E., and McElhinny, M. W. (1982). A 9000-year geomagnetic secular variation record from three Australian Maars. *Geophys. J. R. Astr. Soc.*, **67**, 465–485.
Bloemendal, J., Oldfield, F., and Thompson, R. (1979). Magnetic measurements used to assess sediment influx at Llyn Goddionduon. *Nature*, **280**, 50–53.

Clark, R. M., and Thompson, R., (1979). A new statistical approach to alignment of time series. *Geophys. J. R. Astr. Soc.*, **58**, 593–607.

Colani, C., (1966). A new type of locating device. I — The instrument. *Archaeometry*, **9**, 3–8.

Collinson, D. W., (1983). *Methods in Rock Magnetism and Palaeomagnetism*, Chapman and Hall, London.

Dodson, R., Fuller, M., and Pilant, W., (1974). On the measurement of the remanent magnetism of long cores. *Geophys. Res. Lett.*, **1**, 185–188.

Elner, J. K., and Happey-Wood, C. M., (1980). The history of two linked but contrasting lakes in the North Wales from a study of pollen, diatoms and chemistry in sediment cores. *J. Ecology*, **68**, 95–121.

Horie, S., Yaskawa, K., Yamamoto, A., Yokoyama, T., and Hyodo, M., (1980). Palaeolimnology of lake Kizaki. *Arch. for Hydrobiol.*, **89**, 407–415.

Howell, M. I., (1966). A soil conductivity meter. *Archaeometry*, **9**, 20–23.

Huttonen, P., and Stober, J., (1980). Dating of palaeomagnetic records from Finnish lake sediment cores using pollen analysis. *Boreas*, **9**, 20–23, 193–202.

Kent, D. V., and Opdyke, N. D., (1977). Palaeomagnetic field intensity variation recorded in a Brunhes Epoch deep-sea sediment core. *Nature*, **266**, 156–159.

Kovacheva, M., (1982). Archaeomagnetic investigations of geomagnetic secular variations. *Phil. Trans. R. Soc. Lond.*, **A306**, 79–86.

Levi, S., and Banerjee, S. K. (1976). On the possibility of obtaining relative palaeointensities from lake sediments. *Earth Planet. Sci. Letts.*, **29**, 219–266.

Mackereth, F. J. H., (1971). On the variation in direction of the horizontal component of remanent magnetization in lake sediments. *Earth Planet. Sci. Letts.*, **12**, 332–338.

McWilliams, M. O., Holcolm, R. T., and Champion, D. E. (1982). Geomagnetic secular variation from ^{14}C dated lava flows on Hawaii and the question of the Pacific non-dipole low. *Phil. Trans. R. Soc. Lond.*, **A306**, 211–221.

Molyneux, L., (1971). A complete result magnetometer for measuring the remanent magnetization of rocks. *Geophys. J. R. Astr. Soc.*, **24**, 429–434.

Molyneux, L., and Thompson, R., (1973). Rapid measurement of the magnetic susceptibility of long cores of sediment. *Geophys. J. R. Astr. Soc.*, **32**, 479–481.

Molyneux, L., Thompson, R., Oldfield, F., and McCallan, M. E., (1972). Rapid

Oldfield, F., (1978). Lakes and their drainage basins as units of sediment-based ecological study. *Prog. Phys. Geog.*, **1**, 460–504.

Oldfield, F., Dearing, J., Thompson, R. and Garrett-Jones, S. E., (1978a). Some magnetic properties of lake sediments and their links with erosion rates. *Pol. Arch. Hydrobiol.*, **25**, 321–333.

Oldfield, F., Rummery, T. A., Thompson, R., Walling, D. E., (1979). Identification of suspended sediment sources by means of magnetic measurements: some preliminary results. *Water Resource Research*, **15**, 211–218.

Oldfield, F., Thompson, R. and Barber, K. E., (1978b). Changing atmospheric fallout of magnetic particles recorded in recent ombrotrophic peat sections. *Science*, **199**, 679–680.

Radhakrishnamurty, C., Likhite, S. D., Amin, B. S., and Somayajulu, B. L. K., (1968). Magnetic susceptibility stratigraphy in ocean sediment cores. *Earth Planet. Sci. Letts.*, **4**, 464–468.

Scollar, I., (1968). A simple direct reading susceptibility bridge. *J. Sci. Instr. Ser.*, **2**, 781–782.

Thompson, R., (1975). Long period European geomagnetic secular variation confirmed. *Geophys. J. R. Astr. Soc.*, **43**, 847–859.

Thompson, R. (1977). Stratigraphic consequences of palaeomagnetic studies of Pleistocene and recent sediments. *J. Geol. Soc. Lond.*, **133**, 51–59.

Thompson, R., (1980). Use of the word 'influx' in palaeolimnological studies. *Quaternary Research*, **14**, 269–270.

Thompson, R., and Barraclough, D. R., (1982). Geomagnetic secular variations based on spherical harmonic and cross validation analyses of historical and archaeomagnetic data. *J. Geomagn. Geoelect.*, **34**, 245–263.

Thompson, R., Battarbee, R. W., O'Sullivan, P. E., and Oldfield, F., (1975). Magnetic susceptibility of lake sediments. *Limnology and Oceanography*, **20**, 687–698.

Thompson, R., Turner, G. M., Stiller, M., and Kaufman, A. (1985). Near East palaeomagnetic secular variation recorded in sediments from the Sea of Galilee (Lake Kinneret) *Quaternary Res.*

Thompson, R., and Wain-Hobson, T., (1976). Palaeomagnetic and stratigraphic study of the Loch Shiel marine regression and overlying gyttja. *J. Geol. Soc. Lond.*, **136**, 383–388.

Tolonen, K., Siiriainen, A., and Thompson, R., (1975). Prehistoric field erosion sediment in Lake Lojarvi, S. Finland and its palaeomagnetic dating. *Ann. Bot. Fenn.*, **12**, 161–164.

Turner, G. M., and Thompson, R., (1979). Behaviour of the Earth's magnetic field as recorded in the sediment of Loch Lomond. *Earth Planet. Sci. Letts.*, **42**, 412–426.

Watkins, N. D., (1971). Geomagnetic polarity events and the problem of 'The Reinforcement Syndrome'. Comments on Earth Sci: *Geophys.*, **2**, 36–43.

Handbook of Holocene Palaeoecology and Palaeohydrology
Edited by B. E. Berglund
© 1986 John Wiley & Sons Ltd.

16

Tephrochronology

THORLEIFUR EINARSSON

*Department of Geosciences
University of Iceland,
Reykjavik, Iceland*

INTRODUCTION

Volcanic products are divided into three categories: *lava* flowing from volcanoes and solidifying as lava flows, *volcanic gases* (volatiles), and *tephra* leaving the volcano with the volcanic cloud and often deposited over wide areas. Tephra is a collective term for all airborne pyroclasts, including both air-fall and flow pyroclastic material, and includes ignimbrite, welded tuff, scoria, bombs, pumice (lapilli) and volcanic ash (Thorarinsson, 1974). In tephrochronological studies, pumice and volcanic ash are the most important.

The term *tephrochronology* was coined by Thorarinsson (1944, 1974) as a chronology based on measurements, connections and dating of tephra layers. Tephrochronology is, therefore, concerned with the establishment of chronosequences of geological events based on the unique characteristics of successive tephra layers.

Tephra layers in profiles in Iceland are mentioned in a written account from 1638 and in print from 1749 (Thorarinsson, 1981). The beginning of modern tephrochronology in Holocene and Late Pleistocene deposits lies in the late 1920s in New Zealand (Grange, 1931), in Tierra del Fuego (Auer, 1932), in Iceland in the 1930s (Bjarnason and Thorarinsson, 1940), and in Japan in the early 1930s (Uragami *et al.* 1933; Machida, 1981). After the Second World War Thorarinsson built up the first tephrochronological 'system' in Iceland. Since then studies have been carried out in many volcanic areas.

Tephra layers formed in hours or days are important time signals in different types of sediments, as tephra from a single volcanic eruption often spreads over wide areas and long distances. A few examples can be mentioned. Tephra from Crater Lake, western USA (the Mazama tephra, 6700 B.P.), has been found in Alberta as a distinct layer, 1550 km north-east of the volcano (Westgate and Gorton, 1981). In the Ålesund area, western Norway, a tephra layer (the Vedde ash bed) has been found in lacustrine sediments from the Younger Dryas chronozone, 10,600 B.P. — its probable source is the volcano Katla, southern Iceland, some 1300 km away (Mangerud *et al.*, 1984). Tephra from the Laacher See area in West Germany has been found in a peat bog on the island of Bornholm, Denmark (Usinger, 1978) and near the island of Gotland, Sweden, 1000 km from the source (Risse *et al.*, 1980). In deep-sea sediments in the South Pacific airborne tephra has been found 3000 km from the source (Huang *et al.*, 1973).

It is therefore obvious that tephra layers are excellent time-parallel markers in various sediments formed on land, in lakes and on the sea-bottom. When the age of a tephra layer has been determined in one place, the dating applies to that particular tephra layer wherever it is found.

Very few compilatory publications have appeared on tephra and tephrochronology. A bibliography and index of Quaternary tephrochronology (Westgate and Cold, 1974), spans the literature up to 1973; the proceedings of an international conference held at Laugarvatn, Iceland, in 1980 covers different topics connected with this subject (Self and Sparks, 1981); and a manual of tephrochronology (Steen-McIntyre, 1977) deals with field and laboratory methods.

DISTRIBUTION OF TEPHRA

Tephra is produced by explosive activity in volcanoes and is carried upwards in the eruption cloud and then sideways by wind. The most important factor in the distribution of tephra is, therefore, the wind. In small or medium eruptions the volcanic cloud will rise up to the tropopause and the distribution of the tephra will, therefore, mainly be decided by the wind direction in the troposphere during the eruption. In the numerous eruptions of the volcano Hekla in Iceland during the Holocene, tephra has spread in all directions around the volcano, depending on the wind direction during individual eruptions (Larsen and Thorarinsson, 1977). In powerful eruptions the eruption cloud rises up through the tropopause into the stratosphere. During the initial phase of the 1947–48 Hekla eruption the cloud reached an altitude of 30 km and the tephra was then carried by the prevailing western wind in the stratosphere. Fine-grained tephra from this eruption began to fall on Helsinki 51 hours after the beginning of the eruption (Thorarinsson, 1954).

After the tephra is ejected it will settle through the air and become sorted according to the fall velocity. The grain size of the tephra in each tephra layer decreases systematically with the distance from the volcano (Thorarinsson, 1954; Walker, 1973; Sigurdsson, 1982).

Two Icelandic tephra layers may serve as examples of tephra distribution over long distances, and the speed of transport.

On the night of 29 March 1875 a violent eruption took place in the Askja caldera in the Dyngjufjöll volcano in the eastern highlands of Iceland. Owing to strong westerly winds, the volcanic cloud was transported to the east, and at about 8 a.m on 29 March greyish-white tephra began to fall 80 km away in the village of Seydisfjördur on the east coast of Iceland. In the evening of the same day, at about 7 p.m., tephra began to fall on the west coast of Norway, 1100 km away, and in Stockholm, 1900 km from the volcano, at 11 a.m. next morning (Mohn, 1877; Figure 16.1)

Another example of widespread Icelandic tephra is the rhyolitic-andesitic tephra layer H_4 from the volcano Hekla, southern Iceland. Its age is about 4000 B.P. This tephra layer covers an area about 78,000 km^2 on land in Iceland (Figures 16.2 and 16.3) — the total area of Iceland being 103,000 km^2. The total volume of the freshly fallen tephra on land and sea was about 9 km^3. This tephra layer is divided into four units on land according to differences in colour because the volcanic material became more basic in composition during the eruption: from the bottom, white, greyish-yellow, greyish-brown and, at the top, brownish-black. The SiO_2 content changed from 74% to 57% as the layer was deposited. According to Thorarinsson (1976) and Larsen and Thorarinsson (1977), the thickness of this tephra layer is about 10 cm in the sector 100 km north of Hekla, and 1–2 cm in north-east Iceland 400 km from the volcano (Figure 16.4). According to detailed mapping by Thorarinsson (1976) and Larsen and Thorarinsson (1977), the wind was blowing from the south or south-east during the beginning of the eruption and then shifted towards the south-west, and again towards the south-east and east. This corresponds to the passing of one depression over Iceland which may have taken 24–30 hours. Traces of the tephra layer H_4 have been found in Norway, Sweden and, possibly, on the Faroe Islands (Persson, 1966, 1967, 1971; Waagstein and Johansen, 1968).

FIELD WORK

Investigations of tephra layers in the field are best made in open sections. If open sections are not available the only alternative is drill coring. The open sections have the advantage that thin (1–2 mm) tephra layers, which can be discontinuous, can be discerned and followed in exposures; whereas a drill core from a spot chosen for coring may or may not show the very thin tephra layers.

In the field fresh tephra layers can be divided into three 'petrographical' categories according to colour. Basaltic layers are black, andesitic are brownish-grey, and rhyolitic grey or white.

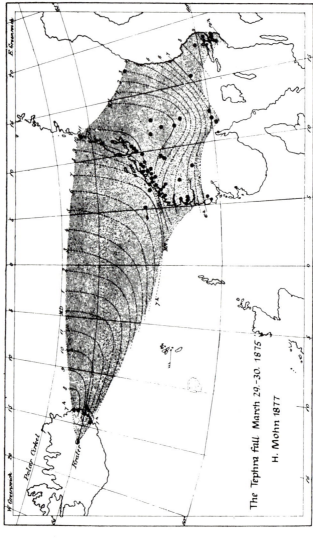

FIGURE 16.1. The tephra sector of the Askja eruption of 29 March 1875. Broken lines are isochrones (GMT) for the beginning of tephra fall (Mohn, 1877)

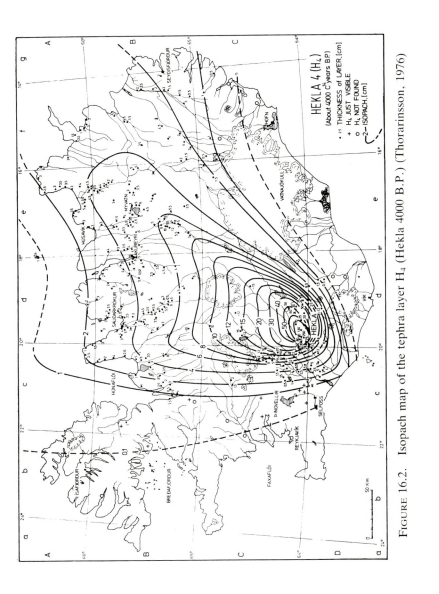

FIGURE 16.2. Isopach map of the tephra layer H₄ (Hekla 4000 B.P.) (Thorarinsson, 1976)

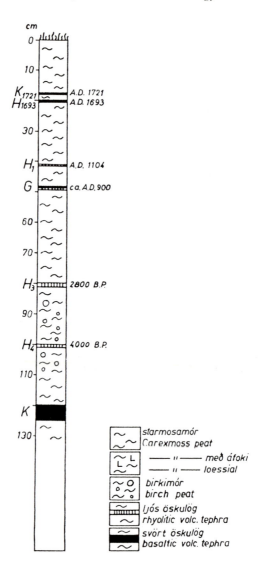

FIGURE 16.3.   A section with tephra layers from the upper part of a peat bog at Skálholt, Southern Iceland. The dates of layers K-1721 (Katla eruption 1721), H-1693 (Hekla 1693) and H-1104 (Hekla 1104) are based on historical records. The tephra layer G is the layer VIIa + b, from Thorarinsson (1944), which, according to pollen-analytical work, is dated to approximately 900 A.D. The tephra layers H_3 (Hekla 3) and H_4 (Hekla 4) have radiocarbon ages of 2800 B.P. and 4000 B.P. respectively. The tephra layer K is from Katla and pollen-analytical work suggests that its age is approximately 5000 B.P. (Einarsson, 1962, 1963)

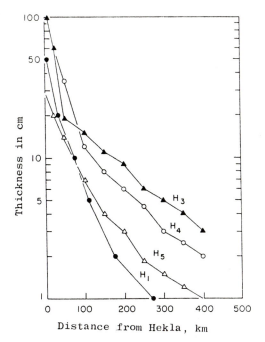

Distance from Hekla, km

FIGURE 16.4.   The change in thickness of the tephra layers Hekla 1 (1104 A.D.), Hekla 3 (2800 B.P.), Hekla 4 (4000 B.P.) and Hekla 5 (6600 B.P.) along the axis of the tephra sectors. From Sigurdsson (1982), compiled from data in Larsen and Thorarinsson (1977)

Preliminary estimates of grain size can also be useful for studies of tephra layers in the field. According to Thorarinsson (1958), a convenient field distinction of grain size of tephra is: (*a*) fine-grained tephra smaller than 0.2 mm, (*b*) medium-grained tephra, 0.2–2 mm, and (*c*) coarse-grained tephra larger than 2 mm.

The colour of tephra layers and a rough estimate of grain size in the field can often facilitate the recognition of individual tephra layers and help build up a preliminary succession of tephra layers for stratigraphical purposes. The field observations must then be followed up by granulometric, petrographical and chemical investigations in the laboratory.

LABORATORY WORK

As tephra layers are important time signals in stratigraphical studies, reliable correlations have to be based on all available means. Among methods for this purpose, granulometric, petrographical and geochemical investigation may be mentioned.

In the proximal areas of a volcano, thickness and grain size analysis, along with petrographical and geochemical analyses of tephra layers, are often sufficient for relating individual tephra layers to the source and for reconstructing maps of tephra sectors. In distal areas, hundreds or thousands of kilometres away from the volcano, chemical analysis of the glass shards may be the safest way for relating tephra layers to the source and for correlations.

In the proximal areas, bulk chemical analysis combined with petrographical studies of minerals (phenocrysts) and xenoliths may be of value if the tephra layers are homogenous or well stratified. In distal areas, both minerals and xenoliths will mostly be absent from the tephra, owing to sorting during the long transport.

In recent years microprobe analysis of fine-grained volcanic shards have made it possible to relate very thin tephra layers, or even traces of tephra in sediments, to volcanoes far away. As an example, glass shards in the 2–3 mm thick Vedde ash-bed in the Ålesund area, western Norway, from Younger Dryas, have an unusually high TiO_2 content. Therefore, it has been suggested that the source of this tephra layer is in the Katla-Eldgjá volcanic system in southern Iceland, which also has erupted volcanics with high TiO_2 content (Mangerud *et al.*, 1984).

More recently, measurements of the largest diameter of tephra grains, i.e. glass shards, in distal areas has been taken to be a reliable indication of the distance from the source.

It has also been suggested that the grain morphology of volcanic glass shards gives an indication as to the character of the eruption: elongated grains from the subaerial eruptions — the so-called Strombolian–Hawaiian eruptions; and spherical grains from subaquatic eruptions — the so-called Surtseyian eruptions.

In tephra studies, both in the field and in a laboratory, one has to be aware of the possibilities of redeposition, weathering, xenoliths etc. which can affect conclusions. (For further reading see: Fisher, 1964; Walker, 1971; Walker and Croasdale, 1972; Walker, 1973; Larsen and Thorarinsson, 1977; Steinthórsson, 1977; Westgate and Gorton, 1981).

ABSOLUTE DATING OF TEPHRA LAYERS

In some volcanic areas the exact date of many historical eruptions and corresponding tephra layers is known. This is the case, for example, in Italy and Iceland. An Italian example is the tephra fall which covered Pompeii and the ash flow which destroyed Herculaneum in the eruption of Vesuvius in the year 79 A.D. Since 1104 A.D., Icelandic written sources have helped to establish the dates of many eruptions in Iceland and, if not the exact date, then at least the year of eruption.

Where historical records are not available the only alternative is absolute (radiometric) dating. Radiocarbon dating, the most common method of dating

Holocene and Upper Pleistocene sequences, however, usually provides only indirect ages because the organic material used for age determination comes from below the tephra layer or is collected from the overlying strata. In a few cases the organic material has been embedded in the tephra layer, or been burnt by it, and, therefore, gives a more accurate age. As an example, it may be mentioned that a birch trunk embedded in the Hekla-tephra H_3, 11 km north of the volcano, gave an age of 2820 ± 70 B.P. (Thorarinsson, 1964). The problem of contamination in radiocarbon dating has been discussed in Chapter 14, but material below a thin tephra layer can easily be contaminated by younger roots.

Other radiometric methods for dating Quaternary tephra layers are fission-track and K–Ar-dating, which both provide direct ages, as the material used is the tephra itself. With further refinement of the apparatus they should provide ages from the Upper Pleistocene and older (Naeser *et al.*, 1981).

RELATIVE DATING OF TEPHRA LAYERS

Among the numerous methods are: the stratigraphical position in relation to dated tephra layers, palaeomagnetic correlations (Chapter 15), annually laminated sediments (Chapter 17), biostratigraphical methods, e.g. pollen analyses (Chapter 22), O^{16}/O^{18}-curves from deep-sea cores (Shackleton, 1977), archaeological correlations and sedimentation rates in dated sections.

Sometimes a combination of some of these methods can yield more accurate dates than those obtained by radiometric dating. The archaeological excavation in Thjórsárdalur, upper southern Iceland, for example, revealed a 3–5 cm tephra layer (VIIa + b, bottom rhyolitic, top basaltic) underneath farm ruins, the farm having been destroyed by tephra fall in the Hekla eruption in 1104 A.D. Pollen analysis of the soils around the farm ruins showed that the tephra layer VIIa + b had fallen shortly before the settlement of the area, i.e. it revealed the characteristic, very abrupt vegetational change from birch to grassland and cultivation that has been seen to have taken place when Iceland was first settled, the Landnám phase. A thin charcoal layer, also characteristic for the settlement horizon in many areas in Iceland, appears just above the layer VIIa + b. From this investigation, Thorarinsson (1944) deduced that the twin tephra layer VIIa + b was a little older than the settlement of Iceland which, according to the Saga literature, occurred in 870–930 A.D. In pollen-analytical research in bogs in the lower part of southern Iceland, Einarsson (1962, 1963) was able to show that the layer had fallen after the settlement had begun, and this has been verified in recent years by pollen-analytical studies carried out by Margrét Hallsdóttir (unpublished). From this it can be concluded that the layer was formed around 900 A.D. This obviously means that the settlement began in the coastal areas, and as the population in Iceland increased the later settlers had to move inland.

Recently Hammer *et al.* (1980), through counting the annual layers in the Greenland ice-sheet, combined with measurements of the acidity caused by aerosols (mainly sulphates) from volcanic eruptions, have shown that a large eruption took place, probably in Iceland, around 900 A.D. (898 A.D.?). It is possible that this is the eruption in which the 'settlement layer' was formed. The volume and distribution of the tephra indicate that the eruption was large and, possibly, a combination of two volcanoes erupting simultaneously (Thorarinsson, 1967, 1976; Larsen, 1982). The counting of annual layers in the ice-sheet, combined with electrical measurements of aerosols, shows possibilities of providing exact dates of volcanic eruptions and tephra layers, even though direct petrological or geochemical analysis could not be applied for the correlations. In the future it may also be possible to trace tephra layers in varved clays, which would be important for correlations (Mangerud *et al.*, 1984).

TEPHROCHRONOLOGICAL 'SYSTEMS'

As mentioned before, systematic tephrochronological research in Holocene deposits, for a given area, was initiated in Iceland by Thorarinsson (1954, 1967, 1976, 1981) in the years after the Second World War. It is now a sophisticated system, well supported by historical records and radiometric dates, and very useful in different subjects in geological research of the Holocene. In the last three decades similar research has been carried out in many volcanic areas around the world.

The Icelandic tephrochronological system is based mainly on the light-coloured rhyolitic tephra layers, which are fairly easily recognized in the field, and not as numerous as the basaltic tephra layers. The most important and widespread acidic tephra layers in Iceland are: Askja (1875 A.D.), Öræfajökull (1362 A.D.), Hekla (1104 A.D.), the settlement layer VIIa + b (about 900 A.D.), and the radiocarbon-dated Hekla$_3$ (2800 B.P.), Hekla$_4$ (4000 B.P.), and Hekla$_5$ (6000 B.P.) (Thorarinsson, 1976).

Traces of all these tephra layers, except VIIa + b, have been found in bog sections in Norway, Sweden and the Faroe Islands (Persson, 1966, 1967, 1971; Waagstein and Johansen, 1968).

It is not within the scope of this chapter to account for other volcanic areas in the world, but a brief mention of western U.S.A. will be given. There, tephrochronology has been used intensively for many aspects of geological research and all means of dating have been applied. The major Late Quaternary tephra layers have been produced by volcanoes in the Cascade Range (Porter, 1981; Naesser *et al.*, 1981). The youngest important tephra layer is the Mazama ash from Crater Lake (radiocarbon age 6700 B.P.); then the Glacier Peak tephra with two widely distributed layers (11,250 B.P. and 12,000–12,700 B.P. respectively); the Pearlette-type tephra layers from the

Yellowstone Park area with fission-track age and K–Ar-age of 0.6–1.9 million years; and the Bishop tephra in California which has a fission-track and palaeomagnetic age of 0.7 million years (Porter, 1981). From Mount St. Helens there are also numerous ash-beds from the last 37,000 years, the last one from 1980, and some of them are known as good marker beds (Naeser *et al.*, 1981).

APPLICATIONS OF TEPHROCHRONOLOGY

Tephra layers and tephrochronology are, as already stated, widely employed for dating purposes in different disciplines of stratigraphy. The study of tephra sequences, for example, reveals the history of individual volcanoes and volcanic areas, and gives information on the eruption history (age of lava flows and tephra layers), geochemistry and volume of the material produced etc. For further reading on this, see: on Iceland (Thorarinsson, 1981); the Eifel area, West Germany (Frechen, 1959); the U.S.A. (Porter, 1981); Japan (Machida, 1981).

Tephra layers are valuable time markers in soil sections, both in loess and peat bogs, and also in lakes and marine sediments for different vegetational and faunal studies. Dated tephra layers in sections in the field give information on hiatuses, and sedimentation rates etc., and are of use for choosing favourable sites for further investigation. For further reading on this, see: Einarsson (1962, 1963), Waagstein and Johansen (1968), and Straka (1971).

In recent years, distinct tephra layers and traces of tephra found in deep-sea sediments have been shown to be of stratigraphical value. This tephra is mostly airborne (Huang *et al.*, 1973; Mangerud *et al.*, 1984), or transported by sea-ice to warmer areas (Ruddiman and Glover, 1975; Sigurdsson and Loebner, 1981; Sigurdsson, 1982).

Tephrochronological studies may also be used for the unravelling of soil formation and erosion rates. Thorarinsson's work (e.g. 1960, 1981) shows that in the 1100 years since the Nordic settlement of Iceland, around 50% of the vegetation cover and loess soils have been stripped off.

Tephrochronology has also been used in studies of fluvial erosion and deposition in some areas around the world, e.g. western U.S.A. (Porter, 1981), Japan (Machida, 1981), and the lower Rhine area (Frechen and Heide, 1969).

The history of the Holocene river erosion in Iceland, evident in the canyons of Jökulsá á Fjöllum (Thorarinsson, 1960, 1981) and Hvítá (Einarsson, 1982), has been elucidated by dated tephra layers. The outer part of the 3 km long Hvítá canyon displays all the white Hekla ash layers, whereas the oldest ash layer found on the terraces adjacent to the waterfall Gullfoss is that of the Hekla eruption in 1693 A.D.

Tephrochronology and archaeology are of mutual use as sometimes archaeological findings can be used for dating of tephra layers along with other dating methods, e.g. in Iceland (Thorarinsson, 1944, 1981), in Japan (Machida, 1981), in the Mediterranean region (Thera, Pompeii) (Keller, 1981) and in America (Steen-McIntyre, 1981).

Tephra layers in glaciers may be used for calculating the accumulation of snow through time. In 1972, a 415 m deep hole was drilled on north-western Vatnajökull ice-sheet at 1800 m above sea level. In this core, 30 tephra layers were found, of which 22 could be attributed to known eruptions; the oldest ones (near the bottom of the hole) from the Grímsvötn eruption in 1659 A.D. and the Katla eruption in 1660 A.D. The mean annual balance was calculated and seems to vary from 3 to 7 m per year (Steinthórsson, 1977). Similar investigations on aerosols from volcanic eruptions have been carried out on drill cores from the Greenland ice-sheet, where the age can be determined through counting of annual layers (Hammer *et al.*, 1980).

In the U.S.A. (Porter, 1981) and in Iceland (Thorarinsson, 1981), glacier oscillations have been dated by tephra layers either underlying or overlying till or glaciofluvial material.

Finally it may be mentioned that drifted pumice has been found at various levels on raised beaches in the North Atlantic, most of which is considered to be of Middle Holocene age (Binns, 1967). The origin of the pumice is still not known with any certainty, but might well be Icelandic.

REFERENCES

Auer, V. (1932). Palaeogeographische Untersuchungen in Feuerland und Patagonien. *Sitz. Berichte Acad. Sci. Fennicae*, 1–17.

Auer, V. (1965). The Pleistocene of Fuego-Patagonia. IV: Bog profiles. *Ann. Acad. Sci. Fennicae A*, III, **80**, 1–80.

Binns, R. E. (1967). Drift pumice on postglacial shorelines of Northern Europe. *Tromsö Museum A. Scientia*, **24**, 1–63.

Bjarnason, H., and S. Thorarinsson (1940). Datering av vulkaniska asklager i isländsk jordmån. *Geogr. Tidskr.*, **43**, 5–30.

Einarsson, Th. (1962). Vitnisburdur frjógreiningar um gródurfar, vedurfar og loftslag á Islandi (Pollen-analytical studies on the vegetation, climate and settlement in Iceland). *Saga*, 442–469.

Einarsson, Th. (1963). Pollen-analytical studies on the vegetation and climate history of Iceland in late and postglacial times. In: *North Atlantic Biota and their History* (Eds. Löve and Löve), Pergamon Press, Oxford, pp. 355–365.

Einarsson, Th. (1982). The history of Hvítárgljúfur and Gullfoss in the light of tephrochronology. In: *Eldur er í nordri* (Eds. H. Thórarinsdóttir *et al.*), Sögufélag, Reykjavík, pp. 443–451.

Fisher, R. V. (1964). Maximum diameter, medium diameter and sorting of tephra. *J. Geophys. Research*, **V**, 341–355.

Frechen, J. (1959). Die Tuffe der Laacher See Vulkangebietes als quartärgeologische Leitgesteine und Zeitmarken. *Fortschr. Geol. Rheinland und Westphalen*, **4**, 301–312.

Frechen, J. and Heide, H. (1969). Tephrostratigraphische Zusammenhänge der

Vulkantätigkeit im Laacher See-Gebiet und der Mineralführung der Terrassenschotter am Unteren Rhein. *Mittel-Rhein. Decheniana*, **122**, 35–44.

Grange, L. I. (1931). Volcanic ash showers: a geological reconnaissance of volcanic showers of the central part of the North Island. *New Zealand J. Sci. Tech.*, **12**, 228–240.

Hammer, C. U., Clausen, H. B., and Dansgaard, W. (1980). Greenland ice-sheet evidence of post glacial volcanism and its climatic impact. *Nature*, **288**, 230–235.

Huang, T. C., Watkins, N. D., Shaw, D. M., and Kennett, J. P. (1973). Atmospherically transported volcanic dust in South Pacific deep sea sedimentary cores at distances over 3000 km from the eruptive source. *Earth and Planetary Sci. Letters*, **20**, 119–124.

Keller, J. (1981). Quaternary tephrochronology in Mediterranean regions. In: *Tephra Studies* (Eds. S. Self and R. S. J. Sparks), Reidel, Dordrecht, pp. 227–244.

Larsen, G. (1981). Tephrochronology by microprobe glass analysis. In: *Tephra Studies* (Eds. S. Self and R. S. J. Sparks), Reidel, Dordrecht, pp. 289–316.

Larsen, G. (1982). Gjóskutímatal Jökuldals (Tephrochronology of Jökuldalur, E-Iceland). In: *Eldur er í nordri* (Eds. H. Thorarinsdóttir *et al.*), Sögufélag, Reykjavík, pp. 51–65.

Larsen, G., and Thorarinsson, S. (1977). H$_4$ and other acid Hekla tephra layers. *Jökull*, **27**, 28–46.

Machida, H. (1981). Tephrochronology and Quaternary studies in Japan. In: *Tephra Studies* (Eds. S. Self and R. S. J. Sparks), Reidel, Dordrecht, pp. 161–191.

Mangerud, J., Lie, S. E., Furnes, H., Kristansen, I. L., and Loemo, L. (1984). A Younger Dryas ash bed in Western Norway and its possible correlations with tephra cores from the Norwegian Sea and the North Atlantic. *Quaternary Research*, **21**, 85–104.

Mohn, M. (1877). Askeregn fra den 29de-30te Marts 1875. *Kria. Vidensk. Selsk. Forhandl.*, **1877**, 1–12.

Naeser, C. W., Briggs, N. D., Obradovich, J. D., and Izett, C. A. (1981). Geochronology of tephra deposits. In: *Tephra Studies* (Eds. S. Self and R. S. J. Sparks), Reidel, Dordrecht, pp. 13–47.

Persson, C. (1966). Försök till tefrokronologisk datering av några svenska torvmossar. *Geol. Fören. Stockh. Förhandl.*, **88**, 361–394.

Persson, C. (1967). Försök till tefrokronologisk datering i tre norska myrar. *Geol. Fören. Stockh. Förhandl.*, **89**, 181–197.

Persson, C. (1971). Tephrochronological investigation of peat deposits in Scandinavia and on the Faroe Islands. *Sveriges Geol. Unders.*, **C565**, 1–33.

Porter, S. C. (1981). Use of tephrochronology in the Quaternary geology of the United States. In: *Tephra Studies* (Eds. S. Self and R. S. J. Sparks), Reidel, Dordrecht, pp. 135–160.

Risse, R., Schminche, H. U., Bogard, P. v. d., and Wörner, G. (1980). *Dispersal of the Laacher See Tephra*, International Association of Sedimentologists, First European Meeting, Bochum, p. 255.

Ruddiman, W. F., and Glover, L. K. (1975). Subpolar North Atlantic circulation at 9300 yr B.P.: faunal evidence. *Quaternary Research*, **5**, 361–389.

Self, S., and Sparks, R. S. J. (Eds.) (1981). *Tephra Studies*, Reidel, Dordrecht.

Shackleton, N. J. (1977). The oxygen stratigraphic record in late Pleistocene. *Phil. Trans. Roy. Soc. London*, **B 280**, 169–182.

Sigurdsson, H. (1982). Útbreidsla íslenzkra gjóskulaga á botni Atlantshafs. (Distribution of Icelandic tephra layers in the North-Atlantic). In: *Eldur er í nordri* (Eds. H. Thórarinsdóttir *et al.*), Sögufélag, Reykjavík, pp. 119–127.

Sigurdsson, H., and Loebner, B. (1981). Deep-sea record of Cenozoic explosive volcanism in the North Atlantic. In: *Tephra Studies* (Eds. S. Self and R. S. J. Sparks), Reidel, Dordrecht, pp. 289–316.

Steen-McIntyre, V. (1977). *A Manual for Tephrochronology*, published by the author, Idaho Springs, Colorado.

Steen-McIntyre, V. (1981). Tephrochronology and its application to New World Archaeology. In: *Tephra Studies* (Eds. S. Self and R. S. J. Sparks), Reidel, Dordrecht, pp. 355–372.

Steinthórsson, S. (1977). Tephra layers in a drill core from the Vatnajökull ice cap. *Jökull*, **27**, 2–27.

Straka, H. (1971). New pollen-analytical and radiocarbon datings of the volcanic 'Maars' in the Eifel Mountains. *Proc. International Conf. on Palynology*, Novosibirsk, 19–25 July 1971.

Thorarinsson, S. (1944). Tefrokronologiska studier på Island. *Geogr. Ann.*, **26**, 1–217.

Thorarinsson, S. (1954). The tephra fall from Hekla on March 29th 1947. In: *The Eruption of Hekla 1947–1948. Soc. Sci. Islandica*, **2**, 1–68.

Thorarinsson, S. (1958). The Öræfajökull eruption of 1362. *Acta Naturalia Islandica*, **22**, 1–99.

Thorarinsson, S. (1960). Der Jökulsá-Canyon und Asbyrgi. *Petermanns Geogr. Mitteilungen*, **104**, 154–162.

Thorarinsson, S. (1961). Uppblástur á Islandi í ljósi öskulagarannsókna (Wind erosion in Iceland: a tephrochronological study). *Arsrit Skógræktarfél. Isl.*, 17–54.

Thorarinsson, S. (1967). The eruptions of Hekla in historical times. In: *The Eruption of Hekla 1947–1948. Soc. Sci. Islandica*, **1**, pp. 1–170.

Thorarinsson, S. (1971). Aldur ljósu gjóskulaganna úr Heklu samkvæmt geislakolstímatali (The age of the light Hekla tephra). *Náttúrufrædingurinn*, **41**, 99–105.

Thorarinsson, S. (1974). The terms tephra and tephrochronology, In: *Quaternary Tephrochronology*, (Eds. J. A. Westgate and C. M. Cold), Univ. of Alberta, pp. 17–18.

Thorarinsson, S. (1976). Gjóskulög (Tephra layers). *Samvinnan*, **70**, 4–9.

Thorarinsson, S. (1981). Greetings from Iceland. Ash-fall and volcanic aerosols in Scandinavia. *Geogr. Ann.*, **63a**, 109–118.

Thorarinsson, S. (1981). The application of tephrochronology in Iceland. In: *Tephra Studies* (Eds. S. Self and R. S. J. Sparks), Reidel, Dordrecht, pp. 109–134.

Thorarinsson, S. (1981). Tephra studies and tephrochronology: a historical review with special reference to Iceland. In: *Tephra Studies* (Eds. S. Self and R. S. J. Sparks), Reidel, Dordrecht, pp. 1–12.

Uragami, K., Yamada, S., and Naganuma, Y. (1933). Studies on the volcanic ashes in Hokkaido. *Bull. Volcanol. Soc. Japan*, **1**, 44–60.

Usinger, H. (1978). Bölling-Interstadial und Laacher Bimstuff in einem neuen Spätglazial-Profil aus dem Vallensgård Mose, Bornholm. Mit pollen-grössenstatistischen Trennung der Birken. *Danm. Geol. Unders. Årbog*, **1977**, 5–29.

Waagstein, R., and Johansen, J. (1968). Tre vulkanske askelag fra Færöerne. *Medd. Dansk Geol. Foren.*, **31**, 257–264.

Walker, G. P. L. (1971). Grain-size characteristics of pyroclastic deposits. *J. Geology*, **79**, 696–714.

Walker, G. P. L. (1973). Explosive volcanic eruptions — a new classification scheme. *Geol. Rundschau*, **62**, 431–446.

Walker, G. P. L., and Croasdale, H. (1972). Characteristics of some basaltic pyroclastics. *Bull. Volc.*, **35**, 303–317.

Westgate, J. A., and Cold, C. M. (1974). *World Bibliography and Index of Tephrochronology*. Univ. of Alberta, Edmonton.

Handbook of Holocene Palaeoecology and Palaeohydrology
Edited by B. E. Berglund
© 1986 John Wiley & Sons Ltd.

17

Annually laminated lake sediments

MATTI SAARNISTO

*Department of Geology, University of Oulu,
Oulu, Finland*

INTRODUCTION

Demands for more accurate palaeoecological reconstructions and correlations have in recent years turned many scholars' attention to annually laminated lake sediments (i.e. seasonal rhythmites or varves), whose presence has been recognized for some time (Heer, 1865; Wesenberg-Lund, 1901). Laminations are composed of thin, alternating, horizontally bedded layers which differ in composition and texture, usually in couplets consisting of a light and dark layer. In principle, annual laminations provide a precise chronological background in terms of years for all stratigraphical data in the sediments, whereas most other dating methods such as radiocarbon analysis have a wide margin of error and thus allow only a broad time-scale to be proposed for the stratigraphical record.

The availability of new sediment sampling methods, especially the freezing technique, has also led to a stimulation of interest in finely laminated sediments. When frozen *in situ*, the fine structures in the sediment can be seen and preserved even in cases where no laminations were recognized in ordinary cores. Annually laminated sediments, which have been found in small and large lakes mainly in various parts of the temperate zone, are more common than was earlier believed, and are not restricted to glaciolacustrine or calcareous sediments, where they were originally described. The longest sequences studied so far are of the order of 9000–10,000 years in duration (Table 17.1).

This chapter describes the types of lakes which tend to have laminations, types of laminations, and field and laboratory techniques for the study of finely

TABLE 17.1. Data from lakes with annually laminated sediments: some published results

Lake	Size (ha)	Greatest depth (m)	Nature of varves	Number of varves	References
Ahvenainen, Finland	7.8	19	D	6174	Tolonen (1978a,b)
Frängsjön, Sweden	6	7	M	8646 ± 190	Renberg and Segerström (1981)
Järlasjön, Sweden	60	24	M		Digerfeldt et al. (1975)
Judesjön, Sweden	15	15.6	M	6130	Renberg et al. (1984)
Kassjön, Sweden	23	12.2	M	6321 ± 700	Renberg and Segerström (1981)
Lampellonjärvi, Finland	9.7	10	D	3372 (+2000)	Tolonen (1980)
Lovojärvi, Finland	4.8	17.5	D	4929	Saarnisto et al. (1977); Huttunen (1980)
Pääjärvi, Finland	1342	87	D		Simola and Uimonen-Simola (1983)
Polvijärvi, Finland	185	35	D		Simola (1983)
Pyhäjärvi, Finland	1150	68	D	3700	Kukkonen and Tynni (1970)
Rudetjärn, Sweden	4	7	M		Renberg (1976)
Sarsjön, Sweden	10	8	M	8909 ± 150	Renberg and Segerström (1981)
Taka-Killo, Finland	2.8	25	D	1506	Huttunen (1980)
Valkiajärvi, Finland	7.8	25	M	9500	Meriläinen (1970); Saarnisto (1985)
Diss Mere, England	3	(7)16	C	3000	Peglar et al. (1984)
Loe Pool, England	57	11	D		Simola et al. (1981)
Zürichsee, Switz.	6730	137	C		Kelts and Hsü (1978)
Lake Van, Turkey	3574km²	457	C	10400	Kempe and Degens (1979)
Crawford Lake, Ontario	2.5	25	C		Boyko-Diakonow (1979)
Fayetteville Green Lake, New York	25.8	52.5	C		Brunnskill and Ludlam (1969); Ludlam (1969)
Greenleaf Lake, Ontario	56	76	(?) I		Cwynar (1978)
Lake of the Clouds, Minnesota	12.5	31	I	9500	Craig (1972); Anthony (1977)
M Lake, N.W.T. (Canada)	7.7	22	C		Ritchie (1977)
Pink Lake, Quebec	14.6	20	I		Dickman (1979)

Nature of varves (mainly according to the original source): D = diatom rich; M = variation in mineral matter; C = calcareous; I = iron-rich.

laminated sediments. It also provides a short review of the applications of annually laminated sediments in stratigraphical work.

CHARACTERISTICS OF LAKES

Sedimentation is in principle rhythmic in all lakes in the temperate zone where the production of the lake and the amount of allochthonous material is controlled by seasonal changes in the climate. However, only in special cases is this rhythmic structure preserved, as the upper loose sediment and its fine structures are easily destroyed and reworked by the benthos fauna and by currents, especially at the spring and autumn overturns. Fine laminations are preserved in lakes with a permanent oxygen deficit in the bottom water layers and thus have a poor bottom fauna. In such lakes (termed meromictic) the vertical water circulation does not extend to the bottom. All meromictic lakes sampled by the author in Finland and in Canada, or those whose sediments are described in the literature, contain fine laminations. Laminated sediments are also found in a number of holomictic lakes which have protracted seasonal anoxia in the hypolimnion and consequently little bioturbation (Ludlam, 1976). All the lakes described by Renberg (1981a, b) in Norrland, northern Sweden, are holomictic.

In some lakes in Finland only the uppermost sediment sequence is laminated. This may be explained by the increased productivity in the lake owing to human influence, which increases the probability of large periodic blooms of diatoms and other algae. This increased productivity also results in a prolonged oxygen deficiency in the bottom water, thus restricting the bottom fauna (Ludlam, 1979).

Perhaps the most important requirement for the occurrence of laminations concerns the morphometry of the lake basin. An ideal lake basin is one that is deep in relation to its surface area and is sheltered from the wind. It is preferably situated in a bedrock basin, has a small drainage area and no significant inflow. Often, such a lake is the highest basin in the drainage area. Kettle-hole lakes in glacial drift have a suitable morphometry, but the groundwater flow may sometimes destroy the laminations. Oxygen-rich groundwater also supports a benthos fauna and thus promotes bioturbation. Lakes surrounded by mires do not normally have laminated sediments.

Although no precise morphometric figures can be given, the lakes where laminated sediments have been described are usually less than 1 km in length, their surface area is less than 20 ha and they are over 15 m deep. Several exceptions occur, however, and Table 17.1 gives morphometric data for selected lakes with laminated sediments.

Annually laminated sediments are also found in isolated, deep basins within large lakes with surface areas up to several hundred square kilometres. These have a reasonably high rate of sedimentation because of the large amount of allochthonous mineral matter, and although holomictic, often have a reduced bottom fauna because of the decomposition of the sparse organic matter before it reaches the bottom. Currents in large lakes vary locally, to the extent that no precise figures can be given in this respect.

The situation can be summed up in the words of Ludlam (1967, 1976) that:

> 'Laminated sediments are found wherever the rate of sediment accumulation is large relative to the rate of disturbance and wherever the proper sampling technique is used.'

TYPES OF LAMINATIONS

The thickness of a single lamination in a rhythmically laminated sediment varies from less than 0.5 mm in some unproductive lakes to some 30–40 cm or more in glacial-lacustrine environments. The normal thickness of organic laminations in small undisturbed lakes is perhaps between 0.5 mm and 1 mm.

The annual character of the laminations must be established individually in each case. This can be studied microscopically by analysing the pollen, diatom and other microfossil content of different layers (e.g. Welten, 1944; Tippett, 1964; Benda, 1974; Simola, 1977; Peglar *et al.*, 1984). Land-use, such as the construction of a road near the lake, is often documented in the sediment and can be verified from historical data. The laminations above such a marker horizon can then be counted and their annual character confirmed (Digerfeldt *et al.*, 1975; Saarnisto *et al.*, 1977). In longer sequences lamination counts and radiocarbon dates can be compared, always bearing in mind, of course, the deviation between radiocarbon and calendar years (Stuiver, 1969). Moreover, radiocarbon analyses from lake sediments involve several sources of error which may well obscure real differences between the two chronologies (Chapter 14; Olsson, 1974; Renberg, 1978; Tolonen, 1978b; Tolonen, 1980).

In general, there are several reasons for the formation of annual laminations superimposed on each other: seasonal, rhythmic changes in biogenic production, water chemistry and the inflow of minerogenic matter. The classification presented here should therefore not be taken too strictly. The following four main varve types have been described in the literature and they are present in lakes throughout the temperate zone. Laminated sediments have also been recognized in the large lakes of tropical Africa (e.g. Richardson and Richardson, 1972) although these latter cases are omitted from consideration here.

Diatom-rich laminations

Diatoms are the most common microalgae found in lake sediments. Seasonal variations in their abundance are recorded in the bottom sediments. In

Lovojärvi, Finland, a comparison of laminations with present-day sedimentation led to the conclusion that the light, diatom-rich layers represent early summer, followed by gradually darker horizons and finally a black humus-rich winter layer also containing chrysophyte cysts (Saarnisto *et al.*, 1977). Simola (1977, 1979) has been able to identify the same diatom succession in the sediments of Lake Lovojärvi as has been observed in the phytoplankton in the water, thus verifying the annual character of the laminations. Superimposed on the variations in diatom sedimentation are seasonal changes in the deposition of minerogenic material. Distinct light-coloured layers composed almost entirely of diatoms may also occur in late summer, causing difficulties in varve identification. In general, the whole diatom-rich summer layer appears as a single light lamina, especially when the sediment accumulation rate is slow. Kukkonen and Tynni (1970) similarly believe that variations in diatom sedimentation explain the annual lamination formations in Lake Pyhäjärvi, Finland. Interglacial annually laminated diatomites are known in Germany (Giesenhagen, 1926; Benda, 1974) and in England (Turner, 1976).

Calcareous laminations

In studying the precipitation of $CaCO_3$ in the meromictic Fayetteville Green Lake, New York State, Brunskill (1969) observed that about 90% of the total sedimentation occurred from June to October, and that 80% of this mass was calcite. This explains the formation of light–dark laminations in similar lakes in calcareous areas. The light–dark couplet is a varve, the light layer representing $CaCO_3$ precipitation in warm water during spring and summer, and the dark layer, with relatively more organic (humus) material, representing the autumn and winter periods. Carbonate deposition occurs either as a result of seasonal temperature changes, as in Fayetteville Green Lake, or as a result of increased photosynthesis, as deduced for Zürichsee (Kelts and Hsü, 1978). These sources of calcium carbonate are interrelated and their relative importance is not easy to assess.

A secondary scanning electron microscope image of the calcareous sediment of Diss Mere, England, is shown in Figure 17.1 (Peglar *et al.*, 1984). The large calcite crystals at the base of the pale laminae grade into smaller crystals at the top. A similar phenomenon, which may reflect the degree of supersaturation of calcium carbonate, has been described in calcareous varves in Faulenseemoos in Switzerland by Welten (1944) and in Zürichsee by Kelts and Hsü (1978).

Calcareous varves have also been described in small lakes in Ontario (Tippet, 1964; Boyko, 1973, 1979). Similarly the formation of laminations in the large Lake Van, Turkey, is said to be due to the seasonal supply of

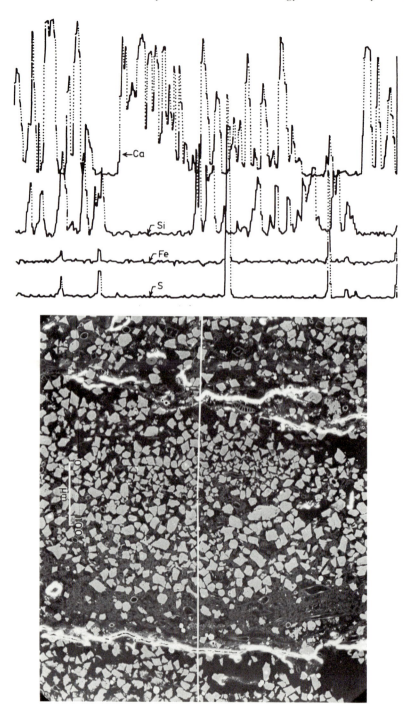

aragonite (calcite) (Kempe and Degens, 1979). Calcareous varves are usually both distinct and reasonably easy to count and handle in the laboratory.

Iron-rich laminations

The sediment of the Lake of the Clouds, Minnesota, has annual laminations covering a period of about 9500 years. According to Anthony (1977), the light-coloured laminae presumably form from iron oxides due to the partial oxygenation of iron-rich bottom water during overturns, mainly in spring. The dark layers would then represent organic remains that settle out during summer.

Renberg (1981b) emphasizes the role of iron in the visual appearance of annual laminations even in holomictic lakes. During and after the overturn periods in spring and autumn the bottom water becomes oxidized and the sediment gains its light colour from iron hydroxides (Figure 17.2), while during the oxygen-poor periods in summer and winter it is coloured black by iron sulphides. The significance of the iron in the appearance of laminations nevertheless depends on the amounts of other components such as diatoms and mineral matter.

Dickman (1979) presents an alternative explanation for the origin of dark laminations. Mass occurrences of dead cells of purple sulphur bacteria, which are intolerant to oxygen, are present in the sediment deposited in the meromictic Pink Lake, Quebec, during the autumn overturn period in October. The pyritization of iron takes place under reducing conditions near the bottom, and separate dark laminae will be formed in iron-rich lakes. The role of anaerobic bacteria may also be of significance in many other lakes with laminated sediments, but this has not yet been adequately investigated.

Variations in mineral matter deposition

Varves in glaciolacustrine environments are the best manifestations of this type which are rich in allochthonous mineral matter (clastic varves by Sturm, 1979). The laminations found in recent, highly minerogenic sediments in some large lakes, such as Lake Päijänne in Finland (Saarnisto, unpublished data) and

---

FIGURE 17.1. Secondary electron image (SEI), taken by a scanning electron microscope in the Department of Electron Optics, University of Oulu, Finland, of a dark-pale lamination sequence in an area 0.5 × 0.7 mm of the calcareous sediments of Diss Mere, England (Peglar *et al.*, 1984). The gradation from large calcite crystals to smaller ones is clearly visible. The upper diagram is the result of digital-line analysis by an energy dispersive spectrometer (EDS) to show the distribution of calcium, iron, silicon and sulphur along the line marked on the SEI. The step intervals along the line are about 2.6 microns. The scales of the elemental distribution curves are arbitrary.
Reproduced by permission of Universitetsforlaget, Oslo

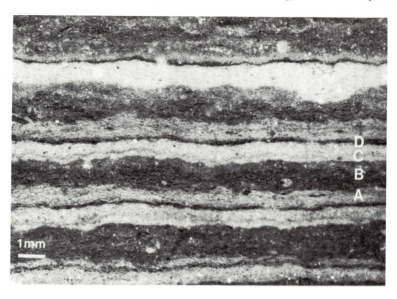

Figure 17.2.   A frozen sample of a varved lake sediment deposited about A.D. 1920 in a holomictic lake in northern Sweden. (Renberg, 1981b). The photograph shows the significance of the form of the iron present for the final appearance of the individual varves in some lakes. During and after the overturn periods in spring and autumn, a sediment coloured brown (light) by iron hydroxides is deposited (A and C, respectively). During the oxygen-poor periods in summer and winter a sediment coloured by black iron sulphide is laid down (B and D). *Photograph previously published in* Forskning och Framsteg *2/1979*

Lake Pääjärvi (Simola and Uimonen-Simola, 1983) are composed of light-coloured, thick summer layer with a high minerogenic content, and a thin, dark winter layer with relatively more organic matter and iron sulphides. Seasonal changes in mineral sedimentation are also important for varve formation in small, unproductive lakes. In the thin laminations, less than 0.5 mm, in Lake Valkiajärvi, Finland (Koivisto and Saarnisto, 1978) the light layer presumably represents minerogenic matter transported by the spring floods after the melting of the ice. The mineral matter and its particle size declines upwards through each lamina while the proportion of dark organic matter increases towards winter. The light layers in some thick varves contain vast quantities of diatoms. Seasonal variation in mineral matter is similarly responsible for varve formation in many lakes in Norrland, Sweden (Renberg, 1976, 1981a) (Figure 17.3).

The problem of massive layers and non-annual laminations

Annual laminations are often found in small, steeply sloping basins where

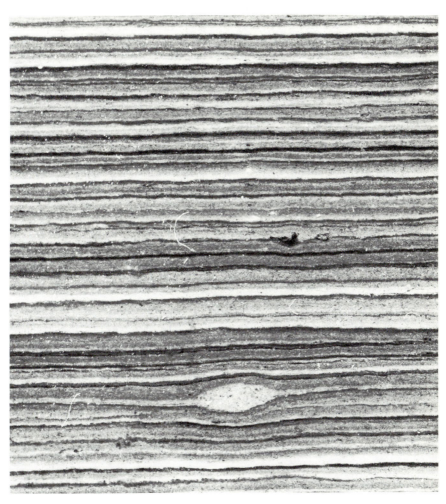

FIGURE 17.3.   An unfrozen sample of the varved sediment of Lake Kassjön, northern
Sweden. Each varve consists of a light layer of mineral grains, deposited during spring,
a more organic summer layer and a thin winter layer with a high organic (humus)
content. The varve thickness is about 0.6 mm. *Photographed and provided by I.
Renberg*

redeposition may occur. In Fayetteville Green Lake about 50% of the
sediment sequence is massive. According to Ludlam (1974) the massive layers
are deposits from turbidite currents and represent instantaneous events. In
Lake Valkiajärvi, 8 cm, i.e. 3.5% of the 2.3 m sediment sequence, appears to
be massive in this sense, the thickness of such layers varying between 2 mm
and 20 mm. Since the requirements for varve preservation in holomictic lakes

are especially sensitive, such sediments may contain both laminated and massive sequences as a result of bioturbation.

Massive layers can severely limit the use of annually laminated sediments for chronological studies, and therefore careful selection of the sampling locality is important. The cores should be taken from a flat area of the lake bottom away from slopes.

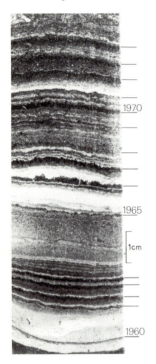

FIGURE 17.4. Inconsistent annual laminations in the upper sediment sequence of Lake Lovojärvi, Finland. The lamination boundaries have been determined by diatom analysis (Simola, 1979). The laminations from 1961 to 1964 appear similar to those found deeper in the sediment, the light layer representing diatom-rich spring layer (Saarnisto *et al.*, 1977). The thick, light layers of 1960 and 1965 are composed mainly of allochtonous mineral matter washed into the lake by human activities. *Photographed by K. Tolonen and provided by H. Simola*

Another problem is the presence of intra-annual laminations (Figure 17.4), as have been recognized in glaciolacustrine sediments in Denmark (Hansen, 1940) and in recent sediments in Lake Lovojärvi, Finland (Simola and Tolonen, 1981). High winds or heavy rains can produce false laminations, and only careful diatom and pollen analyses across such sequences can ensure accurate determination of the varves (Simola, 1977, 1979). Intra-annual laminations originating from human activities in the surroundings of the lake have been shown to lead to erroneous varve counts in the upper sediment sequence of Lake Lovojärvi (Saarnisto *et al.*, 1977; Simola and Tolonen, 1981).

Laminations with a periodicity longer than a year may occur, especially in large lakes, for the same reasons as intra-annual laminations. Thus it should be emphasized that the annual character of the varves must be established in each case.

THE FREEZING TECHNIQUE

Undisturbed samples from the uppermost loose sediment sequence in lakes are difficult to obtain by ordinary piston corers or samplers designed for surface sediments. Fine structural features in such sediments are easily destroyed during the sampling procedure, extrusion, transportation, or gas present in the sediments forming bubbles when the core is brought to the surface, can all destroy the fine structure within minutes. These difficulties can be overcome by freezing the sediment *in situ* and handling it only when it is frozen.

The freezing technique was introduced by Shapiro (1958) and further simplified and developed by H. E. Wright Jr. at the University of Minnesota in the late 1960s (Wright, 1980). The author gained experience of the method from J. H. McAndrews, Royal Ontario Museum, in Ontario, and it was introduced into Finland in 1972. Descriptions have been given by Swain (1973), Saarnisto (1975) and Saarnisto *et al.* (1977). It is simple, inexpensive, and often the only way to obtain undisturbed finely laminated sediments.

The device is composed of a metal tube 1–4 m long, e.g. an iron tube 2.5 m long, 80 mm in diameter and with a wall thickness of 1 mm, tightly closed at the base with a pointed wooden or plastic plug (Figure 17.5). If a tightly packed sediment is to be sampled or a light aluminium tube is used, lead weights may be placed inside the tube to ensure vertical penetration. The tube is then filled alternately with pieces of dry ice (size less than 10 cm^3) and small amounts (less than 0.5 litre altogether) of trichlorethylene (poisonous!) or some other liquid with a low freezing point, e.g. normal butanol or ethanol. The tube should be packed carefully, tapping it on the outside. The temperature of the dry ice is $-79°C$, and the added liquid accelerates its evaporation while cooling it further. A long and narrow polythene tube or rubber pipe is fastened to the top to prevent water from entering, while allowing the release of gas from the tube. A long polythene tube is recommended, as this will lead the gas well above the sediment surface and bottom water layer which would otherwise be disturbed.

The device is then allowed to fall freely through the water on a cable or rope and penetrate the soft sediment. It is often important to regulate the free fall to avoid mixing the uppermost loose sediment and to prevent the tube from descending too far into the sediment. For this purpose the water depth of the site should be carefully measured. In Lake Valkiajärvi, an iron tube 2.5 m long penetrated a little over 1 m into highly organic sediment after a free fall through 25 m of water, while a similar tube 4 m long and 80 mm in diameter penetrated 2.3 m, representing about 7300 years. In the highly minerogenic sediment of Lake Päijänne of Finland the same 2.5 m tube penetrated 0.6 m, representing about 100 years, after a free fall through 70 m of water. Even deeper penetration can be achieved using extension rods, which is possible in shallower lakes.

In 10–15 minutes a crust of sediment 1–3 cm thick freezes onto the tube,

(a)   (b)

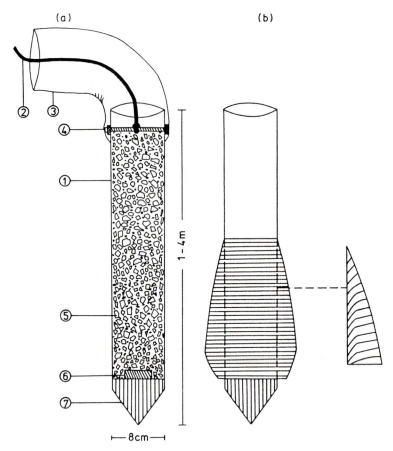

← 8 cm →

FIGURE 17.5. Sketch of dry ice sampler (not to scale). (a) shows technical details: 1 = metal tube; 2 = rope; 3 = polythene tube (10 cm diameter, 1–2 m long) fastened with insulating tape; 4 = bolt; 5 = small pieces of dry-ice plus small amounts of trichlorethylene; 6 = lead weights, 1–2 kg; 7 = wooden or plastic plug (waterproof seal). (b) indicates the ideal shape of the frozen sample and the bending of the sediment layers in contact with the tube

which is then pulled to the surface. The outermost unfrozen sediment should be immediately removed using either a knife or rubber gloves, and the frozen sediment surface may then be washed. The fine laminations, if present, can be seen on the frozen surface. The device can be handled by one person from a small boat provided the sample is not more than about a metre long, but two men standing on the ice could only just pull a 2.3 m long frozen core from the bottom of Lake Valkiajärvi.

The frozen sediment crust is loosened by removing the remaining dry ice and liquid from the tube and pouring warm water into it, preferably introducing it through a pipe to the bottom first, where the crust is thickest. Sometimes it is useful to open the lower, pointed end, for if this is made of wood or plastic it will not be covered by frozen sediment, and extrusion will be easier. Automobile exhaust can also be used for warming the inner surface of the sampler tube. Extrusion is easier if the tube and sediment crust are not too cold, and therefore it is preferable to wait some time after the dry ice has been poured out of the tube. If extrusion is to be carried out in the laboratory, which is often a practical solution, the tube is filled with dry ice and wrapped in newspaper for transportation. The frozen sample can be stored in a freezer, but it is important to wrap it tightly in aluminium foil and polythene to prevent drying. Properly packed, a sample can be kept for several years if necessary, and no oxidation occurs.

Some bending of the sediment layers nearest to the tube occurs during its passage through the sediment, but no mixing, provided that penetration is slow. Laminations of less than 1 mm are preserved even in the topmost loose sediment, although near the sediment–water interface, where usually only a thin frozen crust of sediment is obtained, bending may produce varves which are apparently too thick when measured directly from the outer surface of the core. In general, the sediment nearest to the tube should be discarded because of the downward bending.

Huttunen and Meriläinen (1978) obtained more material from the uppermost sediment layers by using a flat 'letterbox' (Figure 17.6) instead of a tube. In their modification, the dry ice cubes are pressed against one wall of the device (the front panel) with pieces of plywood and strings. Thus the sediment freezes, predominantly to the surface of the front panel. Huttunen and Meriläinen believe that the disturbance of the sediment is smaller on the side of the front panel because of the pointed asymmetrical keel at the lower end of the device. The flatness of the box, however, makes it uncertain whether penetration is absolutely vertical without supporting rods, and therefore the thickness of the laminations may still be measured incorrectly. Another adaptation of the freezing method is that proposed by Renberg (1981a), as illustrated and described in Figure 17.7.

LABORATORY TECHNIQUES

Subsampling the frozen sediment

Sediment structures must be studied and measured from the surface of the frozen sample, as they will be destroyed on melting. If the volume of a subsample for pollen studies etc. is not important, it can be taken with a sharp spatula, knife or saw. Known volumes are more difficult to obtain, but two possible methods can be recommended. Cwynar (1978) used pins and a piece of

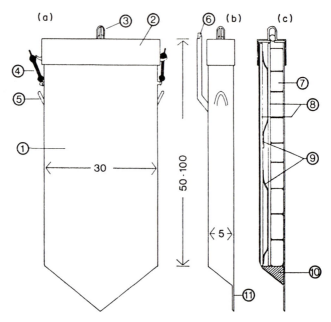

FIGURE 17.6.   Box freezer (from Huttunen and Meriläinen, 1978). (a) Front view: 1 = freezing panel; 2 = cover; 3 = ball valve; 4 = rubber fastener; 5 = wire handle. (b) Side view: 6 = socket. (c) Lateral transsection: 7 = dry ice cube; 8 = tray of double plywood; 9 = spring made of steel plate; 10 = lead weight; 11 = keel. *Reproduced by permission of the Finnish Botanical Publishing Board*

thread to mark every tenth varve and then transferred the length of the 10-year section to millimetre graph paper. The frozen core (or a sawn section of it) was then allowed to thaw and 10-year sequences of sediment packed into a calibrated spoon (0.3 ml). In this way the volume and sedimentation rate are known and the subsample can thus be used for determining pollen influx etc. It is important that the sediment should thaw at a low temperature and that it should be covered with polythene to prevent evaporation of water.

Another useful method is the embedding of the sediment into water-soluble carbowax, polyethyleneglycol (Tippett, 1964). A wax with a melting point of approximately +50°C is suitable. A section of the frozen core, 1–2 cm wide and 15–20 cm long, is placed into warm wax in a container in an oven (at a temperature just above this melting point). In about three days the wax replaces the sediment water without destroying the fine structures, although colour changes may occur. The wax sample is easy to handle after cooling, and samples of known volume can be cut off. Thin subsamples may be obtained by use of a microtome. The disturbed laminations nearest to the original sampling tube can be readily seen and discarded when working with solid wax samples. For additional methods, see Renberg and Segerström (1981b) and Renberg (1981).

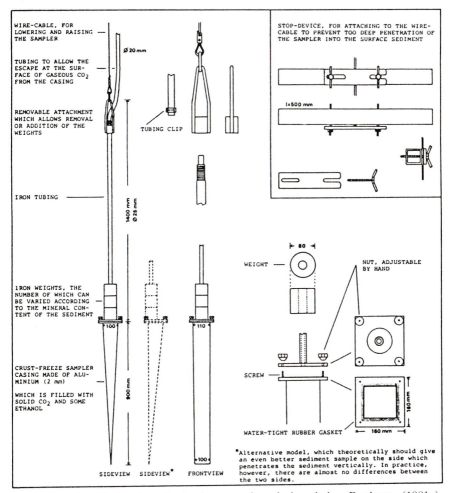

FIGURE 17.7. Wedge-shaped dry-ice sampler designed by Renberg (1981a).
Reproduced by permission of Universitetsforlaget, Oslo

The use of adhesive transparent tape is recommended for subsampling long sequences from the surface of a frozen core (Simola, 1977). The surface is cleaned with a knife and left open in a freezer, or preferably in a freezer room with a fan. A thin layer at the sediment surface freeze-dries within one or two days, and the loose particles are then caught on a tape by pressing it smoothly on to the sediment surface. Excessively thick samples should be avoided as these make the identification of diatoms and other microfossils increasingly difficult. Pieces of the tape can then be used when making microscopic slides for pollen, diatoms etc. simply by mounting them sediment side up under cover glass using the normal procedures (Figure 17.8). Some particles may break

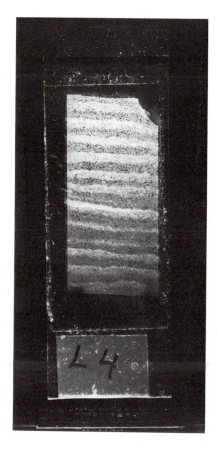

FIGURE 17.8.   Tape-peel preparation of the annually laminated sediments from Lake Lovojärvi representing the years 1907–18. The light component of the laminae correspond to the diatom maxima. The width of the tape is 19 mm. *Photographed and provided by H. Simola*

loose from the thin laminations; but if the varves are at least, say, a millimetre thick, detailed enough information on the fine structures can be obtained by this method. Long tapes, even those greater than 0.5 m in length, give a highly informative picture of the general pattern in sediment structures.

Simola *et al.* (1984) have analysed a 418-year sequence from the period 1539–1956 from such a tape peel preparation. The analysis was continuous, each level corresponding to a visual field 0.1 mm in width. The identification of about 250,000 diatoms and many other particles from 12,000 levels took about 10 months, which shows just how laborious an accurate varve study may be.

X-ray radiography

Unfrozen, normal sediment cores can be successfully used in the study of thinly laminated sediments when the sediment is not so loose that structures are destroyed during sampling or transportation. Subsamples of known volume are also easier to take from such cores than from frozen ones. X-ray radiography of unfrozen cores is recommended for documenting and measuring annually laminated sediments, one of the most valuable advantages being that the measurements are much easier to obtain from X-ray radiographs than from the wet, loose sediment surface, which tends to alter rapidly. The radiographs often render the laminations visible even in apparently massive sequences.

Conventional radiography is recommended as a routine for the study of fine laminations, this method being best suited for studying thin sediment samples of 5–10 mm. Sections of the core can be taken for X-ray radiography using the plexiglass trays designed by G. Digerfeldt, University of Lund (Digerfeldt *et al.*, 1975; Koivisto and Saarnisto, 1978). These are 20 cm long (the length depending on the film size), 3 cm wide and 1 cm deep, with a wall thickness of 1 mm. The edges are sharpened and one end is left open. A tray is carefully pressed into the sediment and the section cut free with a thin steel wire. The sediment surface is cleaned and then wrapped in thin plastic foil. The sooner the samples are exposed for radiography the less the laminations will be disturbed by drying etc.

A medical X-ray apparatus designed for mammography is suitable for sediment study. Mammography film with emulsion on one side only (Kodak PL 4006 or DuPont) gives good results. A number of the 0.3 mm varves in Lake Valkiajärvi were identified and counted from such radiographs. Further technical details, such as exposure time, are daily routine in the radiology laboratory, and can be most easily worked out with a specialist. The method is not expensive, as several sections can be exposed at once on the same film.

Microanalysis

A scanning electron microscope (SEM) equipped with an energy-dispersive spectrometer (EDS), as used routinely in mineralogy, is an accurate tool for investigating the element content within the laminations. This instrument has been used in the study of varved sediments from Lake Valkiajärvi, Finland (Alapieti and Saarnisto, 1981) and Diss Mere, England (Peglar *et al.*, 1984) (Figure 17.1). The measurements are performed from polished thin sections (Moreland, 1968) made from dried sediment samples which are immersed in the EPOFIX impregnating medium under a vacuum. Unfortunately the instrument cannot detect elements lighter than sodium (e.g. carbon, hydrogen and oxygen).

Preliminary qualitative microanalyses are carried out first from a small area (e.g. 1 × 1 mm) to determine what elements are present in the sediment. The detailed distribution of these elements is then studied by means of a digital-line profile analysis in which the characteristic X-ray lines of the elements are simultaneously mapped along a single line (Figure 17.1). The electron beam of the SEM is moved across the sample digitally step by step at intervals of several microns, and data for each element are collected from exactly the same points on the scan line. The data are stored in the EDS memory and then transferred to a graph plotter or to the cathode-ray tube of the SEM, where they can be photographed (Alapieti and Saarnisto, 1981).

In Figure 17.1, for the calcareous sediments of Diss Mere (Peglar *et al.*, 1984), a secondary electron image (SEI) has been produced for an area 0.5 × 0.7 mm, including a pale-dark lamination sequence 0.4–0.5 mm in thickness. The upper diagram is the result of a digital-line analysis by EDS to show the distribution of calcium, iron, silica and sulphur.

Photography

Photography is sometimes the simplest way of documenting both frozen and unfrozen varves (e.g. Saarnisto *et al.*, 1977; Renberg, 1978, 1981a), but it is practicable only when the colour differences between the layers within a varve are clear. In frozen cores the temperature is important. The sample surface can be carefully warmed using an electric lamp. The frozen core can also be left unwrapped for a day or two in a deep-freeze, whereupon the iron sulphide becomes oxidized as the sediment become freeze-dried. The same also holds true for unfrozen sulphide-rich sediments, whose varved structure may not be visible at all before oxidation. Frozen cores can also be submerged in a shallow container filled with 50% glycerine solution cooled to a temperature of −20°C (Renberg, 1981a). Unfrozen samples with a high water content are difficult to photograph because the water causes light reflections. The sediment surface has to be cleaned carefully in the direction of the laminations and then quickly covered with a small amount of water and a high-quality plastic film, which often makes the lamination remain visible for a longer time (Renberg, 1981a). Accurate photographs are very difficult to obtain in the case of very thin laminations, and some information on the structural features is always lost.

Thin sections

Thin sections from highly organic laminated sediments are rather laborious to make. These may reveal useful information in special cases, however, when the origin of the varves is to be studied. The sediment is first allowed to dry. This produces cracks and disturbances in the fine structure, and the original varve thicknesses can no longer be measured. The dry sample is then embedded in epoxy resin (e.g. EPOFIX) in a vacuum and thin sections are made using the same procedures as for mineralogical thin sections (Moreland, 1968). Sections thin enough for mineralogical analysis are difficult to make, but 30–40 μm sections are suitable for the study of micro-organisms and structural features (Koivisto and Saarnisto, 1978) (Figure 17.9). Merkt (1971) describes a procedure for making long, thin sections (10 cm) by freeze-drying the sediment before embedding it into Araldite resin. This preserves the original varve thickness, and the long slides can thus be used for varve counts and measurements.

FIGURE 17.9.   The varved sediment of Lake Valkiajärvi, Finland, as revealed in a microphotograph taken from a thin section. Average varve thickness (light-dark couplet) about 0.3 mm

Counting the varves

The counting of the varves is the most difficult, time-consuming and important step in the study of annually laminated sediments. Reliable counting is imperative before any detailed chronological interpretation can be made. Counting proceeds reasonably easily if the varve thickness is 0.5 mm or more, but thinner laminations are more problematical. Short intervals can be counted from microscope slides made from transparent tape, or from thin sections, but such methods are not practicable for counting sequences of hundreds or thousands of years.

    Counting of the varves from the surface of a frozen sample is often the most practical way. It is most effectively done in a walk-in cold room, using a magnifying glass and cold light whose angle can be determined. The surface of the frozen core is cleaned with a knife and then allowed to warm for 10–20 minutes to ensure that the laminations are visible. The surface of the core can also be warmed by blowing on it gently. The varve counts are made in convenient lengths, e.g. 10 or 30 varves or 1–2 cm, the intervals counted being marked with pins. In about an hour the core becomes so warm that the colours change and the laminations are no longer so clearly visible. The intervals are then marked on millimetre graph paper and the core put back into the freezer. The procedure is repeated until the whole core has been counted, preferably three times along different sides. The deviations from the average number of varves give the accuracy of the varve counts. The numbers of varves and information of structural features are also marked on millimetre graph paper.

Similarly, the laminations can be counted directly from the surface of an unfrozen core on the basis of colour and texture. A large Russian peat sampler (chamber length 100 cm), the principle of which has been described by Dowsey (1966), is suitable for sampling undisturbed varved sediments below loose surface layers. The core can be covered with plastic film, as for photography, and counting may proceed for several hours under a stereo microscope without any significant change in the visibility of the laminations (Renberg, 1981a).

It is often convenient to count from X-ray radiographs on a light table, in which case a microdensitometer is also useful for documenting and measuring the varves (Koivisto and Saarnisto, 1978). Care is needed when interpreting the densitometer curves, however, and therefore the original radiographs should be consulted at the same time.

Prints from the X-ray radiographs, or enlarged black-and-white photographs, can be studied under a binocular microscope and the varve counts can be recorded using devices designed for tree-ring analysis. In very favourable situations varve measurements can be performed directly from the core surface (Renberg *et al.*, 1984). The varves need to be reasonably clear before accurate counts can be made, however. The advantage of both microdensitometry and dendrochronology devices is that the variations in varve thickness can be recorded accurately. This presents a difficulty if the varves are counted directly from the core surface, although the latter is often the only practicable way.

APPLICATIONS OF ANNUALLY LAMINATED SEDIMENTS

The accurate time-scale provided by annually laminated lake sediments is an argument in favour of their extensive use in palaeoecological studies. The applications are numerous, and cover a wide range from the duration of interglacials (Giesenhagen, 1926; Turner, 1970; Benda, 1974; Meyer, 1974; Müller, 1974) to studies on vegetational history, the history of land use, lake development and climatic change. The applications have been reviewed briefly by Saarnisto (1979) and in more detail by O'Sullivan (1983). The following discussion introduces examples which are also of interest from a methodological point of view.

Studies on vegetational history

Closely spaced pollen analyses from laminated sediments furnish an unparalleled opportunity for detailed studies of vegetational history. The 9500 laminations identified in the Lake of the Clouds, Minnesota, provide a detailed time-scale for relative and influx pollen diagrams leading to an accurate reconstruction of the forest history (Craig, 1972). Watts (1973), using Craig's and other data, emphasizes the particular suitability of varved sediments for exact dating of the duration of plant successions and changes in the plant population.

Laminated sediments have one important advantage over massive sediments for pollen studies, namely minimization of the pollen recirculation effect, as pointed out by Ritchie (1977) in his study of M Lake near Inuvik, N.W.T., Canada. Ritchie also suggests that if individual varves could be studied for annual pollen influx, attempts could be made to correlate the results with short-term climatic variations under the critical conditions prevailing near the forest limit.

History of agriculture indicated by pollen and charcoal

Detailed pollen studies on varved sediments from Crawford Lake, near Toronto, Ontario, by Boyko (1973, 1979) and McAndrews (Byrne and McAndrews, 1975; McAndrews, 1976) have elegantly demonstrated the use of varved sediment for the study of human influence on the vegetation and as evidence of the presence of prehistoric man in the neighbourhood of the lake. The presence of agricultural Indians in the area was indicated by the findings of pollen and seeds of maize (*Zea mays*) and sunflower (*Helianthus amuus*) in sediments dating from 1360–1659 A.D. Another important discovery was that of purslane pollen and seeds (*Portulaca oleracea L.*) in sediments dating from 1350–1539 A.D., the highest pollen influx values dating from the period 1430–1449 A.D., which means that purslane was already present in North America in pre-Columbian times. The findings actually led to archaeological excavations and the discovery of an Indian village near the lake (Byrne and McAndrews, 1975).

The effect of fire in natural forest ecosystems in North America has been studied by Swain (1973) in Minnesota and Cwynar (1978) in Ontario. In Finland, and elsewhere in Fennoscandia, fire intervals reflect both the incidence of natural forest fires and the intensity of the widely practiced technique of slash and burn cultivations. Huttunen (1980) studied the influx of pollen and charcoal particles in a varved sequence covering 1506 years from Lake Taka-Killo near Lammi, southern Finland, using a continuous series of one centimetre thick samples (Figure 17.10). The charcoal was analysed from the same slides as pollen. The amounts of pollen and spores indicative of agriculture (e.g. *Rumex acetosa*, *Pteridium aquilinum* and Cerealia) and charcoal vary in a broadly similar way in the sediment. Pollen of *Juniperus* also increases significantly as the burning of the forest intensifies.

The mean interval between fires in the Taka-Killo sediments from 470–1100 A.D. is 95 years, and only weak evidence of agriculture can be seen. By 1450 A.D. charcoal increases permanently, as does the pollen of rye (*Secale*). From 1600 to 1900 A.D. the mean fire return period is only 30 years, and this together with the general use of forest for pasture may be assumed to have effectively prevented the regeneration of the forest, especially the spruce element. Low totals for conifer and birch pollen, and high *Alnus* and *Juniperus* in the middle of the nineteenth century, show that the surroundings of the lake had become practically open land with alder and juniper bushes, as was common around Lammi up until the end of the century. Nowadays spruce is the dominant tree species in the area. Similar studies have been performed on varved sediments from Lake Ahvenainen, also near Lammi, by Tolonen (1978a). There, slash and burn cultivation had begun by the early Iron Age, the 5th century B.C., and indications of still earlier human activity are present. The intensive cultivation of fibre plants, flax (*Linum*) and hemp (*Cannabis*) is also clearly seen in the pollen diagrams from both of these lakes (Figure 17.9).

In addition to pollen and charcoal, magnetic parameters of the sediments, namely saturation isothermal remanent magnetization and suspectibility, seem to be useful indicators of fires in the catchment area of a given lake (Rummery, 1983; see also Chapter 15).

Human influence on a lake

Increased sediment accumulation rates, changes in diatoms and other algae, and changes in sediment chemistry, are the most common indications of human impact on the surroundings of a lake. Increased nutrient loading leads to increased productivity, and varved lake sediments are invaluable archives of such developments because of the exact dating they provide for the course of events. The above mentioned lakes Taka-Killo and Ahvenainen near Lammi

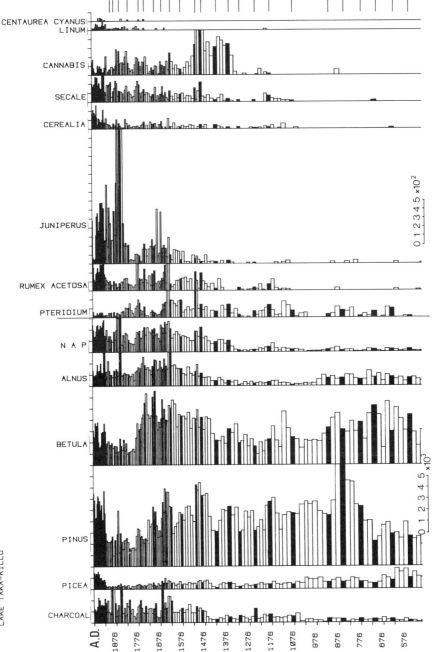

LAKE TAKA-KILLO

FIGURE 17.10. Charcoal and selected pollen influx diagram from the annually laminated sediments of Lake Taka-Killo, Finland (Huttunen, 1980). Forest fires and/or slash-and-burn cultivation are indicated by black lines. The time-scale is based on varve counts. *Reproduced by permission of the Finnish Botanical Publishing Board*

(Huttunen, 1980; Tolonen, 1978a, 1978c) have a long eutrophication history not only because of the practice of agriculture in the surrounding area, but also to a great extent because of the soaking of fibre plants in the lakes themselves.

The extensive peatland drainage undertaken in Finland in recent decades is also recorded in varved sediments. Simola (1983) has demonstrated how sensitively the plankton diatoms react to the increase in allochthonous material from drained peatlands. The eutrophication of Lake Polvijärvi, northern Karolia, is shown by the succession *Tabellaria flocculosa–Asterionella formosa––Melosira ambiqua–Fragilaria crotonensis*, beginning in 1967, when extensive drainage also commenced. The effect of peatland drainage was intensified by fertilization of the drained area in 1971. The increased allochthonous material in the sediment is also shown by rises in the carbon to nitrogen ratio and in magnesium. The values for phosphorous, iron and manganese, on the other hand, do not change significantly. As Simola (1983) points out, many of the significant elements are not bound to the sediment particles but are present in the sediment water, and therefore the chemical stratigraphy is not so sharply defined. The drainage of peatlands and moist forest areas has also led to increased mercury levels in some Finnish lakes (Simola and Lodenius, 1982).

Industrial waste waters can similarly affect the eutrophic status of a lake, as shown by Digerfeldt *et al.* (1975) at Järlasjön, near Stockholm, and at Loe Pool, Cornwall, England, by Simola *et al.* (1981), and this can again be dated by means of varved sediments. The effect of air pollutants such as lead are also recorded in sediments (Renberg and Segerström, 1981). In addition to algae, pigments of carotenids have also been used to study the eutrophication history of a lake (e.g. Zürichsee — (Züllig (1981)). See Chapter 21.

Climatic change

Variations in varve thickness have attracted interpretations of short-term climatic fluctuations. It is often difficult, however, to differentiate between regional factors (e.g. climate) and local factors (forest fires, human activity), both of which can influence the influx of allochthonous material into the sediments and the production of the lake itself. It may be suggested that the influence of local factors upon varve thickness is more pronounced in small lakes than in large basins.

Seibold (1958) made a detailed comparison between climatic variables and variations in the thickness of calcareous varves from the bottom of a closed bay, Malo Jesero, on the island of Mljet in the Adriatic Sea, Yugoslavia. The clearest correlation existed between high summer precipitation records being for Rome, where they go back to 1782 A.D. In Lake Van, Turkey, varve counts cover 10,400 years (Kempe and Degens, 1979). Here, varve thickness was found to vary in sequence with high water level and high precipitation, with a periodicity of 10 years, which is close to the duration of the sunspot cycle.

Renberg *et al.* (1984), who measured about 1300 varves from Lake Judesjön in northern Sweden, found a cyclic variation in varve thickness with a periodicity of 30–40 years which depends on varying amounts of organic material. This may reflect variations in primary production caused by changes in insolation during the summer months. In the study of climatic changes, as in connection with other applications, the crucial importance of the accuracy of the varve count should again be emphasized.

REFERENCES

Alapieti, T., and Saarnisto, M. (1981). Energy-dispersive X-ray microanalysis of laminated sediments from Lake Valkiajärvi, Finland. *Bull. Geol. Soc. Finland*, **53**, 3–9.

Anthony, R. S. (1977). Iron-rich rhythmically laminated sediments in Lake of the Clouds, northeastern Minnesota. *Limnol. Oceanogr.*, **22**, 45–54.

Benda, L. (1974). Die Diatomeen der niedersächsischen Kielselgur-Vorkommen, palökologische Befunde und Nachweis einer Jahresschichtung. *Geol. Jahrb.*, **A 21**, 171–197.

Boyko, M. (1973). *European impact on the vegetation around Crawford Lake in southern Ontario*, M.Sc. thesis, Department of Botany, University of Toronto, Ontario.

Boyko-Diakonow, M. (1979). The laminated sediment of Crawford Lake, southern Ontario, Canada. In: *Moraines and Varves* (Ed. C. Schlüchter), Balkema, Rotterdam, pp. 303–307.

Brunskill, G. J., and Ludlam, S. D. (1969). Fayetteville Green Lake, New York: I. physical and chemical limnology. *Limnol. Oceanogr.*, **14**, 817–829.

Byrne, R., and McAndrews, J. H. (1975). Pre-Columbian purslane *Portulaca oleracea* L. in the New World. *Nature*, **253**, 726–727.

Craig, A. J. (1972). Pollen influx to laminated sediments: a pollen diagram for northeastern Minnesota. *Ecol.*, **53**, 46–57.

Cwynar, L. C. (1978). Recent history of fire and vegetation from laminated sediment of Greenleaf Lake, Algonquin Park, Ontario. *Can. J. of Bot.*, **56**, 10–21.

Dickman, M. D. (1979). A possible varving mechanism for meromictic lakes. *Quat. Res.*, **11**, 113–124.

Digerfeldt, G., Battarbee, R. W., and Bengtsson, L. (1975). Report on annually laminated sediment in lake Järlasjön, Nacka, Stockholm. *Geol. Fören. Stockholm Förhandl.*, **97**, 29–40.

Giesenhagen, K. (1926). Kieselgur als Zeitmass für eine Interglazialzeit. *Z. Gletscherkunde*, **14**, 1–10.

Hansen, S. (1940). Varvighed i danske og skaanske senglaciale Aflejringer, med saerlig Hensyntagen til Egernsund Issøsystemet; med et atlas (Varvity in Danish and Scanian late-glacial deposits, with special reference to the system of ice-lakes at Egernsund in Jutland; with Atlas). *Danm. Geol. Unders.*, **II 63**.

Heer, O. (1865). *Die Urwelt der Schweiz*, F. Schulthess, Zürich.

Huttunen, P. (1980). Early land-use, especially the slash and burn cultivation in the commune of Lammi, Southern Finland, interpreted mainly using pollen and charcoal analysis. *Acta Bot. Fennica*, **113**, 1–45.

Huttunen, P., and Meriläinen, J. (1978). New freezing device providing large unmixed sediment samples from lakes. *Ann. Bot. Fenn.*, **15**, 128–130.

Jowsey, P. C. (1966). An improved peat sampler. *New Phytol.*, **65**, 245–248.

Kelts, K., and Hsü, K. J. (1978). Freshwater carbonate sedimentation. In: *Lakes: Geology, Chemistry, Physics* (Ed. A. Lerman), Springer Verlag, New York, pp. 295–323.

Kempe, S., and Degens, E. G. (1979). Varves in the Black Sea and Lake Van (Turkey). In: *Moraines and Varves* (Ed. C. Schlüchter), Balkema, Rotterdam, pp. 309–318.

Koivisto, E., and Saarnisto, M. (1978). Conventional radiography, xeroradiography, tomography, and contrast enhancement in the study of laminated sediments. *Geogr. Ann.*, **A 60**, 55–61.

Kukkonen, E., and Tynni, R. (1970). Die Entwicklung des Sees Pyhäjärvi in Südfinnland im Lichte von Sediment und Diatomeenuntersuchungen. *Acta Bot. Fenn.*, **90**, 1–30.

Ludlam, S. D. (1967). Fayetteville Green Lake, New York. 3: The laminated sediments. *Limnol. Oceanogr.*, **14**, 849–857.

Ludlam, S. D. (1976). Laminated sediments in holomictic Berkshire lakes. *Limnol. Oceanogr.*, **21**, 743–746.

Ludlam, S. D. (1979). Rhythmite deposition in lakes of the northeastern United States. In: *Moraines and Varves* (Ed. C. Schlüchter), Balkema, Rotterdam, pp. 287–294.

McAndrews, J. H. (1976). Fossil history of man's impact on the Canadian flora: an example from southern Ontario. *Can. Bot. Assoc. Bull. Suppl.*, **9:1**, 1–6.

Meriläinen, J. (1970). On the limnology of the meromictic Lake Valkiajärvi in the Finnish Lake District. *Ann. Bot. Fenn.*, **7**, 29–51.

Merkt, J. (1971). Zuverlässige Auszählungen von Jahresschichten in Seesedimenten mit Hilfe von Gross-Dünnschliffen (Reliable counting of annually laminated layers of lake sediments by means of long-sized thin sections). *Arch. Hydrobiol.*, **69**, 145–154.

Meyer, K. J. (1974). Pollenanalytische Untersuchungen und Jahresschichtenzählungen an der holsteinzeitlichen Kieselgur von Hetendorf. *Geol. Jahrb.*, **A 21**, 107–140.

Moreland, G. C. (1968). Preparation of polished thin sections. *Amer. Mineralogist*, **53**, 2070–2074.

Müller, H. (1974). Pollenanalytische Untersuchungen und Jahresschichtenzählungen an der Eem-zeitlichen Kieselgur von Bispingen/Luhe. *Geol. Jahrb.*, **A 21**, 87–105.

Nipkow, F. (1920). Vorläuge Mitteilungen über untersuchungen des Schlammbatzes im Zürichsee. *Zeitschrift für Hydrologie*, **1**, 100–122.

Olsson, I. U. (1974). Some problems in connection with the evaluation of C^{14} dates. *Geol. Fören. Stockholm Förhandl.*, **96**, 311–320.

O'Sullivan, P. E. (1983). Annually-laminated lake sediments and the study of Quaternary environmental changes — a review. *Quat. Sci. Rev.*, **1**, 245–313.

Peglar, S., Fritz, S. C., Alapieti, T., Saarnisto, M., and Birks, H. J. B. (1984). The composition and formation of laminated lake sediments in Diss Mere, Norfolk, England. *Boreas*, **13**, 13–28.

Renberg, I. (1976). Annually laminated sediments in Lake Rudetjärn, Medelpad province, northern Sweden. *Geol. Fören. Stockholm Förhandl.*, **98**, 335–360.

Renberg, I. (1978). Palaeolimnology and varve counts of the annually laminated sediment of Lake Rudetjärn, northern Sweden. *Early Norrland*, **11**, 63–92.

Renberg, I. (1981a). Improved methods for sampling, photographing and varve-counting of varved lake sediments. *Boreas*, **10**, 255–258.

Renberg, I. (1981b). Formation, structure and visual appearance of iron-rich, varved lake sediments. *Verh. Int. Ver. Limnol.*, **21**, 94–101.

Renberg, I., and Segerström, U. (1981a). The initial points on a shoreline displacement curve for southern Västerbotten, dated by varve counts. *Striae*, **14**, 174–176.

Renberg, I., and Segerström, U. (1981b). Application of varved lake sediments in palaeoenvironmental studies. *Wahlenbergia*, 7, 125–133.

Renberg, I., Segerström, U., and Wallin, J.-E. (1984). Climatic reflection in varved lake sediments. In: *Climatic Changes on a Yearly to Millennial Basis* (Eds. N.-A. Mörner and W. Karlén). D. Reidel Publishing Company, pp. 249–256.

Richardson, J. L., and Richardson, A. E. (1972). History of an African Rift Lake, and its climatic implications. *Ecol. Monographs*, 42, 499–534.

Ritchie, J. C. (1977). The modern and late Quaternary vegetation of the Campbell-Dolomite uplands, near Inuvik, N. W. T. Canada. *Ecol. Monographs*, 47, 401–423.

Rummery, T. (1983). The use of magnetic measurements in interpreting the fire histories of lake drainage basins. In: *Palaeolimnology*, Proc. 3rd Inter. Symp. Palaeolimn., Joensuu, Finland (Eds. J. Meriläinen, P. Huttunen and R. W. Battarbee); also *Hydrobiologia*, 103, pp. 53–58, Dr. W. Junk Publishers, The Hague.

Saarnisto, M. (1975). Pehmeiden järvisedimenttien näytteenottoon soveltuva jäähdytysmenetelmä (A freezing method for sampling soft lake sediments). *Geologi*, 26, 37–39.

Saarnisto, M. (1979). Applications of annually-laminated lake sediments: a review. *Acta Universitas Ouluensis*, A 82, Geology 3, 97–108.

Saarnisto, M. (1985). Long varve series in Finland. *Boreas*, 14, 133–137.

Saarnisto, M., Huttunen, P., and Tolonen, K. (1977). Annual lamination of sediments in Lake Lovojärvi, southern Finland, during the past 600 years. *Ann. Bot. Fenn.*, 14, 35–45.

Seibold, E. (1958). Jahreslagen in Sedimenten der mittleren Adria. *Geol. Rundschau*, 47, 100–117.

Shapiro, J. (1958). The core-freezer — a new sampler for lake sediments. *Ecol.*, 39, 748.

Simola, H. (1977). Diatom succession in the formation of annually laminated sediment in Lovojärvi, a small eutrophicated lake. *Ann. Bot. Fenn.*, 14, 143–148.

Simola, H. (1979). Micro-stratigraphy of sediment laminations deposited in a chemically-stratifying eutrophic lake during the years 1913–1976. *Holarctic Ecol.*, 2, 160–168.

Simola, H. (1982). Limnological effects of peatland drainage and fertilisation as reflected in the varved sediment of a deep lake. In: *Palaeolimnology*, Proc. 3rd Inter. Symp. Palaeolimn., Joensuu, Finland (Eds. J. Meriläinen, P. Huttunen and R. W. Battarbee); also *Hydrobiologia*, 103, pp. 43–57, Dr. W. Junk Publishers, The Hague.

Simola, H., Coard, M. A., and O'Sullivan, P. E. (1981). Annual laminations in the sediment of Loe Pool, Cornwall. *Nature*, 290, 238–241.

Simola, H., Hanski, I., and Liukkonen, M. (1984). History of diatom populations during 418 years in the laminated sediment of Lake Lovojärvi. *Proc. Nordic Meeting Diatomologist*, Bergen, May 1983, Rep. Bot. Inst. Univ. Bergen. (in press).

Simola, H., and Lodenius, M. (1982). Recent increase in mercury sedimentation in a forest lake attributable to peatland drainage. *Bull. Environ. Contam. and Toxicol.*, 29, 298–305.

Simola, H., and Tolonen, K. (1981). Diurnal laminations in the varved sediment of Lake Lovojärvi, South Finland. *Boreas*, 10, 19–26.

Simola, H., and Uimonen-Simola, P. (1983). Recent stratigraphy and accumulation of sediment in an oligotrophic, deep lake in South Finland. In: *Palaeolimnology*, Proc. 3rd Inter. Symp. Palaeolimn., Joensuu, Finland (Eds. J. Meriläinen, P. Huttunen and R. W. Battarbee); also *Hydrobiologia*, 103, pp. 287–293, Dr. W. Junk Publishers, The Hague.

Stuiver, M. (1969). Long term C14 variations. In: *Radiocarbon Variations and Absolute Chronology* (Ed. I. U. Olsson), Almqvist & Wiksell, Stockholm, pp. 197–213.

Sturm, M. (1979). Origin and composition of clastic varves. In: *Moraines and Varves* (Ed. C. Schlüchter), Balkema, Rotterdam, pp. 281–285.

Swain, A. M. (1973). A history of fire and vegetation in north-eastern Minnesota as recorded in lake sediments. *Quat. Res.*, **3**, 383–396.

Tippett, R. (1964). An investigation into the nature of the layering of deep-water sediments in two eastern Ontario lakes. *Can. J. of Bot.*, **42**, 1693–1704.

Tolonen, K. (1980). Comparison between radiocarbon and varve dating in lake Lampellonjärvi, South Finland. *Boreas*, **9**, 11–19.

Tolonen, M. (1978a). Palaeoecology of annually laminated sediments in Lake Ahvenainen, S. Finland. I: Pollen and charcoal analyses and their relation to human impact. *Ann. Bot. Fennici*, **15**, 177–208.

Tolonen, M. (1978b). Palaeoecology of annually-laminated sediments in Lake Ahvenainen, South Finland. II: Comparison of dating methods. *Ann. Bot. Fennici*, **15**, 209–222.

Tolonen, M. (1978c). Palaeoecology of annually-laminated sediments in Lake Ahvenainen, South Finland. III: Human influence on lake development. *Ann. Bot. Fennici*, **15**, 223–240.

Turner, C. (1975). The correlation and duration of middle Pleistocene interglacial periods in northwest Europe. In: *After the Australopithecines* (Eds. K. Butzer and G. Isaac), Mounton, The Hague, pp. 259–308.

Watts, W. A. (1973). Rates of change and stability in vegetation in the perspective of long periods of time. In: *Quaternary Plant Ecology* (Eds. H. J. B. Birks and R. G. West), Blackwell, Oxford, pp. 195–206.

Welten, M. (1944). Pollenanalytische, stratigraphische und geochronologische Untersuchungen aus dem Faulenseemoos bei Spiez. *Veröff. Geobot. Inst. Rübel*, **21**, 1–201.

Wesenberg-Lund, C. (1901). Studier over søkalk, bønnemalm og søgytje i danske indsøer (Lake-lime, pea-ore, lake-gytje). *Medd. Dansk. Geol. Foren.*, **2**, 1–180.

Wright, H. E. (1980). Cores of soft lake sediments. *Boreas*, **9**, 107–114.

Züllig, H. (1981). On the use of carotenoid stratigraphy in lake sediments for detecting past developments of phytoplankton. *Limnol. and Oceanog.*, **26**, 970–976.

Handbook of Holocene Palaeoecology and Palaeohydrology
Edited by B. E. Berglund
© 1986 John Wiley & Sons Ltd.

18

Dendrochronology applied to mire environments

André-V. Munaut

Laboratory of Palynology and Dendrochronology,
Louvain-la-Neuve, Belgium

INTRODUCTION

In certain conditions, stumps or trunks are buried in sediments and the wood is preserved without structural deformation. These subfossil woods can be used as living trees or historical beams in dendrochronological research. The main purpose of these studies is to collect information concerning the past environment.

In areas where only one growth season occurs, the annual increment of wood is apparent on the cross-section of the tree stem as one circumcentric layer, the so-called tree-ring (Figure 18.1). The ring-width is limited by several genetic or external factors (fertility of the soil, water supply, competition, climate etc.). If a climatic parameter has a limiting effect on the tree, and if this parameter fluctuates from one year to the next, the trees belonging to the same botanical species and living in the same general conditions will suffer these limitations in the same way. This phenomenon gives a fluctuating time series of ring-widths which can be measured (Figure 18.1). The chronological sequences are often very similar in contemporary trees. Graphic representations of synchronous sequences show characteristic patterns which can be matched between many trees (Figure 18.2). This matching process, or cross-dating, is the basic principle of dendrochronology.

When species of various ages are cross-dated, and if the cutting date of at least one tree is known, the absolute dating of each ring from each sample is

371

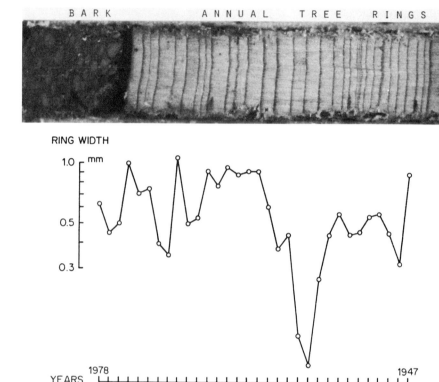

FIGURE 18.1.   Tree-rings of *Cedrus atlantica* and measured ring-widths
along a logarithmic scale

possible (Figure 18.3). Otherwise only relative dates can be assigned. Besides its use as a chronological scale, tree-ring analysis is able to provide ecological information.

FIELD WORK

Field observation

Stumps and trunks in situ

When the material is preserved in the place where it grew, various informations concerning the dynamics of the vegetation and the evolution of the environment can be collected in the field. The following features must be observed:

(1) the stratigraphical position of stumps, roots and prostrate trunks;
(2) the topographical situation of each element and the spatial extension of possible wood layers;
(3) the presence of uprooted trees and the orientation of the prostrate trunks;

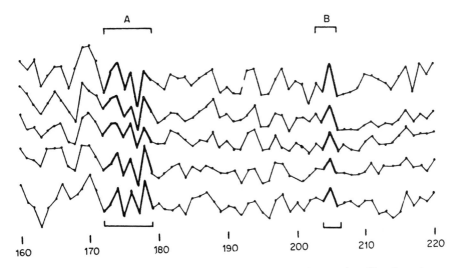

FIGURE 18.2. Cross-dating between archaeological oak samples (floating chronology)

(4) the dimensions (diameter and length) of the wood;
(5) the relative position of stumps and trunks: isolated, scattered, clustered;
(6) the presence of bark and branches on stumps and trunks;
(7) the shape of the stubs and stems: sickle, straight, low-branched;
(8) the type of rooting: deep, superficial;
(9) the state of preservation of the wood, and the nature of the degradation
     (anaerobic organisms, fungi, insects etc.).

Various inferences can be made from these observations. For instance, the rooting gives information about the water level in the soil. Roots growing downwards indicate a more-or-less deep water level, while plank-shaped ones can indicate very moist soil. When the roots are growing upwards, or if multilevel rooting is observed, we can conclude that the water level has risen. When the sediments are heterogeneous, special attention must be paid to the stratigraphical level of the rootlets in order to determine the origin of the nutrients. The vitality of the trees is revealed by their length and the shape of the trunk. Deformed and low-branched stems give evidence of very difficult growth conditions. A sickle-shaped stump is often the result of a geotropic correction of the trunk by a tree growing on soft and unstable subsoil.

Uprooting gives evidence of sudden and violent action such as wind and erosion causing the death of the trees, while broken stems, without any bark on branches, indicate that the trees died standing up. In this case death often results from slow processes such as physiological degradation due to age, asphyxiation by a rising water level, destruction by climatic change, etc.

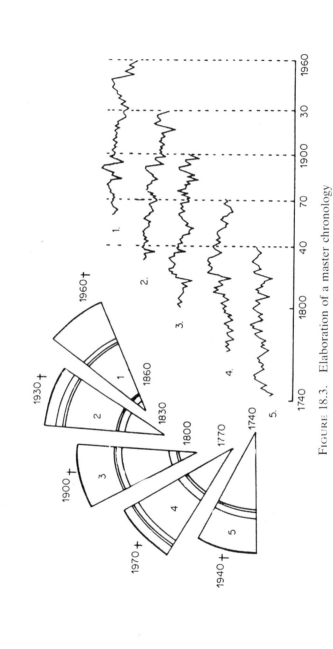

FIGURE 18.3.   Elaboration of a master chronology

The state of preservation depends on the changes in the environment, after the trees died. High, well-preserved stumps, covered with some bark, are the result of a very fast accumulation process.

Driftwoods

Driftwoods buried in lake or river sediments very often result from a disrupted equilibrium in the drainage basin, for example by flood, erosion, landslide etc. Nevertheless, it can provide chronological information about these phenomena. The direction of the tree (the roots are generally upstream), the presence of bark and branches, and the corrosion affecting the surface of the wood, are useful for further dendrochronological interpretation.

Sampling

Selection of samples

In order to obtain accurate statistical information, all categories of material must be sampled according to the stratigraphical position, the spatial dispersion, the dimensions, and the number of rings. If the material is abundant and easily sampled, special attention must be paid to the conservation of the wood and to the presence of the outermost ring.

A general rule concerning the number of samples does not exist, but to obtain a reliable chronology about ten specimens is desirable. Depending on the qualities of the wood, two or three times this number must be collected in the field. A low number of rings (less than 100) makes cross-datings difficult.

Technical work

If possible, cross-sections are cut off near the base of stumps and trunks. The disc must be at least 5 cm thick to avoid breakage and to keep material for future needs (^{14}C dating, isotope analysis etc.).

When the preservation is good, the sections are allowed to dry slowly. Otherwise, to conserve the wood moisture, they need to be wrapped closely in plastic bags.

If sawing a cross-section is impossible and if the wood is hard, it is possible to bore cylindrical cores along radii. Equipment for this has been described by Eckstein and Bauch (1969), but other types are available.

LABORATORY WORK

Information sheet and identification

A sheet must be filled in for each sample and each site, using field data and close observations of the specimen (presence of bark, pith, growth anomalies etc.).

An identification must be attached to each specimen. The most convenient method for future computer processing is a six-digit number system. Each number gives, according to a code, the species, the site, the tree in the site, the radius etc.

Surfacing

To be observed and measured with precision, the raw surfaces have to be smoothed. If the wood is very well preserved and can be submitted to desiccation without major deformataions, shrinkage or crumbling, the method used for living trees gives the best results. After desiccation the surface is polished with a mechanical sander using, in succession, different grades of abrasive papers (for instance 220 and 400×). Sanding is especially recommended for Gymnospermae.

When the specimens are soft and must be kept moist, it is necessary to cut a surface of observation along one or several radii. On hard wood a strong blade is necessary, but on soft wood only a thin razor blade will allow preparation of a smooth surface without scratching and crushing.

Measuring

Most of the measuring machines are fitted out with a low-power binocular (magnification ×20). This magnification is adequate for a precise view of the wood structure and to read the width to the nearest 0.01 mm. Although mechanical or electrical devices are still useful in some conditions, many tree-ring laboratories have now developed more sophisticated electronic equipment, built with compatible elements: an incremental measuring machine and a microcomputer linked to a high-power computer and various peripherals. Various types of equipment exist on the market, but care should be taken that all the interfaces are compatible.

When the tree is asymmetrical, the measured radius is not necessarily equal to the geometric one. The ring-width must always be measured perpendicularly to the ring's limit. This requirement is especially necessary for the measuring method using X-rays (Chapter 19).

Dendrochronological curves

The measurements are plotted on semi-transparent paper: the years against the *x*-axis, the widths against the *y*-axis along a logarithmic scale. This scale reduces the dimensions of the graph when the variability of the rings is very high and transfers the graphic representation of an absolute difference into a relative one (e.g. Hoëg, 1956). This transformation makes it possible to compare dendrochronological curves which exhibit very different ring-width means.

Various types of curves may be drawn. The absolute curves have every ring dated; the floating ones have not (for instance, subfossil or prehistoric ones). A single radius can give a curve as well as the mean of several curves. The master chronology is an absolutely accurate mean curve used as a reference to cross-date other curves of unknown age.

Cross-dating

Most dendrochronological curves are cross-dated by ocular comparison. This process consists of identifying the same sequences of width variations in different curves. The identification is made easier when many years are characteristic. (A characteristic year is defined as a year for which at least 80% of the compared curves show a width variation in the same direction.) When many characteristic years are succeeding each other they constitute a so-called *signature* (Figure 18.4).

Another role for ocular cross-dating is to detect anomalies affecting the growth of some trees. These anomalies can be very severe. For instance, when a tree has been wounded, several rings can be missing and several others deformed by abnormal physiological processes. Such specimens must be discarded. When the anomaly is weak (the absence or duplication of only one ring) this procedure allows for the identification of the missing or false ring. This identification is absolutely necessary before it is possible to combine individual curves in a mean chronology.

A technique has been described using the coefficient of correlation to locate false or missing rings (Wendland, 1975). It should be stressed, however, that a coefficient of correlation is not a definite proof of a cross-dating, but a help for the final ocular decision. Edaphic factors also have an effect on the cross-dating. In some environments (for instance on fertile soils) the limiting effect of the climate is sometimes so much reduced that the rings have almost the same widths; these are called complacent rings. In harsher environments (for instance on drained slopes) the reactions to a fluctuating climate are stronger and registered in terms of widely varying ring-widths; these are called sensitive rings.

Other difficulties occur in some climates, especially in oceanic ones; the response of the trees can be different according to the microenvironment, and some specimens are impossible to cross-date (Munaut and Casparie, 1971).

Indices

Absolute tree-ring values are not always suitable when computing a mean chronology. Indeed, the mean value of the tree-ring varies during the life of the

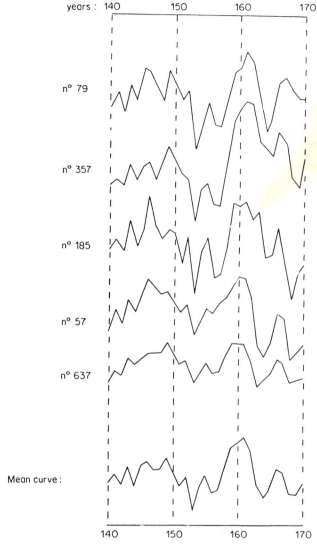

FIGURE 18.4.   Signature: a very characteristic pattern in five
individual curves and preserved in the mean of twenty years

tree. Moreover, trees growing in the same site often have a different mean
growth, and their weight in the mean curve is consequently different.

For these reasons, it is necessary to remove the trend of the growth while
keeping the annual variations. This effect can be obtained by computing the
trend curve for each sample and by dividing each absolute width by its
corresponding trend value. The resulting indices form stationary time-series

and their means are equal to 1 (Matalas, 1962). Thus, each sample has an equal weight which avoids introducing artificial variations into the mean chronology (Figure 18.5).

Dendrochronological parameters

The chronologies can be characterized by statistical parameters: the standard deviations, the serial correlation of 1st to n^{th} order, the mean sensitivity, and by a variance analysis (Fritts, 1976).

The standard deviation (classic definition) gives the dispersion of the indices from the mean.

The serial correlation is the correlation between a chronological sequence and itself after a shifting of 1 to n years. If significant, it shows that tree-ring sequences are not generated by a random process, the growth of a given year being influenced by phenomena generated during the previous one — for instance, storage of food products from one year to the next (Matalas, 1962).

The mean sensitivity *ms* is given by the following equation:

$$MS = \frac{1}{n-1} \sum_{t=1}^{t=n} \left| \frac{2(x_{i-1} - x_i)}{(x_{i+1} + x_i)} \right|$$

This coefficient depends on the variability between two successive rings. It expresses the reaction of the tree to the variations of climatic factors. The value 0.20 constitutes the limit under which the tree can be considered as 'complacent'.

A special analysis of variance was described by Fritts (1976) in order to express as a percentage the part of the variance attached to the cores, to the trees and to the mean chronology of a site. The magnitude of this last variance can show how strong the effect of the climate is.

Presentation of results

Tables

All the information collected in the field, during laboratory processing or by computation should be tabulated.

For each sample, a table should show:

(1) the length and orientation of the core;
(2) the presence of bark pith;
(3) the number of rings;
(4) the dates of the first and the last ring related to the master chronology;
(5) the mean ring-width;
(6) the equation used to fit the growth trend.

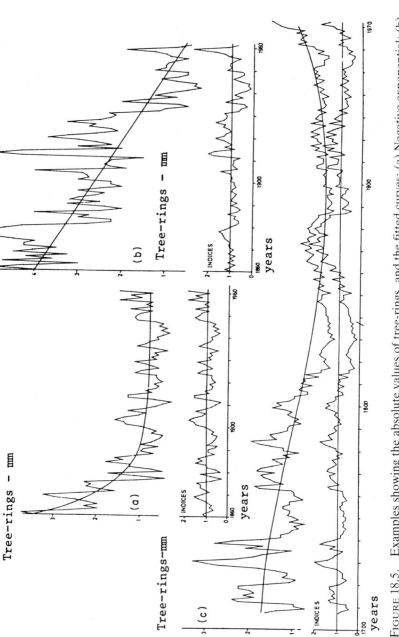

FIGURE 18.5. Examples showing the absolute values of tree-rings, and the fitted curves: (a) Negative exponential; (b) straight line; (c) polynomial curve. Also shown are the indices after removing the trend (from Munaut, 1978)

For each sample and each chronology another table should show:

(1)   the identification number;
(2)   the standard deviation;
(3)   the mean sensitivity;
(4)   the serial correlation coefficient;
(5)   if possible, the matrix of the cross-correlation coefficients among samples and among tree chronologies.

For the master chronology only there should be a table with the annual ring-width or indices; the signatures are pointed out by special typographical marks. This table can also give the results of the variance analysis. Note, however, that as the variance analysis requires some peculiar conditions (equal number of rings in each sample, even number of samples in each chronology) it is often necessary to be precise concerning the period and the samples used to compute the variance analysis.

Graphic representation

Though attractive, the diagrammatic representation of dendrochronological curves is useless; only numerical values should be published for future use. Nevertheless, it is sometimes interesting to give details as illustrations (signatures). A more useful way to summarize the cross-dating is as a graph showing the relative position of each cross-dated sample against the master chronology (Figure 18.6).

APPLICATIONS

The application of tree-ring analysis to palaeohydrological or palaeoecological studies is not very common. The major reason is the scarcity of favourable sites.

The first attempt to use palynology, dendrochronology, ^{14}C, stratigraphy and macrofossil analyses together was carried out in the early sixties at Terneuzen (Munaut, 1966) and Emmen (Munaut and Casparie, 1971) in the Netherlands. On both the sites studied, *Pinus sylvestris* grew on peaty soils in large bogs. It was demonstrated that such an environment was suitable for dendrochronological work.

At Terneuzen, the forest of *Pinus sylvestris* was preserved in a peat layer of Late Atlantic age. On a surface 2 ha in area, 722 *Pinus* stumps with a mean diameter of 30 cm were observed. Of these, 56 samples were selected and 49 were cross-dated in a floating chronology 242 years long. This work showed the gradual occupation (at least 125 years) of the surface by a Pine forest. This occupation was longer than 200 years and the disappearance took at least 40 years. Taking these results into consideration, the author concluded that a gradual asphyxiation of the forest, by a rising water level, had occurred.

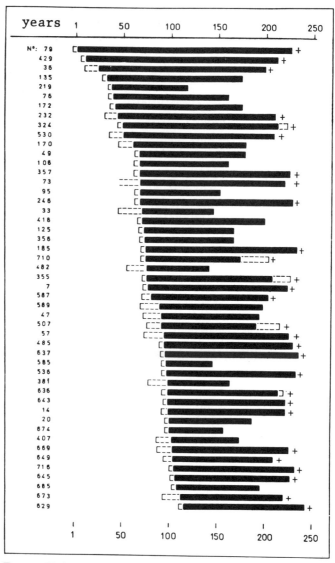

FIGURE 18.6. Table showing the cross-dating of subfossil trees. The trees covered by bark on the last ring are marked +

At Emmen, 52 samples were taken out at seven different places dispersed over 3 km along an edge of the bog, and in three levels of Early, Middle and Late Atlantic age. The difficulties in cross-dating individual or group curves of the same age, and some characteristics of the samples (for instance 42% of the variance related to differences between trees), were attributed to the effect of the local and very peculiar type of environment. Nevertheless, this study demonstrated the complementary character of dendrochronological and ^{14}C datings and the usefulness of a floating chronology to study the trend in the ^{14}C content of the atmosphere between 6000 and 7000 B.P. (Vogel *et al.*, 1969). Such a calibration using a low-altitude European chronology was described by Pearson *et al.* (1977). This work was carried out on a 1200-year section of a 2990-year-old floating chronology derived from the north of Ireland. The authors hope to use this chronology to make both stable isotope measurements for palaeotemperature estimation, and climatic interpretations based on ring widths (Pilcher *et al.*, 1977). The same kind of study was undertaken in Germany by Becker *et al.* (1977), using subfossil oaks preserved in river sediments. These studies also gave ecological information concerning the erosion or aggradation processes, since 9000 B.P.

A multidisciplinary investigation has been carried out in gravel pits of the Main river (southern Germany) by the two authors. Stratigraphical and pedological studies have shown that several episodes of gravel deposition occurred during the Holocene in a valley formed within Würmian gravels. Many subfossil oak trunks were lying in each gravel layer, provided by the destruction of a riverine forest. At the end of the accumulation flood loamy sediments were deposited in which specific soils developed under a new oak forest. Stratigraphical, pedological and dendrochronological studies show alternate increased and reduced fluvial activity since the early Holocene. Indeed, after cross-dating, it was possible to point out that trunks were not dispersed in time but clustered in groups dated by the ^{14}C, or since the Iron-Romano time by an absolute master chronology. These groups of trunks show that the culminations of the fluvial activities dated from the Middle Atlantic, the subboreal, the Iron-Romano age, the Main Middle Ages, until earliest modern time.

Scrutinizing the age of the subfossil trunks, the authors observe that 91% of the individuals lie within a 100–300 year-old group, clearly below the maximum age of living oaks (up to 600 years). This fact gives an indication that the riverine forests did not reach maturity and were subjected to periodic destruction. No climatic explanation is given by the authors, but such a cause cannot be discarded. Indeed, in the Danube Valley, where dendrochronological studies have been carried out, most of the chronologies have the same age as in the Main river.

In Italy, Corona (1972, 1973) has observed huge accumulations of trunks submerged in two lakes located in the Alps (Rovino and Tovel). After a dendrochronological study, the author concluded that the trees were destroyed by catastrophic landslides resulting from geomorphological perturbations occurring during the 'little ice age'.

A long-term reconstruction of the water-level changes for Lake Athabasca,

Canada, was proposed by Stockton and Fritts (1973). Indeed the ecology of a very large area is controlled by the lake level, the fluctuations of which have been affected by the building of a dam. By using tree-ring analysis of spruce growing on soil adjacent to the lake, the authors were able to calibrate width variations of the trees with the fluctuations in the water level during a 33-year period. Hence by using the tree-ring indices from six sites as predictors they reconstructed the water-level fluctuations during a period of 158 years. This reconstruction was checked by a comparison between estimated values and actual values observed during the last 33 years. The work shows that dendrochronology can be used as a tool to aid water management.

Analysis of reaction wood for indications of slope processes is described in the U.S.S.R. by Turmanina (1968). A reaction wood quotient is computed (equal to 1 for a vertical tree). This method is of practical interest in dating stages of landslide processes (avalanches, mudflows etc.). The work of Lamarche (1968) must also be mentioned, concerning the erosion of slopes in the White Mountains of California and based on the study of the age and the rooting of over-aged *Pinus longeava*.

A combination of tree-ring analysis and wood anatomy using archaeological remains has provided Bartholin (1978a, b) with information concerning the landscape development in southern Sweden during the Neolithic and the Medieval periods. The composition of the vegetation was calculated by identifying all wood remains, while the structure, density and utilization of the forest was deduced from the age, mean ring-width and growth trend of the trees. Such reconstructions, taken over a very short period, provide much more detailed information than palynology alone.

In 1976, Marshall-Libby *et al.* published results concerning the ratios of the stable isotopes of hydrogen and oxygen in tree-rings. The authors assumed that these ratios are dependent on the air temperature prevailing when the ring was formed. The use of such a method in reconstructing the palaeotemperatures is fascinating, but there is some controversy concerning how stable isotopes are incorporated into the wood and how the resulting ratios may be climatically interpreted (Wigley *et al.*, 1978).

It can be concluded from these examples that in favourable circumstances tree-ring analysis can be a very valuable method for reconstructing past events affecting the vegetation, the soil, the climate and the hydrology of an area.

REFERENCES

Bartholin, T. S. (1978a). Dendrochronology, wood anatomy and landscape development in southern Sweden. In: *Dendrochronology in Europe, BAR Int. Series,* **59**, pp. 125–130.

Bartholin, T. S. (1978b). Alvastra pile dwelling: tree studies, the dating and the landscape. *Fornvännen,* **73**, 213–219.

Becker, B., and Schirmer, W. (1977). Palaeoecological study on the Holocene valley development of the River Main, southern Germany. *Boreas*, **6**, 303–321.

Corona, E. (1972). Indagini sui tronchi sommersi del Lago delle Rovine nel bacino del Gesso (Cuneo). *L'Italia Forestale e Montana*, **1**, 7.

Corona, E. (1973). I tronchi sommersi nel lago di Tovel. *Esperienze e Ricerche*, **4**, 333–343.

Eckstein, D., and Bauch, J. (1969). Beitrag zu Rationalisierung eines dendrochronologischerverfahrens und zu Analyse seiner Aussagesicherheit. *Fortwis. Centralbl.*, **88**, 230–250.

Fritts, H. C. (1976). *Tree-Rings and Climate*, Academic Press, London, p. 567.

Høeg, O. A. (1956). Growth ring research in Norway. *Tree-Ring Bull.*, **21**, 1–15.

Lamarche, V. C. (1968). Rates of slope degradation as determined from botanical evidence White Mountains. Erosion and Sedimentation in a semiarid environment. *Geol. Surv. Prof. Paper 352–1*, 341–377.

Marshall Libby, L., Pandolfi, L. J., Payton, P. N., Marshall, J., Becker, B., and Giertz-Siebenlist, V. (1976). Isotopic tree thermometers. *Nature*, **261**, 284–288.

Matalas, M. C. (1962). Statistical properties of tree-ring data. *Int. Ass. Sci. Hydrol. Publ.*, **7**, 32–47.

Munaut, A. V. (1966). Recherches dendrochronologiques sur Pinus sylvestris. II: Première application des méthodes dendrochronologiques à l'étude de pins sylvestres sub-fossiles (Terneuzen-Pays-Bas). *Agricultura*, **14** (2e série-3), 361–389.

Munaut, A. V. (1978). La dendrochronologie: une synthèse de ses méthodes et applications. *Lejeunia*, **91**, 1–47.

Munaut, A. V., and Casparie, W. A. (1971). Etude dendrochronologique des *Pinus silvestris L.* subfossiles provenant de la tourbière d'Emmen (Drenthe, Pays-Bas). *Rev. Palaeobotan. Palynol.*, **11**, 201–226.

Pearson, G. W., Pilcher, J. R., Baillie, M. G. L., and Hillam, J. (1977). Absolute radiocarbon dating using a low altitude, European tree-ring calibration. *Nature*, **270**, 25–28.

Pilcher, J. R., Hillam, J., Baille, M. G. L., and Pearson, G. W. (1977). A long sub-fossil oak tree-ring chronology from the North of Ireland. *New Phytol.*, **79**, 713–729.

Stockton, Ch.W., and Fritts, H. C. (1973). Long-term reconstruction of water level changes for Lake Athabasca by analysis of tree-rings. *Water resources Bull. American Water Resources Ass.*, **9**, 1006–1026.

Stokes, M. A., and Smiley, T. L. (1968). *An Introduction to Tree-Ring Dating*, University of Chicago Press, p. 73.

Turmania, V. I. (1968). Analysis of reaction wood for indication of slope processes (All Union conferences on dendrochronology and dendroclimatology: first conference, Vilnius, Lithuania, S.S.R.). In: *Russian Papers on Dendrochronology and Dendroclimatology*, 1962–1968–1970–1972. Translated and edited by J. M. Fletcher and W. Linnard, Research Lab. for Archaeology and History of Art, Oxford University, pp. 29–32.

Vogel, J. C., Casparie, W. A., and Munaut, A. V. (1969). Carbon 14 trends in subfossil pine stubs. *Science*, **166**, 1143–1145.

Wendland, W. M. (1975). An objective method to identify missing or false rings. *Tree-Ring Bull.*, **35**, 41–48.

Wigley, T. M. L., Gray, B. M., and Kelly, P. M. (1978). Climatic interpretation of $\delta^{18}O$ and δD in tree-rings. *Nature*, **271**, 92–94.

19

Dendrochronology applied in mountain regions

WALTER BIRCHER

*Swiss Federal Institute of Forestry Research,
Birmensdorf, Switzerland*

INTRODUCTION

One special application of dendrochronology and radiodensitometry is in the analysis of material from alpine and subalpine sites. While traditional methods of dendrochronology yield information on annual ring width only, radiodensitometry permits the measurement of five additional parameters: earlywood width, latewood width, latewood as a percentage of total ring width, minimum density and maximum density (Figure 19.1).

The measurement of maximum density allows inferences to be made about the climatic conditions existing in earlier times. In the alpine timberline zone (cold, moist sites), temperatures in late summer (July and August) exert a decisive influence on wood production, thus affecting overall latewood and maximum density (Schweingruber, 1978, 1979). Consequently, density curves from such sites can be employed not only for dating but also for climatic studies. Radiodensitometric research on fossil wood has two main objectives:

(1) *synchronization* of the existing floating chronologies (Röthlisberger *et al.*, 1980) by means of cross-dating to obtain a single absolute chronology for the absolute dating of geomorphological, archeological and historical events, as well as of pollution damage to trees;

(2) *reconstruction* of climatic conditions over long periods for correlation and detailed analysis of results obtained through other palaeoecological methods (palynology, analysis of laminated lake sediments, chemical analysis of material).

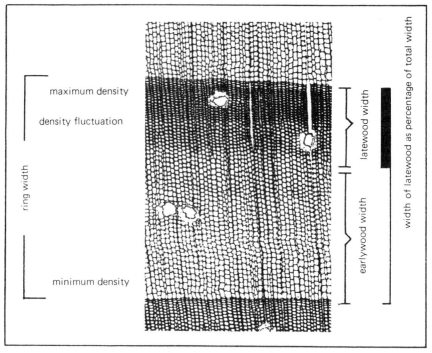

FIGURE 19.1. Annual ring parameters measurable with radiodensitometry (Schweingruber, 1983)

Research on material from living trees aims at (a) analysing climatic conditions in the site area, and (b) determining large-scale climatic changes by measurement and comparison of local conditions at numerous sites over a large area.

The climatic interpretation of annual ring curves obtained from trees of the alpine timberline zone is only possible where the following conditions are met:

(1) The growth of the tree must be predominantly influenced by one climatic parameter (temperature, precipitation).
(2) The growth site must be homogeneous in terms of internal and external factors.
(3) The chronology must be constructed from samples of the same tree species, since each species reacts differently to the same climatic factor.
(4) Climatic data appropriate to the growth site must be available to permit comparison and extrapolation.
(5) A thorough knowledge of the growth characteristics of the species is essential.

OBTAINING THE MATERIAL

Suitable tree species

The following species shown in Table 19.1 are influenced by climatic conditions in such a way as to be suitable for radiodensitometric analysis.

TABLE 19.1. Suitable species for analysis

Species	Suitability	Durability of fossil material
Larix decidua Mill.	Good. In modern and relatively young fossil trunks, damage due to attacks of the larch bud moth (*Zeiraphera dimiana Gn.*) may hinder analysis and prohibit climatic interpretation.	Good
Picea abies Karst.	Good. This species, however, seldom reaches the alpine timberline.	Fair/good
Pinus cembra L.	Poor. Large resin ducts in the latewood and enlarged cells round them hinder analysis. This species displays great variation in density (Bircher, 1982).	Fair/poor
Pinus mugo Turra	Fair. Better than *P. cembra*, as cells round resin ducts are not enlarged.	Fair/good

Collection of samples

Fossil stems from the alpine timberline can be found in bogs, lateral moraines and areas of glacier fronts (Vorfelder). Tree trunks found in such places date from 10,000 B.P. to the present. Fossil stems in bogs or soft sediments are located by means of a probe-pole and exposed by digging a trench. Samples are taken either by sawing off discs with a power saw or by coring with a cutting borer of 10 mm diameter (Figure 19.2; Schweingruber, 1983). An exact record of the stratigraphical positions of the stems must be made, and careful labelling is essential.

Historic material may also be found in old wooden buildings of the timberline zone. The survival of suitable material to the present is largely

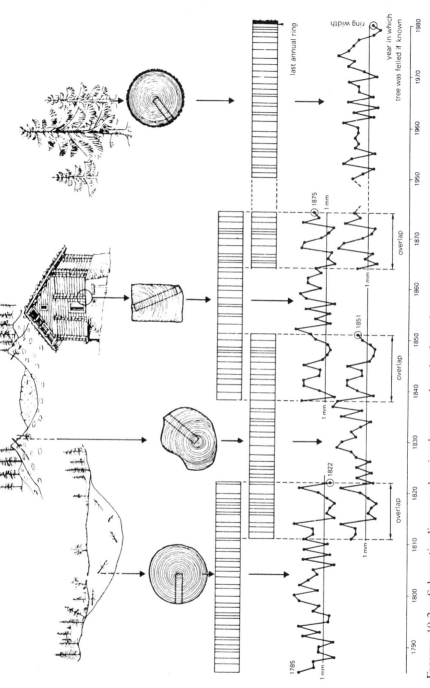

FIGURE 19.2. Schematic diagram showing how synchronization of annual ring sequences from different sources is used to construct an absolute chronology (Schweingruber, 1983).

determined by the type of construction. Discs may be sawn from projecting beams or core samples taken as described above. Samples must be collected from living trees on comparable sites for the purpose of comparison.

In the case of living trees, samples are obtained with an increment borer (diameter 5 mm, length 40 cm). Orientation to the trunk axis is facilitated by the use of a holding and orientation device (Schweingruber, 1983). The samples are stored in strong, permeable containers.

Assessment of material

Fossil material must be inspected and assessed with regard to site reconstruction (Figure 19.3). This involves the following questions:

(1) Does the find site correspond to the growth site, or was the tree transported there after its death?
(2) Does the species composition suggest a realistic vegetation pattern?
(3) Do sediment and pollen analyses indicate that the wood is of the same age as the stratum in which it was found?

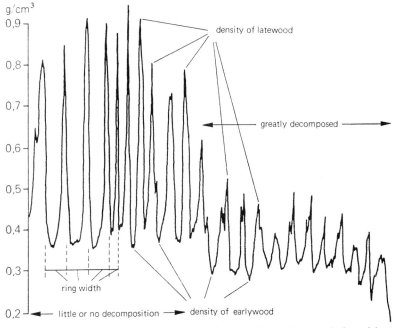

FIGURE 19.3. Density diagram from a late Ice Age pine trunk found in a clay pit near Dättnau, Switzerland (13,000 years old). In the well-preserved section on the left, both density and ring width can be used for analysis, but in the decomposed part on the right only the ring widths give valid information (Schweingruber, 1983, after Kaiser, 1979)

(4)   Is there any evidence of external factors which may have disturbed the growth of the tree?
(5)   Has the wood structure been affected by decomposition by micro-organisms or fungi, pressure or fire? Various rot fungi may cause extensive decomposition of cell walls in both earlywood and latewood. In such cases, density measurements are pointless, though the annual ring widths may still be recorded accurately (Fig. 19.3).

LABORATORY WORK

Preparation of samples

In the case of disc samples from fossil or historical wood, two or three radial samples are taken and allowed to dry at a constant relative humidity for 2–3 weeks. The transverse sections are polished with sandpaper in such a way that the annual ring boundaries are perpendicular to the radial surface. Smaller, rectangular samples are then cut from these sections. Under no circumstances should the residual material be thrown away; it is best kept stored under water. For core samples these initial steps are unnecessary. For both types of sample, the subsequent steps are as follows:

(1)   The samples are glued on to wooden supports and inspected through a binocular microscope to determine the deviation of the fibre orientation from the perpendicular.
(2)   By means of a twin-bladed circular saw, laths 1.25 mm thick are cut out of the samples.
(3)   Extractives such as resin are removed from the laths by distillation in alcohol, acetone or water for several hours. This step can also be carried out before glueing to the wooden supports.
(4)   Finally, the samples are acclimatized for several hours at 20°C and 50% relative humidity. This is necessary to ensure that the absolute values obtained from all the samples will be comparable.

Radiography

The samples are arranged in a cellophane carrier, which is then placed directly on the film (Figure 19.4). To ensure deep contrast on the film, the samples are irradiated for 90 minutes at 11 kV and 20 mA (Schweingruber, 1983). The film used may be Kodak RP/M, and development is carried out in an automatic processor.

Basic principles of densitometric measurement

The densitometric measurement of wood density, shown in schematic form in

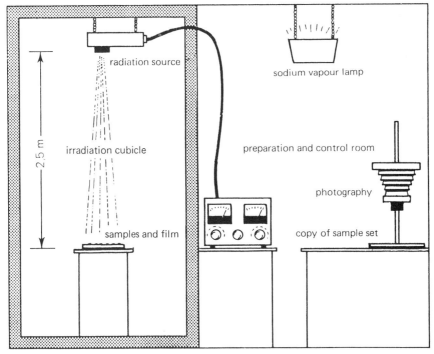

FIGURE 19.4. Plan of the radiography unit; the irridiation cubicle is separated from the preparation and control room (Schweingruber, 1983)

Figure 19.5, may be hindered by technical errors and biological defects which can render synchronization of the curves impossible. Some examples of these hazards are shown in Table 19.2.

DATA PROCESSING

Preprocessing

Preparation of the data for analysis is carried out with the help of a mainframe computer and comprises the following steps (Schweingruber, 1983):

(1) transcription and listing of values registered by the densitometric unit; detection and correction of certain operational and mechanical errors;
(2) change of format and further corrections; calculation of ring width and latewood percentage; listing of values for all parameters; determination and correction of remaining errors;
(3) printing of values in graph format and manual connection of graph points to produce curves for ring width and maximum density;

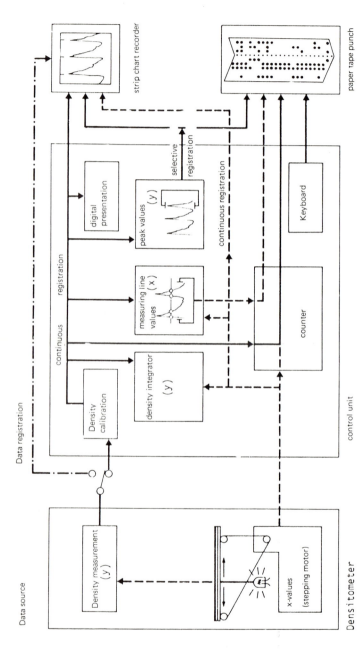

FIGURE 19.5.   Schematic diagram of the data-acquisition system. The densitometer sends impulses to the control unit, the strip-chart recorder and the tape punch. The control unit permits continuous, periodic or selective registration of data. The strip-chart recorder continuously reproduces all signals from the densitometer in the form of a 2-dimensional graph (analogue values). The tape punch registers data in digital form (Schweingruber, 1983)

TABLE 19.2. Hazards of densitometric measurements

Technical errors	Biological defects
Cracks in samples	Missing annual rings
Wrong matching of ring sequence in overlapping samples	(climatological or pathological reasons)
Faulty orientation in preparing samples, leading to blurring of ring boundaries on the film and consequent impossibility of measurement	Double annual rings
	Non-extractable, highly reflecting substances, e.g. fungal hyphae
	Traumatic resin ducts
	Large resin ducts surrounded by enlarged tracheids
Non-registration of narrow rings due to too wide a measuring slit	Reaction wood, e.g. compression wood
Errors in operation of the densitometer	Irregularities in growth, e.g. hazel growth

(4) visual checking of the curves on a light table; comparison with X-ray film and strip chart diagram in case of doubt;
(5) calculation and printing of the raw data mean curve;
(6) visual comparison of this with the individual curves; calculation of percentage of agreement and *t*-values for each curve in relation to the mean curve; correction of remaining errors.

The corrected data are then ready for further processing.

Dendrochronological investigation is based on the analysis of chronological sequences of cambial activity. These exhibit fluctuations caused by both internal and external factors. According to the research objectives, interest centres on the short-, medium- or long-term fluctuations within the sequence. Long-term fluctuations can be eliminated by the process of indexing, while short-term ones can be damped by means of filters and smoothing functions, which may also be used to determine and represent the pattern and wavelength of these biological–ecological variations (Schweingruber, 1983; LaMarche, 1974; Fritts, 1976).

Indexing and age trend

With increasing age of a tree, the width and maximum density of the annual rings decreases. This phenomenon is known as age trend (Bräker, 1981).

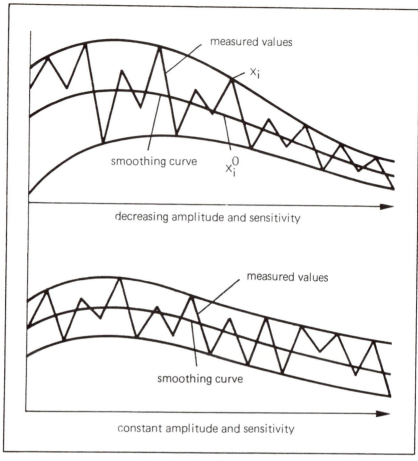

FIGURE 19.6.   Indexing (standardization) (Schweingruber, 1983)

Climatic interpretation is only possible if this trend is first removed from the data. In order to achieve this, a smoothing curve is superimposed on the raw data, giving a set of smoothing values (Figure 19.6).

Depending on the annual ring parameter under investigation, the smoothing value is either divided into or subtracted from the measured one to give a new set of indexed values:

For width: $I_t = \dfrac{x_t}{x_t^0}$

For density: $I_t = x_t - x_t^0$

where x_t is the measured value and

x_t^0 is the smoothing value.

Although the absolute values are no longer given, the indexed values are free of age trend. Indexing (standardization) consequently permits the valid comparison of annual ring data from different trees and the construction of a mean curve representative of a particular site.

Synchronization of curves

The correlation coefficient for each given pair of curves is calculated. The more homogeneous the curves, the greater the correlation between them. The mean correlation between the curves from one site is thus a measure of site homogeneity. The significance of the correlation can be determined by a *t*-test, which can also be applied to check the accuracy of synchronization.

This can also be tested by calculating the percentage of agreement between curves. The increase or decrease in value between any pair of adjacent points on one curve is compared with the trend in the corresponding interval on the other. Calculation of the percentage of agreement permits the synchronization of annual ring sequences of known data with undated ones.

Modelling and reconstruction

When annual ring sequences from homogeneous sites are available together with appropriate, checked meteorological data, response functions (multiple linear regressions) can be employed to test for relationships between *annual ring parameters* and *climatic factors* (temperature, precipitation). Linear regressions may be used to compare annual ring density curves with changes in glacier mass balance or tongue length, as well as with grain harvests, vine yields etc. (Pfister, 1979).

The reconstruction process comprises three phases (Figure 19.7):

(1) In the *calibration phase*, dependency models between maximum density and meteorological parameters are calculated. The influential variables (predictands) are the indexed maximum density values, while the target variables (predictors) are mean temperature and precipitation.

(2) In the *testing phase*, the quality of the residuals is compared with that of the calibration phase: if it remains constant, the model is appropriate.

(3) In the *reconstruction phase*, climatic parameters are extrapolated on the basis of the mean maximum density values for the relevant period. The accuracy of the reconstruction depends greatly on the homogeneity of the tree samples (site quality) and of the climatic data.

Composites of floating chronologies from fossil material permit the reconstruction of climatic parameters in relative terms only, since the absolute density

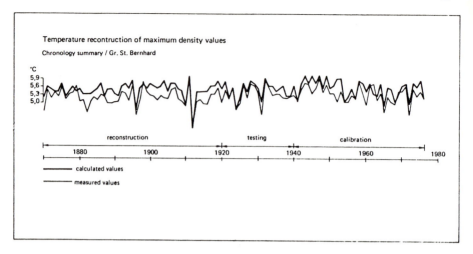

FIGURE 19.7.   Temperature reconstruction on the basis of maximum density. The predictands are the indexed maximum density values of chronologies from 4 sites in the Alps and pre-Alps (Rigi, Aletschwald, Cortina d'Ampezzo, Lauenen). The predictors are the mean temperatures for August–September at the Great St. Bernard meteorological station, 2479 m above sea-level (Bircher, 1982)

values of the raw data vary considerably from curve to curve. Each floating chronology is the mean of several individual ones, with short-term fluctuations suppressed by means of low-pass filters. The chronologies are arranged on a common time axis. Figure 19.8 shows clearly that an analysis of annual rings reveals greater detail than that of known glacier peak periods or of pollen analysis.

APPLICATIONS OF DENDROCHRONOLOGY AND RADIODENSITOMETRY

Table 19.3 shows some applications, mostly referring to the analysis of annual ring width. The table is not a complete catalogue of possible applications.

TABLE 19.3. Applications of dendrochronology and radiodensitometry

Application	Parameter	References
Climatology:		
History of glaciers	Ring width, maximum density	Röthlisberger (1976), Bircher (1982), Renner (1982), Röthlisberger et al. (1980)
History of rivers	Ring width	Becker and Frenzel (1977)
History of sea floods	Ring width	Munaut (1967), Heyworth (1978)
History of forest fires	Ring width	Arno and Sneck (1977)
Dating for historical studies:		
Architecture	Ring width	Eidem (1955), Stahle (1979)
History of settlements	Ring width	Hollstein (1980), Ruoff (1979)
Art history	Ring width	Robinson (1976), Bauch et al. (1972)
Criminal investigations	Ring width	Fletcher (1978), Liese and Eckstein (1971)
Air pollution:		
Evidence for and dating of pollution damage to trees	Ring width	Kienast et al. (1981), Eckstein et al. (1981)
Isotope physics:		
Calibration of ^{14}C decay sequences	Ring width	Mook and van der Straaten (1980), Stuiver and Quay (1980), Becker (1980)
Relationships between stable isotopes in annual rings and climatic factors	Ring width, maximum density	White and Lawerence (1980)

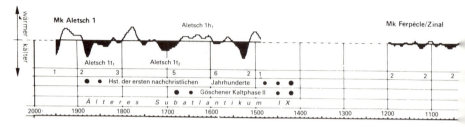

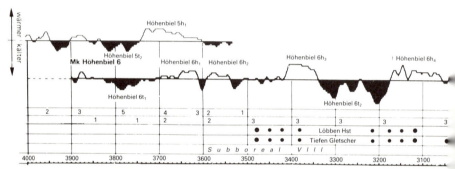

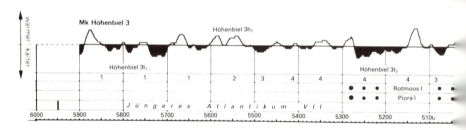

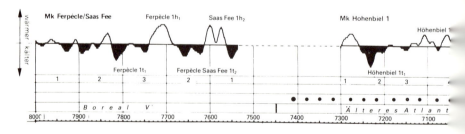

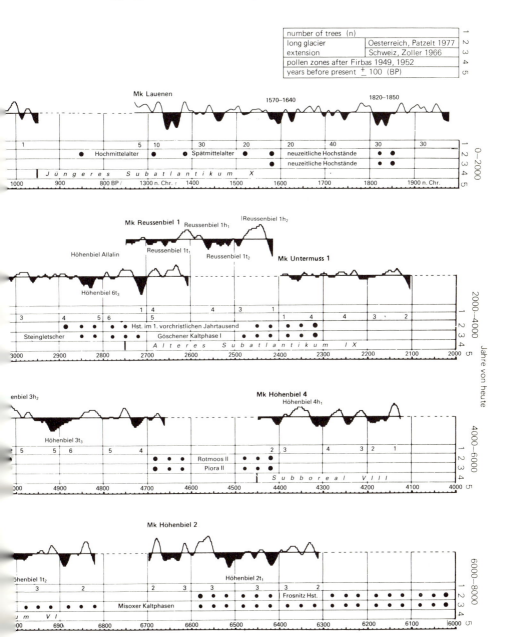

FIGURE 19.8.   Temperature reconstruction for the period 8000 B.P. to the present in the subalpine zone. Comparison of results obtained through analysis of annual rings, pollen and history of glaciers (Bircher 1982, Renner 1982)

REFERENCES

Arno, S. F., and Sneck, K. M. (1977). *A Method for Determining Fire History in Coniferous Forest of the Mountain West*, USDA Forest Serv. Gen. Tech. Report INT-12.

Baillie, M. G. L., and Pilcher, J. R. (1973). A simple cross-dating programme for tree-ring research. *Tree-Ring Bull.*, **33**, 7–14.

Bauch, J., Eckstein, D., and Meier-Siem, M. (1972). Dating the wood panels by a dendrochronological analysis of tree rings. *Nederl. Kunsth.* **JB 23**.

Becker, B. (1980). Tree ring dating and radiocarbon calibration in South-Central Europe. *Radiocarbon*, **22**, 219–226.

Becker, B., and Frenzel, B. (1977). Paläoökologische Befunde zur Geschichte postglazialer Flussauen im südlichen Mitteleuropa. In: *Dendrochronologie und postglaziale Klimaschwankungen in Europa. Verhandlungen des Symposiums über die Dendrochronologie des Postglazials, Grundlagen und Ergebnisse, 13–16 June 1974*, Steiner, Weisbaden, pp. 43–61.

Bircher, W. (1982). Zur Gletscher- und Klimageschichte des Saastales. Glazial-morphologische und dendroklimatologische Untersuchungen. *Physische Geographie, Universität Zürich*, **9**.

Bräker, O. U. (1981). Der Alterstrend bei Jahrringdichten und Jahrringbreiten von Nadelhölzern und sein Ausgleich. *Proc. Int. Symp. on Radial Growth in Trees*, Vienna, IUFRO Working Party S1. 01-04, pp. 75–102.

Eckstein, D. (1972). Tree-ring research in Europe. *Tree-Ring Bull.*, **32**, 1–18.

Eckstein, D., *et al.* (1981). Dendroklimatologische Untersuchungen zur Entwicklung von Strassenbäumen. *Forstw. Cbl.*, **100**, 381–396.

Eidem, P. (1955). Dendrochronological dating of a cast-house from Istad, Slidre, Valdres. *Blyttia*, **13**, 65–70.

Fletcher, J. M. (1978). Tree ring analysis of panel paintings. In: *Dendrochronology in Europe, BAR Int. Series* 51, pp. 303–306.

Fritts, H. C. (1963). Computer programs for tree-ring research. *Tree-Ring Bull.*, **25**, 2–7.

Fritts, H. C. (1976). *Tree Rings and Climate*, Academic Press, London.

Heyworth, A. (1978). Submerged forests around the British Isles: their dating and relevance as indicators of postglacial land and sea-level changes. In: *Dendrochronology in Europe, BAR Int. Series* 51, pp. 279–288.

Hollstein, E. (1980). Mitteleuropäische Eichenchronologie: Trierer dendrochronologische Forschungen zur Archäologie und Kunstschichte. *Trierer Grabungen und Forschungen*, **11**, Mainz am Rhein.

Kaiser, N. F. J. (1979). *Ein späteiszeitlicher Wald im Dättnau bei Winterthur, Schweiz*, Ziegler, Winterthur.

Kienast, F., *et al.* (1981). Jahrringanalysen an Föhren (*Pinus silvestris L.*) an immissionsgefährdeten Waldbestände des unteren Wallis (Saxon, Schweiz). *Mitt. schweiz. Anst. forst. Vers'wes.*, **57**, 415–432.

LaMarche, V. C. (1974). Palaeoclimatic inferences from long tree-ring records. *Science*, **183**, 1043–1048.

Liese, W., and Eckstein, D. (1971). Die Jahrringchronologie in der Kriminalistik. In: *Grundlagen der Kriminalistik* (Ed. H. Schäfer), Steintor, Hamburg, pp. 395–422.

Mook, W. G., and van der Straaten, C. M. (1980). Preliminary D/H results on tree-ring cellulose from oak in the province of Drente, the Netherlands. In: *Edit. Carbon Dioxide Effects: Research and Assessment Program, Proc. Int. Meeting on*

Stable Isotopes in Tree-Ring Res. (Ed. G. C. Jacoby), Lamont-Doherty Geol. Observ., Columbia Univ., New York, pp. 56–57.

Munaut, A. (1967). La forêt ensevelie de Ternuzen. *Revue Industrie.*

Pfister, C. (1979). Reconstruction of past climate: the example of Swiss historical weather documentation project (16th to early 19th century). *Proc. Int. Conf. on Climate and History*, 8–14 July 1979, pp. 128–147.

Renner, F. (1982). Beiträge zur Gletschergeschichte des Gotthardgebietes und dendroklimatologische Analysen an fossilen Hölzern. *Physische Geographie, Universität Zürich*, **8**.

Robinson, W. H. (1976). Tree-ring dating and archeology in the American South-West. *Tree Ring Bull.*, **36**, 9–20.

Röthlisberger, F. (1976). 8000 Jahre Walliser Gletschergeschichte: ein Beitrag zur Erforschung des Klimaverlaufs in der Nacheiszeit. *Die Alpen*, **52**, 3/4, Pt II, 59–152.

Röthlisberger, F., Bircher, W., and Renner, F. (1980). Holocene climatic fluctuations: radiocarbon dating of fossil soils and woods from moraines and glaciers in the Alps. *Geographica Helvetica*, **35**, special issue.

Ruoff, U. (1979). Neue dendrologische Daten aus der Ostschweiz. *Zeitschr. Schweiz. Arch. und Kunstgesch.*, **36**, 94–96.

Schweingruber, F. H., *et al.* (1979). Dendroclimatic studies on conifers from Central Europe and Great Britain. *Boreas*, **8**, 427–452.

Schweingruber, F. H. (1983). *Der Jahrring:Standort, Methodik, Zeit und Klima in der Dendrochronologie*, Haupt, Bern.

Stahle, D. W. (1979). Tree-ring dating of historic buildings in Arkansas. *Tree-Ring Bull.*, **39**, 1–28.

Stuiver, M., and Quay, P. D. (1980). Changes in atmospheric carbon-14 attributed to a variable sun. *Science*, **207**, 11–19.

Stuiver, M., and Quay, P. D. (1981). Atmospheric ^{14}C changes resulting from fossil fuel CO_2 release and cosmic ray flux variability. *Earth and Planetary Science Letters*, **53**, 349–362.

White, J. W. C., and Lawrence, J. R. (1980). The relationship between the non-exchangeable hydrogens of tree-ring cellulose and the source waters for trees. In: *Carbon Dioxide Effects: Research and Assessment Program, Proc. Int. Meeting on Stable Isotopes in Tree-Ring Res.*, (Ed. C. G. Jacoby), Lamont-Doherty Geol. Observ., Columbia University, New York, pp. 58–65.

Physical and Chemical Methods

20

Stable oxygen and carbon isotope analyses

Ulrich Siegenthaler and Ulrich Eicher

Physics Institute, University of Bern, Bern, Switzerland

PRINCIPLE OF THE METHOD

In natural oxygen, the ratio of the (stable) isotopes with masses 18 and 16 ($^{18}O/^{16}O$) is subject to small variations owing to slight isotopic differences in the physico-chemical behaviour. The use of oxygen isotopes in palaeoclimatic studies of the continents is based on the observation that the $^{18}O/^{16}O$ ratio of precipitation — and consequently of glacier ice, groundwater and lakes — depends on climate, especially on temperature. In general, in precipitation the ratio increases with increasing temperature. Well-known examples of palaeoclimatic records based on this fact are polar ice cores, such as that of Camp Century, Greenland, which reflect the whole Würm glacial period and the Holocene with their climatic variations (Dansgaard *et al.*, 1982).

This chapter considers mainly studies performed on authigenic lake carbonate. The carbonate may be precipitated chemically because of biological withdrawal of CO_2 for assimilation (lake marl), or may be in mollusc shells originating in the lake. This authigenic material essentially records the $^{18}O/^{16}O$ ratio of the lake water, as discussed by Stuiver (1970).

Lake marl (Seekreide, craie lacustre) is formed when submerged aquatic plants and algae use dissolved CO_2 for photosynthesis, so that a CO_2 deficit is created. In consequence, bicarbonate is decomposed into CO_2 and insoluble carbonate, which precipitates:

$$Ca^{++} + 2HCO_3^- = CO_2 + CaCO_3 + H_2O$$

The $^{18}O/^{16}O$ ratios in water, bicarbonate and carbonate are related by equilibrium constants (isotope fractionation factors) which are functions of temperature. Therefore, the $^{18}O/^{16}O$ variations measured in carbonate are in general not identical with the original variations in the water, a fact which must be considered in the interpretation of ^{18}O results (see below). For lakes in temperate and cold regions, however, the isotope signal of the water in general prevails over the temperature effect of this isotopic fractionation.

Carbon has two stable isotopes with masses 12 and 13. The $^{13}C/^{12}C$ ratio of authigenic lake carbonate depends mainly on local factors. Under certain conditions it is a measure of biological productivity in the lake, but in general it is not possible to interpret $^{13}C/^{12}C$ results in such a general way as $^{18}O/^{16}O$ results.

The following conditions must be met for applying the stable isotope method:

(1) The sediment must contain a sufficient amount of carbonate formed in the lake itself. This means that there is a good chance of finding suitable material only in regions with rocks containing carbonate, but not, for instance, in purely granitic areas.

(2) The relative amount of allochthonous carbonate must be small. This may sometimes be problematic for lake marl, since there is no simple way of identifying allochthonous carbonate.

(3) A large fraction of organic material can lead to analytical problems (because of volatile components that interfere with the mass-spectrometric measurement). The best materials are pure lake marl or mollusc shells.

(4) Shells of different mollusc species are not necessarily isotopically identical. If working with shells, therefore, only samples of the same species must be compared.

Pollen analysis, in parallel with oxygen and carbon isotope analyses, is highly desirable, enabling the synchronization of isotope curves with pollen chronology. Another valuable parameter which is easy to measure is the carbonate content of the sediment, giving an indication of the amount of authigenic carbonate and thus of biological productivity.

$^{18}O/^{16}O$ and D/H ratios in organic matter (e.g. tree-rings or peat bogs) should in principle also provide records of variations in precipitation. However, their interpretation is not easy, and results obtained so far are equivocal. A brief discussion is given later.

EXPERIMENTAL METHODS

Notation and standards

By convention, oxygen isotope ratios are not given as absolute values (since these would mostly be numbers very close to 0.200%), but in the delta notation as relative deviation (in per mil) from the $^{18}O/^{16}O$ ratio of a standard substance:

$$\delta^{18}O = \frac{R(\text{sample}) - R(\text{standard})}{R(\text{standard})} \times 1000$$

where R is the absolute ratio $^{18}O/^{16}O$. Ratios $^{13}C/^{12}C$ and $^{2}H/^{1}H$ are also indicated analogously in the delta notation.

Commonly used standards are SMOW (Standard Mean Ocean Water — Craig, 1961) for $^{18}O/^{16}O$ in water and D/H, and PDB (from a belemnite of the Peedee formation — Craig, 1957) for $^{18}O/^{16}O$ and $^{13}C/^{12}C$ in carbonate and other materials. Standard materials are distributed by the International Atomic Energy Agency, Vienna.

Preparation and measurement of carbonate samples

Sample collection in the field is the same as for pollen analysis. One sample should contain about 50 mg of carbonate, preferably more for eventual replicate analysis. Macroscopic plant remains and shell fragments are taken out from lake marl samples. In the laboratory, CO_2 gas is prepared by treating the samples with 100% or 95% phosphoric acid; CO_2 is then analysed by mass spectrometry for its $^{18}O/^{16}O$ and $^{13}C/^{12}C$ ratios. The whole procedure is similar to that for deep-sea sediments, as described by McCrea (1950). The molecules $^{12}C^{16}O_2$, $^{13}C^{16}O_2$ and $^{12}C^{16}O^{18}O$ have masses 44, 45 and 46, so that the isotope ratios $^{13}C/^{12}C$ and $^{18}O/^{16}O$ are obtained essentially by measuring the mass ratios 45/44 and 46/44. The mass spectrometric measurement technique has been thoroughly discussed by Craig (1957) and by Mook and Grootes (1973).

The measurements are performed on specially equipped mass spectrometers, which means that they can be carried out only in specialized laboratories.

Organic impurities may lead to erroneous results, since organic vapours can yield molecular fragments in the mass range 44–46. Therefore, carbonate samples are sometimes pretreated before acidification in order to purify them. One such method which is frequently used for deep-sea carbonates is roasting the samples at about 450°C for one hour or less, either in a vacuum or in a helium stream which removes the volatile products more thoroughly. Another method consists of treating the samples with a chemical oxidant such as a solution of KClO (Eau de Javelle), NaClO (Fritz *et al.*, 1975) or H_2O_2, by which the reactive organic matter is oxidized.

Eicher (1979) compared samples pretreated in different ways with untreated samples. He found that lake marl samples yielded, after heating in a vacuum, systematically lower $\delta^{18}O$ values than those without pretreatment; the difference was typically 0.2–0.5 per mil, but in a few cases it was greater. Changes of similar magnitude, of either sign, were also observed in $\delta^{13}C$ after heating. Clearly too high $\delta^{13}C$ values were observed for some samples with low carbonate content, probably because of an impurity of mass 45 produced or released by the heating. Pretreatment by chemical oxidation yielded smaller deviations than through heating. It is not clear whether or not the isotopic results from treated or untreated samples are nearer the true value of the pure carbonate. Since erroneous results have been observed from pretreated samples, lake marl samples are now prepared without pretreatment in the authors' laboratory. Mollusc shells are pretreated with KClO solution.

In order to produce CO_2, a sample of 5–40 mg of carbonate is put in the side-arm of a special glass vessel containing 1–2 ml of orthophosphoric acid; the air is then pumped away. In order to start the reaction, the sample is transferred to the acid by tilting the vessel. The reaction takes place at a constant temperature, which is important because there is an oxygen isotope fractionation between the reaction products CO_2 and H_2O. Temperatures between 25°C and 50°C are used; at higher temperatures, the reaction proceeds faster. The usual reaction time is one hour. The carbon dioxide produced is passed over a dry-ice cooled trap, which removes traces of water vapour, and frozen with liquid nitrogen in a gas-sample container.

Phosphoric acid is used because there is no interaction between its oxygen atoms and those of the evolving CO_2. Usually 100% phosphoric acid is used; however, Mook (1968) found that the reaction is more rapid and more reliable with 95% acid. The sample size should not be too large. Samples exceeding about 50 mg react incompletely, even if stoichiometrically there is a large excess of acid. Incomplete reaction is accompanied by poorly reproducible $\delta^{18}O$ results (usually too low values); $\delta^{13}C$ is either not effected or to a lesser extent.

The $\delta^{18}O$ and $\delta^{13}C$ results have, for well-prepared and pure CO_2 samples, an error comparable to the precision of the mass spectrometer itself, which is about ± 0.02 per mil for modern machines. It is generally observed that $\delta^{18}O$ values are more susceptible to impurities etc. than are $\delta^{13}C$ values, which can at least partly be explained by the lower $^{18}O/^{16}O$ ratio (2×10^{-3}, compared with 1.1×10^{-2} for $^{13}C/^{12}C$).

RESULTS FROM LAKE CARBONATES

Oxygen isotopes

Interesting results have been obtained mainly for the transition Late Glacial–Early Postglacial (i.e. about 14,000–9000 B.P.) in Europe, when drastic climatic changes took place. Figures 20.1–20.4 show results from several sites in Switzerland and France, taken from published work. In all cases, pollen analysis had been performed first, so that it was possible to identify the different pollen zones in the profiles and thus to date them indirectly. Furthermore, carbonate content was measured, except in the profiles analysed first.

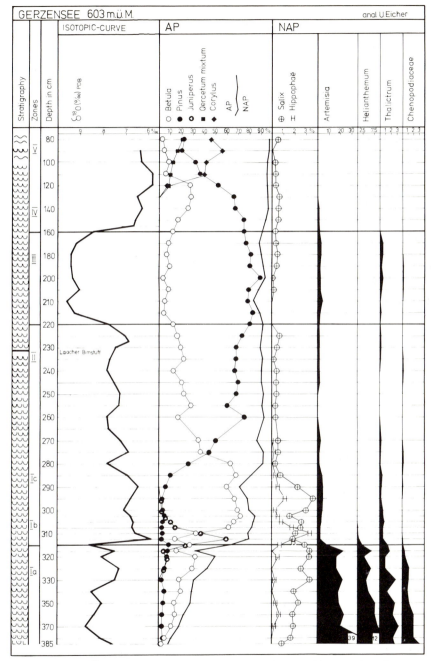

FIGURE 20.1. $\delta^{18}O$ and pollen diagram of profile II from Lake Gerzensee (Switzerland). $\delta^{18}O$ changes abruptly from pollen zone $1a$ to zone $1b$, parallel to the percentage of arboreal pollen (AP). The Younger Dryas period (III) clearly shows lower values than the adjacent periods. Note the change of depth scale at 200 cm depth (from Eicher and Siegenthaler, 1976). *Reproduced by permission of Universitetsforlaget, Oslo*

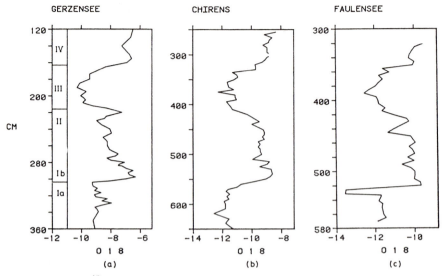

FIGURE 20.2. $\delta^{18}O$ in lake marl profiles from:
(a) Gerzensee III (Switzerland; Eicher, 1980)
(b) Tourbière de Chirens (France; Eicher *et al.*, 1981)
(c) Faulensee (Switzerland; Eicher and Siegenthaler, 1976)
Horizontal axes: $\delta^{18}O$ in per mil; vertical axes: depth in cm. Left column in (*a*) are
pollen zones. *Reproduced by permission of the International Glaciological Society*

Inspection of the figures shows that the same pattern of variations is
recorded in all $\delta^{18}O$ profiles. Three rapid changes in $\delta^{18}O$ are observed,
corresponding to the transitions between the pollen zones Oldest Dry-
as/Bølling (Ia/Ib), Allerød/Younger Dryas (II/III) and Younger Dry-
as/Preboreal (III/IV). Obviously, $\delta^{18}O$ in precipitation shifted by 2–3 per
mil during these transitions within relatively short time spans (less than about
100 years). The reasons for these shifts must have been drastic climatic changes
which affected at least the whole of Central Europe. $\delta^{18}O$ in precipitation is
correlated with temperature, so we may consider $\delta^{18}O$ changes in lake
carbonate as qualitative indicators of temperature variations. Comments on
the detailed interpretation are given below.
    Judging from the values of $\delta^{18}O$ only, the pollen zones Oldest Dryas (1a) and
Younger Dryas (III) appear as cold periods, which agrees with palynological
results. There is little or no indication in the $\delta^{18}O$ values of an Older Dryas cool
phase. The zones Bølling, Older Dryas and Allerød (1b, 1c, II) appear as one
relatively uniform warm period, with only minor fluctuations. This view is
supported by Berglund (1979), Gray and Lowe (1977) and other authors, based
on non-isotopic evidence. A slow but definite decline of $\delta^{18}O$ from Bølling to
Allerød is observed. It can probably be explained partly by a slow decrease of
$\delta^{18}O$ in the ocean, which forms the source of the atmospheric water vapour,

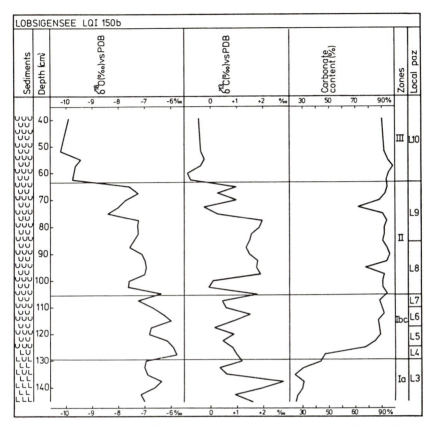

FIGURE 20.3. $\delta^{18}O$ and $\delta^{13}C$ in lake marl and carbonate content of the sediment from Lobsigensee (Switzerland; Eicher and Siegenthaler, 1984). *Reproduced by permission of Museum d'histoire naturelle de Genève*

caused by the melting of large continental ice masses strongly depleted in ^{18}O. After the transition from the Younger Dryas to the Preboreal period, which marks the boundary between Pleistocene and Holocene, similar values as observed during the Bølling are found. The similarities between the different profiles also include minor fluctuations. Thus, before the end of the Allerød pollen zone, a $\delta^{18}O$ minimum is observed which has also been found in marl profiles from other sites and which has been termed the Gerzensee fluctuation by Eicher (1980). At Gerzensee and Faulensee there is a small minimum after the rise at the transition Younger Dryas/Preboreal; it is also observed in several other profiles not shown here. Thus, we conclude that lake carbonate sediments form a reliable record of even minor variations of $\delta^{18}O$ in precipitation. The variations observed in marl may be different in amplitude from those in precipitation — in general slightly smaller owing to the

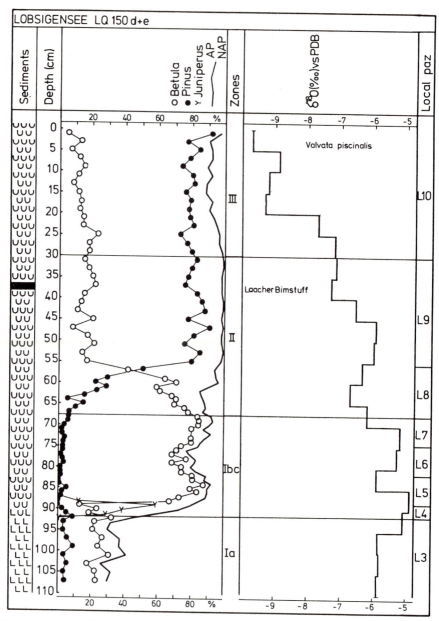

FIGURE 20.4. $\delta^{18}O$ for mollusc shells, and simplified pollen diagram of Ammann and Tobolski (1984) (from Eicher and Siegenthaler, 1984). *Reproduced by permission of Museum d'histoire naturelle de Genève*

temperature-dependent isotope fractionation between water and carbonate — but the relative pattern of changes is preserved.

Studies on *mollusc shells* have the advantage that there are fewer problems of possible allochthonous carbonate than for lake marl. In Figure 20.4, results obtained on snail shells of *Valvata piscinalis* from Lobsigensee are shown (Eicher and Siegenthaler, 1984). Comparison with the marl results of Figure 20.3 show that $\delta^{18}O$ in the shells is systematically higher by about 1 per mil. A similar difference between marl (calcite) and aragonitic shells was observed by Stuiver (1970) and by Blanc *et al.* (1977). The difference can be explained partly by the fact that the isotopic fractionation between water and $CaCO_3$ is different for aragonite and calcite, such that $\delta^{18}O$ in aragonite is about 0.6 per mil higher than in calcite in isotopic equilibrium. In addition, a specific metabolic fractionation seems to play a role (Lemeille, 1980).

The $\delta^{18}O$ profiles from Lobsigensee do not show the low values during the Oldest Dryas as observed elsewhere. This must be caused by a local phenomenon, for instance changes in the water regime of the lake.

Figure 20.3 also shows that the carbonate content of the sediment was considerably lower during pollen zone Ia (Oldest Dryas) than during all later periods. Obviously biological productivity — and therefore formation of lake marl — was rather low before the first warming. Thus, carbonate content analyses, which can be performed without additional effort during CO_2 sample preparation by simply measuring the CO_2 pressure, give information on the lake's palaeoproductivity.

Comparison with other studies

Other authors have found similar $\delta^{18}O$ variations in lake carbonates. The Younger Dryas $\delta^{18}O$ minimum was also found by Staesche in Lake Mindelsee in southern Germany (results presented by Lang, 1970). Lemeille *et al.* (1983) observed the same pattern as reported here in mollusc shells from an ancient lake sediment core in south-western Switzerland, covering the period from Bølling to Preboreal. Mörner and Wallin (1977) measured stable isotopes in marl from a lake on the island of Gotland (Sweden). In its oldest part (clay sediments), the profile probably contains mainly allochthonous carbonate, followed by lake marl with low $\delta^{18}O$ and relatively high $\delta^{13}C$ values. A sharp $\delta^{18}O$ increase by 3.5 per mil possibly marks the transition Younger Dryas–Preboreal. (Our interpretation differs from that of Mörner and Wallin who translated the isotopic changes in an uncritical way into temperature changes.) A $\delta^{18}O$ record on lake marl from a site in Denmark (Kolstrup and Buchardt, 1982) exhibits a continuous decrease during the Allerød, as observed in central Europe, but higher values during the Younger Dryas instead of the expected decrease. This may be due to a large content of allochthonous carbonate with relatively high $\delta^{18}O$, up to 50% according to the authors' estimate, while during the preceding and following periods it was less than 10%.

The abrupt climatic variations reflected in our $\delta^{18}O$ profiles are also well known from other palaeoclimatic studies (e.g. Watts, 1980; Coope, 1975). An important question for our understanding of climatic change is whether these rapid changes were a hemispheric or even global phenomenon or if they were restricted to Europe. In order to determine if similar $\delta^{18}O$ variations in precipitation also occurred in North America, as in central Europe, Eicher *et al.* (in preparation; see also Eicher and Siegenthaler, 1982) have analysed Late Glacial carbonates from the Great Lakes region. The results obtained so far do not show indications of abrupt changes. This is consistent with the findings from other evidence that in North America there was no climatic fluctuation in the Late Wisconsin comparable to the Younger Dryas period (Wright, 1977). Results obtained by Stuiver (1970) from North American lakes also do not show clear evidence for such fluctuations.

The same sequence of $\delta^{18}O$ variations as observed in the European lake sediments is also present in ice cores from Greenland (Oeschger *et al.* 1984). As discussed by Siegenthaler *et al.* (1984), the reason for the climatic changes recorded in the deposits from Greenland and Europe were probably dramatic changes in the surface conditions of the North Atlantic Ocean. There is convincing evidence from deep-sea sediments for retreats and readvances of cold, ice-laden polar water during the Late Glacial (Ruddiman and McIntyre, 1981), which were contemporaneous with the continental climatic events.

The postglacial period has so far not been investigated by $\delta^{18}O$ studies as thoroughly as has the Late Glacial. On an average, there seems to exist a decreasing $\delta^{18}O$ trend toward the present. Besides that, however, fluctuations observed in individual profiles cannot yet be synchronized.

Climatic interpretation of $\delta^{18}O$ results

In order to be certain that $\delta^{18}O$ variations in carbonate do not merely reflect local processes or even some irregularity, it is necessary that the same pattern of change be observed in profiles from more than one site. If this is the case —as for the Late-Glacial variations discussed in the last sections — one may conclude that they are indeed a record of large-scale environmental changes.

The isotopic variations in meteoric precipitation are transferred to the water of a lake and then to sedimentary carbonate. Thus, the following phenomena are of influence:

(1) $\delta^{18}O$ changes in sea water;
(2) $\delta^{18}O$ difference between mean annual precipitation in the study area and sea water in the vapour source region;
(3) isotopic modification of lake water compared with mean meteoric precipitation;

(4)   isotopic fractionation during the precipitation of carbonate, depending on water temperature.

Before going into details, we must note that the dominating effect is point (2), the positive correlation of $\delta^{18}O$ in precipitation with temperature. The next-important effect is point (4), which has a negative temperature coefficient of about -0.25 per mil/°C (Craig, 1965). Point (4) thus tends to attenuate the variations in precipitation, which means that the $\delta^{18}O$ variations in carbonate represent a lower limit for the variations that occurred in mean annual precipitation.

The Ice Age ocean was enriched in ^{18}O by about 1 per mil, compared with the present-day ocean, because large amounts of water were stored in continental ice masses with low $\delta^{18}O$ values. These changes occurred rather slowly. A relatively rapid process affecting $\delta^{18}O$ in sea water was the input of ^{18}O-depleted melt-water to the North Atlantic Ocean during the cold periods Oldest Dryas and Younger Dryas (Ruddiman and McIntyre, 1981). The corresponding decrease in the ocean surface is, however, difficult to estimate.

Dansgaard (1964) studied the world-wide variations of $\delta^{18}O$ in precipitation and found that on a global scale there is a linear correlation between mean annual values of $\delta^{18}O$ and surface air temperature for marine and polar stations with mean annual temperatures below 10°C, with a slope of 0.7 per mil/°C. On a regional scale, deviations do exist. When comparing the seasonal cycles of $\delta^{18}O$ and temperature at some fixed station, a lower temperature coefficient is obtained, typically 0.2–0.5 per mil/°C (Siegenthaler and Oeschger, 1980). These figures include the effects of the rainout history of air masses (point (2) above) and of differing vapour source conditions (point (1)). It is not easy to discern the various effects, but the indicated figures give an idea of the influence of temperature. For point (2) alone, a reasonable range for the temperature coefficient of $\delta^{18}O$ is, based on model calculations, 0.5–0.7 per mil/°C for temperate latitudes (Siegenthaler and Matter, 1983).

Lake water is often enriched in ^{18}O, compared with mean precipitation, because of evaporation. Extensive evaporation is probably the reason why the absolute values for Gerzensee are several per mil higher than for nearby Faulensee (Figure 20.2). Evaporation, and therefore ^{18}O enrichment, increases with rising temperature. For Lake Gerzensee this effect was estimated at 0.2 per mil/°C (Eicher and Siegenthaler, 1976); for most other lakes it will be smaller.

Neglecting evaporative enrichment and changes in sea water, the combined effect of points (2) and (4) can be estimated to yield a $\delta^{18}O$ change in carbonate of 0.25–0.45 per mil/°C change in annual mean temperature. The three major changes observed in the Late Glacial amount to 2–3 per mil, which therefore would correspond to temperature steps of 4–12°C. If it is assumed that $\delta^{18}O$ of sea water in the vapour source region was 1 per mil lower in cold than in warm periods, the climate-induced isotopic changes would be only 1–2 per mil, corresponding to temperature changes between 2 and 8°C. These temperature results do not refer to the absolute temperature changes at the sites studied, but rather to the differences between these sites and the oceanic vapour source region.

The above discussion shows that the relationship between environmental changes and $\delta^{18}O$ in precipitation or lake carbonate are quite complex. Until these are better understood, a quantitative interpretation of continental $\delta^{18}O$ records is affected by considerable uncertainties. While it is obvious that $\delta^{18}O$ variations do reflect climatic changes, it appears to be necessary, at present, to consider them mainly as isotope records of palaeoprecipitation and to restrict oneself to a qualitative interpretation.

Stable carbon isotopes

The $^{13}C/^{12}C$ ratio of lake carbonate depends on many factors: $\delta^{13}C$ in the water feeding the lake; exchange of the lake water with atmospheric CO_2; biological productivity (since plants assimilate $^{12}CO_2$ in preference to $^{13}CO_2$, the remaining carbonate is enriched in ^{13}C); and residence time of the water in the lake. Groundwater and river water in general have $\delta^{13}C$ values between -10 and -15 per mil. Equilibrium with atmospheric CO_2 leads to higher values (about $+2$ per mil at final equilibrium) and in addition biological activity of plants which preferentially withdraw ^{13}C-depleted carbon involves a ^{13}C enrichment in the dissolved bicarbonate. The latter effect obviously depends, among other factors, on productivity. Therefore, Stiller and Hutchinson (1980) could discuss $\delta^{13}C$ in sediments of Lake Huleh (Israel) in terms of productivity.

$\delta^{13}C$ values of lake marl in Lobsigensee are plotted in Figure 20.3. In pollen zone Ia the carbonate content is low, indicating a low productivity, and a considerable fraction of the carbonate may be allochthonous. It is therefore difficult to interpret the $\delta^{13}C$ results of this section. In the rest of the profile, allochthonous contributions are probably small. The values range between -1 and $+2$ per mil, clearly higher than typical values for groundwater, which indicates that they were modified by exchange with air-CO_2 and/or by biological activity in the lake. $\delta^{13}C$ variations are remarkably parallel to those of $\delta^{18}O$. Since there is no direct influence of climate on $\delta^{13}C$, this suggests an indirect connection. A likely explanation is that the productivity of aquatic vegetation responded to climatic change, with relatively strong biological activity in warm phases leading to enhanced ^{13}C enrichment in the water. In this way, $\delta^{13}C$ may under some conditions be an indicator of palaeoproductivity.

This need not always be the case, however, as demonstrated by results from Lake Gerzensee (not reproduced in Figure 20.1), where $\delta^{13}C$ values similar to those in Lobsigensee are observed (0 to $+3$ per mil). (This is not so everywhere; the Faulensee profile, cf. Figure 20.2 for $\delta^{18}O$, exhibits values mainly in the range −6 to −2 per mil.) There is no clear correlation between $\delta^{13}C$ and $\delta^{18}O$ at Gerzensee, as would be expected if ^{13}C was enriched due to productivity. Rather, these values are presumably a result of isotopic equilibration with air-CO_2, implying a relatively long water residence time in the lake. This interpretation is supported by the absolute $\delta^{18}O$ values which are significantly higher than at Faulensee (cf. Figure 20.2) — although both sites are in the same region and at the same altitude — probably because of relatively strong ^{18}O enrichment due to evaporation, which again points to a long residence time.

$\delta^{18}O$ AND δD IN ORGANIC MATTER

A related subject, which is therefore briefly mentioned here, is stable isotope analysis of organic matter, especially peat or tree-rings. A large number of studies exist, but so far with restricted success as to providing palaeoclimatic information. The basis of the method is that $\delta^{18}O$ and δD in organic matter ultimately depend on the isotopic composition of local precipitation, so that variations in the latter should be recorded.

For sample preparation, cellulose is often extracted from the organic matter; then H_2 or CO_2 are prepared for δD or $\delta^{18}O$ analysis, respectively. Sample preparation has been described for δD, by, for example, Epstein *et al.* (1976) and Yapp and Epstein (1982), and for $\delta^{18}O$ by, for example, Thompson and Gray (1977) and Brenningkmeijer and Mook (1981).

The isotopic composition of plant material does not directly reflect the composition of precipitation. Rather it is influenced by evapotranspiration which leads to an enrichment of $\delta^{18}O$ and D in the leaf water used for photosynthesis. This enrichment depends on the relative humidity of the ambient air (Dongman *et al.*, 1974). Furthermore, there is a biochemical fractionation involving a considerable enrichment in $\delta^{18}O$ and a slight depletion in D in the cellulose, compared with the leaf water (DeNiro and Epstein, 1979). The overall fractionation is thus a complex function of climatic and plant-physiological parameters. Correlations of isotopic composition of plant material with air temperature (Schiegl, 1974; Gray and Thompson, 1977), or with temperature and relative humidity (Burk and Stuiver, 1981), have been demonstrated empirically.

Yapp and Epstein (1977) measured δD in North American trees from the period 9500–22,000 years B.P. and inferred that glacial age water (from 14,000–22,000 B.P.) in ice-free regions of North America was on average 19 per mil heavier in δD (corresponding to 2–2.5 per mil in $\delta^{18}O$) than modern water. About half of the difference may be explained by the isotopic shift in sea water; but this still leaves the surprising result that precipitation during the Ice Age seems to have been isotopically heavier than now, in contrast to what is known in Greenland and Europe. While the translation of δD values from cellulose back to precipitation involves some uncertainty, these results do indicate that the δD of precipitation in ice-free North America probably cannot have been much lower than the δD in modern precipitation.

Brenningkeijmer *et al.* (1982) report $\delta^{18}O$ and δD results from peat from a bog in the Netherlands, covering the period 3100–2400 years B.P., and include a thorough discussion of the method. Their results qualitatively agree with the pollen-analytical evidence for the climatic deterioration at the subboreal–subatlantic transition.

The subject of stable isotope variations in plants is quite complex, and the above discussion, as well as the selected references, are necessarily incomplete. Readers interested in the method will find more information in the cited literature.

REFERENCES

Ammann, B., and Tobolski, K. (1984). Vegetational development during the Late Lateglacial at Lobsigensee, Swiss Plateau. *Revue de Paléobiol.*, **2**, 163–180.

Berglund, B. E. (1979). The deglaciation of southern Sweden 13,500–10,000 B.P. *Boreas*, **8**, 89–118.

Blanc, Ph., Chaix, L., Fontes, J.-Ch. Letolle, R., Olive, Ph., and Sauvage, J. (1977). Etude isotopique préliminaire de la craie lacustre des grands marais de Genève. *Arch. Sc. Genève*, **30**, 421–431.

Brenningkmeijer, C. A. M., and Mook, W. G. (1981). A batch process for direct conversion of organic oxygen and water to CO_2 for $^{18}O/^{16}O$ analysis. *Int. J. Appl. Radiat. Isot.*, **32**, 137-141.

Brenningkmeijer, C. A. M., van Geel, B., and Mook, W. G. (1982). Variations in the D/H and $^{18}O/^{16}O$ ratio in cellulose extracted from a peat bog core. *Earth Planet. Sci. Lett.*, **61**, 283–290.

Burk, R. L., and Stuiver, M. (1981). Oxygen isotope ratios in trees reflect mean annual temperature and humidity. *Science*, **211**, 1417–1419.

Coope, G. R. (1975). Climatic fluctuations in NW Europe since the last interglacial, indicated by fossil assemblages of Coleoptera. In: *Ice Ages, Ancient and Modern*, (Eds. A. E. Wright *et al.*), Seel House Press, Liverpool, pp. 154–168.

Craig, H. (1957). Isotopic standards for carbon and oxygen and correction factors for mass-spectrometric analysis of carbon dioxide. *Geochim. Cosmochim. Acta*, **12**, 133–149.

Craig, H. (1961). Standard for reporting concentrations of D and ^{18}O in natural waters. *Science*, **133**, 1833–1834.

Craig, H. (1965). The measurement of oxygen isotope palaeotemperatures. In: *Stable Isotopes in Oceanographic Studies and Paleotemperatures*, (Ed. E. Tongiorgi), Pisa, Consiglio Nazionale delle Ricerche, pp. 161–182.

Dansgaard, W. (1964). Stable isotopes in precipitation. *Tellus*, **16**, 436–463.

Dansgaard, W., Clausen, H. B., Gundestrup, N., Hammer, C. U., Johnsen, S. E., Kristindottir, P. M., and Reeh, N. (1982). A new Greenland deep ice core. *Science*, **218**, 1273–1277.

DeNiro, M. J., and Epstein, S. (1979). Relationship between the oxygen isotope ratios of terrestrial plant cellulose, carbon dioxide, and water. *Science*, **204**, 51–53.

Dongman, G., Förstel, H., Nürnberg, H. W., and Wagener, K. (1974). On the enrichment of $H_2^{18}O$ in the leaves of transpiring plants. *Radiat. Environ. Biophys.*, **11**, 41–52.

Eicher, U. (1979). Ph.D. thesis, University of Bern.

Eicher, U. (1980). Pollen- und Sauerstoffisotopenanalysen an spätglazialen Profilen vom Gerzensee, Faulensee und Regenmoos ob Boltigen. Mitteilungen Naturforsch. *Gesellschaft Bern*, **37**, 6 5–80.

Eicher, U., Siegenthaler, U. (1976). Palynological and oxygen isotope investigations on Late-Glacial sediment cores from Switzerland. *Boreas*, **5**, 109–117.

Eicher, U., Siegenthaler, U., and Wegmüller, S. (1981). Pollen and oxygen isotope analyses on Late- and Post-Glacial sediments of the Tourbière de Chirens (Dauphine, France). *Quat. Res.*, **15**, 160–170.

Eicher, U., and Siegenthaler, U. (1982). Klimatische Informationen aus Sauerstoff-Isotopenverhältnissen in Seesedimenten. *Physische Geographie*, **1**, 103–110 (Inst. of Geography, Univ. of Zurich).

Eicher, U., and Siegenthaler, U. (1984). Stable isotopes in lake marl and mollusc shells from Lobsigensee (Swiss Plateau). *Revue de Paléobiol.*, **2**, 217–220.

Epstein, S., Yapp, C. J., and Hall, J. H. (1976). The determination of the D/H ratio of non-exchangeable hydrogen in cellulose. *Earth Planet. Sci. Lett.*, **30**, 241–251.

Fritz, P., Anderson, T. W., and Lewis, C. F. M. (1975). Late quaternary climatic trends and history of Lake Erie from stable isotope studies. *Science*, **190**, 267–269.

Gray, J. M., and Lowe, J. J. (1977). *Studies in the Scottish Late-Glacial Environment*, Pergamon, New York.

Gray, J., and Thompson, P. (1977). Climatic information from $^{18}O/^{16}O$ analysis of cellulose, lignin and whole wood from tree rings. *Nature*, **270**, 708–709.

Kolstrup, E., and Buchardt, B. (1982). A pollen analytical investigation supported by an ^{18}O-record of a Late Glacial lake deposit at Graenge (Denmark). *Rev. Palaeobot. Palyn.*, **36**, 205–230.

Lang, G. (1970). Florengeschichte und mediterran-mitteleuropäische Florenbeziehungen. *Feddes Reportorium*, **81**, 315–335.

Lemeille, E. (1980). Ph.D. thesis, Université de Paris XI, Centre de Recherches Géodynamiques, 74203 Thonon-les Bains, France.

Lemeille, E., Létolle, R., Melière, F., and Olive, P. (1983). Isotope and other physico-chemical parameters of palaeolake carbonates. In: *Palaeoclimates and Palaeowaters*, IAEA, Vienna (STI/PUB/621), pp. 135–150.

McCrea, J. M. (1950). On the isotopic chemistry of carbonates and a palaeotemperature scale. *J. Chem. Physics*, **18**, 849–857.

Mook, W. G. (1968). Ph.D. thesis, University of Groningen.

Mook, W. G., and Grootes, P. M. (1973). The measuring procedure and corrections for the high-precision mass-spectrometric analyses of isotopic abundance ratios. *Int. J. Mass Spectrom. Ion Physics*, **12**, 273–298.

Mörner, N.-A., and Wallin, B. (1977). A 10,000-year temperature record from Gotland, Sweden. *Palaeogeogr, Palaeoclimat. Palaeoecol.*, **21**, 113–138.

Oeschger, H., Beer, J., Siegenthaler, U., Stauffer, B., Dansgaard, W., and Langway, C. C. (1984). Late-Glacial climatic history from ice cores. In: *Climate Processes and Climate Sensitivity* (Eds. J. Hansen and T. Takahashi) Am. Geophys. Union 299–306.

Ruddiman, W. F., and McIntyre, A. (1981). The North Atlantic Ocean during the last deglaciation. *Palaeogeogr. Palaeoclimatol. Palaeoecol.*, **35**, 145–214.

Schiegl, W. E. (1974). Climatic significance of deuterium abundance in growth rings of *Picea*. *Nature*, **251**, 582–584.

Siegenthaler, U., Eicher, U., and Oeschger, H. (1984). Lake sediments as continental ^{18}O records from the transition Glacial-Post-glacial. *Annals of Glaciol.*, **5**, 149–152.

Siegenthaler, U., and Oeschger, H. (1980). Correlation of ^{18}O in precipitation with temperature and altitude. *Nature*, **285**, 314–317.

Siegenthaler, U., and Matter, H. A. (1983). Dependence of δ^{18}O and δD in precipitation on climate. In: *Palaeoclimates and Palaeowaters*, IAEA, Vienna, (STI/PUB/621), pp. 37–52.

Stiller, M., and Hutchinson, G. E. (1980). The waters of Merom: a study of Lake Huleh. *Arch. Hydrobiol.*, **89**, 275–302.

Stuiver, M. (1970). Oxygen and carbon isotope ratios of fresh water carbonates as climatic indicators. *J. Geophys. Res.*, **75**, 5247–5257.

Thompson, P., and Gray, J. (1977). Determination of δ^{18}O/^{16}O ratios in compounds containing C, H and O. *Int. J. Appl. Rad. Isot.*, **28**, 411–415.

Watts, W. A. (1980). Regional variations in the response of vegetation to Late-Glacial climatic events in Europe. In: *The Late-Glacial of North-West Europe* (Eds. J. J. Lowe *et al.*), Pergamon, Oxford, pp. 1–21.

Wright, H. E. (1977). Quaternary vegetation history — some comparisons between Europe and North America. *Ann. Rev. Earth Planet. Sci.*, **5**, 123–158.

Yapp, C. J., and Epstein, S. (1977). Climatic implications of D/H ratios of meteoric water over North America (9500–22,000 B.P.) as inferred from ancient wood cellulose C–H hydrogen. *Earth Planet. Sci. Lett.*, **34**, 333–350.

Yapp, C. J., and Epstein, S. (1982). A reexamination of cellulose carbon-bound hydrogen δD measurements and some factors affecting plant-water D/H relationships. *Geochim. Cosmochim. Acta*, **46**, 955–965.

Handbook of Holocene Palaeoecology and Palaeohydrology
Edited by B. E. Berglund
© 1986 John Wiley & Sons Ltd.

21

Chemical analysis

LARS BENGTSSON AND MAGNUS ENELL

Institute of Limnology, University of Lund, Lund, Sweden

INTRODUCTION

Palaeoecological studies include chemical and physical analyses of different materials, such as sediments, peats and soils, and the contents of constituents vary within wide ranges. Usually the physical and chemical methods used vary according to the material analysed. In the interpretation of the results this is unsatisfactory, as in many cases the comparison of results obtained by different methods is unreliable. Besides, it is impracticable to work with quite different chemical methods depending on the material being analysed.

The need for an analytical reference work to meet the requirements of basic chemical analysis within palaeoecological studies is obvious. But it must also be realized that some workers do not have laboratories well equipped for chemical analysis. For this reason the idea here is to present a sampling treatment scheme in which most of the work can be done in laboratories without expensive equipment. The final analytical step presupposes access to a spectrophotometer (or auto analyser) and an atomic absorption spectrophotometer.

Sediments, peats and soils can be regarded as mirrors of past conditions in ecosystems and in the surrounding land. This was concluded by Lundqvist (1942), who wrote: 'Sediment is a product of the life-activity in lakes and the development of the catchment area, and can therefore provide information on the trophic state'. Sedimentological investigations offer the possibility of environmental control and can perhaps be better than other investigation methods in terms of reaching conclusions about which areas are or have been polluted and which are not. They also inform as to whether chemical constituents are present in increased amounts compared with the natural state, and which are not, or how the eutrophication and pollution situation has developed in the ecosystem (Håkansson, 1981). A chemical investigation of the historical development of the ecosystem can be summarized as a chemostratigraphic study; this, together with, for example, a biostratigraphical study, can very well often describe the former ecosystem. Engstrom & Wright (1984) have written an excellent chapter about 'Chemical stratigraphy of lake sediments as record of environmental change'.

SAMPLING, SUBSAMPLING AND STORING

Usually, stratigraphical samples of sediment, peat and soil are taken with a corer, which gives a certain restricted volume of sample. For the physical and chemical analysis of sediment a sample volume of about 50 cm^3 is usually enough. Recommended sampling equipment, sampling density and temporal resolution are described in Chapters 4 and 8. Reviews of different samplers and sampling techniques are given by, for example, Hopkins (1964), Wright *et al.* (1965), Bouma (1969), Sly (1969) and Håkansson and Jansson (1983). The lack of uniform methods of sampling and preparation of samples for chemical analysis has impeded interpretations and comparisons of results from ecological and sedimentological studies (Håkansson and Jansson, 1983).

The subsampling and sample preparation must be tailored to the subsequent analysis of the sediment, and hence to the aim of the investigation. There are two subsampling principles:

(1) *structural subsampling*, in which the samples are taken according to the sediment structure, lamination, strata or varves. Some existing structures may not be visible without X-ray techniques;
(2) *uniform subsampling*, in which the samples are taken at certain given levels determined before the survey, e.g. every 2 centimeters (0–2, 2–4, 4–6 cm, etc.) or in layers of greater thickness. The uniform subsampling technique ignores the sedimentary structures, which implies that important information may be lost.

To avoid contamination of the samples, corers and slicing equipment made of stainless steel and plastic should be used. Brass should be avoided as contamination by copper and zinc can result. Soil samples should be sieved through a 2 mm mesh to separate the fractions of gravel and stones. Brass sieves should be avoided for the reason mentioned above. Stainless steel or nylon sieves are recommended. Usually large stones and roots are picked out before sieving.

Sediments do not usually contain particles greater than 2 mm (especially organic sediments) and sieving is not usually necessary. Plant and animal fragments like leaves, small branches, shells, living benthic organisms etc. may be picked out.

It is recommended that the cores be subsampled in the field and as a general rule the period of storage minimized. The subsamples can be put in watertight plastic bags or jars. These should then be transported and stored in 'cold-boxes' with prefrozen 'cold-packs'. It is important to keep the samples cold and dark to avoid (or reduce) microbial activity, which will influence determinations of, especially, organic compounds such as nitrogen, amino acids, carbohydrates, lipids and pigments. Care must also be taken regarding the fact that water diffuses through most plastic materials and that prolonged storage might alter the water content. For storage during longer periods deep-freezing may be recommended. The water content of the sample should be analysed as promptly as possible.

PHYSICAL ANALYSIS

Physical analysis in this context refers to parameters such as bulk density, and water, organic and carbonate content. Accurate determination of these parameters is essential as the chemical results should be expressed per unit volume of fresh sediment (FW) and per dry-weight of sediment (DW), or as sedimentation per m^2 and time. It is important to remember that sediment, peat and soil are complex materials and that the vertical stratification is closely connected with the biology and chemistry of the material.

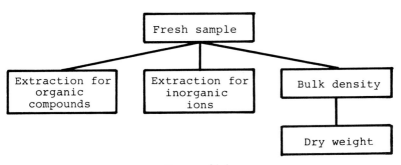

FIGURE 21.1.

For many chemical analyses drying of the sample is both possible and convenient. The temperature and conditions for drying are critical, however. (Some analyses must be carried out as soon as possible after sampling, and on the fresh material.) The general scheme given in Figure 21.1 can be followed. The dried samples may be stored for long periods and used for determinations of the total content of individual elements.

If pH and redox measurements are to be carried out they should be done as soon as possible after sampling, preferably in the sampler out in the field or on the subsamples directly after subsampling. These measurements will not be dealt with further here; those who are interested are recommended to use the methods outlined by the Baltic Marine Biologists (1976).

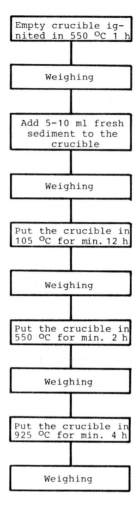

FIGURE 21.2.

It is recommended that bulk density (*D*), and water, organic and carbonate content, be determined together in the following manner (summarized in Figure 21.2):

(1) Heat a porcelain crucible (about 30 ml) for 1 hour at 550°C in a muffle furnace, cool it to room temperature in a desiccator, and determine *the weight of the crucible* (*A*) to an accuracy of ±0.1 mg.

(2) Transfer a fixed volume of the fresh sample (5–10 ml, with a 'cut' plastic syringe) to the crucible and immediately determine *the weight of sample + crucible* (*B*).

(3) Place the crucible in an air-circulation oven at 105 ± 2°C and dry to constant weight (overnight for about 12 hours).

(4) Let the crucible and sample cool to room temperature in a desiccator and determine *the weight of the dry sample + crucible* (*C*).
(5) Place the crucible with the dry sample in a muffle furnace for 2 hours at 550°C. If the sample has a high organic content, put a porcelain lid on the crucible and start at a lower temperature. By this procedure ash losses by violent burning of the sample can be prevented. If the ignition is incomplete (black with carbon) after 2 hours, the residue can be treated after cooling with 1 or 2 drops of NH_4NO_3 solution (20 g/100ml) and ignited for a further 30 minutes.
(6) Let the crucible with the ignited sample cool to room temperature in a desiccator and determine *the weight of ash + crucible* (*D*).
(7) Place the crucible with the ignited sample in a muffle furnace for 4 hours at 925°C.
(8) Let the crucible with the ignited sample cool to room temperature in a desiccator and determine *the weight of ash + crucible* (*E*).

The water content is distributed in a typical way in sediments: low values in shallow waters where coarser materials predominate, and high values in deeper parts of the lake. The decrease in water content with depth in sediment cores is due to compaction. A vertical variation in the water content depends on many factors, such as the rate of sedimentation, the quality and character of the deposits, and the degrees of compaction and bioturbation. The loss on ignition is usually used as a measure of the organic content of the material. Most of the organic substances, but also some of the chemically bound water, will be included in this analysis. The loss on ignition may, under certain conditions, be used to estimate the content of organic carbon in the sample (Mackereth, 1966; Digerfeldt, 1972; Cato, 1977; Håkansson, 1983). The organic carbon then represents 12/30 of the organic content ($(CH_2O)_n$). The correlation between the content of organic carbon and the loss on ignition is generally very good (Håkansson and Jansson, 1983). The carbonate content in the sample is caused by allogenic, endogenic and authigenic minerals (sources).

Calculations

$$\text{Bulk density } D = \frac{B - A}{\text{ml fresh sample}} \text{ kg/dm}^3$$

$$\text{Dry weight } DW = \frac{C - A}{B - A} \text{ g DW/kg FW}$$

$$\text{Loss on ignition } IG = \frac{C - D}{B - A} \text{ g IG/kg DW}$$

$$\text{Carbonate content } CO_3 = \frac{D - E}{B - A} \times f_1 \text{ g } CO_3/\text{kg DW}$$

$$\text{Conversion factor (dry weight} \rightarrow \text{volume)} = D \times DW(\text{g/dm}^3 \text{ FW})$$

$$f_1 = \frac{60(\text{molecular weight for } CO_3)}{44(\text{molecular weight for } CO_2)} = 1.36$$

Loss on ignition is, as already mentioned, a rough indication of the amount of organic matter in the sample, and usually the correlation with organic carbon is acceptable for non-calcareous samples (Allen *et al.*, 1974). Various conversion factors have been reported for converting loss on ignition to organic carbon. As the factor depends on the material being analysed the use of conversion factors should be handled with care even within a single sediment profile (Digerfeldt, 1972). Usually the organic carbon is 40–60% of the loss on ignition.

Depending on the ignition temperature, various losses of volatile salts, structural water and ammonia can occur, and therefore it is essential to check the temperature carefully.

As carbonate compounds become volatile at different temperatures, the carbonate content estimation is only a rough determination. Various conversion factors have been given by different authors, but as a general rule the weight loss obtained between 550 and 925°C should be multiplied by 1.36 to obtain the carbonate content.

As the dried sediment takes up water from the air the samples should be kept in a desiccator or be redried before further use.

Special care must be taken with iron-rich sediments, while weight E (above) can be greater than weight D.

SOLUTION PREPARATION

The elements in sediments, peats and soils can be grouped into five categories, according to Kemp *et al.* (1976):

(1) *major elements:* Si, Al, K, Na and Mg which constitute the main group in lake sediments;
(2) *carbonate elements:* Ca, Mg and CO_3-C which constitute the second most important group in the sediments, about 15% of the sediment;
(3) *nutrient elements:* organic-C, N and P, constituting about 10% of recent lake sediments;
(4) *mobile elements:* Mn, Fe and S, especially mobile elements, which react rapidly in changes in the oxidation–reduction state of the sediment — they contribute about 5%;

(5) *trace elements:* Hg, Cd, Pb, Zn, Cu, Cr, Ni, Ag, V, etc. which contribute about 0.1% of the sediment — to this group belongs the so-called heavy metals and toxic metals, which usually show the influence of pollution.

Sediment, peat and soil materials consist of a highly complex mixture of minerals and organic compounds. The number of elements studied are limited, naturally depending on the aim of the investigation.

Many different techniques have been developed for determining the total and extractable amounts of elements bound to organic matter etc. in sediment, peat and soil. An introduction of existing procedures has been given by Engstrom & Wright (1984). As these methods are usually time-consuming and hazardous, a simplified treatment of the samples is suggested here for determinations of organic and acid-soluble amounts of various elements. For palaeoecological work it seems more useful to analyse a variety of elements at close intervals than a few elements at long intervals. A solution preparation technique, including an autoclaving step, is under development and this technique will reduce the time for preparation.

The two solution-preparation techniques described below can be used in any laboratory with ordinary equipment. In order to check the exact exchange in different sediment, peat and soil it is recommended that some separate checks with sodium carbonate fusion, hydrofluoric–perchloric digestion or an X-ray fluorescence technique be carried out (Allen *et al.*, 1974) (see Figure 21.3).

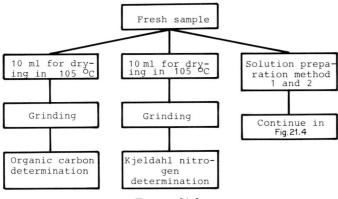

FIGURE 21.3.

Solution-preparation method 1

As the solution preparation proceeds from the dried sample stage, it is convenient to start with the drying of a fixed volume (5–10 ml) in a 30 ml porcelain crucible and to determine bulk density, dry weight and loss on ignition according to the procedure mentioned above. The carbonate content determination is excluded. It is recommended that the determinations are made in duplicate. After the determinations of loss on ignition, proceed as follows (summarized in Figure 21.4):

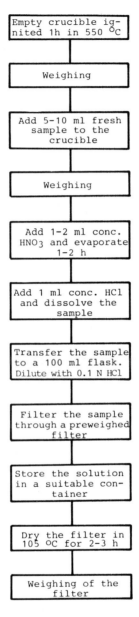

FIGURE 21.4.

(1) Add 1–2 ml (the sample should be soaked) of concentrated nitric acid and evaporate over a water bath to dryness (1–2 hours) after crushing the ignited crust with a glass rod. Clean the rod with a few drops of distilled water.

(2) Add 1 ml of concentrated hydrochloric acid and dissolve the sample (if

necessary add a few drops of 0.1N hydrochloric acid). The solution will now be more or less yellow and clear.

(3)   Transfer the sample quantitatively to a 100 ml volumetric flask. Use a glass rod to rinse the crucible. Dilute with 0.1N hydrochloric acid to 100 ml.

(4)   Shake vigorously and filter through a glass-fibre filter (Whatman GF/C, preweighed if the acid insoluble residue is to be determined).

(5)   Reject the first 10–15 ml of the filtrate and collect the rest in a suitable glass or plastic container, in which this stock solution can be stored for further analyses of phosphorus and metals.

(6)   If the volumetric flask is well rinsed with several portions of 0.1N hydrochloric acid which is put through the filter, the acid-insoluble residue can be determined by drying (105°C for 2–3 hours) and weighing the filter, and the volumetric flask (now well rinsed) can be used for the next sample.

Carry out blank and standard determinations in the same way.

As the phosphorus determination is sensitive to the acid concentration, keep to the same acid solutions for each series and treat at least three phosphorus standards in the same way as the samples. With the treatment described above the final acid concentration in the stock solution will be 0.2 mole/litre.

A thorough washing with hydrochloric acid (about 5–6) NClH is necessary for all glassware and crucibles being used, to avoid contamination.

Solution preparation method 2

With a wet digestion of the dried material in a strong acid solution with an oxidizing reaction, all organic compounds are mineralized, which means that the bound elements such as P, Ca, Mg, Na, K, heavy metals etc. are liberated from the particulate matter and transferred into solution.

After drying the sample, proceed as follows (summarized in Figure 21.5):

(1)   Homogenize the dried sample by grinding in a porcelain mortar with a pestle. A fine-grained powder will facilitate the following digestion. Vacuum-drying of the fresh sediment will directly result in a fine-grained dried sample.

(2)   Re dry the sample for 2 hours at 105°C. Store the sample in a desiccator.

(3)   Transfer 1.000 g of dried homogenized sample to a wide-necked Ehrlenmeyer flask (size 250–300 ml).

(4)   Add 15 ml of acid solution. The acid solution consists of one part concentrated perchloric acid and four parts concentrated nitric acid. The acid solution must be handled with great care since there is a risk of explosion if instructions are not correctly followed.

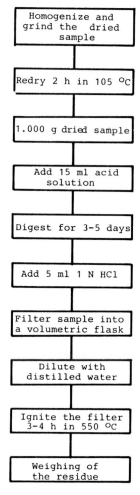

FIGURE 21.5.

(5) Place the Ehrlenmeyer flask on a heating plate. Put a watch-glass on the flask. Nitric gases are produced which means that the digestion must be made in an evacuation chamber (fume cupboard).

(6) Increase the temperature on the heating plate slowly. The acid solution must not evaporate totally. After 3–4 days the gas production decreases and the organic matter is mineralized. Slightly increase the temperature on the heating plate.

(7) When the residue in the flask is whitish-yellow, take away the watch-glass. Let the acid solution 'smoke off', but dryness must not occur.

(8) Take the flask from the heating plate and add 5 ml of 1N hydrochloric acid. Use a glass rod to scrape all minerogenic matter from the walls of the flask.

(9)   Transfer the sample quantitatively to a 200–300 ml volumetric flask. Use a glass rod to rinse the Ehrlenmeyer flask.

(10) Shake vigorously and filter through an ash-free filter (Munktell 00). Add 10–15 ml portions of distilled water to the Ehrlenmeyer flask, rinse carefully and put it through the filter. At least 3–4 portions of distilled water should be used to transfer the sample quantitatively to the volumetric flask.

(11) Fill up to the mark in the volumetric flask with distilled water. This stock solution can be stored for further analyses for phosphorus and metals.

(12) If the Ehrlenmeyer flask is well rinsed with several portions of distilled water which is put through the filter, the acid-insoluble residue can be determined by ignition (at 550°C) and weighing the minerogenic residue.

Carry out blank and standard solutions in the same way. The notes following method 1 apply here too.

PHOSPHORUS CONTENT

Total phosphorus

Phosphorus exists in both organic and mineral forms. Much attention has been paid to the various mineral forms in order to determine the available or exchangeable forms in soils and sediments (Dean, 1938; Chang and Jackson, 1957; Williams *et al.*, 1967, 1971, 1976; Hieltje and Lijklema, 1980). However, the methods developed for the determinations have serious limitations according to the extractants being used (Allen *et al.*, 1974).

Phosphorus is essential to all living organisms and occurs in, for example, phytates, nucleic acids and phospholipids. It is also important in the ADP-ATP cell energy and as a structural element in bones and shells.

The main sources of phosphorus for an ecosystem are:

(1)   weathering of bedrock, mainly apatite ($Ca_{10}(PO_4)_6(OH)_2$);
(2)   arable land: drainage water from agricultural land may be enriched with phosphorus from manure and fertilizers;
(3)   point sources, like municipal and industrial waste water.

The build-up of phosphorus in sediments, peats and soils resulting from excessive use of fertilizers and detergents, and the disposal of various industrial wastes, can often increase the concentrations by up to one order of magnitude. The supply of phosphorus from farmland is about ten times higher than the background.

Phosphorus is deposited in nature as (Håkansson and Jansson, 1983):

(1)   allogenic apatite minerals;

(2) organic associates, partly as structural elements of settling dead organisms;
(3) precipitates together with inorganic complexes, like iron or aluminium hydroxides or as coprecipitates together with calcite.

The ranges of concentration of phosphorus normally encountered are given below in grams per kilogram dry weight:

Mineral soils	0.2–2	gP/kg DW
Organic soils	0.1–2	gP/kg DW
Lake sediments	0.5–3	gP/kg DW
Plant materials	0.5–3	gP/kg DW
Animal tissue	3–40	gP/kg DW

The molybdenum blue method

In this procedure (Murphy and Riley, 1962), a blue phosphomolybdate compound is developed in aqueous solution. The method has been adapted for an automatic procedure which is recommended if large numbers of samples are to be processed.

Reagents

(1) *H_2SO_4, 5N:* 70 ml of concentrated H_2SO_4 is mixed into deionized H_2O to make 500 ml. This solution can be stored.
(2) *Ammonium molybdate, 4%:* 2 g $(NH_4)_6Mo_7O_{24}\cdot4H_2O$ is dissolved in deionized H_2O in a volumetric flask and diluted to 50 ml. Prepare a new solution before each period of analysis.
(3) *Ascorbic acid, 0.1 mole/l:* 1.75 g ascorbic acid is dissolved in deionized water in a volumetric flask and diluted to 100 ml. Prepare a new solution before each period of analysis.
(4) *Potassium antimony tartrate, 1 mg Sb/ml:* 0.158 g $K(SbO)C_4H_4O_6\cdot$ $\frac{1}{2}H_2O$ is dissolved in deionized water in a volumetric flask and diluted to 50 ml. The solution can be stored if kept in the dark and cold.
(5) *Phosphorus stock solution, 100 mg P/l:* 0.4392 g KH_2PO_4 *pro analysi* dried in 105°C for 2 hours is dissolved in deionized H_2O in a volumetric flask and dilute to 1000 ml after the addition of 1 ml of chloroform for preservation. The solution should be stored in the dark.

A reagent mixture is prepared of (1), (2), (3) and (4) before each analysis in the proportions given in Table 21.1. The reagents should be mixed in the order indicated.

Procedure (see Figure 21.6)

(1) Add a suitable aliquot (0.5 ml) of the sample stock solution and deionized

TABLE 21.1. Reagent requirements for molybdenum blue method

Number of samples	H_2SO_4 (ml)	Ammonium molybdate (ml)	Ascorbic acid (ml)	Potassium antimony tartrate (ml)
10	50	15	30	5
20	100	30	60	10
30	150	45	90	15
40	200	60	120	20
50	250	75	150	25
100	500	150	300	50

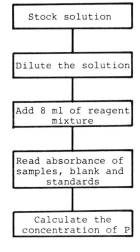

FIGURE 21.6.

H$_2$O to a 50 ml volumetric flask. Add 8 ml of the reagent mixture whilst agitating, dilute to 50 ml and mix carefully.
(2) Include a blank and at least three standards (0.25, 0.5 and 1.0 mg P/l in the final solution), which have been treated in the same way as the samples, for each batch of samples.
(3) Read the absorbance after at least 10 minutes at 882 nm, with deionized water as reference.
(4) Calculate the concentration of phosphorus of the sample by using the standard curve.

If the colour of a sample is darker than the highest standard, a pre-dilution must be done prior to step (1).

The acid treated standards should be checked against standards made up directly from the phosphorus stock solution. Dilute 10 ml to 1000 ml giving 10 μg P/ml. 10 μg P per sample will give an absorbance of about 0.150 in a 1 cm cuvette.

The colour absorption follows the law of Lambert-Beer up to 1.25 mg P/l (50 μg P per sample) with 1 cm cuvettes. The reproducibility is as good as ±1%.

At further dilution of more than 100 times for phosphorus determinations the residual acidity will not influence the results. If less dilution is needed the residual acidity may require neutralization of the sample by adding an appropriate amount of sodium hydroxide in order to keep the final acidity in the phosphorus determination at 0.40 M (Olsen, 1967).

Extractable phosphorus

As a consequence of many pitfalls in past techniques, simplified extraction schemes have been developed. One of the most relevant fractionation schemes has been developed by Williams *et al.* (1976). This distinguishes between organic phosphorus, apatite phosphorus and non-apatite inorganic phosphorus. The extraction procedure described by Hieltje and Lijklema (1980) suggests a fractionation scheme that includes subsequent extractions with NH_4Cl; giving loosely bound phosphorus, $NaOH$; giving iron- and aluminium-bound phosphorus and HCl; giving calcium-bound phosphorus. For detailed descriptions of the above fractionation methods, refer to the original articles. Recent research, described in a review article by Boström *et al.* (1982), indicates that even the simplified approaches described by Williams *et al.* and Hieltje and Lijklema cannot be used uncritically, since this type of extraction may underestimate the organic phosphorus fraction and the calcium-bound phosphorus.

NITROGEN CONTENT

Nitrogen is essential to all organisms as a structural component, mainly of proteins. As most of the nitrogen in sediments, peats and soils is bound in organic matter, total organic nitrogen is usually determined as Kjeldahl-N, which includes ammonia and organically bound nitrogen. Nitrates are not included and are usually present in relatively small amounts in these types of materials. In waters, however, nitrates can represent the main part of the total nitrogen. Usually the inorganic soluble fractions NH_4-N and NO_3-N are determined separately in waters (Ahlgren and Ahlgren, 1975).

The ranges of concentration of nitrogen normally encountered are given below:

Mineral soils	1–5	(g Kj-N/kg DW)
Organic soils	5–15	
Sediments	5–20	
Plant material	10–30	
Animal tissue	40–100	
Waters	0.2 –2 mg Kj-N/l	
	0.01–1 mg NH_4-N/l	
	0.01–2 mg NO_3-N/l	

The Kjeldahl nitrogen method

In this method organic nitrogen is converted into ammonia. The sample is digested with sulphuric acid. Potassium sulphate is added to raise the temperature sufficiently to break down most organic compounds present. Copper sulphate is added as a catalyst. Mercury oxide is generally preferred for soils and gives a better gain (Allen *et al.*, 1974); but being a serious pollutant, mercury is not recommended.

After digestion a distillation or a colorimetric method follows. It is more convenient to use the colorimetric method of determining ammonia with the indophenol blue method in conjunction with an auto-analyser. The digestion procedure is given below.

Reagents

(1) *Potassium sulphate–copper sulphate mixture*: mix K_2SO_4 and $CuSO_4$ (*pro analysi*) in the ratio of 3:1 by weight.
(2) H_2SO_4, *concentrated, p.a.*

Make a digestion solution by dissolving 1 g of the K_2SO_4 + $CuSO_4$ mixture in 50 ml of freshly prepared deionized H_2O, add 50 ml concentrated H_2SO_4 and make up to 100 ml in a volumetric flask. This solution is freshly made for each batch of samples.

Procedure (Figure 21.7)

(1) Weigh about 25 mg of dried (105°C) and finely ground sample (0.1 mg weighing precision) and put the sample in a special digestion tube (for example, Tecator).
(2) Add 1 ml of the digestion mixture while slowly rotating the tube, thus washing the sample down. Residues on the glass walls can be washed down with a few drops of freshly prepared deionized water.
(3) Put the sample tubes on the heating rack at 100°C overnight. Increase the heat to 365°C for 4 hours or until the samples are colourless or pale green. The H_2SO_4 will reflux down the neck of the tube.

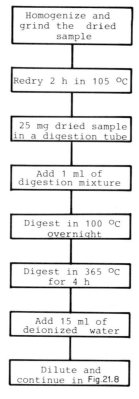

FIGURE 21.7.

(4) After cooling (put Al-foil on the top of the tubes to prevent contamination from NH_3 in the air), add 15 ml of freshly prepared deionized water.
(5) Mix the sample carefully (with a tube vibrator) and let the sample stand until the solution is clear (about 1 hour).
(6) Take 1 ml of the clear solution, dilute to 20 ml, and store in clean glass tubes with lids until NH_4 analysis (Chaney and Marbach, 1962).

The indophenol blue method for ammonium analysis

In this procedure a blue indophenol complex is developed when ammonium reacts with hypochlorite in a weak alkaline solution. Monochloramin is formed. In the presence of phenol and excess hypochlorite the indophenol blue complex is produced. The exact reaction mechanism is not fully known.

Reagents

(1) *Phenol-nitroprussid solution*: 5.0 g phenol (*p.a.*) and 0.025 g sodium nitroprussid $(Na_2(Fe(CN)_5NO) \cdot 2H_2O)$ is dissolved in deionized H_2O in a volumetric flask and diluted to 100 ml. The solution is stable for about 2 months when stored in cold and dark conditions.

(2) *Alkalihypochlorite solution*: 2.5 g NaOH (*p.a.*) is dissolved in about 90 ml deionized H_2O. Cool the solution to room temperature. Add 0.21 g active chlorine in the form of a hypochlorite solution (about 2.2 ml of the commercial sodium hypochlorite solution). Dilute to 100 ml and store in a refrigerator. The solution is stable for about 2 months.

(3) *Ammonium stock solution*, 100 mg NH_4^+/l: 0.3819 g NH_4Cl (*p.a.*) dried at 100°C for 2 hours is dissolved in deionized water in a volumetric flask and diluted to 1000 ml. Store the solution in a dark bottle and in cold and dark conditions.

Procedure (Figure 21.8)

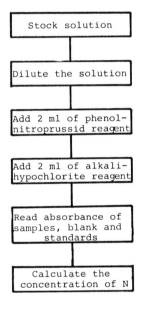

Stock solution

Dilute the solution

Add 2 ml of phenol-nitroprussid reagent

Add 2 ml of alkali-hypochlorite reagent

Read absorbance of samples, blank and standards

Calculate the concentration of N

FIGURE 21.8.

(1) The analysis is continued from (6) above, i.e. the glass tubes with 20 ml of diluted sample.
(2) Add 2 ml of the phenol-nitroprussid solution. Cover the tube with para-film and mix carefully.
(3) Add 2 ml alkali-hypochlorite solution. Cover the tube with para-film and mix carefully.
(4) Let the samples stand for at least 30 minutes, maximum 24 hours, and read the absorbance at 635 nm with deionized water as reference.
(5) Calculate the concentration of NH_4-N of the sample by using the standard curve.

Prepare blank (with reagents alone) and at least three standard digests (between 2 and 6 mg N/l in the final solution).

The digested blank and standards are checked against undigested NH_4-N standards containing the same amount of digestion solution (about 3 ml/l).

The special digestion rack, for example from Tecator, is recommended as all samples can be kept at the same, well controlled temperature and the risk of samples spitting is minimized.

SEDIMENTARY PLANT PIGMENTS

Together with phosphorus, nitrogen and loss on ignition analyses, sediment-ary plant pigment analyses can give information on changes in trophic conditions (Vallentyne, 1960). The relationship between the commonly most dominant pigment groups, chlorophyll derivatives and carotenoids, have also been used in interpretations of changes in the balance between allochthonous and autochthonous organic contributions to sediments and changed oxygen conditions at the sediment surface (Gorham and Sanger, 1975; Sanger and Crowl, 1979).

Although specific carotenoids have sometimes been used to establish the occurrence and abundance of blue-green algae (Vallentyne, 1956; Züllig, 1961; Griffiths and Edmondson, 1975), only the analyses of chlorophyll derivatives and total carotenoids will be dealt with here as specific analyses are too tedious. The method is modified after Fogg and Belcher (1961).

The concentration ranges normally encountered are shown in Table 21.2.

TABLE 21.2. Normal concentration ranges per gram of organic matter

Sediment type	Chlorophylls	Carotenoids
Oligotrophic lakes	1–5	1–10
Eutrophic lakes	10–16	25–60
Meromictic lakes	20–50	95–135

Reagents

(1) 90% (v/v) acetone. Freshly distilled acetone is mixed with distilled water 9:1.
(2) 20% (w/v) methylalcoholic KOH. Dissolve 200 g KOH (*p.a.*) in 1000 ml CH_3OH (*p.a.*).
(3) Petroleum ether (40–60°C).

Procedure

A. Extraction (Figure 21.9)

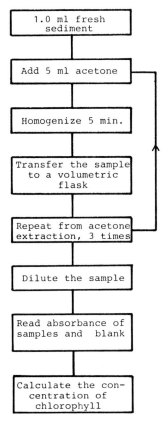

FIGURE 21.9.

(1) Put 1 ml (about 1 g) well-mixed fresh sediment into a centrifuge tube with a rubber stopper with the help of a 'cut' syringe.
(2) Add 5.0 ml 90% acetone and homogenize with a motor-driven Teflon-coated pestle for 1 minute.
(3) Centrifuge the suspension for 5 minutes at 3700 rev/min and decant into a 50 ml volumetric flask.
(4) Repeat steps (2) and (3) three times.
(5) Dilute to 50 ml with 90% acetone. Store in cold and dark conditions.
(6) The chlorophyll derivatives are determined spectrophotometrically in the acetone extract with 90% acetone as blank using a slitband less than 0.06 mm and at the 664–667 nm peak.

B. Separation (Figure 21.10)

(1) Shake 25 ml of the acetone extract, containing the chlorophyll derivatives, with 25 ml 20% CH_3OH-KOH in a 100 ml separating funnel on a shaking table for 2 hours at room temperature.

FIGURE 21.10.

(2) Add 25 ml petroleum ether and 25 ml distilled water and shake vigorously for 2 minutes.

(3) After separation, the water phase is thrown away and the petroleum ether phase is washed with another 25 ml distilled water.

(4) Throw away the water phase and the first 5–10 ml of the petroleum ether phase. The rest is stored for carotenoid determinations.

(5) The carotenoids are determined in the petroleum ether extract with petroleum ether as blank at the 445–450 nm peak.

C. Calculation

Calculate the pigment concentrations expressed as spectrophotometric units (PU) per gram of organic matter, one unit being equivalent to an absorbance of 0.1 in a 1 cm quartz cuvette when dissolved in 100 ml of solvent:

$$\text{Pigment units } (PU) = \frac{10 \times (abs_x - abs_0) \times V_{extr}}{(V_x \cdot D \cdot DW \cdot IG \cdot 10^{-6}) \times 100 \times cell}$$

where abs_x = absorbance of sample
abs_o = absorbance of blank
V_{extr} = volume of acetone extract (50 ml)
V_x = volume of sample (about 1 cm^3)
D = density of sample (g/cm^3)
DW = dry-weight of sample (mg/g)
IG = loss on ignition of sample (mg/g DW)
$V_x \cdot D \cdot DW \cdot IG \cdot 10^{-6}$ = organic matter (g)
$cell$ = length of the cuvette (cm)

To minimize degradation of the pigments, the samples should be stored in cold and dark conditions in gas-tight jars or bottles.

As the wavelength scale of the spectrophotometer can be inaccurate, the absorption peak should be checked before measurements are started.

METALS

The determination of different metals by palaeoecologists often depends on the fact that the metals are easy to analyse by flame techniques.

Today the most-used technique is the atomic absorption method (the flame AA method). Most elements can be analysed by this technique (including Na, K, Ca, Mg, Fe, Mn, Cu, Zn, Cr, Pb, Co, Ni, Al, Mo and Cs). The element is dissolved in a solution which is sprayed into a flame. Radiation of a characteristic wavelength passes the flame and the atoms of the element absorb the element's specific radiation in relation to the concentration of the element in the solution. The decrease in intensity of the characteristic wavelength is measured by a detector system, and by using standard solutions the concentration in an unknown sample can be determined.

The use of a graphite furnace, where a portion of the sample is placed in the light path and heated quickly to a very high temperature, increases the detection limit a thousand-fold, and with automatic sampling good reproducibility is achieved.

Using the digestion solution techniques described previously and appropriate dilution (0–1000 times, depending on the element being analysed), the element can be analysed according to the instrument's manual. Some elements of palaeoecological interest are summarized in Table 21.3.

TABLE 21.3. Summary of some elements of interest for palaeoecological investigations

Element	Suitable range in sample solution[1] (mg/l)	Interferences	Sediment[4] range (g/kg DW)	Interpretation possibilities
Na	15– 1000	+ CsCl[2]	0.2 – 2	Na:K salt-water influence
K	40– 2000	+ CsCl[2]	0.5 – 5	Illite indication
Ca	80– 5000	+ La$_2$O$_3$[3]	5 – 20	CaCO$_3$ indication
Mg	7– 500	+ La$_2$O$_3$[3]	0.5 – 20	Ca:Mg salt-water influence
Fe	120– 5000	Ni + HNO$_3$	5 –300	Redox conditions
Mn	55– 3000		0.2 – 3	Redox conditions
Al	1– 500	Fe	10 – 40	Acidification indication
Cu	90– 5000		0.005– 0.1	Redox conditions
Zn	18– 1000		0.05– 0.5	Pollution indication
Cd	25– 2000		0.2 – 5[5]	Pollution indication
Cr	100– 5000		0.01– 0.3[5]	Pollution indication
Pb	150–20000		10 –500[5]	Pollution indication
Co	150– 5000		0.2 – 20[5]	N$_2$ fixation
Ni	150– 5000		5 –100[5]	
Cs	200–15000		0.3 – 25[5]	
Mo	500–40000		0.3 – 3[5]	

[1] The lowest figure is the concentration in the sample solution giving 1% absorption with air-acetylene flame according to Perkin–Elmer manual. Concerning graphite tube see the text.

[2] To 9 ml sample solution 1 ml CsCl solution is added (6 g CsCl p.a. dissolved in 1 litre deionized water).

[3] To 10 ml sample solution 0.5 ml La$_2$O$_3$ solution is added (dissolve 29.3 g La$_2$O$_3$ in 125 ml concentrated HCl, dilute to 500 ml with deionized water).

[4] Concentration ranges generally encountered. The figures are taken from the literature and our own experiences.

[5] Ranges expressed in mg/kg DW instead of g/kg DW.

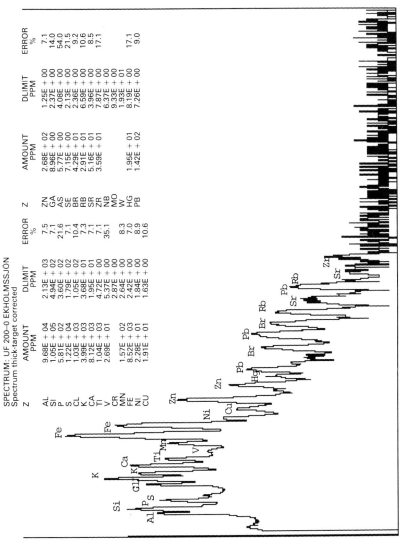

FIGURE 21.11. Results from a PIXE-analysis of a lake sediment sample

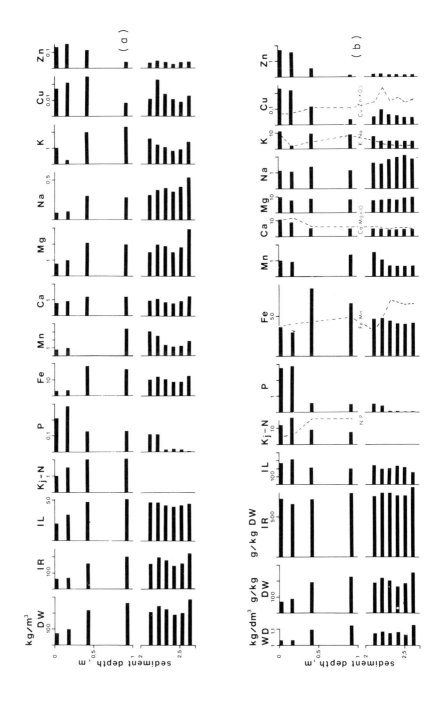

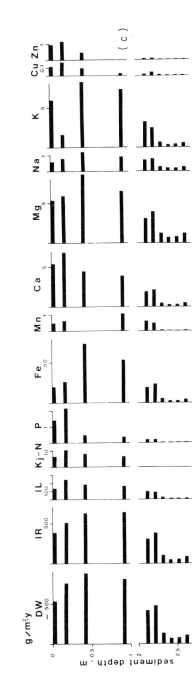

FIGURE 21.12

(a) Wet density (WD), dry weight per wet sediment (DW), ignition residue at 550°C (IR), organic matter (IL), Kjeldahl-N (K_j-N), P, Fe, Mn, Ca, Mg, Na, K, Cu and Zn per dry weight. Sediment core from Lake Järlasjön, Stockholm (Digerfeldt et al., 1975).

(b) Dry weight (DW), ignition residue at 550°C (IR), organic matter (IL), Kjeldahl-N (K_j-N), P, Fe, Mn, Ca, Mg, Na, K, Cu and Zn per volume wet sediment.

(c) Annual accumulation per m^2. Dry weight (DW), ignition residue (IR), organic matter (IL), Kjeldahl-N (K_j-N), P, Fe, Mn, Ca, Mg, Na, K, Cu and Zn.

In more recent years instruments for multi-element analysis have been developed for routine analysis of water, sediment, peat and soil as well as for biological materials. Although these instruments are quite expensive, the cost for analysing several elements can be cheaper than with the AA-technique. The PIXE (Particle Induced X-ray Emission) and OES-ICP (Optical Emission Spectroscopy-Inductive Coupled Plasma) methods give for most elements a sensitivity as good as or better than that of the flame AA-method.

In Figure 21.11 the results from a PIXE analysis of lake sediment are illustrated. Except for freeze-drying and homogenizing no sample preparation is required, and in 1–3 minutes the analysis of, for example, pelletized material, 20 to 30 elements heavier than aluminium can be determined simultaneously. Despite the requirements for expensive experimental equipment (a particle accelerator) these properties make the technique quite economical. The PIXE method is a non-destructive analytical technique described by Johnsson and Johnsson (1976) and Mangelson and Hill (1981). Soil and plant analyses have been described by Stanford *et al.* (1975) and Jahnke *et al.* (1981).

PRESENTATION OF RESULTS

The results of chemical analyses should be presented in such a manner that both qualitative and quantitative differences can be easily examined. Usually the results are presented either per dry weight or per wet volume. If so, bulk density and dry weight should be presented, so that the reader can convert the results properly.

For sediment, peat and soil profiles it is recommended that the results be presented as in Figure 21.12. This figure shows the effect of expressing the results in different ways. For use in palaeoecological interpretations the presentation of annual accumulation per square metre is most valuable, but a figure or a table with the results expressed per dry weight should be included in the presentation. Usually results expressed per fresh weight or per volume give insufficient information.

For elements mainly related to either organic or inorganic matter, it can also be useful to express these on an ignition-loss basis (for example Kj-N, chlorophyll and carotenoids etc.) or on ash-content basis (for example Ca, Mg and Fe). Heavy metals and P can also be usefully expressed in relation to organic matter.

Sometimes it can be valuable to express one element in relation to another by the quotient. For example Fe:Mn changes, as well as Cu:Zn changes, in a sediment profile can give information on changed redox conditions in the near-bottom water (Hallberg, 1972). Changes in the quotients of Na:K and Ca:Mg can give information on salt-water influence, and N:P relations can give information on the nutrient status (Digerfeldt *et al.*, 1975; Bengtsson and Persson, 1978; Figure 21.13).

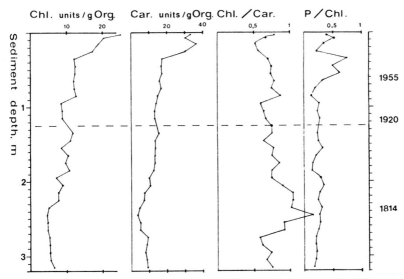

FIGURE 21.13.   Chlorophyll derivatives (Chl.) and charotenoids (Car.) in the sediment profile of Lake Södra Bergundasjön, Växjö (Bengtsson and Persson, 1978)

Finally, the conclusions drawn from the results of chemical analyses must, of course, be checked and confirmed by pollen and microfossil analyses as well as by comparison with known historical events.

REFERENCES

Ahlgren, I., and Ahlgren, G. (1975). *Methods of Water Chemical Analyses compiled for Instruction in Limnology*. University of Uppsala, Dept. of Limnology, mimeographed.

Allen, S. E., Grimshaw, H. M., Parkinson, J. A., and Quarmby, C. (1974). *Chemical Analysis of Ecological Materials* (Ed. S. E. Allen), Blackwell, Oxford.

Baltic Marine Biologist. (1976). *Recommendations on Methods for Marine Biological Studies in the Baltic Sea* (Eds. B. I. Dybern, H. Ackefors, and R. Elmgren), University of Stockholm, Dept. of Zoology.

Bengtsson, L., and Persson, T. (1978). Sediment changes in a lake used for sewage reception. *Pol. Arch. Hydrobiol*, **25**, 17–33.

Boström, B., Jansson, M., and Forsberg, C. (1982). Phosphorus release from lake sediments. *Arch. Hydrobiol. beih. Ergebn. Limnol.*, **18**, 5–59.

Bouma, A. H. (1969). *Methods for the study of sedimentary structures*, Wiley, New York.

Cato, I. (1977). Recent sedimentological and geochemical conditions and pollution problems in two marine areas in south western Sweden. *STRIAE*, University of Uppsala, **6**, 1–158.

Chaney, A., and Marbach, E. P. (1962). Modified reagents for determination of urea and ammoinia. *Clin. Chem.*, **8**, 130–132.

Chang, S. C., and Jackson, M. L. (1957). Fractionation of soil phosphorus. *Soil Sci.*, **84**, 133–144.

Dean, L. A. (1938). An attempted fractionation of the soil phosphorus. *J. Agr. Sci.*, **28**, 234–246.

Digerfeldt, G. (1972). The post-glacial development of Lake Trummen. Regional vegetation history, water level changes and palaeolimnology. *Folia Limnol. Scand.*, **16**.

Digerfeldt, G., Battarbee, R. W., and Bengtsson, L. (1975). Report on annually laminated sediment in Lake Järlasjön, Nacka, Stockholm. *Geol. Fören. Stockh. Förh.*, **97**, 29–40.

Engstrom, D. R., and Wright, Jr., H. E. (1984). Chemical stratigraphy of lake sediments as a record of environmental change. In: *Lake Sediments and Environmental History*. (Eds. E. Y. Haworth and J. W. G. Lund), Leicester University Press, pp. 11–67.

Fogg, G. E., and Belcher, J. H. (1961). Pigments from the bottom deposits of an English lake. *New Phytol.*, **60**, 129–142.

Gorham, E., and Sanger, J. E. (1975). Fossil pigments in Minnesota lake sediments and their bearing upon the balance between terrestrial and aquatic inputs to sedimentary organic matter. *Verh. Internat. Verein. Limnol.*, **19**, 2267–2273.

Griffiths, M., and Edmondson, W. T. (1975). Burial of oscillaxanthin in the sediment of Lake Washington. *Limnol. Oceanogr.*, **14**, 317–326.

Hallberg, R. O. (1972). Sedimentary sulfid mineral formation: an energy circuit system approach. *Miner. Deposita*, **7**, 189–201.

Hieltje, A. H. M., and Lijklema, L. (1980). Fractionation of inorganic phosphorus in calcareous sediments. *J. Environ. Qual.*, **9**, 405–407.

Hopkins, T. L. (1964). A survey of marine bottom samplers. In: *Progress in Oceanography*, Vol. *II* (Ed. M. Sears), Pergamon-Macmillan, New York, pp. 213–256.

Håkansson, L. (1981). *A Manual of Lake Morphometry*, Springer, New York.

Håkansson, L. (1983). *On the Relationship between Lake Trophic Level and Lake Sediments*. Manuscript, Natl. Swed. Environ. Prot. Board, Uppsala.

Håkansson, L., and Jansson, M. (1983). *Principles of Lake Sedimentology*. Springer Verlag, New York.

Jahnke, A., Shimmen, T., Koyama-Ito, H., and Yamazaki, T. (1981). PIXE trace element analysis of aquatic plant as indicator of heavy metal pollution. *Chemosph.*, **10**, 303–312.

Johnsson, S. A. E., and Johnsson, T. B. (1976). Analytical applications of particle induced X-ray emission. *Nucl. Instr. and Meth.*, **137**, 473–516.

Kemp, A. L. W., Thomas, R. L., Dell, C. I., and Jaquet, J.-M. (1976). Cultural impact on the geochemistry of sediments in Lake Erie. *J. Fish. Res. Board Can.*, special issue, **33**, 440–462.

Lundqvist, G. (1942). Sjösediment och deras bildningsmiljö. *Sver. Geol. Unders.*, Series C, **444**, 1–126.

Mackereth, F. J. J. (1966). Some chemical observations on post-glacial lake sediments. *Phil. Trans. R. Soc. London*, **250**, 167–213.

Mangelson, N. F., and Hill, M. W. (1981). Recent advances in particle induced X-ray emission analysis applied to biological samples. *Nucl. Instr. and Meth.*, **181**, 243–254.

Murphy, J., and Riley, J. P. (1962). A modified single solution method for the determination of phosphate in natural waters. *Anal. Chim. Acta.*, **27**, 31–36.

Olsen, S. (1967). Recent trends in the determination of orthophosphate in water. In: Proceedings of an I.B.P. symposium held in Amsterdam and Niewersluis, 10–16 October 1966. N. V. NoordHollandsche uitgevers maatschappij, Amsterdam, pp. 63–105.

Sanger, J. E., and Crowl, G. H. (1979). Fossil pigments as a guide to the palaeolimnology of Browns Lake, Ohio. *Quat. Res.*, **11**, 342–352.

Sly, P. G. (1969). Bottom sediment sampling. In: *Proc. 12th Conf. Great Lakes Res.*, Ann Arbor, pp. 883–898.

Stanford, J. M., Willis, R. O., Walter, R. L., Gutkneckt, W. F ., and Antonivics, J. (1975). Proton-induced X-ray emission analysis — a promising technique for studying the metal content of plants and soils. *Rad. and Environm. Biophys.*, **12**, 175–180.

Vallentyne, J. R. (1956). Epiphasic carotenoids in post-glacial lake sediments. *Limnol. Oceanogr.*, **1**, 252–262.

Vallentyne, J. R. (1960). Fossil pigments. In: *Comparative Biochemistry of Photoreactive Systems* (Ed. M. B. Allen), Academic Press, New York, pp. 83–105.

Williams, J. D. H., Syers, J. K., and Walker, T. W. (1967). Fractionation of soil inorganic phosphate by a modified Chang and Jackson's procedure. *Proc. Soil. Sci. Soc. Amer.*, **31**, 736–739.

Williams, J. D. H., Syers, J. K., Harris, R. F., and Armstrong, D. E. (1971). Fractionation of inorganic phosphate in calcareous lake sediments. *Proc. Soil Sci. Soc. Amer.*, **35**, 250–255.

Williams, J. D. H., Jaquet, J.-M., and Thomas, R. L. (1976). Forms of phosphorus in the surficial sediments of Lake Erie. *J. Fish. Res. Board Can.*, **33**, 413–429.

Wright, H. E., Cushing, E. J., and Livingstone, D. A. (1965). Coring device for lake sediments. In: *Handbook of Palaeontological Techniques* (Eds. B. Kummel and D. Raup), Freeman, San Francisco, pp. 494–520.

Züllig, H. (1961). Die Bestimmung von Myxoxanthophyll in Bohrprofilen zum Nachweis vergangener Blaualgenentfaltung. *Verh. Internat. Verein. Limnol.*, **14**, 263–270.

Biological methods

Handbook of Holocene Palaeoecology and Palaeohydrology
Edited by B. E. Berglund
© 1986 John Wiley & Sons Ltd.

22

Pollen analysis and pollen diagrams

Björn E. Berglund

*Department of Quaternary Geology,
Tornavägen 13, Lund, Sweden*

and

Magdalena Ralska-Jasiewiczowa

Institute of Botany, Polish Academy of Science, Krakow, Poland

INTRODUCTION

Pollen analysis is one of the most important palaeoecological techniques. Pollen diagrams contribute to the knowledge of biostratigraphy and chronology as well as local and regional environments. For a general description of the method, its application and errors, reference may be made to several textbooks (e.g. Faegri and Iversen, 1964, 1975, new edition in prep; Moore and Webb, 1978; Birks and Birks, 1980). For pollen morphology, see also Punt and Clarke (1976–).

This chapter presents some recommendations which are of importance to achieve reliable and uniform material for biostratigraphical and environmental correlations. The pollen-analytical information from a palaeoecological reference site will be of fundamental importance to specialists within different branches, and it should therefore be treated and presented in a simple and uniform way. Also, for correlations using time–space transects or a mapping technique a uniform treatment of pollen data is important.

Selection of reference sites has been discussed in Chapter 4, where a pollen-analytical representativity test has been proposed. Field sampling has been described in Chapter 8. General subsampling and dating strategies have been described in Chapters 4–6.

LABORATORY TECHNIQUE

The laboratory technique will not be discussed here in detail. It normally includes the following main steps:

(1) subsampling of a standard volume for later calculation of concentration/influx values;
(2) dispersion of sediment (deflocculation) by mechanical or chemical treatment. Most common chemical dispersants are hydroxides (KOH, NaOH), or sodium pyrophosphate;
(3) chemical removal of extraneous matter, such as calcium carbonate (by using HCl), humic acids (KOH), pyrite (HNO_3), silica (HF);
(4) if needed, mechanical removal of coarse particles, by sieving, or by gravity separation with heavy liquids (most common are $ZnCl_2$, KJ with CdJ_2, but others are in use);
(5) removal of cellulose (acetolysis);
(6) staining (e.g. with safranin, basic fuchsin);
(7) mounting (glycerine or silicone oil).

Alternative laboratory techniques are described by the flowcharts in Figure 22.1. These are abbreviated descriptions. More complete versions can be obtained from the authors.

Alternative A

The general treatment used in Lund follows Faegri and Iversen (1964), later slightly revised by Berglund (1966 a, p. 31). Sediments very rich in minerogenic matter are treated by the $ZnCl_2$ concentration method as described by Björck *et al.* (1978). Such treatment is often needed when dealing with late-glacial deposits.

Pollen concentration values are obtained by adding *Lycopodium* tablets to a specific volume of sediment. This technique was developed by Stockmarr (1971, 1973). These tablets are distributed for sale by the laboratory in Lund (orders may be sent to B. E. Berglund). The tablets contain about 10,000 spores. The number of tablets needed for each sample depends on the concentration of fossil grains. Our experience is that the relation of counted fossil pollen/counted exotic spores should be about 1:1 (Regal and Cushing, 1979). This means that 1–4 tablets may be used. Normally our sediment volume is 1–2 cm^3. The *Lycopodium* tablets contain spores of *Lycopodium clavatum*. Although they are dark-coloured, it may be difficult to separate them from fossil spores of this species. Other exotic particles may, therefore, be preferred. One alternative would be the use of plastic microspheres of suitable size and density (see Chapter 26).

Alternative B

This description following a report by H. J. B. Birks also includes the possibility of calculating pollen concentration, but the method of adding a known volume of suspended exotic pollen is applied here (Benninghoff, 1962). The description includes a treatment to remove pyrite. The mounting medium used is silicone oil according to the method described by Andersen (1960).

IDENTIFICATIONS — NOMENCLATURE AND ACCURACY

The taxonomic nomenclature in pollen diagrams should follow international standards, e.g. in Europe the Flora Europaea nomenclature is to be recommended. Palaeoecological interpretations are dependent on the accuracy of palynological identifications. This accuracy should be stated by applying a standardized nomenclature. The same set of conventions described by Birks (1973, pp. 225–226) is recommended, and his definitions are therefore cited here:

Gramineae (=*Poaceae*)	Family determination certain, types or subgroups undetermined or indeterminable.
Thalictrum	Genus determination certain, types or subgroups undetermined or indeterminable.
Plantago lanceolata	Species determination certain.
Sedum cf. *S. rosea*	Genus determination certain, species identification less certain because of imperfect preservation of fossil grain or spore, inadequate reference material, or close morphological similarity of the grain or spore with those of other taxa. In each case, the reason should be explained in the notes on the determination.
Plantago major/P. media	One fossil type present; only two taxa are considered probable alternatives, but further distinctions are not possible on the basis of pollen or spore morphology alone. In view of their modern ecology and/or distribution, the occurrence of both taxa is considered equally likely.
Angelica type	One fossil type present, three or more taxa are possible alternatives, but further distinctions are not possible on the basis of pollen or spore morphology alone. The selection of

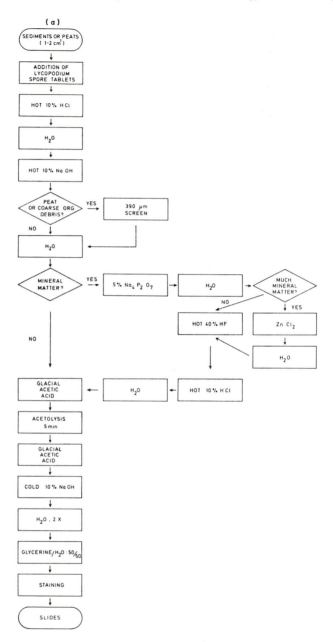

FIGURE 22.1.   Flowcharts showing the laboratory technique for pollen preparation. Two alternatives are presented and discussed in the text

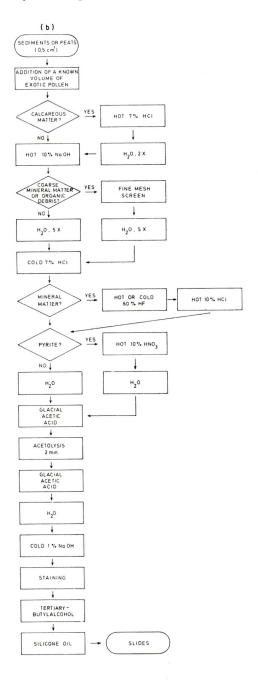

the taxon name is based on modern ecological and/or phytogeographical criteria. The notes on the determination should list all the known possibilities.

Rosaceae undifferentiated (undiff.)

Family determination certain, some morphological types distinguished and presented separately. Curve represents fossil grains or spores that were not or could not be separated beyond family level.

Stellaria undiff.

Genus determination certain, some morphological types distinguished and represented separately. Curve represents fossil grains or spores that were not or could not be separated beyond genus level.

The basis for identification down to the level of species or other taxonomic units within problematic genera should be explained by a synthetic pollen-morphological discussion (Andersen, 1961; Berglund, 1966a; Birks, 1973). In some cases the species identification by size measurements of pollen grains is possible. In such situations size–frequency curves should be presented and their interpretation discussed (Andersen, 1961; Birks, 1973). Modern statistical methods for size differentiation of species has been developed by e.g. Usinger (1978) and Prentice (1981), who both applied these on *Betula* (cf. also Birks and Gordon, 1985). — The accuracy of identifications is also dependent on the reference collection of recent pollen.

For judging the accuracy of a pollen spectrum it is recommended that unidentified pollen grains are also classified. This information also has a bearing on the depositional conditions, in lacustrine as well as terrestrial environments (Lowe, 1982; see also Chapter 7). Unidentified pollen grains may be classified into the following categories (Cushing, 1967a; Delcourt and Delcourt, 1980):

Unknown

Indeterminable
$\left\{ \begin{array}{l} \text{Corroded} \\ \text{Degraded} \\ \text{Broken or crumpled} \\ \text{Concealed} \end{array} \right.$

Definitions of the deterioration classes are given in Table 22.1. However, the values of deterioration obtained in this way refer to indeterminable grains only. In some cases it is useful to have an estimate of the corrosion effect, even though it is quite possible to identify the grains. The frequency of corroded pollen grains with smooth, psilate surfaces may then be counted (mainly following Jörgensen, 1963) — e.g. number of corroded *Alnus*, *Betula*, *Corylus*, Gramineae.

TABLE 22.1. Classes of deteriorated pollen according to Delcourt and Delcourt (1980). *Reproduced with permission of authors*

List of major deterioration classes	Description of major classes	Primary process(es) responsible for pollen deterioration
Corrosion	Exine locally etched, pitted or perforated	Biochemical oxidation related to localized fungal and bacterial activity (*e.g.* exine perforation by fungal rhizoids; Goldstein, 1960); secondary, pitting can be caused by chemical oxidation (Brooks and Elsik, 1974)
Degradation	Exine thinned, fusion of sculptural features, or fusion of structural elements forming the wall layers	Chemical oxidation within aerial and subaerial environments
Mechanical Damage	Grain broken or crumpled	A. Physical transport of the pollen grains, and B. syn-depositional and post-depositional compaction of pollen grains within the sediments (particularly during the interval of increased sediment packing resulting in the progressive extrusion of water from
Concealment by Authigenic Minerals	Grain infilled with crystals of pyrite or marcasite	Diagenesis with precipitation of authigenic minerals (*e.g.* precipitation of pyrite and marcasite within chemically-reducing alluvial environments)
Concealment by Detritus	Grain hidden by mineral or organic debris	Detrital concealment of pollen grains is related to: A. mineralogical composition of the sediment (*e.g.* relative abundance of chemically- resistant grains of minerals such as zircon, tourmaline, and garnet), and B. the chemical schedule used to extract the pollen residue in the laboratory

CALCULATION OF RESULTS

The following terminological definitions follow Davis (1969; cf. Birks and Birks, 1980, p. 207).

Pollen sum. To reduce the statistical errors in the calculation of pollen spectra the minimum pollen sum should be 500 tree pollen in forested areas and 500 tree and non-tree pollen in open areas (see Chapter 4). However, when dealing with human-influenced landscapes it is recommended that the analyses be extended to at least 1000 pollen per sample; otherwise important but rare indicator species will not be represented in sufficient numbers. Our experience of Late Holocene pollen diagrams is that a pollen sum of 2000 will facilitate the identification of human impact (cf. 1) Digerfeldt, 1972.

Pollen percentages. The traditional procedure is to express the pollen countings in percentages. This is also the most reliable method when the sediment cores represent irregular deposition, as, for example, at sites close to lake shores, or in fen deposits. The calculation sum for different parts of the pollen diagram is stated in the section on pollen diagrams.

Pollen concentration (P_{conc}) *and pollen influx* (P_{influx}). The development of pollen analysis in recent times includes the calculation of pollen concentration per unit volume of wet sediment and pollen accumulation rate (pollen influx) per unit area of sediment surface per unit time. The standard units are grains/cm^3 and grains/cm^2 per year respectively. This presupposes an exact determination of the sediment accumulation rate (v) in cm per year. However, it is not always possible to obtain reliable datings. But pollen concentration values alone will also give some additional palaeoecological information, for example, about redeposition and sedimentological changes in a basin. The absolute pollen values are calculated in the following way when using exotic spores:

$$P_{conc} = \frac{\text{Spores added}}{\text{Spores counted}} \times \frac{\text{Fossil } P}{\text{Volume cm}^{-3}, \text{grains, cm}^{-3}}$$

$$P_{influx} = P_{conc} \times v(\text{grains/cm}^2/\text{year})$$

These calculations are made automatically when using the POLLDATA program (see Chapter 37). Errors in the absolute calculations when using exotic spores have been estimated by Regal and Cushing (1979) and Maher (1981).

Sediment accumulation rate. The calculation of the sediment accumulation rate is fundamental for the construction of a pollen influx diagram. This has to be based on a series of radiocarbon dates (or another dating method). The dates should be plotted against depth in a diagram like that of Figure 22.2. When sediments younger than about 7000 ^{14}C years are dealt with, the dates may be calibrated to calendar years according to methods described in Chapter 14. The actual calculation of the accumulation rate can be performed in two main ways:

(1) by joining the points for the mean age and the mean depth of each dated sample. This was done in the example of Figure 22.2(*a*). This may be the most accurate method when only a few dates are available. However, this assumes a high degree of accuracy in each date. When the density of dates is greater, irregularities in the constructed line may occur and a second method will give more reliable values for the accumulation rate;
(2) by constructing a curve which fits to the dated levels. This can be done mathematically by constructing a polynomial curve (Davis, 1967; Digerfeldt, 1972) or graphically (Digerfeldt, 1972). A graphically constructed curve is demonstrated in Figure 22.2(*b*).

POLLEN DIAGRAMS

Pollen-analytical information may be presented and ordered with regard to several factors, i.e. the local or regional origin of pollen grains, the different ecology of the pollen producers, the deposition environment (lake or bog), or the immigration order. The aim of a study will influence the way of presenting the information. It is extremely difficult to construct a diagram type which meets these different demands and which is applicable in different geographic regions. Many pollen analysts therefore prefer the more neutral way of ordering taxa after immigration. The proposals here are based on experience in Sweden (cf. papers by Berglund, Digerfeldt, Königsson, Engelmark and others), which was originally influenced by the Danish system (Iversen, Troels-Smith, Andersen, Jörgensen). Ecological grouping of plants is based on the life-form system described by Iversen (1936) and applied by Troels-Smith (1954). This method follows the indicator species approach (Birks and Birks, 1980) which means that modern ecological and sociological preferences are traced backwards in time.

As an example we have chosen a diagram from a late-glacial lake sequence in southern Sweden (Berglund *et al.*, 1984; cf. Berglund, 1971). However, we have been inspired by the diagram technique of, for example, Birks and Mathewes (1978), Birks (1981), and others. Our example in Figures 22.3 and 22.4 illustrates 'main pollen diagrams' from a reference site (cf. Chapter 4). Later on we will briefly mention some examples of 'special pollen diagrams'.

Pollen percentage diagram

Normally pollen percentages are plotted against depth. However, for reference sites it may be appropriate to plot against time. When reliable sediment accumulation rates have been calculated (see above) it is an advantage for correlations between different diagrams to calibrate the layer sequence against the time-scale (as is done in Figures 22.3 and 22.4).

The pollen analyses will include a lot of information concerning the occurrence of pollen and other microfossils. When reproduction permits, rare elements should also be included, or at least listed in a special column (Ralska-Jasiewiczowa, 1980). The rare taxa are often of great palaeoecological importance. For publishing in some journals a selection of taxa is sometimes needed for the diagram. All raw data should be stored in tables and later transferred to computer cards (see Chapter 37).

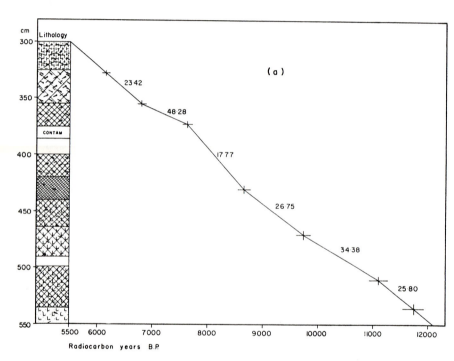

FIGURE 22.2. (a) Sediment accumulation rate curve where radiocarbon dates are plotted against depth for a profile covering the time-span 12,000–5500 B.P. Accumulation rates are calculated and indicated along the line (from H. H. Birks and Mathewes, 1978). *Reproduced by permission of The New Phytologist.* (b) Sediment accumulation rate curve where radiocarbon dates are plotted against depth for a profile covering the time span 12,700–9300 B.P. The thin line runs through all radiocarbon dates. It demonstrates different dating errors. The heavy line corresponds to a more probable sediment accumulation rate curve and the estimated time-scale in the pollen diagrams of Figures 22.3 and 22.4 is based on this curve (from Berglund *et al.*, in preparation). Arrows indicate levels where the accumulation rate changes

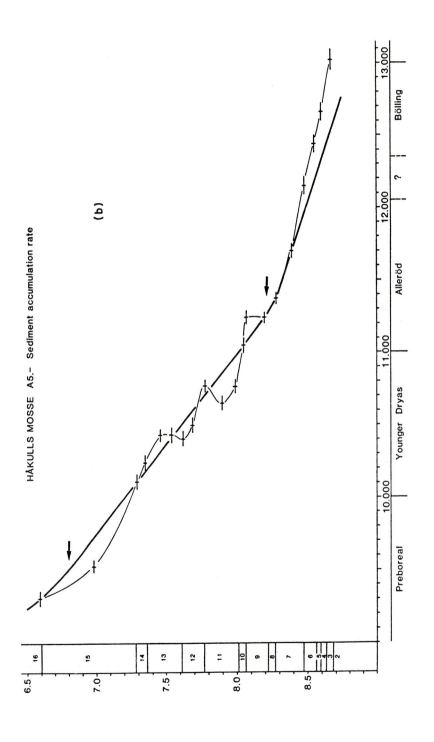

HÅKULLS MOSSE A5.- Sediment accumulation rate

(b)

The description of the pollen diagram below refers to sites within temperate forest regions. Referring to Figure 22.3, it is proposed that the information be presented in columns following the same order as below:

(1) Pollen influx curve, i.e. total amount of terriphytic spermatophytes used for the main pollen sum (ΣA), expressed as grains/cm^2 per year. If this curve shows little change it suggests that it is not worth presenting a separate pollen influx diagram.

(2) Pollen concentration curve, i.e. ΣA expressed as grains/cm^3.

(3) Radiocarbon dates, i.e. dates given in uncorrected radiocarbon years B.P.

(4) Estimated time-scale, i.e. radiocarbon years interpolated from dated levels — when the diagram is drawn against time-scale.

(5) Core segments.

(6) Depth (cm) below surface.

(7) Layer numbers, i.e. numbers of each layer referring to the stratigraphic description.

(8) Lithology, i.e. sediment column described by using the symbols of the Troels-Smith system (1955; see Chapter 12). When specific layers, such as tephra layers, occur, these ought to be indicated in an extra column and given names (and possibly also dates).

(9) Pollen analysed levels indicated by horizontal lines.

(10) Pollen diagram, divided into the main parts A–K, which are further subdivided. Here, at least, it is convenient to use these letter designations.

Diagram A. Terriphytic spermatophytes. Pollen diagram illustrating the occurrences of terrestric higher plants, primarily grouped according to main life forms. Pollen sum, ΣA = pollen of all trees, shrubs, dwarf-shrubs, herbs and graminids — without transformations for different pollen production and dispersal. Pollen of plants with local over-representation may be excluded from the pollen sum. This is valid for e.g. Cyperaceae in overgrown lakes and in fens and bogs, Ericaceae in bogs. Values for such plants are given in Diagram D.

A1: A total diagram, a cumulative area diagram, showing the proportions between the following plant groups:

   (*a*) Trees. Within this diagram area some tree pollen curves may be drawn such as *Betula* undiff., *Pinus, Quercus, Fagus* and *Picea.* Conventional symbols are to be found in pollen-analytical textbooks.

   (*b*) Shrubs such as *Salix* undiff., *Betula nana, Hippophaë, Rubus* undiff., *Corylus, Viburnum, Rhamnus, Myrica.*

   (*c*) Dwarf-shrubs, i.e. mainly dwarf-shrubs of open heaths such as *Calluna, Empetrum,* Ericaceae undiff., sometimes *Salix herbacea* type (and other indentified dwarf-willows).

(*d*) Herbs and graminids, i.e. non-lignose plants excl. limnophytes, telmatophytes and amphiphytes (cf. definitions by Iversen, 1936). Pollen identified to taxa (family, genus), where the species have different demands in relation to the water-level, are normally included in the pollen sum, e.g. Cyperaceae, Ranunculaceae. When these are distinctly over-represented they may be excluded and referred to diagram D.

A2: A series of silhouette curves, both per cent and per mil, for important species included in pollen sum *A*, grouped according to the subdivision mentioned above.

(*a*) Trees with silhouette curves for species not shown in the total diagram A1.

(*b*) Shrubs.

(*c*) Dwarf-shrubs.

(*d*) Herbs and graminids, mentioned in the following order:

  (1) Plants which cannot with accuracy be placed into any definite ecological group, such as Cyperaceae, Gramineae undiff., Compositae undiff., Cruciferae undiff., *Ranunculus* undiff., Rosaceae undiff., Umbelliferae undiff., Rubiaceae, *Thalictrum*.

  (2) Indigenous species, not influenced by man, such as heliophilous, arctic–alpine tundra elements.

  (3) Apophyte species, such as *Plantago lanceolata*, *Rumex acetosella*, *Artemisia*, Chenopodiaceae.

  (4) Anthropochor species, such as Cerealea, *Plantago major*, *Centaurea cyanus* (though some of these may be indigenous).

  (5) Forest plants, such as *Anemone*, *Melampyrum*.

  (6) Plants which prefer rather moist habitats, occurring in tall herb meadows, upper zones of shore meadows or forest edges. They are named poly- and mesohygrobic terriphytes by Iversen (1936, p. 61). This group may be exemplified by *Filipendula ulmaria*, *Geum*, *Urtica*, *Potentilla palustris*, *Caltha*, *Valeriana*.

Within each group the taxa are presented in stratigraphical immigration order, so that pure tundra species will be found together etc. However, many apophytes frequently occur in a juvenile, arctic–subarctic landscape as well as in an agrarian cultural landscape. Alternative orders may be applied depending on the geographical area and time-span covered by the diagram. This applies, for example, to the forest plants which in a Holocene diagram may be placed close to group (2).

This diagram is terminated by two columns of figures, one for the pollen sum (ΣA, or in POLLDATA called ΣP) and one for the number of pollen types in the pollen sum (Σ Taxa). Just before the pollen sum column a Varia column may be inserted.

Diagram B. Pteridophytes. Calculation sum: $\Sigma A + \Sigma$ Pteridophytes. The diagram is composed of silhouette curves for pteridophytes such as *Equisetum*, Polypodiaceae undiff., *Gymnocarpium*, Lycopodium species etc. In late-glacial conditions it may, however, be of interest to include terrestrial pteridophytes in the total diagram (Berglund, in prep.).

Diagram C. Limnophytes, i.e. aquatics. Pollen sum: $\Sigma A + \Sigma$ Limnophytes. This group is defined by Iversen (1936). The diagram is composed of silhouette curves for the limnophytes and possibly a summation curve for all limnic spermatophytes. The individual taxa may be ordered in life-form groups: isoëtiden, elodeiden and nymphaeiden (Du Rietz, 1931; Thunmark, 1931).

Diagram D. Telmatophytes and amphiphytes. Pollen sum: $\Sigma A + \Sigma$ Telmatophytes + amphiphytes. This group is defined by Iversen (1936) and corresponds to the so-called helophytic plants (Thunmark, 1931). It includes plants of lake shores, fens and bogs. The diagram is composed of silhouette curves for the telmatophytes and amphiphytes and possibly a summation curve for all these plants. In some cases Cyperaceae pollen may derive mainly from telmatophytic sedges and then they should be included in this diagram. When *Phragmites* is identified it ought to be included in the telmatophytes.

Diagram E. Bryophytes. Calculation sum: $\Sigma A + \Sigma$ Bryophytes. The diagram is composed of silhouette curves for some spore types, such as *Sphagnum*, Bryophyta undiff.

Diagram F. Fungi. Calculation sum: $\Sigma A + \Sigma$ Fungi. The diagram is composed of silhouette curves for *inter alia* Fungi undiff. However, when fungi spores are studied in more detail, e.g. for bog-stratigraphical purposes, a specific fungi diagram is needed (see Chapter 24).

Diagram G. Unidentified pollen grains. Calculation sum: $\Sigma A + \Sigma$ Unident. p.gr. The diagram is composed of silhouette curves for unknown and undeterminable pollen grains (see above).

Diagram H. Rebedded sporomorphs. Calculation sum: $\Sigma A + \Sigma$ Rebedded sp. The diagram may be composed of silhouette curves for *inter alia* Pre-Quaternary sporomorphs and trilete spores, possibly also rebedded Quaternary pollen (warmth-demanding taxa).

Diagram I. Charcoal dust. Calculation sum: $\Sigma A + \Sigma$ Charcoal. Charcoal dust particles >10 μm may be further size-classified and different silhouette curves presented. For detailed charcoal dust analysis, see Chapter 23.

Diagram K. Algae. Calculation sum: $\Sigma A + \Sigma$ Algae. Routine identifications of green algae, e.g. *Pediastrum* species, *Botryococcus* etc., may be presented as separate silhouette curves. For detailed algae analysis, see Chapter 25.

(11) Local pollen zones, named by using a letter code for the site name followed by a number in a sequence from the base upwards. Zone boundaries may be indicated by thick broken lines throughout the pollen diagram. The pollen zones are defined as site pollen assemblage zones (PAZ). The zonation can be performed in the traditional way or by using numerical methods (Chapter 37). When pollen zones from different sites have been correlated, regional pollen assemblage zones may be defined and also indicated in a separate column.

(12) Depth (cm) below surface (cf. column (6) which is repeated here).

(13) Estimated time-scale (cf. column (4) which is repeated here). A column for chronozones may be added here, but there are no general agreements concerning definitions of Late-Quaternary chronozones (Mangerud *et al.* 1982).

Pollen diagrams from arctic–subarctic regions may quite well be subdivided into two main parts: (1) local taxa, excluding aquatics and exotic plants, forming the main pollen sum; and (2) aquatics, exotic plants and other microfossils (Fredskild, 1973, 1983).

Pollen influx diagram

When reliable sediment accumulation rates have been calculated (see above), pollen influx diagrams may be constructed. For this diagram only the most abundant taxa selected and curves drawn in a manner similar to that in the percentage diagram. The subdivision and the order of curves should follow the same pattern as earlier described.

It is recommended that a specific pollen influx diagram be presented, as shown in Figure 22.4. Sometimes it is an advantage to present percentage and influx diagrams for selected taxa, side by side, to illustrate the information from the two different calculation methods. This was done when the first influx diagram was presented, namely the Rogers Lake diagrams from Connecticut, U.S.A. (Davis, 1967).

When a very close correlation between percentage values and influx values is of interest, the two series of information can be combined into one diagram. This has been done, for example, for late-glacial sites in Britain

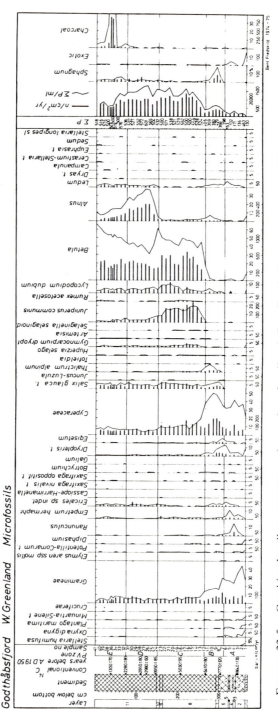

FIGURE 22.5. Combined pollen percentage (curves) and pollen influx (bars) diagram covering a Holocene sequence on western Greenland (from Fredskild, 1983). *Reproduced by permission of Kommissionen for Videnskabelige Undersøgelser i Grönland*

(Bonny, 1972; Pennington, 1977), and in the Holocene lake sequences of Greenland (Fredskild, 1983) (see Figure 22.5).

Survey cumulative pollen diagrams

The general features of the vegetational development, according to the information from one or several main pollen diagrams, may be summarized in a survey diagram where pollen of terrestrial plants are grouped in three or more ecological groups and the values plotted against a radiocarbon time-scale. Pollen-analytical information from Quaternary interglacials was treated in this way by Andersen (1964) and from Late-Quaternary sediments by Berglund (1966b) and others. The diagram from Berglund is reproduced in Figure 22.6. Another form of presenting the vegetational changes in a survey diagram was introduced by Fries (1958). It is a composite total diagram with surface symbols for different ecological plant groups. It has also been applied by others (e.g. Engelmark, 1978).

Schematic pollen diagrams

For complete Holocene or Late-Quaternary sequences schematic pollen diagrams, with silhouette curves for a selected group of taxa, are useful for general correlations between regions (e.g. Iversen, 1973; Hafsten, 1960; Fredskild, 1977). They may even be compiled from several pollen diagrams within the region in order to obtain an average picture (Iversen, 1973; see Figure 22.7). Several diagrams can also be compiled and reproduced on a smaller scale. This may be useful for survey correlations (Wright, 1971; Ritchie, 1979). In schematic diagrams it is an advantage to smooth the pollen curves. A simple method for statistical smoothing is described and applied by Usinger (1978).

Transformed tree pollen diagrams

The differences in pollen productivity and dispersal for different tree species will give a misleading picture of their abundance in a pollen diagram. Hence, transformed pollen diagrams calculated by using special representation factors for unequal pollen representation are sometimes constructed. Such representation factors have been proposed by different authors (Birks and Birks, 1980; and Chapter 39). One example is shown in Figure 22.8. They may appear useful especially where the over-representation of pollen coming from the high pollen producers provides no opportunity to reconstruct the actual development of deciduous forests. However, such a transformation can only be done accurately where careful research on present-day pollen deposition in relation to vegetation has been done in the investigated area, as in some Danish and

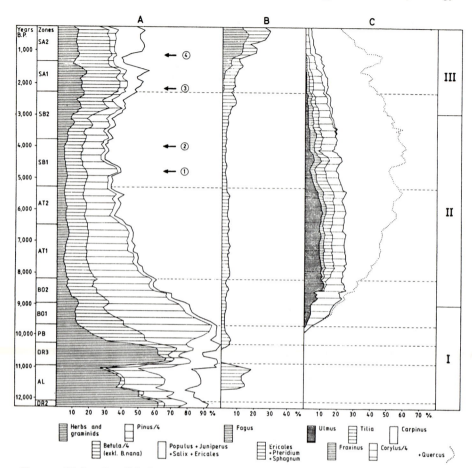

FIGURE 22.6. Simplified survey pollen diagram showing the main features in the vegetational development of the coast area in eastern Blekinge, south-eastern Sweden (Berglund, 1966b). The diagram is compiled from two sites. The terrestric spermatophytes form the basic sum. *Alnus, Betula, Corylus* and *Pinus* are divided by four. The percentages given are means of three samples. Otherwise the diagram is constructed according to the scheme used by Andersen (1964) for interglacial sequences in Jutland. Part A comprises plants which do not tolerate well competition and shade, part B comprises plants indicative of acid humus or peat (raw-humus), part C comprises plants characteristic of fertile soils (mull). I = protocratic stage; II = mesocratic stage; III = oligocratic stage. Arrows in part A indicate the prehistoric stages of increased human influence. This should be compared with Figure 22.9

Polish forests (Andersen 1970, 1973; Dąbrowski, 1975). A simplified transformation has earlier been made in some schematic diagrams like those of Figures 22.6 and 22.7, but this procedure is not accurate enough (see Chapter 39).

Combined local and regional pollen diagrams

Following the research strategy of Chapter 4 it is very important to obtain local and regional pollen diagrams from a palaeoecological reference area. One way of illustrating similarities and differences between two such diagrams is shown in a study by Andersen (1978), where the values for the dominant trees at two sites are plotted together. That diagram is here reproduced in Figure 22.8. Another method when comparing paired sites was applied by Jacobson (1979). By matching two diagrams and subtracting the pollen influx at one site from the corresponding influx at another site, a difference diagram was constructed. It is also possible to make such numerical correlations when only percentage values are available.

Human impact diagrams

For correlation of changes in the cultural landscape between different regions, a specially designed pollen diagram type is needed. When discussing the human impact on the Scandinavian landscape, Berglund (1969) used a simplified diagram type where selected taxa were subdivided into six main ecological groups (Figure 22.9):

(a)   trees on damp soils, e.g. *Alnus*;
(b)   high-competitive and shade-tolerant trees, ordered according to their light demand;
(c)   Trees of the same ecological group but immigrating during late Holocene, e.g. *Fagus, Picea*;
(d)   Low-competitive and light-demanding trees, often belonging to the primary stage in Holocene forest successions;
(e)   apophytes, i.e. shrubs, herbs and graminids favoured by man;
(f)   anthropochors, i.e. herbs and graminids introduced by man.

A similar diagram type has been used by other authors (e.g. Ralska-Jasiewiczowa, 1977). Stages of human expansion can be interpreted from such diagrams and indicated in a specific column or diagram (Berglund, 1969; Königsson, 1968; Hjelmroos-Ericsson, 1981).

Classifying the plants into these groups has to be discussed carefully for each region and related to the knowedge of indicator species for different land use in the present or historical landscape. This was done for the central European lowland by Behre (1981), which made it possible to propose pollen indicators for different kinds of pastures, open fields etc. The authors of this chapter have proposed a modified grouping (Berglund and Ralska-Jasiewiczowa, in prep.):

(a)–(d)   trees and shrubs ordered from shade-tolerant taxa on more or less damp soils to less competitive, light-demanding taxa on more or less dry soils (as above);

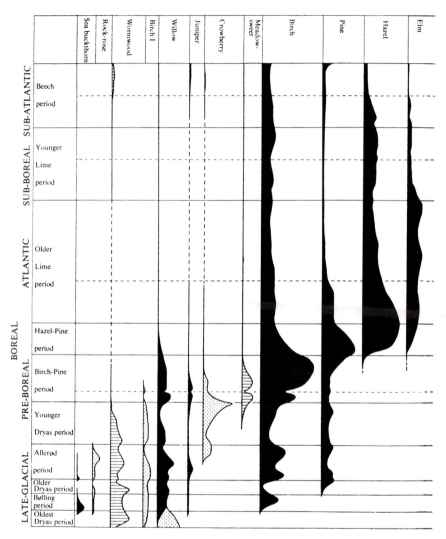

FIGURE 22.7.   Schematic pollen diagram where the curves are smoothed. Pollen data
        from Lake Bølling in West Jutland, Denmark (from Iversen, 1973)

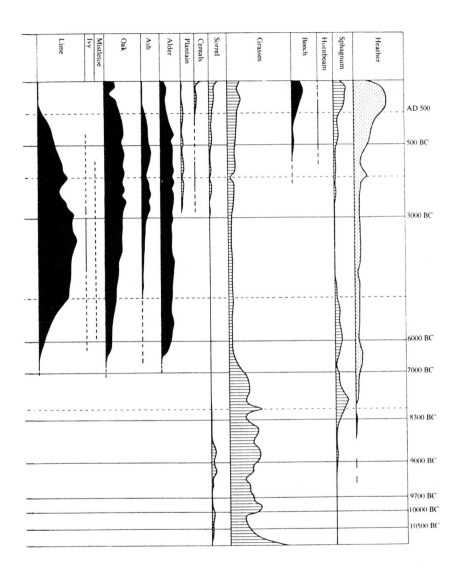

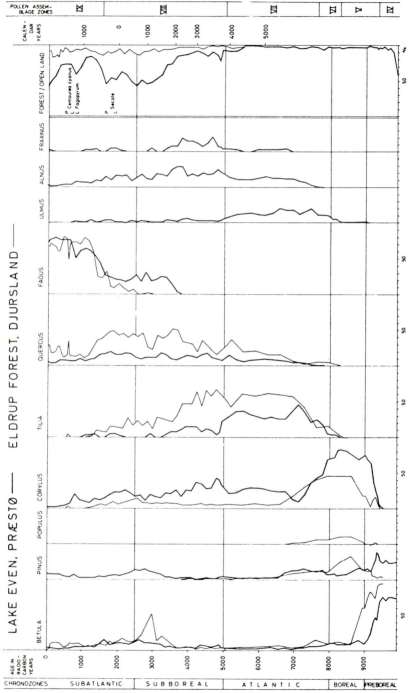

FIGURE 22.8. Combined local forest pollen diagram (Eldrup Forest, Jutland) and regional lake pollen diagram (Lake Even, Praestø, Zealand). Tree pollen frequencies are corrected according to Andersen (1970, 1973). Timescale calibrated (from Andersen, 1978)

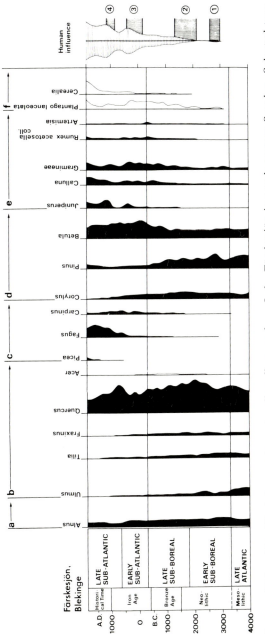

FIGURE 22.9. Simplified, human-impact pollen diagram from Lake Färskesjön in south-eastern Sweden. Selected taxa subdivided into six ecological groups of importance for the interpretation of the human impact. Human expansion phases indicated in the column to the right (from Berglund, 1969). *Reproduced by permission of Oikos*

WORYTY 80, NE Poland (Ralska – Jasiewiczowa)

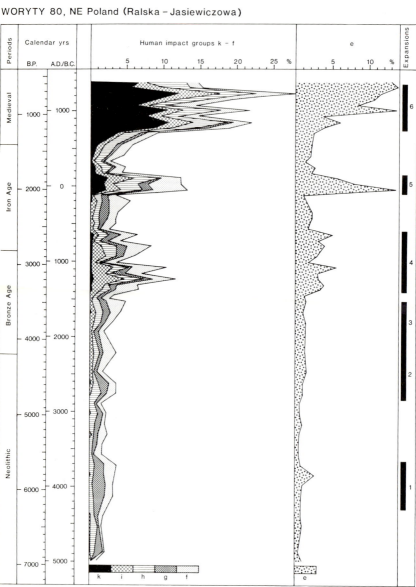

FIGURE 22.10. Synthetic human impact pollen diagram from Lake Woryty in north-eastern Poland. Apophytes and anthropochors grouped according to Behre (1981): k = cultivated land; i = fresh meadows; h = dry pastures; g = grazed woodlands; f = ruderal communities; e = general apophytes. Time-scale calibrated.

(*e*)   general apophytes;
(*f*)   plants of ruderal communities;
(*g*)   plants of grazed woodlands;
(*h*)   plants of dry pastures;
(*i*)   plants of fresh meadows;
(*k*)   plants of cultivated land. This group was subdivided by Behre into three subgroups: winter cereals, summer cereals and root crops, plants of fallow land.

An example of a synthetic, cumulative diagram for the apophytes and anthropochors of the groups (*e*)–(*k*) is given in Figure 22.10. When possible, the pollen values of all these human impact diagrams should be plotted against a calibrated radiocarbon time-scale (see Chapter 14).

POLLEN-ANALYTICAL CORRELATIONS AND SYNTHESES

Pollen-analytical correlation within basins in order to correlate stratigraphical columns is discussed in Chapter 13. Correlations of local pollen diagrams from different sites to build up regional pollen assemblage zones is discussed in Chapter 37. When several diagrams from one region are compared, a correlation may be synthesized into a table with the following headings (Birks, 1976, 1981):

Age B.P.	Local PAZ	Regional PAZ	Physiognomy of vegetation	Suggested modern analogue

Correlations between different regions can be outlined in tables similar to those proposed above (e.g. Hyvärinen, 1975) or in time–space diagrams often used as palaeovegetation cross-sections (Cushing, 1967b; Wright and Watts, 1969; Birks, 1980; see also Figure 37.4). However, the most powerful technique for correlation of numerous pollen diagrams is producing isopollen maps, and related maps, for individual taxa (e.g. Bernabo and Webb, 1977, for eastern North America; Huntley and Birks, 1983, for Europe; Ralska-Jasiewiczowa, 1983, for Poland). By defining vegetation units in pollen percentage terms it is even possible to construct palaeovegetation maps for the past (Delcourt and Delcourt, 1981; Huntley and Birks, 1983). This kind of map is here illustrated by the map sequence of Figure 22.11.

Acknowledgements

We wish to thank Dr. H. J. B. Birks for valuable comments on this chapter and for putting his pollen preparation report at our disposal. We are also grateful to

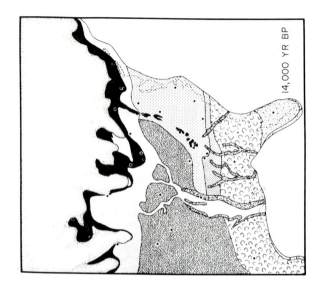

14,000 YR BP

SOUTHEASTERN EVERGREEN FORESTS

OAK-HICKORY-SOUTHERN PINE

SOUTHERN PINE

CYPRESS - GUM

SUBTROPICAL HARDWOODS

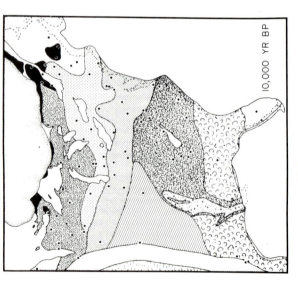

10,000 YR BP

LAURENTIDE ICE SHEET

TUNDRA

BOREAL FORESTS

SPRUCE

SPRUCE-JACK PINE

JACK PINE - SPRUCE

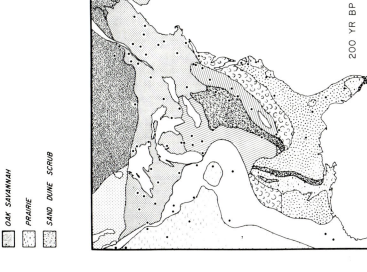
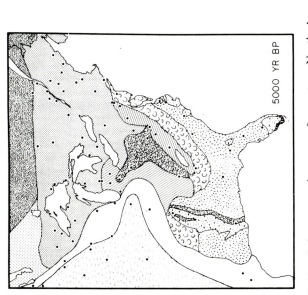

FIGURE 22.11. Palaeovegetation maps for eastern North America from 14,000 to 200 B.P. This map sequence illustrates the climatically conditioned displacements of vegetation belts (from Delcourt and Delcourt, 1981). *Reproduced by permission of Plenum Press*

Drs. H. H. Birks, B. Fredskild, S. T. Andersen, P. Delcourt and H. Delcourt who permitted us to reproduce material from their publications. Mr. T. Persson has assisted in the preparation of the pollen diagrams from Håkulls Mosse, and Miss I. Lander has redrawn some figures. We are grateful for their skilful work.

REFERENCES

Andersen, S. T. (1960). Silicone oil as a mounting medium for pollen grains. *Danm. Geol. Unders.* IV, **4**, 1–24.

Andersen, S. T. (1961). Vegetation and its environment in Denmark in the Early Weichselian Glacial (Last Glacial). *Danm. Geol. Unders.* II, **75**, 1–175.

Andersen, S. T. (1964). Interglacial plant succession in the light of environmental changes, *Rep. VI. Int. Congr. Quat. Warsaw 1961*, **2**, 359–368.

Andersen, S. T. (1970). The relative pollen productivity and pollen representation of north European trees, and correction factors for tree pollen spectra. *Danm. Geol. Unders.* II, **96**, 1–99.

Andersen, S. T. (1973). The differential pollen productivity of trees and its significance for the interpretation of a pollen diagram from a forested region. In: *Quaternary Plant Ecology* (Eds. H. J. B. Birks and R. West) Blackwell, Oxford, pp. 109–115.

Andersen, S. T. (1978). Local and regional vegetational development in eastern Denmark in the Holocene. *Danm. Geol. Unders., Årbog 1976*, 5–27.

Behre, K. E. (1981). The interpretation of anthropogenic indicators in pollen diagrams. *Pollen et Spores*, **23**, 225–245.

Benninghoff, W. S. (1962). Calculation of pollen and spore density in sediments by addition of exotic pollen in known quantities. *Pollen et Spores*, **4**, 332–333.

Berglund, B. E. (1966a). Late-Quaternary vegetation in eastern Blekinge, southeastern Sweden: a pollen-analytical study. I: Late-Glacial time. *Opera Botanica*, **12**, 1–180.

Berglund, B. E. (1966b). Late-Quaternary vegetation in eastern Blekinge, southeastern Sweden: a pollen-analytical study. II: Post-Glacial time. *Opera Botanica*, **12**, 1–190.

Berglund, B. E. (1969). Vegetation and human influence in south Scandinavia during Prehistoric time. *Oikos Supp.*, **12**, 9–28.

Berglund, B. E. (1971). Late-Glacial stratigraphy and chronology in south Sweden in the light of biostratigraphic studies on Mt. Kullen, Scania. *Geol. Fören. i Stockholm Förh.*, **93**, 11–45.

Berglund, B. E., Lemdahl, G., Liedberg-Jönsson, B., and Persson, T. (1984). Biotic response to climatic changes during the time span 13,000–10,000 B.P. — A case study from SW Sweden. In *Climatic Changes on a Yearly to Millenial Basis* (Eds. N. A. Mörner and W. Karlén). D. Reidel Publ. Comp., pp. 25–36.

Bernabo, J. C., and Webb, T., III (1977). Changing patterns in the Holocene pollen record of northeastern North America: a mapped summary. *Quat. Res.*, **8**, 64–96.

Birks, H. H., and Mathewes, R. W. (1978). Studies in the vegetational history of Scotland. V: Late Devensian and Early Flandrian pollen and macrofossil stratigraphy at Abernethy Forest, Inverness-Shire. *New Phytol.*, **80**, 455–484.

Birks, H. J. B. (1973). *Past and Present Vegetation of the Isle of Skye: a Palaeoecological Study*, Cambridge University Press, London. 415 pp.

Birks, H. J. B. (1976). Late-Wisconsinan vegetational history at Wolf Creek, central Minnesota. *Ecol. Monographs*, **46**, 395–429.

Birks, H. J. B. (1980). *Quaternary Vegetational History of West Scotland*, Guidebook for excursion C8, 5. Intern. Palynol. Conf., Cambridge.

Birks, H. J. B. (1981). Late Wisconsin vegetational and climatic history at Kylen Lake, northeastern Minnesota. *Quat. Res.*, **16**, 322–355.

Birks, H. J. B., and Birks, H. H. (1980). *Quaternary Palaeoecology*, Edward Arnold, London. 289 pp.

Birks, H. J. B., and Gordon, A. D. (1985). *Numerical methods in Quaternary pollen analysis*, Academic Press, London (in press).

Björck, S., Persson, T., and Kristersson, I. (1978). Comparison of two concentration methods for pollen in minerogenic sediments. *Geol. Fören. i Stockholm Förh.*, **100**, 107–111.

Bonny, A. P. (1972). A method for determining absolute pollen frequencies in lake sediments. *New Phytol.*, **71**, 393–405.

Cushing, E. J. (1967a). Evidence for differential pollen preservation in late Quaternary sediments in Minnesota. *Rev. Palaeobot. Palynol.*, **4**, 87–101.

Cushing, E. J. (1967b). Late-Wisconsin pollen stratigraphy and the glacial sequence in Minnesota. In: *Quaternary Palaeoecology* (Eds. E. J. Cushing and H. E. Wright) Yale University Press, New Haven, pp. 59–88.

Dąbrowski, M. J. (1975). Tree pollen rain and the vegetation of the Bialowieza National Park. *Biul. Geolog. Polon.*, **19**, 157–172.

Davis, M. P. (1967). Pollen accumulation rates at Rogers Lake, Connecticut, during Late and Postglacial time. *Rev. Palaeobotan. Palynol.*, **2**, 219–230.

Davis, M. B. (1969). Palynology and environmental history during the Quaternary Period. *Amer. Sci.*, **57**, 317–332.

Delcourt, P. A., and Delcourt, H. R. (1980). Pollen preservation and Quaternary environmental history in the southeastern United States. *Palynology*, **4**, 215–231.

Delcourt, P. A., and Delcourt, H. R. (1981). Vegetation maps for eastern North America: 40,000 B.P. to the present. In: *Geobotany II* (Ed. R. C. Romans) Plenum Publ. Corp., 123–165.

Digerfeldt, G. (1972). The post-glacial development of Lake Trummen: regional vegetation history, water level changes and palaeolimnology. *Fol. Limnol. Scand.*, **16**, 1–104.

Du Rietz, G. E. (1931). Life-forms of terrestrial flowering plants, I. *Acta Phytogeogr. Suec.*, **3**, 1–95.

Engelmark, R. (1978). The comparative vegetational history of inland and coastal sites in Medelpad, N Sweden, during the Iron Age. *Early Norrland*, **11**, 25–62.

Faegri, K., and Iversen, J. (1964, 1975). *Textbook of pollen analysis*, Munksgaard, Copenhagen.

Fredskild, B. (1973). Studies in the vegetational history of Greenland: palaeobotanical investigations of some Holocene lake and bog deposits. *Medd. om Grønland*, **198**, 1–245.

Fredskild, B. (1977). The development of the Greenland lakes since the last glaciation. *Fol. Limnol. Scand.*, **17**, 101–106.

Fredskild, B. (1983). The Holocene vegetational development of the Godthåbsfjord area, West Greenland. *Medd. om Grønland, Geoscience*, **10**, 1–28.

Fries, M. (1958). Vegetationsutveckling och odlingshistoria i Varnhemstrakten. En pollenanalytisk undersökning i Västergötland (Zusammenfassung). *Acta Phytogeogr. Suec.*, **39**, 1–64.

Hafsten, U. (1960). Pollen-analytical investigations in South Norway. In: *Geology of Norway* (Ed. O. Holtedahl), *Nor. geol. unders.*, **208**, 434–462.

Hjelmroos-Ericsson, M. (1981). Holocene development of Lake Wielkie Gacno area, northwestern Poland. *LUNDQUA Thesis*, **10**, Dept. of Quaternary Geology, Univ. of Lund, pp. 1–101.

Huntley, B., and Birks, H. J. B. (1983). *An atlas of past and present pollen maps for Europe: 0–13,000 years ago*, Cambridge University Press. 667 pp.

Hyvärinen, H. (1975). Absolute and relative pollen diagrams from northernmost Fennoscandia. *Fennia*, **142**, 1–23.

Iversen, J. (1936). *Biologische Pflanzentypen als Hilfsmittel in der Vegetationsforschung*, Munksgaard, Copenhagen.

Iversen, J. (1973). The development of Denmark's nature since the Last Glacial. *Danm. Geol. Unders.*, **V**, 7-C, 1–126.

Jacobson, G. L. (1979). The palaeoecology of white pine (*Pinus strobus*) in Minnesota. *J. Ecology*, **67**, 697–726.

Jörgensen, S. (1963). Early post-glacial in Aamosen. Geological and pollen-analytical investigations of Maglemosian settlements in the West-Zealand bog Aamosen, I *Danm. Geolog. Unders.* II, 87, 1–79.

Königsson, L.-K. (1968). The Holocene history of the Great Alvar of Öland. *Acta Phytogeogr. Suec.*, 55, 1–172.

Lowe, J. J. (1982). Three Flandrian pollen profiles from the Teith Valley, Perthshire, Scotland. II: Analysis of deteriorated pollen. *New Phytol.*, **90**, 371–385.

Maher, L. J. (1981). Statistics for microfossil concentration measurements employing samples spiked with marker grains. *Review of Palaeobotany and Palynology*, **32**, 153–191.

Mangerud, J., Birks, H. J. B., and Jäger, K.-D., Eds. (1982). Chronostratigraphical subdivision of the Holocene. *Striae* **16**.

Moore, P. D., and Webb, J. A. (1978). *An Illustrated Guide to Pollen Analysis*, Hodder and Stoughton, London. 133 pp.

Pennington, W. (1977). The Late Devensian flora and vegetation of Britain. *Phil. Trans. R. Soc. Lond.*, **B 280**, 247–271.

Prentice, I. C. (1981). Quantitative birch (*Betula L.*) pollen separation by analysis of size frequency data. *New Phytol.*, **89**, 145–157.

Punt, W., and Clarke, G. C. S. (1976–). *The Northwest European Pollen Flora*, I–. Elsevier Scient. Publ. Comp., Amsterdam.

Ralska-Jasiewiczowa, M. (1977). Impact of prehistoric man on natural vegetation recorded in pollen diagrams from different regions of Poland. *Folia Quaternaria*, **49**, 75–91.

Ralska-Jasiewiczowa, M. (1980). *Late-Glacial and Holocene Vegetation of the Bieszczady Mts. (Polish Eastern Carpathians)*, Polska Akad. Nauk, Instytut Botaniki Krakow. 202 pp.

Ralska-Jasiewiczowa, M. (1983). Isopollen maps for Poland: 0–11,000 years B.P. *New Phytol.*, **94**, 133–175.

Regal, R. R., and Cushing, E. J. (1979). Confidence intervals for absolute pollen counts *Biometrics*, **35**, 557–565.

Ritchie, J. C. (1979). Towards a Late-Quaternary palaeoecology of the ice-free corridor. In: *The Ice-Free Corridor*, AMQUA Vth Biennial Meeting Symposium, Univ. of Alberta Press, Edmonton.

Stockmarr, J. (1971). Tablets with spores used in absolute pollen analysis. *Pollen et Spores*, **13**, 615–621.

Stockmarr, J. (1973). Determination of spore concentration with an electronic particle counter. *Danm. Geol. Unders.*, *Årbog 1972*, 87–89.

Thunmark, S. (1931). Der See Fiolen und seine Vegetation. *Acta Phytogeogr. Suec.*, **II**, 1–198.

Troels-Smith, J. (1954). *Pollenanalytische Untersuchungen zu einigen Schweizerischen Pfahlbauproblemen*, København. 64 pp.

Troels-Smith, J. (1955). Characterization of unconsolidated sediments. *Danm. Geol. Unders.*, **IV**, 1–73.

Usinger, H. (1978). Bölling-Interstadial und Lacher Bimstuff in einem neuen Spätglazial-Profil uas dem Vallensgård Mose, Bornholm. Mit pollengrössenstatistischer Trennung der Birken. *Danm. Geol. Unders.*, *Årbog 1977*, 5–29.

Wright, H. E. (1971). Late Quaternary vegetational history of North America. In: *The Late Cenozoic Glacial Ages* (Ed. K. K. Turekian), Yale Univ. Press, pp. 425–464.

Wright, H. E., and Watts, W. A. (1969). *Glacial and Vegetational History of North Eastern Minnesota*, Minnesota Geological Survey Special Publ. 11. 59 pp.

23

Charred particle analysis

KIMMO TOLONEN

Department of Botany, University of Helsinki, Helsinki, Finland

INTRODUCTION

The burning of any carbonaceous material (wood, coal, oil etc.) produces a high number of carbon particles, which after being deposited can in favourable conditions be preserved in sediments throughout geological periods.

On the basis of a microscopic morphology of the preserved sedimentary charred particles, it is possible to distinguish the combustion products of fossil fuels from particles derived from forest fires (Griffin and Goldberg, 1979, 1981, 1983). Carbonaceous particles originating from oil and coal combustion may be called 'soot' (Renberg and Wik, 1983, 1984) to distinguish them from charcoal of burned biospheric material, here called 'charcoal particles'. Identification and counting of these charcoal occurrences in sediment or peat columns assists in tracing the history of air pollution and forest fires. The palaeoecological contribution of both these areas of research is very valuable owing to the lack of sufficient written documents over a long period.

Charred particle analysis when combined with a pollen record leads to an understanding of vegetational successions and their causes, the developmental history of lakes and mires and other ecological problems, both in the long term and the short term, depending on the resolution of the methods used. Under certain conditions, fires are climate-controlled because certain weather conditions initiate lightning, and fuel is especially available through biomass production (Payette, 1980). Consequently, 'natural' fires may be used as an index of palaeoclimate (e.g. Terasmae and Weeks, 1979; Singh *et al.*, 1981). At many sites, however, both historic and prehistoric man are potential causes of ignition, as already demonstrated by Iversen (1941) in his quantitative estimations of charred particles in pollen preparations from Ordrup Mose, Denmark. Later, Iversen successfully applied charcoal analyses in several studies, but surprisingly it was not until the 1960s that other scientists followed his initiative (e.g. Hutchinson and Goulden, 1966; Davis, 1967; Fredskild, 1967; Tsukada and Deevey, 1967; Waddington, 1969).

The next important advance in developing the method was the quantitative analysis of charred particles plus pollen grains, in annually laminated sediments of the Lake of the Clouds, Minnesota. The fossil record significantly correlated with known fires around the lake basin (Swain, 1973). Numerous studies in different areas carried out after this example include, among others: Bradbury and Waddington (1973), Hope and Peterson (1976), Green (1976), Mehringer *et al* (1977), Cwynar (1978), Tolonen (1978, 1985), Amundson and Wright (1979), Anderson (1979) and Huttunen (1980).

Although annual laminae were not always present, the charcoal analyses gave the most important information about the past fire regime in each area.

The essentially uniform counting methodology applied to four varved lakes (one in Minnesota, one in Ontario and two in Finland) enabled some comparison to be made (Table 23.1). For comparison, one lake from Quebec was also considered although it was not dated by varves but by means of ^{14}C datings. The differences in the fire frequency and interval figures between the localities corroborate with the present-day and dendrochronologically obtained fire information about the susceptibilty of the prevailing forest vegetation to fires (Tolonen 1983).

Unlike pollen grains, which are almost constant in size, the diameter of charred particles in smoke varies from submicroscopic to several centimetres (Schaefer, 1976). Similarly, production and transport of particles from different fires varies under different meteorological conditions (Burzynski, 1982). The charcoal influx also varies with the morphology of the basin and the surrounding landscape (Swain, 1980). All these facts mean that there are special requirements for relevant collection and interpretation of data. This is further supported by the fact that charred particle production is not an annual feature as is the flowering of plant species, but an irregular one. Therefore, the prevailing local physiographic and other conditions must be taken into consideration. In addition, as many other palaeoecological analyses as possible, from the same cores, are recommended.

CHOICE OF STUDY SITES

The analysis of charred particles from sediment is of maximum use when it is finely resolved, i.e. when carried out continuously using sufficiently thin layers (containing some 20 years of deposits or less). This means that annually laminated sediments, where bioturbation and other sediment disturbances are minimal, are desirable materials. In non-varved sediments, tubificids and

TABLE 23.1. Comparison of charred particle analyses from some lake sediments (Tolonen, 1983)

Site reference	Lac Louis, Quebec: 47°15'N, 79°07'W (Terasmae and Weeks, 1979)	Lake of Clouds, Minnesota: 48°N, 91°07'W (Swain, 1973)	Greenleaf Lake, Ontario: 45°50'N, 77°50'W (Cwynar, 1978)	Lake Ahvenainen, South Finland: 61°02'N, 25°07'E (Tolonen, 1978)	Lake Laukunlampi, East Finland: 62°40'N, 29°10'E (Tolonen, 1980)
Elevation (m)	300	462	?	122	82
Surface area (ha)	10	12.5	56.9	7.6	8.4
Depth (max.) (m)	7.6	>30	76	18.7	28
Drainage area (ha)	~50	~68	~460	42.5	21
Charcoal influx, mean ± S.E. (10^6 $\mu m^2/cm^2/yr$)	?	6.6	3.98 ± 1.51	2.83 ± 2.08	2.22 ± 2.01
Charcoal influx, range (10^6 $\mu m^2/cm^2/yr$)	?	1–17[3]	1.8–9.4	0.02–13.1	0.3–11.7
Period analysed	7000 B.C.–recent	A.D. 1000–1750	A.D. 770–1270	~5000 B.C.–A.D. 1976	A.D. 1001–1859
Fire frequency, mean ± S.E. (yr)[1]	95–100[2]	~60–70	83	62 ± 8/158 ± 12	71 ± 21/78 ± 26
Fire interval, range (yr)	?	20–100	?	12–336	26–146
Prevailing vegetation type on the catchment	Picea mariana forests and muskeg (Populus tremuloides + Betula papyrifera)	Jack pine forests (Pinus banksiana, B. papyrifera, Populus tremuloides, Picea mariana etc.)	Mosaic of conifer and hardwood forests (Pinus strobus, P. resinosa, Populus spp. etc.)	Mesic spruce forests (Picea abies)	Dry pine heath forest (Pinus sylvestris)

[1] For the 'culture period'/for the natural period.
[2] During the pine zone (7000–4000 B.P.) the fire frequency was 48–50 yr.
[3] During the period 7500 B.C. to 1750 A.D. charcoal influx ranges 2–11 × 10^6 $\mu m^2/cm^2yr$ (Swain, 1973).

other organisms can seriously obscure the charcoal and pollen record (Davis, 1967, 1974). There is no point in carrying out historical studies of fire from lake sediments if the sediment quality and coring sites do not fulfil the criteria for finely resolved pollen analysis. An example of an unsuccessful charred particle analysis is that by Bettson and Cawker (1983) from Mashanga Lake, Ontario.

Small, closed, deep lakes with steep slopes and without a wide filtering vegetation zone on the shore are ideal for collecting larger charred plant fragments entering the lake by surface runoff and, from a shorter distance, via air currents. The subsequent interpretations might be of 'local fires' (Tolonen, 1978). The most appropriate lakes for a reliable detection of 'regional' atmospheric charcoal input might be lakes with similar basins but with gentle slopes and, around the lakes, a wetland collar which effectively filters the long-term input of particles inwashed by surface flow (Terasmae and Weeks, 1979).

Peat deposits have an excellent and continuous record of airborne charred particle input (Mehringer *et al.*, 1977; Burzynski, 1982, Tolonen, 1985), and they are superior to unlaminated lake muds since post-depositional mixing and transport is generally absent.

Even local fire history is detectable from charred dust in peat if small mire basins are chosen. This is because larger (>237 µm diameter) charred fragments can readily be distributed at least 0.8 km from the place of the fire, as evidenced from two experimental burnings (Burzynski, 1982) where the burning of clearcut conifer and mixed wood areas (slash left on the site) was carried out under prevailing wind speeds of 5 m/s and greater than 20 m/s, respectively. Generally, the only problem with peat is that there is no very precise dating method, except in volcanic areas (tephrachronology).

LABORATORY WORK

Charred particles can be counted from microscope preparations made according to standard pollen slide techniques for sediment and peat (see Chapter 22).

A 'charcoal verification test' can be made by boiling the sample in nitric acid (Singh *et al.*, 1981). Nevertheless, decomposed plant debris (*Substantia humosa* of Troels-Smith) which cause confusion can normally easily be identified from standard pollen slides, and be differentiated from charred plant material. Reference slides of burned plant structures can be used for comparison. A detailed description of pyrite spherules (5–70 µm in diameter) was given by Dell (1975). Usually, these spherules are a tarnished to brownish yellow colour, while the charred particles are black.

The widely used procedure for counting charred particles has been to measure the particle area for different size classes, as introduced by Waddington (1969). He used eleven size classes, from 56 µm^2 to 6187 µm^2.

Larger fragments were measured and counted separately. Particles smaller than 56 μm² were ignored. His eyepiece graticule had a 15 × 15 μm grid. More recently authors have divided the continuum of the size of charred particles into different size classes, the number of which varies from 1 to 20.

Some authors have counted even the smallest particles, whereas the minimum diameter of the grains included by some others has varied from 5 μm to 50 μm. Usually the total 'charcoal area' of the sample has been computed as the sum over all the size classes of the product of the arithmetic mean and number of fragments in each size class. When inner markers (e.g. *Lycopodium* spores, see Chapter 22) have been counted, together with charred particles, the 'charcoal area' is expressed per cm³ of sediment and finally converted to flux values/cm²/year (assuming that the rate of sedimentation is known). Usually charred particles have been recorded in 10 preset cover-slip traverses. Good descriptions of this method are given by Swain (1973) and Cwynar (1977). Both these authors and Waddington (1969) made statistical tests to determine the differences in charcoal (area) influx between replicate slides from the same depth. Reasonably satisfactory results could be obtained from single countings.

It is unnecessary, and in routine analysis too time-consuming, to use more than two or three size classes. This has been shown by several authors. The setting of the lower limit has no basis for many reasons. The input of small charred particles less than 25 μm in diameter from the atmosphere is apparently continuous in most areas. However, the smallest size class always strongly correlates with the larger size classes. It is difficult to give an exact universal recommendation, but the methodology of Mehringer *et al.* (1977) seems to hold well in the light of experience. Only two size classes were enough. The first one includes the charred fragment, longer axis ≤25 μm, and the second one >25 μm, because most of the larger particles fell into the 25–50 μm class. The larger fraction is then most indicative of (more or less) past fires in the local area.

A preferable alternative for counting charred particles in pollen preparations and thin sections of sediment has been suggested by Clark (1983). The method is simple and rapid: it takes 10–30 minutes per sample as against 1–3 hours using a square eyepiece grid (Clark, 1983). This point count method gives only the total particle area but not numbers of particles or size-classing. The basic principle of the method is given in Figure 23.1. The number of points on the reticle, and the magnification, should be chosen so that only one point falls on the majority of individual particles. The transects across the slide must be placed so that a representative sample is achieved.

The area of charred particles on the slide can then be estimated when the total number of points used is known using the nomogram and formulae given by Clark (1983). If the volume of a subsample on the slide is known or the markers have been used, the charred particle unit area can be expressed per

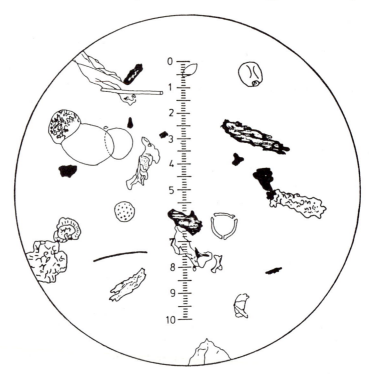

FIGURE 23.1.   Principal of the point-count method for measuring the total area of charred particles in pollen slides. Of the eleven points defined by the upper ends of lines nearest the numbers of a standard eyepiece micrometer, only one 'touches' a charred particle. From the ratio of charred particles 'touched' to the number of points within the fields viewed, the area of charred particles can be calculated using the nomogram and formulae given in Clark (1983). *Redrawn with permission from Pollen et Spores*

cm^3 of sediment or as annual influx per unit area (if the sedimentation rate is known).

From polished or microtome thin sections, the volume of charred particles per unit volume of sediment (V_c) can be estimated using:

$$V_c = C/N$$

where C is the number of points falling on charred particles and N the total number of points used (Clark, 1983). Correspondingly, the annual input volume can be computed if the number of years is known.

INTERPRETATION, APPLICATION AND RECOMMENDATIONS

Very little is known about the relationship between the charred particle curves and the intensity and/or frequency of fires, but it is usually assumed that relatively high frequencies and/or intensities of fire are expressed by the charred particle peaks in sediments (Singh *et al.*, 1981). Several attempts have been made to test this hypothesis.

The charcoal:pollen ratio (C:P), as introduced by Swain (1983), has been for him, and for many other authors, an extremely useful and recommendable index in tracing 'true' fires. Its advantages over charred particle area influx include: (1) it is not sensitive to the errors involved in determining the rate of sedimentation; (2) false peaks resulting from the redeposition of charred particles can be minimized assuming that pollen is deposited along with charcoal; and (3) charcoal:pollen ratios may also emphasize charred particle peaks associated with actual fires, because increased charcoal influx combined with depressed post-fire pollen production would result in a maximum value for this index. Using the C:P ratio, Swain (1973) discovered a historically known fire from 1910 around the Lake of the Clouds that was not apparent on the charred particle influx diagram.

Other parameters used in searching for indirect indicators of fires in sediments are the conifer:sprouter ratio (Swain, 1973); varve thickness (Swain, 1973; Cwynar, 1977); influx of pollen, aluminium and vanadium (Cwynar, 1977); measurement of abrupt rises in ash and mineral content; declines in *Picea* pollen and changes in diatom assemblages (Tolonen, 1978; Tolonen, 1983); and correlations between influx of charred particles and magnetic minerals (Rummery *et al.*, 1981). Attempts to apply statistical techniques (including time series analysis) to the interpretation of finely resolved data are also noteworthy (Green, 1981, 1983).

As emphasized by Cwynar (1977), the most useful approach for identifying fires from sediment records is to observe several indices expressing landscape changes which are a result of burning .

Based on published and unpublished historical fire studies, it seems likely that a palaeoecological analysis can effectively provide information in terms of fire frequencies of different forest types in different areas and through different time periods. It is apparent, given the present state of the art, that charred particle analyses performed in very different ways and expressed in different units cannot be completely comparable. But this is only natural when all the differences in the production, dispersion and sedimentation of charred particles in different areas are considered. Only results obtained from samples having distinct varves are comparable to some degree (Table 23.1). Many authors have examined the usefulness of simple relative figures like charred particle percentage (of the pollen total plus charcoal total), or charcoal number per pollen number ratio, etc. These figures well indicate large changes in the ancient fire regime and strongly correlate with more sophisticated indices (Mehringer *et al.*, 1977; Amundson and Wright, 1979; Pattersson and Nordheim, unpublished).

Therefore, the easy and rapid point count method developed by Clark for a routine count in every pollen analysis can, to date, be recommended as being as good as more complicated techniques. If the size distribution and the number of charred particles remains unknown, the basis for estimating the proximity of the fire is less sound than if the parameters are available. But a more serious point is the non-random occurrence of very large charred particles in microscope slides which have very few medium and small particles. In such a case, a total charred particle area gives a strongly distorted record of the 'fire activity' in the area, as examined in detail by Battson and Cawker (1983). Further studies, especially on material from varved lakes and accurately dated peat cores (sectioned into increments of a few millimetres) relating to inferred fires from fire scars, will allow a more realistic interpretation.

The main results from Laukunlampi, a varved lake in eastern Finland (Table 23.1, Figure 23.2) are given here as an example of charred particle analysis (Tolonen, 1983). The 42 cm sediment section represents the period A.D. 1001–1659. The year can be dated back to A.D. 1300, the onset of permanent settlement, by means of *Triticum* plus *Rumex acetosa–acetosella* pollen. This roughly correlates with the historical data.

The three lowermost 'fire peaks' result in a 'natural' fire frequency of 78±28 years and fire intervals with a range of 36–124 years. This is in agreement with the fire scar information (for years 1712–1969) for a rather similar landscape type in Ulvinsalo nature reserve, where Haapanen and Siitonen (1978) found the average time between fires to be 82±43 years, with a range of 18–219 years.

The varve thickness curve and dominant diatom changes were evidence that all charred particle influx (and charcoal:pollen ratio) peaks except two (from *c.* 1158 and 1282) had reactions traceable in the lake basins. Thus the charred particle peaks indicate local fires, as further evidenced by the palaeomagnetic measurements from the same core (Rummery, 1983). Excluding the lowermost charred particle peak, where there was a concurrent pollen influx peak, these influx curves exhibit an alternating pattern. This is in contrast to the pattern observed in the material from Greenleaf Lake (Ontario) sediments (Cwynar, 1978). The reason is unknown, but can be speculated to be due to the great differences in the geomorphology of the two landscapes: the Canadian lake has very steep and long slopes, the opposite to the landscape around Lake Laukunlampi. Correspondingly, the erosion from the shores of Laukunlampi might always have been low. The mass occurrence of the alkaliphilous diatom *Synedra acus* in A.D. 1515 and its almost complete absence in the varve of A.D. 1516 (Dr. Rick Battarbee, personal communication) and its absence thereafter for the next 400 years, strongly supports this hypothesis. But differences can also result from a long-lasting burning tradition (slash-and-burn cultivation) in the catchment of Lake Laukunlampi. This is different from the likely occurrence of wild-fires around Greenleaf Lake because of the

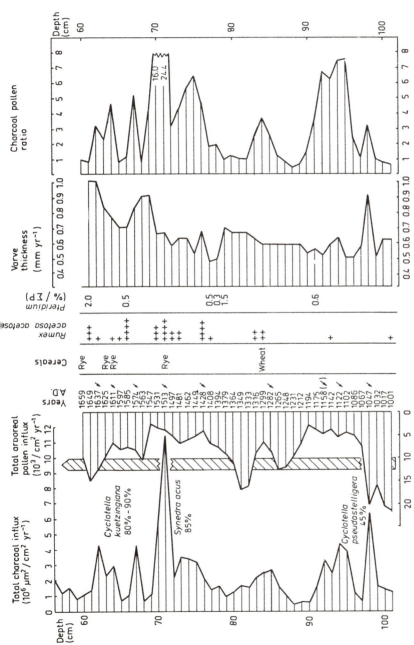

FIGURE 23.2. Example of charred particle analysis in varved sediments of Lake Laukunlampi, a small forest lake in Eastern Finland. The sediment was analysed in 1 cm sections and the inferred fire occurrences are indicated by arrows. The thick vertical cross-hatched column denotes *Cyclotella kuetzingiana*, which comprised 80–90% of the total diatoms, except at two depths where the other species indicated predominated. *Redrawn from Tolonen (1983) by permission of SCOPE*

slashed areas of Lake Laukunlampi were strictly bordered. The size of each individual fire remains undetermined in this study, as it does in all sediment studies.

For environmental application, and for dating, counting of 'soot particles' seems to be a valuable approach (Renberg and Wik, 1983, 1984).

REFERENCES

Amundson, D. C., and Wright, H. E. (1979). Forest changes in Minnesota at the end of the pleistocene. *Ecological Monographs*, **49**, 1–16.

Anderson, S. R. (1979). A holocene record of vegetation and fire at Upper South Branch Pond in northern Maine. M.Sci. Thesis, University of Maine at Orono.

Battson, R. A., and Cawker, K. B. (1983). Methodology to determine long term fire history: an examination of charcoal and pollen from Mashagama Lake, Ontario. In: *Resources and Dynamics of the Boreal Zones* (Eds. R. W. Wein, R. R. Ricwe and J. R. Metheven), Proc. of a conference held at Thunder Bay, Ontario, August 1982, Acuns. Ottawa, pp. 226–248.

Bradbury, J. P., and Waddington, J. C. B. (1973). The impact of European settlement on Shagawa Lake Northeastern Minnesota, U.S.A. In: *Quaternary Plant Ecology* (Eds. H. J. B. Birks and R. G. West), Blackwell, pp. 109–115.

Burzynski, M. P. (1982). Prehistoric fires and forest development as recorded in peat from New Brunswick. M.Sci. Thesis, Dept. of Biology, University of New Brunswick, Fredericton, Canada.

Byrne, R., Michaelson, J., and Souter, A. (1977). Fossil charcoal as a measure of wildfire frequency in southern California: preliminary analysis. Proceedings of the Symposium on the Environmental Consequences of Fire and Fuel Management in Mediterranean Ecosystems, 1–5 August 1977, Palo Alto, California, USDA Forest Service General Technical Report WO-3, Washington, pp. 361–367.

Clark, R. L. (1982). Point count estimation of charcoal in pollen preparations and thin sections of sediments. *Pollen et Spores*, **24**, 523–535.

Cwynar, L. C. (1978). Recent history of fire and vegetation from laminated sediment of Greenleaf Lake, Algonquin Park, Ontario. *Can. J. Botany*, **56**, 10–21.

Davis, R. B. (1967). Pollen studies of near-surface sediments in Maine lakes. In: *Quaternary Palaeoecology* (Eds. E. J. Cushing and H. E. Wright) Yale University Press, New Haven, pp. 143–174.

Davis, R. B. (1974). Stratigraphic effects of tubificids in profundal lake sediments. *Limnol. Oceanogr.*, **19**, 466–487.

Dell, C. J. (1975). Pyrite concretions in sediment from South Bay, Lake Huron. *Can. J. Earth Sci.*, **12**, 1077–1083.

Fredskild, B. (1967). Palaeobotanical investigations at Sermermuit, Jakobshavn, West Greenland. *Medd. Grønland*, **178**, 1–54.

Green, D. G. (1976). *Nova Scotian Forest History — Evidence from Statistical Analysis of Pollen Data*. Ph.D. Thesis, Dalhousie University, Halifax.

Green, D. G. (1981). Time series and postglacial forest ecology. *Quat. Res.*, **15**, 265–277.

Green, D. G. (1983). The ecological interpretation of fine resolution pollen records. *New Phytol.*, **94**, 459–477.

Goldberg, E. D., Hodge, V. F., Griffin, J. J., and Koide, M. (1981). Impact of fossil fuel combustion on the sediments of Lake Michigan. *Env. Sci & Technology*, **15**, 466–471.

Griffin, J. J., and Goldberg, E. D. (1979). Morphologies and origin of elementary carbon in the environment. *Science*, **206**, 563–565.

Griffin, J. J., and Goldberg, E. D. (1981). Sphericity as a characteristic of solids from fossil fuel burning in a Lake Michigan sediment. *Geochimica et Cosmochimica Acta*, **45**, 763–769.

Griffin, J. J., and Goldberg, E. D. (1983). Impact of fossil fuel combustion on sediments of Lake Michigan: a reprise. *Environ. Sci. Technol.*, **17**, 244–245.

Haapanen, A., and Siitonen, P. (1978). Kulojen esiintyminen Ulvinsalon Luonnon-puistossa (Forest fires in Ulvinsalo strict nature reserve). *Silva Fennica*, **12**, 187–200.

Hutchinson, G. E., and Goulden, C. E. (1966). The history of Laguna de Petenxil: the plant microfossils. *Mem. Connecticcut. Acad. Arts Sci.*, **17**, 67–73.

Huttunen, P. (1980). Early land use, especially the slash-and-burn cultivation in the commune of Lammi, southern Finland, interpreted mainly using pollen and charcoal analyses. *Acta Bot. Fennica*, **113**, 1–45.

Iversen, J. (1941). Land occupation in Denmark's Stone Age. *Danmarks Geologiske Undersøgelse*, II Raekke Nr. 66, 1–68.

Mehringer, P. J., Arno, S. F., and Petersen, K. L. (1977). Postglacial history of Lost Trail Pass Bog, Bitterroot Mountains, Montana. *Arctic and Alpine Research*, **9**, 345–368.

Moore, P. D. (1978). Forest fires. *Nature*, **272**, 754.

Patterson, W. A., and Nordheim, E. V. (unpublished). A statistical analysis of sedimentary charcoal data. Dept. of Forestry and Wildlife Management, University of Massachusetts, Amferst, MA, U.S.A., unpublished manuscript.

Payette, S. (1980). Fire history at the treeline in northern Quebec: a paleoclimatic tool. Proceedings of the Fire History Workshop, 20–24 October 1980, Tucson, Arizona, Rocky Mountain Forest and Range Exp. Sta. Forest Service, U.S. Dept. of Agriculture, General Techn. Rept. RM-81, pp. 126–131.

Renberg, I., and Wik, M. (1983). Soot particle counting in recent lake sediments: an indirect dating method. *Ecol. Bulletins*, 36.

Renberg, I., and Wik, M. (1984). Dating recent lake sediments by soot particle counting. *Verh. Internat. Verein. Limnol.*, **22**, 712–718.

Rummery, T. A. (1983). The use of magnetic measurements in interpreting the fire histories of lake drainage basins. *Hydrobiologia*, **103**, 53–58.

Rummery, T. A., Bloemendal, J., Dearing, J., and Oldfield, F. (1979). The persistence of fire-induced magnetic oxides in soils and lake sediments. *Annales de Geophysique*, **35**, 103–107.

Schaefer, V. J. (1976). The production of Optirasun particle smoke in forest fires. In: *Proc. International Symp. Air Quality and Smoke from Urban and Forest Fires*, Colorado State Univ., Fort Collins, Colorado, pp. 27–29.

Singh, G., Kershaw, A. P., and Clark, R. (1981). Quaternary vegetation and fire history in Australia. In: *Fire and the Australian Biota* (Eds. A. M. Gill, R. H. Groves and J. R. Noble), Australian Academy of Science, Canberra, pp. 23–54.

Smith, D. M., Griffin, J. C., and Goldberg, E. D. (1973). Elemental carbon in marine sediments: a baseline for burning. *Nature*, **241**, 268–270.

Swain, A. M. (1973). A history of fire and vegetation in northeastern Minnesota as recorded in lake and sediments. *Quaternary Research*, **3**, 383–396.

Swain, A. M. (1978). Environmental changes during the past 2000 years in North-Central Wisconsin: analysis of pollen, charcoal and seeds from varved lake sediments. *Quaternary Research*, **10**, 55–68.

Swain, A. M. (1980). Landscape patterns and forest history in the Boundary Waters Canoe area, Minnesota: a pollen study from Hug Lake. *Ecology*, **61**, 747–754.

Terasmae, J., and Weeks, N. C. (1979). Natural fires as an index of paleoclimate. *Can. Field Naturalist*, **93**, 116–125.

Tolonen, M. (1978). Palaeoecology of annually laminated sediments in Lake Ahvenainen, South Finland. I: Pollen and charcoal analyses and their relation to human impact. *Ann. Bot. Fennici*, **15**, 155–208.

Tolonen, M. (1985). Palaeoecological record of local fire history from a peat deposit in SW Finland. *Ann. Bot. Fennici*, **22**, 15–29.

Tolonen, K. (1983). The post-glacial fire record. In: *The Role of Fire in Northern Circumpolar Ecosystems* (Eds. R. W. Wein and D. A. MacLean), Scope 18, John Wiley, pp. 21–44.

Tsukada, M., and Deevey, E. S. (1967). Pollen analysis from four lakes in the southern Maya area of Quatemala and El Salvador. In: *Quaternary Palaeoecology* (Eds. E. J. Cushing and H. E. Wright), Yale Press, New Haven, pp. 303–331.

Waddington, J. C. B. (1969). A stratigraphic record of the pollen influx to a lake in the Big woods of Minnesota. *Geol. Soc. America*, special paper, **123**, 263–282.

Handbook of Holocene Palaeoecology and Palaeohydrology
Edited by B. E. Berglund
© 1986 John Wiley & Sons Ltd.

24

Application of fungal and algal remains and other microfossils in palynological analyses

BAS VAN GEEL

Hugo de Vries-Laboratory, University of Amsterdam, Amsterdam, The Netherlands

INTRODUCTION

Apart from pollen records, most palynologists use evidence provided by fossilized colonies of *Pediastrum*, *Botryococcus* and *Scenedesmus*. However, there are often many more recognizable microfossils present in pollen slides. During the last decade efforts have been made at the University of Amsterdam to identify all the microfossils (van Geel 1972, 1976, 1978, 1979, in prep; van Geel *et al.*, 1981, 1983, 1984; van der Wiel, 1982; Pals *et al.*, 1980; Ellis-Adam and van Geel, 1978; Bakker and van Smeerdijk, 1982). In these articles many additional records of microfossils have been reported and illustrated.

Although there is as yet no complete inventory of such fossils from all sediment types, results from studies of lake deposits, eutrophic to mesotrophic bogs, carr peat and raised bogs are available. In many cases it has not yet been possible to identify all microfossil types encountered, but their indicator value could nevertheless be deduced indirectly by comparing the frequency curves in the diagrams, and by looking for positive or negative correlations with other data. Even in those cases where identification to the species level was available in the form of published descriptions, detailed data concerning the ecology of the organism were often found wanting. By comparing the fluctuation in representation of an unknown microfossil with the fluctuation of known pollen types and of certain macrofossils or other recognized taxa, one can obtain an indication of the ecological significance of the incidence of an unknown and newly encountered microfossil. The representatives of the Zygnemataceae (see below) may be mentioned as an example. It appeared that almost all fungi and algae (also when only their spores were conserved) strictly occurred *in situ*. As a consequence, the information deduced from the curves also pertains to local conditions.

Palynological studies with short sample distances (i.e. with a high resolution of time) using records of all microfossils, sometimes enable a very detailed reconstruction of the local vegetation development and of changes in the local hydrology. In this way it is possible to connect the local history with the regional development so as to obtain a better insight into the impact of climatic changes on the vegetation cover (van Geel *et al.*, 1981). Minor climatic changes and their effect on pollen production in regional forests and on the local developments of peat bogs can also be traced in this way (van Geel, 1978).

The usefulness of the very time-consuming research into all micro- and macrofossils will increase as more results become available, so that it will suffice to make a selection on only those microfossils which provide relevant information. Up to now the advantage of recording all microfossils in pollen slides (in combination with the identification of macrofossils) is no greater than that of the conventional method, because only a selection of curves is used for interpretation. It is quite clear that only a selection of the species of fungi, algae, small animals etc. originally present become fossilized, but any improvement in the still inadequate knowledge concerning the vegetational succession deducible from pollen analysis is most welcome.

LABORATORY STUDIES

Up to now the normal preparation procedure for the samples has been followed (see Chapter 22). No differences in preservation have been observed between samples treated with HF and those treated with bromoform-alcohol. A special procedure is required for the study of diatoms and rhizopods (see Chapters 26 and 31).

During the analysis every distinctive type of microfossil has to be recorded and counted. It is of great importance to keep photographic records of all the types encountered, augmented by a concise description. A number code system has been used for identification of the newly recognized microfossils. So far over 350 different type numbers have been recorded at the Hugo de Vries-Laboratory.

Normally palynologists disregard microfossils other than pollen. It is not necessary to be specially trained in order to study other microfossils, since all palynologists are experienced in observing morphological differences and only need to extend their interest. A word of warning before any literature references are given is not inappropriate at this point. One must bear in mind that the assistance of trained mycologists, algologists, zoologists and other specialists is necessary to avoid errors of identification. One cannot always expect to elicit a taxonomic identification from a specialist, because they are often hampered by the incompleteness of the fossil remains. The asci of Ascomycetes, for instance, are not preserved, and only the spores of filamentous algae are preserved — the filaments disappear. This is the reason why keys for the identification of living organisms are often inadequate. In spite of this handicap many specialists are able to provide the literature to be used for identification. In some cases it is necessary to compare the fossils with recent material.

It appears that among the fossil fungi the dematiaceous hyphomycetes play an important role. For the preliminary guidance of palynologists the treatments by Ellis (1971, 1976) are recommended. For the relatively important, thick-walled Ascomycetes, that by Dennis (1968) can be used. For more detailed descriptions of taxa it is necessary to consult the relevant literature. The spores of Sordariaceae normally become fossilized. Lundqvist (1972) has made a detailed study of the representatives of that family; many species are coprophilous, and often so specialized that they only grow on the dung of a single or of only a few species of animals.

Apart from algae such as *Pediastrum*, *Botryococcus* and *Scenedesmus*, well-known to palynologists, the representatives of the Zygnemataceae and Oedogoniaceae form an interesting group. Spores of these algae are often very well-preserved (van Geel, 1976, 1978) and their morphology often permits identification to the genus or even the species level. For purposes of identification the publications of Transeau (1951), Randhawa (1959), Czurda (1932) and Kolkwitz and Krieger (1941) are very useful. In order to gain an impression of the morphological diversity of the spores in the genera *Mougeotia*, *Spirogyra*, *Debarya* and *Bulbochaete*, the illustrations of van Geel and van der Hammen (1978) may be mentioned. In the genus *Zygnema*, but also in some other taxa, lenticular spores occur whose morphology is so similar that the literature pertaining to recent algae is insufficient for a reliable identification of this type of spore. However, the fossil spores of this 'Zygnema-type' often show enough detail to permit the recognition of distinct morphological subtypes (see also Chapter 25).

Representatives of blue-green algae (*Gloeotrichia*, *Rivularia*) are preserved in gyttjas and their presence permits conclusions concerning the conditions prevailing during sedimentation.

Other than fungi and algae, a complete inventory always includes a number of microfossil types of animal origin (e.g. rhizopods such as *Amphitrema flavum*, Rotatoria, Cladocera) but remains of phanerogams also occur (e.g. hairs of *Ceratophyllum* and Nymphaeaceae). A number of microfossils are of unknown origin, but in this early phase of inquiry this can hardly be an obstacle to recording such fossils and for drawing curves of representation.

PRESENTATION OF RESULTS

The first decision to be made after the recording of fossils other than pollen grains concerns the calculation of a percentage. Up to now it has been deemed meaningful to calculate the frequencies as percentages of Σ pollen (for instance, for Holocene material as ΣAP, and for late-glacial material as $\Sigma AP +$ dry herbs). Probably the best way to obtain a better understanding of the changes of frequency is to construct an absolute diagram in addition to the percentage diagram (see Chapter 22).

The sequence of the microfossil Type numbers is, of course, arbitrary because it depends solely on the moment when a new microfossil became recognized. Meanwhile the total number of microfossil Types has grown so large, and the requirement of readily locating the curves in the diagrams without problems is such a primary condition, that of necessity the curves are arranged according to the sequence of Type numbers. The best method of summarizing the results is probably the separate presentation of the diagrams of the fungi, the algae, the animal remains and the unidentified microfossils.

Some additional diagrams showing a selection of curves (and including both micro- and macrofossils) may be used to show, among other things, changes in local moisture conditions, changes in trophic conditions, relations of fungi with their respective host plants, or other examples of ecological mutualisms (e.g. van Geel, 1978).

INTERPRETATIONS

It is not possible to present here the indicative ecological information nor the usefulness for palaeoecological and palaeoclimatic interpretations of all newly established microfossil taxa or form taxa. The reader is referred to the above-mentioned literature. Some examples are given below which show the additional information that can be obtained by the recording of all microfossils present.

The numbers in square brackets refer to the labels in Figures 24.1 and 24.2.

(1) The curve of conidia (or chlamydospores), [1–3] indicated as Type 10, appears to be indicative of changes in local moisture conditions at an early stage of an investigation into all microfossils in raised bogs (van Geel, 1972). High percentages of Type 10 are always concomitant with a relatively high degree of decomposition of the peat, and with relatively high percentages of Ericales pollen (i.e. with relatively dry conditions during the formation of the peat). Peat formed under relatively wet conditions contains low percentages of Type 10, and Type 10 may often be lacking altogether. We can interpret the behaviour of the curve of Type 10 in the diagrams: Type 10 occurs on roots of *Calluna vulgaris* which occurs in raised bogs under relatively dry conditions.

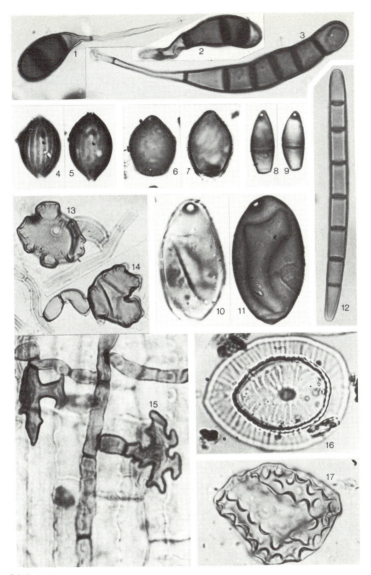

FIGURE 24.1.

1–3 : fungal spores of Type 10 (×1000)
4, 5 : ascospores of *Neurospora* sp. (Type 55C) in high and low focus (×1000)
6, 7 : sordariaceous ascospores (Type 169: cf. *Tripterospora*; ×1000)
8, 9 : ascospores of *Cercophora*-type (Type 112; ×1000)
10, 11 : sordariaceous ascospores (Type 368; ×1000)
12 : ascospore of *Geoglossum sphagnophilum* (Type 77A; ×1000)
13–15 : hyphopodia and mycelium of *Gaeumannomyces* cf. *caricis* (Type 126; ×1000). 13 and 14 from microfossil samples; 15 showing the fungus still in organic contact with the epidermis of a *Carex* species in a macrofossil sample
16 : zygospore of *Debarya glytosperma* (×1000)
17 : zygospore of aplanospore of the *Zygnema*-type (Type 314; ×1000).

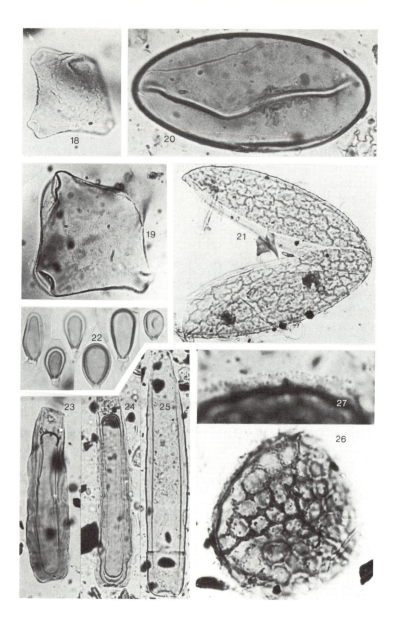

FIGURE 24.2.

18   : zygospore of *Mougeotia* cf. *gracillima* (Type 61; ×1000)
19   : zygospore of *Mougeotia* cf. *punctata* (Type 313F; ×1000)
20   : psilate zygospore or aplanospore of *Spirogyra* sp. (Type 315; ×1000)
21   : reticulate zygospore or aplanospore of *Spirogyra* sp. (Type 132; ×500)
22   : heterocysts of *Rivularia*-type (Type 170; ×1000)
23–25 : sheaths of *Gloeotrichia*-type (Type 146; ×500); 23 and 24 with a resting spore
26   : spore of *Riccia* cf. *sorocarpa* (Type 165; ×750)
27   : detail of wall of *Riccia* cf. *sorocarpa* (×1000).

(2)   The ascospores of *Neurospora* sp. (Type 55C; [4 and 5]) indicate the incidence of a local bog fire.

(3)   Ascospores of coprophilous Sordariaceae [6–11] are often restricted to one or a few kinds of dung. This renders the Sordariaceae interesting in palynological inquiry, especially when applied to archaeology.

(4)   The ascospores of *Geoglossum sphagnophilum* (Type 77A; [12]) occur in the remains of hummock vegetation types, often just before the overgrowing phase containing *Scheuchzeria palustris*. Their presence in the Netherlands during the Atlantic and Subboreal periods and their absence during the Subatlantic period may indicate that the climatic changes that enabled the growth of *Sphagnum imbricatum* may have been unfavourable for *G. sphagnophilum*. There are no recent records of *G. sphagnophilum* in the Netherlands.

(5)   There is a correlation between the presence of hyphopodia of the fungus *Gaeumannomyces* (Type 126; [13–15]) and the local appearance of Cyperaceae in peat. This has interesting implications for the palynological analysis of, for example, late-glacial deposits. Since Cyperaceae are usually included in the pollen sum as a basis for calculations, the answer to the question whether or not Cyperaceae constituted an element of the local vegetation is a crucial piece of information (Pals *et al.*, 1980; Van Geel, in prep.).

(6)   Fossil spores of Zygnemataceae [16–21] are not rare in pollen diagrams, but have usually been disregarded by palynologists. Within the group of Zygnemataceae hardly any appreciable difference in ecology is known, but different zygnemataceous spore types do not occur at the same stratigraphical levels. The curves of the spores at the very least show that there are ecological differences in habitat preference of the various Zygnemataceae, so far as the conditions for spore production are concerned. Most representatives of the Zygnemataceae need a relatively high water temperature to sporulate. The presence of, for example, zygnemataceous spore types in sediment formed during the so-called Rammelbeek phase could indicate that this was not a cold oscillation during the Preboreal as has sometimes been suggested (van Geel *et al.*, 1981).

In pre-Quaternary deposits the spores of Zygnemataceae have often been described as 'form taxa' and they have so far not been recognized as early representatives of Zygnemataceae (van Geel, 1979). According to Rich *et al.* (1982), spores of *Spirogyra*, which are known as the form taxon *Ovoidites* in pre-Quaternary palynology, may be a valuable indicator of ancient freshwater coal deposits. Details of the ecology and information regarding local conditions yielded by the recording of spores of Zygnemataceae was given by van Geel (1978, 1979), van Geel and van der Hammen (1978), Ellis-Adam and van Geel (1978) and van Geel *et al.* (1981, 1983).

(7)   Fossil sheaths and heterocysts of Cyanophyceae [22–25] occur in lake deposits and indicate an alkaline environment (pH=7–8.5). The pioneer role of the *Gloeotrichia*-type in a late-glacial pool deposit may be mentioned. In

an environment which was originally poor in nutrients and where nitrogen was almost completely lacking, the nitrogen-fixing capacity of these pioneer algae opened up the site for other aquatic plants (van Geel *et al.*, 1984).

(8) Spores of representatives of the genera *Riccia* [26 and 27] and *Anthoceros* (Hepaticae) may indicate agricultural activities: some species occur as pioneers on lime-rich arable land (see also Chapter 30).

REFERENCES

Bakker, R., and van Smeerdijk, D. G. (1982). A palaeoecological study of a Late Holocene section from 'Het Ilperveld', W. Netherlands. *Rev. Palaeobot. Palynol.* **36**, 95–163.

Czurda, V. (1932). *Zygnemales* (2nd edn), Gustav Fischer Verlag, Jena.

Dennis, R. W. G . (1968). *British Ascomycetes*, Lehre.

Ellis-Adam, A. C., and van Geel, B. (1978). Fossil zygospores of *Debarya glyptosperma* (De Bary) Wittr. (Zygnemataceae) in Holocene sandy soils. *Acta Bot. Neerl.*, 389–396.

Ellis, M. B. (1971). *Dematiaceous Hyphomycetes*, Commonwealth Mycol. Inst., Kew.

Ellis, M. B. (1976). *More Dematiaceous Hyphomycetes*, Commonwealth Mycol. Inst., Kew.

Kolkwitz, R., and Krieger, H. (1941). Zygnemales. In: *Rabenhorst's Kryptogamen-Flora*, **13**(2), pp. 1–449.

Lundqvist, N. (1972). Nordic Sordariaceae s. lat. *Symbolae Bot. Upsal.*, **20**, 1–374.

Pals, J. P., van Geel, B., and Delfos, A. (1980). Palaeoecological studies in the Klokkeweel bog near Hoogkarspel (Prov. of Noord-Holland). *Rev. Palaeobot. Palynol.*, **30**, 371–418.

Randhawa, M. S. (1959). *Zygnemaceae*, Indian Council Agric. Res., New Delhi.

Rich, F. J., Kuehn, D., and Davies, T. D. (1982). The palaeoecological significance of Ovoidites. *Palynology*, **6**, 19–28.

Transeau, E. N. (1951). *The Zygnemataceae*, Columbus Graduate School Monographs, Contrib. Bot. 1.

van der Wiel, A. M. (1982). A palaeoecological study of a section from the foot of the Hazendonk (Zuid-Holland, the Netherlands), based on the analysis of pollen, spores and macroscopic remains. *Rev. Palaeobot. Palynol.*, **38**, 35–90.

van Geel, B. (1972). Palynology of a section from the raised peat bog 'Wietmarscher Moor', with special reference to fungal remains. *Acta Bot. Neerl.*, **21**, 261–284.

van Geel, B. (1976). Fossil spores of Zygnemataceae in ditches of a prehistoric settlement in Hoogkarspel (the Netherlands). *Rev. Palaeobot. Palynol.*, **22**, 337–344.

van Geel, B. (1978). A palaeoecological study of Holocene peat bog sections in Germany and the Netherlands. *Rev. Palaeobot. Palynol.*, **25**, 1–120.

van Geel, B. (1979). Preliminary report on the history of Zygnemataceae and the use of their spores as ecological markers. *Proc. IV Int. Palynol. Conf.*, Lucknow (1976–77), **1**, pp. 467–469.

van Geel, B. (in prep.). A palaeoecological study of a Late Glacial sequence near Usselo (the Netherlands).

van Geel, B., and van der Hammen, T. (1978). Zygnemataceae in Quaternary Colombian sediments. *Rev. Palaeobot. Palynol.*, **25**, 377–392.

van Geel, B., Bohncke, S. J. P., and Dee, H. (1981). A palaeoecological study of an upper Late Glacial and Holocene sequence from 'De Borchert', the Netherlands. *Rev. Palaeobot. Palynol.*, **31**, 367–448.

van Geel, B., Hallewas, D. P., and Pals, J. P. (1983). A Late Holocene deposit under the Westfriese Zeedijk near Enkhuizen (Prov. of Noord-Holland, the Netherlands): palaeoecological and archaeological aspects. *Rev. Palaeobot. Palynol.*, **38**, 269–335.

van Geel, B., de Lange, L., and Wiegers, J. (1984). Reconstruction and interpretation of the local vegetational succession of a late-glacial deposit from Usselo (the Netherlands), based on the analysis of micro-and macrofossils. *Acta. Bot. Neerl.*, **33**, 535–546.

Handbook of Holocene Palaeoecology and Palaeohydrology
Edited by B. E. Berglund

25

Blue–green algae, green algae and chrysophyceae in sediments

GERTRUD CRONBERG

Institute of Limnology, University of Lund, Lund, Sweden

REVIEW

As long ago as 1854, Ehrenberg recorded algal remains in soil samples from different parts of the world (Ehrenberg, 1854). Von Post (1860, 1862) noticed that the gyttja of Swedish lakes was often very rich in algal remains, and therefore called it 'algal gyttja'. Borge (1892) studied what he referred to as subfossil algae from Gotland, Sweden; by this term he meant algae deposited in post-glacial sediments (Messikommer, 1938). Lagerheim (1902) investigated sediments from 35 localities in Switzerland, Germany, Finland, Denmark and various parts of Sweden. Krieger (1927) investigated subfossil algae in Diebelsee, Germany, both out in the open water and on the high moor. He found great differences in these biocoenoses indicating different ecological demands. He pictured both chlorococcal green algae and cysts of Chrysophyceae. Messikommer (1938) studied sediments collected by himself or received from other scientists in Europe. He was, however, mainly interested in the desmid flora which he investigated thoroughly with ecological comments. Steinecke (1927) investigated microfossils from Zehlau high moor, Germany and compared the fossil algal flora with that of the more recent.

In these pioneering works, the algal remains found in sediments were identified to genus or species, but this mostly involved qualitative analyses and no real sediment profiles were investigated. Attempts were also made to classify the past trophy of lakes, moors, swamps and other wetlands with the aid of the algal composition in their sediments. However, when making pollen analyses of sediments taken from different cores, algal remains were often

TABLE 25.1. Algal species recorded from sediments

Species	Ahonen and Ristiluoma (1973)	Borge (1982)	Borge and Erdtman (1954)	Cookson (1953)	Crisman (1978)	Cronberg (1982a,b this paper)	Fjerdingstad (1954)	Frederick (1977)	Fredskild (1973, 1975, 1983)	van Geel (1976)	Goulden (1970)	Jankovska and Komárek (1982)	Korde (1966)	Krieger (1929)	Lagerheim (1902)	Lundqvist (1927)	Messikommer (1938)	Salmi (1963)	Sebestyen (1969)	Simola (1979)	Simola et al. (1981)	Steinecke (1927)	Whiteside (1965)	Wilson and Hoffmeister (1953)
CYANOPHYTA																								
Anabaena	4					1							1		1	spp.								
Aphanizomenon							1									1				1				
Aphanocapsa							1						1			spp.								
Aphanothece					1		1						1			spp.								
Chroococcus					1	5	3	5								spp.								
Gloeotrichia					1	1	1	2			1		1		1									
Gloeocapsa								1					1			1								
Gomphosphaeria																1								
Merismopedia							1	1					1		1									
Microcystis							1									1								
Lyngbya							1				1					spp.								
Pleurocapsa								1																
CHLOROPHYTA																								
Volvocales																								
Phacotus						1		1			1		1			1								
(Chlorococcales)																								
Botryococcus	1			1	1	1						1	1			1								
Coelastrum	1					1					1		1	1	1	1								
Oocystis	1														1							1		
Pediastrum	4			1	7	5	5	7	7		3	10	1	5	14			7	5		1	1	4	4
Scenedesmus	1				7	ssp.	2	12				ssp.	1	1	4	spp.					spp.	1		

(Zygnematales)																		
Bambusina	–	–	–	–	–	–	–	–	–	–	–	–	–	–	–	–	–	–
Closterium	–	–	–	–	–	–	–	–	–	–	–	–	–	–	–	–	2	–
Cosmarium	8	–	55	–	–	7	5	146	3	–	1	–	–	–	–	–	–	–
Cylindrocystis	–	–	1	–	–	–	1	1	–	–	–	–	–	–	–	–	–	–
Desmidium	–	–	–	–	–	1	–	2	–	–	–	–	–	–	–	–	–	–
Docidium	–	–	–	–	–	–	–	–	–	–	–	–	–	–	–	–	–	–
Euastrum	1	–	10	–	2	2	1	23	–	–	–	–	–	–	–	–	–	–
Gonatozygon	–	–	–	–	1	1	1	2	–	–	–	–	–	–	–	–	–	–
Hyalotheca	–	–	–	–	–	–	–	1	–	–	–	–	–	–	–	–	–	–
Mesotaenium	–	–	–	–	–	–	–	1	–	–	–	–	–	–	–	–	–	–
Mougeotia	–	–	–	–	spp.	–	–	–	–	–	–	–	–	–	–	–	–	–
Netrium	–	–	–	–	–	–	–	1	1	–	–	–	–	–	–	–	–	–
Penium	–	–	–	–	–	–	–	–	–	–	–	–	–	–	–	–	1	–
Pleurotaenium	–	–	1	–	–	–	–	–	–	–	–	–	–	–	–	–	–	–
Roya	–	–	15	–	3	2	spp.	1	–	–	–	–	–	–	–	–	–	–
Staurastrum	–	–	2	–	–	–	–	46	1	–	–	–	–	–	–	–	–	–
Staurodesmus	–	–	–	–	–	–	–	3	–	–	–	–	–	–	–	–	–	–
Tetmemorus	–	–	–	–	1	1	–	2	–	–	–	–	–	–	–	–	–	–
Xanthidium	–	–	1	–	–	–	–	1	1	–	–	–	–	–	–	–	–	–
Zygnema	–	ssp.	–	–	–	–	–	–	–	–	–	–	–	–	–	–	1	–
(Ulothricales)																		
Binuclearia	–	–	–	–	–	1	–	1	–	–	–	–	–	–	–	–	–	–
(Charales)																		
Chara	–	ssp.	–	–	–	spp.	spp.	–	–	–	–	–	–	–	–	–	–	–
CHROMOPHYTA																		
Chrysophyceae	6	ssp.	ssp.	ssp.	spp.	spp.	spp.	–	spp.	spp.	–	–	–	–	–	–	–	–

recorded and counted together with pollen (Cookson, 1953; Fredskild, 1983, 1975, 1983; Livingstone *et al.*, 1958; Salmi, 1963; Whiteside, 1965; Wilson and Hoffmeister, 1953). The algae identified in such investigations have been summarized in Table 25.1. *Pediastrum* seemed to be the most recorded and best preserved algal genus in the sediments (excluding diatoms). They can evidently stand the hard acetolysis method (Faegri and Iversen, 1974) used for cleaning the pollen.

Pediastrum species have usually been counted as a group, and the percentage distribution of total *Pediastrum* presented. However, in sediments *Pediastrum* is well preserved and can easily be identified to the species level. The frequency (in numbers per cm^3) of *Pediastrum* species has been calculated and compared with chemical–physical conditions or other organisms in the sediments (Alhonen and Ristilouma, 1973; Crisman, 1978; Cronberg, 1982a, b; Jankovska and Komárek, 1982; Goulden, 1970; Korde, 1966; Salmi, 1963; Sebestyen, 1969) — see Table 25.1. The ecological significance of *Pediastrum* species has also been discussed (Crisman, 1978; Cronberg, 1982a, b; Jankovska and Komárek, 1982; Salmi, 1963).

Fjerdingstad (1954) investigated the subfossil algal flora of Lake Bølling Sø, Denmark, and Frederick (1977) studied three sediment cores from central Ohio lakes, U.S.A. Both authors discussed the environmental significance of different algal species deposited in the sediments. Frederick studied all algal remains except diatoms, while Fjerdingstad included these species in his study.

Green algae deposited in the sediments resist dissolution and their cells can be found as whole coenobia; only the cell contents disappear. Blue–green algal remains can be found as empty cells (*Chroococcus*, *Gloeocapsa*), but mostly it is only the spores that are left (*Aphanizomenon*, *Anabaena*). In the *Gloeothrichia* it is the sheath around the spores that is preserved, whereas the rest of the trichomes are lost. In the sediments from Rostherne mere, England, Livingstone and Jaworski (1980) studied the viability of akinets from species of blue–green algae. They found that *Aphanizomenon* and *Anabaena* spores cultured from sediments up to 18 and 64 years old respectively were still viable and could develop gas-vacuolated populations in cultures.

Silica scales and cysts of Chrysophyceae are sometimes well preserved in the sediments. However, the acetolysis method combined with HF treatment for making pollen preparations (Faegri and Iversen, 1974) dissolves the silica of diatom frustules and chrysophycean scales and cysts, and possibly destroys other algae too. For these reasons, other preparation methods have been applied for studying algae partly made of silica. Such treatments have included strong acids (Patrick and Reimer, 1966) or hydrogen peroxide (see Chapter 26).

Deflandre (1932) investigated diatom-rich sediments (diatomite) from many parts of the world and found, besides diatoms, many round silica cysts with characteristic structure and form. He classified them as fossil cysts belonging to marine chrysophytes and constructed the class of Archaeomonadaceae for them. In 1936 Deflandre summarized the biological and palaeontological value of these cysts and stated that if their biological affinity and ecology was known, they could serve as indicators in palaeoliminological investigations. Much research has been devoted to describing and depicting the chrysophyte cysts (Cronberg, 1985), first with light microscopy (LM) and since the beginning of 1970 with scanning electron microscopy (SEM) (Cornell, 1972; Gritten, 1977). With SEM it was revealed that the cysts often had a characteristic ultrastructure of the cell wall, even though they look quite smooth when studying them with LM. During the last decade many new studies on the ultrastructure of the cysts have been published (Gritten, 1977; Cronberg, 1980, 1982b; Carney, 1982; Smol, 1983; Carney and Sandgren, 1983; Sandgren and Carney, 1983). In all these works, characteristic cysts have been counted in sediment cores and related to changes in the surroundings.

As the biological affinities of most cysts are unknown, different systems have been used for naming them. Deflandre (1932, 1936) classified them according to size, form and structure into different genera, but emphasized that it was a completely artificial system only devised as a help in cataloguing the different morphotypes. Nygaard (1956) presented another system for naming the cysts, and used *nomina tempora*. He wrote 'cysta' and a characteristic epithet such as *cysta longispina*, meaning a cyst having long spines. Nygaard grouped the cysts he found in Lake Gribsø according to characteristic features and made a key for the species found in the lake sediments. Nygaard's system has been much used for describing chrysophycean cysts (Gritten, 1977; Elner and Happey-Wood, 1978; Carney, 1982; Carney and Sandgren, 1983; Sandgren and Carney, 1983). Nygaard found no chrysophycean scales in the deeper sediments of Lake Gribsø, although they were rich in cysts. In surficial sediments, however, scales were found, representing remains of the recent chrysophycean flora in the lake. The solubility of the scales must consequently be much greater than that of the cysts.

Fott (1966), on the other hand, found that chrysophycean scales were sometimes well preserved in sediments, and he stated that these scales could be useful palaeoindicators, if the scales could be related to the species from which they were derived. Chrysophycean scales are small and possess a species-specific ultrastructure. To get an accurate identification of the scaled chrysophytes they have to be studied with EM, preferably transmission electron microscopy (TEM). Since the 1960s about 130 species of scale-bearing chrysophytes have been identified and described with EM (Kristiansen, 1982).

In 1980, three papers appeared simultaneously in which chrysophycean scales were identified, counted and used for interpretation of the lake's earlier history (Battarbee *et al.*, 1980; Munch, 1980; Smol, 1980). In these studies it was clearly demonstrated that the scales were well preserved in the sediments and could be identified to species.

In many lakes, laminated sediments have been deposited (Nipkow, 1927;

Tippett, 1964; Renberg, 1976; Simola, 1977, 1979, 1983; Battarbee *et al.*, 1980, Simola *et al.*, 1981). The lamina very often consist of chrysophycean scales and cysts, and may be deposited annually. Battarbee *et al.* (1980) showed that the different lamina (white and dark) in the sediments from Laukunlampi, Finland, consisted of chrysophycean cysts and scales deposited during different seasons of the year. The white lamina consisted partly of *Mallomonas* scales and bristles corresponding to the summer or early autumn, and were followed by darker lamina of chrysophycean cysts and organic material deposited during the winter. Before a stratigraphical change in the sediment core at 58 cm, the dominant species in the sediments was *Mallomonas crassisquama* (Asmund) Fott. The change reflected an influence from human activities. In the upper sediments the laminations were not developed and here *M. crassisquama* was substituted by the more eutrophic species *M. elongata* Reverdin. The use of chrysophycean scales as palaeoindicators is, however, very promising, especially as the scales can be identified to species or even subspecies levels. Unfortunately, the dissolution of scales is greater than that of the cysts, and hence sometimes they are completely missing in the sediments (Nygaard, 1956; Cronberg, 1980), although the cysts are present.

From the above-mentioned papers it is obvious that algal remains other than diatoms can be of much use for palaeolimnological reconstructions.

METHODS

It has long been recognized that, in pollen preparations, algae can be identified and counted, particularly the green algae (Table 25.1). In a few investigations the main interest was the study of algae deposited in lake sediments (Korde, 1961, 1966; Fjerdingstad, 1954; Frederick, 1977; Cronberg, 1982a; Jankovska and Komárek, 1982). The study of algae in sediments was usually a by-product of the pollen investigations. During more recent years it has now been realized that the study of algal remains can give extra information for palaeoecological reconstructions. However, before now no standard methods for algal investigation in sediments (excluding diatoms) have been recommended.

Field methods

The same coring technique is applied as for the investigation of pollen and diatoms (Chapter 8). However, special sampling techniques have to be used if the investigation involves laminated sediments (Chapter 17). It is strongly recommended that, if lake sediments are being studied, plankton samples be taken. Such sampling ought to be performed during different seasons and at several places in the lake. It is then possible to compare the recent algal flora with that of past times.

Laboratory methods

The following methods have been used:

(1) No treatment. Fresh sediment has been analysed directly under the microscope, and the degree of cover on a slide estimated (Lundqvist, 1927; Korde, 1961, 1966).
(2) The acetolysis method (e.g. Faegri and Iversen, 1974) combined with HF treatment. This is a strong treatment and many algae are dissolved (e.g. diatoms, chrysophytes, and algae with calcareous scales (*Phacotus*)).
(3) The acid cleaning method (e.g. Patrick and Reimer, 1966). With this method organic material is removed but diatoms, scales and cysts of Chrysophyceae are preserved.
(4) The hydrogen peroxide cleaning method (Chapter 26). This also removes organic material, but diatoms and chrysophytes are left. The advantage of this technique over the acid cleaning method is that it is not so time-consuming.

The following methods (labelled A–D) are recommended.

Samples for algal analyses should be taken out of the core at 0.5–10 cm intervals, or from separate layers in laminated sediments (Battarbee, 1980) depending on the aim of the investigation. If samples are not going to be treated immediately it is advisable to preserve the subsample with 37% formalin. A defined volume of sediments (0.5–1 cm^3) should be taken from each subsample/level and transferred to centrifuge tubes. If preserved samples are used, they must be cleaned with distilled water before further treatment.

A. Determination of the presence of algal remains in the sediments (e.g. presence of calcareous algae)

(1) Place small amounts of the sediment on a slide.
(2) Add 1–2 drops of glycerin.
(3) Stir carefully until lumps are no longer visible.
(4) Carefully place a coverslip on top of the sediment, ensuring that no air bubbles appear underneath the coverslip.
(5) Seal the coverslip to the slide with glycel or nailpolish.
(6) Investigate with LM (phase-contrast).

Comments. In slides prepared according to this method it is possible to judge the quality of sediment; if many algae are present; if they have viable chloroplast; and if calcareous algae, e.g. *Phacotus*, are present.

B. Determination of the presence of chrysophycean scales in the sediments

(1) Place 1–2 drops of fresh sediment on a coverslip.
(2) Add 1–2 drops of distilled water.

(3) Stir the sediment carefully and spread it out to a very fine layer on the coverslip.
(4) Let it dry at room temperature for a minimum of 12 hours.
(5) Place the coverslip with the sediment turned towards the slide.
(6) Seal the coverslip with hot stearin from a candle.
(7) Investigate with LM (phase-contrast).

Comments. In these dry preparations the presence of chrysophycean scales can be investigated. The scales are best observed in a dry condition without any embedding. For identification of the scales see method D.

C. *Blue–green and green algal analyses*

(1) Place 0.5–1 ml of sediment into a centrifuge tube.
(2) Add standard microsphere solution or *Lycopodium* spore tablets.
(3) Add 5–10 ml of 10% NaOH solution.
(4) Boil on a waterbath for 3–10 minutes, until lumps are no longer visible.
(5) Rinse with distilled water. Repeat several times until a neutral reaction is obtained.
(6) Add a volume of glycerin equal to that of the sediment.
(7) Stir the mixture carefully with a vortex mixer.
(8) Take 1–2 drops of treated sediment solution and place on a slide.
(9) Add 1 drop Fuchsin B-glycerin solution.
(10) Mix well with a glass rod.
(11) Add a coverslip, and seal with glycel or nailpolish.
(12) Investigate with LM (phase-contrast).

Comments. Investigation of algal remains in sediments are very time-consuming. However, with the addition of external markers, the investigation time can be reduced. Without markers added, microfossils have to be counted in a defined volume of sediments (0.5–1 ml) to make it possible to calculate the absolute number of algal remains in sediments. When using external markers, the counting can be interrupted when enough microfossils and markers have been counted. With this method it is possible to count the number of microfossils per cm^2 per year (Kaland and Stabell, 1981; Battarbee, 1982).

D. *Chrysophycean analyses*

(1) Place 0.5–1 ml of sediment into a centrifuge tube.
(2) Add microsphere solution (not for EM, or procedures D.*b* and D.*c*)
(3) Add 5–10 ml of 37% H_2O_2.
(4) Boil on a waterbath for about half an hour.
(5) Rinse the solution several times with distilled water.

D.*a*
  (6) Spread 1–2 drops evenly on a coverslip
  (7) Dry for a minimum of 12 hours at room temperature.
  (8) Embed with Hyrax or Naphrax.
  (9) Investigate with LM (phase-contrast).

or
D.*b*
  (6) Place 1–2 drops on a round coverslip.
  (7) Dry for a minimum of 12 hours at room temperature.
  (8) Glue the coverslip on to an SEM specimen stub.
  (9) Coat the preparation with gold or gold/palladium.
 (10) Investigate with SEM.

or
D.*c*
  (6) Place 1 drop of solution on formvar coated, shadowcast coppergrid.
  (7) Dry for a minimum of 12 hours at room temperature.
  (8) Investigate with TEM.

Comments on method D.a. With this method the number of chrysophycean scales and cysts (diatoms) can be calculated together with the external markers, preferably microspheres (see method D.*c*). The chrysophycean scales are often difficult or impossible to identify to species levels with LM. However, they can be counted to genus levels as *Synura, Mallomonas, Chrysophaerella* and *Paraphysomonas*. Some scales are characteristic and can be identified in the embedded preparations (Smol, 1980; Munch, 1980; Battarbee *et al.*, 1980), but the identifications must be checked with EM (TEM) according to method D.*c*. Scale-bearing Chrysophyceae have variable numbers of scales covering their cells and it is therefore difficult or impossible to use the number of scales counted per cm^3 sediment for calculating the number of cells deposited per cm^3. The number of different scales per species is at present unknown.

Comments on method D.b. With this method chrysophycean cysts can be identified. In SEM it is possible to study in detail the ultrastructure of the cysts, since the description of the cysts, the pore, the pore region and the cell wall structure are of important taxonomic value (Cronberg, 1980; Sandgren and Carney, 1983). Micrographs of whole specimens showing these features must be taken. Such photographic documentation is important. Much has been written about chrysophycean cysts (Cronberg, 1985), but few have been identified to species levels. When a cyst has been well-described with SEM, it is often possible to find characteristic features that can also be found in the preparations, embedded according to D.*b*, and thus the cyst can also be identified under LM. By EM examination, characteristics are recognized that are important for the identification of a specific cyst and one can then look for these in LM preparations.

Comments on method D. c. This method is used for an accurate identification of chrysophycean scales belonging to the Mallomonadaceae. It is also possible to scan the grids and to obtain the percentage distribution of the scales on the grids (=sediments). This method is important as it is possible to make identifications to species levels. The chrysophycean species are mostly members of the planktonic community, and therefore reflect the conditions in the water body during earlier periods of development.

Preparation of microsphere solution

Concentrated solutions of microspheres (diameter 5–20 μm) can be bought.* A standard solution is prepared through dilution of the concentrated solution. The number of microspheres per millilitre is calculated in the prepared standard solution through counting the microspheres in a hematocytometer, or with an electronic particle counter. Tests have to be performed to get the appropriate number of microspheres in the standard solution (Kaland and Stabell, 1981).

Fuchsin-B solution for staining

0.5 ml fuchsin-B crystals are dissolved in 50 ml of absolute alcohol. 1–3 drops of the fuchsin solution are then added to the amount of glycerin needed for the preparations. Fuchsin stains the cell walls red, and hence the cells are more easily observed.

CASE STUDY

Subfossil algae from Lake Växjösjön, south central Småland, Sweden, were studied in a sediment core taken from the surface down to a depth of 3 m, covering the period A.D. 1200–1970. The deposition of blue–green algae, green algae and cysts of Chrysophyceae was recorded (Cronberg, 1980, 1982a, b).

The lake was originally oligotrophic but has been influenced by human activities since A.D. 600. (Digerfeldt, 1977; Battarbee *et al.*, 1980). In recent times the lake has been utilized as a recipient for sewage water from the surrounding town. In about 1860 the pollution started and was maintained at high levels throughout the period 1880–1930 (Battarbee and Digerfeldt, 1976). Diversion of sewage started in 1930, but the lake did not recover. During the

*Microspheres can be obtained from: Christison (Scientific Equipment) Ltd., Albany Road, Gateshead NE8 3AT, England; and from Nuclear Products, 3M Company, 3M Center, Saint Paul, Minnesota 55101, U.S.A.

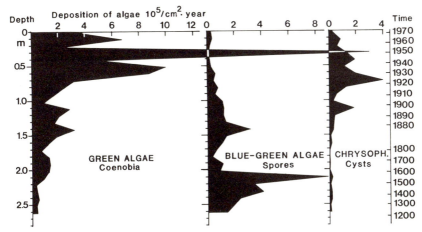

FIGURE 25.1. Deposition of green algae, blue-green algae and Chrysophyceae cysts per cm^2 per year in Lake Växjösjön, 1200–1970 A.D.

worst pollution period black anaerobic gyttja developed. The sediments were rich in nutrients and algal remains.

For quantitative evaluation of algal remains in the sediments, method C described above was applied, and for the identification of chrysophycean cysts method D.*b* was used. Subsamples were taken at 10 cm intervals from the surface to a depth of 2.6 m.

In the lake sediments most blue–green algal spores were found between depths of 2.5 m and 2 m, corresponding to the period 1300–1600. From 1.3 m to the sediment surface (1890–1970) the number of spores decreased. This means that most *Anabaena* spores were found during the oligotrophic period of the lake (Figure 25.1).

Chrysophycean cysts were rare at depths from 2.5 m to 1.3 m, corresponding to the period 1200–1890. When pollution began in 1890, the number of cysts increased and reached a maximum at 0.7 m corresponding to the year 1925. Thereafter the number of cysts deposited decreased again and stabilized at a lower level.

The deposition of green algae increased slowly with depths from 2.6 m to 1 m, but thereafter a pronounced development of green algae occurred. The maximum density was reached at 0.35 m, corresponding to 1950. They then decreased again.

The blue–green algal remains consisted of *Anabaena* spores. The dominant species was *Anabaena lemmermannii* P. Richt. This alga was dominant during the oligotrophic period. *A. lemmermannii* is also common today in oligotrophic lakes in the Växjö region. This alga can sometimes form short blooms during the summer. The spores are gathered in the centre of the colonies, and when they are deposited on the sediment surface the vegetative cells dissolve,

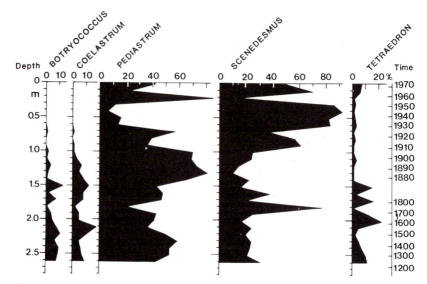

FIGURE 25.2.   Percentage distribution of different green algae in the
sediments of Lake Växjösjön, 1200–1970 A.D

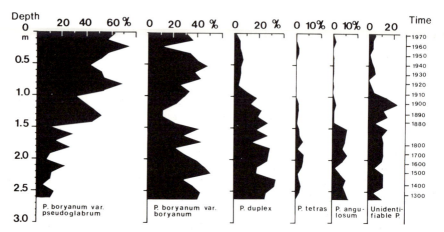

FIGURE 25.3.   Percentage distribution of the different species of *Pediastrum* in the
sediments of Lake Växjösjön, 1200–1970 A.D *Reproduced by permission of E.
Schweizerbart'sche Verlagsbuchhandlung*

leaving the spores in their typical position. This simplifies the identification of
these spores in the sediment. The reduction in *Anabaena* spores from the
beginning of 1900 is a result of changes in the plankton community and the
blue–green algal flora from a dominance of *Anabaena* to a dominance of
Microcystis spp. (Thunmark, 1945).

The green algae found in the sediments were *Botryococcus braunii* Kütz.,

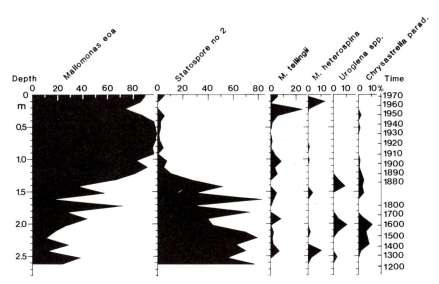

FIGURE 25.4. Percentage distribution of chrysophycean cysts in the sediments of
Lake Växjösjön, 1200–1970 A.D

Coelastrum reticulatum (Danng.) Senn., *Tetraedron minimum* (Brunnt.)
Hansg., *Staurastrum tetracerum* Ralfs, *S. paradoxum* var. *parvum* W. West, *S.
uplandicum* Teil., *Pediastrum* spp. and *Scenedesmus* spp. As few cells of
Staurastrum were found in the sediments, they have been omitted in the
figures.

Botryococcus, *Coelastrum* and *Tetraedron* had a similar distribution in the
sediments (Figure 25.2). They were frequent from 2.6 m to 1.2 m and
decreased from there upwards. Thus they were common during the oligotrophic period, but decreased from about 1880 onwards. *Tetraedron*, however,
showed an increase again after 1960.

Pediastrum spp. were the best preserved green algae in the sediments and
increased in the whole core from 1200 to 1950. A slight decrease in number
was observed from 1950 onwards. The following *Pediastrum* species were
identified: *P. boryanum* var. *boryanum* (Turp.) Menegh., *P. boryanum* var.
pseudoglabrum Parra, *P. duplex* var. *duplex* Meyen, *P. duplex* var. *punctatum*
(Krieg.) Parra, *P. angulosum* (Ehrenb.) Menegh. and *P. tetras* (Ehrenb.)
Ralfs. The percentage distribution of the *Pediastrum* species was about
constant during the period 1200–1850. More recently (1880–1910) changes in
the distribution occurred. *P. boryanum* var. *pseudoglabrum* increased while *P.
angulosum* and *P. tetras* nearly disappeared. *P. duplex* and unidentified
Pediastrum diminished from about 1920. The number of *P. boryanum* var.
boryanum fluctuated but no real change was observed between 1200 and
1970.

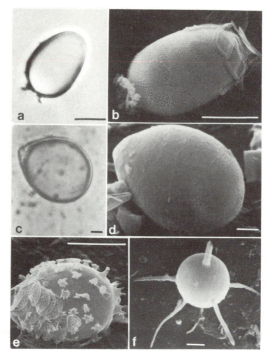

Figure 25.5.   Chrysophycean cysts found in the sediments of Lake Växjösjön — 5 μm bars indicated:
(*a*) *Mallomonas eoa* (LM)
(*b*) *M. eoa* with remaining scales (SEM)
(*c*) *M. heterospina* (LM, photo R. Battarbee)
(*d*) *M. heterospina* (SEM)
(*e*) *M. teilingii* with remaining scales (SEM)
(*f*) *Chrysastrella paradoxa* (SEM).

Scenedesmus species were recorded throughout the whole core (Figure 25.3). Most coenobia were damaged and nearly impossible to identify, so the *Scenedesmus* were counted as total *Scenedesmus* coenobia per cm^2 per year. Common species were, however: *S. acuminatus* (Lagerh.) Chod., *S. armatus* Chod. *S. oahuensis* (Lemm.) G. M. Smith, *S. opoliensis* P. Richt., *S. quadricauda* (Turp.) Bréb. The number of *Scenedesmus* coenobia increased slowly from 1200 to 1900. Thereafter, a drastic increase in number was recorded with maximal deposition between 1940 and 1950. The maximal deposition of *Pediastrum* and *Scenedesmus* occurred between 1930 and 1950, the period when the lake was most heavily loaded with pollution.

Most chrysophycean cysts were deposited during the period 1890–1970, with a maximum occurring around 1925 (Figure 25.1). Six different morphotypes were identified: *M. eoa* Takah., *M. teilingii* (Teil.) Conr., *M. heterospina*

Lund, *Chrysastrella* cf *paradoxa* Chod. and *Statospore no. 2* Cronberg, which is probably a cyst of a *Mallomonas* species. These cysts were found in the whole sediment core (*Uroglena* spp. excluded) but had divergent distributions. *M. eoa* increased in connection with the increase in nutrients, and reached its maximal development during the worst polluted years 1910–1950 (Figure 25.4). *Statospore no 2* was reduced in number drastically around 1880, and thereafter stabilized at a lower level. *M. heterospina* showed two maxima around 1400 and 1960. *M. teilingii* was evenly distributed in the core until 1940 when a maximum developed, after which it decreased in number again. *Chrysastrella* cf. *paradoxa* diminished in number from 1890, and *Uroglena* spp. disappeared completely at the same time. With the increasing pollution of Lake Växjösjön from around 1890, the algal composition in the sediments changed greatly. Oligotrophic species were reduced in numbers or even disappeared completely at that time, while eutrophic species increased drastically.

Chrysophycean cysts found in the lake are illustrated in Figure 25.5.

CONCLUSIONS

It is evident that algae can be good indicators, and hence useful tools, in palaeolimnological reconstructions. More information about earlier conditions in a water body can be gained if the algae are identified to species and not only to genus levels. Different species belonging to the same genus may have quite dissimilar ecological demands. The Chrysophyceae are often looked on as typical oligotrophic algae; but when the ecology of the separate species is studied, it is apparent that they can have completely contrasting demands for requirements such as nutrients and temperature (e.g. *Mallomonas eoa* and *Statospore no. 2*).

Even though it is not possible to identify the algae deposited in the sediments, it is important to record them with drawings or micrographs. If the organisms found are illustrated, it may be possible to identify them later, when more information has been acquired. It is possible that another scientist has found the same algae, and has more knowledge about it.

The blue–green and green algae can be identified with the standard algological literature (see below). The common scale-bearing Chrysophyceae have been identified with EM, and good EM illustrations of the scales can be found in the literature cited.

However, the identification of the chrysophycean cysts is more difficult. The literature is spread over several hundred publications, most of which are very difficult to obtain (Cronberg, 1985). The situation is made even more complicated as most cysts described have unknown biological affinities, and most of them are only illustrated with LM drawings.

During the last decade the cysts have been studied with SEM, and through this technique higher magnification and resolution can be obtained. Furthermore the EM pictures give a stereometric image of the specimen and it can be photographed from different angles. The cysts have a specific ultrastructure that mostly cannot be seen with LM. So the old LM drawings are difficult to compare with the modern SEM micrographs.

However, cysts have been shown to be good palaeoindicators and, to allow more use of them to be made in future investigations, a proposal for the standardization of statospore nomenclature was outlined with recommendations at the First International Chrysophyte Symposium, held in August 1983 at Grand Forks, North Dakota, U.S.A. (Cronberg and Sandgren, 1985).

Thus, with more standardized methods for identifying, counting and recording algae remains from the sediments of different water bodies, it is expected that additional helpful information will be gained from future palaeolimnological investigations and reconstructions.

RECOMMENDED TAXONOMIC LITERATURE

Blue–green algae	Bourrelly (1970)
	Huber-Pestalozzi (1938)
Green algae	Bourrelly (1966)
	Förster (1982)
	Komárek and Fott (1983)
	Růžička (1977, 1981)
	West and West (1904, 1905, 1908, 1911, 1922)
Chrysophyceae: cysts	Carney and Sandgren (1983)
	Cronberg (1980)
	Cronberg (1985)
	Cronberg and Sandgren (1985)
	Deflandre (1932, 1936)
	Frenguelli (1936)
	Huber-Pestalozzi (1941)
	Nygaard (1956)
Chrysophyceae: scales	Asmund and Kristiansen (1985)
	Takahashi (1978)
	Wee (1982)
	Bourrelly (1957, 1968)

Otherwise the standard taxonomic algal literature can be used.

REFERENCES

Alhonen, P., and Ristiluoma, S. (1973). On the occurrence of subfossil *Pediastrum* algae in a Flandrian core at Kirkkonummi, southern Finland. *Bull. Geol. Soc. Finland*, **45**, 73–77.

Asmund, B., and Kristiansen, J. (1985). The genus Mallomonas: a monograph based on the ultrastructure of scales and bristles. *Opera Botanica* (in press).

Battarbee, R. W., and Kneen, M. J. (1982). The use of electronically counted microspheres in absolute diatom analysis. *Limnol. Oceanogr.*, **27**, 184–188.

Battarbee, R. W., Cronberg, G., and Lowry, S. (1980). Observations on the occurrence of scales and bristles of *Mallomonas* spp. (Chrysophyceae) in the micro-laminated sediments of a small lake in Finnish North Karelia. *Hydrobiologia*, **71**, 225–232.

Battarbee, R. W., and Digerfelt, G. (1976). Palaeoecological studies of the recent development of Lake Växjösjön. I: Introduction and chronology. *Arch. Hydrobiol.*, **77**, 330–346.

Battarbee, R. W., Digerfelt, G., Appleby, P., and Oldfield, F. (1980). Palaeoecological studies of the recent development of Lake Växjösjön. III: Reassessment of recent chronology on the basis of modified 210Pb dates. *Arch. Hydrobiol.*, **89**, 440–446.

Borge, O. (1892). Subfossila sötvattensalger från Gotland. *Bot. Notiser 1892*, 55–69.

Borge, O., and Erdtman, G. (1954). On the occurrence in Tertiary Strata in the Isle of Wight. *Bot. Notiser* **107**, 112–113.

Bourrelly, P. (1957). Recherches sur les Chrysophycées, *Rev. Algol. Mem. Hors.*, série 1.

Bourrelly, P. (1966). *Les Algues d'Eau Douce. 1: Les Algues Vertes*, Boubée & Cie, Paris.

Bourrelly, P. (1968). *Les Algues d'Eau Douce. 2: Les Algues Jaunes et Brunes*, Boubée & Cie, Paris.

Bourrelly, P. (1970). *Les Algues d'Eau Douce. 3: Les Algues Bleues et Rouges*, Boubée & Cie, Paris.

Brugam, R. B. (1980). Postglacial diatom stratigraphy of Kirchner Marsh, Minnesota. *Quaternary Research*, **13**, 133–146.

Carney, H. J. (1982). Algal dynamics and trophic interactions in the recent history of Frains Lake, Michigan. *Ecology*, **63**, 1814–1816.

Carney, H. J., and Sandgren, C. D. (1983). Chrysophycean cysts: indicators of eutrophication in recent sediments of Frains Lake, Michigan. *Hydrobiologia*, **101**, 195–202.

Cookson, I. C. (1953). Records of the occurrence of *Botryococcus braunii*, *Pediastrum* and Hystrichosphaerideae in cainzoic deposits of Australia. *Mem. Nat. Mus.*, **18**, 107–123.

Cornell, W. C. (1972). Late cretaceous chrysomonad cysts. *Palaeogeogr., Palaeoclimatol., Palaeoecol.*, **12**, 33–47.

Craig, A. J. (1972). Pollen influx to laminated sediments: a pollen diagram from northeastern Minnesota. *Ecology*, **53**, 46–57.

Crisman, T. (1978). Algal remains in Minnesota lake types: a comparison of modern and late-glacial distributions. *Verh. Internat. Verein. Limnol.*, **20**, 445–451.

Cronberg, G. (1980). Cyst development in different species of Mallomonas (Chrysophyceae) studied by scanning electron microscopy. *Algological Studies*, **25**, 421–434.

Cronberg, G. (1982a). *Pediastrum* and *Scendesmus* (Chlorococcales) in sediments from Lake Växjösjön, Sweden. *Arch. Hydrobiol. Suppl.*, **60**, 500–507 (Algological studies 29).

Cronberg, G. (1982b). Phytoplankton changes in Lake Trummen induced by restoration. *Folia Limnol. Scand.*, **18**.

Cronberg, G. (1985). Chrysophycean cysts and scales in sediments: a review. In: *Chrysophytes — Aspects and Problems* (Eds. J. Kristiansen and R. A. Andersen), Cambridge University Press.

Cronberg, G., and Sandgren, C. (1985). A proposal for the development of standardized nomenclature and terminology for chrysophycean statospores. In: *Chrysophytes —*

Aspects and Problems (Eds. J. Kristiansen and R. A. Andersen), Cambridge University Press.

Deflandre, G. (1932). Archaeomonadaceae, une famille nouvelle de Protistes fossiles marins a loge siliceuse. *C. R. Acad. Sci.*, **194**, 1859–1861.

Deflandre, G. (1936). *Les Flagellés Fossiles. Apercu Biologique et Paleontologique Role geologique, Actualités Scientifiques et Industrielles* 335, Exposés de Geologique, Paris, Hermann & Cie Editeurs, pp. 8–97.

Digerfelt, D. (1977). Palaeoecological studies of the recent development of Lake Växjösjön. II: Settlement and landscape development. *Arch. Hydrobiol.*, **79**, 465–477.

Ehrenberg, C. G. (1854). *Microgeologie. Das Erden und Felsen schaffenden Wirken des unsichtbaren kleinen selbständigen Lebens auf der Erde*, Leipzig. L. Voss.

Elner, J. K., and Happey-Wood, G. M. (1978). Diatom and Chrysophyceae cyst profiles in sediment cores from two linked but contrasting Welsh lakes. *Br. Phycol. J.*, **13**, 341–360.

Faegri, K., and Iversen, J. (1974). *Textbook of Pollen Analysis*, Munksgaard, Copenhagen.

Fjerdingstad, E. (1954). The subfossil algal flora of the Lake Bölling Sö and its limnological interpretation, *Det Konglige Danske Videnskapernes Selskab. Biol. Skrifter*, **7**, 1–35.

Fott, B. (1966). Elektronenmikroskopischer Nachweis von Mallomonas-Schuppen in Seeablagerungen. *Int. Revue Ges. Hydrobiol.*, **51**, 787–790.

Frederick, R. (1977). *The Environmental Significance of the Floras from Three Central Ohio Sediment Profiles*, Ph.D. Thesis, Ohio State University.

Fredskild, B. (1973). Studies in the vegetational history of Greenland: palaeobotanical investigations of some holocene lake and bog deposits. *Meddel. Grönland*, **198**, 1–245.

Fredskild, B. (1975). A late-glacial and early post-glacial pollen-concentration diagram from Langeland, Denmark. *Geol. Fören. Stockh. Förh.*, **97**, 151–161.

Fredskild, B. (1983). The Holocene development of some low and high arctic Greenland lakes. *Hydrobiologia*, **103**, 217–224.

Frenguelli, J. (1936). Crisostomataceas del Neuquén. *Notas del Museo de la Plata*, **1**, 247–275.

Förster, K. (1982). Das Phytoplankton des Süsswassers. 8. Teil, 1. Hälfte. Conjugatophyceae. Zygnematales und Desmidiales (excl. Zygnemataceae). In: *Die Binnengewässer*, XVI (Ed. G. Huber-Pestalozzi).

Geel, B. van (1976). Fossil spores of Zygnemataceae in ditches of a prehistoric settlement in Hoogkarspel (the Netherlands). *Rev. Palaeobot. Palynol.*, **22**, 337–344.

Goulden, C. E. (1970). The fossil flora and fauna (other than siliceous fossils, pollen and chrionomid head capsules). In: (Hutchinson, G. E.) Ianula: an account of the history and development of the Lago di Monterosi, Latium, Italy. *Trans. Am. Phil. Soc.*, New Series, **60**, 102–111.

Gritten, M. M. (1977). On the fine structure of some chrysophycean cysts. *Hydrobiologia*, **53**, 239–252.

Hayworth, E. Y. (1983). Diatom and chrysophyte relict assemblages in the sediments of Blelham Tarn in the English Lake District. *Hydrobiologia*, **103**, 131–134.

Hayworth, E. (1984). Stratigraphic changes in algal remains (diatoms and chrysophytes) in the recent sediments of Blelham Tarn, English Lake District. In: *Lake Sediments and Environmental History*, (Eds. E. Hayworth and J. W. G. Lund) Leicester University Press.

Huber-Pestalozzi, G. (1938). Das Phytoplankton des Süsswassers. 1. Teil, Allgemeiner Teil. Blaualgen. Bakterien. Pilze. In: *Die Binnengewässer*, XVI, (Ed. G. Huber-Pestalozzi).

Huber-Pestalozzi, G. (1941). Das Phytoplankton des Süsswassers. 2. Teil, 1. Hälfte. Chrysophyceen. Farblose Flagellaten. Heterokonten. In: *Die Binnengewässer*, XVI (Ed. G. Huber-Pestalozzi).

Jankovska, V., and Komárek, J. (1982). Das Vorkommen einiger Chlorokokkalalgen in bömischen Spätglazial und Postglazial. *Folia Geobot. Phytotax.*, **17**, 165–196.

Kaland, P. E., and Stabell, B. (1981). Methods for absolute diatom frequency analysis and combined diatom and pollen analysis in sediments. *Nord. J. Bot.*, **1**, 697–700.

Komárek, J., and Fott, B. (1983). Das Phytoplankton des Süsswassers. 7. Teil, 1. Hälfte. Chlorophyceae (Grünalgen), Ordnung: Chlorococcales. In: *Die Binnengewässer*, XVI (Ed. G. Huber-Pestalozzi).

Korde, N. W. (1961). Characteristische Merkmale der Stratifikation der Bodenablagerungen in Seen mit verschiedenartigem Zufluss. *Verh. Int. Verein. Limnol.* **14**, 524–532.

Korde, N. W. (1966). Algenreste in Seesedimenten. Zur Entwicklungsgeschichte der Seen und umliegenden Landschaften. *Ergebn. Limnol.*, **3**, 1–38.

Krieger, W. (1929). Algologisch-monographische Untersuchungen über das Hochmoor am Diebelsee. *Beitrag z. Naturdenkmalpflege*, **13**, 235–300.

Lagerheim, G. (1902). Untersuchungen über fossile Algen: I, II. *Geol. Fören. Förhandl.*, **24**, 475–500.

Livingstone, D., and Jaworski, G. H. M. (1980). The viability of akinets of blue–green algae recovered from the sediments of Rostherne Mere. *Br. Phycol. J.*, **15**, 357–364.

Livingstone, D. A., Bryan, K., and Leahy, R. G. (1958). Effects of an arctic environment on the origin and development of freshwater lakes. *Limnol. & Oceanogr.*, **3**, 192–214.

Lundqvist, G. (1927). Bodenablagerungen und Entwicklungstypen der Seen. In: *Die Binnengewässer*, II. (Ed. A. Thienemann).

Messikommer, V. E. (1938). Beitrag zur Kenntnis der fossilen und subfossilen Desmidiaceen. *Hedwigia*, **78**, 107–201.

Moss, B. (1979). Algal and other fossil evidence for major changes in Strumpshaw Broad, Norfolk, England, in the last two centuries. *Br. Phycol. J.*, **14**, 263–283.

Munch, C. S. (1980). Fossil diatoms and scales of Chrysophyceae in the recent history of Hall Lake, Washington. *Freshwater Biology*, **10**, 61–66.

Nipkow, F. (1927). *Über das Verhalten der Skelette planktischer Kieselalgen im geschichteten Tiefenschlamm des Zürich- und Baldeggersees*, Ph.D. Thesis, Eidgenössische Technische Hochscule, Zürich, pp. 1–49.

Nygaard, G. (1956). Ancient and recent flora of diatoms and Chrysophyceae in Lake Gribsø. *Folia limnol. Scand.*, **8**, 1–99 and 254–262.

Patrick, R., and Reimer, C. W. (1966). *The Diatoms of the United States exclusive of Alaska and Hawaii*, Vol. I, Acad. Nat. Sci. Philadelphia Monograph 13.

Post, H. (1860). Resultater af en undersökning utaf gyttja, dy, torf och mylla samt deras huvudbeståndsdelar. *K. Vet.-Akad. Förhandl.*, **17**, 41–57.

Post, H. (1862). Studier öfver Nutidens koprogena jordbildningar, gyttja, dy, torf och mylla. *K. Svenska Vetenskapsakademien Handlingar*, **4**, 1–59.

Renberg, I. (1976). Annually laminated sediments in Lake Rudetjärn, Medelpad Province, northern Sweden. *Geol. Fören. Stockh. Förh.*, **98**, 355–360.

Růžička, J. (1977). *Die Desmidiaceen Mitteleuropas: 1(1)*, E. Schweizerbart'sche Verlagsbuchhandlung, Stuttgart.

Růžička, J. (1981). *Die Desmidiaceen Mitteleuropas:* 1(2), E. Schweizerbart'sche Verlagsbuchhandlung, Stuttgart.

Salmi, M. (1963). On the subfossil Pediastrum algae and molluscs in the Late-Quaternary Sediments of Finnish Lapland. *Arch. Soc. Vanamo*, **18**, 105–120.

Sebestyen, O. (1969). Studies on Pediastrum and cladoceran remains in the sediments of Lake Balaton, with reference to lake history. *Mitt. Int. Verein. Limnol.*, **17**, 292–300.

Simola, H. (1977). Diatom succession in the formation of annually laminated sediment in Lovojärvi, a small eutrophicated lake. *Ann. Bot. Fennici*, **14**, 143–148.

Simola, H. (1979). Micro-stratigraphy of sediment laminations deposited in a chemically stratifying eutrophic lake during the years 1913–1976. *Holarctic Ecology*, **2**, 160–168.

Simola, H., Coard, M. A., and O'Sullivan, P. E. (1981). Annual laminations in the sediments of Loe Pool, Cornwall. *Nature*, **290**, 238–241.

Simola, H., and Uimonen-Simola, P. (1983). Recent stratigraphy and accumulation of sediment in the deep oligotrophic Lake Pääjärvi in south Finland. *Hydrobiologia*, **103**, 287–293.

Smol, J. P. (1980). Fossil synuracean (Chrysophyceae) scales in lake sediments: a new group of palaeoindicators. *Can. J. Bot.*, **58**, 458–465.

Smol, J. P. (1983). Palaeophycology of a high arctic lake near Cape Herschel, Ellesmere Island. *Can. J. Bot.*, **61**, 2195–2204.

Steinecke, F. (1927). Leitformen und Leitfossilien des Zehlaubruches. Die Bedeutung der fossilen Microorganismen für die Erkenntnis der Nekrozönosen einer Moors. *Bot. Archiv*, **19**, 327–344.

Stockmarr, J. (1971). Tablets with spores used in absolute pollen analysis. *Pollen et Spores*, **13**, 615–621.

Takahashi, E. (1978). *Electron Microscopical Studies of the Synuraceae (Chrysophyceae) in Japan*, Tokai University Press, Tokyo.

Tippett, R. (1964). An investigation into the nature of the layering of deep-water sediments in two Eastern Ontario Lakes. *Can. J. Bot.*, **42**, 1693–1709.

Wee, J. L. (1982). Studies on the Synuraceae (Chrysophyceae) of Iowa. *Bibliotheca Phycologica*, **62**, 1–183.

West, W., and West, G. S. (1904). *The British Desmidiaceae*, Vol. I, Monogr. Ray Soc. 82.

West, W., and West, G. S. (1905). *The British Desmidiaceae*, Vol. II, Monogr. Ray Soc. 83.

West, W., and West, G. S. (1908). *The British Desmidiaceae*, Vol III, Monogr. Ray Soc. 88.

West, W., and West, G. S. (1911). *The British Desmidiaceae*, Vol IV, Monogr. Ray Soc. 92.

West, W., and West, G. S. (1922). *The British Desmidiaceae*, Vol V, Monogr. Ray Soc. 108.

Whiteside, M. C. (1965). On the occurrence of *Pediastrum* in lake sediments. *J. Arizona Acad. Sci.*, **3**, 144–146.

Wilson, L. R., and Hoffmeister, W. S. (1953). Four new species of fossil Pediastrum. *Am. J. Sci.*, **251**, 753–760.

Handbook of Holocene Palaeoecology and Palaeohydrology
Edited by B. E. Berglund

26

Diatom analysis

Richard W. Battarbee

Palaeoecology Research Unit, Department of Geography, University College London, London, U.K.

INTRODUCTION

The assemblages of diatoms preserved in lake sediments can directly reflect the floristic composition and productivity of lake diatom communities, and can indirectly reflect lake water quality, especially pH and alkalinity, nutrient status, and salinity. In most lakes diatom communities are diverse, they occupy a wide range of habitats, and the siliceous frustules of the diatom cell, that possess most of the characters needed for precise identification, accumulate in high concentrations in sediments. The potential of diatoms as palaeoecological indicators is clear, but it should also be noted that the quality of preservation varies from site to site, many aspects of the ecology of diatoms are, as yet, poorly understood, identification can often be difficult, and palaeoecological interpretation must take into account the complexities of the sedimentary environment in which the frustules accumulate.

Diatoms have been recorded and classified for over two centuries. In the late nineteenth century the systematic and taxonomic investigations of modern and fossil diatoms began to be complemented by attention to aspects of distributional ecology (Cleve, 1894–95), but it was not until the second decade of the twentieth century that the palaeoecological value of diatoms in lake sediments began to be recognized (e.g. Cleve-Euler, 1922; Lundqvist, 1924; Nipkow, 1920, 1927).

Since then a body of information on the relation of diatom taxa to habitat, salinity, pH and nutrient content has been accumulated (Kolbe, 1927; Hustedt, 1937–39; Jørgensen, 1948; Cholnoky, 1968). There is a growing literature on diatom autecology and resource competition (Lund, 1949, 1950,

1954, 1955; Lund *et al.*, 1963; Shear *et al.*, 1976; Haffner, 1977; Kilham, 1971; Tilman, 1977), and aspects of the role of diatoms in contemporary lake ecosystems has been evaluated (e.g. Lund, 1949; Jørgensen, 1957; Gibson, 1981; Bailey-Watts, 1976; Reynolds, 1973).

A more recent element has been the use of multivariate statistical techniques to relate surface sediment diatom assemblages to environmental parameters in a way that allows downcore changes to be evaluated in a more quantitative way (e.g. Brugam, 1980, 1983; Gasse and Tekaia, 1983; Huttunen and Meriläinen, 1983; Davis and Anderson, 1985).

On the basis of these ecological approaches and advances, and with the taxonomic help of general floras (Schmidt, 1874–1959; Hustedt, 1930, 1930–66; Cleve-Euler, 1951–55; Patrick and Reimer, 1966, 1975) diatomists have been able to address themselves to a wide range of palaeoenvironmental questions. In lacustrine environments these include the nature of lakes in the late glacial period (e.g. Fjerdingstad, 1954; Florin, 1970; Haworth, 1976), the ontogeny of lakes (e.g. Pennington, 1943; Nygaard, 1956; Round, 1957; Haworth, 1969; Digerfeldt, 1972), water level and associated climatic change (Ehrlich, 1973; Bradbury *et al.*, 1981; Gasse and Street, 1978; Richardson and Richardson, 1972; Stoffers and Hecky, 1978) and the disturbance of lake systems as a result of human activity (e.g. Bradbury, 1975; Battarbee, 1978; 1984a; Brugam, 1978; Moss, 1978, 1980).

In addition to these freshwater studies, diatom analysis is widely used at the interface of fresh and saline environments to identify lake isolation from the sea in areas of land uplift (Halden, 1929, 1931; Fromm, 1938; Florin, 1946; Miller, 1964; Alhonen, 1971; Eronen, 1974; Kjemperud, 1981a; Renberg, 1976), to indicate marine and brackish water transgressions (Florin, 1944; Digerfeldt, 1975), and to locate the positions of past shorelines (e.g. Florin, 1944; Miller, 1977; Miller and Robertsson, 1979).

This chapter reviews the methods used in analysing and interpreting the diatom content of sediments. Most emphasis is placed on lacustrine sediments.

METHODS

Laboratory techniques

There are a number of ways of extracting diatoms from sediments for slide preparation. The precise combination of steps must be determined experimentally in relation to the individual characteristics of the sediment being analysed. Care should be taken at all stages not to lose or damage valves, and occasional checks for such events should be made by examining treated material before, during and after the various stages of preparation. It should be noted that strong acids can destroy delicate spines and processes and that vigorous stirring and rapid centrifugation can break fragile diatoms. At all times care should be taken to avoid contamination of samples. It is especially important in laboratories where diatoms from different sites are being prepared simultaneously. Clean glassware is essential and if glassware is to be reused it should be cleaned with hot 10% Na_2CO_3.

Three kinds of sedimentary material may require removal in order to obtain a satisfactory slide preparation, and the techniques commonly used for these procedures are as follows (see Figure 26.1):

(1) *Salts soluble in hydrochloric acid.* Carbonate and many metal salts and oxides can be removed by treatment with dilute hydrochloric acid. For non-calcareous sediments this stage can be omitted. Place a small quantity of sediment into a beaker and cover with 10% HCl. Agitate and heat gently for about 15 minutes. Centrifuge or settle, and wash in distilled water.

(2) *Organic matter.* Organic matter can be removed by oxidation. Add 30% H_2O_2 and heat gently in a water-bath or on a hotplate until all organic matter has been removed. If there is a coarse organic residue sieve using a 0.5 mm screen. For some sediments removal of the organic matter is the only step required, in which case a sample residue should be thoroughly washed (centrifuge or settle and wash with distilled water at least three times) and suspended in distilled water ready for slide preparation.

(3) *Minerogenic matter.* Where minerogenic material prevents good slide preparation, attempts can be made to reduce the mineral material by physical methods. Coarse-grained mineral matter may be removed by sieving (mesh size not less than 0.5 mm) or by gentle swilling in a beaker followed by decanting the diatoms and the finer mineral fractions (Brander, 1936). Clays can be partially removed in a related manner, allowing diatoms and coarser mineral particles to sediment before decanting and discarding suspended clay. This technique may be improved by allowing the sedimentation to take place in dilute ammonium hydroxide solution.

Diatoms may also be separated from mineral particles by flotation in heavy liquids (Jouse, 1966; Knox, 1942). However, these methods are tedious, use hazardous chemicals, and can be ineffecient in total diatom recovery. It is probably wiser to make dilute preparations and inspect a greater coverslip area than to risk diatom loss.

After treatment and the final wash the preparation should be checked to ensure adequate removal of unwanted material and to ensure that most frustules and colonies have been separated into single valves. If not, and it appears that identification and counting might be impaired by valves lying in girdle view, a subsample of the final suspension can be sonicated in an ultrasonic bath. The treatment is effective, but since it also leads to fracturing

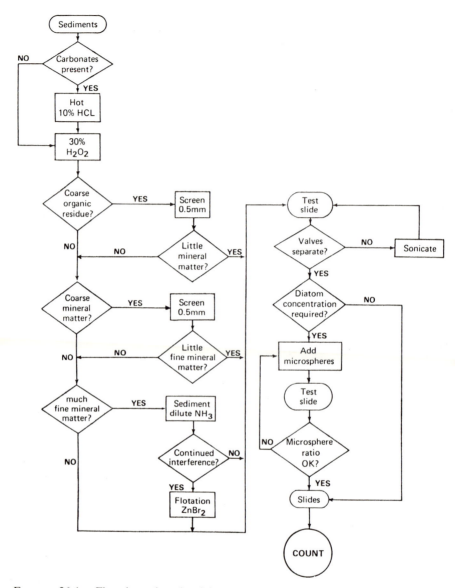

FIGURE 26.1. Flowchart for the laboratory preparation of diatoms from lake sediments, based on the procedure used at the Palaeoecology Research Unit, University College London

of valves it should only be used in a qualitative way, as a guide to identification and relative frequency.

The prepared suspension of diatoms in distilled water is diluted to an appropriate concentration and thoroughly mixed. If diatom concentration is required marker microspheres can be added at this stage. Using a clean pipette about 0.2 ml of suspension is dropped carefully on to a coverslip and the diatoms are allowed to settle and the water evaporate at room temperature. When dry the coverslip is mounted using a resin of high refractive index (e.g. Hyrax, Naphrax).

Light microscopy

Almost all diatom examination needs to be carried out using oil immersion objectives and magnifications of ×750 and over. A phase contrast condenser is recommended and a powerful source of illumination should be used. An eyepiece graticule or a micrometer eyepiece is needed to measure valve dimensions, and the graticule should be calibrated using a micrometer slide with a scale marked off in 0.01 mm divisions. A camera lucida attachment is useful, and microphotography should be available for use during routine counting.

Identification and taxonomy

Diatom taxonomy is not easy. Most sediments contain well in excess of 100 taxa and each taxon is usually present in a range of different morphological forms. Morphological variation is caused especially by changes in size and shape associated with both vegetative and sexual reproduction (McDonald, 1869; Pfitzer, 1869; Nipkow, 1927; Geitler, 1932) and by changing environmental conditions (e.g. Margalef, 1969; Kilham and Kilham, 1975; Tropper, 1975). In the analysis of sedimentary sequences it is wise to separate all forms that appear to have morphological unity since their behaviour through a core may have stratigraphic interest. Reference collections of drawings and photographs should be made of the material from each site studied. Standard floras are Hustedt (1930, 1930–66), Cleve-Euler (1951–55), Huber-Pestalozzi (1942), Patrick and Reimer (1966, 1975), Molder and Tynni (1967–80), and Schmidt's *Atlas* (1874–1959). Terminology should be in accordance with the proposals outlined in Anon. (1975) and Ross *et al.* (1979).

Electron microscopy

While traditional floras (with the exception of Helmcke and Krieger, 1953–64) are based on light microscope taxonomy, modern systematic research is conducted using electron microscopy (e.g. Ross and Sims, 1972; Florin, 1970; Lowe, 1975; Round, 1972; Håkansson, 1976; Schoeman and Archibald, 1976; Hasle and Heimdal, 1970), and palaeoecologists have used electron microscopes extensively in routine analysis (e.g. Miller, 1969; Gasse, 1975; Haworth, 1975; Renberg, 1976; Battarbee, 1978a).

The preparation of diatom material for the scanning electron microscope is straightforward. A small drop of prepared suspension is allowed to evaporate directly on a specimen stub which is then coated with gold, in a sputter-coater apparatus. The sample is then ready for examination. The transmission electron microscope (TEM) lacks many of the advantages of the SEM for diatom taxonomy, but its resolving power is much greater and it is especially appropriate for the inspection of thin-walled specimens. Samples are prepared by drying a drop of clean suspension on to formvar-coated copper grids. As diatom taxonomy is progressively being re-evaluated using electron microscopy, use of such instruments is becoming essential in routine palaeoecological work both for assessing specific taxonomic problems and for aiding the interpretation of images seen in the light microscope.

Percentage counting

The techniques and statistics of percentage diatom counting are analogous to those for pollen analysis in which the number of individuals of a particular taxon are expressed as a percentage of a total sum. Most often the sum used in diatom analysis comprises all taxa. However, as in pollen analysis, there are sometimes considerable advantages in either excluding some taxa from the sum (e.g. diatoms suspected of being reworked from older deposits or derived from an upstream source), and in constructing special sums, such as a plankton sum, in which unwanted variation caused by the more irregular occurrence of periphytic taxa can be eliminated (Battarbee, 1978b, 1984b). The usefulness of such manipulation is dependent on the ease of separating taxa into groups of different provenance, habitat or ecology.

The total number of valves to be counted for each sample varies according to the purpose of the analysis and according to the need to produce statistically good results. The statistical precision of percentage counting depends on the frequency of the taxon in the sample count in relation to the size of the sample count. Mosimann (1965) and Maher (1972) have discussed these relationships with regard to pollen analysis. Their remarks and conclusions can be applied to diatom analysis. An illustration of the way in which percentages change as the total counted increases is shown in Figure 26.2. It can be seen that there are marked differences in the percentages between a count of 100 and 200 while there is little change between 400 and 500. A count of 300 to 600 may therefore be recommended for purposes of routine analysis.

In some cases this total may be inappropriate. A lower count is sufficient if it is only necessary to locate biostratigraphic boundaries or events in replicate cores that are to be used for stratigraphic correlation (Battarbee, 1978c), although more detailed counting is necessary to fix stratigraphic changes precisely. On the other hand, a higher count may be necessary if the complete floristic composition of the sample is to be recorded. While most of the taxa in a

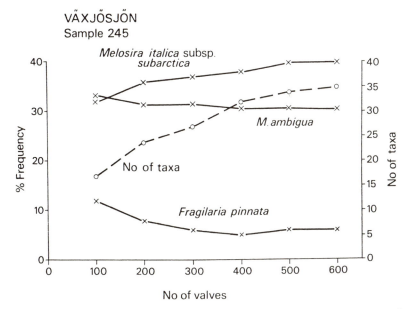

VÄXJÖSJÖN
Sample 245

Melosira italica subsp. *subarctica*

M. ambigua

No of taxa

Fragilaria pinnata

% Frequency

No of taxa

No of valves

FIGURE 26.2. Relationship between percentage of various taxa, and number of taxa, as the sample count is increased

sample may be encountered after a count of about 400 valves (cf. Figure 26.2) additional taxa are likely to be encountered if greater total numbers are counted. For example, by counting over 2000 individuals at each level Miller recorded a number of rare and interesting forms that may otherwise have passed unnoticed (Miller, 1964). A large count is also necessary if the fluctuations of interesting or ecologically important taxa through a core are obscured by the mass occurrence of more common taxa. This can occur especially in large eutrophic lakes where the periphyton may be swamped by the quantity of planktonic individuals present, or in shallow alkaline lakes where massive dominance by *Fragilaria* species is not unusual. In such cases special sums excluding these groups could be constructed.

When counting it is necessary to ensure that a representative proportion of the coverslip is examined. This can be achieved by counting from a number of randomly selected fields of view, or, more usually, by counting along continuous traverses. The traverses should include equal proportions of edge and centre to override any sorting that may have taken place on evaporation.

The basic counting units should be the single valve, so that complete frustules are counted as two. Where chains of frustules remain intact each valve should be counted individually. When a long chain of an infrequent diatom is encountered this may lead to statistical bias, but in this case the sample count can be increased so that a more accurate estimate of the true relative frequency of the taxon is obtained.

Fragments of diatom valves are included in the counts as long as a system is adopted that excludes the possibility of double or multiple counting. The best ways of doing this partly depends on the taxon concerned. In most cases the most satisfactory solution is to count all those fragments that include the valve centre (e.g. central nodule in *Pinnularia*, central inflation in *Tabellaria*, central porate area in *Stephanodiscus*), or a single characteristic feature of the valve, e.g. the larger apical inflation in *Asterionella formosa*. Some taxa (e.g. some *Synedra* species) do not possess recognizable centres, and an alternative approach is to count only fragments that include a valve end, and divide the total by 2.

The counting process itself can be facilitated by restricting the width of a traverse with parallel hairs or a micrometer eyepiece, thus obviating the problem of identifying taxa on the fringes of the field of vision, and allowing attention to be concentrated on a smaller area. Where it is uncertain whether a valve lies in or outside a line of traverse a convention can be adopted whereby valves on one side are included and those on the other excluded.

Diatom concentration

It is often useful to estimate the concentration of diatoms per unit weight or unit volume of sediment. Given the rate of sediment accumulation, diatom concentration values can be used to calculate the diatom accumulation rate per cm^2 per year.

There are three main ways of carrying out counts for diatom concentration calculations, all of which begin with the laboratory preparation of known quantities of sediment.

Aliquot method

In this method a measured amount of suspension is pipetted on to a circular coverslip and a slide prepared as described above. Because of the shape of the meniscus of the initial drop of suspension, and because of convection and surface tension effects on evaporation, the distribution of valves on coverslips prepared in this way is non-random (Eaton and Moss, 1966; Battarbee, 1973b). It is invalid, therefore, to calculate a total for the coverslip on the basis of counts over a sample area (e.g. a diameter traverse or randomly selected fields of vision), and it is extremely time-consuming to count the total number of valves over the whole coverslip under oil immersion (see Table 26.1). Instead the following procedure can be adopted. First, carry out a detailed percentage count of the coverslip; second, select a fairly large, distinctive and numerically frequent diatom as an internal marker; third, count the total number of these

TABLE 26.1.   Diatom concentration of a sediment sample according to the method of estimation. The counting time for each method is also indicated

	Valves counted	Sample total ($\times 10^6$)	Time (h)
Microspheres	1000	9.7416	1
Aliquot	4145	9.270	10
Evaporation tray	1193	8.589	0.5

individuals over the whole coverslip at low magnification ($\times 400$); and fourth, calculate the total number of individuals on the coverslip according to the relationship between the total number of marker valves and the percentage representation of the marker valves in a sample count according to:

$$N = 100\,x/a$$

where N is the total of individuals on the coverslip, x is the total of counted marker valves, and a is the percentage frequency of the marker valve in the relative count. While the slides are easy to prepare this technique is laborious, especially if replicate slides are counted; moreover, there are inaccuracies in pipetting small quantities of suspension.

Some of these problems can be solved by using the technique employed by Renberg (1976) in which marker valves were counted using an inverted microscope. In this way counts can be made more rapidly and errors in measuring subsample volumes are reduced.

Evaporation tray method

This technique has been described by Battarbee (1973b) and Battarbee and McCallan (1974) and is designed specifically for estimating microfossil concentrations. It exploits the principles used in the preparation of samples for inverted microscope counting. The technique involves the use of a circular tray. Four evenly spaced circular wells in the floor of the tray exactly accommodate circular coverslips, so that when in position the coverslips do not project above the floor of the tray. 25 ml of a thoroughly mixed sample suspension is gently poured into the centre of the tray. The water is left to evaporate at room temperature, after which the coverslips can be picked out and mounted in the normal way.

There may be two problems with this method. First, occasional redistribution of valves as the water finally evaporates may disturb the pattern of valves on the coverslip; and second, since a large volume (25 ml) of water is evaporated the suspension needs to be as free as possible from dissolved salts.

If the sample is not thoroughly washed crystallization of salts on the coverslip may occur. This problem can be solved by increasing the number of washes or by carefully removing about 15 ml of the water in the tray by pipette after the diatoms have had sufficient time to settle.

The advantage of the method lies in its statistical precision. Sedimentation from a well-mixed suspension leads to a random distribution of valves on the coverslips, and totals can be estimated from sample counts either along diameter traverses or in selected fields of view. The total of valves in the subsample can be easily calculated after computing the appropriate factors:

$$\text{Total} = \text{count} \times \frac{\text{coverslip area}}{\text{area counted}} \times \frac{\text{sample volume}}{\text{subsample volume}}$$

The total should then be divided by 2 to represent the number of cells rather than valves. Since the counting procedure involves sample counts from the four coverslips in each of two trays, 95% confidence intervals are attached on the basis of 8 sample counts and 6 degrees of freedom.

Microsphere markers

The most convenient method of estimating diatom concentration is to use external markers added to the sample in known quantities. This is a standard technique in pollen analysis (Benninghoff, 1962; Matthews, 1969; Bonny, 1972; Stockmarr, 1971), and the principle can be applied to diatom analysis (Battarbee and Kneen, 1982). Some diatomists have used pollen (Brugam, 1978; Donner *et al.*, 1978) but Battarbee and Kneen (1982) recommend the use of microspheres that are commercially available, are easy to count using a particle counter, and are easy to use in the high concentration necessary for much diatom analysis. Figure 26.3 shows an example of results obtained using this method. Details of the method are described by Battarbee and Kneen (1982), who also compare the counting times for the three methods described here (see Table 26.1).

Diatom concentration is calculated from:

$$\Sigma = \frac{\text{microspheres introduced} \times \text{diatoms counted}}{\text{microspheres counted}}$$

Diatom biovolume and accumulation rate

The concentration values can often have little palaeoecological significance, but they are values from which other data can be calculated. Taking into account the rate of sediment accumulation, diatom accumulation rate (D.A.R.) can be calculated for individual taxa and for all taxa combined.

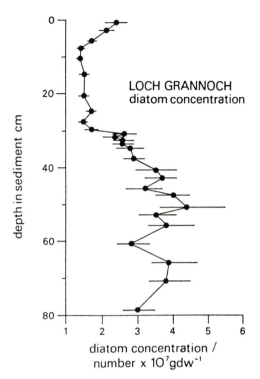

FIGURE 26.3.   Diatom concentration values for a short core from Loch Grannoch, south-western Scotland, with 95% confidence intervals for the counting error. Counts are obtained using the microsphere method (Battarbee and Kneen, 1982)

However, since diatoms vary considerably in size this latter statistic may not be very meaningful in palaeoproductivity terms, unless some conversion from cell number to biomass or biovolume is undertaken (Paasche, 1960). Mean diatom volumes can be computed and used to convert cell number data to cell volume data for individual taxa (e.g. Table 26.2), and then the totals for individual taxa can be added, corrected for the rate of sediment accumulation, and expressed as diatom biovolume per cm^2 per year (Figure 26.4). This can be plotted both against depth and, more realistically, since the depth axis does not often represent a linear timescale, against time (Battarbee, 1978a, b).

Measurements of total amorphous silica

In most freshwater sediments diatoms account for most of the amorphous silica present, although in some sediments the siliceous cysts of chrysophytes may also be important, and sponge spicules are sometimes a minor component. An estimate of the amount of biogenic amorphous silica in lake sediments has been

TABLE 26.2. Diatom cell volumes of selected taxa
from Lough Neagh

Taxon	μm^3
Melosira italica subsp. *subarctica*	494
M. ambigua	280
M. islandica subsp. *helvetica*	3825
Cyclotella comta	1700
C. comensis	100
C. ocellata	534
Stephanodiscus astraea	11,536
S. astraea var. *minutula*	695
S. dubius	511
Tabellaria flocculosa	2137
Asterionella formosa	408

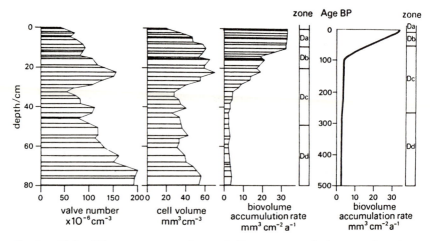

FIGURE 26.4. Diatom concentration and diatom accumulation rate data for a
core from Lough Neagh, N. Ireland, showing the conversion of valve
concentration versus depth values to biovolume accumulation rate versus time
(from Battarbee, 1978). *Reproduced by permission of The Royal Society*

used as a substitute for diatom concentration (e.g. Digerfeldt, 1972; Renberg,
1976). The most commonly used technique involves the measurement of
weight loss after dissolution of the amorphous silica with sodium carbonate.
The good relationship between values of biogenic silica and diatom biovolume
based on sediment trap samples has been shown by Flower (1980). It is clear,
however, from the relationship shown in Figure 26.5 that the regression line
does not pass through the origin, indicating the presence of a small systematic
error in the method equal to an overestimate of about 7.5 mg/g in this case.

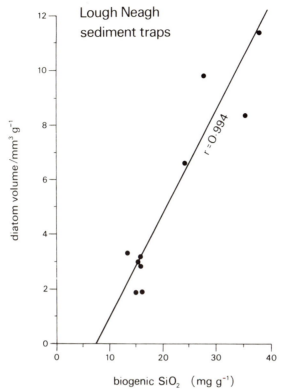

FIGURE 26.5.   Correlation and linear regression of diatom volume and biogenic SiO_2 for sediment trap data from Lough Neagh (from Flower, 1980). *Reproduced by permission of The New University of Ulster*

Data manipulation and presentation

Manipulation of the data from diatom counts requires simple but repetitive calculations that are best carried out by a computer. Computer handling also facilitates calculating percentages according to alternative sums, combining percentages of taxa into ecological or other groupings, statistical analysis, zonation or other purposes (Gordon and Birks, 1972), and rapid plotting and diagram construction (Vuorinen and Huttunen, 1981). Diagrams are usually drawn using bar histograms to represent the percentages of all or selected taxa at each level sampled down the core.

REPRESENTATIVITY

Before discussing the various ways in which diatoms in lake sediments can be used to indicate ecological and environmental change, it is necessary to consider how faithfully the diatom assemblages in the sediment represent the source communities from which the assemblages are derived, and how the spatial and temporal variability of the sediment can affect the diatom record. Questions of representativity embrace a number of issues.

If the diatom concentration and accumulation rate of the sediment is to be used to indicate productivity changes, it is essential to realize that the diatom response to nutrient changes may be non-linear, especially at high levels of productivity where silica is a limiting nutrient (Lund, 1950; Jørgensen, 1957), and increased production is accounted for by, for example, blue–green algae (Gibson *et al.*, 1971). A similar problem is likely to occur in acid waters where diatom plankton is replaced by *Cryptophyceae* and other algal groups without any necessary change in primary production.

Second, the individual valves and frustules comprising the diatom assemblage of a subsample of lake mud are derived from a variety of different source communities. While the plankton communities are usually well represented (Battarbee, 1979, 1981; Simola, 1981; Haworth, 1980), periphytic communities are likely to be less well represented at the core site, and taxa from external sources (e.g. upstream habitats, catchment soils, reworked sediments) may also be present in the sediment assemblage (Battarbee and Flower, 1984).

Third, not all diatoms produced in the lake are incorporated in the sediment. Many, depending on the standing crop and the water-retention period, will be lost down the outflow; others will be lost in the guts of zooplankton and fish down the outflow; and grazing by herbivores may fracture frustules to such an extent that fragments become unidentifiable and dissolution is enhanced (Cooper, 1962).

Fourth, in some cases partial or complete dissolution of frustules may also occur in the water column and in the sediments. Preservation is sometimes poor in calcareous sediments, although this is not always the case (e.g. Hecky and Kilham, 1973; Evans, 1970), and poor preservation can occur in acidic lakes (Round, 1964; Eronen in Donner *et al.*, 1978) and in meromictic lakes (Merilainen, 1971, 1973). Many factors are important (Hurd, 1972, 1973), chief of which may be the silica content of the cell wall (Jouse, 1966); temperature and pH (Rippey, 1977, 1983); the concentration gradient of the dissolved SiO_2 between mud and water (Tessenow, 1966); benthic macroinvertebrate activity (Tessenow, 1964); water depth (Jouse, 1966); sediment accumulation rate (Bradbury and Winter, 1976); and the availability of polyvalent cations for adsorption to the cell wall to provide a protective coating (Lewin, 1961).

Fifth, while in studies of long-term change the temporal resolution of a sample may not be important, where it is necessary to examine rapid stratigraphic transitions or short time periods this can be a limiting factor. The period of time represented by a sediment sample depends on the thickness of the sample, sediment mixing and the rate of sediment accumulation; and these latter two factors can vary considerably from lake to lake. In laminated sediments seasonal diatom changes can be reconstructed if the tape-peel method of Simola is used (Simola, 1977, 1979; Battarbee, 1981; and Figure 26.6). In other lakes slow accumulation rates and bioturbation may place the limit of resolution greater than 10 years.

Lastly, it is usually assumed that the central deepest part of the lake provides a representative location for the analysis of diatom assemblage changes through time. This assumption has rarely been explicitly tested, although there is considerable evidence to suggest that it is a valid assumption at least if peripheral littoral zones are excluded. For Lough Neagh, 14 replicate cores from three main widely separated localities on the lake gave almost identical percentages and sequences of the major planktonic taxa (Battarbee, 1978a, b), and in Augher Lough a transect of cores including shallow and deep-water sites showed very good replication of trends (Anderson, personal communication; Figure 26.7), despite variations in sediment accumulation rates, temporal resolution and percentages. Diatom concentration values, however, may vary more from site to site, and attempts to assess lake-wide diatom accumulation rates are likely to require a multiple-core approach (Moss, 1980).

ENVIRONMENTAL RECONSTRUCTION

Autecological and resource competition studies of diatoms (e.g. Lund, 1949, 1950, 1954, 1955; Knudson, 1954; Kilham and Kilham, 1975; Tilman, 1977) provide useful information for palaeoecological interpretation. However, the number of taxa studied in this way is small and palaeoecologists are more concerned with the behaviour of diatom communities than with single species. Consequently palaeoecologists tend to use classification schemes of various kinds in order to group diatoms into categories reflecting life-form, pH, salinity, and other major variables. In addition there is an increasing attempt to identify the ranges and optima of diatoms with respect to these variables by relating surface sediment diatom assemblages to water-quality characteristics (Brugam, 1980; Huttunen and Meriläinen, 1983; Charles, 1982).

Life-form

While it is often difficult to subdivide periphytic taxa in sediment assemblages into detailed life-form groups, since few taxa are found exclusively in any one of these habitats, a general distinction between plankton and periphyton can usually be made, especially in large productive lakes where euplanktonic taxa dominate. Care is essential, however, since some taxa often found in the plankton can also occur in the periphyton (e.g. *Tabellaria flocculosa*), and

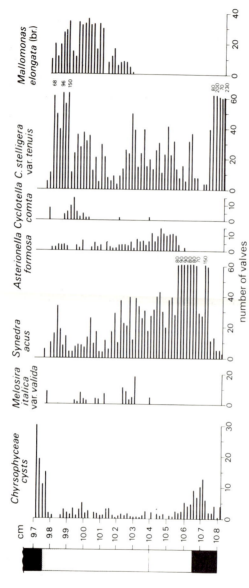

FIGURE 26.6.   Diatom analysis of a single year of sediment accumulation in Laukunlampi, eastern Finland, using the adhesive-tape-peel method of Simola (1977, 1979). The data record the seasonal succession of planktonic diatoms in the lake for the year 1964 (from Battarbee, 1981). *Reproduced by permission of Societas Uppsaliensis pro Geol. Quat.*

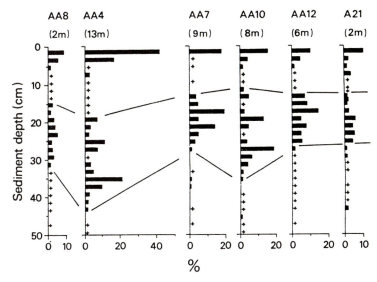

FIGURE 26.7. The record of *Stephanodiscus tenuis* for Augher Lough, N. Ireland, as represented by six cores aligned along a transect of water depth from 2 m to 13 m (from Anderson, unpublished)

Stockner and Armstrong (1971) record *Cyclotella stelligera* beginning its seasonal development in the littoral zone but achieving its maximum population after being washed into the plankton.

Changes in the plankton/periphyton ratio can be interpreted in two main ways: either as due to a change in the morphology of the lake basin as a lake fills in, and/or to a change in productivity (Battarbee, 1978b).

The presence of aerophilous diatoms in lake sediments can also be used to advantage. These taxa can form distinct and diverse communities (Krasske, 1932; Petersen, 1935; Hustedt, 1942; Lowe and Collins, 1973), and while many of the constituent taxa (*Eunotia* spp., *Cymbella* spp., *Pinnularia* spp.) occur both in moist terrestrial and aquatic habitats, some (e.g. *Hantzschia amphioxys, Navicula contenta* f. *biceps, Melosira roseana, Pinnularia borealis*) are aerobiontic (Hustedt, 1957). Haworth (1976) has recorded some of these taxa in the late-glacial sediments of Lochs Cam and Borralan in northern Scotland, and she suggested that they were transported to the lake by solifluction processes. Florin in her study of the history of Kirchner March categorized the recorded taxa according to life-form and thereby showed how a small lake basin had developed following the collapse of the forest floor as underlying dead ice melted. Early aerophilous (eu-terrestrial) forms character-

Kirchner Marsh Minnesota core K-6

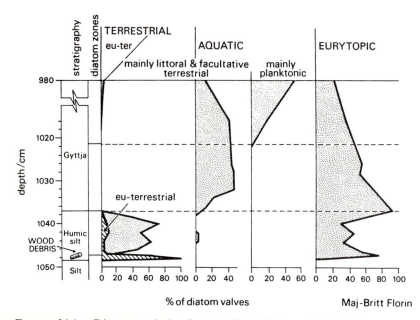

FIGURE 26.8. Diatom analysis of a core from Kirchner Marsh, Minnesota, showing the dominance of eu-terrestrial taxa in the basal sediments (from Florin, 1970b). *Reproduced by permission of J. Cramer*

istic of the forest floor moss flora dominated the lowest levels of the sediment (Figure 26.8; Florin and Wright, 1969).

pH

One of the most useful and commonly used systems for grouping diatoms is the pH system set up initially by Hustedt in his classic paper on the diatom flora of Java, Bali and Sumatra (Hustedt, 1937–39). The diatoms were divided into five categories according to their individual pH tolerances as follows:

(1) *alkalibiontic*, occurring at pH values over 7;
(2) *alkaliphilous*, occurring at pH values about 7 and with widest distribution at pH>7;
(3) *indifferent*, occurring equally on both sidfes of pH = 7;
(4) *acidophilous*: occurring at pH values about 7 with widest distribution at pH<7;
(5) *acidobiontic*: occurring at pH values under 7, with optimum distribution at pH = 5.5 and under.

Nygaard (1956) attempted to develop Hustedt's system by introducing a quantitative element. To do this, instead of taking only the dominant or most frequent forms ('die Massenformen') into consideration he included all taxa, as well as their relative frequencies, arguing that rare species may be as informative as common ones. He categorized his taxa according to the pH preferences reported in the literature, and, as many later authors have found, noted that in some cases the same species had a different placing.

On the assumption that acidobiontic and alkalibiontic forms were much better ecological indicators than their respective acidophilous and alkaliphilous counterparts, Nygaard suggested that in any statistical calculation these groups should be awarded more significance. He therefore weighted the relative frequencies of each acidobiontic and alkalibiontic taxon by an arbitrary factor of five. By adding the totals of alkalibiontic and alkaliphilous taxa and the totals of 'alkaline' units and 'acidic' units (the indifferent taxa were not included), he proposed three indices as follows:

$$\alpha = \frac{\text{acid units}}{\text{alkaline units}}$$

$$\omega = \frac{\text{acid units}}{\text{number of acid species}}$$

$$\varepsilon = \frac{\text{alkaline units}}{\text{number of alkaline species}}$$

Some of the limitations of Nygaard's system have been pointed out by more recent authors (e.g. Meriläinen, 1967; Digerfeldt, 1972; Renberg, 1976). Specific criticisms of these indices can be made. First, since acidobiontic and alkalibiontic taxa are accorded such weight, it is even more critical that the pH range of individual taxa be known. Second, the exclusion of indifferent (or circumneutral) taxa from the indices can lead to large fluctuations in the index unrelated to any real corresponding change in nature (Renberg, 1976). And third, both ω and ε are related to the number of taxa identified per sample. Since this is dependent on the total number of individuals counted at any one level (see above) and on taxonomic conventions varying from worker to worker, fluctuations in these indices are likely to be less clearly related to pH changes than is α (Digerfeldt, 1972; Renberg, 1976).

Despite these limitations Nygaard's system (especially index α) has proved of value, although its present popularity is mainly due to work by Meriläinen (1967) in which the applicability of the system to Finnish lakes was tested. Meriläinen refined the method by measuring pH values only at the time of the autumnal circulation, thereby making the values from each lake more comparable. As a result he was able to produce calibration curves for the various indices against measured pH values (Figure 26.9). The values of α

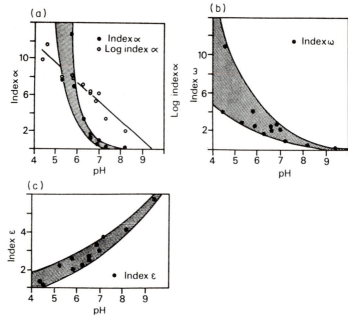

FIGURE 26.9. Measured pH versus index α, index ω and index ε, for a series of Finnish lakes (after Meriläinen, 1967). *Reproduced by permission of Societas Vanamo*

ranged from almost zero to infinity (because of the exclusion of circumneutral taxa), with very high and very variable levels for the most acidic lakes. When transformed to a logarithmic scale the values were grouped along a straight line determined by regression analysis (Figure 26.9(a)).

A further improvement of the Nygaard/Meriläinen system was proposed by Renberg and Hellberg (1982). In order to avoid values of infinity they modified the Nygaard index α by including indifferent (circumneutral) taxa in the equation. They then calculated the coefficient for each group using multiple regression analysis, formulating an index B as follows:

$$B = \frac{\% \text{ ind} + (5 \times \% \text{ acp}) + (40 \times \% \text{ acb})}{\% \text{ ind} + (3.5 \times \% \text{ alk}) + (108 \times \% \text{ alb})}$$

When plotted against observed pH values from 30 lakes in Sweden, Finland and Norway, the computed linear regression equation was as shown in Figure 26.10.

The index has been widely used (Flower and Battarbee, 1983; Tolonen and Jaakkola, 1983) and is especially useful in situations where local surface sediment diatom assemblage data are absent since the data classifications used in the equation are based on information available in the literature.

If local surface sediment diatom assemblage data sets are available it is

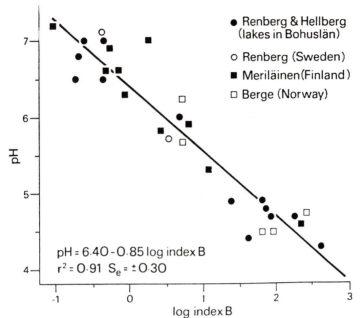

FIGURE 26.10. Measured pH versus log (index B) for lakes from Sweden, Norway and Finland (from Renberg and Hellberg, 1981). *Reproduced by permission of The Royal Swedish Acad. of Sciences*

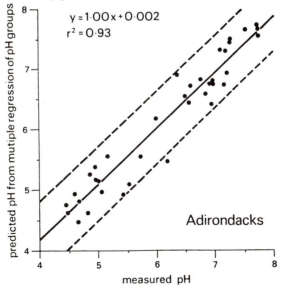

FIGURE 26.11. Measured pH versus predicted pH from a multiple regression analysis of pH groupings for Adirondack lakes, New York State (from Charles, 1982). *Reproduced by permission of Ecological Soc. of America*

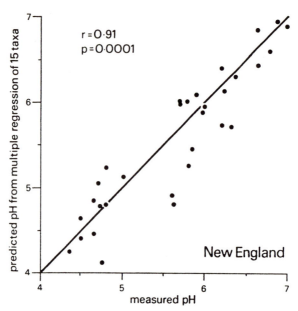

FIGURE 26.12.   Measured pH versus predicted pH from a
multiple regression analysis of 15 taxa for New England
lakes, U.S.A. (from Davis and Anderson, 1985). *Reproduced by permission of Hydrobiologia*

possible to produce regression equations to predict pH in a way that is
demonstrably valid for the region under consideration (see Figure 26.11;
Charles, 1982); and if individual taxa are used as predictors of pH it is possible
to produce pH reconstructions without prior classification of the taxa into the
artificial groupings of Hustedt (see Figure 26.12; Davis and Anderson, 1985;
Battarbee, 1984a).

In using any system based on the Hustedt classification the most important
factors are taxonomy and the correct allocation of taxa to pH groups, especially
where taxa of high percentages are concerned and when they belong to groups
that have large weighting coefficients. Other problems have been set out by
Battarbee (1984a).

The increase in studies of lake acidification has been a driving force in the
development of methods of pH reconstruction. Most have been based on one
or more of the methods presented here, and these have also been reviewed by
Battarbee (1984a). An example of the use of the Hustedt classification and
index B is shown in Figure 26.13.

Productivity

Many studies of lake development are concerned with changes in productivity
through time (trophic history), either in the long-term context of lake ontogeny

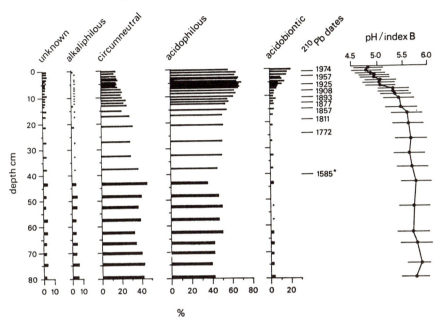

FIGURE 26.13. Diatom evidence for the acidification of Round Loch of Glenhead, south-western Scotland. A decline in pH of about 1 unit since approximately 1850 is indicated by the index B pH reconstruction (from Battarbee, 1984 *extrapolated date). *Reproduced by permission of The Royal Society*

or the shorter-term context of lake eutrophication (Whiteside, 1983). A number of productivity indices have been developed by various authors but most have limitations of one kind or another.

Plankton:periphyton ratio

The values and limitations of this ratio have been discussed above. It is more useful in indicating productivity changes associated with cultural eutrophication than productivity changes associated with post-glacial lake development since, over a long period of time, the ratio is also likely to be sensitive to morphometric changes as the lake fills in.

Centrales:Pennales (C:P) ratio

Nygaard proposed that the ratio of the number of species of Centrales to the number of species of Pennales could be used as an index of productivity (Nygaard, 1949). He recognized that some Pennales were undoubtedly of a eutrophic character and that some Centrales (e.g. some *Cyclotella*) were typical of oligotrophic waters. Nevertheless, on the whole he considered Centrales to indicate eutrophic conditions and Pennales to be more or less eurytrophically disposed.

Foged (1954b, 1960, 1965, 1968, 1969) has been the main user of the C:P idea, applying it to a variety of freshwater sediments. He suggested that the number of valves rather than species should be used in the quotient (Foged, 1954). The index (*sensu* Foged) has also been used and discussed by Kaczmarska (1976) and Mannion (1978).

By applying the ratio to diatom assemblages in sediments it seems that Nygaard's original intention has been distorted. The original quotient referred only to those taxa found in the plankton (Centrales and mainly Araphidinate forms). The sediment assemblages usually contain a much greater quantity of periphytic Pennales washed in from the littoral, and the influence of these not only affects the numerical quantity of the quotient but also introduces and includes a very large group of taxa in the denominator of very mixed ecological preference. The limited success of the ratio as applied to sediments does not stem from the difference between the trophic preferences of the Centrales (which span a wide range of trophic conditions), and the Pennales group, but probably from the fact that, as used by Foged, it approximately reflects the partition of taxa into planktonic and periphytic forms (cf. (i) above).

Araphidineae:Centrales (A:C) ratio

In 1967 Stockner and Benson proposed that increases in the A:C ratio indicates increasing eutrophication. Since, in Nygaard's original proposal, the Pennales were almost all araphidinate taxa, Stockner's suggestion inverts Nygaard's original C:P scheme. Nygaard regarded Centrales as more or less eutrophic indicators while Stockner regards them as oligotrophic indicators. Because of the increase of *Fragilaria crotonensis* and decrease of *Stephanodiscus astraea* var. *minutula* and *Melosira italica* in the uppermost sediment of Lake Washington, Stockner and Benson suggested that the A:C ratio might be used as an indicator of cultural enrichment. Stockner developed the idea further after studying lakes in the English Lake District and in the Experimental Lakes Area (E.L.A.) in Canada, and proposed (Stockner, 1971, 1972) that trophic status could be simply assessed by using the A:C ratio (on the basis of percentage counts) as follows:

A:C	Trophic status
0–1.0	Oligotrophic
1.0–2.0	Mesotrophic
>2.0	Eutrophic

In a response to doubts about the universality of the scheme he limited the use of it to temperate dimictic lakes, excluding shallow lakes, bog lakes, rivers and

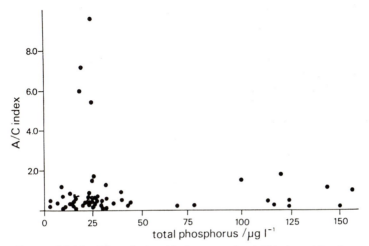

FIGURE 26.14. The relationship between the A/C index of Stockner and total phosphorus values for Minnesota lakes showing that the most productive lakes do not have a high A/C index (from Brugam, 1979). *Reproduced by permission of Blackwell Scientific Publishers Ltd*

dams. He later excluded all lakes which were 'meso-eutrophic' for most of their post-glacial history!

Brugam (1979) has thoroughly evaluated the index by comparing the diatom floras and A:C ratios of 80 surface sediment samples along a gradient of water chemistry values.

The data (Figure 26.14) showed that high A:C ratios were not characteristic of the most productive lakes where small *Stephanodiscus* taxa tended to dominate, but were more characteristic of lakes with medium total phosphorus levels and very low alkalinities. The subsequent survey to test this conclusion was carried out by Brugam and Patterson (1983) by comparing lakes of similar productivity (TP) but different alkalinities. As predicted, lakes with low alkalinities generally had higher A:C ratios, although this was not always the case. The use of this index is limited and certainly contrary to the use suggested of it by Stockner. If anything it might be used as an indicator of eutrophication of low alkalinity lakes (Brugam and Patterson, 1983). In the light of these problems it is unfortunate that this index has been popularized by textbooks (e.g. Wetzel, 1975) and in review articles (e.g. Frey, 1969, 1974). Contrary to the A:C ratio the diatom populations of some of the most eutrophic lakes in the world are often dominated by centric taxa as demonstrated by Brugam (e.g. Florin, 1946; Bradbury and Waddington, 1973; Bradbury and Winter, 1976; Haworth, 1972; Digerfeldt, 1972; Battarbee, 1973a, 1978a, 1978b).

Indicator species

General productivity levels can sometimes be inferred from the presence and relative abundance in sediments of stenotrophic taxa. However, there are often considerable differences of opinion between authors on the correct designation of particular taxa. This may be because some taxa have wider ranges than supposed, because of taxonomic confusion, or because trophic categories have been differently defined by the various authors. It is wise, therefore, to base interpretations on the behaviour of groups of taxa rather than on individual species. Of these, the most faithful indicators of high productivity associated with culturally enriched lakes is the small *Stephanodiscus* group including *Stephanodiscus hantzschii*, *Stephanodiscus astraea* var. *minutula*, *Stephanodiscus tenuis*, and *Stephanodiscus* (now *Cyclostephanos*) *dubius*. The clear association of these taxa with culturally enriched lakes can be seen in Figure 26.15, which shows the sudden appearance of *Stephanodiscus* in lake Trummen following the introduction of sewage effluent after 10,000 years of dominance in the lake by *Melosira* and *Cyclotella* taxa.

Diatom accumulation rate

Methods of estimating the annual accumulation rate of diatoms in the sediments have been discussed above. Many factors influence diatom accumulation rates, but in circumstances where preservation is good and sediment accumulation is continuous the most important factor is the level of diatom production in the water column. If diatom production is related to primary production some indication of change in primary productivity may thereby be obtained (see Figure 26.4). However, a number of considerations, in addition to general questions of representativity outlined above, should be taken into account in the interpretation of such data:

(1) Where silica limitation restricts crop size, increases in primary production may not be reflected by accumulation rate values.

(2) Where internal silica recycling occurs, the diatom crop may continue to increase but accumulation rate values may be reduced if recycling leads to the complete dissolution of some frustules in the sediment.

(3) If accumulation rate is not expressed in biomass/biovolume terms then the interpretation should allow for conspicuous differences between the relative weights or volumes of the most important constituent taxa.

(4) Accumulation rates are sensitive to counting errors and to errors in the estimation of the rate of sediment accumulation. As a result the curves produced should not be over-interpreted. Peaks and troughs represented by only single sample points may have little real significance.

(5) Accumulation rate values can be disturbed as a result of changing patterns of sediment redistribution within a lake basin through time. This can affect valve concentration and the rate of sediment accumulation, both of which influence the calculation of accumulation rate values (Lehman, 1975).

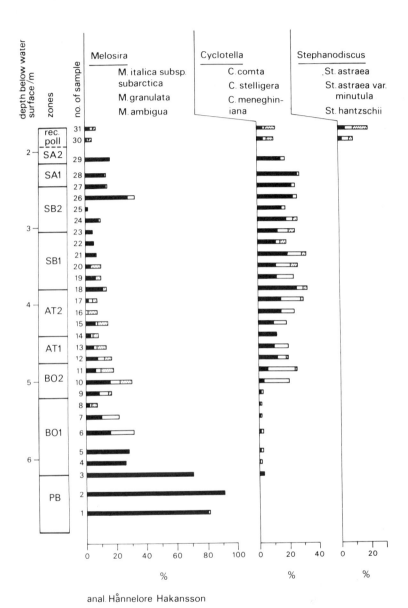

anal. Hånnelore Hakansson

FIGURE 26.15. Diatom diagram from Lake Trummen, southern Sweden, from preboreal to the present indicating the development of a *Stephanodiscus* flora during the recent pollution period (from Digerfeldt, 1972)

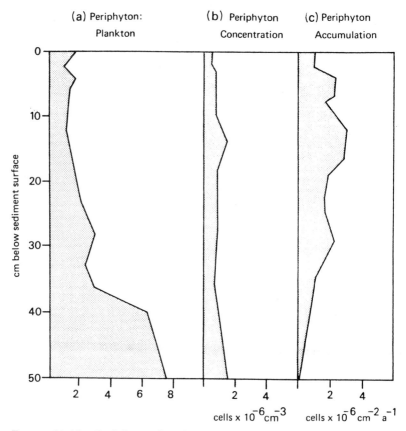

FIGURE 26.16.   Periphyton data from a core from Lough Erne, N. Ireland, showing the response of periphyton to enrichment. Although there is a decline in the proportion of periphyton following enrichment (a), there is no decline in concentration (b), and there is an increase in periphyton accumulation rate (c) (from Oldfield *et al.*, 1983). *Reproduced by permission of The Geographical Journal*

Despite these limitations there are considerable advantages in this kind of analysis (especially if it is used in conjunction with percentage counting techniques), since it is the only direct method of assessing past productivity changes. Most importantly it is independent of floristic changes and thus allows possible changes in productivity to be identified within periods of more or less constant floristic composition. The value of this approach also lies in its ability to help the interpretation of relative diagrams. From the concentration and accumulation rate values of individual taxa it is possible to judge whether relative increases or decreases are real (see Figure 26.16; Battarbee, 1978b). This approach also allowed Battarbee and Flower (1984) to identify inwashed diatoms from the catchment (see Figure 26.17).

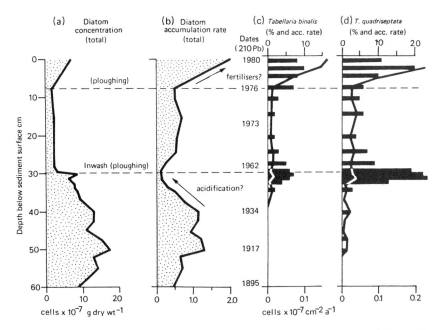

FIGURE 26.17. Diatom data from a core from Loch Grannoch showing: (a) the effect of catchment ploughing and an increase in sediment accumulation rate on diatom concentration; (b) changes in diatom accumulation rate indicating a decrease before ploughing and an increase after ploughing as diatoms from peats in the catchment are washed into the lake; (c) and (d) a comparison of percentage and accumulation rate values for two acidobiontic taxa, *Tabellaria binalis* and *T. quadriseptata*. The lower peaks are characterized by high percentages but unchanging accumulation rates (from Battarbee and Flower, 1984). *Reproduced by permission of Amer. Soc. of Limnol. and Oceanogr.*

Salinity

Of the environmental factors that influence the distribution of diatoms salinity is probably the strongest. Consequently diatoms are of great use to palaeoecologists interested in coastal (thalassic) areas and in inland salt lake (athalassic) environments.

Historically, most emphasis has been placed on the thalassic areas (Halden, 1929; Florin, 1946; Miller, 1964), but in recent decades there has been a growing interest in salt-lake diatoms (Hustedt, 1959; Ehrlich, 1973; Richardson, 1968; Gasse, 1974; Bradbury *et al.*, 1981). Although many saline lake diatoms are also found in brackish environments in coastal areas, it is necessary to deal with these two different environments separately since the major ion concentrations are different and since salinity variations in both long and short terms tend to be more frequent and more extreme in saline lakes.

Coastal environments

The halobian system of classifying diatoms according to salinity stems from Kolbe's (1927) work in the Sperenberg region near Berlin. Consequently, his scheme and following modifications were based on waters where the dominant anion was Cl⁻. He divided the diatoms into three major groups: euhalobous (30–40 per mil S), mesohalobous (5–20 per mil S) and oligohalobous (<5 per mil S), and subdivided the last group into halophilous, indifferent and halophobous forms. Halophilous forms were those that grew well in slightly brackish conditions, indifferent forms were defined as those that tolerated brackish water but found their optimum growth in freshwater, while halophobous taxa were those that did not withstand even small concentrations of salt.

Hustedt (1953, 1957) modified Kolbe's system largely from his experiences with the diatom flora of the River Weser. He preferred to use the term polyhalobous for stenohaline marine diatoms and used the term 'euhalobous' as a more general term to include polyhalobous and mesohalobous diatoms. He also argued that the 5 per mil salinity lower limit for the mesohalobous category was far too high since mesohalobous diatoms are found in the Weser at salinities as low as 0.2 per mil.

A further change was the subdivision of the mesohalobous category into three, mainly to separate diatoms found in the upper water of a stratified estuary (α − mesohalobe) from those found mainly in the lower water (β − mesohalobe). The third category included euryhaline mesohalobous diatoms.

The Hustedt system, then, is as follows:

(1) *polyhalobous:* salt concentrations 30 per mil and above, but some euryhaline forms tolerate lower salinities;
(2) *mesohalobous:*
    (a) euryhaline mesohalobous: 0.2–3 per mil
    (b) α−mesohalobous: species found in the lower brackish water minimum 10 per mil
    (c) β−mesohalobous: species found in the upper brackish water, about 0.2–10 per mil;
(3) *oligohalobous:* as in the Kolbe system
    (a) halophilous
    (b) indifferent;
(4) *halophobous*.

Following on from Hustedt, Simonsen (1962) pointed out that within each of these major groups there were taxa with greatly varying tolerance limits and that an understanding of the tolerance limits of the various taxa was especially important if such diatoms were to be used in geological–palaeoecological studies. He argued that brackish water taxa (mesohalobes) were all euryhaline to some degree but that marine and freshwater diatoms could be divided into

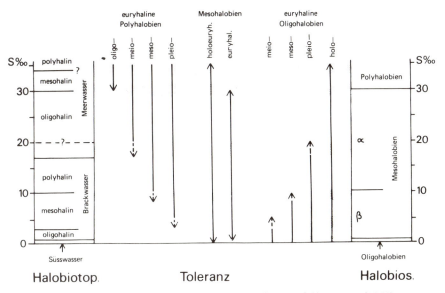

FIGURE 26.18.   The salinity tolerance scheme of Simonsen (1962)

tolerance groups. He maintained that this procedure linked together the Kolbe/Hustedt system with the biotope or saltwater zonation scheme of Ekman (Figure 26.18), thus enabling the salt concentration of the environment and past environments to be reconstructed from the diatom data.

The earliest use of diatoms in palaeoecology as indicators of former salinity was in Scandinavia in the debate on land uplift, shore displacement and the isolation of lakes (Cleve-Euler, 1923, 1944; Lundqvist and Thomasson, 1923; Halden, 1929; Florin, 1946). One of the clearest illustrations of this application was a detailed analysis by Florin of the sediments of Myskasjön, a lake now 10 m above sea-level. The analysis showed a progression from marine diatoms, through brackish-water diatoms to freshwater forms, indicating that the site had once been an open bay in the Baltic and had gradually passed through a series of coastal configurations: from an inlet, to a lagoon, before final isolation as a freshwater lake (Figure 26.19). Many further examples of this kind of sequence have been described, some from the Baltic (Eronen, 1974; Renberg, 1976; Alhonen, 1971; Alhonen *et al.*, 1978) and others from the west coast of Sweden and Norway (Miller, 1964; Kjemperud, 1981a, b).

Closed basin lakes

Closed-basin lakes are of interest to the palaeoecologist since the salinity of such lakes is strongly related to hydrological and climatic conditions, and the diatom record in the sediments of these lakes can be used as a record of climatic

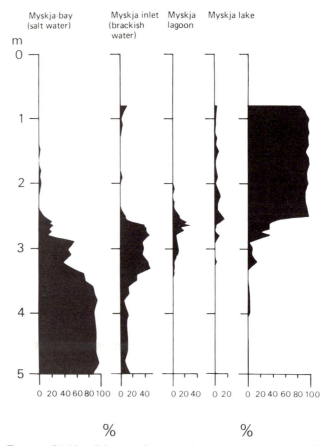

FIGURE 26.19.   Diatom diagram showing the isolation of Myskjasjön from the Baltic Sea. Polyhalobous diatoms are progressively replaced by mesohalobous and then oligohalobous forms (from Florin, 1946)

change (e.g. Begin, *et al.*, 1974; Bradbury *et al.*, 1981; Gasse, 1974). Since the dominant anion in saline lakes can be chloride, carbonate or sulphate, the chemistry and range of variation of these systems are quite different from the water chemistry of thalassic waters (Eugster and Hardie, 1978). Although many of the brackish taxa found in coastal environments are the same, this, and the lack of a fully marine component in saline lakes, makes the Kolbe–Hustedt halobian system not directly applicable for salinity reconstruction from sediment cores. In the Ethiopian part of Africa, Gasse used certain aspects of the halobian system (Gasse, 1974), while further south, in East and Central Africa, Richardson (1968) relied on information from modern diatom distributions and water chemistry. This approach has been recently pursued by

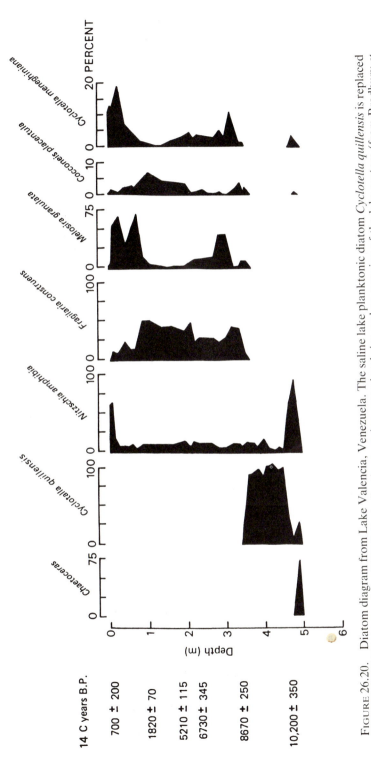

FIGURE 26.20. Diatom diagram from Lake Valencia, Venezuela. The saline lake planktonic diatom *Cyclotella quillensis* is replaced by freshwater diatoms at about 8500 B.P., indicating a major water-level rise and an opening of the lake system (from Bradbury *et al.*, 1981). *Reproduced by permission of Science*

Gasse (1983). In studies on Lake Patzcuaro, Mexico (Watts and Bradbury, 1982) and of Lake Valencia, Venezuela (Bradbury *et al.*, 1981) Bradbury makes use of autoecological information. In the L. Valencia study the transition from a saline closed basin to an alkaline freshwater lake is clearly shown by the replacement in the plankton of *Cyclotella quillensis* by *Melosira granulata* and *Cyclotella meneghiniana* about 8700 B.P. (see Figure 26.20).

Acknowledgements

I should like to thank Urve Miller, Maj-Britt Florin and Jouko Meriläinen for their useful comments on an earlier version of this paper, and Roger Flower and John Anderson for their help with this revised edition. I am also grateful to Claudette John, Alick Newman and Chris Cromarty for their technical services.

REFERENCES

Alhonen, P. (1968). On the late-glacial and early post-glacial diatom succession in Loch of Park, Aberdeenshire, Scotland. *Memor. Soc. F. Fl. Fenn*, **44**, 13–20.

Alhonen, P. (1971). The stages of the Baltic Sea as indicated by the diatom stratigraphy. *Acta Bot. Fennica*, **92**, 1–18.

Alhonen, P., Eronen, M., Nuñez, M., Salomaa, R., and Uusinoka, R. (1978). A contribution to Holocene shore displacement and environmental development in Vantaa, South Finland: the stratigraphy of Lake Lammaslampi. *Bull. Geol. Soc. Finland*, **50**, 69–79.

Anonymous (1975). Proposals for a standardisation of diatom terminology and diagnoses. *Nova Hedwigia Supp.*, **53**, 223–254.

Bailey-Watts, A. E. (1976). Planktonic diatoms and silica in Loch Leven, Kinross, Scotland: a one month silica budget. *Freshw. Biol.*, **6**, 203–213.

Battarbee, R. W. (1973a). Preliminary studies of Lough Neagh sediments. II: Diatoms from the uppermost sediment. In: *Quaternary Plant Ecology* (Eds. H. J. B. Birks and R. G. West), Oxford, Blackwell, pp. 279–289.

Battarbee, R. W. (1973b). A new method for estimating absolute microfossil numbers with special reference to diatoms. *Limnol. Oceanogr.*, **18**, 647–653.

Battarbee, R. W. (1977). *Lough Erne Survey: Sediments, Dating and Diatom Analysis*, Traad Point Limnology Laboratory Report, The New University of Ulster, pp. 1–78.

Battarbee, R. W. (1978a). Observations on the recent history of Lough Neagh and its drainage basin. *Phil. Trans. R. Soc.*, **B 281**, 303–345.

Battarbee, R. W. (1978b). Relative composition, concentration and calculated influx of diatoms from a sediment core from Lough Erne, Northern Ireland. *Pol. Arch. Hydrobiol.*, **25**, 9–16.

Battarbee, R. W. (1978c). Biostratigraphical evidence for variations in the recent pattern of sediment accumulation in Lough Neagh, Northern Ireland. *Verh. Int. Verein. Limnol.*, **20**, 625–629.

Battarbee, R. W. (1979). Early algological records — help or hindrance to palaeolimnology? *Nova Hedwigia Beih.*, **64**, 379–394.

Battarbee, R. W. (1981a). Changes in the diatom microflora of a eutrophic lake since 1900 from a comparison of old algal samples and the sedimentary record. *Holarctic Ecology*, **4**, 73–81.

Battarbee, R. W. (1981b). Diatom and chrysophyceae microstratigraphy of the annually laminated sediments of a small meromictic lake. In: *Florilegium Florinis Dedicatum* (Eds. L.-K. Königsson and K. Paabo), Striae 14, Uppsala, pp. 105–109.

Battarbee, R. W. (1984a). Diatom analysis and the acidification of lakes. *Phil. Trans. R. Soc.* B305, 451–477.

Battarbee, R. W. (1984b). Spatial variations in the water quality of Lough Erne, Northern Ireland, on the basis of surface sediment diatom analysis. *Freshwater Biol.*, **14**, 539–545.

Battarbee, R. W., and Flower, R. J. (1984). The inwash of catchment diatoms as a source of error in the sediment-based reconstruction of pH in an acid lake. *Limnol. Oceanogr.* **29**, 1325–1329.

Battarbee, R. W., and Kneen, M. J. (1982). The use of electronically counted microspheres in absolute diatom analysis. *Limnol. Oceanogr.*, **27**, 184–188.

Battarbee, R. W., and McCallan, M. E. (1974). An evaporation tray technique for estimating absolute pollen numbers. *Pollen et Spores*, **16**, 143–150.

Begin, Z. B., Ehrlich, A., and Nathan, Y. (1974). Lake Lisan, the Pleistocene precursor of the Dead Sea. *Geol. Surv. Isr. Bull.*, **63**, 1–30.

Bengtsson, L., and Persson, T. (1978). Sediment changes in a lake used for sewage reception. *Pol. Arch. Hydrobiol.*, **25**, 17–34.

Benninghoff, W. S. (1962). Calculation of pollen and spore density in sediments by addition of exotic pollen in known quantities. *Pollen et Spores*, **4**, 332–333.

Berge, F. (1975). pH-forandringer og sedimentasjon av diatomeer i Langtjern. *Norges Teknisk-Naturvitenskapelige Forskningsrad.* Internal Report, pp. 1–18.

Berge, F. (1976). Kiselalger og pH i noen elver og insjoer i Agder og Telemark. En Sammenlikning mellom arene 1949 og 1975. *SNSF — prosjektet IR 18/76.*

Berge, F. (1979). Kiselalger og pH i noen innsjoer i Agder og Hordaland. *SNSF — prosjektet IR 42/79.*

Birks, H. H., Whiteside, M. C., Stark, D. M., and Bright, R. C. (1976). Recent palaeolimnology of three lakes in Northwestern Minnesota. *Quat. Res.*, **6**, 249–272.

Bonny, A. P. (1972). A method for determining absolute pollen frequencies in lake sediments. *New Phytol.*, **71**, 393–405.

Bradbury, J. P., and Megard, R. O. (1972). Stratigraphic record of pollution in Shagawa Lake, Northeastern Minnesota. *Geol. Soc. America Bull.*, **83**, 2639–2648.

Bradbury, J. P. (1975). Diatom stratigraphy and human settlement in Minnesota. *Geol. Soc. Am. Special Paper* 171.

Bradbury, J. P., Leyden, B., Salgado-Labouriav, M., Lewis, W. M., Scubert, C., Binford, M. W., Frey, D. G., Whitehead, D. R., and Weibezahn, F. H. (1981). Late-Quaternary environmental history of Lake Valencia, Venezuela. *Science*, **214**, 1299–1305.

Bradbury, J. P., and Waddington, J. C. B. (1973). The impact of European settlement on Shagawa Lake, northeastern Minnesota, USA. In: *Quaternary Plant Ecology* (Eds. H. J. B. Birks and R. G. West), Oxford, Blackwell, pp. 289–307.

Bradbury, J. P., and Winter, T. C. (1976). Areal distribution and stratigraphy of diatoms in the sediments of Lake Sallie, Minnesota. *Ecology*, **57**, 1005–1014.

Brander, G. (1936). Uber das Einsammeln von Erdproben und ihre Praparation fur die qualitative und quantitative Diatomeenanalyse. *Bull. Comm. Geol. Finlande*, **115**, 131–144.

Brugam, R. B. (1978). The human disturbance history of Linsley Pond, North Branford, Connecticut. *Ecology*, **59**, 19–36.

Brugam, R. B. (1979). A re-evaluation of the Araphidineae/Centrales index as an indicator of lake trophic status. *Freshw. Biol.*, **9**, 451–460.

Brugam, R. B. (1980). Postglacial diatom stratigraphy of Kirchner Marsh, Minnesota. *Quat. Res.*, **13**, 133–146.

Brugam, R. B. (1983). The relationship between fossil diatom assemblages and limnological conditions. *Hydrobiol.*, **98**, 223–235.

Brugam, R. B., and Patterson, C. (1983). The A/C (Araphidineae/Centrales) ratio in high and low alkalinity lakes in eastern Minnesota. *Freshw. Biol.*, **13**, 47–55.

Charles, D. F. (1982). *Studies of Adirondack Mountain (N.Y.) Lakes: Limnological Characteristics and Sediment Diatom–Water Chemistry Relationships*, Ph.D. Thesis, Indiana University, Bloomington, U.S.A.

Cholnoky, B. J. (1968a). The relationship between algae and the chemistry of natural waters. *CSIR Reprint 129*, 215–225.

Cholnoky, B. J. (1968b). *Die Okologie der Diatomeen*, Weinheim.

Cleve, P. T. (1894–95). Synopsis of the naviculoid diatoms. *Kgl. Sven. Vet. Akad. Handl.*, **26**, 1–194; **27**, 1–219.

Cleve-Euler, A. (1922). Om diatomacevegetationen och dess förandringar i Sabysjön, Uppland samt några dämda sjöar i Salatrakten. *Sver. Geol. Unders.*, C 309, 1–76.

Cleve-Euler, A. (1923). Forsök til analys av Nordens senkvartära nivåförandringar jämte några konsekvenser. *Geol. Fören. Förhandl.*, **45**, 19–107.

Cleve-Euler, A. (1944). Die diatomeen als quartärgeologische Indikatoren. *Geol. Fören. Förhandl.*, **66**, 383–410.

Cleve-Euler, A. (1951–55). Die Diatomeen von Schweden und Finland. *Kungl. Svenska Vetensk. Handl. Ser. 4*, **2**(1), 3–163; **4**(1), 3–158; **4**(5), 3–255; **5**(4), 3–231; **3**(3), 3–153.

Cooper, L. H. N. (1952). Factors affecting the distribution of silicate in the N. Atlantic Ocean and the formation of N. Atlantic deep water. *J. Mar. Biol. Assoc.*, **30**, 511–526.

Crabtree, K. (1969). Post-glacial diatom zonation of limnic deposits in North Wales. *Mit. Int. Ver. Limnol.*, **17**, 165–171.

Dahm, H-D. (1959). Diatomeen aus spätglazialen Ablagerungen der Eichholz-Neiderung bei Heiligenhaften (Holstein). *Zeitschr. Deutsch Geol. Gesellsch.*, **11**, 8–12.

Davis, R. B., and Anderson, D. S. (1985). Methods of pH calibration of sedimentary diatom remains for reconstructing recent history of pH in lakes. *Hydrobiologia*, **120**, 69–87.

Davis, R. B., and Norton, S. A. (1978). Palaeolimnological studies of human impact on lakes in the United States, with emphasis on recent research in New England. *Pol. Arch. Hydrobiol.*, **25**, 99–115.

Digerfeldt, G. (1972). The post-glacial development of Lake Trummen: regional vegetation history, water level changes and palaeolimnology. *Fol. Lim. Scand.*, **16**, 1–104.

Digerfeldt, G. (1975a). The post-glacial development of Ranviken Bay in Lake Immeln. III: Palaeolimnology. *Geol. Fören. Förh.*, **97**, 13–28.

Digerfeldt, G. (1975b). A standard profile for Littorina transgressions in western Skåne, South Sweden. *Boreas*, **4**, 125–142.

Digerfeldt, G. (1977). *The Flandrian Development of Lake Flarken: Regional Vegetation History and Palaeolimnology. Univ. of Lund, Dept. of Quat. Geol. Report*, 13.

Donner, J. J., Alhonen, P., Eronen, M., Jungner, H., and Vuorela, J. (1978). Biostratigraphy and radiocarbon dating of the Holocene lake sediments of Tyotjarvi and the peats in the adjoining bog Varrossuo west of Lahti in southern Finland. *Ann. Bot. Fennici*, **15**, 258–280.

Duthie, H. C., and Sreenivasa (1971). Evidence for the eutrophication of Lake Ontario from the sedimentary diatom succession. *Proc. 14th Conf. Great Lakes Res.* Ann. Arbor, Michigan, pp. 1–13.

Eaton, J. W., and Moss, B. (1966). The estimation of numbers and pigment content in epipelic algal populations. *Limnol. Oceanogr.*, **11**, 584–595.

Ehrlich, A. (1973). Quaternary diatoms of the Hula basin (Northern Israel). *Geol. Survey of Israel Bull.*, **58**, 1–39.

eugster, H. P., and Hardie, L. A. (1978). Saline lakes. In: *Lakes: Chemistry, Geolog y, Physics* (Ed. A. Lerman), Springer-Verlag, New York, 237–294.

Engstrom, D., and Wright, H. E. (1984). Chemical stratigraphy of lake sediments as a record of environmental change. In: *Lake sediments and environmental history* (Eds. E.Y. Haworth and J. W. G. Lund) Leicester University Press pp. 11–67.

Eronen, M. (1974). The history of the Litorina Seà and associated Holocene events. *Comment Physico-Math.*, **44**, 79–195.

Evans, G. H. (1970). Pollen and diatom analyses of late Quaternary deposits in the Blelham Basin, North Lancashire. *New Phytol.*, **69**, 821–874.

Evans, G. H., and Walker, R. (1977). The late Quaternary history of the diatom flora of Llyn Clyd and Llyn Glas, two small oligotrophic high mountain tarns in Snowdonia (Wales). *New Phytol.*, **78**, 221–236.

Faegri, K., and Iversen, J. (1964). *Textbook of Pollen Analysis*, Blackwell Sci. Pubs., Oxford.

Fjerdingstad, F. (1954). The subfossil algal flora of the Lake Bölling Sö and its limnological interpretation. *Kgl Da. Vid. Selsk. Biol. Skr.*, **7**, (6), 1–56.

Florin, M-B. (1944). En sensubarktisk transgression i trakten av Södra Kilsbergen enligt diatomacé-succession i områdets högre belägna fornsjölagerföljder. *Geol. Fören. Förh.*, **66**, 417–488.

Florin, M-B. (1946). Clypeusfloran i postglaciala fornsjölagerföljder i östra Mellansverige. *Geol. Fören. Förh.*, **68**, 429–458.

Florin, M-B. (1970a). The fine structure of some pelagic freshwater diatom species under the scanning electron microscope: I. *Svensk Bot. Tidskr.*, **64**, 51–64.

Florin, M-B. (1970b). Late-glacial diatoms of Kirchner Marsh, S. E. Minnesota. *Nova Hedwigia*, **31**, 667–756.

Florin, M-B. (1973). Ekologiska Salthaltsfrågor med Tyngdpunkt på Diatomeer inom Östersjöområdet, en Historisk Översikt, *Univ. of Lund, Dept. of Quat. Geol. Report*, **3**, pp. 12–38.

Florin, M-B. (1977). Late-glacial and pre-boreal vegetation in Southern Central Sweden. II: Pollen, spore, and diatom analysis. *Striae, Soc. Upps. Geol. Quat.*, 60 pp.

Florin, M-B., and Wright, H. E. (1969). Diatom evidence for the persistence of stagnant glacial ice in Minnesota. *Bull. Geol. Soc. Am.*, **80**, 695–704.

Florin, S. (1944). Havstrandens forskjutningar och bebyggelseutvecklingen i östra Mellansverige under senkvartär tid. I: Allmän översikt. *Geol. Fören. Förh.*, **66**, 1–80.

Flower, R. J. L. (1980). *A Study of Sediment Formation, Transport and Deposition in Lough Neagh, Northern Ireland, with Special Reference to Diatoms*, D.Phil. Thesis, The New University of Ulster.

Flower, R. J., and Battarbee, R. W. (1983). Diatom evidence for recent acidification of two Scottish lochs. *Nature*, **305**, 130–133.

Foged, N. (1948). Diatoms in water-courses in Funen 1-6. *Dansk Bot. Arkiv*, **12**, 5, 6, 12. Copenhagen.

Foged, N. (1953). Diatoms from West Greenland. *Meddr. Gronland*, **147**, (10), 1–86.

Foged, N. (1954a). On the diatom flora of some Funen lakes. *Folia limnol. Scand.*, **6**, 1–75.

Foged, N. (1954b). En interglacial diatomejordaflejring i Ost-Fyn. *Medd. Dan. Geol. For.*, **12**, 541–547.

Foged, N. (1960). Diatomefloran i en interglacial kiselguraflejring ved Rands fjord i Ostjylland. *Medd. Dan. Geol. For.*, **14**, 197–211.

Foged, N. (1964). Freshwater diatoms from Spitzbergen. *Tromso Museums Skr.*, **11**, 205.

Foged, N. (1965). En senglacial ferskvands-diatomeflora fra Fyn. *Medd. Dan. Geol. For.*, **15**, 459–469.

Foged, N. (1968). Diatomeerne i en postglacial Boreprove fra Bunden af Esrom So, Danmark. *Medd. Dan. Geol. For.*, **18**, 161–183.

Foged, N. (1969). Diatoms in a postglacial core from the bottom of the lake Grane Langso, Denmark. *Bull. Geol. Soc. Denmark*, **19**, 237–256.

Foged, N. (1972). The diatoms in four post-glacial deposits in Greenland. *Meddr. Gronland*, **194**, 66 pp.

Frey, D. G. (1955). Langsee: a history of Meromixis. *Mem. Ist. Ital. Idrob. Suppl.* 8, 141–164.

Frey, D. G. (1969). Evidence for eutrophication from remains of organisms in sediments. In: *Eutrophication: Causes, Consequences, Correctives.* Washington D.C., pp. 594–613.

Frey, D. G. (1974). Palaeolimnology. *Mitt. Int. Ver. Limnol.*, **10**, 95–123.

Fromm, E. (1938). Geokronologisch datierte Pollendiagramme und Diatomeenanalysen aus Ångermanland. *Geol. Fören. Förh.*, **60**, 365–381.

Gasse, F. (1974). *Les Diatomées des Sediments Holocènes du Bassin du Lac Afrera (Giulietti) (Afar Septentrional, Ethiopie)*, Essai de Reconstitution de l'Evolution du Milieu. *Int. Revue ges. Hydrobiol.*, **59**, 95–122.

Gasse, F. (1975). L'évolution des lacs de l'Afar Central (Ethiopie et T.F.A.I.) du Plio-Pleistocene a l'Actuel. *Thesis D.Sci.Nat., University of Paris.* — I. Text 1–406, II. Annexes 1–103, III. Planches photographiques.

Gasse, F., and Street, F. A. (1978). Late Quaternary lake-level fluctuations and environments of the northern rift valley and Afar region (Ethiopia and Djibouti). *Palaeogeogr. Palaeoclim. Palaeocol.*, **24**, 279–325.

Gasse, F., and Tekaia, F. (1983). Transfer functions for estimating palaeoecological conditions (pH) from East African diatoms. *Hydrobiologia*, **103**, 85–90.

Geitler, L. (1932). Der Formwechsel der pennaten Diatomeen (Kieselalgen). *Arch. Protistenk.*, **78.**, 1–226.

Gibson, C. E. (1981). Silica budgets and the ecology of planktonic diatoms in an unstratified lake (Lough Neagh, N. Ireland). *Int. Revue ges. Hydrobiol.*, **66**, 641–664.

Gibson, C. E., Wood, R. B., Dickson, E. L., and Jewson, D. H. (1971). The succession of phytoplankton in Lough Neagh 1968–70. *Mitt. Int. Ver. Limnol.*, **19**, 146–160.

Gordon, A. D., and Birks, H. J. B. (1972). Numerical methods in Quaternary palaeoecology. I: Zonation of pollen diagrams. *New Phytologist*, **71**, 961–979.

Håkansson, H. (1976). Die Struktur und Taxonómie einiger *Stephanodiscus* Arten aus eutrophen Seen Sudschwedens. *Bot. Notiser*, **129**, 25–34.

Halden, B. (1929). Kvartärgeologiska diatomacéestudier belysande den postglaciala transgressionen å Svenska Västkusten. *Geol. Fören. Förh.*, **51**, 311–366.

Halden, B. (1931). Diatomaceers succession i deltasediment. *Geol. Fören. Förh.*, **53**, 150–158.

Hasle, G. R., and Heimdal, B. R. (1970). Some species of the centric diatom genus *Thalassiosira* studied in the light and electron microscopes. *Nova Hedwigia, Beih.*, **31**, 543–581.

Haworth, E. Y. (1969). The diatoms of a sediment core from Blea Tarn, Langdale. *J. Ecol.*, **57**, 429–441.

Haworth, E. Y. (1972). The recent diatom history of Loch Leven, Kinross. *Freshw. Biol.*, **2**, 131–141.

Haworth, E. Y. (1975). A scanning electron microscope study of some different frustule forms of the genus *Fragilaria* found in Scottish late-glacial sediments. *Br. Phycol. J.*, **10**, 73–80.

Haworth, E. Y. (1976). Two late-glacial (late Devensian) diatom assemblage profiles from northern Scotland. *New Phytol.*, **77**, 227–256.

Haworth, E. Y. (1980). Comparison of continuous phytoplankton records with the diatom stratigraphy in the recent sediments of Blelham Tarn. *Limnol. Oceanogr.*, **25**, 1093–1103.

Hecky, R. E., and Kilham, P. (1973). Diatoms in alkaline saline lakes: Ecology and geochemical implications. *Limnol. Oceanogr.*, **18**, 53–71.

Helmcke, J. G., and Krieger, W. (1953–64). *Diatomeenschalen im Elektronmikrosko-pischen Bild*, Teil I–X, Cramer Verlag, Weinheim.

Huber-Pestalozzi, G. (1942). *Das Phytoplankton des Susswassers: Systematik und Biologie Diatomeen Binnengewasser* **16**(2), 183 pp.

Hurd, D. C. (1972). Factors affecting solution rate of biogenic opal in seawater. *Earth Planet. Sci. Lett.*, **15**, 411–417.

Hurd, D. C. (1973). Interactions of biogenic opal, sediment and seawater in the Central Equatorial Pacific. *Geochim. Cosmochim. Acta*, **37**, 2257–2283.

Hustedt, F. (1930). *Die Susswasser flora Mitteleuropas. Heft 10*, 1–466. Bacillariophyta *(Diatomeae)*. Jena. Mitteleuropas.

Hustedt, F. (1930–66). *Die Kieselalgen Deutschlands, Osterreich's und der Schweiz unter Besucksichtigung der ubrigen Lander Europas sowie der angrensenden Meeresgebiete*. In Dr. L. Rabenhorts Kryptogramen. Flora von Deutschland, Osterreich, und der Schweiz. VII: 1–3.

Hustedt, F. (1937–39). Systematische und okologische Untersuchungen uber den Diatomeen-Flora von Java, Bali, Sumatra. *Arch. Hydrobiol.* (Suppl.), 15 & 16.

Hustedt, F . (1942). Aerophile Diatomeen in der nordwestdeutschen Flora. *Ber. Deutsch Bot. Ges.*, **60**, 55–73.

Hustedt, F. (1957). Die Diatomeenflora des Fluss-systems der Weser im Gebiet der Hansestadt Bremen. *Ab. Naturw. Ver. Bremen*, **34**, 181–440.

Hutchinson, G. E. *et al.* (1970). Ianula: an account of the history and development of the Lago di Monterosi, Latium, Italy. *Trans. Amer. Phil. Soc.*, **60**, 1–178.

Hutchinson, G. E., and Wollack, A. (1940). Studies on Connecticut lake sediments. II: Chemical analyses of a core from Linsley Pond, North Branford. *Amer. J. Sci.*, **238**, 492–517.

Huttunen, P., and Meriläinen, J. (1983). Interpretation of lake quality from contemporary diatom assemblages. *Hydrobiologia*, **103**, 91–98.

Huttunen, P., Meriläinen, J., and Tolonen, K. (1978). The history of a small dystrophied forest lake, Southern Finland. *Pol. Arch. Hydrobiol.*, **25**, 189–202.

Huttunen, P., and Tolonen, K. (1976). Human influence in the history of Lake Lovojärvi, S. Finland. *Finskt Museum 1975*, 68–117.

Jouse, A. (1966). Diatomeen in Seesedimenten. *Arch. Hydrobiol. Beih. Ergebn. Limnol.*, **4**, 1–32.

Jørgensen, E. G. (1948). Diatom communities in some Danish lakes and ponds. *Det Kong. Danske Vidensk. Selskab Biol. Sk.*, **V**(2), 1–140.

Jørgensen, E. G. (1957). Diatom periodicity and silicon assimilation. *Dan. Bot. Ark.*, **18**(1), 1–54.

Kaczmarska, I. (1973). Late-glacial diatom flora at Knapowka near Wloszczowa (South Poland). *Acta Palaeobotanica*, **14**, 179–193.

Kaczmarska, I. (1976). Diatom analyses of Eemian profile in freshwater deposits at Imbramowice near Wroclaw. *Acta Palaeobotanica*, **17**, 3–34.

Kalbe, L., and Werner, H. (1974). Das Sediment des Kummerrower Sees. Untersuchungen des Chemismus und der Diatomeenflora. *Int. Revue ges. Hydrobiol.*, **59**, 755–782.

Kilham, P. (1971). A hypothesis concerning silica and the freshwater planktonic diatoms. *Limnol. Oceanogr.*, **16**, 10–18.

Kilham, S. S., and Kilham, P. (1975). *Melosira granulata* (Ehr.) Ralfs: morphology and ecology of a cosmopolitan freshwater diatom. *Verh. Internat. Verein. Limnol.*, **19**, 2716–2721.

Kjemperud, A. (1981a). Diatom changes in sediments of basins possessing marine/lacustrine transitions in Frosta, Nord-Trondelag, Norway. *Boreas*, **10**, 27–38.

Kjemperud, A. (1981b). A shoreline displacement investigation from Frosta in Trondheimsfjorden, Nord-Trondelag, Norway. *Norsk. Geol. Tidskr.*, **61**, 1–15.

Knox, A. S. (1942). The use of bromoform in the separation of non-calcareous microfossils. *Science*, 95, 307.

Knudson, B. M. (1954). The ecology of the diatom genus *Tabellaria* in the English Lake District. *J. Ecol.*, **42**, 345–358.

Koivo, L. K. (1976). Species diversity in post-glacial diatom lake communities of Finland. *Palaeogeogr., Palaeoclimatol., Palaeoecol.*, **19**, 165–190.

Kolbe, R. W. (1927). Zur Okologie, Morphologie und Systematik der Brackwasser-Diatomeen. *Pflanzenforschung*, **7**, 1–146.

Kolbe, R. W. (1932). Grundlinien einer allegemeinen Okologie der Diatomeen. *Ergebn. Biol.*, **8**, 221–348.

Kolkwitz, R., and Marsson, M. (1908). Okologie der pflanzlichen Saprobien. *Ber. Deutsch. Bot. Ges.*, **26a**, 505–519.

Kolkwitz, R. (1950). Okologie der Saprobien. Uber die Beziehungen der Wasser-organismen zur Umwelt. *Schriftenreihe des Vereins fur Wasser-, und Lufthygiene*, **4**, 1–64.

Krasske, G. (1932). Beitrage zur Kenntniss der Diatomeenflora der Alpen Dresden. *Hedwigia*, **72**, 92–134.

Krishnaswamy, S., Lal, D., Martin, J. M., and Meybeck, M. (1971). Geochronology of lake sediments. *Earth Plan. Sci. Lett.*, **11**, 407–414.

Kukkonen, E., and Tynni, R. (1972). A sediment core from Lake Lovojärvi, a former meromictic lake (Lammi, S. Finland). *Aqua Fennica*, 70–74.

Lange-Przybylowska, W. (1976). Diatoms of lake deposits from the Polish Baltic Coast. I: Lake Drnzno. *Acta Palaeobotanica*, **17**, 35–74.

Lehman, J. T. (1975). Reconstructing the rate of accumulation of lake sediment: the effect of sediment focusing. *Quat. Res.*, **5**, 541–550.

Lewin, J. C. (1961). The dissolution of silica from diatom walls. *Geochim. Cosmochim. Acta*, **21**, 182–198.

Lowe, R. L. (1975). Comparative ultrastructure of the valves of some *Cyclotella* species (Bacillariophyceae). *J. Phycol.*, **11**, 415–424.

Lowe, R. L., and Collins, G. B. (1973). An aerophilous diatom community from Hocking County, Ohio. *Trans. Am. Micr. Soc.*, **92**, 492–496.

Lowe, R. L., and Crang, R. E. (1972). The ultrastructure and morphological variability of the frustule of *Stephanodiscus invisitatus* Hohn and Hellerman. *J. Phycol.*, **8**, 256–259.

Lund, J. W. G. (1949). Studies on *Asterionella*. I: The origin and nature of the cells producing seasonal maxima. *J. Ecol.*, **37**, 389–419.

Lund, J. W. G. (1950). Studies on *Asterionelia formosa* Hass II. Nutrient depletion and the spring maximum. *J. Ecol.*, **38**, 1–35.

Lund, J. W. G. (1954). The seasonal cycle of the plankton diatom, *Melosira italica* (Ehr.) Kutz. subsp. *subarctica* O. Mull. *J. Ecol.*, **42**, 151–179.

Lund, J. W. G. (1955). Further observations on the seasonal cycle of *Melosira italica* (Ehr.) Kutz. subsp. *subarctica* O. Mull. *J. Ecol.*, **43**, 90–102.

Lund, J. W. G., Mackereth, F. J. H., and Mortimer, C. H. (1963). Changes in depth and time of certain chemical and physical conditions and of the standing crop of *Asterionella formosa* Hass. in the North Basin of Windermere in 1947. *Phil. Trans. R. Soc.*, **B246**, 255–290.

Lundqvist, G. (1924). Utvecklingshistoriska insjötudier i Sydsverige. *Sver. Geol. Unders.*, Series C, **330**, 1–129.

Lundqvist, G. (1927). Bodenablagerungen und Entwicklungstypen der Seen. *Die Binnengewasser*, **2**, 124 pp.

Lundqvist, G., and Thomasson, H. (1923). Diatomacéekologien och kvartärgeologien. *Geol. Fören. Förh.*, **45**, 379–385.

Lundqvist, J. (1971). The interglacial deposits at the Leveaniemi Mine, Svappavaara, Swedish Lapland. *Sver. Geol. Unders.*, Series C, **658**, 1–163.

McDonald, J. D. (1869). On the structure of the diatomaceous frustule and its genetic cycle. *Ann. Mag. Hist.*, Series 4, **13**, 1–8.

Maher, L. J. (1972). Nomograms for computing 0.95 confidence limits of pollen data. *Rev. Palaeobotan. Palynol.*, **13**, 85–93.

Mannion, A. (1978). Late Quaternary deposits from Southeast Scotland. II: The diatom assemblage of a marl core. *J. of Biogeog.*, **5**, 301–318.

Marciniak, B. (1969). Die ersten Ergebnisse der Diatomeenanalyse der Spatglazialen Sedimente des Mikolajkisees (No-Polen). *Mitt. Int. Ver. Limnol.*, **17**, 344–350.

Marciniak, B. (1973). The application of the diatomological analysis in the stratigraphy of the late-glacial deposits of the Mikolajki Lake. *Studia Geologica Polonica*, **39**, 1–159.

Margalef, R. (1969). Size of centric diatoms as an ecological indicator. *Mitt. Internat. Verein. Limnol.*, **17**, 202–210.

Matthews, J. (1969). The assessment of a method for the determination of absolute pollen frequencies. *New Phytol.*, **68**, 161–166.

Meriläinen, J. (1967). The diatom flora and the hydrogen ion concentration of water. *Ann. Bot. Fenn.*, **4**, 51–58.

Meriläinen, J. (1969). The diatoms of the meromictic lake Valkiajärvi, in the Finnish Lake District. *Ann. Bot. Fenn.*, **6**, 77–104.

Meriläinen, J. (1971). The recent sedimentation of diatom frustules in four meromictic lakes. *Ann. Bot. Fenn.*, **8**, 160–176.

Meriläinen, J. (1973). *The Dissolution of Diatom Frustules and its Palaeoecological Interpretation*, Univ. of Lund, Dept. of Quat. Geol. Report 3, pp. 91–95.

Miller, U. (1964). Diatom floras in the Quaternary of the Göta River Valley (W. Sweden). *Sver. Geol. Unders.*, **44**, 1–67.

Miller, U. (1969). Fossil diatoms under the scanning electron microscope — a preliminary report. *Sver. Geol. Unders.*, C **624**, 1–65.

Miller, U. (1971). Diatom floras in the sediments at Leveaniemi. App. In: Lundqvist, 1971 (q.v.).

Miller, U. (1977). Pleistocene Deposits of the Alnarp Valley, Southern Sweden: Microfossils and their Stratigraphic Application, *Univ. of Lund, Dept. of Quat. Geol. Thesis*, **4**, 125 pp.

Miller, U., and Robertsson, A-M. (1979). Biostratigraphical Investigations in the Anundsjö region, Ångermanland, N. Sweden. *Early Norrland*, **12**, 1–76.

Molder, K., and Tynni, R. (1967–80). Uber Finnlands rezente und subfossile Diatomeen I–XI. *Comptes Rendus de la Societé Géologique de Finlande.*

Mölder, K. (1944). Die Entwicklungsgeschichte des Sees Vieljärvi in Ostkarelien und die Klimaschwankung im Lichte der Fossilen Diatomeenfunde aus den Seesedimenten. *Co. Ren. de la Soc. Geol. de Finl.*, **16**, 101–146.

Mosimann, J. E. (1965). Statistical methods for the pollen analyst: multinomial and negative multinomial techniques. In: *Handbook of Palaentological Techniques* (Eds. B. Kummel and D. Raup), Freeman, San Francisco, Calif.

Moss, B. (1978). The ecological history of a medieval man-made lake, Hickling Broad, Norfolk, United Kingdom. *Hydrobiol.*, **60**, 23–32.

Moss, B. (1980). Further studies on the palaeolimnology and changes in the phosphorus budget of Barton Broad, Norfolk. *Freshw. Biol.*, **10**, 261–279.

Nipkow, F. (1920). Vorlaufige Mitteilung uber Untersuchungen des Schlammabsatzes im Zurichsee. *Schweiz. Z. Hydrobiol.*, **1**, 100–122.

Nipkow, F. (1927). Uber das Verhalten der skelette planktischer Kieselalgen in geschichteten Tiefenschlamm des Zurich- und Baldeggersees. *Schweiz. Z. Hydrobiol.*, **4**, 112–120.

Nygaard, G. (1949). Hydrobiological studies on some Danish ponds and lakes. II: The Quotient hypothesis and some new or little known phytoplankton organisms. *Det Kongel. Dansk. Vid. Selsk. Biol. Skr.*, **7**, 1–293.

Nygaard, G. (1956). Ancient and recent flora of diatoms and chrysophyceae in Lake Gribso. In: (Kaj Berg and I. B. Clemens Petersen) Studies on the Humic Acid Lake Gribso. *Fol. Limnol. Scand.*, **8**, 32–94.

Paasche, E. (1960). On the relationship between primary production and standing stock of phytoplankton. *J. Cons. Int. Explor. Mer.*, **26**(1).

Patrick, R. (1977). Ecology of freshwater diatoms — diatom communities. In: *The Biology of Diatoms* (Ed. D. Werner), Botanical Monographs Vol. 13, Blackwell.

Patrick, R., Hohn, M. H., and Wallace, J. H. (1954). A new method for determining the pattern of the diatom flora. *Notulae Naturae Acad. Nat. Sci. Philadelphia*, **259**, 1–12.

Patrick, R., and Reimer, C. (1966). *The Diatoms of the United States, Exclusive of Alaska and Hawaii. I: Fragilariaceae, Eunotiaceae, Achnanthaceae, Naviculaceae*, Acad. Nat. Sci. Philadelphia Monograph 13, pp. 1–688.

Patrick, R., and Reimer, C. (1975). *The Diatoms of the United States Exclusive of Alaska and Hawaii. II: Part I.*, Acad. Nat. Sci. Philadelphia Monograph 13, pp. 1–213.

Pennington, W. (1943). Lake sediments: the bottom deposits of the N. Basin of Windermere with special reference to the diatom succession. *New Phytol.*, **43**, 1–27.

Pennington, W. (1973). The recent sediments of Windermere. *Freshwat. Biol.*, **3**, 363–382.

Pennington, W., Cambray, R. S., and Fisher, E. M. R. (1973). Observations on lake sediments using fallout 137Cs as a tracer. *Nature*, **242**, 343–346.

Petersen, B. J. (1935). Studies on the biology and taxonomy of soil algae. *Dan. Bot. Ark.*, **8**, 183 pp.

Petersen, B. J. (1943). Some halobian spectra (Diatoms). *Kgl. Dansk Vidensk. Selsk. Biol. Medd.*, **17**, 1–95.

Pfitzer, E. (1869). Uber Bau und Zellteilung der Diatomeen. *Sitzungsber. niederrhein. Ges. Nat. Heilk.*, **26**, 86–89.

Quennerstedt, N. (1955). Diatoméerna i Långans sjövegetation. *Acta phytogeogr. Suec.*, **36**, 1–208.

Renberg, I. (1976). Palaeolimnological investigations in Lake Prastsjon. *Early Norrland*, 9, 113–160.

Renberg, I., and Hellberg, T. (1982). The pH history of lakes in south-western Sweden, as calculated from the subfossil diatom flora of the sediments. *Ambio*, 11, 30–33.

Reynolds, C. S. (1973). The seasonal periodicity of planktonic diatoms in a shallow eutrophic lake. *Freshw. Biol.*, 3, 89–110.

Richardson, J. L. (1968). Diatoms and lake typology in East and Central Africa. *Int. Rev. Ges. Hydrobiol.*, 53, 299–338.

Richardson, J. L., and Richardson, A. E. (1972). History of an African rift lake and its climatic implications. *Ecol. Monogr.*, 42, 499–534.

Rippey, B. (1977). The behaviour of phosphorus and silicon in undisturbed cores of Lough Neagh sediments . In: *Interactions between Sediments and Freshwater*, (Ed. H. L. Golterman), Dr. W. Junk B. V. Publ., pp. 348–353.

Rippey, B. (1983). A laboratory study of the silicon release process from a lake sediment (Lough Neagh, Northern Ireland). *Arch. Hydrobiol.*, 96, 417–433.

Ritchie, J. C., and Koivo, L. K. (1975). Postglacial diatom stratigraphy in relation to the recession of Glacial Lake Agassiz. *Quat. Res.*, 5, 529–540.

Robertsson, A-M. (1973). Late-glacial and pre-boreal pollen and diatom diagrams from Skurup, Southern Scania. *Sver. Geol. Unders.*, C 679, 1–75.

Rodhe, W. (1968). Crystallization of eutrophic concepts in Northern Europe. In: *Eutrophication: Causes, Consequences, Correctives*, Washington, D. C., pp. 50–64.

Ross, R., and Sims, P. A. (1972). The fine structure of the frustule in centric diatoms: a suggested terminology. *Br. Phycol. J.*, 7, 139–163.

Ross, R., Cox, E. J., Karayeva, N. I., Mann, D. G., Paddock, T. B. B., Simonsen, R., and Sims, P. A. (1979). 'An amended terminology for the siliceous components of the diatom cell', *Nova Hedwigia Beih.* 64, 513–533.

Round, F. E. (1957). The late-glacial and post-glacial diatom succession in the Kentmere Valley deposit. I: Introduction, methods & flora. *New Phytologist*, 56, 98–126.

Round, F. E. (1961). Diatoms from Esthwaite. *New Phytol.*, 60, 43–59.

Round, F. E. (1964). The diatom sequence in lake deposits; some problems of interpretation. *Verh. Int. Verein. Limnol.*, 15, 1012–1020.

Round, F. E. (1972). *Stephanodiscus binderanus* (Kutz) Krieger or *Melosira binderana* Kutz (Bacillariophyta Centrales). *Phycologia*, 11, 109–119.

Schmidt, A. (1874–1959). *Atlas der Diatomaceenkunde*, Leipzig.

Schoeman, F. R., and Archibald, R. E. M. (1976). *The Diatom Flora of Southern Africa*, CSIR Special Report Wat. 50.

Schofield, C. L., and Galloway, J. N. (1977). The utility of palaeolimnological analyses of lake sediments for evaluating acid precipitation effects on dilute lakes. *Technical Completion Report*. Cornell University, pp. 1–26.

Shear, H., Nalewajko, C., and Bacchus, H. M. (1976). Some aspects of the ecology of *Melosira* spp. in Ontario lakes. *Hydrobiologia*, 50, 173–176.

Simola, H. (1977). Diatom succession in the formation of annually laminated sediment in Lovojärvi, a small eutrophicated lake. *Ann. Bot. Fenn.*, 14, 143–148.

Simola, H. (1979). Micro-stratigraphy of sediment laminations deposited in a chemically stratifying eutrophic lake during the years 1913–1976. *Hol. Ecol.*, 2, 160–168.

Simola, H. (1981). Sedimentation in a eutrophic stratified lake in S. Finland. *Ann. Bot. Fenn.*, 18, 23–36.

Simonsen, R. (1957). Spatglaziale Diatomeen aus Holstein. *Arch. Hydrobiol.*, 53, 337–349.

Simonsen, R. (1962). Untersuchungen zur Systematik und Okologie der Bodendiatomeen der Westlichen Ostsee. *Int. Rev. Hydrobiol. Syst. Beih.*, 1, 144.

Staub, E. (1977). Stratigraphy of diatoms and chlorophyll derivatives in the uppermost sediment of a Swiss lake. In: *Interactions between Sediment and Freshwater*, (Ed. H. L. Golterman), Dr. W. Junk B.V. Publishers, The Hague, pp. 161–164.

Stockner, J. G. (1971). Preliminary characteristics of lakes in the Experimental Lakes Area, Northwestern Ontario, using diatom occurrences in the sediment. *J. Fish. Res. Bd. Canada*, **28**, 265–275.

Stockner, J. G. (1972). Palaeolimnology as a means of assessing eutrophication. *Verh. Int. Verein. Limnol.*, **18**, 1018–1030.

Stockner, J. G., and Benson, W. W. (1967). The succession of diatom assemblages in the recent sediments of Lake Washington. *Limnol. & Oceanogr.*, **12**, 513–532.

Stoffers, P., and Hecky, R. E. (1978). Late Pleistocene–Holocene evolution in the Kivu-Tanganyika basin. *Spec. Publs. int. Ass. Sediment.*, **2**, 43–55.

Tessenow, U. (1964). Experimentaluntersuchungen zur Kieselsaureruckfuhrung aus dem Schlamm der Seen durch Chironomidenlarven (Plumosus-Gruppe). *Arch. Hydrobiol.*, **60**, 497–504.

Tessenow, U. (1966). Untersuchungen über den Kieselsaurehaushalt der Binnengewasser. *Arch. Hydrobiol. Suppl.*, **32**, 1–136.

Thomas, S. R., and Soltero, R. A. (1977). Recent sedimentary history of a eutrophic reservoir: Long Lake, Washington. *J. Fish. Res. Bd. Canada*, **34**, 669–676.

Tilman, D. (1977). Resource competition between planktonic algae: an experimental and theoretical approach. *Ecology*, **58**, 338–348.

Tolonen, K. (1978). Effects of prehistoric man on Finnish lakes. *Pol. Arch. Hydrobiol.*, **25**, 419–422.

Tolonen, K., and Jaakkola, T. (1983). History of lake acidification and air pollution studied on sediments in South Finland. *Ann. Bot. Fenn.*, **20**, 57–78.

Tropper, C. B. (1975). Morphological variation of *Achnanthes haukiana* (Bacillariophyceae) in the field. *J. Phycol.*, **11**, 297–302.

Vuorinen, J., and Huttunen, P. (1981). The desktop computer in processing biostratigraphic data. *Pollen et Spores*, **23**, 165–172.

Walker, R. (1978). Diatom and pollen studies of a sediment profile from Melynllyn, a mountain tarn in Snowdonia, North Wales. *New Phytol.*, **81**, 791–804.

Watts, W. A., and Bradbury, J. P. (1982). Palaeoecological studies at Lake Patzcuaro on the west-central Mexican Plateau and at Chalco in the Basin of Mexico. *Quat. Res.*, **17**, 56–70.

Wetzel, R. G. (1975). *Limnology*. W. B. Saunders, Philadelphia.

Whiteside, M. (1983). The mythical concept of eutrophication. *Hydrobiologia*, **103**, 107–112.

Handbook of Holocene Palaeoecology and Palaeohydrology
Edited by B. E. Berglund
© 1986 John Wiley & Sons Ltd.

27

Analysis of fossil fruits and seeds

Krystyna Wasylikowa

Institute of Botany, Polish Academy of Sciences, Krakow, Poland

INTRODUCTION

Macrofossil plant remains preserved in Quaternary deposits are fruits, seeds, fragments of wood and other parts of vascular plants (leaves, buds, scales, spines and so on), mosses and some algae (oospores of Charales). Most of them are found in an uncharred condition. Only wood occurs commonly as charcoal, while charred fruits and seeds are less common in natural deposits. Carbonization may result from natural or man-made fires and the accumulation of charred plant remains is usually connected with archaeological sites. Other types of fossilization are seldom encountered among the Quaternary fossils (e.g. impressions, mineralization). A special case are impressions on potsherds and daub preserved in archaeological sites.

This chapter concerns the analysis of fossil fruits and seeds; for the other plant remains the reader is referred to Chapters 28, 29 and 30. The intention here is to touch on only the main principles which should be recognized and observed at every stage of macrofossil investigation (see Figure 27.1), in order that reliable results might be obtained. More information can be found in the literature. Growing interest in plant macrofossils, over the last twenty years, is reflected in the accumulation of valuable data presented in numerous papers devoted to individual sites. Some more comprehensive publications on this subject are also available (e.g. Birks, 1980; Dickson, 1970; Godwin, 1975; Nikitin, 1969; Wasylikowa, 1973; Watts, 1978; West, 1969).

FIELDWORK

In Quaternary studies macrofossil analysis is usually part of wider palaeoeco-

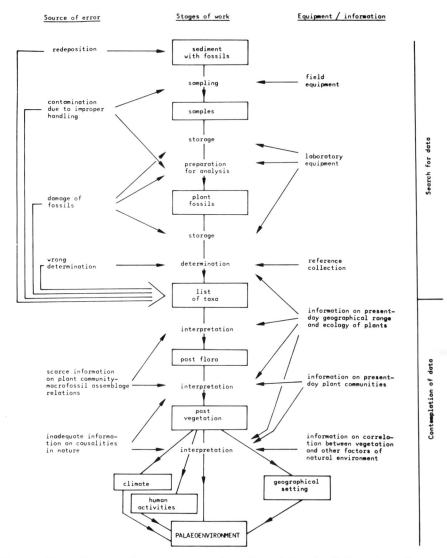

FIGURE 27.1.   Scheme of successive steps in studies of plant fossils from the sediment
          sample to the reconstruction of palaeoenvironment

logical research, and the selection of sampling site depends on the general aim of
the investigation. It is not normally possible to find a place equally suited for all
kinds of analyses. If plant macrofossils are to contribute substantially to
information about the past environment, they should be abundant (not less than 10
per 100 cm^3, see Watts, 1978), well-preserved (identifiable) and their stratigraphic
position must be well correlated with sequences obtained from other analyses.

In water basins the best material for macrofossil analysis may be obtained from the near-shore sediments where most plant remains accumulate. If several cores are to be used for different analyses, it is advisable to sample at least two cores for macrofossils — one in the centre and one near the shore of a lake. Their correlation should be checked by pollen analysis or other methods (Chapter 13). For the detailed reconstruction of local changes in plant communities of a lake or a peat-bog a series of borings should be made at least along two transects of a basin (see Chapters 8 and 28).

Sampling from archaeological sites poses special questions which cannot be discussed in detail. For the study of vegetational changes connected with the settlement of a particular culture, a series of samples should be taken in places where the culture layer is included in organic sediments. Otherwise it may be difficult to correlate exactly environmental and cultural phenomena. If an investigation of human husbandry in a particular site is of special interest, several sample series should be collected at various points within a site. The number of samples, and their distribution, on the plane of excavation must be decided for each site individually.

General principles for sample collection are the same as in other analyses. Samples may be taken from exposures or borings. In both cases two main conditions must be fulfilled: (1) the precise stratigraphic position of each sample must be known in relation to both the whole sequence of sediments and to the samples used for other analyses; and (2) samples must be clean, which means that they must not be contaminated by plant material older or younger than the layer from which a particular sample is taken.

The density of sampling should be adjusted according to the character of the deposits and to the purpose of the study. Samples should be taken at closer intervals for periods when vegetational changes were rapid or when the sedimentation was slow. For periods characterized by more stable vegetation or very fast sedimentation, the sampling intervals may be wider. In Quaternary studies a continuous series of samples for macrofossil analysis is usually taken in such a way that each macrosample corresponds to only one or two samples for pollen analysis. In stratified deposits sampling intervals should be adapted to the thickness of layers in order to obtain samples from pure layers. Mixing of material from two neighbouring horizons should be avoided.

LABORATORY TECHNIQUES

Storage of samples

Samples for macrofossil analysis should not be exposed to rapid drying, which may damage plant remains. Slow drying is not harmful and material may be used for macrofossil analysis as long as it does not become air dry, which can make subsequent disintegration impossible. This may happen particularly with highly humified organic sediments (dy, decomposed peat or gyttja). Properly wrapped samples kept in cool conditions will neither dry completely nor get mouldy during prolonged storage.

Dispersion of sediment

Samples collected in plastic bags in the field are ready for further treatment. Core segments stored as such must be sampled in the laboratory. Each segment is unwrapped, its outer surface is cleaned with a knife or spatula and the sediment is described. Pollen samples are taken in glass vials, macrofossil samples in plastic bags. If the main pollen study is being undertaken on another core, additional samples for pollen analysis should be taken from the same core as the macrofossils, for future correlation. All utensils must be thoroughly cleaned before each sampling.

The preparation of samples for analysis includes the dispersion of sediment and the separation of identifiable plant remains from indeterminable residue. The procedures employed should be as gentle as possible because certain plant remains may be very fragile. Accelerated dispersion by mechanical means (shakers, ultrasonics) or chemical reagents (strong acids or bases) should be undertaken with great caution.

Each sample which is large enough should be divided into two parts. One portion (usually the smaller) is left untouched for future reference. The size of the portion taken for analysis should be measured for volume or weight. It is convenient to work with samples of constant size, but in many cases it is not possible (e.g. stratification of a deposit may require varying sampling intervals). In profiles composed of more or less homogenous sediments both the weight and volume of samples may be used. If organic and mineral sediments occur in one section it seems more appropriate to use volume as a more or less comparable measure of sample size. Sample volume is measured by displacement of water in a measuring cylinder.

Certain sediments (e.g. sand) may disperse in pure water, but most of them require the addition of dissolving reagents. Peats and most other sediments are soaked in water with a small amount of Na_2CO_3 or 10% KOH for one to several days, and stirred occasionally. Some heating or boiling of the sample is usually required. When complete dispersion is achieved, the sample is washed through two sieves with meshes 0.5 and 0.2 (or 0.1) mm. Tap water (warm or cold) is used for washing. The stream of water should be strong enough to cause small mineral particles to go through the sieves, but not too strong, because fragile plant remains may become damaged. When the water passing through the sieves is clean, washing is finished, and the material from both sieves is transferred separately to Petri dishes, beakers or jars. For deposits containing larger plant fragments more sieves of various size may be used. Care should be taken that no plant material is left on the sieves or in any dish used during the preparation of samples.

Completely dried dy, decomposed gyttja or decomposed peat disperse with great difficulty. Boiling in pure 5–10% KOH may be necessary, which may be harmful to plant tissues.

Highly calcareous sediments are treated with 5–10% cold or warm HCl until effervescence stops. This may be the only way to disperse some sediments, but HCl dissolves all calcareous fossils (e.g. envelope of lime deposited around *Chara* oospores), and other plant fragments become brighter and transparent.

In some cases flotation methods are useful. They enable the separation of organic matter, which floats on the surface of the liquid, from mineral matter which sinks and accumulates at the bottom. The simplest way is to pour water on to a dry sample, stir, wait a while until plant remnants float to the surface and pour the water with plant remains through a sieve. The procedure should be repeated several times. The method gives good results with dry sand containing charred material. HNO_3 or H_2O_2 may be used for flotation of samples with uncharred fossils (Watts, 1959; Schwarzenholz, 1961). Different methods of froth flotation have been tested mainly for archaeological purposes, but they are of no importance for the study of non-anthropogenic sediments, where only small samples are available. In every case, washing in the field should be undertaken with care, to ensure that the device and water used for washing are clean and contain no living or fossil plant remains.

Separation of fossils

For the separation of fossils, a small portion of washed sample is placed on a Petri dish or small plate, and a small amount of water is poured into it (enough to cover the dish with 2–3 mm depth of water). Under a low-power microscope, at a magnification of 10–20 times, the material is pushed from one end of the dish to the other, and all identifiable plant remains are picked out with a delicate brush and placed in another dish. Uncharred fossils are put in a small amount of glycerol mixed with 96% alcohol and distilled water, in the proportions 1:1:1, with addition of thymol; or in an equal mixture of glycerol and 2% formalin (Watts, 1978). Charred fragments are best preserved in their dry state.

When the preliminary segregation of fossils is accomplished the material is ready for determination. In rare cases the bleaching of fossils or the cleaning of their surface is necessary (Barghoorn, 1948; Nikitin, 1969). Any bleaching or cleaning method should be used with great caution and should be tested for each material.

IDENTIFICATION OF FRUITS AND SEEDS

The determination of fossils is based on comparisons with modern plants. A reference collection of fruits and seeds is indispensable for work with fossil material. A good reference collection should include at least all species which might appear in fossil material from a given area, on the basis of their present-day geographical distributions and ecology. However, the investigator should always be prepared to find an unexpected taxon, not known within the study area at present and not previously described in the fossil state. In addition to seeds and fruits, it is also convenient to obtain some vegetative parts of plants as reference material (needles of conifers, small Ericaceae leaves, spines, buds, scales etc.). Any addition to the reference collection leads to potential improvement in the comprehensive analysis of fossil floras.

To identify a taxon it is not sufficient to state that the fossil is similar to extant specimens of a taxon, but it is necessary to check that it differs from all closely related taxa (all species within a genus or related genera) having similar diaspores. There are no general keys for fossil seeds and fruits, and nothing can substitute for a reference collection. There are, however, several publications which may greatly help in the work. Atlases with photographs and drawings facilitate a first approximation (e.g. Beijerinck, 1947; Berggren, 1969, 1981; Bertsch, 1941; Brouwer and Stählin, 1955; Grosse-Brauckmann, 1972, 1974; Kats *et al.*, 1965). Publications on seed and fruit morphology of selected families and genera are very useful for specific determinations. A comprehensive bibliography was published by Delcourt *et al.* (1979). Valuable information may also be found in regional floras.

Seeds and fruits preserved in middle and late Quaternary sediments usually do not show great differences in shape and sculpture compared with their modern counterparts. In compact sediments, buried under a thick cover of mineral deposits, they may become compressed. The sculpture of fossils may be more pronounced than in modern specimens because the outer walls of the epidermic cells are destroyed and the pattern formed by lateral (vertical) walls is clearly visible. In some cases, characters not visible on modern dry specimens may be seen on fossil seeds or fruits. For instance, seeds of *Lythrum*, which are more or less smooth when dry and covered with elongated cells adhering tightly to the seed surface, are often found with very small papillae all over their surface. This appearance is caused by swelling of the epidermic cells. Delicate swelling of epidermis cells may also occur in seeds of other genera (e.g. in the family Cruciferae). Frequently, modern specimens must be subjected to special preparation in order to reveal the characters observed in the fossils; artificial swelling of the epidermis cells may be produced by heating modern seeds in 10% KOH. It may be necessary to remove outer parts of present-day fruits or seeds: cuticule, complete epidermis or the outer walls of epidermis cells (*Ranunculus* fruits, *Hypericum* seeds), the entire pericarp (seeds of *Chenopodium*) or just part of it (*Potamogeton* fruit-stones). To do this, modern specimens are boiled for a short time in water with the addition of Na_2CO_3 or 5–10% KOH and left in this solution for a few minutes or up to several hours. Then they are washed in pure water, and the soft outer layers are removed with a needle under the microscope. Artificial fossilization of *Juncus* seeds and caryopses of Gramineae was described by Körber-Grohne (1964). This procedure leads to the decomposition of the inner part of seeds without destroying their outer shape and thus enables the comparison of fossils with present-day specimens. Special changes of shape are observed in charred specimens. Experimental carbonization of modern material helps towards a better understanding of deformations caused by high temperature (Wilson, 1984).

The size of fossil seeds or fruits is usually given with their descriptions, but compared with pollen analysis little attention is paid to the change in size due to fossilization and sample preparation. Fossilization may cause the diminution of fossil diaspores owing to dehydration and compression. Secondary shrinkage occurs when the specimens dry completely after being isolated from the sediment, and may often be irreversible. On the other hand, boiling of a sample in Na_2CO_3 causes an increase in size (swelling) of fossil fruits and seeds (Bialobrzeska, 1964). The same effect may be produced by heating in KOH followed by storage in a glycerine mixture. Carbonization (by heat) causes a change in size and shape which depends on the temperature of combustion, and on the anatomical and chemical structure of fossils. Cereal grains usually become shorter but thicker and broader. Many other seeds and fruits show a similar tendency to become more or less round (e.g. *Polygonum convolvulus*). In some cases no clear changes of shape are apparent, only the size of charred specimens is smaller when compared with uncharred material.

The change of size in the fossil state does not cause much trouble in routine work but may become important in all cases when size is a decisive feature for the identification of a species. If data on dimensions, as published by various authors, should be comparable, information should be given concerning the state of preservation (charred–uncharred), preparation technique, the approximate time of storage before the measuring was done, and whether or not dry or wet specimens were measured.

For routine work on the identification of seeds and fruits, a low-power binocular microscope with magnification 5–100 times is used. More detailed observations may be done with a high-power microscope using transmitted or reflected light. The scanning electron microscope (SEM) is of great help in studies of seed and fruit morphology.

The glossy surface of fruits and seeds soaked with glycerine mixture obscure the fine details of the surface sculpture. To observe them better, it is necessary to wash the specimens with alcohol and dry them on blotting-paper. Desiccation should be controlled under the microscope and should be stopped at the right moment. Some fossils split, roll or shrink very easily when drying. To avoid this, the observation of sculpture may also be carried out on fossil specimens immersed in pure water.

PRESENTATION AND INTERPRETATION OF RESULTS

Qualitative results

A list of identified taxa is a qualitative result of an investigation. For each taxon the degree of reliability of the determination should be indicated. Identification of a species is certain when a fossil specimen is only comparable with one species on the basis of its morphology or anatomy. Very often, however, two different criteria are combined, that of the structure and of the present-day occurrence in nature (geographical range and more seldom ecological requirements). For instance, working with Holocene sediments in central Europe our experience is to eliminate from consideration those species which are not native to Europe or which are native plants of rather limited geographical range (e.g. growing in the extreme North or South or in high mountains). In some cases phytogeographical and ecological criteria may be misleading; well-known examples are modern American species found in interglacial floras of Europe (*Brasenia schreberi, Dulichium spathaceum*). This problem is less likely to arise in Holocene floras, but *Xanthium strumarium* may be an example. In Europe it was considered by plant geographers as naturalized, probably from North America, until the finding of its fruits from the medieval (Lange, 1968) and then from the Roman periods (Willerding, 1983) proved its presence on this continent at least since that time.

Uncertainty of determination may be indicated with plant name by the addition of cf. or type (or other marks). These qualifications are applied with different meanings by various authors and it is advisable to state in what sense they are used. It is very useful to a reader if the author mentions all species with which a particular fossil was compared. Identified fossils should be stored in such a way that they are easily accessible for future reference.

Species identifications should be well-founded and well-documented because they are the basis for ecological interpretations concerning the site in question and for broader considerations of the vegetational and climatic history.

Quantitative results

All fossils which are found in a sample of known size and which represent more or less complete plant organs should be counted. These are seeds, fruits, and all other plant fragments which can be counted, such as parts of flowers, spines, needles or bud scales. While counting fragments a reasonable and constant principle should be adopted; for instance 2 seed halves or 4 smaller fragments make one specimen, the number of needles corresponds to the number of fragments having tips or complete basal parts, etc. Chapters 23, 28, 29 and 30 may be referred to for quantitative assessment of vegetative plant fragments.

Some fossils may occur so abundantly (e.g. *Juncus* or *Typha* seeds) that it is not possible to count them all. Their number may be counted in a smaller portion of the sample and estimated for the whole sample, or an approximate number may be given, for instance 'over 500'.

Quantitative results may be presented in tables and diagrams. The simplest table includes plant names, sample numbers, depths and sizes and the number of specimens of each taxon in each sample. For special purposes different schemes of tables have been developed by various authors (e.g. Behre, 1976; Grosse-Brauckmann, 1979; Wasylikowa, 1978; van Zeist, 1974). Synthetic tables (Table 27.1) that combine pollen and macrofossil data were produced by Rybníček (1973) and Rybníček and Rybníčkova (1974).

Macrofossil concentration diagrams represent an absolute number of specimens in samples of constant size (Figures 27.2 and 27.3). In profiles having a sufficient number of radiocarbon dates, macrofossil influx may be calculated and presented in a diagram to show the changes in yearly deposition of macrofossils per cm^2. An alternative is to draw percentage curves or to combine both percentage and absolute curves in one diagram (Watts and Winter, 1966). Most often, however, diagrams are based on macrofossil concentration numbers.

One point should be stressed. Macrofossil diagrams, at first sight resembling pollen diagrams, have a different significance because numerical changes in macrofossil spectra are based on a much smaller number of records than are pollen spectra, and because seeds are produced in smaller quantities than are pollen grains.

Studies of macrofossil plant remains have a longer tradition than pollen analysis but, until now, methods of quantitative interpretation have been much less refined and their improvement is a matter for future investigation.

Sources of errors

Errors may result from contamination of material, wrong determination and misinterpretation. The two latter questions are obvious and only the problem of contamination, seldom considered in connection with the investigation of plant macrofossils, will be discussed here.

The deposit which is intended for sampling may be contaminated by plant material redeposited from older sediments or by younger material introduced to older layers by animals (burrowers, worms) or by mechanical forces. Redeposition of older fossils is most likely to occur in rivers where sediments could have been exposed to a fairly strong flow of water for a certain period of time. Older sediments eroded at one place could have been deposited downstream on top of younger sediments. Nikitin (1969) quotes 200 species of Tertiary plants which were found secondarily deposited in Holocene sediments of western Siberia. Redeposition may also take place in lakes subjected to the fluctuations of water level (West, 1969). Watts and Bright (1968) described a lake in North Dakota where a lowering of the water table during the warmest

TABLE 27.1. Local plant succession from an alder stand to a waterlogged cultivated meadow during the subatlantic period, at Nahořany in south Bohemia, Czechoslovakia (sporadically occurring species are omitted) (after Rybníček and Rybníčková, 1974, simplified)

Taxon		Depth (cm)										
		97–100	85–97	75–85	65–75	55–65	45–55	37–45	25–37	15–25	8–16	2–8
Cyperaceae	t,p	5	5		20	20	50	50	60	50	50	30
Caltha palustris agg.	p		(+)	(+)	(+)	(+)	(+)	(+)	(+)	[+]	(+)	(+)
Cirsium sp.	p				(+)	(+)	(+)	(+)	(+)	(+)	(+)	(+)
Galium sp.	p				(+)	(+)	(+)	(+)	(+)	(+)	(+)	(+)
Epilobium sp.	p		(+)							(+)		
Campanula sp.	p		(+)							(+)		
Alnus glutinosa (L.) Gaertn.	p,w	10	[+]	[+]	[+]	[+]	(+)	(+)	[+]	(+)		
Polypodiaceae	p		30	20	30	25	25	10				
Filipendula cf. ulmaria (L.) Maxim.	p				(+)		(+)	(+)	(+)			
Coenococcum geophilum Fr.	scl	1	1			3	[+]	[+]				
Rubus cf. idaeus L.	s		1	1	1	1	2	1				
Ranunculus repens L.	s		2	2	1	1		1				
Sambucus cf. nigra L.	s			1								
Carex echinata Ehrh.	s								2			2
Carex fusca All.	s								1	1	2	
Carex sp.	s								1	1		1
Viola palustris L.	s,p									(2)	1	4
Menyanthes trifoliata L.	s,p									1	(+)	
Ajuga reptans L.	s										1	
Cardamine cf. pratensis L.	p					(+)			(+)	(+)	(+)	(+)
Trifolium t. pratense L.	p							(+)	(+)	(+)	(+)	(+)
Rumex cf. acetosa L.	s,p							(+)	1	(+)	1	(+)
Ranunculus acer L.	s								1		3	10
Lotus sp.	p					(+)				(+)	(+)	
Succisa pratensis Moench.	p									(+)	(+)	(+)

Explanations: n = pollen or spores; s = seeds, fruits (absolute number per 100 ml); scl = sclerotia (absolute number per 100 ml); t = plant tissues and
... %; (+) = important and [+] = maximum occurrance of pollen and spores; (1) = number of macrofossils and

part of the Holocene caused the erosion of shore sediments, resulting in the incorporation of late-glacial pollen and macrofossils into the mid-Holocene sediments.

If the age of redeposited fossils is distinctly different from the age of the sediment, the elimination of contaminating material may be easy (e.g. Tertiary plants in Holocene sediments of central Europe). If, however, redeposited taxa occur at present in or near the study area the discovery of contamination may be difficult or impossible. In such cases the detailed field observations of the physical properties of the sediment (particularly of the boundaries between layers) may be helpful, together with observation of the state of preservation of fossils.

Transfer of present-day fruits and seeds into older layers by animals may occur particularly in aerated, neutral or slightly basic soils which support a rich fauna. Cracking of sediment due to drying of its surface may cause the penetration of the soil from the top to the deeper layers, and frost action may result in various disturbances. Modern seeds brought down in this way may still have viable embryos and may be easy to distinguish from fossils, but if their inner parts have already rotted they may be very similar to fossil specimens.

Contamination may also be caused by the careless handling of samples: taking them from sections which were not properly cleaned, use of dirty utensils, exposing samples to contact with living plants bearing fruits or with the surface of soil rich in fruits and seeds, or with other sediments which contain plant remains of various ages.

Advantages and limitations of plant macrofossil studies

The main question in palaeoecological studies of plant fossils is how does a fossil assemblage reflect the composition of plant communities from which the fossils were derived? Many studies have been carried out from this point of view on pollen and on autochthonous plant tissues in peats, but much less is known in this respect about seeds, fruits and other detached plant fragments.

Factors which influence the composition of fossil seed and fruit assemblages are connected, on the one hand, with the structure and biology of plants, and on the other with the external agents regulating the spread, deposition and preservation of diaspores. An extensive description of the biology of dispersal was presented by Harper (1977). Several questions important for the interpretation of fossil spectra and waiting for elucidation in further studies arise from already known facts. These facts are as follows: (1) plant species differ greatly in their seed production and in the adaptation to their dispersal; (2) species composition of diaspores in the soil, the seed bank, differs to a variable extent from that of the local vegetation — in general, the pioneer species may lie dormant for a long time in the soil; (3) there exists some vertical movement of seeds in the soil, usually down the soil profile but exceptionally in the opposite direction (by some burrowing animals); and (4) only a small fraction of seeds produced has a chance to become fossilized (Figure 27.4).

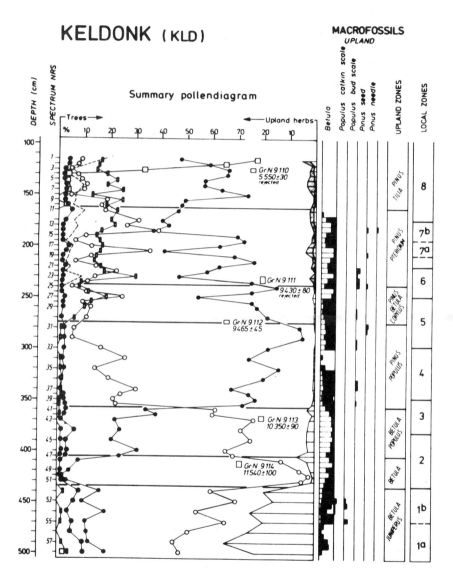

FIGURE 27.2. Macrofossil diagram from Keldong, Netherlands, combined with a shortened pollen diagram. Bars for individual species, arranged according to their appearance, show the local succession from open-water communities, through reed-swamps to alder woods (from Leeuwaarden, 1982). *Reproduced by permission of Dr W. van Leeuwaarden*

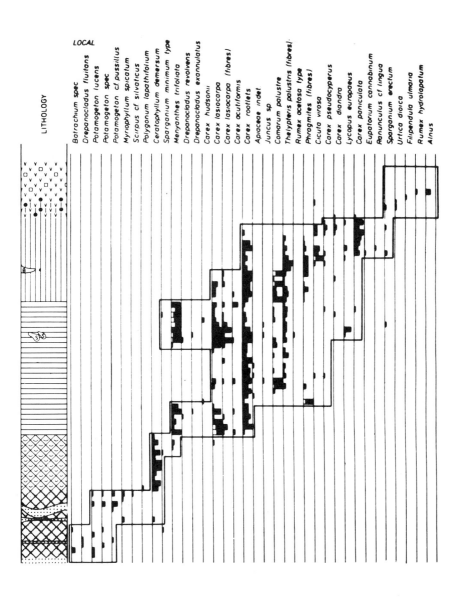

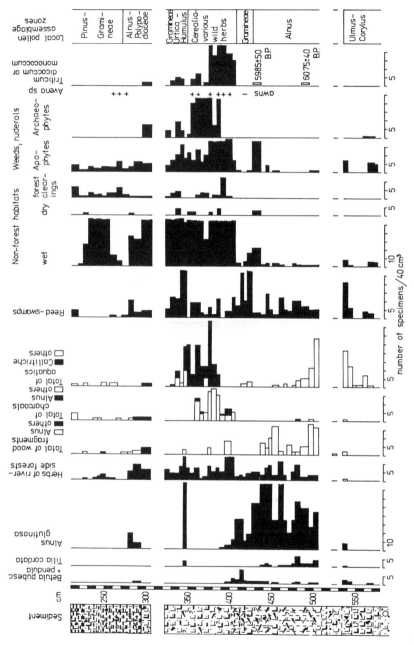

FIGURE 27.3. Macrofossil diagram from the Neolithic site Pleszów in Cracow, Poland. Bars for selected species and ecological groups illustrate vegetational succession in a river valley from natural alder stands to wet meadow communities

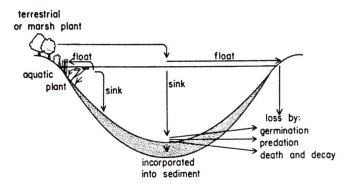

FIGURE 27.4. Schematic presentation of the deposition of macroscopic plant remains in a lake (from Birks, 1980). *Reproduced by permission of E. Schweizerbart'sche Verlagsbuchhandlung*

Some useful information to palaeoecologists may be found in publications on the seed bank in arable, meadow and forest soils (for reference see Harper, 1977). Old records of weed content in cereal grain stored for sowing may be useful for reconstructing weed flora based on the samples of prehistoric grain. Unfortunately, studies of surface samples undertaken with the aim of aiding the interpretation of fossil assemblages, which are the most important source of data, are not common.

Investigations of the transportation and accumulation of diaspores (and vegetative parts) on snow surfaces and along a stream in alpine zones were carried on in Norway (Ryvarden, 1971, 1975) and in Alaska (Glaser, 1981). They have shown that those plants most common in the area, and which produce the largest number of particles capable of being transported, were most abundant in surface samples. The effect of long distance transport, marked by the presence of species represented by a small number of specimens, was of minor importance in samples collected during summer and was better manifested in winter samples (transport up to 5 km and up and down the mountain slope). Surface samples taken along the Dunajec River in the Pieniny Mountains, Poland, also contained mostly diaspores of plants growing near the river channel. Plants of gravel banks, marshy areas, riverside forests and waters constituted over 50% of all species, and many of them were very abundant (Pelc, 1983).

Wider studies carried out by Birks (1973) in Minnesota lakes have shown that diaspores which fall on to the surface of water and sink do not spread uniformly over a lake bottom. They concentrate in a belt along the shore and come mostly from plants which grow on the site (Figure 27.5).

All these observations, fragmentary as they are, indicate that the numerical relations between species in a fossil assemblage may not properly reflect their quantitative occurrence in past communities. The actual number of specimens

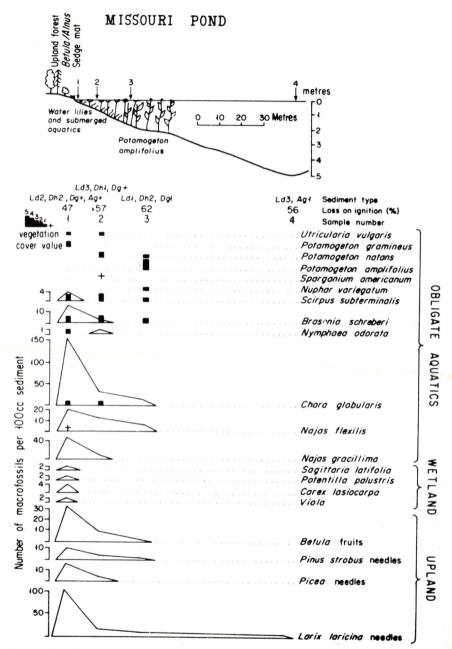

FIGURE 27.5. Missouri Pond, Minnesota. Present-day vegetation belts compared with the distribution of plant macrofossils in surface samples. Bars denote the cover value of the plant in the vegetation plots, silhouettes denote the number of macrofossils in 100 cm³ of the surface sediment (from Birks, 1973). *Reproduced by permission of Blackwell Scientific Publications Ltd*

should never be taken literally. In general, the abundant occurrence of a species, particularly repeated in several samples, is a fairly good indication of its derivation from local population of plants living locally. In some cases, however, the occurrence of single diaspores is important, for instance with species which reproduce vegetatively for long periods of time and are under-represented in seed spectra. It seems that the interpretation of fossil spectra should be based on the observation of both presence and abundance of a taxon corrected on the basis of its known ecology.

In spite of our incomplete knowledge of the relationship between living communities and macrofossil assemblages, the analysis of macroscopic plant remains is a very important source of data in palaeoecological studies. Its great advantage lies in the possibility of determining the species with frequently greater accuracy than in pollen analysis. This, in turn, enables one to draw more precise ecological conclusions. In order to minimize errors resulting from unpredictable changes in the quantitative appearance of macrofossils, investigators should attempt to base palaeoecological reconstructions on the behaviour — in stratigraphic sequence — of the entire groups of those taxa having similar ecological requirements (Table 27.1, Figure 27.3).

Another specific feature of macrofossils is their local character. Most seeds are dispersed close to the parent plant, and macrofossil assemblages are composed mostly of plants growing near the sampling site. These are, first of all, aquatic and bog herbs. Trees and shrubs from nearby places may also be well-represented, while plants of upland habitats may be brought to the sediment only occasionally by wind or water (rain or streams). Upland plants have a better chance to move into the sediment of lakes having poorly developed marginal zones of swamp or bog vegetation (Watts, 1978). The identification of a taxon among macrofossils is usually better evidence for its presence in the study area in the past than that provided by pollen analysis. For this reason, the occurrence of macrofossils is of special value in periods when new species immigrate to an area (e.g. the immigration of trees in late-glacial times).

Local phenomena, if they occur in similar sequences in several localities, may have the same regional cause. It has been shown by Watts (1967) that there are several macrofossil assemblages of regional importance, characteristic of particular pollen assemblage zones in late- and post-glacial times in Minnesota (Figure 27.6).

To conclude, it should be emphasized that fruits and seeds are only parts of plants. Other remnants of these same plants — pollen, wood, leaves etc. — are scrutinized by other specialists. These artificial divisions, induced by the nature of data, should not obscure the fact that the common aim of study, namely the reconstruction of past vegetation, can be achieved only by a combination of various methods.

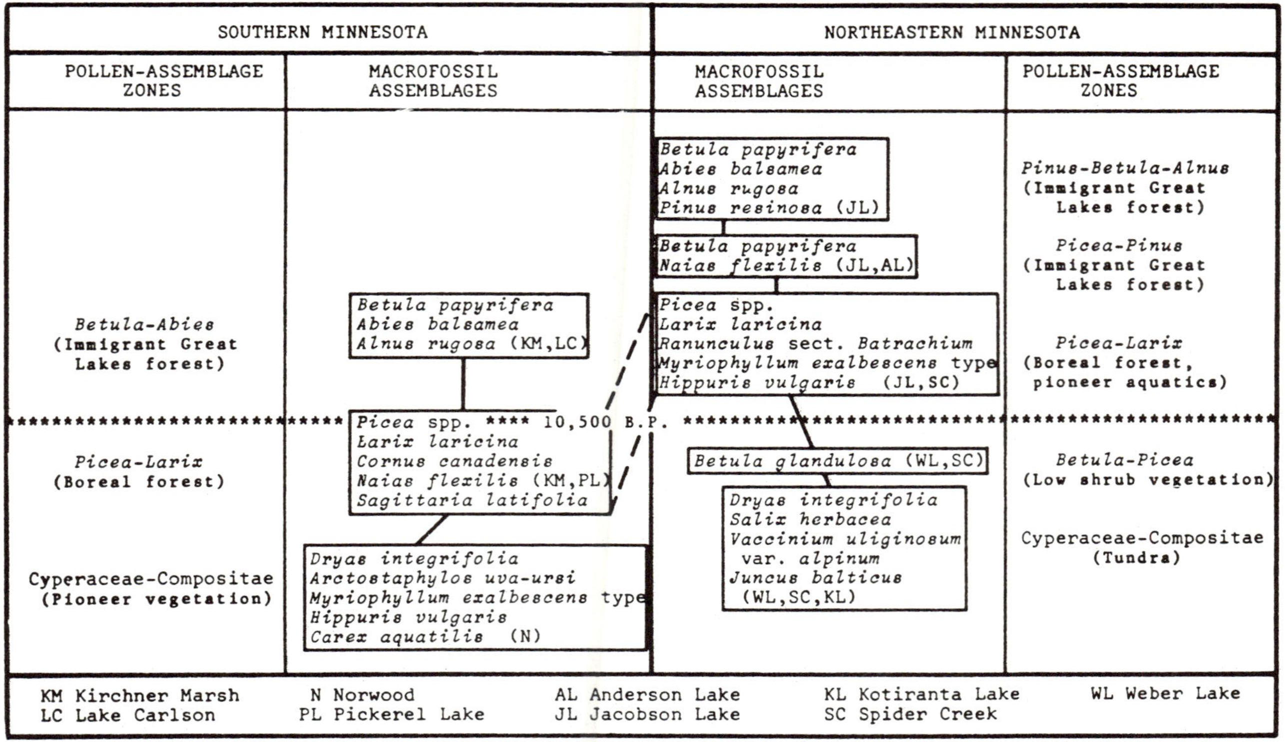

FIGURE 27.6. Comparison of pollen and macrofossil assemblages from the late-glacial and early postglacial sediments in Minnesota (from Watts, 1967). *Reproduced by permission of Yale University Press*

REFERENCES

Barghoorn, E. S. (1948). Sodium chlorite as an aid in palaeobotanical and anatomical study of plant tissue. *Science*, **107**, 480–481.

Behre, K.-E. (1976). Die Pflanzenreste aus der frühgeschichtlichen Wurt Elisenhof. In: *Studien zur Küstenarchäologie Schleswig-Holsteins*, vol. 2, Herbert Lang, Bern.

Beijerinck, W. (1947). *Zadenatlas der Nederlandsche Flora*, H. Veenman & Zonen, Wageningen.

Berggren, G. (1969). *Atlas of Seeds. 2: Cyperaceae*, Swedish Natural Science Research Council, Stockholm.

Berggren, G. (1981). *Atlas of seeds. 3: Salicaceae to Cruciferae*, Swedish Natural Science Research Council, Stockholm.

Bertsch, K. (1941). *Früchte und Samen: Handbücher der Praktischen Vorgeschichtsforschung*, vol. 1, Ferdinand Enke, Stuttgart.

Białobrzeska, M. (1964). Wpływ różnych czynników na wielkość i kształt kopalnych owoców graba. *Acta Palaeobot.*, **5**, 3–21.

Birks, H. H. (1973). Modern macrofossil assemblages in lake sediments in Minnesota. In: *Quaternary Plant Ecology* (Eds. H. J. B. Birks and R. G. West), Blackwell Scientific Publications, Oxford, pp. 173–189.

Birks, H. H. (1980). Plant macrofossils in Quaternary lake sediments. *Arch. Hydrobiol. Beih., Ergebn. Limnol.*, **15**, 1–60.

Brouwer, W. and Stählin, A., (1955). *Handbuch der Samenkunde*, DLG-Verlags-GmbH, Frankfurt.

Delcourt, P. A., Davis, O. K., and Bright, R. C. (1979). *Bibliography of Taxonomic Literature for the Identification of Fruits, Seeds, and Vegetative Plant Fragments*, Environmental Science Division, Publ. 1328, Oak Ridge National Laboratory.

Dickson, C. A. (1970). The study of plant macrofossils in British Quaternary deposits. In: *Studies in the Vegetational History of the British Isles* (Eds. D. Walker and R. G. West), Cambridge University Press, pp. 233–254.

Glaser, P. H. (1981). Transport and deposition of leaves and seeds on tundra: a late-glacial analog. *Arctic and Alpine Research*, **13**, 173–182.

Godwin, H. (1975). *The History of the British Flora*, Cambridge University Press.

Grosse-Brauckmann, G. (1972). Über pflanzliche Makrofossilien mitteleuropäischer Torfe. I: Gewebereste krautiger Pflanzen und ihre Merkmale. *Telma*, **2**, 19–55.

Grosse-Brauckmann, G. (1974). Über pflanzliche Makrofossilien mitteleuropäischer Torfe. II: Weitere Reste (Früchte und Samen, Moose u.a.) und ihre Bestimmungsmöglichkeiten. *Telma*, **4**, 51–117.

Grosse-Brauckmann, G. (1979). Sukzessionen bei einigen torfbildenden Pflanzengesellschaften nach Ergebnissen von Grossrest-Untersuchungen an Torfen. In: *Int. Symposien Int. Verein. Vegetationskunde, 1967: Gesellschaftsentwicklung Syndynamik* (Eds. R. Tüxen and W.-H. Sommer), J. Cramer, Vaduz, pp. 393–412.

Harper, J. L. (1977). *Population Biology of Plants*, Academic Press, London.

Kats, V. Ya., Kats, S. V., and Kipiani, M. G. (1965). *Atlas i opredelitel piodov i semyan vstrechayushchikhsya v chetvertichnych otlozheniyakh SSSR*, Nauka, Moskva.

Körber-Grohne, U. (1964). Bestimmungsschlüssel für subfossile Juncus-Samen und Gramineen-Früchte. In: *Probleme der Küstenforschung im Südlichen Nordseegebiet*, vol. 7, August Lax Verlagsbuchhandlung, Hildesheim.

Lange, E. (1968). Zum Vorkommen von Xanthium strumarium L. in Mitteleuropa. *Feddes Repertorium*, **77**, 57–60.

Leeuwaarden, W., van (1982). *Palynological and Macropalaeobotanical Studies in the Development of the Vegetation Mosaic in Eastern Noord-Brabant, the Netherlands, during Late-Glacial and Early Holocene Times*, Jumbo-offset B. V.'s Gravenpolder.

Nikitin, V. P. (1969). *Paleokarpologicheski Metod*, Izd. Tomskovo Univer., Tomsk.

Pelc, S. (1983). Owoce i nasiona we współczesnych osadach Dunajca w rejonie Pienin i przełomu beskidzkiego. *Prace Monogr. Wyższej Szkoły Pedagogicznnej w Krakowie*, **59**, Kraków.

Rybníček, K. (1973). A comparison of the present and past mire communities of Central Europe. In: *Quaternary Plant Ecology* (Eds. H. J. B. Birks and R. G. West), Blackwell Scientific Publications, Oxford, pp. 237–261.

Rybníček, K., and Rybníčkova, E. (1974). The origin and development of waterlogged meadows in the central part of the Šumava Foothills. *Folia Geobot. Phytotax.*, **9**, 45–70.

Ryvarden, L. (1971). Studies in seed dispersal. I: Trapping of diaspores in the alpine zone at Finse, Norway. *Norw. J. Bot.*, **18**, 215–226.

Ryvarden, L. (1975). Studies in seed dispersal. II: Winter-dispersed species at Finse, Norway. *Norw. J. Bot.*, **22**, 21–24.

Schwarzenholz, W. (1961). Isolierung von Azolla-Sporen aus erdigen und sandigen Sedimenten. *Geologie*, **10**, Beiheft 32, 5–9.

Wasylikowa, K. (1973). Badanie kopalnych szczątków roślin wyższych. In: *Metodyka Badań Osadów Czwartorzędowych* (Ed. E. Rühle), Wyd. Geolog., Warszawa, pp. 161–210.

Wasylikowa, K. (1978). Plant remains from Early and Late Medieval time found on the Wawel Hill in Cracow. *Acta Palaeobot.*, **19**, 115–198.

Wasylikowa, K. (1982). Pollen diagram from the vicinity of the linear pottery culture site in Cracow. In: *Siedlungen der Kultur mit Linearkeramik in Europa* (Ed. J. Pavuk), Archäolog. Inst. Slovak. Akad. Wissenschaften, Nitra, pp. 285–290.

Watts, W. A. (1959). Interglacial deposits at Kilbeg and Newton Co. Waterford. *Proceedings of the Royal Irish Academy*, **B 60**, 79–134.

Watts, W. A. (1967). Late-glacial plant macrofossils from Minnesota. In: *Quaternary Palaeoecology* (Eds. E. J. Cushing and H. E. Wright), Yale University Press, New Haven, pp. 89–97.

Watts, W. A. (1978). Plant macrofossils and Quaternary palaeoecology. In: *Biology and Quaternary Environments* (Eds. D. Walker and J. C. Guppy), Australian Acad. of Science, pp. 53–67.

Watts, W. A., and Bright, R. C. (1968). Pollen, seed and mollusk analysis of a sediment core from Pickerel Lake, northeastern North Dakota. *Geol. Soc. Am. Bull.*, **79**, 855–876.

Watts, W. A., and Winter, T. C. (1966). Plant macrofossils from Kirchner Marsh, Minnesota. *Geol. Soc. Am. Bull.*, **7**, 1339–1359.

West, R. G. (1969). *Pleistocene Geology and Biology*, Longman, London.

Willerding, U. (1983). Paläo-Ethnobotanik und Ökologie. *Verhandlungen Gesellsch. Ökologie (Festschrift Ellenberg)*, **11**, 489–503.

Wilson, G. (1984). The carbonisation of weed seeds and their representation in macrofossil assemblages. In: *Plants and Ancient Man* (Eds. W. van Zeist and W. A. Casparie), A. A. Balkema, Rotterdam, pp. 201–206.

Zeist, W. van (1974). Palaeobotanical studies of settlement sites in coastal area of the Netherlands. *Palaeohistoria*, **16**, 223–371.

28

Analysis of vegetative plant macrofossils

GISBERT GROSSE-BRAUCKMANN

Botanical Institute, Technical University, Darmstadt, F.R.G.

INTRODUCTION

The interpretation of macrofossil records differs greatly from that of results of pollen-analytical investigations. The pollen contents of peats and sediments primarily give information on the vegetational character of the surroundings of the place of deposition; the macrofossil content of the peat profiles, on the other hand, as a rule only gives information about the plant cover of the successional stages of the mires themselves.

The term 'macrofossil' covers everything that cannot be considered as microfossil. It means all plant remains which can be recognized by the naked eye or at least with the aid of a strong lens, though an exact identification is sometimes possible only by microscopy using a magnification of × 100–200 (rarely × 500).

Fruits and seeds, wood remains and bryophytes, which also belong to macrofossils by this definition, are dealt with in Chapters 27, 29 and 30. Here, the subject is the remains of the vegetative organs of the kormophytes, in particular their rhizomes and stems, their leaves (mainly rhizome leaves or other lower leaves as well as leaf sheaths), and their roots. In the sections on preparation of the peats and on presentation and evaluation of macrofossil records, the other kinds of macrofossils, however, are of course not excluded.

When peat research in Quaternary botany began, macrofossils gave rise to the first attempts at interpretation with respect to vegetation history of the peat sequences. In view of this the classical publications of Nathorst (e.g. 1872, 1892), Schröter (1882), Weber (1897, 1902, 1905, 1930), Andersson (1898), Holmboe (1903), Früh and Schröter (1904), Lewis (1905–7) and some others may be mentioned. Furthermore, Nathorst and, above all,

Andersson were the first to make use of the washing technique, which has turned out to be essential for peat research. The extensive and detailed description of the methods of macrofossil investigation given by Andersson (1892a, b, 1883) has remained valid in many respects.

Many years ago pollen-analytical investigation techniques developed rapidly, and consideration of macrofossil records became less popular. Only in the sixties and seventies has more frequent use of them been made. This was done with regard to investigations of general vegetation history, in order to take into consideration as many single results as possible and so to enable a maximum amount of data for interpretation.

Beyond its importance in Quaternary botany, knowledge of plant macrofossils is a prerequisite for mire mappings or surveys (see later). It should be borne in mind that the classical work on Swedish mire inventory by von Post and Granlund (1926) included some very clear statements on macroscopic characteristics of important plant remains in peat.

DEPOSITION AND PRESERVATION OF MACROFOSSILS IN PEAT

The main cause of the different frequencies in peat of identifiable plant remains lies in the original conditions of peat formation (Grosse-Brauckmann, 1964b, 1980). High pH values, greater base content and, above all, temporarily low water levels which enable aeration of the upper layers of the mire, result in the formation of strongly humified peats containing a few preserved macrofossils. The opposite is true in conditions of high and scarcely varying water levels, especially if nutrient supply and pH are low — in which case peats of slight humification are formed, the plant material of which looks nearly unaltered compared with its appearance when alive.

It must be borne in mind, however, that also in slightly humified peats a large proportion of the original organic matter production of the mire vegetation has disappeared by decomposition, as shown by Holling and Overbeck (1960), Grosse-Brauckmann and Puffe (1964), Overbeck (1975) and Grosse-Brauck-mann (1980). The main reason is that above-ground parts of the mire plants (with the exception of fruits and seeds, and mosses for which the older and therefore deeper parts are 'continuously' dying) are preserved only in special cases when, by chance, after dying, they fall into more or less anaerobic environmental conditions (e.g. into water-filled hollows or luxuriantly growing moss carpets). Accordingly it is, with the exception of mosses, fruits and seeds, in general, only the subterranean parts of the mire plants which form the peat macrofossils. But not all of them do so, as is well-known. There are two individual characteristics which determine which subterranean organs are transformed to (identifiable) macrofossils in peat. The first one is their 'specific decomposition resistance', as determined by certain histological and chemical properties (e.g. pine wood and *Eriophorum vaginatum* sheaths disintegrate only slowly); and the second one is the depth of their penetration into the ground, where a longer phase of aerobic decomposition cannot take place (extreme examples are rhizomes of *Phragmites* or *Equisetum fluviatile*, and roots of species of *Eriophorum*).

Thus normally subterraneous organs of mire plants are growing within a peat matrix that is as a rule somewhat older than themselves, as shown by Figure 28.1. If these peat layers have an origin very different from the peat-forming canopy of that time, a mixed macrofossil composition must be the consequence. Such peats have been called by Weber (1930) 'displacement' peats ('Verdrängungstorfe').

The vegetational changes which are the preconditions for such effects can be caused, for example, by changes in hydrology, as shown in detail in Figure 28.2. Displacement peats with a very heterogeneous macrofossil mixture, however, are not found very often. Evidently sharp vegetational changes, caused by sudden successional transitions, were rather infrequent in the vegetational history of the mires. Exceptions to this general rule are possible only in regions where hydrological changes were relatively major and/or more frequent, as was the situation during the sea-level changes in the past.

FIGURE 28.1.   Diagrammatic representation of the growing mechanism of *Eriophorum vaginatum* (roots and the upper parts of living leaves omitted). The shoot bases are always several years younger than the *Sphagnum* material surrounding them (from Grosse-Brauckmann, 1964b). *Reproduced by permission of Deutsche Botanische Gesellschaft*

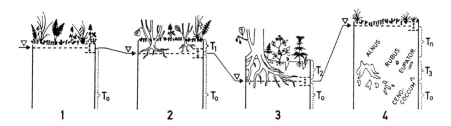

FIGURE 28.2.   Diagrammatic illustration of the formation of a 'displacement peat'
rich in wood after a temporary lowering of the water table. The broken lines symbolize
the groundwater tables, the double arrows the range of their variation. The four stages
are as follows:

(1) In an actively growing mire a slightly humified peat T_0 has been deposited below a
     peat-forming plant cover of herbs and mosses.
(2) By a slight lowering of the water table woody species can invade; however, the
     shrubs and trees are, throughout, dying very early because of the rather high water
     table. The peat T_1 has been originated partly by a further additional humification
     of the uppermost layers of the initial peat T_0, partly by deposition of remains of
     the scrub-like plant cover, which still possesses a low capacity of peat formation.
(3) A deeper water lowering has enabled the trees to grow up into a closed forest; the
     peat T_1 has more or less disappeared by mineralization. The remaining peat layer
     above the water table (originally T_0) has been transformed by total humification
     into the peat T_2; here the trees are rooting and some durable seeds and fruits of
     the forest vegetation may be preserved.
(4) The water table has risen again in the course of time, the forest has been replaced
     by herbaceous vegetation; instead of peat mineralization, a new peat formation
     has begun again (peat T_n). The 'displacement' peat T_3 now contains within a
     completely humified peat matrix the root wood and some other durable remains of
     the (not peat-forming) plant cover of stage 3 (from Grosse-Brauckmann, 1979d).

Reproduced by permission of J. Cramer, Braunschweig.

Apart from these cases, the peat-forming vegetation cover has persisted
unchanged for rather long periods of time, often for centuries. So the
deep-growing plant organs as a rule enter a peat matrix of the same
composition as the canopy to which they themselves belong, and no
heterogeneous mixtures of macrofossils are established.

The statement made before regarding the relationship between the degree of
decomposition of a peat and its number of plant species (or, generally
speaking, plant taxa) as represented by macrofossils, applies only for peats
derived from similar vegetation types. So the number of taxa in ombrogenous
peats of slight decomposition is sometimes hardly greater than that of, for
example, strongly decomposed *Phragmites* peats. The greatest number of
species is found, generally speaking, in fairly acid, fairly nutrient-rich
'transitional' peats which are not strongly decomposed. This can be seen also in
Table 28.1, even though it only covers vegetative plant remains.

TABLE 28.1. Identification in the field of the most important vegetative plant remains (with a few mosses included) in central European peats. The entries are generalized and thus necessarily simplified. So the general occurrence of some macrofossils or their greater abundance is restricted only to parts of Europe (e.g. *Myrica*, *Pinus mugo*, presumably also *Vaccinium uliginosum*), or to certain periods of the past (*Cladium*). The occurrence in certain types of mires, too, can be restricted to certain regions (e.g. *Eriophorum angustifolium* in ombrotrophic mires). The differences in macrofossil number between the single mire types are clearly seen; the oligo- and mesotrophic peats would have proved, when considering the whole of the mosses, to be much richer in species

General abundance in central Europe	Identification in the field	Macrofossils	Trophic character of mires in which peats containing remains are found			
			Ombrotrophic	Minerotrophic		
				Oligotrophic	Mesotrophic	Eutrophic
		HERBACEOUS PLANTS				
A	ea	*Eriophorum vaginatum*	V	V		
r	po	*Rhynchospora alba*	V	V		
r	ea	*Trichophorum cespitosum*	V	V		
A	ea	*Scheuchzeria palustris*	V	V	(v)	
r	ea	*Eriophorum angustifolium*	V	V	(v)	
a	ea	*Carex limosa*		V	V	
r	po	*Molinia caerulea*		V	V	
A	po	*Menyanthes trifoliata*		(v)	V	(v)
A	ea	*Equisetum fluviatile*			V	V
a	ea	*Cladium mariscus*			(v)	V
A	ea	*Phragmites australis*			(v)	V
r	po	*Thelypteris palustris*			(v)	V
		DWARF SHRUBS				
a	po	*Andromeda polifolia*	V	V		
a	ea	*Erica tetralix*	V	V		
a	ea	*Oxycoccus palustris*	V	V	(v)	
A	ea	*Calluna vulgaris*	V	V	(v)	
r	po	*Vaccinium uliginosum*		V	V	
		TREES AND SHRUBS				
A	po	*Pinus mugo/sylvestris*	(v)	V	(v)	
a	ea	*Myrica gale*		V	V	
A	ea	*Betula (pubescens)*		V	V	
A	po	*Salix (aurita/cinerea)*		(v)	V	V
A	po	*Alnus (glutinosa)*			(v)	V
		SOME MOSSES				
A	·	ombrotrophic *Sphagna*	V			
a	ea	*Polytrichum strictum*	V	V		
a	po	*Aulacomnium palustre*	V	V	V	
a	·	*Sphagnum palustre*		V	V	
r	·	*Sphagnum teres*			V	(v)

General abundance: r = rare, a = rather abundant and frequent, A = very abundant and frequent.
Changes of identification: ea = easy and with good certainty, po = in many cases possible but not always with the highest degree of certainty.
Comment on the occurrence: V = typical, widespread, more-or-less abundant, (v) = rarer or more local.

LABORATORY TECHNIQUES

For macrofossil analyses, samples of a volume of 50 ml have proved adequate. Greater volumes can sometimes be useful in peats which are very strongly decomposed or in sediments. In slightly humified peats, however, greater quantities should be avoided since in such peats it is already very difficult to determine the whole content of vegetative macrofossils even where a 50 ml sample is used. According to the present author's experience it seems, moreover, that doubling the size of the samples from 50 to 100 ml yields on average an increase in taxa by no more than one-fifth or one-quarter. For the sake of comparability, in each investigation only samples of the same size should be used, as stressed already by Andersson (1893), who proposed, however, much bigger samples.

The following procedure of peat preparation (see also the flow chart in Figure 28.3) has proved effective in the work of the author and his students in the Botanical Institute in Darmstadt. It does not differ very much from the Polish example described in the preceding chapter, and it naturally enables us to discover more or less quantitatively not only the vegetative plant remains but also the mosses as well as the fruits and seeds.

From the original sample, as collected in the field, a piece of the layer concerned, which amounts to somewhat more than 50 ml, is horizontally cut out using a very sharp knife. The layer taken should be about 1 cm thick in the case of a monolith sample, but may be up to 5 cm in the case of a borer sample. Its marginal parts are cut off and removed because of possible contamination by material of other layers. From the remaining peat the sample needed for pollen analysis is taken and, additionally, a small sample is kept without any preparation for the purpose of a later investigation of general peat characteristics (in particular, quantity and kind of mineral components) and the presence of conspicuous microfossils (e.g. diatoms, sponge needles, phytoliths). The remaining peat should now have a volume of 50 ml. If the peat sticks together sufficiently, the volume is given easily by a shape of appropriate dimensions. If it crumbles to pieces, one may measure the volume required by pressing it by hand into a porcelain dish of the right size.

The dissolution of humic substances is brought about by KOH, of which a concentration of 5% has proved to be sufficient. Before adding the KOH solution the sample is broken up by hand into small pieces. In doing so all macroscopic plant remains which might be destroyed by the KOH treatment or brought into conditions unfavourable for microscopical determination are collected for later separate investigation. Presumably this concerns all wood pieces, leaves, greater stem bases (as those of *Cladium* or *Eriophorum angustifolium*) and rhizomes, especially those with delicate leaves. If the peat shows marked layers, the breaking up is done according to its natural splitting, since certain plant remains are mainly found lying on the split surfaces and can

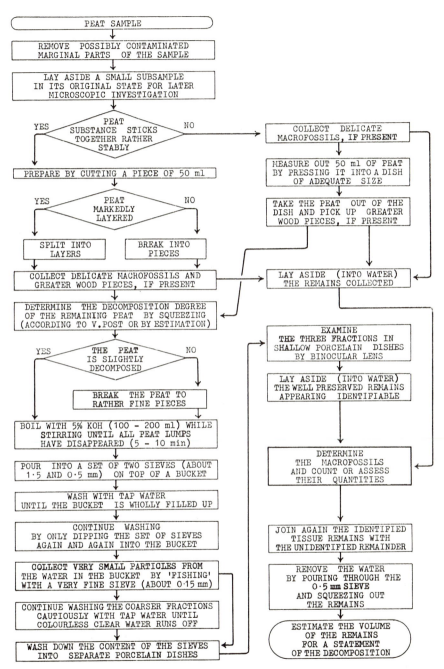

FIGURE 28.3. Flowchart showing the laboratory technique for macrofossil preparation, as applied in Darmstadt

be collected without difficulty in this way. This is true of leaves of, for example, *Betula, Salix* or *Myrica*, the typical margin and venation of which can easily be recognized on split surfaces; these characteristics might be destroyed, however, by the KOH treatment, because the mesophyll is usually completely macerated by the humification and only the conducting bundles and the epidermis are left. Sometimes larger parts of plants are found on such split surfaces (e.g. shoot systems of *Andromeda* or *Erica* as a whole, or fruiting stems of *Scheuchzeria*).

After the sample is broken and placed in a porcelain dish of sufficient size, a good quantity of KOH solution (100–200 ml) is added and then the sample is cautiously boiled on an asbestos mat for about five minutes, occasionally stirring it and, if necessary, crushing remaining lumps of peat.

After that the sample is poured into two sieves of differing mesh size (1.2–1.8 mm and about 0.5 mm) lying one on top of the other on a bucket (common kitchen sieves made of metal and not plastic are very suitable). It is then washed by a slight jet of tap water from a hose until the bucket is filled up. The washing is henceforth continued only by repeated dipping of the set of two sieves into the water.

The fluid washed down through the finer of the sieves, lying below, still contains some small plant remains (e.g. small moss leaflets, seeds of *Juncus* and some other species). To collect these, as well, a very fine sieve (0.15–0.18 mm mesh size) is used. It would, however, be impossible to pass the whole volume of fluid through the sieve as its mesh would soon be blocked totally by the many fine tissue fragments present, and furthermore an exact and thorough examination of all the fine detritus would hardly be possible. So it is advisable to take, and thereafter examine, only a small proportion of the fine tissue fraction. This can be done by sieving off a small quantity of the water in the bucket (after stirring well) or by 'fishing' with the small, fine sieve inside the bucket. In the bucket now the amount of sand and silt admixed to the peat can easily be recognized and assessed after sedimentation. Finally, in order to remove the remaining alkali and humic substances, the coarser fractions lying in the wide-meshed sieves continue to be washed by more tap water. (For a simple procedure, also included in the flow chart, which gives a rough idea on the degree of decomposition, see the next section.)

The three fractions are examined in small quantities by a binocular lens, lying in sufficient water in white shallow porcelain dishes. In many cases certain histological identification is only possible by microscopic investigation. For remains of a very dark coloration, bleaching can sometimes be useful in order to make the recognition of characteristics of high diagnostic importance easier. This can be done very quickly and easily by concentrated oxalic acid. In doing so it is important that the water the remains are lying in be absolutely free of Ca, as otherwise oxalate precipitations would result, so impeding the investigation. For an even more effective bleaching the use of $KMnO_4$

solution, preceding the application of the oxalic acid, has been recommended (Lagerheim, 1902).

For the sake of completeness it should be added that in the past entirely different methods of preparation have been employed as well. Thus the use of nitric acid should be mentioned, as described by Andersson (1892a, b, 1893). The reagent used was crude HNO_3, diluted by twice its own volume of water. (Later on von Bülow (1929) recommended concentrations of only 10–20%.) The treatment, carried out in the cold, takes 1–2 days. The effect is a gentle loosening of the peat material caused by a weak gas formation (nitric oxides) and, additionally, a bleaching of dark tissues. Compared with the KOH method, the advantages of this method are better preservation of certain more-or-less delicate tissues, and somewhat easier collection of certain fruits etc. By gas formation in their interior these plant remains accumulate swimming on the surface of the fluid, which allows them to be decanted easily. A disadvantage to be considered in practice, however, is the aggressiveness of the nitric acid and the formation of dangerous gas.

To avoid these unpleasant effects, Lagerheim (1902) proposed that the nitric acid be replaced by a solution of oxalic acid (3%). This forms gas in contact with the peat as well; however, the influence of light (preferably sunlight) is absolutely necessary for the process. In this case the gas is CO_2, originating from oxidation of the oxalic acid by the peat. According to Lagerheim, the treatment takes only a few hours.

Both methods have certain disadvantages in common. By the treatment the humified material is not dissolved but only disintegrated into crumbs. Very small remains (e.g. seeds or moss leaflets), if hidden in larger crumbs, can therefore easily be overlooked. Furthermore, the lengthy pretreatment of the single samples must be taken into account, being inconvenient at least for investigations of more extensive sets of peat material. Therefore, these methods should be considered only when the main concern is the preservation of the most delicate tissues, and the loss of some small seeds etc. can be neglected.

Under certain circumstances it may even be favourable to work without any chemicals at all. Klinger (1968, 1984) showed that the remains of hepatics are completely destroyed by any kind of chemical treatment; whereas without this treatment he found several species, some in great quantities, in certain ombro- to mesotrophic peats.

QUANTITATIVE ASSESSMENTS IN MACROFOSSIL INVESTIGATIONS

The identifiable tissue remains of the various species, including mosses, are best estimated as a percentage of the whole washing residue (excepting the finest fraction, 0.5 mm), even if this is only roughly possible. Usually there is a remainder of plant material which cannot be identified in detail. This may be further classified as rhizomes, other shoot organs, and roots (sometimes wood and bark too, as well as mosses) which can also be estimated as a percentage. In this way an idea may be established as to the type of the peat as a whole. In this respect an additional statement concerning the decomposition of the peat sample is essential. If no other, more precise, determinations of the degree of decomposition or humification are made, at least the H-values (according to von Post, see Figure 28.4, cf. also Chapter 12) should be given. Another possibility is to give a statement on the decomposition, as follows: Having investigated the sample with respect to macrofossils, and estimated the quantities of its components, the volume of all the remains together is estimated after the water is sieved off and squeezed out. The relation of this estimated volume to the initial peat volume is a rough measure of the degree of decomposition with respect to its macrofossil content.

GENERAL REMARKS ON IDENTIFICATION OF PLANT REMAINS

In general terms the identification of plants or parts of them is more reliable the greater the number of basic characteristics. This is also true for the identification of plant remains in peat. Good identification is achieved where not only characteristics in anatomy and histology but also in macromorphology can be taken as its basis. Well-known examples in this respect are the rhizomes of *Phragmites*, *Equisetum fluviatile*, *Scheuchzeria* and *Carex limosa*, the subterranean axes of *Cladium* and *Eriophorum angustifolium*, the stem bases with basal leaves as in *Trichophorum cespitosum* or *Rhynchospora alba*, or the naked ones as those of *Molinia*, the leaves of trees and shrubs (e.g. *Betula*, *Salix*, *Myrica*) as well as leaves or axes of dwarf shrubs (*Calluna*, *Erica*, *Andromeda*, *Oxycoccus* and *Vaccinium uliginosum*).

Sometimes characteristics of colouring of the remains can give useful additional indication, such as in the case of *Equisetum fluviatile*, the roots and rootlets of which can easily be distinguished from the roots of all flowering plants (only the roots of *Dryopteris* s.l. are also black). Some mosses, too, are marked by garish colourings (cf. the blackish-brown leaflets of *Polytrichum strictum*). Occasionally colourings in peat in its fresh, unoxidized state can be of particular interest. The most striking example in this respect is the bright brownish-red colouring of the inner parts of the subterranean axes of *Cladium*. In *Sphagnum* peats the axes of *Eriophorum angustifolium* are very conspicuous by their dark reddish-brown colour contrasting well against lighter *Sphagnum* material. The colourings of fresh wood remains should be mentioned in this connection. *Betula* and *Salix* sometimes exhibit a striking pinkish colour whilst *Alnus*, *Myrica* and *Pinus* never do.

On the whole the occurrence of a species has proved to be more reliable where different remains of it have been found (e.g. vegetative remains and seeds of *Menyanthes*, rhizomes and roots of *Carex limosa*, epidermis, fibres

"H"-VALUE ACCORDING TO v.POST	CHARACTERISTICS OF MOIST, FRESH PEATS			CHARACTERISTICS OF DRY OR BY AERATION DISINTEGRATED CRUMBLY PEATS	
	THE PLANT TISSUE STRUCTURES IN THE PEAT ARE	SQUEEZING BY HAND MAKES TO PASS BETWEEN THE FINGERS	APPEARANCE OF THE REMAINDER AFTER SQUEEZING	PRIMARY PLANT STRUCTURES IN THE PEAT	COLOUR OF THE PEAT
1	DISTINCT	COLOURLESS, CLEAR	NOT PULPY	THE ONLY VISIBLE PEAT COMPONENT	WHITISH TO YELLOW
2	DISTINCT	WEAKLY YELLOWISH-BROWN, NEARLY CLEAR (WATER)	NOT PULPY	THE ONLY VISIBLE PEAT COMPONENT	RATHER LIGHT BROWN
3	DISTINCT	BROWN, EVIDENTLY SOMEWHAT TURBID (WATER)	NOT PULPY	THE ONLY VISIBLE PEAT COMPONENT	DARKER BROWN
4	DISTINCT	BROWN, VERY TURBID (WATER)	NOT PULPY	THE ONLY VISIBLE PEAT COMPONENT	
5		VERY TURBID, BESIDES THAT A LITTLE (WATER)	A LITTLE PULPY	NEARLY THE ONLY PEAT COMPONENT	
6	SOMEWHAT INDISTINCT	UP TO 1/3 OF THE	VERY PULPY	MORE THAN 2/3	
7	RECOGNIZABLE TO A CERTAIN DEGREE	ABOUT 1/2 OF THE (PEAT SUBSTANCE, MORE OR LESS WET AND PULPY)	STRUCTURES OF PLANT TISSUES MORE DISTINCT THAN BEFORE SQUEEZING	ABOUT 1/2 OF THE WHOLE PEAT MATERIAL	RATHER DARK TO BLACK
8	VERY INDISTINCT	ABOUT 2/3 OF THE	IN PARTICULAR CONSISTING OF MORE RESISTENT COMPONENTS (FIBERS, WOOD)	ABOUT 1/3	
9	HARDLY DISCERNIBLE	NEARLY THE WHOLE		ONLY A VERY LITTLE PART	
10	NO LONGER DISCERNIBLE	THE WHOLE	NO REMAINDER	WITHOUT ANY PLANT STRUCTURES	

FIGURE 28.4. The ten-grade scale by von Post of the decomposition of peats: tabular representation of the definitions, complemented in respect to dry and drained peats, respectively (adapted from Grosse-Brauckmann *et al.*, 1977)

and sclerenchyma-spindles of *Eriophorum vaginatum*). Concerning the microscopic characteristics of the shoot remains, which are mainly histological features of epidermal (possibly also subepidermal) layers, a few remarks should be made. It must not be presupposed that tissues, particularly epidermises, of a single species or even of the same plant specimen, are totally uniform. For there are often clear differences between the internodes of rhizomes and of bases of aerial shoots with respect to the size and proportions of the cells, and even in a single internode the cells in the middle of the internode differ from those near the nodes. Still greater differences are known between the epidermises of rhizomes and leaves, and in the leaves the characteristics of rhizome leaves are again different from those of the sheaths of basal foliage leaves. Furthermore there are often strong differences between inner and outer faces of the leaves, between leaf margin and centre, and even between the cells on top of the subepidermal sclerenchyme skeins and those in the middle between two neighbouring skeins (in monocotyledons). These differences have not yet been studied to the last detail, and so attempts to determine a species only on the basis of a few tissue fragments alone and without the combined consideration of further macroscopic or microscopic characteristics will probably be very problematic.

Similar statements are true with respect to the roots. The characteristics of the rhizodermis of one and the same species are possibly very different between the larger roots and the rootlets of last rank (radicells). On the whole the roots have, compared with the shoots, far fewer histological characteristics which might be used for an identification of the species or genus. So it has hardly ever been possible, up to now, to identify fossil roots or rootlets of herbaceous plants with complete certainty, with the exception of only a few special cases (*Carex limosa*, *Equisetum*, *Thelypteris*, *Eriophorum*, and perhaps also *Cladium*, *Schoenoplectus* and some others).

SOME NOTES ON PHYTOLITHS

Additionally, types of plant remains should here be mentioned, which, though of vegetative origin, are nevertheless microfossils with respect to their size: the phytoliths or phytolitharia (also called opal phytoliths or plant opal). These terms mean very small silica bodies of a typical and often very striking shape (as dumb-bells, short bones, pieces of jigsaw puzzles, denticulate or smooth rods or trapezoids etc.). Their size is mostly between 25 and 250 μm; including the extremes, their dimensions range from 10 up to 500 μm. They originate by deposition of silica in various plant tissues (Netolitzky, 1929) of various members of the Spermatophytes, conifers as well as mono- and dicotyledons, herbs as well as trees. After the dying of the plant organs and the decomposition of the organic matter they become incorporated into the soil as isolated bodies. By their origin in plant tissues they reflect the morphological features of the cells typical for those plants, and so it is possible to assign the phytoliths to certain broad groups of plants. As shown by Rovner (1971) the dicotyledons (especially the deciduous trees) have phytoliths very different from those of the monocotyledons, and within this group, particularly in the grass family, some further differentiations are possible, to subfamily level, or even to tribe or genus. This possibility of a crude identification of the plant group from which the phytoliths originate brought about pedological interest, in particular in North America and Australia, into this group of very durable plant remains, which can be expected to elucidate some pedogenetic features of generally dry soils where they are found, sometimes in great quantities. Since, in the author's experience, phytoliths also occur in waterlogged alluvial soils, they may also be found during mire investigations, and so some attention should be paid to them by Quaternary botanists. Our knowledge of the distribution among the plant taxa of the morphological types of phytoliths (or of the characteristic 'type-sets', as Rovner points out in detail) has, however, been rather incomplete until now, especially with regard to mire plants, and so at first further basic comparative studies are needed; the methods of which are dealt with by Rovner.

DETERMINATION OF VEGETATIVE MACROFOSSILS: LITERATURE, REFERENCE COLLECTIONS

It is not the aim of this chapter to describe in detail the single fossils found in sediments and peats. So here only a literature review is given, covering those publications which deal with, by means of pictures, descriptions or keys, the morphological and histological features of macrofossils.

Among the few books covering the complete field of macrofossils, the most comprehensive is by N. Y. Katz, S. V. Katz and E. I. Skobeyeva (1977), written in Russian. Its very early precursor was the booklet by the Katzs (1933) written not only in Russian but also in German; it was the very first contribution exclusively on macrofossils, and though in many respects only of very preliminary character it was rather useful in those days. Two further Russian books to be mentioned here are those of Istomina, Korenyeva and Turemnov (1938) and of Dombrovskaya, Korenyeva and Turemnov (1959), both covering not only vegetative macrofossils, but several other remains too. The same applies to the excellent compendium of Nilsson (1961) with very good line drawings and descriptions (in Swedish).

Some very useful details on macrofossils are furnished in some publications of more extensive character dealing also with other topics. Here the book of Früh and Schröter (1904), for example, should be mentioned, and above all the excellent descriptions of Rudolph (1917, 1935). The book by Bertsch (1942) includes several drawings of vegetative remains too, taken, however, mostly from earlier publications of other authors. The thesis work of Klinger (1968), though unfortunately unpublished, should also be referred to.

The most common vegetative plant remains, as far as they can easily be identified, are surveyed by means of photographs, descriptions and some keys by the present author (1972, 1974a); a few selected macrofossils have also been dealt with (1964a, b). In this connection one must also refer to the extensive descriptions of the subterranean parts of *Narthecium* and *Scheuchzeria* made by Schumacker (1961), and to the short monograph on the rootlets ('radicells') of *Carex* by Matjuschenko (1924), the statements of which need, however, further confirmation and completion in some respect. Furthermore some special papers should be mentioned: on Conifer stomata by Trautmann (1953) and on bud scales of trees and shrubs by Schumann (1889), Hesmer (1935) and Rabien (1953). For the sake of completeness, and supplementary to Chapter 29, some references should be added with respect to the barks of trees and shrubs (Moeller, 1882; Holdheide, 1951) and concerning wood keys with regard to root wood (Riedl, 1937) and dwarf shrubs (Hiller, 1922/23; Greguss, 1959); for dwarf shrubs see also Rudolph (1917), Klinger (1968) and Grosse-Brauckmann (1974).

Supplementary to Chapter 24 an additional reference to Characeae should be given here: an extensive treatment of their oospores, at present under preparation by the German botanist Dr. W. Krause, is expected to appear in 1986.

Finally, it should be stressed that a knowledge of macrofossils cannot be gained if guided only by literature. In particular the knowledge of their macroscopic appearance under various conditions of preservation and kinds of extraction out of the mire (Hiller borer versus the Russian or other borers — see Chapter 8) is broadly a matter of experience in the field, which is learned most easily with the help of a practised teacher.

For a correct identification of vegetative macrofossils by means of histological characteristics a good reference collection is a very essential requisite. The reference samples should ideally not be made of recent material, as its characteristics, even if 'fossilized' in any way (e.g. by cooking with alkali or by short treatment with Schulze's maceration mixture, $KClO_3 + HNO_3$ — see Andersson (1892a)) are not necessarily the same as in the real fossil state. So the material used for reference slides should be derived, if possible, from fossil remains, for which the correct identification has been made, according to the statements given above. The use of such real fossil reference samples also provides some experience in differentiating between the various appearances of the remains, in various degrees of decomposition.

Reference slides are most easily made using glycerol jelly or an analogous medium (e.g. Gelvatol, manufactured by Burkard Manufacturing Co. Ltd., Rickmansworth, Hertfordshire, England) in the usual way.

PRESENTATION OF MACROFOSSIL RECORDS

Presentation of the results of macrofossil investigations, including mosses wood and charcoal, fruits and seeds, is possible in tabular form and by graphs. Each method has its own advantages and problems.

Graphs can be constructed analogously to pollen diagrams, so the macrofossil results are immediately comparable with pollen results in the same profile.

The main disadvantage of graphical presentations of macrofossil results is the difficulty in recognizing at a glance the typical macrofossil assemblages, since the presentation stresses the single species and its more or less abundant occurrence. So the characteristic species combinations and their changes, reflecting the successions in the peat profile, can less easily be made clear, unless the arrangement of the species is well adapted to their typical groupings. Examples of macrofossil diagrams of this type were published by, among others, Marek (1965), Rybníček and Rybníčková (1968), Casparie (1969), van Geel (1978), and Hölzer and Schloss (1981).

Sometimes the same arrangement as in pollen diagrams, i.e. the depths along the vertical axis and the single species side-by-side, is not only used for graphic but also for tabular presentations. Examples were published by, among others, Marek (1965), Pacowski (1967) and Palczynski (1975).

The character of the peat-forming vegetation as a whole is expressed better by a more 'phytosociological' tabular presentation of macrofossil records. It is clear that because of the quite different arrangement, this kind of presentation makes immediate comparisons with pollen records of the same profile more difficult.

The tabular arrangement preferred by the present author is in detail made analogous to tables with vegetation relevés. An example of this is given in Table 28.2. The plant species appear one on another and the results of the single samples of the profile are arranged in their natural sequence from left to right. To elucidate the relationships of macrofossil assemblages to the present-day vegetational units, it is essential to arrange the species in appropriate groups of more-or-less the same ecological 'constitution' and phytosociological character. Several very good examples of this kind of presentation were published by Obidowicz (1975), Oświt (1973, 1976), Rybníček and Rybníčková (1968, 1974, 1977), Rybníčková and Rybníček (1972) and Rybníčková *et al.* (1975).

The phytosociological groupings can be checked and adapted to the specific character of macrofossil assemblages, too, by a comparative 'working-up' of the macrofossil record of all samples from one investigation area; here the principles and methods are the same as those applied to the present-day vegetation relevés (cf. e.g. Ellenberg, 1956). For examples see the publications of Schwaar (e.g. 1976 and 1978) and of the present author (1962c, 1963, 1968, 1973, 1974b, 1985).

On macrofossil tables it is an advantage to use a simple scale for the quantities of macrofossils (including fruits and seeds) found in the single

TABLE 28.2. Example of a 'phytosociological' representation of a macrofossil record. For definitions of the figures and letters, see the text. For further details of the mire, see Grosse-Brauckmann (1974b, 1976)

Dreckmoor (Steinhuder Meer, NW-Germany)
Macrofossil record of profile Q[1]

Stage of development[2]	S	a				b			c			d
Depth[3] below floor (cm)	130	110	90	75	65	55	45	35	25	15	8	1
Decomposition[4] (H. v. Post)	.	3	3	3	.	4	5	5	7	7	7	4
Number of taxa	11	11	10	7	10	9	11	12	16	16	11	9
SPECIES OF OLIGOTROPHIC HUMMOCKS												
Sphagnum magellanicum	.	.	.	.	.	.	.	.	+	+	.	.
Polytrichum strictum	.	.	.	.	.	.	.	.	+	+	1	.
Aulacomnium palustre	.	.	.	.	.	.	.	.	1	+	1	.
Calluna vulgaris	.	.	.	.	.	.	.	.	1	1	.	.
Eriophorum vaginatum	.	.	.	.	.	1	3	.	2	+	+	.
Oxycoccus palustris	+	.	2	1	+	1	1	1	+	1	1	.
SPECIES OF OLIGO- TO MESOTROPHIC HOLLOWS												
Sphagnum sect. *Cuspidata*	2	3	1	1	1	1	2	.	1	1	1	4
Carex rostrata	s	h	.	.	.	.	.	h	m	h	h	.
Scheuchzeria palustris	2	3	4	5	2	3	+	1	.	.	.	.
MESOTROPHIC BRYOPHYTES												
Sphagnum palustre	.	.	.	2	1	1	5	5	2	+	.	1
Calliergon stramineum	.	1	+	.	1	1	2	1	2	+	.	1
MESO- TO EUTROPHIC HERBACEOUS SPECIES[5]												
Juncus effusus	.	.	.	.	.	.	.	.	.	.	.	.
Juncus type *articulatus*	.	.	.	.	.	.	.	.	.	s	.	s
Hydrocotyle vulgaris	.	.	.	.	.	.	.	.	m	s	s	H
Viola palustris	.	s	.	.	.	.	.	.	m	m	s	s
Carex canescens	s	.	.	.	.	.	.	.	h	m	h	H
Potentilla erecta	.	.	.	.	.	.	.	.	s	h	s	s

SPECIES OF VERY WET MIRES, MESO- TO EUTROPHIC

Species	1	2	3	4	5	6	7	8	9	10	11
Equisetum fluviatile	4	.	.	.	3	2	3	2	.	3	1
Menyanthes trifoliata	.	3	.	.	3	1	3	1	.	2	.
MORE-OR-LESS EUTROPHIC SPECIES											
Sphagnum teres	.	.	.	.	.	.	1	.	1	.	2
Acrocladium cuspidatum	.	.	.	.	1	.	.	.	.	.	.
Typha latifolia	3	1	.	.	s	.	.	.	.	.	.
Thelypteris palustris	.	2	.	.	.	2	1	.	.	.	.
Lycopus europaeus	.	m	.	.	.	.	.	.	.	.	.
WOODY SPECIES											
Betula alba s.l.	s	.	.	.	.	h	2	1	.	s	s
Alnus glutinosa	s	m	.	.	.	.	.	.	.	.	.
TOTAL PERCENTAGES											
Wood, bark	-	35	20	-	8	1	3	2	-	2	75
Herbaceous roots	10	43	77	72	60	50	-	-	35	85	8
Other herbaceous remains	80	22	3	8	27	45	17	10	10	13	17
Bryophytes totally[6]	10	.	.	20	5	4	80	88	55	v	v
hereof *Sphagna*	5	3	.	.	2	3	75	80	53	2	2

Additional species occurring once: 130 cm: *Drepanocladus* cf. *exannulatus* 2, *Carex limosa* 3; 110 cm: *Potentilla palustris* s; 90 cm: *Rhynchospora alba* +; 55 cm: *Erica tetralix* +; 45 cm: *Drosera rotundifolia* s; 35 cm: *Carex* type *paniculata* s; 25 cm: *Hypnum cupressiforme* +; 15 cm: *Salix* sp.

[1] For details see Grosse-Brauckmann (1974b, 1976); the period of peat formation does not exceed the last 7 centuries.
[2] Characterized by rather uniform macrofossil content; S = allochthonous organic lake sediment.
[3] Depth of the upper surface of each sample; samples 1 cm thick (except sediment sample: 20 cm).
[4] In the layer 30–5 cm with some admixture of sediments.
[5] In the present case evidently favoured by human influence (perhaps grazing of cattle).
[6] v = less than 0.5%.

samples. In this way similarities with actual plant communities can be recognized particularly well. The author uses a scale analogous to the Braun–Blanquet scale, the definitions of which are as follows:

Fruits and seeds, insofar as no other remains of the same taxon were found in the whole sample of 50 ml:

  s = 1–2 specimens ('seldom')
  m = 3–5 specimens ('moderately frequent')
  h = 6–14 specimens ('higher quantities')
  H = more than 14 specimens ('very high quantities')

If the fruits and seeds are found in the finest fraction of the washing preparation which was obtained by 'fishing', it must be taken into account that only a small part of the whole sample was analysed. In this case, 's' is used for only 1, 'm' for 2, 'h' for 3–5 and 'H' for more than 5 specimens.

Tissue remains of mosses and higher plants (in the main, roots and rootlets, rhizomes, stalks, leafs, wood and bark):

  + = only a few pieces of tissue remains, forming, at the same time, much less than 1% of the washing residue. Fruits or seeds of the same taxon are not present or do not exceed the frequency 'm';
  1 = *either* the same as before, fruits or seeds of the same taxon more frequent than 'm',
      or tissue remains represented by many pieces, even if less than 1% of the washing residue, fruits or seeds present or not,
      or tissue remains in any number and 1–3% of the washing residue, etc.
  2 = tissue remains to 4–10% of the washing residue, etc.;
  3 = tissue remains to 10–25% of the washing residue, etc.;
  4 = tissue remains to 25–50% of the washing residue, etc.;
  5 = tissue remains to >50% of the washing residue, etc.;

SIGNIFICANCE OF MACROFOSSIL INVESTIGATIONS

Macrofossil records of Quaternary peat deposits might be compared with results of pre-Quaternary palaeobotanical research. But besides the different ages of the deposits investigated, there is a great difference in the aims of the investigations; while in palaeobotany the single species found and their geographical and historical relations are the matters of main interest, the investigations of Quaternary plant remains, at least those of late- or postglacial origin, refer much more to the assemblage of the macrofossils, which furnishes the chance of, by 'extrapolation', a reconstruction of the original peat-forming plant community of the past.

There are several disciplines in science which can make use of such reconstructions: phytosociology and ecology, vegetation- and landscape-history, prehistory, and soil science.

One of the ecological and phytosociological questions asked in this connection is which of the vegetation units known on the mires of today are not only peat-inhabiting but also peat-forming plant communities (e.g. Grosse-Brauckmann, 1962a, b)? This question arises particularly in the central and western European mire regions, where most of the mires, if not exploited totally, were drained or otherwise influenced by man. In answering this question, however, the existence of 'displacement peats' must be borne in mind, though they seem not to be very frequent, as mentioned above.

An example of a phytosociological evaluation of a greater number of macrofossil records, exclusively comprising, however, north-west German peats containing macrofossils of *Myrica,* is given in Table 28.3. It shows a clear division into four distinct plant communities. Phytosociological synopses of the whole inventory of 'subfossil plant communities' of single mires or smaller areas have been made by Grosse-Brauckmann (1962c, 1963, 1967, 1973, 1974b, 1979c), Pacowski (1967), and others; phytosociological records of Polish peats are given by Jasnowski (1959, 1962), Tołpa *et al.* (1967), Palczynski (1969), and Oświt *et al.* (1975b). An attempt to compile from several publications a list of all European vegetation units as represented by macrofossil records was made by Rybníček (1973).

It is essential to stress that, strictly speaking, vegetation units of today are of course comparable only to macrofossil records of peats of subatlantic age. But also peats of late- or older postglacial age can be compared with actual mire vegetation types, and until now only macrofossil assemblages have been found, which have proved to be more or less analogous to any vegetational units of appropriate parts of Europe. The absence of fundamental phytosociological divergences maintained herewith, applies also to the strongly humified 'older Sphagnum peat' of northern central Europe, the macrofossil content of which is not generally different from what we know about certain ombrogenous mires. (Concerning this problem, see particularly Casparie (1972) and van Geel (1978).)

Further phytosociological considerations relate to the distribution of macrofossil assemblages in all dimensions of a mire. So the (vertical) distribution within one profile depicts the course of the successions of mire vegetation at that place (cf. the example in Table 28.1), and differences in horizontal direction reflect the vegetational zonations in the various developmental stages of the whole mire (Grosse-Brauckmann, 1969, 1976, 1979c; Oświt *et al.*, 1975a).

Knowledge of the ecology of the various plant communities found in such investigations also enables us to reconstruct the changes in time and the differences in space of the ecological conditions of a mire in the past. One of the

TABLE 28.3.   Example of a phytosociological synopsis of several macrofossil records (adapted from Grosse-Brauckmann, 1979d). The plant species are arranged in groups which were found by phytosociological 'working-up' of the single investigation results. The figures are the frequencies (in %) of the single species in each of the four types, the 'parent communities' of which are characterized as follows:

(1)   Most clearly oligotrophic and not extremely wet, characterized by the ombrotrophic hummock plants; minerotrophic species are, however, not wholly absent.
(2)   Comparatively poor in nutrients and very wet, characterized by absence of the hummock plants of type (1) and by the highest frequencies of *Carex limosa*, *Menyanthes*, *Sphagna* of the *Subsecunda* group and other mosses.
(3)   Comparatively rich in nutrients and very wet, characterized by high frequencies of the reed species (Phragmitetalia) and near absence of the oligotrophic hollow species (Rhynchosporion).
(4)   A 'mixed' type, appearing as a combination of types (1) and (3), which cannot be discussed here.

The mires that the records are derived from are situated in the lowlands between the Weser and Elbe rivers. The peats are of subatlantic age on the basis of pollen analysis. Some rarer species have been omitted. *Reproduced by permission of J. Cramer, Braunschweig*

PHYTOSOCIOLOGICAL TYPES OF PEATS CONTAINING REMAINS OF MYRICA GALE: FREQUENCIES OF THE SPECIES FOUND

Type	(1)	(2)	(3)	(4)
Samples investigated	21	17	19	13
Mean number of taxa	8.2	7.5	6.8	9.9
± FREQUENT IN ALL TYPES				
Myrica gale	100	100	100	100
Sphagnum paluste	67	35	21	77
Molinia caerulea	24	12	5	62
OMBROTROPHIC SPECIES (MOSTLY OXYCOCCO-SPHAGNETEA)				
Calluna vulgaris	86	.	.	85
Eriophorum vaginatum	81	.	.	54
Andromeda polifolia	29	6	.	8
Erica tetralix	24	.	.	31
Trichophorum caespitosum	15	.	.	.
Oxycoccus palustris	10	.	.	8
Sphagnum papillosum	38	.	.	23
Sphagnum magellanicum	10	.	.	8
Sphagnum imbricatum	5	.	.	8
Aulacomnium palustre	28	.	.	23
Polytrichum strictum	24	.	.	.
Dicranum bonjeani	5	.	.	8
Pohlia nutans	5	.	.	8
Rhynchospora alba	48	24	5	15
Carex limosa	10	47	.	.
Scheuchzeria palustris	10	6	.	8

PHYTOSOCIOLOGICAL TYPES OF PEATS CONTAINING REMAINS OF MYRICA GALE: FREQUENCIES OF THE SPECIES FOUND

SCHEUCHZERIA-CARICETEA FUSCAE:
REMAINING SPECIES

Menyanthes trifoliata	10	47	11	.
Hydrocotyle vulgaris	10	6	16	23
Ranunculus flammula	.	6	11	8
Carex nigra	5	.	5	.
Potentilla palustris	5	6	.	.
Sphagnum sect. *Cuspidata*	38	41	11	46
Sphagnum sect. *Subsecunda*	24	41	5	15
Calliergon stramineum	5	12	21	15
Sphagnum teres	5	6	11	8
Calliergon giganteum	.	12	5	.
Tomenthypnum nitens	.	.	5	8

PHRAGMITION OR PHRAGMITETALIA (IN PART)

Phragmites australis	5	.	74	77
Cladium mariscus	.	.	53	54
Schoenoplectus lacustris	5	.	21	.

VARIOUS SPECIES

Lychnis flos-cuculi	10	41	37	23
Thelypteris palustris	5	35	37	.
Eleocharis palustris	.	6	32	31
Salix spec.	.	.	16	8
Potentilla erecta	5	12	5	.
Drepanocladus fluitans	5	24	21	8
Acrocladium cuspidatum	10	24	11	8
Scorpidium scorpioides	.	18	.	.
Bryum pseudotriquetrum	.	12	.	8

IDENTIFIED ONLY TO GENUS

Carex sp.	10	71	68	46
Drepanocladus sp.	.	12	11	23
Calliergon sp.	5	24	5	31

points to note in this respect is the water relations, particularly the changes from wet to very wet or to only moist conditions, which are then characterized by a cessation in peat formation. Another point of view is the mineral nutrition of the mire. A predominantly significant step in the trophic conditions of a mire is the transition from minerotrophic (or 'rheophilic' as defined by Kulczyński (1949)) to ombrotrophic conditions, which can be well-defined by phytosociological methods according to the concept of the 'fen plant limit' (Sjörs, 1948) or 'Mineralbodenwasserzeigergrenze' (Du Rietz, 1954). A phytosociological definition with the help of macrofossils is, it should be noted, the only correct way to study this change from minerotrophy to true ombrotrophy in peat profiles, since chemically no sharp limit is possible to find. Moreover, some chemical characteristics of the peats might have changed after deposition by outwash of minerals or, in certain cases, also by mineral enrichment because of the high cation exchange capacity of peats (Grosse-Brauckmann, 1980, 1976).

The preceding phytosociological and ecological discussion, which started with the significance of macrofossil records to the present vegetation, has led us, via vegetational successions, to the historical questions quoted above, since mire development is part (and a very essential one in some lowland or oceanic regions) of the landscape- and vegetation-history. It should also be noted that at times macrofossil records can give some information on the environmental conditions of prehistoric man (e.g. Grosse-Brauckmann, 1979a).

The importance of macrofossil records with respect to pollen analyses must also be mentioned. Self-evidently, a critical interpretation of pollen records must consider the characteristics of the place of catchment, deposition and embedding of pollen, and the best idea on this is given by the peat-forming plant community at the times in question. In addition, indications are given as to which part of the pollen content of the sample analysed can perhaps be of local origin. Here it should be noted, in passing, that herb pollen can contribute to our ideas on the composition of the peat-forming plant communities, as shown, for example, in several publications of Rybníček and Rybníčková.

Finally, the significance of macrofossils for geological or soil mapping in mire areas should be mentioned. The macrofossil content of the uppermost peat layers gives (sometimes in connection with certain general geomorphological and hydrological facts) a means to place mires (even if they no longer have their original plant cover because they have been drained and forested, or are in agricultural use) in the right place in a mire classification, which usually comprises only a few units. So the German soil classification has only the three mire types Niedermoor (eutrophic mire, fen), Übergangsmoor (mesotrophic, transitional mire), and Hochmoor (ombrogenous mire, bog). Insertion of the single peats into this classification scheme is mostly possible by means of the macrofossil content, as can be seen in Table 28.1; here only those macrofossils which can already be identified, with some experience, macroscopically (or by hand lens) in the field are included. Other peat characteristics, as are the degree of humification, the mineral content, and characteristics of colour and fabric (this in the pedological meaning) are likewise of interest for mapping, but they are, compared with the macrofossil content of the peats, of lower rank with respect to the major classification.

In certain mire mappings or surveys, statements concerning the composition of the whole of the profiles are also asked for, which can only be made on the basis of a field identification of macrofossils. For further details of the pedological approach to mires refer to the German soil mapping instruction (Arbeitsgruppe Bodenkunde, 1982), in which the chapters on peats and mires are based to a great extent on proposals or publications of the present author (see also Grosse-Brauckmann *et al.*, 1977).

REFERENCES

Andersson, G. (1892a). Om metoden för växtpaleontologiska undersökningar af torfmossar. *Geol. Fören. Förhandlingar*, **14**, 165–175.

Andersson, G. (1892b). Om slamning af torf. *Geol. Fören. Förhandlingar*, **14**, 506–508.

Andersson, G. (1893). Om metoden för botanisk undersökning af olika torfslag. *Svenska Mosskulturfören. Tidskr.*, **6**, 526–530.

Andersson, G. (1898). Studier öfver Finlands torfmossar och fossila kvartärflora. *Bull. Comm. Géol. Finlande*, **8**, 210 pp.

Arbeitsgruppe Bodenkunde (1982). *Bodenkundliche Kartieranleitung* (3rd edn), Schweizerbartsche Verlagsbuchhandlung, Stuttgart.

Bertsch, K. (1942). Lehrbuch der Pollenanalyse. In: *Handbücher der praktischen Vorgeschichtsforschung*, vol. 3, F. Enke, Stuttgart.

Bülow, K.v. (1929). *Allgemeine Moorgeologie, Einführung in das Gesamtgebiet der Moorkunde*, Gebr. Borntraeger, Berlin.

Casparie, W. A. (1969). Bult- und Schlenkenbildung in Hochmoortorf (Zur Frage des Moorwachstums-Mechanismus). *Vegetatio Acta Geobotanica*, **19**, 146–180.

Casparie, W. A. (1972). *Bog development in south-eastern Drenthe (the Netherlands)*, Dr. W. Junk N.V., the Hague.

Dombrovskaya, A. V., Korenyeva, M. M., and Turemnov, S. N. (1959). *Atlas of the Plant Remains Occurring in Peat*, Gosenergoizdat, Moscow and Leningrad, (in Russian).

Du Rietz, G. E. (1954). Die Mineralbodenwasserzeigergrenze als Grundlage einer natürlichen Zweigliederung der nord- und mitteleuropäischen Moore. *Vegetatio Acta Geobotanica*, **5/6**, 571–585.

Ellenberg, H. (1956). Grundlagen der Vegetationsgliederung. I: Aufgaben und Methoden der Vegetationskunde. In: *Einführung in die Phytologie* (Ed. H. Walter), vol. 4/1, E. Ulmer, Stuttgart.

Früh, J., and Schröter, C. (1904). Die Moore der Schweiz, mit Berücksichtigung der gesamten Moorfrage. In: *Beiträge zur Geologie der Schweiz*, geotechn. Ser., 3. Lief., Bern.

Greguss, P. (1959). *Holzanatomie der europäischen Laubhölzer und Sträucher*, Hungarian Museum of Natural History, Budapest.

Grosse-Brauckmann, G. (1962a). Zur Moorgliederung und -ansprache. *Zeitschrift für Kulturtechnik*, **3**, 6–29.

Grosse-Brauckmann, G. (1962b). Torfe und torfbildende Pflanzengesellschaften. *Zeitschrift für Kulturtechnik*, **3**, 205–225.

Grosse-Brauckmann, G. (1962c). Moorstratigraphische Untersuchungen im Nieder-wesergebiet (Über Moorbildungen am Geestrand und ihre Torfe). *Veröffent-lichungen des Geobotanischen Institutes der Eidgenössischen Technischen Hochschule, Stiftung Rübel*, Zürich, **37**, 100–119.

Grosse-Brauckmann, G. (1963). Über die Artenzusammensetzung von Torfen aus dem nordwestdeutschen Marschen-Randgebiet (eine pflanzensoziologische Auswertung von Grossrestuntersuchungen). *Vegetatio Acta Geobotanica*, **11**, 325–341.

Grosse-Brauckmann, G. (1964a). Einige wenig beachtete Pflanzenreste in nordwest-deutschen Torfen und die Art ihres Vorkommens. *Geol. Jb.*, **81**, 621–644.

Grosse-Brauckmann, G. (1964b). Zur Artenzusammensetzung von Torfen. (Einige Befunde und Überlegungen zur Frage der Zersetzlichkeit und Erhaltungsfähig-keit von Pflanzenresten). *Ber. Dtsch. Botan. Ges.*, **26**, (22)–(37).

Grosse-Brauckmann, G. (1967). Über die Artenzusammensetzung einiger nordwest-deutscher Torfe. In: *Pflanzensoziologie und Palynologie, Bericht über das internationale Symposium in Stolzenau/Weser 1962 der Internationalen Vereini-gung für Vegetationskunde* (Ed. R. Tüxen), W. Junk, the Hague, pp. 160–180.

Grosse-Brauckmann, G. (1968). Einige Ergebnisse einer vegetationskundlichen Auswertung botanischer Torfuntersuchungen, besonders im Hinblick auf Sukzes-sionsfragen. *Acta Botanica Neerlandica*, **17**, 59–69.

Grosse-Brauckmann, G . (1969). Zur Zonierung und Sukzession im Randgebiet eines Hochmoores (nach Torfuntersuchungen im Teufelsmoor bei Bremen). *Vegetatio*, **17**, 33–49.

Grosse-Brauckmann, G. (1972). Über pflanzliche Makrofossilien mitteleuropä-ischer Torfe. I: Gewebereste krautiger Pflanzen und ihre Merkmale. *Telma*, **2**, 19–55.

Grosse-Brauckmann, G. (1973). Grossrest- und pollenanalytische Ergebnisse zur Vegetationsentwicklung im Poggenpohlsmoor. In: Zur historischen und aktuellen Vegetation im Poggenpohlsmoor bei Dötlingen (Oldenburg) (G. Grosse-Brauckmann and K. Dierssen, *Mitteilungen der Floristisch-soziologischen Arbeits-gemeinschaft N.F.*, **15/16**, 123–145.

Grosse-Brauckmann, G. (1974a) Über pflanzliche Makrofossilien mitteleuropä-ischer Torfe. II: Weitere Reste (Früchte und Samen, Moose u.a.) und ihre Bestimmungsmöglichkeiten. *Telma*, **4**, 51–117.

Grosse-Brauckmann, G. (1974b). Zum Verlauf der Verlandung bei einem eutrophen Flachsee (nach quartärbotanischen Untersuchungen am Steinhuder Meer). I: Heutige Vegetationszonierung, torfbildende Pflanzengesellschaften der Ver-gangenheit. *Flora*, **163**, 179–229.

Grosse-Brauckmann, G. (1976). Zum Verlauf der Verlandung bei einem eutrophen Flachsee (nach quartärbotanischen Untersuchungen am Steinhuder Meer). II: Die Sukzessionen, ihr Ablauf und ihre Bedingungen. *Flora*, **165**, 415–455.

Grosse-Brauckmann, G. (1979a). Pflanzliche Grossreste von Moorprofilen aus dem Bereich einer steinzeitlichen Seeufer-Siedlung am Dümmer. *Phytocoenologia*, **6**, 106–117.

Grosse-Brauckmann, G. (1979c). Sukzessionen bei einigen torfbildenden Pflan-zengesellschaften (nach Ergebnissen von Grossrest-Untersuchungen an Torfen). In: *Gesellschaftsentwicklung (Syndynamik)*, Ber. Int. Symp. Ver. Vegetationskunde Rinteln, 1967 (Ed. R. Tüxen), Vaduz, pp. 393–408.

Grosse-Brauckmann, G. (1979d). Zur Deutung einiger Makrofossil-Vergesellschaf-tungen unter dem Gesichtspunkt der Torfbildung. In: *Werden und Vergehen von Pflanzengesellschaften*, Berichte der Int. Symp. der Int. Vereinigung für Vegeta-tionskunde, Rinteln, 20–23 März 1978 (Eds. O. Wilmanns and R. Tüxen), J. Cramer, Vaduz, pp. 111–131.

Grosse-Brauckmann, G. (1980). Ablagerungen der Moore. In: *Moor-und Torfkunde* (2nd edn.) (Ed. K. Göttlich), E. Schweizerbart, Stuttgart, pp. 130–173.

Grosse-Brauckmann, G. (1985). Über einige torfbildende Pflanzengesellschaften der Vergangenheit in der Rhön und auf dem Vogelsberg. *Tuexenia, Mitteilungen der Floristisch-soziologischen Arbeitsgemeinschaft*, N.S. **5**, 191–206.

Grosse-Brauckmann, G., Hacker, E., and Tüxen, J. (1977). Moore in der boden-kundlichen Kartierung — ein Vorschlag zur Diskussion. *Telma*, **7**, 39–54.

Grosse-Brauckmann, G., and Puffe, D. (1964). Über Zersetzungsprozesse und Stoffbilanz im wachsenden Moor. In: *Trans. Int. Congress of Soil Science, Bucharest, Romania, 1964*, **5**, 635–648. Publishing House of the Academy of the Socialist Republic of Romania, Bucharest.

Hesmer, H. (1935). Samen- und Knospenschuppenanalyse in Mooren. *Zeitschr. f. Forst- u. Jagdwesen*, **67**, 600–621.

Hiller, W. (1922/23). Das Bestimmen von Hölzern nach mikroskopischen Merkmalen. *Mikrokosmos, Zeitschr. Angewandte Mikroskopie*, **16**, 179–182 and 193–197.

Holdheide, W. (1951). Anatomie mitteleuropäischer Gehölzrinden (mit mikrofotografischem Atlas). In: *Handbuch der Mikroskopie in der Technik. 5: Mikroskopie des Holzes und des Papiers*, Pt. 1 (Ed. H. Freund), Umschau-Verlag, Frankfurt am Main, pp. 193–368.

Holling, R., and Overbeck, F. (1960). Über die Grösse der Stoffverluste bei der Genese von Sphagnumtorfen. *Flora*, **150**, 191–208.

Holmboe, J. (1903). Planterester i norske torvmyrer. Ett bidrag til den norske vegetationshistorie efter den sidste istid. *Videns.-selsk. Skr. I, math.*-nat. Kl., (2).

Hölzer, A., and Schloss, S. (1981). Paläoökologische Studien an der Hornisgrinde (Nordschwarzwald) auf der Grundlage von chemischer Analyse, Pollen- und Grossrestuntersuchung. *Telma*, **11**, 17–30.

Istomina, E. S., Korenyeva, M. M., and Turemnov, S. N. (1938). *Atlas of the Plant Remains Occurring in Peat*. Academy of Science, Moscow and Leningrad, (in Russian).

Jasnowski, M. (1959). Klassifizierung von Moostorfarten quartärer Niedermoore und ihre Entstehung (in Polish, with German summary). *Acta Societatis Botanicorum Poloniae*, **28**, 319–364.

Jasnowski, M. (1962). Über die Klassifizierung der Torfarten. *Freiberger Forschungshefte*, **A 254**, 13–26.

Katz, N. Y., and Katz, S. W. (1933). *Atlas der Pflanzenreste im Torf*, Selchozgiz, Moscow and Leningrad.

Katz, N. Y., Katz, S. V., and Skobeyeva, E. I. (1977). In: *Atlas of the Plant Remains in Peats*, Nyedra, Moscow, (in Russian).

Klinger, P. U. (1968). *Feinstratigraphische Untersuchungen an Hochmooren. Mit Hinweisen zur Bestimmung der wichtigsten Grossreste in nordwestdeutschen Hochmoortorfen und einer gesonderten Bearbeitung der mitteleuropäischen Sphagna cuspidata*, unpublished thesis.

Klinger, P. U. (1984). Einige Bemerkungen zu Lebermoosfunden bei stratigraphischen Torfuntersuchungen und zur Aufbereitung von Torfproben. *Telma*, **14** 81–88.

Krause, W. (1986). Die Bestimmung der Characeen-Oosporen nebst ihrer Bedeutung für die Vegetationskunde und Vorgeschichte. *Phytocoenologia* (in preparation).

Kulczyński, S. (1949). Peat bogs of Polesie. *Mém. Acad. Polon. Sci. Lettr., Cl. Sci. math. et nat.*, Sér B, **15**, Kraków, 356 pp.

Lagerheim, G. (1902). Torftekniska notiser. *Geol. Fören. Förhandlingar*, **24**, 407–411.

Lewis, F. J. (1905–7). The plant remains in the Scottish peat mosses, I, II and III. *Trans Roy. Soc. Edinburgh*, **41**, (28); **45** (13); **46** (2).

Marek, S. (1965). Biology and stratigraphy of the alder bogs in Poland (in Polish, with English summary). *Zeszyty problemowe postępów nauk rolnyczych*, **57**, 5–303.

Marek, S. (1974). Plant succession on the peat bog near Kłodawa in the Gorzów Wielkopolski district (in Polish, with English summary). *Badania Fizjograficzne nad Polską Zachodnią*, **26**, ser. B. — Biologia, 195–207.

Marek, S. and Palczyński, A. (1959). Raised bogs in the western Bieszczady region (in Polish, with English summary). *Zeszyty problemowe postępów nauk rolniczych*, **34**, 255–299.

Matjuschenko, W. (1924). Schlüssel zur Bestimmung der in den Mooren vorkommenden Carexarten (Übersetzung von S. Ruoff). *Geolog. Archiv, Zeitschr. Gesamtgebiet d. Geol.*, **3**, 183–188 and 192–193.

Moeller, J. (1882). *Anatomie der Baumrinden, vergleichende Studien*, J. Springer, Berlin.

Nathorst, A. G. (1872). Om arktiska växtlemningar i Skånes sötvattensbildningar. *Vetensk.-Akad. Förh.*, **29** (2).

Nathorst, A. G. (1892). Über den gegenwärtigen Standpunkt unserer Kenntnis von dem Vorkommen fossiler Glacialpflanzen. *Bih. till Kungl. Svenska Vetenskapsakademiens Handlingar* 17/III, **5**, Stockholm.

Netolitzky, F. (1929). Die Kieselkörper. In: *Handbuch der Pflanzenanatomie* (Ed. K. Linsbauer), **III**/1a, Gebr. Borntraeger, Berlin, pp. 1–19.

Nilsson, T. (1961). *Kompendium i kvartärpaleontologi och kvartärpaleontologiska undersökningsmetoder* (2nd edn), vols. 1 and 2, Lund University, Lund.

Obidowicz, A. (1975). Entstehung und Alter einiger Moore im nördlichen Teil der Hohen Tatra. *Fragmenta Floristica et Geobotanica*, **21**, 289–323.

Oświt, J. (1973). Warunki rozwoju torfowisk w dolinie dolnej Biebrzy na tle stosunków wodnych. *Roczniki Nauk Rolniczych*, **D 143**, 5–80.

Oświt, J., Pacowski, R., and Zurek, S. (1975a). Succession of peat-forming vegetation on the peatlands of Poland. In: *Peatlands and their Utilization in Poland*, Wydawnictwa Czasopicz Techniczne Not, Warszawa, pp. 29–35.

Oświt, J., Pacowski, R., and Zurek, S. (1975). Characteristics of more important peat species in Poland. In: *Peatlands and their Utilization in Poland*, Wydawnictwa Czasopicz Techniczne Not, Warszawa, pp. 51–60.

Overbeck, F. (1975). *Botanisch-geologische Moorkunde unter besonderer Berücksichtigung der Moore Nordwestdeutschlands als Quellen zur Vegetations-, Klima- und Siedlungsgeschichte*, K. Wachholtz Verlag, Neumünster.

Pacowski, R. (1967). Biologie und Stratigraphie des Hochmoores Wieliszewo (in Polish, with German summary). *Zeszyty Problemowe Postępów Nauk Rolniczych*, **76**, 101–196.

Palczynski, A. (1969). An outline of phytocoecology of peat-bogs in Poland and the genetic classification of peats based on the ecologo-phytocoenologic principle (in Russian, with English summary). *Botanicheskii Journal*, **54**, 1921–1938.

Palczyński, A. (1975). Die Jaćwieskie-Sümpfe (Biebrza-Urstromtal), geobotanische, palaeophytosoziologische und wirtschaftliche Probleme (in Polish, with German summary). *Roczniki Nauk Rolniczych*, **D 145**, 1–232.

Post, L. von, and Granlund, E. (1926). Södra Sveriges Torvtillgångar I. *Sveriges Geologiska Undersökning, Årsbok*, **19/2**, 1–127.

Rabien, I. (1953). Zur Bestimmung fossiler Knospenschuppen. *Paläontolog. Zeitschr.*, **27**, 57–66.

Riedl, H. (1937). Bau und Leistungen des Wurzelholzes. *Jahrb. wiss. Botanik*, **85**, 1–75.

Rovner, I. (1971). Potential of opal phytoliths for use in palaeoecological reconstruction. *Quaternary research*, **1**, 343–359.

Rudolph, K. (1917). Untersuchungen über den Aufbau böhmischer Moore. I: Aufbau und Entwicklungsgeschichte südböhmischer Hochmoore. *Abh. Zool.-Botan. Ges. Wien*, **9**, (4).

Rudolph, K. (1935). Mikrofloristische Untersuchung tertiärer Ablagerungen im nördlichen Böhmen. *Beihefte Botan. Centralblatt*, **54**, Abt. B, 244–328.

Rybníček, K. (1973). A comparison of the present and past mire communities of Central Europe. In: *Quaternary Plant Ecology* (Eds. H. J. B. Birks and R. G. West), Blackwell, Oxford, pp. 237–261.

Rybníček, K., and Rybníčková, E. (1968). The history of flora and vegetation on the Bláto mire in southeastern Bohemia, Czechoslovakia (palaeoecological study). *Folia Geobotanica et Phytotaxonomica*, **3**, 117–142.

Rybníček, K., and Rybníčková, E. (1974). The origin and development of waterlogged meadows in the central part of the Šumava foothills. *Folia Geobotanica et Phytotaxonomica*, **9**, 45–70.

Rybníček, K., and Rybníčková, E. (1977). Mooruntersuchungen im oberen Gurgltal, Ötztaler Alpen. *Folia Geobotanica et Phytotaxonomica*, **12**, 245–291.

Rybníčková, E., and Rybníček, K. (1972). Erste Ergebnisse paläogeobotanischer Untersuchungen des Moores bei Vracov, Südmähren. *Folia Geobotanica et Phytotaxonomica*, **7**, 285–308.

Rybníčková, E., Rybníček, K., and Jankovská (1975). Palaeoecological investigations of buried peat profiles from the Zbudovská blata marshes, southern Bohemia. *Folia Geobotanica et Phytotaxonomica*, **10**, 157–178.

Schröter, C. (1882). Die Flora der Eiszeit. In: *Neujahrsblatt d. naturforsch. Gesellsch. Zürich auf d.J. 1883*, pp. 1–41.

Schumacker, R. (1961). Étude d'une tourbe à Scheuchzeria palustris dans les couches inférieures des dépots de la Fagne Wallonne. *Bull. Soc. Royale Sci. Liège*, **30**, 496–511.

Schumann, C. R. G. (1889). Anatomische Studien über die Knospenschuppen von Coniferen und dicotylen Holzgewächsen. *Bibliotheca Botanica*, **15**.

Schwaar, J. (1976). Paläogeobotanische Untersuchungen im Belmer Bruch bei Osnabrück. *Abhandlungen Naturwiss. Verein Bremen*, **38**, 207–257.

Schwaar, J. (1978). Frühere Pflanzengesellschaften küstennaher nordwestdeutscher Moore. *Telma*, **8**, 107–121.

Sjörs, H., (1948). Myrvegetation i Bergslagen (Mire vegetation in Bergslagen, Sweden). *Acta Phytogeographica Suecica*, **21**, Uppsala.

Tołpa, S., Jasnowski, M., and Pałczyński, A. (1967). System der genetischen Klassifizierung der Torfe Mitteleuropas. *Zeszyty Problemowe Postępów Nauk Rolniczych*, **76**, 9–99.

Trautmann, W. (1953). Zur Unterscheidung fossiler Spaltöffnungen der mitteleuropäischen Coniferen. *Flora*, **140**, 523–533.

van Geel, B. (1978). A paleoecological study of Holocene peat bog sections in Germany and the Netherlands. *Rev. Palaeobot. Palynol.*, **25**, 1–120.

Weber, C. A. (1897). Über die Vegetation zweier Moore bei Sassenberg in Westfalen (Ein Beitrag zur Kenntnis der Moore Nordwestdeutschlands). *Abhandl. Naturwiss. Verein Bremen*, **14**, 305–321.

Weber, C. A. (1902). *Über die Vegetation und Entstehung des Hochmoores von Augstumal im Memeldelta, mit vergleichenden Ausblicken auf andere Hochmoore der Erde*, Berlin.

Weber, C. A. (1905). Über Litorina- und Prälitorinabildungen der Kieler Förde. *Englers Botan. Jahrbücher*, **35**, 1–54.

Weber, C. A. (1930). Grenzhorizont und Älterer Sphagnumtorf. *Abhandl. Naturwiss. Verein Bremen*, **28**, 57–65.

Handbook of Holocene Palaeoecology and Palaeohydrology
Edited by B. E. Berglund

29

Wood and charcoal analysis

Werner Schoch

Swiss Federal Institute of Forestry Research
Birmensdorf, Switzerland

INTRODUCTION

Wood has been universally employed as an energy source since earliest times; charcoal remains in fireplaces of bivouac and dwelling sites from all periods of human history bear witness to this fact. Later, wood was also used for tools, weapons and buildings; but this cannot be traced in as much detail since such objects have only survived to the present-day where ambient conditions have precluded biological decomposition: in perpetually arid zones, in permafrost soil layers, or in sediments where continuous moisture has prohibited fungal growth. Nevertheless, such material does exist and can be analysed to obtain information on the type of vegetation present around earlier settlements and on prehistoric man's relationship with his environment.

FIELDWORK

Collection of samples

The collection and subsequent treatment of find material demands great care, since faulty handling can greatly hinder or even prohibit analysis. Samples may comprise isolated finds, material from single sediment strata or whole profile series. The notes below refer to the numbers in Figure 29.1.

(1)   Sediment from a profile section. Such samples should preferably be collected by the laboratory which is to conduct the analysis and are usually taken in conjunction with material for sediment and pollen analysis.

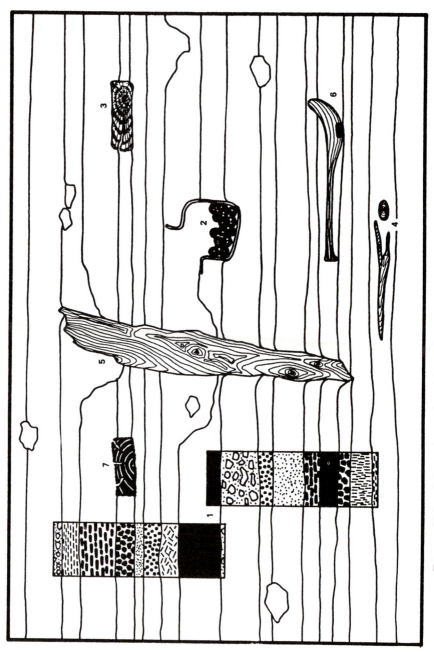

FIGURE 29.1. Schematic stratigraphic section. Numbers refer to the description in the text

Where the sediment contains few stones, samples can be taken by pressing a stainless-steel container into the profile, loosening the surrounding soil with a trowel and lifting out after labelling and/or photographing. Airtight packing in plastic film follows.

(2) Conspicuous accumulations of organic material in various strata: twigs, wood chips, charcoal, leaves, seeds, remains of fruits, contents of vessels, remnants of fabric and cords. Such organic material is packed in plastic bags immediately after collection.

(3) Prostrate pieces of wood, hewn, may be part of a building.

(4) Prostrate pieces of wood, unhewn, may be logs or driftwood.

(5) Upright posts may be part of a building. Analysis of such finds is only profitable if their relative positions are indicated on a plan. This often permits the identification of building phases or ground plans of buildings on the basis of wood species alone.

(6) Wooden artefacts. It is desirable to identify the wood species before preserving the find, especially in the case of artefacts, since the preserving media usually obscure the anatomical structure to a greater or lesser extent.

(7) Carbonized pieces of wood.

Sediments in both caves and open country often contain only charcoal. This can be picked directly out of the soil or separated by washing and sieving. Care must be taken to note whether the find occurs in a fireplace or a post hole, or has no apparent function.

Quantity

The quantity of material collected depends on the richness of the find and the objectives of the study. Where there is much material, 100–200 ml may suffice; but for meagre finds it may be necessary to collect several litres of the sediment. If the deposit contains only little wood, as many fragments as possible should be collected. Even the smallest scraps (1 mm^2) can provide valuable information on the history of vegetation.

Labelling

Careful labelling is essential. Both labels and inscriptions should be waterproof, and the latter should also be smudge-proof. Suitable labels are of PVC with a matt finish, such as those used in gardening, and can be inscribed with pencil or waterproof felt pen. In order to permit interpretation of the finds, general information must be recorded for every sample. This is best achieved by means of specially designed record sheets, as shown in Figure 29.2.

Begleitformular zu Proben an das Labor für quartäre Hölzer, Eidgenössische Anstalt für das forstliche Versuchswesen, CH–8903 Birmensdorf.

Einsender und Adresse _____

Datum _____

Fundortbeschreibung (Koordinaten, Planeintragung, Lage in der Stratigraphie, Fotos, etc.) _____

Alter der Probe _____ | Kulturgruppe _____

Bezeichnung der Probe

☐ Holz   ☐ Kunstgegenstand   ☐ Artefakt   ☐ Konstruktion   ☐ bearbeitet   ☐ unbearbeitet ☐ _____

☐ Holzkohle   ☐ von Feuerstelle, Brandgrube   ☐ ohne erkennbare Funktion ☐ _____

☐ Sediment ☐ _____

Untersuchungsziel: _____

Nach der Untersuchung   ☐ kann über das Material verfügt werden   ☐ ist das Material dem Absender zurückzugeben

☐ ist das Material weiterzuleiten an _____

Für die Analyse   ☐ steht kein Geld zur Verfügung

☐ kann Rechnung gestellt werden an _____

NICHT AUSFÜLLEN	Analysiert	Bericht	Rechnung	Publikation	Analysen–Nr.	Bemerkungen:

FIGURE 29.2. Example of record sheet for wood and charcoal finds

Packing and storage

Non-carbonized wood and other botanical remains are best analysed in a moist condition, since otherwise the work becomes prohibitively laborious. Drying out of the material leads to shrinkage, sometimes to the point of disintegration, and greatly hinders analysis. All samples should therefore be packed in airtight plastic bags as soon as possible after collection, while they are still moist. Fungal attack is best prevented by storage at low temperatures, if possible in a refrigerator. If a long period is to elapse before analysis, a fungicide such as thymol or phenol crystals should be added, but only if the analysis does not include ^{14}C dating.

Charcoal, on the other hand, is investigated in the dry state and should be allowed to dry out slowly after collection. Care must be taken not to subject it to any mechanical stress.

LABORATORY WORK

The species of wood from living trees can often be determined on the basis of macroscopic characteristics, particularly colour. In fossil, historical or carbonized material, however, the original colour has usually disappeared. Consequently, very few indigenous species or anatomical types can be definitely identified from their macroscopic features, although the following types of wood are easily distinguished:

coniferous
ring-porous broadleaf
diffuse-porous broadleaf

An exact species identification requires the inspection of microscopic features (Figure 29.3). To this end, microsections are cut by hand with a razor blade. With some practice it is possible to produce excellent results, and sectioning with a microtome is much more time-consuming and costly. Definite identification, however, is only possible through comparison with preparations from living material or with excellent photographs. A useful aid in such work is *Microscopic Wood Anatomy* by F. H. Schweingruber (1978). Using this book, which includes a description of how to prepare sections, it is possible to identify almost all indigenous species with ease.

For charcoal, microscopic examination is carried out on fracture surfaces instead of sections, with surface illumination.

TYPES OF ANALYSIS AND FINDINGS

The various ways in which wood and charcoal finds can be analysed and the information revealed can be divided into the following main groups.

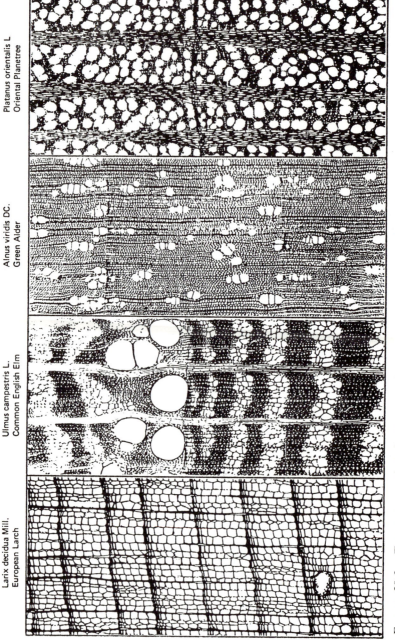

Larix decidua Mill.
European Larch

Ulmus campestris L.
Common English Elm

Alnus viridis DC.
Green Alder

Platanus orientalis L
Oriental Planetree

FIGURE 29.3. Transverse sections from one coniferous, one ring-porous and two diffuse-porous broadleaved trees (all at the same magnification)

Species identification

This forms the basis of all investigations of carbonized and non-carbonized wood, whatever its state of preservation, and permits inferences about vegetation patterns and selection by man.

Type of material

Each type of find provides a particular type of information:

    artefacts
    posts and stakes
    large pieces of wood in deposits
    charcoal, wood chips, twigs, bark, leaves, needles, seeds, remains of fruits.

Time of felling

The annual ring directly beneath the bark reveals the time of year at which the wood was cut from the living tree.

Occurrence of fungal hyphae and state of preservation

These factors provide information on the conditions to which the material has been exposed and the quality of both the deposit and the find site.

Form of wood and charcoal finds

In the case of settlements in lakes or rivers, or loess areas, the form of the find can supply information on stratigraphical development (e.g. transposal of sediments at the waterline).

Technological studies

A valuable supplement to typological studies on wooden utensils, technological analysis also helps trace the development of manual skills.

Vegetation analysis

Existing vegetation patterns can provide useful information for the reconstruction of the prehistoric environment.

Place of investigation

According to the aims of the study, investigations can be conducted at the find site, in a museum or, which is the normal case, in a laboratory.

APPLICATIONS

Wood and charcoal analysis yields valuable, indeed often the only, information on the vegetation growing around settlements and the general environment in prehistoric times. Further, it often indicates how prehistoric man related to his surroundings in his daily life.

Since wood and charcoal samples are not contaminated with organic substances during analysis, they can afterwards be used for ^{14}C dating. If the material is intended for dendrochronological or densitometric studies, it is advisable first to identify the wood species and, by microscopic examination, to determine the state of preservation and degree of decomposition of the material. With valuable samples such as artefacts or art objects, a section of 1 × 3 × 0.1 mm is sufficient. If possible, species identification should precede preservation, since the preserving media may obscure or even destroy the anatomical structure.

INTERPRETATION

In most cases it is necessary to inspect the find site in order to determine whether analysis of the finds will be profitable, and if so, which method will yield most information and how detailed the investigation should be. Further, the costs of the study can be estimated. At the same time, the existing vegetation, topography and geology of the location should be recorded to provide a basis for interpretation of the wood analysis results.

The full ecological or archaeological information contained in the find, however, can only be revealed by comparisons with the findings from other types of investigation, such as analyses of pollen, molluscs and bone deposits. Only through such interdisciplinary studies can the prehistoric environment be reconstructed as accurately as possible.

REFERENCES

Detailed information on wood structure, analysis of prehistoric wood, findings in the region of the Alps and current research can be found in:

Schweingruber, F. H. (1976a). Prähistorisches Holz: Die Bedeutung von Holzfunden aus Mitteleuropa für die Lösung archäologischer und vegetationskundlicher Probleme. *Academica Helvetica*, **2**, 1–106.
Schweingruber, F. H. (1976b). Veröffentlichungen über Untersuchungen prähistorischer Hölzer und Holzkohlen. *Courier Forschungsinstitut Senkenberg*.
Schweingruber, F. H. (1978). *Microscopic Wood Anatomy: Structural Variability of Stems and Twigs in Recent and Subfossil Woods from Central Europe*, Kommissionsverlag Zücher AG, Zug.
Schweingruber, F. H. (1983). *Der Jahrring: Standort, Methodik, Zeit und Klima in der Dendrochronologie*, Haupt, Bern. English edition in preparation.

30

Bryophyte analysis

J. H. DICKSON

Botany Department, University of Glasgow, Glasgow, U.K.

INTRODUCTION

Plants of the Bryophyta comprise the mosses (Musci), liverworts (Hepaticae) and hornworts (Anthocerotae). Bryophytes are abundant Quaternary subfossils existing as both dispersed spores and as macroscopic remains, fragments of leafy shoots (gametophytes) rarely with attached capsules (sporophytes). With the exception of *Sphagnum*, bryophyte spores are little known to palynologists, but macroscopic remains are reported by many workers. Remains of liverworts are much less frequent than those of mosses. Hornworts are only known by their distinctive spores, records of which have been increasing. Some mosses, especially *Sphagnum* but also several other genera, are major peat formers. Lake and river deposits, and even estuarine sediments, can be rich in moss remains not only of species which grew as aquatics but also of species transported from terrestrial habitats. Waterlogged archaeological layers often contain plentiful moss remains.

Abundant data from the British Isles, mostly of macroscopic remains but also of spores, have been summarized by Dickson (1973), who also mentioned results from other areas of Europe including Poland, Scandinavia and the Soviet Union. Odgaard (1981) has catalogued Danish Quaternary bryophytes. The North American evidence has been assessed by Miller (1976a, 1980a, b, c), who also listed results from Greenland (1980b).

SPORES

Dickson (1973) emphasized the difficulties in attempting to identify bryophyte spores, especially those of mosses. Even if the spores are well-preserved in large numbers, small size combined with minute granular sculpturing and lack of apertures militate against ease of recognition. The difficulty confronting the palynologist can be realized by examining the illustrations, especially the photomicrographs of exine surfaces, the spores of Mniaceae in Sorsa and Koponen (1973). Though there is no key, the atlas by Boros and Jarai-Komlodi (1975) is a compendium of mosses, liverworts and hornworts intended for anyone attempting this demanding task of identification, for which a reference collection of spores is indispensable. The relevant descriptions and illustrations in Erdtman (1957, 1965) are useful. Clarke (1979) has provided a summary of spore morphology in relation to systematics.

As anticipated in 1973, large and coarsely sculptured spores have been found and recognized, while the generality of moss spores has been overlooked or merely claimed without precise identification. Curves of unexplained 'Bryales' are common in the Soviet literature. However, Ukraintseva *et al.* (1978) have provided illustrations of five subfossil spores claimed as moss taxa: Bryales, *Dicranum*, *Polytrichum*, cf. *Pottia* and *Sphagnum*. Pennington *et al.* (1972) have curves for spores tentatively called 'bryophyte'; the spores are described as small, spherical, unornamented objects 5 μm or less in diameter, having lost their outer wall. While many bryophytes have small spores, few have spores less than 10 μm diameter and do not have a thick outer wall to lose. Publication of stereoscan electron-micrographs of an unknown spore (Pilcher, 1968) from early Flandrian peat led to the suggested identity of *Polytrichum* (Pilcher and Smith, 1979; Dickson, 1973).

Spores of the coprophilous moss *Aplodon wormskioldii* have been recorded by Brassard and Blake (1978). However, they were recognized not as dispersed spores but as part of a mass of the moss (gametophytes and sporophytes) which persisted for some 2000–2500 radiocarbon years in the mid postglacial. The large, strongly heteropolar, verrucose spores of the moss *Encalpta (cf. rhabdocarpa)* have been recovered from three late-glacial sites in Scotland (Dickson *et al.*, 1976). The considerable range of spore morphology in the Encalyptaceae has been well-illustrated by both Vitt and Hamilton (1974) and Jarai-Komlodi and Orban (1975). *Encalypta rhabdocarpa* and other members of the genus such as *E. vulgaris* and *E. alpina* are calcicoles of bare rock and soil. Many species are well-known in arctic and alpine habitats. The occurrence of *Encalypta* spores in many more glacial deposits is to be expected.

The greater variety of liverwort and hornwort spores compared with moss spores is readily apparent in many recent taxonomic articles illustrated with stereoscan electron-micrographs. However, this has not led to a diversity of Quaternary records as yet. There appear to be no records of taxa addition to those listed in 1973. Postglacial records of the large, trilete spores of *Anthoceros* and *Phaeoceros* (the principal genera of hornworts) have increased. Spores of *Anthoceros punctatus* agg. have been found in Czechoslovakia (Rybnickova, 1973), the Netherlands (Casparie, 1976; van Geel *et al.*, 1981) and north-western Germany (Behre, 1976) as well as at five Scottish sites including the Orkney Islands (Dickson, unpublished). *Phaeoceros laevis* spores have been found in Czechoslovakia (Rybnickova, 1973) and the three sites in the Netherlands (Casparie, 1976; van Geel *et al.*, 1981). The prostrate thalli of *Anthoceros* and *Phaeoceros* occur on bare soil in such habitats as stubble fields, marshy ground, clayey banks and dune slacks. Being associated with appropriate pollen types the spores from the German, Dutch and Czechoslovakian sites can be taken as indicative of agricultural activity. By contrast, coming from dune slack deposits, two of the Scottish occurrences need not have been derived from vegetation influenced by human activity. Hornworts are not plants of arctic or alpine vegetation; the subfossil spores, all of postglacial and interglacial age, are indicative of temperate climates.

MACROSCOPIC REMAINS

Laboratory work

The methods of sampling and extraction are similar to those employed in the study of vascular plant remains. Leafy shoots or detached leaves can be readily mounted in gum chloral (Hoyer's solution — Anderson, 1954) or gelvatol on microscope slides. Cavity slides are useful for thick specimens. Material can also be kept in liquid preservatives in tubes. In these ways subfossils can be stored satisfactorily for many years.

Identification

Standard manuals of bryophyte taxonomy are very useful in identifying macroscopic remains (see Appendix). However, a good bryophyte herbarium is essential for reference. Fragments of well-authenticated herbarium material can be kept on slides like subfossils for easy consultation.

The identification of bryophyte remains, especially those retrieved from lake deposits, is beset with a difficulty not met in the study of other types of plant remains. Growth may have continued on the lake bed before preservation. Bryophytes, including inwashed terrestrial species, are now very well-known as components of the vegetation deep in lakes, as in Scotland (Light and Lewis-Smith, 1976) and in Norway (Malme, 1978). Growth in such habitats can produce atypical shoots with the usual highly recognizable characters reduced or absent. Priddle (1979) has described the great plasticity of submerged *Calliergon sarmentosum*, a not infrequent Quaternary subfossil.

Apart from this special difficulty which may limit identification, in studying subfossil material which may be tiny scraps the investigator does not have at his disposal details of habitat, colour and habit, all of which aid the identification of living bryophytes. Moreover, sporophytes are seldom found. Furthermore, the state of preservation may be poor with few or no intact leaves. The index of reliability of identification concerning preservation, devised by Dickson (1973), has been followed by Janssens (1977) and by Odgaard (1982), and later elaborated (Janssens, 1983a; Janssens and Zander, 1980; Schweger and Janssens, 1980).

Few Quaternary bryophytes have been described as extinct species and many are questionable, perhaps based on ignorance of the great phenotypic variability of many mosses (Dickson, 1973; Kuc, 1974; Miller, 1980a). There are no reliably founded extinct mosses from the late-glacial and postglacial of Europe. Therefore, this is not likely to be a difficulty unless the investigator is studying early Quaternary or Tertiary material.

Despite these pitfalls, the careful investigator, prepared to gain a working knowledge of the living bryophytes of his area, has little difficulty in identifying most Quaternary bryophytes.

Any investigator who finds macroscopic remains of a liverwort is experiencing a rare, happy accident which may well have added to the scant knowledge of Quaternary Hepaticae (see Figure 30.1). In more than 20 years of work on Quaternary bryophytes the author has encountered less than 10 species of liverworts (out of 285 British species). This compares with about 220 out of 675 species of British mosses. The liverwort most recently found by the author was a well-preserved but small and very dark fragment of *Frullania*. It was extracted from mid-postglacial sediments of the north basin of Loch Lomond, north-west of Glasgow, but it completely disintegrated on transfer from one microscope slide to another! No other subfossil of *Frullania* has ever been recovered from the British Quaternary. Well-preserved material of *F. tamarisci* has been extracted recently from the postglacial of north-eastern Spain (Cartañá, 1983, Cartañá and Casas, 1984).

This grossly disproportionate lack of liverworts has been emphasized by Miller (1980a) in discussing North American data. For the Pleistocene he gives 132 Musci and 4 Hepaticae and for the Holocene 24 Musci and 2 Hepaticae. Whereas mosses are abundant in late-glacial sediments and are major components of postglacial peats, it is impossible to predict the occurrence of liverworts in Quaternary deposits of particular types or ages. Even the liverworts characteristic of mire surfaces have very few records; they might have been expected to be frequent but not necessarily abundant components of *Sphagnum* peats. However, it has recently become likely that peat stratigraphers may have overlooked the very small liverworts of the genus *Kurzia* von Martens. [*Lepidozia* Dum; *Microlepidozia* (Spruce) Joerg; *Telaranea* K. Mull.]. Overbeck (1975), quoting the work of Klinger, has recorderd *Kurzia pauciflora* (Dicks) Grolle (as *T. setacea*) from raised bog peat at Dätgen, Schleswig-Holstein, Germany, and van Geel (1976) has recovered the same species (as *Microlepidezia setacea*) from many samples of a peat deposit at

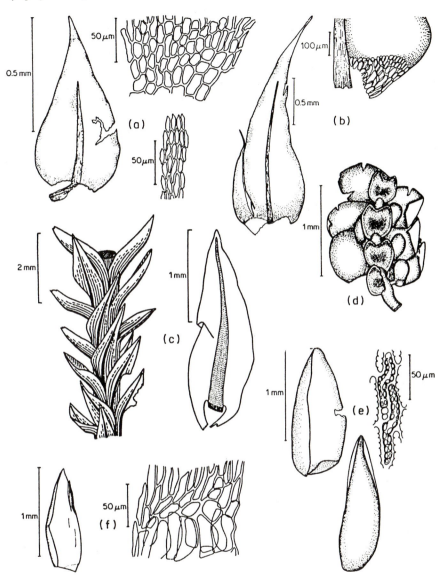

FIGURE 30.1. Macroscopic remains of bryophytes from the postglacial period of Pla de L'Estany, Catalonia, Spain (Cartana, 1983):

(a) *Amblystegium humile*, leaf and leaf base and mid-leaf cell details
(b) *Drepanocladus aduncus*, leaf and details of leaf base
(c) *Meesia longiseta*, leafy stem fragment and single leaf
(d) *Frullamia tamarisci*, leafy stem fragment
(e) *Sphagnum* section *Subsecunda*, two leaves and cell details
(f) *Calliergon cuspidatum*, leaf and details of leaf base.

Engbertsdijksveen in the Netherlands. The author has recently recovered a *Kurzia* species from peat at Carstairs Kames, Strathclyde, Scotland. On ecological grounds the most likely species is *K. pauciflora.* However, both *K. trichoclados* (K. Mull.) Grolle and *K. sylvatica* Evans can also grow in peat-forming vegetation, sometimes with *K. pauciflora.* Separation of these three species may well be difficult if not impossible on subfossil material which lacks inflorescence characters. *Kurzia* undiff. or K. cf. *pauciflora* would be a safe determination.

Tables 30.1, 30.2 and 30.3 show the mosses which have been most commonly recovered from Quaternary deposits in the British Isles, Denmark and North America (with Greenland). Since Miller's compilation, a thousand new records of North American subfossil mosses have been made by Janssens (1983b) who, following many previous workers, stressed the great frequency of *Drepanocladus.* These lists give a guide to the species with which the beginner must become familiar. Amblystegiaceae and Brachytheciaceae will give the most trouble. Peat stratigraphers will most often encounter *Aulacomnium palustre,* Amblystegiaceae, *Homalothecium (Camptothecium, Tomenthypnum) nitens, Dicranum,* Hypnaceae, Meesiaceae, Polytrichaceae, Thuidiaceae including *Helodium,* and *Sphagnum.* Palaeolimnologists should recognize the aquatic genus *Fontinalis* with fifteen Quaternary localities in the

TABLE 30.1.   Mosses with most (>20) Quaternary localities in the British Isles

BRYOPSIDA

Amblystegiaceae

Calliergen cuspidatum	52	*Hypnum cupressiforme*	52
Calliergon giganteum	59	*Pleurozium schreberi*	32
Calliergon stramineum	25	*Rhytidiadelphus squarrosus*	32
Campylium stellatum	36	Leucodontaceae	
Cratoneuron filicinum	31	*Antitrichia curtipendula*	55
Drepanocladus fluitans	28	Meesiaceae	
Drepanocladus revolvens	36	*Paludella squarrosa*	24
Scorpidium scorpioides	45	Neckeraceae	
Aulacomniaceae		*Neckera complanata*	36
Aulacomnium palustre	71	Polytrichaeceae	
Brachytheciaceae		*Polytrichum commune*	39
Eurhynchium praelongum	24	*Polytrichum alpinum*	37
Homalothecium nitens	45	Thuidaceae	
Climaciaceae		*Thuidium tamariscinum*	
Climacium dendroides	24		41
Dicranaceae		*SPHAGNOPSIDA*	
Dicranum scoparium		Sphagnaceae	
Grimmiaceae		*Sphagnum cuspidatum*	32
Racomitrium lanuginosum	34	*Sphagnum imbricatum*	56
Hypnaceae		*Sphagnum palustre*	41
Hylocomium splendens	66	*Sphagnum papillosum*	43

TABLE 30.2.   Bryophytes recorded from eight or more chronostratigraphic divisions of the Danish Quaternary (Odgaard, 1981)

BRYOPSIDA		Climaciaceae	
Amblystegiaceae		*Climacium dendroides*	9
Calliergon cordifolium	8	Hypnaceae	
Calliergon cuspidatum	8	*Hylocomium splendens*	9
Calliergon giganteum	12	Mniaceae	
Campylium stellatum	8	*Plagiomnium affine*	9
Drepanocladus aduncus	12	Polytrichaeceae	
Drepanocladus exannulatus	17	*Polytrichum alpestre*	10
Drepanocladus intermedius	10	*Polytrichum juniperinum*	9
Scorpidium scorpioides	17		
Aulacomniaceae		**SPAGNOPSIDA**	
Aulacomnium palustre	17	Sphagnaceae	
Brachytheciaceae		*Sphagnum plaustre*	11
Homalothecium nitens	10		

TABLE 30.3.   Bryopsida with most (>10) Quaternary localities in North America (Miller, 1980b)

Amblystegiaceae		Brachytheciaceae	
Calliergon giganteum	18	*Homalothecium nitens*	16
Calliergon trifarium	10	Bryaceae	
Campylium stellatum	20	*Bryum pseudotriquetrum*	19
Drepanocladus aduncus	16	Ditrichaceae	
Drepanocladus exannulatus	26	*Distichium capillaceum*	10
Drepanocladus fluitans	14	*Districhum flexicaule*	14
Drepanocladus revolvens	24	Hypnaceae	
Scorpidium scorpioides	20	*Hylocomium splendens*	10
Scorpidium turgescens	14	Meesiaceae	
		Meesia triquetra	11

British Isles and three in North America, but perhaps often overlooked in both areas.

Presentation of results

Counts of leafy shoots and/or detached leaves can be made per unit volume, as in the studies of allochthonous (inwashed) material in postglacial sediments by Birks (1970) and of rich allochthonous assemblages in late-glacial sediments by Burrows (1974). Care should be taken in gentle sieving to minimize the breakup of shoots and the tearing off of leaves. Mörnsjö (1969), working on autochthonous deposits (preservation *in situ*, mostly peats), and Hall (1980) working on last interglacial peat and mud, used a six-point scale of abundance.

Birks and Mathewes (1978) constructed a moss diagram through muds and silts of late-glacial and early postglacial age (Figure 30.2). They used a five-point scale of abundance, as did Aaby and Jacobsen (1979) in a study of *Sphagnum* and other moss leaves in raised bog peat. Using the most elaborate methods yet devised to estimate numbers of fragments per 100 ml of sediment, and to evaluate preservation indices, Janssens (1983a) has produced moss diagrams from postglacial peat. Working on archaeological assemblages, Seaward and Williams (1976) estimated species volumes, as did Dickson (1981). For an exceptional occurrence Miller (1976b) was able to estimate areas (percentage cover) in a study of *in situ* subfossil mossy forest litter.

BRYOGEOGRAPHICAL INFERENCES

Many studies of Quaternary bryophytes have concentrated on understanding bryophyte distribution patterns (Dickson, 1973; Miller, 1976a). Climatic and vegetational changes have been postulated to explain the disjunct patterns shown by many bryophytes, some of which are well-known as Quaternary subfossils. Though bryophyte spores are microscopic, long-range dispersal need not be invoked to explain disjunctions which occurred in the Quaternary; perhaps, however, it needs consideration to explain disjunctions thought to have occurred in pre-Quaternary times (Miller, 1982; van Zanten and Pocs, 1981). This interpretation of Quaternary disjunct distributions follows exactly the interpretation of similar vascular plant patterns (Godwin, 1975). Dickson (1973) drew particular attention to the Tertiary-Quaternary transition, the glacial–interglacial cycle, and especially the late-glacial–postglacial transition in changing bryophyte distributions.

The disjunction and extinction of peat-forming mosses such as *Paludella, Meesia* and *Homalothecium nitens*, the northern continental rich-fen relicts of Dickson (1973), have been interpreted in terms of changing trophic state and vegetation of the mires mostly in the early and middle postglacial period with recent human destruction of the mires as the final blow (Warncke, 1980).

PALAEOCLIMATIC AND PALAEOENVIRONMENTAL INFERENCES

Many bryophytes have very extensive geographical ranges and wide ecological amplitudes, and many of the Quaternary subfossils most commonly found are examples. A particularly good instance is *Hylocomium splendens,* a bipolar species very extensively distributed in the circumboreal and circumpolar regions of the Northern Hemisphere. It reaches the highest northern latitudes and inhabits coniferous and birch forests, *Dryas* and *Calluna* heath, calcicolous grassland in southern Britain and glacial outwash in central Alaska (Dickson, 1973). Accordingly this species and many others are poor indicators of past environments and macroclimates (Miller, 1980c).

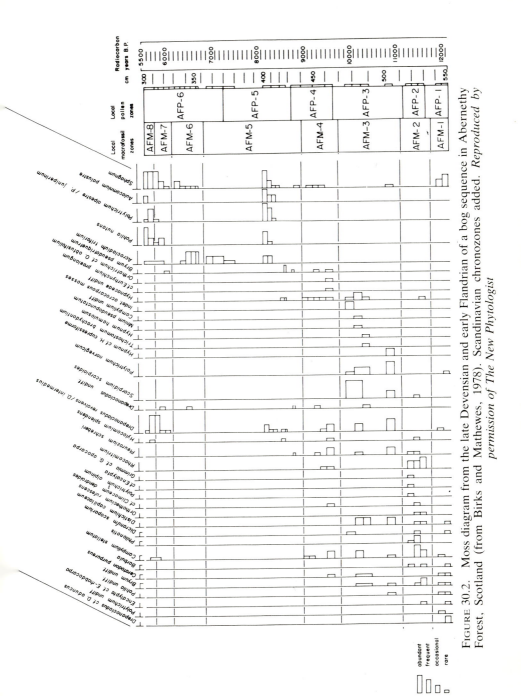

FIGURE 30.2. Moss diagram from the late Devensian and early Flandrian of a bog sequence in Abernethy Forest, Scotland (from Birks and Mathewes, 1978). Scandinavian chronozones added. *Reproduced by permission of The New Phytologist*

Odgaard (1980) has claimed a correlation between low summer temperatures (16°C July isotherm or less) and the present occurrence of *Polytrichum alpinum*. This is one of the commonest late-glacial mosses in Britain, with 37 localities mostly at low altitudes and very widespread but mainly in the highland zone. This moss may be a good indicator of late-glacial summer temperatures.

However, on the basis of assemblages rather than single species climatic deductions are more convincing, as has been well-demonstrated by Odgaard (1982). The site in question is Hirtshals, northern Denmark. Odgaard dissected macroscopic subfossils from 15 cm^3 of slightly calcareous sand from a layer of radiocarbon dated at 47,300 B.P. This allochthonous assemblage used for climatic inference totalled 29 mosses and one vascular plant (*Betula nana*). He deduced that the climate was not oceanic because of the recovery of several species absent from the north-western fringes of Europe. Then from the assessment of the present frequencies of these 30 taxa in the vegetation regions of non-oceanic Fennoscandia he considered the assemblage to be subarctic/low-arctic (subalpine/low-alpine) in character. He concluded that the mean July temperatures were between 8°C and 10°C.

Some bryophytes have very restricted habitats and are therefore good indicator species. *Polytrichum sexangulare* (Florke ex Hoppe) Brid. (*P. norvegicum* auct. non Hedw.) is a cogent example. This species is strictly confined throughout its range to base-poor soils where snow persists late into summer (usually over 1000 m, mostly on slopes of the north and east aspects in the Scottish Highlands). It has been recorded at nine British last-glacial sites, including five in the Scottish Highlands and lowland sites in Cornwall, the Isle of Man, Northumberland and south-east Scotland (Dickson, 1973, 1984; Birks and Mathewes, 1978; Webb and Moore, 1982). On the assumption that the ecological preferences have not changed, the inference is reasonable that acidic late snow bed vegetation was widespread in Britain towards the end of the last glaciation.

Miller (1980c), on analysing the growth-form spectra of three Late Wisconsin moss assemblages, has made deductions regarding substrata and climate. He concludes:

'The moss species involved are almost all characteristic of soils of low organic content. The predominance of short-turf species in continental areas of present-day boreal America and the prevalence of such species in the fossil assemblages add support for a continental late-glacial climate in the Great Lakes–New England area of North America.'

Referring to Hirtshals, Odgaard (1982) used growth-form analyses to stress the importance of dry exposed habitats. He pointedly refrained from climatic deductions from this data.

The richness of the British bryoflora is in large part accounted for by the large number of Atantic (or oceanic) species flourishing in the west while being absent or little known from the rest of Europe. The distribution patterns of these species have been correlated with wetness of climate, in particular the number of wet days (Ratcliffe, 1968). The subfossil history of these Atlantic species is meagre but a few interglacial and postglacial discoveries point to mild, moist climates as at present.

Though moss assemblages are of limited use in reconstructing past macroclimates, they convey cogently details of the microhabitat (Birks, 1980; Miller, 1980c; Odgaard, 1982). Mosses are good indicators of edaphic conditions, often giving clear indications of the trophic state of soil and water and of the mineral-organic content. Dickson (1973) and Miller (1980c) have drawn attention to strong representations of calcicolous species of mineral soil and rock and also of rich fens in last-glacial assemblages. However, calcifuge species are far from unknown in last-glacial deposits of Europe and *Polytrichum, Pleurozium schreberi* and a few species of *Sphagnum* are examples. However, not all *Sphagnum* species indicate oligotrophic conditions. Under present-day conditions in Britain (Hill, in Smith, 1978), *Sphagnum palustre*, section *Squarrosa, S. girgensohnii, S. warnstorfii*, most species of section *Subsecunda* and *S. riparium* grow in mesotrophic even eutrophic mires. The only *Sphagnum* species recorded from Middle Devensian Britain is *S. squarrosum* (Dickson, unpublished).

Moss assemblages can indicate the open or closed nature of past vegetation because some mosses are heliophiles and others are shade-tolerant. Last-glacial assemblages are often rich in the former and interglacial and postglacial assemblages in the latter (Dickson, 1973). The discovery of subfossil *Aulacomnium turgidum*, an arctic–alpine species abundant in the Arctic, has been used by Odgaard (1982) to postulate that climatic severity rather than the impossibility of immigration from too distant refugia accounts for the treelessness of the vegetation at Hirtshals.

Archaeological layers often yield informative moss assemblages. Viking and medieval towns were very mossy, mosses having been used for a variety of purposes including toilet paper (Dublin — Dickson 1973; Bergen — Krzywinski and Faegri, 1979; Aberdeen — Fraser and Dickson, 1982; Perth — D. E. Robinson, 1983, unpublished). Mosses extracted from well-preserved turves cut for constructional purposes can be very instructive. Störmer (1949) inferred a grass vegetation or grazing ground (Dark Age burial mound at Raknehaugen, near Oslo), and Dickson (1973) inferred a damp, weedy turf (Neolithic burial mound at Newgrange, Ireland). Very striking results have been published recently by Fritz and Wilmanns (1982), who recorded 30 moss taxa and deduced 'extensively grazed dry grassland (Mesobromion)' around a Celtic royal burial mound at Villingen, Germany.

The detailed analysis of moss assemblages, both allochthonous and autochthonous, in long sequences of Quaternary sediments can be used to great effect in revealing vegetational and environmental change (Birks, 1982). Birks and Mathewes (1978) have published a moss diagram from the late-glacial and early postglacial of Abernethy forest in the Grampian Highlands of Scotland (Figure 30.2). The allochthonous assemblages from the late-glacial sediments are the richer in species (31 taxa) and the autochthonous Flandrian assemblages the poorer (17 taxa). (Several mire species in AFM-5 may in part or all be contaminants from AFM-7 and 8.) Open conditions prevailed in late-glacial times with indications of base-rich flushes throughout. In Older Dryas times there were late snow beds and associated melt-water runnels close to the site. In Allerød times the snow beds receded (*Polytrichum sexangulare* absent) and shrub tundra developed as well as grass/sedge heath and sedge swamps. In Younger Dryas times the sedge swamps persisted (*Scorpidium scorpioides*) and the snow beds returned, to disappear finally in AFM-4 times with the closing of the vegetation and growth of *Betula* forest with *Populus tremula* on which *Orthotrichum obtusifolium* often grows. Later in the Flandrian mires, both rich (*Calliergon trifarium*) and poor (*Polytrichum alpestre*) developed on the site.

APPENDIX

Manuals for identification of bryophytes

It is impossible to list the great multiplicity of regional and national bryophyte manuals. The following is a selection of mainly modern books concerned with areas where most work is carried out.

Europe

Monkemeyer, W. (1972). *Die Laubmoose Europas. IV: Andreales–Bryales*, Akademische Verlagsgesellschaft, Leipzig.
Muller, K. (1954) and (1957). *Die Lebermoose Europas* (two volumes), Akademische Verlagsgesellschaft, Leipzig.

Great Britain and Ireland

Smith, A. J. E. (1978). *The Moss Flora of Britain and Ireland*, Cambridge University Press. Hardback and paperback editions are available. This is the standard British Moss Flora covering one of the richest bryofloras in Europe and hence applicable to more than the British Isles.
Watson, E. V. (1981). *British Mossese and Liverworts* (3rd edn), Cambridge University Press. Hardback and paperback editions are available. An excellent introductory book for beginners.
MacVicar, S. M. (1962). *The Student's Handbook of British Hepatics* (2nd edn),

Sumfield, Eastbourne. This authoritative work is now very out of date, but there is as yet no modern replacement for the study of the richest hepatic flora of Europe.

Scandinavia

Nyholm, E. (1954–69). *II Musci. Moss Flora of Fennoscandia* (six volumes), Gleerup, Lund. This is the standard authoritative work on moss flora covering the Scandinavian countries and Finland.

Arnell, S. (1965). *Illustrated Moss Flora of Fennoscandia. I: Hepaticae.* Gleerup, Lund. This is the standard work on hepatic flora covering the Scandinavian countries and Finland.

Denmark

Andersen, A. G., *et al.* (1976). *Den Dansk Mossflora. I: Bladmosser.* Glydendal, Copenhagen. This is a useful little book but does not include *Sphagnum.* In Danish.

The Netherlands

Landwher, J. (1974). *Atlas van der Nederland Bladmossen* (2nd edn), K.N.N.V., Amsterdam. In Dutch.

Landwher, J., Gradstein, S. R., and van Melick, H. (1980). *Atlas Nederlandse Levermossen*, K.N.N.V., Amsterdam. In Dutch.

Poland

Szfran, B. (1957) and (1961). *Flora Polska, Musci* (volumes 1 and 2), Polish Academy, Warsaw. In Polish.

Soviet Union

Abramova, A. L., Savicz-Ljubitzkaya, K. I., and Smirnova, S. N. (1961). *Moss Flora of the Arctic U.S.S.R.,* Academia Nauk, Moscow. In Russian.

Kats, N., Kats, S. V., and Skobeyeva, E. I. (1977). *Atlas of Plant Remains in Peat,* Nedra, Moscow. In Russian.

Savicz-Ljubitskaya, L. I., and Smirnova, Z. N. (1970). *Handbook of the Mosses of the U.S.S.R.: The Mosses Acrocarpous,* Nauka, Leningrad. In Russian.

North America

Crum, H. A., and Anderson, L. E. (1981). *Mosses of Eastern Northern America* (two volumes), Columbia University Press, New York.

Ireland, R. R. (1982). *Moss Flora of the Maritime Provinces,* Ottawa National Museum of Canada.

Lawton, E. (1971). *Moss Flora of the Pacific Northwest.* Nichiman; Hattori Botanical Laboratory.

Schuster, R. U. (1966–1980). *The Hepaticae and Anthrocerotae of North America East of the Hundredth Meridian* (four volumes), Columbia University Press, New York.

Sphagnum

Because of the importance of this genus in peat stratigraphy a list of reference works is appended. Most of the moss floras listed above deal with *Sphagnum*.

Coster, I., and Pankow, H. (1968). Illustrierter Schlussel zur Bestimmung einiger Mitteleuropaischer *Sphagnum*-Arten. *Wiss. Z. Univ. Rostock*, **415**, 286–323. This is very well-illustrated with many photomicrographs of leaves and stem sections.
Grosse-Brauckmann, G. (1974). Uber pflanzenliche Makrofossilien mitteleuropaischer Torfe. II: Weitere Reste (Fruchte und Samen, Moose u.a.) und ihre Bestimmungsmoglichkeiten. *Telma*, **4**, 51–117. This deals with some *Sphagnum* species and some other mosses. It is one of the very few works that has a key to peat-forming mosses.
Hill, M. O. (1978). Sphagnopsida. In: *The Moss Flora of Britain and Ireland* (Ed. A. J. E. Smith), Cambridge University Press, pp. 30–78. This is a modern, very useful, well-illustrated account.
Isoviita, P. (1966). Studies on *Sphagnum* L.: nomenclatural revision of the European taxa. *Ann. Bot. Fenn.*, **3**, 199–264.
Isoviita, P. (1970). Studies on *Sphagnum* L. II: Synopsis of the distribution in Finland and adjacent parts of Norway and the U.S.S.R. *Ann. Bot. Fenn.*, **7**, 157–162. This and the earlier paper is good for information on nomenclature and chorology.
Sonesson, M. (1966). *De Svenska Sphagnum-arternas systematik och ekologi.* (revision of duplicated booklet by N. Malmer, 1959), Lund University. In Swedish.

REFERENCES

Aaby, B., and Jacobson, J. (1979). Changes in biotic conditions and metal deposition in the last millenium as reflected in ombrotrophic peat in Draved Mose, Denmark. *Danm. Geol. Arbog 1978*, 5–43.
Anderson, L. E. (1954). Hoyer's solution as a rapid permanent mounting medium for bryophytes. *Bryologist*, **57**, 242–244.
Behre, K-E. (1976). Beginn und Form der plaggenwirtschaft in Nordwest-deutschland nach Pollenanalytishen Untersuchungen in Ostfriesland. *Neue Ausgrabungen und Forschungen in Nierdersachen*, **10**, 197–224.
Birks, H. H. (1980). Plant macrofossils in Quaternary lake sediments. *Ergebnisse der Limnologie*, **15**, 1–60.
Birks, H. H., and Mathewes, R. W. (1978). Studies in the vegetational history of Scotland. V: Late Devensian and early Flandrian pollen and macrofossil stratigraphy at Abernethy Forest, Inverness-shire. *New Phytol.*, **80**, 455–484.
Birks, H. J. B. (1970). Inwashed pollen spectra at Loch Fada, Isle of Skye. *New Phytol.*, **69**, 807–820.
Birks, H. J. B. (1982). Quaternary bryophyte palaeo-ecology. In: *Bryophyte Ecology* (Ed. A. J. E. Smith).
Boros, A., and Jarai-Komlodi, M. (1975). *An Atlas of Recent European Moss Spores*, Akademiai Kiado, Budapest.
Brassard, G., and Blake, W. (1978). An extensive subfossil deposit of the arctic moss Aplodon wormskioldii. *Can. J. Bot.*, **56**, 1852–1859.
Burrows, C. J. (1974). Plant macrofossils from Late-Devensian deposits at Nant Ffrancon, Caernarvonshire. *New Phytol.*, **73**, 1003–1033.
Cartañá, M. (1983). *Estudi dels briofits i alteres macrorestes semifossils al pla de L'Estany (Garrotxa)*, M.Sc. Thesis, Universitat Autonoma de Barcelona.
Cartañá, M. and Casas, C. (1984). *Meesia longiseta* Hedw. en una turbera del cuaternario superior en el Pla de L'Estany (Garrotxa, Girond). *Cryptogamie, Bryof. Lichénol.* **5**, 127–134.

Casparie, W. A. (1976). Palynological investigation of the Celtic field near Vaassen, the Netherlands. In: *Air Photography and Celtic Field Research in the Netherlands* (Ed. J. A. Brongers), Nederlands & Oudhenen, pp. 105–113.

Clarke, G. C. S. (1979). Spore morphology and bryophyte systematics. In: *Bryophyte Systematics* (Eds. G. C. S. Clarke and J. G. Duckett), Academic Press, London. pp. 231–280.

Dickson, J. H. (1973). *Bryophytes of the Pleistocene*, Cambridge University Press.

Dickson, J. H. (1981). Mosses from a Roman well at Abingdon. *J. Bryol*, **11**, 559–560.

Dickson, J. H. (1984). A pollen diagram from Straloch, Strathardle, South-east Scottish Highlands. *Scottish Field Studies 1984*, 33–38.

Dickson, J. H., Jardine, W. G., and Price, R. J. (1976). Three Late-Devensian sites in West-Central Scotland. *Nature*, **262**, 43–44.

Erdtman, G. (1965). *Pollen and Spore Morphology and Plant Taxonomy. Gymnosperms, Pteridophyta, Bryophyta*, Almquist and Wiksell, Stockholm.

Fraser, M., and Dickson, J. H. (1982). Plant remains. In: *Excavations in the Medieval Burgh of Aberdeen 1973–81* (Ed. J. C. Murray), Society of Antiquaries of Scotland Monograph, Edinburgh, pp. 239–243.

Fritz, W., and Wilmanns, O. (1982). Die Aussage raft subfossiler Moos-Synusien bei der Rekonstruktion eines keltischen Lebensraumes — Das Beispiel des Furstengrabhugles Magdaleneberg bei Villingen. *Ber, Deutsch. Bot. Ges.*, **95**, 1–18.

Godwin, H. (1975). *History of the British Flora*, 2nd edn, Cambridge University Press.

Hall, A. R. (1980). Late Pleistocene deposits at Wing, Rutland. *Phil. Trans. R. Soc. B.*, **298**, 135–164.

Janssens, J. A. (1977). Bryophytes from the Pleistocene of Belgium and France. *J. Bryol.*, 349–360.

Janssens, J. A. (1983a). A quantitative method for stratigraphic analysis of bryophytes in Holocene peat. *J. Ecol.*, **71**, 189–196.

Janssens, J. A. (1983b). Past and extant distribution of *Drepanocladus* in North America, with notes on the differentiation of fossil fragments. *J. Hattori Bot. Lab.*, **54**, 251–298.

Janssens, J. A. (1983c). Past and present record of *Drepanocladus crassiocostatus* sp. nov. (Musci : Ambylstegiaceae) and the status of *D. trichophyllus* in North America. *The Bryologist*, **86**, 44–53.

Janssens, J. A. P., and Zander (1980). *Lepodontium flexifolium* and *Pseudocrossidium revolutum* as 60,000 year-old subfossils from the Yukon Territory, Canada.

Jarai-Komlodi, M., and Orban, S. (1975). Spore morphological studies on recent European *Encalypta* species. *Acta Bot. Hung.*, **21**, 305–345.

Krzywinski, K., and Faegri, K. (1979). Etno-botanisk bidrag til funksjonsanalyse Eksempler fra middelalderundersøkelser i Bergen. *Arkeo*, **1**, 33–40.

Kuc, M. (1974). *Calliergon aftonianum* Steere in Late Tertiary and Pleistocene deposits in Canada. *Geol. Surv. Can Paper* 74–24, 1–8.

Light, J. J., and Lewis-Smith, R. I. (1976). Deep-water bryophytes from the highest Scottish locks. *J. Bryol.*, **9**, 55–62.

Malme, L. (1978). Floristic and ecological studies of bryophytes in some Norwegian inland lakes. *Norw. J. Bot.*, **25**, 271–279.

Miller, H. A. (1982). Bryophyte evolution and geography. *Biol. J. Linn Soc.*, **18**, 145–196.

Miller, N. G. (1976a). Quaternary fossil bryophytes in North America: a synopsis of the record and some phytogeographic implications. *J. Hattori Bot. Lab.*, **41**, 73–85.

Miller, N. G. (1976b). Studies in North American Quaternary bryophytes. 1: New moss assemblages from the Two Creeks forest bed of Wisconsin (June 1976), *Occ. Pap. Farlow Herb. Harvard Univ.* 9, 21–42.

Miller, N. G. (1980a). Fossil mosses of North America and their significance. In: *The Mosses of North America*, American Association for the Advancement Science, San Francisco, pp. 9–36.

Miller, N. G. (1980b). Quaternary fossil bryophytes in North America: catalogue and annotated bibliography. *J. Hattori Bot. Lab.*, **47**, 1–34.

Miller, N. G. (1980c). Mosses as palaeoecological indicators of late-glacial ʼerrestrial environments: some North American studies. *Bull. Torrey Bot. Club*, **107**, 373–391.

Mörnsjö, T. (1969). Studies on vegetation and development of a peatland in Scania, south Sweden. *Opera Botanica*, **24**, 1–187.

Odgaard, B. (1980). Ecology, distribution and late Quaternary history of *Polytrichastrum alpinum* (Hedw.) G. L. Smith in Denmark. *Lindbergia*, **6**, 155–158.

Odgaard, B. V. (1981). The Quaternary bryoflora of Denmark. I: Species list. *Danm. Geol. Unders. Arbog. 1980*, 45–74.

Odgaard, B. V. (1982). A Middle Weichselian moss assemblage from Hirtshals, Denmark, and some remarks on the environment 47,000 B.P. *Danm. geol. Unders Arbog 1981*, 5–45.

Overbeck, F. (1975). *Botanisich-Geologische Moorkunde*. Karl Wachholtz Verlag, Neumunster.

Pennington, W., Haworth, E. Y., Bonny, A. P., and Lishman, J. P. (1972). Lake sediments in northern Scotland. *Phil. Trans. R. Soc. Ser. B.*, **264**, 191–244.

Pilcher, J. R. (1968). Some applications of scanning electron microscopy to the study of modern and fossil pollen. *Ulster J. Archaeol.*, **31**, 87–91.

Pilcher, J. R., and Smith A. G. (1979). Palaeoecological investigations at Ballynagilly, a Neolithic and Bronze Age settlement in County Tyrone, Northern Ireland. *Phil. Trans. R. Soc. B.*, **226**, 345–369.

Priddle, J. (1979). Morphology and adaptation of aquatic mosses in an Antarctic lake. *J. Bryol*, **10**, 317–529.

Ratcliffe, D. A. (1968). An ecological account of Atlantic Bryophytes in the British Isles. *New Phytol.*, **67**, 265–439.

Rybnickova, E. (1973). Pollenanalytische Unterlagen fur die Rekonstruktion der ursprunglichen Waldvegetation im mittleren Teil des Ostava Bohmerwald-vorgebirges. *Folia Geobot. Phytotax.*, **8**, 117–142.

Schweger, C. E., and Janssens, J. A. P. (1980). Palaeoecology of the Boutellier Nonglacial Interval, St. Elias Mountains, Yukon Territory, Canada. *Arctic Alpine Res.*, **12**, 309–317.

Seward, M. R. D., and Williams, D. (1976). An Interpretation of mosses found in recent archaeological excavations. *J. Archaeol. Sci.*, **3**, 173–177.

Smith, A. J. E. (1978). *The Moss Flora of Britain and Ireland*. Cambridge University Press.

Sorsa, P., and Koponen, T. (1973). Spore morphology of Mniaceae (Bryophyta) and the taxonomic significance. *Ann. Bot. Fenia*, **10**, 187–200.

Störmer, P., (1949). Moser funnet i Raknehaugen ved utgraviningen 1939–40. *Blyttia*, **7**, 92–95.

Ukraintseva, V. V., Flerov, K. K., and Solonevich, N. G. (1978). Analysis of plant remains from the alimentary tract of Mylakhchinsk bison (Yakutia). *Botanical Journal*, **63**, 1001–1004 (in Russian).

Van Geel, B. (1976). *A Palaeoecological Study of Holocene Peat Bog Sections, based on Analysis of Pollen, Spores and Macroremains of Fungi, Algae, Cormophytes and Animals*. University of Amsterdam.

Van Geel, B., Bohncke, S. J. P., and Dee, H. (1981). A palaeoecological study of an upper Late Glacial and Holocene sequence from 'De Borchert', the Netherlands. *Rev. Palaeobot. Palyn.*, **31**, 367–448.

Van Zanten, B. O., and Pocs, T. (1981). Distribution and dispersal of Bryophytes. *Advances in Bryology*, 479–562.

Vitt, D., and Hamilton, C. D. (1974). A scanning electron microscope study of the spores and selected peristomes of the north American Encalyptaceae (Musci). *Can. J. Bot.*, **52**, 1973–1981.

Warncke, E. (1980). Spring areas: ecology, vegetation, and comments on similarity coefficients applied to plant communities. *Holarctic Ecology*, **3**, 233–333.

Webb, J. A., and Moore, P. D. (1982). The Late Devesian vegetational history of the Whitlaw Moses, south-east Scotland. *New Phytol*, **91**, 341–399.

Handbook of Holocene Palaeoecology and Palaeohydrology
Edited by B. E. Berglund
© 1986 John Wiley & Sons Ltd.

31

Rhizopod analysis

KIMMO TOLONEN

Department of Botany, University of Helsinki, Helsinki, Finland

INTRODUCTION

Rhizopod analysis refers to the use of subfossil shells (tests) of freshwater amoebae (Rhizopoda: Testacea (Protozoa)) as an index of palaeohydrological changes in peatlands and lakes. The application of rhizopod analysis to soil profiles seems difficult owing to the uneven and usually poor preservation of testacean remains in the prevailing conditions. The first investigations of subfossil testaceans published in Finland (Lindberg, 1899) and Sweden (Lagerheim, 1902) dealt with lacustrine deposits; it was not until recent years that the method was actually examined within the field of the palaeolimnology (Schönborn, 1973; Ruzicka, 1982). In mire stratigraphical studies, rhizopod analysis has a long tradition, and since its foundation (Steinecke, 1927) it has been very useful in clarifying the moisture changes during peatland history. The method was further developed by Harnisch (1927, 1951) and Grospietsch (1953) into a tool which has elucidated essential problems of Quaternary geology such as boundary horizons of raised bogs ('Grenzhorizont') and other recurrence surfaces as described by Frey (1964), and the regenerative growth of *Sphagnum* peat (Tolonen, 1971, 1979).

The real potential of the method (regarding peat) lies in the fact that many species of good indicator value are almost quantitatively fossilized and preserved, as well as easily identified. By examining qualitative and quantitative changes in peat deposits whose matrix is rich in rhizopod remains (e.g. Steinecke, 1927; Hoogenraad, 1935, 1936; Schmeidl, 1940), conclusions can be drawn about the character of the past environment. The method is most useful in cases where the parent plant community is not recognizable owing to the presence of highly decomposed peat (Steinecke, 1929). For further history of rhizopod analysis see Frey (1964), who also gives a summary of rhizopod finds from Pleistocene and Tertiary deposits (see also Martin, 1971).

Schönborn (1962a, 1964, 1967) put forward an interesting hypothesis about the evolution and natural history of Testacea. He postulated that the primary habitat of this group of organisms was the depths of great oligotrophic lakes of the Ice Age. In the course of their development, they occupied secondary biotopes (submersed vegetation of the littoral zone, mosses and mires, including raised bogs, soils etc.). In this primary environment, the evolution and diverse morphoecological adaptations of the protozoa took place.

As in the case of diatoms, many contradictory records and opinions have been presented about the ecology of lake and peatland rhizopods. Only a few measurements obtained by means of field studies have so far been published (Heal, 1964; Meisterfeld, 1977). Difficulties in applying controlled culture experiments have not been overcome (see, however, Heal, 1964a). All rhizopod specialists agree that the main ecological factor explaining variations in the distribution and abundance of Testacea outside aquatic habitats is moisture, the secondary one being the availability of humus (detritus). In lakes, however, the nutritional level (trophy) and the humosity grade (dystrophy) seem to be the properties which mostly determine their recent rhizopod assemblages (Schönborn et al., 1965; Schönborn, 1966a, 1967). Ruzicka (1982) demonstrated that in certain circumstances the sedimentary rhizopod remains provide further valuable information about the water-level fluctuations and the occurrence of 'peak rains' around the lake.

Unfortunately, ecological interpretations of rhizopods are hampered by confusions in the taxonomy of several species (e.g. de Graaf, 1956; Heal 1963; Hoogenraad and Groot, 1940; Paulson, 1954; Meisterfeld, 1979a), but also by the fact that there are too few similarly collected data on the occurrence of Testacea along measured ecological gradients in different areas (Meisterfeld, 1979b).

Rhizopod analysis of mires and lakes will be discussed separately owing to the abovementioned fundamental ecological differences and to the different preservations of rhizopod shells in peat and sediment.

RHIZOPOD ANALYSIS OF MIRE DEPOSITS

General

Testaceans are the most common group of one-celled organisms living in surface peat, and according to Heal (1962) their numbers are about 16×10^6 individuals per m^2, which corresponds to a biomass of about $1g/m^2$. (Meisterfeld (1977) gives even greater figures.)

The use of rhizopod analysis is most productive when combined with microscopic analyses of plant remains of peat, such as the identification and numerical estimation of *Sphagnum* leaves and the counting of fungal hyphae (Olausson, 1957; Steinecke, 1927; Tolonen, 1966, 1968, 1971). Thus rhizopod analysis can help to construct a reliable and much more detailed picture of the successional changes of mire surfaces than can a peat analysis alone, for example, in a case when a single *Sphagnum* species (often *S. fuscum*) has been responsible for the peat growth throughout several millennia (Tolonen, 1971). Thus one might not recommend the use of rhizopod diagrams separately as an independent study or as one addition to pollen counts, without a simultaneous analysis of plant remains.

Successful rhizopod analysis is quite often restricted to *Sphagnum* peat (or *Sphagnum*-containing mixed peats). The abundant occurrence of 'useful' rhizopod species is concentrated in ombrotrophic mires (= bogs). This species group contains most of the 'pseudochitineous' tests resistant enough to be preserved against humification and mineralization over thousands of years. Many genera whose tests are mainly composed of siliceous plates and mineral particles are largely restricted to minerotrophic mires (= fens), where they dissolve or break down during the decomposition process.

Montane mires with exceptional amounts of melt-waters obviously deviate from this pattern, as seen from the study by Laminger *et al.* (1981) on the testacean necrocoenoses in diverse minerotrophic peat layers of a small alpine mire in Tyrol.

Ecological factors

Water content

All testaceans are small (10–250 μm) water organisms able to live only in an aquatic environment. The water bodies which they inhabit can be very small and, in extreme conditions such as in mineral soils or in lichen carpets, can be dry even for half a year. Many morphological adaptations can be seen in species found in the driest habitats. Small size and the development of a flat 'ventral' surface enable movement in restricted water films, while, for example, cryptostomy and plagiostomy belong to adaptations which prevent desiccation (Heal, 1963; Schönborn, 1967). The importance of the water factor in testacean ecology is emphasized by the diminishing content of algae suitable as their food within, for example, the bog pool to hummock succession (Heal, 1964).

The different moisture requirements of certain peatland-inhabiting Testacea form the foundation of rhizopod analysis of peat. Nevertheless, most of the testacean species of the secondary biotopes (for example, of mires) are still widely euryplastic organisms in relation to moisture. Within their tolerance range, however, many of the 'peatland species' clearly exhibit optimal to maximal occurrence within fairly narrow limits. Accordingly, many species are stenotope enough to be placed into certain 'moisture' groups. In recent ecological studies, a I–VIII scale (Jung, 1936) has been commonly used for practical estimation of the water content of the moss layer. The corresponding rhizopod groups of de Graaf (1956) and the average water content of the substrate (percentage of wet-weight) according to Meisterfeld (1977) are as follows:

	I	Open water or submerged vegetation, >95%
Hygrophilous	II	Floating vegetation, partly submerged, partly at the surface, >95%
	III	Emerged vegetation, very wet, water drops out without pressure, >95%
α-hydrophilous	IV	Wet, water drops out with weak pressure, ~95%
	V	Half-wet, water drops out with moderate pressure, 95–85%
β-hydrophilous	VI	Moist, water drops out with strong pressure, 85–90%
Xerophilous	VII	Half-dry, a few drops with strong pressure, <80%
	VIII	Dry, no water drops with strong pressure, <50%

Figure 31.1 shows a scheme of a distinct horizontal variation of testacean fauna at different microsites within a small bog, according to Schönborn (1962b).

The application of this classification in the interpretation of rhizopod necrocoenoses ('the community of the dead', i.e. the assemblages of dead individuals) might be more preferable than the use of, ecologically speaking, rather inhomogenous and 'broad' rhizopod association types of Harnisch and Grospietsch (Harnisch, 1927, 1951; Grospietsch 1953). (See also the discussion in Meisterfeld (1979b).)

Nutrient status

The commonly used peatland rhizopod association types of Harnisch include '*Amphitrema flavum* type' and '*A. wrightianum* type' which were described as present 'most widely in bogs' (Hochmoore) or 'only in old, well-developed raised bogs', respectively (Grospietsch, 1953, 1958; Harnisch, 1927, 1951). In the light of later studies, it is evident that no species are confined only to ombrotrophic sites (raised bogs — Hochmoore), even though *Bullinularia indica* (Pen.) Defl. and a couple of associated testaceans from about a hundred

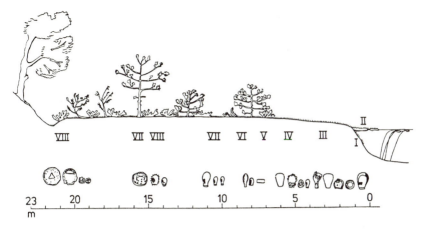

FIGURE 31.1. Horizontal distribution of Testacea within a small forested bog along the moisture gradient corresponding to the moisture classes I–VIII. The species from left to right: *Trigonopyxis arcula, Centropyxis orbicularis, Phryganella hemisphaerica* (lateral and frontal view), *Bullinula indica, Euglypha compressa, Corythion dubium, Nebela collaris, N. minor, N. militaris, Hyalosphenia elegans, Assulina muscorum, Amphitrema flavum, Hyalosphenia papilio, Heleopera sphagni, Centropyxis minuta, Nebela minor, Difflugia bacillifera, Hyalosphenia papilio, Difflugia globulosa* (lateral and frontal view), *Nebela carinata. Redrawn from Schönborn (1962b) by permission of Akademie Verlag*

rhizopod taxa were met only in bog biotopes in two English studies (Heal, 1961, 1962). On the other hand, altogether 10 species from the long lists of mire rhizopods occurred only in fen sites. In the same studies, the species group which was largely restricted to sites with fen character consisted of 26 taxa, and the assemblage primarily occurring in bog sites consisted of 13 taxa. Where *Sphagnum* grows above water level in fen vegetation, a 'typical bog hummock' testacean fauna was found (Heal, 1961, 1962).

The few laboratory experiments so far published about the ecology of Testacea suggest that the chemical composition of mire water, and particularly its pH value (plus, in some cases at least, the silica concentration) are more essential in separating fen from bog-pool species than is the fulfilment of their food requirements or the presence of suitable test material.

In some extreme cases, the absence of detritus from a habitat might prevent species which unambiguously need detritus for their tests from occupying such sites; but these conditions probably prevail only at the microhabitat level, according to Meisterfeld (1977). The fundamental explanation for the superabundance of such important 'autochthonous peatland Testacea' as *Amphitrema* (Schönborn, 1962b) present in bog hollows as opposed to their presence in fen sites with similar moisture conditions to the bogs, is still open (see, however, Schönborn, 1966b, 1967). The decreased composition from the usually larger and more aquatic fen rhizopods and other related organisms obviously provides a partial explanation.

Vertical distribution

In very wet *Sphagnum* carpets, no living Testacea can be found below 10–15 cm from the surface (Meisterfeld, 1977); in drier and denser hummocks the limit rises still higher, approximately corresponding to the thickness of the brown, undecomposed part of the peat mosses (Schönborn, 1963). The lack of rhizopods in the deeper peat layers is in part due to the unfavourable oxygen conditions in the interspace and in interstitial water. The species with symbiotic zoochloroellae have their vertical maximum in the topmost centimetres, probably because of light requirements, while most of the other species have their maxima at a depth of 6–12 cm (Meisterfeld, 1977). No seasonal differences in the vertical distribution of Testacea have so far been observed.

Below the 15 cm level, the surface testacean biocoenoses gradually change to necrocoenoses, which more or less agree with the composition of the living assemblages. To what extent this correspondence is valid depends on:

(1) the resistance of the tests against decomposition. The resistance varies from 0 to 100%, depending on the test material of the organisms (tests of siliceous plates are very soluble in peats);

(2) the amount of down-washing of empty shells. Few quantitative estimates have been made of this process in recent ecological studies (Heal, 1961). There is a significant horizontal gradient both in the species composition and in the abundance of Testacea from hollows to adjacent hummocks. This is valid for recent biocoenoses and for fossilized necrocoenoses (Tolonen, 1968). Therefore, it is evident that this down-core filtration error cannot be very great. Jung (1936) suggested that *Phryganella hemisphaerica* and *Assulina muscorum* are enriched in peat.

The superimposition of the verticality in the testacean taxocoenoses of *Sphagnum* must be emphasized (Figure 31.2; Schönborn, 1963; Meisterfeld, 1979b), and kept in mind when collecting material for recent ecological testacean associations. Most studies have neglected the fundamental difference between the upper and lower horizons, resulting in strongly misleading and often contradictory results in terms of the association types (Meisterfeld, 1979b). Nevertheless, there have been no attempts so far to apply the six Schönborn's associations (Figure 31.2), nor to apply the 25 Meisterfeld's (1979b) taxocoenoses for subfossil rhizopod assemblages. Meisterfeld's (1979b) result is interesting in that his cluster-analytically differentiated testacean associations (characterized by their dominants) could not be

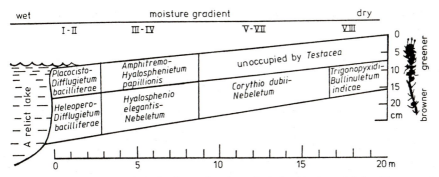

FIGURE 31.2.   Scheme of the horizontal and vertical distribution of living Testacea in *Sphagnum* sward of a raised bog. *Redrawn from Schönborn (1963) by permission of Academie Verlag*

explained by the variability of abiotic habitat conditions owing to the high coenotic variability.

'Evolution'

Harnisch (1951; see also Frey, 1964) has suggested that the biotope dependence of peatland Testacea is not likely to have developed until about the middle Atlantic time, and therefore it would not be possible to interpret faunal lists from older periods in the light of the present ecology of Testacea. This idea has been supported by Schönborn in his numerous studies (e.g. 1962b, 1966b, 1967). A simple explanation of this problem may be that there has been a long-term 'natural' development of mire habitats since the last glaciation, as clarified by many detailed peat-stratigraphical studies. The moisture requirements of '*Amphitrema wrightianum*-type', which nowadays in central Europe is restricted to well-developed raised bogs (according to Harnisch), were earlier met in the wet minerotrophic fen stages both in central Europe and in Finland. (Note the discussion on *A. wrightianum* later in this chapter; also Schmeidl (1940), Meisterfeld (1977) and Laminger *et al.* (1981).)

Sampling, storage and laboratory work

It is best to use fresh peat material and the following recommendations are for such samples. If open peat sections are not available, peat monoliths obtained either by excavation or by a large-capacity peat sampler (for example, a Russian peat sampler) are preferred. Frozen storage is preferable.

In the laboratory, subsamples of about 3 cm^3 are taken from uncontaminated fresh peat samples for humification and peat constituent determinations. The sample interval should be adjusted according to peat quality and may vary depending on the purpose of the study (e.g. Grospietsch, 1953; Tolonen, 1971).

Preparation of slides can be done according to the processing steps shown in the flowchart of Figure 31.3.

A new pipette is preferred for each sample, as it is often difficult to wash testaceans well enough to avoid contamination. From each sample, three replicate slides (= cover glasses) are made using cover glasses of 18 × 18 mm. The total number of rhizopod shells (with magnification of × 150–200) should be expressed per 5 slides (Harnisch and Grospietsch) or per 2.5 slides (Tolonen).

This type of rhizopod analysis is to a large degree only semi-quantitative. However, experience has shown that the method is satisfactory, despite errors in numeric values due to technical inaccuracy. A simple explanation for this conclusion lies in the enormous primary differences between the Testacea found in different microhabitats. On the other hand, it is extremely difficult to elaborate on a truly absolute analysis for any kinds of microfossils in peat strata owing to the fact that it is impossible to determine the exact accumulation rate of peat.

A rather common practice is to count rhizopod frequences as percentages from some basic pollen sums in connection with pollen analysis. This has, however, a limited value for two reasons. These are the selective destruction of the tests in the chemical treatment necessary for pollen preparations (when only a couple of species remain), and the fact that the origin, distribution and accumulation of pollen grains are quite different from that of the extremely 'local' moss-inhabiting Testacea (Grospietsch, 1965a; and own observations). The latter error can be avoided when exotic markers (e.g. Stockmarr's *Lycopodium* spores) are used (Tolonen, 1979).

Construction and interpretation of diagrams

Owing to the paucity of reliable quantitative data about the relation of Testacea to different ecological factors (Meisterfeld, 1977), the vertical distribution of rhizopod remains has usually been presented alphabetically in rhizopod diagrams. One exception is the group of 'tyrphoxene' taxa, which is often shown on the extreme right. This group consists of the species *Bullinularia indica, Hyalosphenia subflava, Trigonopyxis arcula* and *Phryganella hemisphaerica* (Grospietsch, 1953). They are not all indigenous to bogs, but are, instead, characteristic of drained peatlands and those overgrown with heather, and of peatland marginal areas (Grospietsch, 1953).

This kind of rhizopod diagram greatly resembles a pollen diagram; examples are shown in Grospietsch (1953, 1958) and Tolonen (1966, 1967, 1968, 1971). Stratigraphical column(s) showing the peat type and the degree of decomposition should be included. In addition, graphs denoting the relative abundance of fungal hyphae and microscopically unidentifiable organic debris (humus) are preferable, since all this information may assist in a correct interpretation.

In samples from open peat faces of raised bog in Finland, the author (Tolonen, 1971) found a cyclic vertical alteration of the abovementioned 'tyrphoxene' testacean association and 'wet' rhizopod species (like *Amphitrema flavum, Arcella artocrea, Hyalosphenia ovalis* and *H. papilio*). Correspondingly, these ecologically different necrocoenoses, in relation to moisture, exhibit a more or less close correlation with the peat humification as expressed in two independent ways: van Post's and Pyavchenko's method (see Tolonen, 1971). Despite the fact that many of the differentially decomposed layers were only a few millimetres thick, the following correlations were observed (Table 31.1).

TABLE 31.1. Correlation of certain testacean remains and the peat humification by two methods (von Post and Pyavchenko) in peat profiles of a Finnish raised bog (Tolonen, 1971)

	Post	Pyavchenko
Amphitrema flavum	-0.353^3	-0.241^1
Hyalosphenia papilio	-0.318^2	-0.215^1
Trigonopyxis arcula	0.205^1	-0.032
Bullinularia indica	0.496^3	0.217^1

Significance levels: $^1p < 0.05$, $^2p < 0.01$, $^3p < 0.001$ ($n = 95$ samples)

Hence, rhizopod analysis resulted in an interpretation for the outstanding striated structure of the *Sphagnum* peats in the given peatland as follows: the thin, black, strongly decomposed peat layers originate from stages when the peat surface was relatively dry and the interspacing weakly decomposed *Sphagnum* sect. Acutifolia peat from moister stages, respectively. An analogous streaked pattern is widely seen, at least in *Sphagnum* peats of the northern amphiatlantic peatlands (author's observations). In coastal Maine (USA) and New Brunswick (Canada), rhizopod analysis evidenced the generality of the short-term regeneration features of *Sphagnum* peat, which explain these great successive moisture variations (Tolonen, *et al.*, 1985). The basis, once again, was the present-day distribution of the peat rhizopods versus measurable ecological properties — and this appeared to be essentially the same as in Europe. In fact, rhizopod analysis remained the only practical method for solving (and measuring) in detail the palaeohydrological changes of the mires because of the impossibility of identifying *Sphagnum* species in the abovementioned American peat deposits (the peat being almost totally of *Sphagnum* sect. Acutifolia).

It may have to be emphasized that only when point profiles have been connected by means of open peat sections can more general conclusions about changes in humidity and about the possible linkage with the development of climate be considered (Aaby, 1975). In the case of the short-term bog

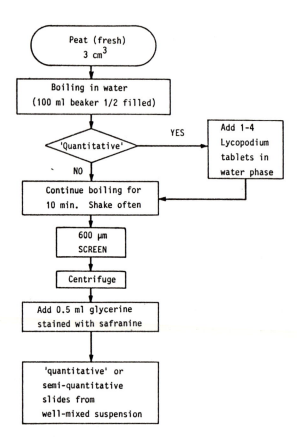

FIGURE 31.3. Flowcharts showing sequential steps in processing peats and sediment for recovery of testacean remains as used for peat by Grospietsch (1953, 1968, 1971 and for sediment by Ruzicka (1982). *Reproduced by permission of Akademie Verla*

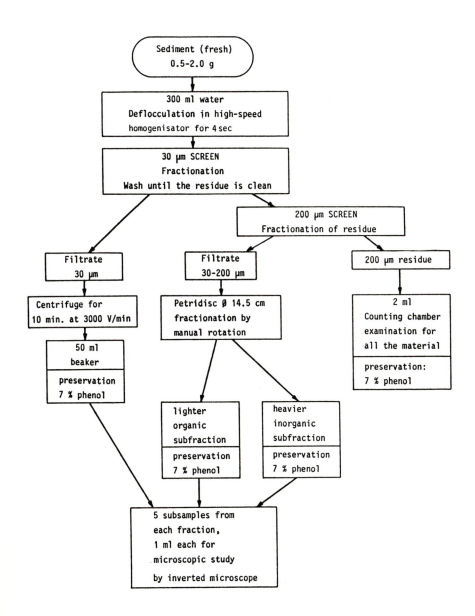

regeneration as discussed above which, contrary to Barber (1981), is obviously a very common feature in our peatlands, there is no indication of involved climatic changes.

Only the zoochlorellae-symbiotic species (e.g. *Amphitrema flavum*, *Hyalosphenia papilio* and *Heleopera sphagni*) are able to inhabit the upper horizon of *Sphagnum*. That is why they are typical of raised bogs (Heal, 1962, 1964b; Schönborn, 1965). Based on the grouping of testacean species in terms of the presence or absence of symbiotic zoochlorellae, the s.c. Meisterfeld's symbiont quotient of subfossil testacean remains (Laminger *et al.*, 1981) throws light on the plant composition of previous mire vegetation. Laminger *et al.* (1981) examined the usefulness of this and some other species indexes in their quantitative rhizopod analysis of an alpine mire in Tyrol.

RHIZOPOD ANALYSIS OF LAKE DEPOSITS

General

Advances in rhizopod analysis of sediments were summarized by Frey (1964): 'The rhizopods give promise of being one of the important groups of organisms in interpreting past limnological conditions.'

Later studies indicate that in addition to the changes in trophy, etc. in the lake basins, the sedimentary testacean remains may contribute to tracing climatically controlled development of watersheds. Durability of shells varies within different lakes and species but its role can be estimated by quantitative analysis, at least in certain circumstances (Ruzicka, 1982).

Ecological factors

Most of the lake rhizopods prefer oligotrophic and acidic waters. Accordingly, Schönborn (1962) stated that the number of testacean individuals of different biotopes is from three to eight times smaller in an eutrophic lake than in an oligotrophic one (Table 31.2).

TABLE 31.2.   Abundance of Testacea in different associations of Lakes Stechlinsee (oligotrophic), Nehmitzsee (oligotrophic–eutrophic) and Dagowsee (eutrophic) (Schönborn, 1962a): figures given per 225 cm^2 bottom area

Biotope	Oligotrophic	Oligotrophic–eutrophic	Eutrophic
Typha-Phragmites	13,000	50,550	24,000
Potamogeton	8,200	5,400	1,000
Submersed meadows	9,450	2,100	5,850
Sediment	6,000	4,000	2,800

The principal ecological factor explaining this distribution pattern is obviously the dissolved oxygen content (Stepanek, 1953; Schönborn, 1962a; Schönborn *et al.*, 1965; Laminger, 1972, 1973b, c). Additionally, factors involved include: (1) the dystrophy-grade of the lake as illustrated by the C/N ratio in its sediment, i.e. the existence of biologically effective humus compounds; (2) the grain size of sediment; and (3) the existence of *Sphagnum* carpet(s) around the lake (Schönborn, 1967). Figure 31.4 illustrates the biotope preference of nineteen lakes inhabited by testacean species.

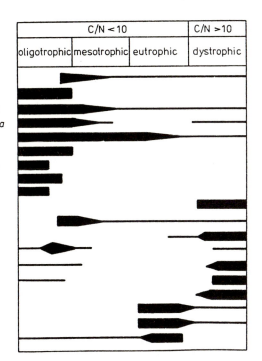

FIGURE 31.4.   Relative abundance of certain Testacea species in lakes of different trophic grade, according to Schönborn (1967). *Reproduced by permission of Akademie Verlag*

Schönborn *et al.* (1965) developed an ecological classification system for the testacean taxocoenoses in lakes. They found that in addition to a large number of ubiquitous species inhabiting sediments of different types of lakes, there are other species which are good indicators. On the basis of those more stenotope species, three associations were suggested, characterizing the oligotrophic, eutrophic and dystrophic lake type.

After only slight revisions by Laminger (1973b), the most useful ecological testacean associations are as follows:

(1) eutrophic: *Difflugia urceolata*-association (besides the nominal species: *Difflugia amphora, Centropyxis ecornis, C. austrica* and *Difflugia labiosa*);
(2) dystrophic: *Arcella gibbosa*-association (besides the nominal species: *Heleopera petricola, Hyalosphenia cuneata, Lesquereusia spiralis, Nebela collaris, N. dentistoma* and *N. vitraea*);
(3) oligotrophic: *Arcella hemisphaerica*-association (besides the nominal species: *Centropyxis aerophila, C. platystoma, Cyclopyxis kahlii, Difflugia lemani* and *D. finstertaliensis*).

The associations seem to be ecologically consistent in both European and North American lakes (Schönborn, 1966a, b, c, 1967, 1968, 1973, 1975; Laminger, 1972, 1973b; Laminger *et al.*, 1979; Ruzicka, 1982).

Sampling, storage and laboratory work

In general, the principles applied in obtaining sediment cores for diatom analysis should be followed. There is, however, no experience of the durability of sedimentary testacean remains if freezing techniques are used, since only fresh sediments have been used so far. No standard method for the rhizopod analysis of sediment has been established (Laminger, 1972, 1973b; Schönborn, 1973; Ruzicka, 1982).

This last-mentioned study (Ruzicka, 1982) is by now the most complete attempt for the quantitative testacean analysis of sediment cores. Therefore, her methods are briefly described below. Fresh subsamples 4.50 ml in size were used, the pretreatment and processing which took place being shown in the flowchart of Figure 31.3. The final counting of the individuals was with an inverted microscope using the smallest suspension volume possible (1 ml) which was totally scanned. The 'absolute' number of each taxa was expressed per gram dry-weight of sediment.

Construction and interpretation of diagrams

Schönborn (1973) presented the relative abundance (4-grade scale for 0.05 cm^3 sample volume per one microscope cover slip) of all the twelve rhizopod species he found during the late and postglacial history of subarctic lake Latnjajaure in Swedish Lapland (Figure 31.5). He stressed that rhizopod analysis of lake sediments has not yet matured enough for tracing detailed limnological changes. He also emphasized that the ecological power of most testacean species is much higher than the ecological valence which has been obtained by recent ecological analyses. According to Schönborn (1973), lake Latnjajaure maintained its oligotrophic character through the ages (*Centropyxis aërophila*), but in addition there were indications of eutrophication in a couple of samples.

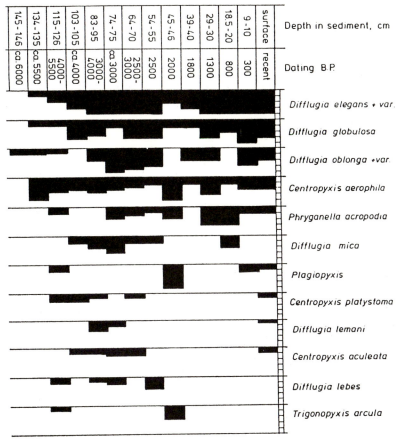

Depth in sediment, cm	Dating B.P.

FIGURE 31.5. Testacean remains encountered in a sediment profile from Lake Latnjajaure, Swedish Lapland (elevation 978 m above sea-level). The abundance of each taxon expressed on a 4-grade scale per 0.05 cm³ sample. The lake maintained its oligotrophic character through the ages as inferred from the abundant occurrence of *Centropyxis aërophila* in all samples. Periods of initial eutrophication are indicated by a decrease of *Centropyxis aerophila* and an increase of *Difflugia* species, but typical eutrophic species are absent. The severe decline in the abundance of *Difflugia* and the increase in *Plagiopyxis* and *Trigonopyxis* at about 2000 B.P. clearly reflected the strongest deterioration of the climate, when the lake might have been covered by ice throughout summer seasons (Schönborn 1973). *Redrawn from Schönborn (1973) by permission of W. Junk Publ*

The number of testacean species encountered in sediment cores can be so high that the presentation of records for individual taxa is impractical. Therefore, Ruzicka (1982) constructed 'absolute' rhizopod diagrams for the total number of shells and for certain trophic groups and other associations. She encountered no less than 85 species and 9 variations in the sediments of meromictic lake Krottensee in Austria. Because the lake has probably been meromictic since the preboreal period, the testacean necrocoenosis of its oxygen-free deep water (43 m) was compiled by using allochthonous individuals (from shallower water and from the surroundings of the lake). But being able to put about 60% of the encountered rhizopod taxa into four main habitat groups — (1) lake, (2) *Sphagnum*, (3) Bryales, and (4) soil — Ruzicka constructed an informative vertical diagram for rhizopod associations during the past 17,000 years or so. She argued that certain combinations of species in her lake could be ascribed to typical climatic conditions in the lake catchment, especially in respect to the amount of rainfall.

Ruzicka examined carefully, for the first time, the differential durability of various rhizopod taxa in sediments. In addition, she gave valuable taxonomical and autoecological information on many species.

IDENTIFICATION OF SPECIES

The number of species which are preserved in peats is small (about 30 in Steinecke, 1927) and shows little geographical variation. Unfortunately, no easily obtainable identification manuals for peat rhizopods exist which contain all species. Grospietsch's textbook (1965) gives keys for (and has pictures of) most of the peat-living Testacea, and it also contains much general information, as well as references. Hoogenraad and de Groot (1940) is a useful source for information on peat rhizopods but is not easily available. Keys and very short descriptions of most species are found also in Harnisch (1959), but the taxonomical notes stated below have to be taken into consideration. Regarding the genera *Arcella*, *Centropyxis*, *Cryptodifflugia*, *Hyalosphenia* and *Nebela*, the monographs mentioned in the reference list have to be borne in mind (Deflandre, 1928, 1929, 1936; Grospietsch, 1958, 1965b).

For lake rhizopods, see Schönborn (1966d) and Ruzicka (1982) and references therein.

AUTECOLOGICAL AND TAXONOMICAL NOTES

This section contains notes on the most important Testacea for rhizopod analysis of peat. The following abbreviations are used according to Heal (1961):

F   = fen species
E   = eurytope (in terms of bog/fen)
E(b) = eurytope, mainly in bogs
BP   = bog pool species
BH   = bog hummock species

The different moisture classes (I–VIII) are explained earlier in this chapter.

Only those species most frequently occurring as subfossils in peat deposits are discussed here. Forms with non-resistant delicate tests, usually not encountered in deep peat, include such genera as *Euglypha, Placocista, Tracheuglypha,*

Cyphoderia, Quadranqula, Pontigulasia and *Pyxidula*, as well as most of the *Difflugia* and *Centropyxis* species.

Amphitrema flavum Archer: BH according to Heal (1964b), but according to Meisterfeld (1977) it occurs in moisture classes II–V (optimum in III) both in bog and fen vegetation and also in isolated *Sphagnum* stands. In Lapland, also in Bryales (Schönborn, 1966b, and literature therein). According to Schönborn (1962) in III–IV.

   A. wrightianum Archer: According to Heal (1964) and to a large number of earlier rhizopod experts, it is mostly restricted to bog pools, occurring in moisture classes II–III (Meisterfeld, 1977). Tolonen (1967, 1968) found the species as subfossil in mesotrophic fen stages in Finland from Early Atlantic to recent times. Several subfossil finds of the species from Oberbayern in Bryales-sedge peat, since Atlantic time.

   A. stenostoma Nüsslin: A bog pool species (Steinecke, 1927; Heal, 1964b) occurring in moisture classes I–III (optimum in II — Meisterfeld, 1977). Schönborn (1966b) found it in submerged and wet *Sphagnum*, together with *A. flavum* at a pH value of 8.0.

   Arcella: Both systematically and ecologically, the genus is insufficiently known and the author's name after each species name is indispensable. Many intermediate forms are impossible to identify; often the lateral view is diagnostic (Deflandre, 1928). Several authors (de Graaf, 1956) consider *A. rotundata aplanata* synonymous with *A. discoides scutelliformis*. The type usually occurs abundantly in floating, submerged or very wet *Sphagnum* swards. But in a relatively wide subfossil material (232 samples) taken from raised bogs in eastern N. America (Tolonen, *et al.*, 1985) the taxon, together with *Assulina muscorum*, belongs to *Trigonopyxis arcula-Hyalosphenia subflava* necrocoenosis (cluster analysis). The whole genus is essentially hydrophilous, with the exception of *A. arenaria* Greef (xerophilous) and *A. catinus catinus* Pen. (hygrophilous) (de Graaf, 1956).

   A. artocrea Leidy: An indicator species for regeneration and stationary complexes of raised bogs, i.e. for moisture classes III–VII, but the species avoids open pools (Jung, 1936). Hygrophilous (Laminger, 1973a).

   Assulina muscorum Greeff: Acidophilous, with a distinct xerophilous tendency (de Graaf, 1956), from II–VIII with a very clear maximum in VI–VIII (Laminger, 1973a; Meisterfeld, 1977). E(b) (Heal, 1964b).

   A. seminulum Ehr.: Hygrophilous (de Graaf, 1956; Schönborn, 1962; Laminger, 1973a). BH (Heal, 1964b).

   Bullinularia indica (Pen.) Defl.: Typical xerophilous species with an optimum in dry hummock *Sphagnum*, especially in forested bog (Jung, 1936) and in other moss carpets, particularly those growing in woods (e.g. Bonnet and Thomas, 1960; Grospietsch, 1953; Schönborn, 1962); occasionally in open water habitats as single individuals only (de Graaf, 1956). BH (Heal, 1964).

Centropyxis aculeata Ehr. and *C. hirsuta* Defl.: These are difficult to identify separately and according to Heal (1961) are regarded as the species complex E. Mostly in aquatic habitats (de Graaf, 1956; Schönborn, 1962).

C. cassis (Wallich) complex: There are considerable difficulties with at least *C. cassis* s. str., *C. aerophila* Defl., *C. constricta* Ehr. and *C. kolkwitzi* van Oye (Paulson, 1954; Heal, 1961). All more or less hygrophilous.

Corythion dubium Taranek: Here discussed with *Trinema lineare* Pen and *T. enchelys* Ehr. because these three species are difficult to separate if the component plates are not easily visible (Paulson, 1954; Heal, 1961). The first species is distinctly xerophilous: IV–VIII (optimum at VI) according to Meisterfeld (1977), while the two latter ones are rather eurytopic in this respect, avoiding, however, the driest sites (Laminger, 1973).

Difflugiella oviformis (Pen.) Bonnet et Thomas Syn., *Cryptodifflugia oviformis* Pen. 1890 and *Difflugiella oviformis* var. *fusca* Pen.: Detailed descriptions of these very small species (13–26 μm) and related taxa are in Bonnet and Thomas (1960) and Grospietsch (1964). Regarding nutrient ecology, it is a eurytope (Heal, 1961). It has its optimum in transitional peat formations and occasionally also in non-peat-forming Sphagneta (de Graaf, 1956). According to my own observations (Tolonen, 1967) the species reaches its maximal abundance on the top of dry bog hummock (E. Finland); but Heal's material (1964b) from England gave quite an opposite picture: all occurrences are in saturated-submerged to wet *Sphagnum* in early minerotrophic mires.

Heleopera petricola Leidy: The ecology is quite variable and disputed. The species has been found in the driest sites of a bog (Jung, 1936), but also in very wet or fairly wet *Sphagnum* (α-hydrophilous according to de Graaf, 1956), and often in bog hummocks (Heal, 1961; 1964b). A typical component in the fauna of central European lake muds (Schönborn, 1966b).

H. sylvatica Pen.: Restricted to drier mosses (Jung, 1936). A bog hummock species (Heal, 1961).

H. sphagni Leidy: Acidobiontic and β-hygrophilous (de Graaf, 1956) which agrees well with Jung (1936).

H. rosea Pen.: In bog hummocks and drier *Sphagnum* swards (Jung, 1936). A eurytopic fen species (Heal, 1964b).

Hyalosphenia elegans Leidy: For taxonomy and identification, see Grospietsch (1965). A typical α-hygrophilous species, in peat and brown mosses (de Graaf, 1956), as well as in oligotrophic lakes (Schönborn, 1966a). BH (Heal, 1961) agrees with Meisterfeld (1977): a distinct optimum in IV, absent in I and VIII.

H. insecta Harnisch: Taxonomy (Grospietsch, 1965). Encountered in Obernbayern from very early (preboreal) sedge peat in several profiles, as well as in later ombrogenous *Sphagnum* peat.

H. papilio Leidy: BH (Heal, 1961), reaches its optimum in wet *Sphagnum* III–IV (Schönborn, 1962; Meisterfeld, 1977) and disappears when moisture falls to lower levels (absent in VII and VIII). See also the next species.

H. ovalis Wailes: The most characteristic indicator species of the rhizopod association of pools of raised bogs (Jung, 1936). It has occurred abundantly in wet minerotrophic habitats (sedge fens) in Finland since Atlantic time (author's observations).

H. subflava Cash: In terrestrial Bryales and *Sphagnum*, also in soils (Grospietsch, 1965), belonging to the so-called tyrfoxene type (Jung, 1936; Grospietsch, 1953), i.e. not indigenous to living mires, but characteristic of drained peatlands.

Nebela carinata Archer: A species of wet or very wet *Sphagnum* (Jung, 1936; de Graaf, 1956); bog pool species (Heal, 1964b) with an optimum in II (Meisterfeld, 1977).

N. collaris Leidy incl. *N. bohemica* Taranek: β-hygrophilous (de Graaf, 1956). II–VII with an optimum in V (Meisterfeld, 1977). For taxonomic difficulties see e.g. Heal (1963).

N. griseola Pen. incl. *N. tenella* Pen.: Very common within the moisture range (II) III–VI (Jung, 1936) with an optimum in III-IV (de Graaf, 1956; Meisterfeld, 1977); in very wet to aquatic *Sphagnum*, both in bogs and fens (Heal, 1964b).

N. marginata Pen.: α-hygrophilous (de Graaf, 1956). In very wet *Sphagnum*: I–IV with an optimum in III (Meisterfeld, 1977).

N. militaris Pen.: BH (Heal, 1964), very common in fens, too. In V (Schönborn, 1962) or VII (Meisterfeld, 1977). One must bear in mind that only the latter author emphasizes that he counted only living individuals.

N. tincta Leidy: A xerophilous species (e.g. Jung, 1936) with an optimum in the half-dry mosses of forests (de Graaf, 1956). II–VII with an optimum in V (Schönborn, 1962; Meisterfeld, 1977). For taxonomy see Heal (1963).

Phryganella hemisphaerica Pen. Syn. *Phryganella acropodia* (Hertw. et Lesser) Hopk.: This species has been commonly used in rhizopod analysis as an index of more or less dry conditions within the 'tyrfoxene type' (see *Hyal. subflava* above); but many experts (since Hoogenraad, 1935) are of the opinion that empty theca of the species cannot be distinguished with certainty from, for example, *Centrophyxis eurystoma* and *Difflugia globulosa* (see also discussion by Meisterfeld (1979) about *Cyclopyxis arcelloides*). Heal (1964) found the species in very different moisture conditions both in fen and bog sites. Therefore, ecological interpretations of its subfossil occurrences are doubtful.

Trigonopyxis arcula Leidy: According to de Graaf (1956) and many earlier scientists, this is a xerophilous species which may be regarded as a guide species for the last dry stages in the raised bog formation. BH (Jung, 1936) but found in wet fen sites, too. Very abundant in moisture class VIII (and only there) according to Schönborn (1963). Positively correlated with the highly humified dwarf shrub peat streaks in the 'short cyclic' regeneration of *Sphagnum fuscum* peat (Tolonen, 1971).

Trinema: See discussion under *Corythion dubium*.

REFERENCES

Aaby, B. (1975). Cykliske klimavariationer i de sidste 7500 år påvist ved undersøgelser af højmoser og marine transgressionsfaser. *Danm. Geol. Unders. Åarbok 1974*, 91–104.

Barber, K. E. (1981). *Peat Stratigraphy and Climatic Change*, A. A. Balkema, Rotterdam.

Bonnet, L., and Thomas, R. (1960). Thecamoebiens du Sol. *Vie et Milieu IX*: 4 Suppl. Faune terrestre et d'eau douce des Pyrenees-Orientales 5, pp. 1–103.

Deflandre, G. (1928). Le gengre Arcella Ahrenberg. *Arch. Protistenk.*, **64**, 152–267.

Deflandre, G. (1929). Le gengre Centropyxis Stein. *Arch. Protistenk*, **67**, 322–375.

Deflandre, G. (1936). Etude monographique sur le gengre Nebela Leidy (Rhizopoda-Testacea). *Ann. Protistol.*, **5**, 20–286.

Frey, D. G. (1964). Remains of animals in Quaternary lake and bog sediments and their interpretation. *Arch. Hydrobiol. Beih. Ergebn. Limnol.*, **2**, I–II, 1–114.

Graaf, F. de (1956). Studies on Rotatoria and Rhizopoda from the Netherlands. *Biol. Jaarb. Dodonaea*, **23**, 145–217.

Grospietsch, Th. (1953). Rhizopodenanalytische Untersuchungen an Mooren Ostholsteins. *Arch. Hydrobiol.*, **47**, 321–452.

Grospietsch, Th. (1958). *Wechseltierchen (Rhizopoden)*, Kosmos, Stuttgart.

Grospietsch, Th. (1964). Die Gattungen Cryptodifflugia und Difflugiella (Rhizopoda testacea). *Zoologisches Anzeiger Leipzig*, **172**(4), 243–257.

Grospietsch, Th. (1965a). Rhizopodenanalytische Untersuchungen im Naturschutzgebiet Bernrieder Filz (Oberbayern). *Arch. Hydrobiol.*, **61**, 100–115.

Grospietsch, Th. (1965b). Monographische Studie der Gattung *Hyalosphenia* Stein. *Hydrobiologia*, **26**(1–2), 211–241.

Grospietsch, Th. (1954). Studien über die Rhizopodenfauna von Schwedisch-Lappland. *Arch. Hydrobiol.*, **49**, 546–580.

Harnisch, O. (1927). Einige Daten zur rezenten und fossilen testaceen Rhizopodenfauna der Sphagnen. *Arch. Hydrobiol.*, **18**(3), 245–360.

Harnisch, O. (1951). Daten zur Gestaltung der ökologischen Valenz der sphagnikolen Rhizopoden im Verlauf des Postglazials. *Deutsch. Zool. Z.*, **1**, 222–233.

Harnisch, O. (1959). Rhizopoda. In: *Die Tierwelt Mitteleuropas* I:1b, 1–75, 1–26.

Heal, O. W. (1961). The distribution of testate amoebae (Rhizopoda: Testacea) in some fens and bogs in northern England. *J. Linn. Soc. (Zool.)*, **44**, 369–382.

Heal, O. W. (1962). The abundance and micro-distribution of testate amoebae (Rhizopoda: Testacea) Sphagnum. *Oikos*, 35–47.

Heal, O. W. (1963). Morphological variation in certain Testacea (Protozoa: Rhizopoda). *Arch. Protistenk.*, **106**, 351–368.

Heal, O. W. (1964a). The use of cultures for studying Testacea (Protozoa: Rhizopoda) in soil. *Pedobiologia*, **4**, 1–7.

Heal, O. W. (1964b). Observations on the seasonal and spatial distribution of Testacea (Protozoa: Rhizopoda) in Sphagnum. *J. Anim. Ecol.*, **33**, 395–412.

Hoogenraad, H. R. (1935). Studien über die sphagnicolen Rhizopoden der niederländischen Fauna. *Arch. Protistenk.*, **84**, 1–100.

Hoogenraad, H. R. (1936). Zusammenstellung der fossilen Süsswasserrhizopoden aus postglazialen Sapropelium- und Torfablagerungen Europas. *Arch. Protistenk.*, **87**, 402–416.

Hoogenraad, H. R., and de Groot, A. A. (1940). Zoetwaterrhizopoden en Heliozoen (Ala). In: *Fauna van Nederland* 9, Leiden.

Jung, W. (1936). Thekamöben ursprünglicher, lebender deutscher Hochmoore.

Abh. Landesmus. Prov. Wstf, Museum f. Naturkd, **7** (4), 1–87.

Lagerheim, G. (1902). Om lämningar af rhizopoder, heliozoer och tintinnider i Sveriges och Finlands lakustrina kvartäraflagringar. *Geol. Fören. Förhandl.*, **23**, 469–520.

Laminger, H. (1972). Die profundale Testaceenfauna (Protozoa, Rhizopoda) älterer und jüngerer Bodenseesedimente. *Arch. Hydrobiol.*, **70**, 108–129.

Laminger, H. (1973a). Die Testaceen in der Umgebung der Station Büschelbach (Spessart/BRD). *Hydrobiologia*, **41** 501–513.

Laminger, H. (1973b). Quantitative Untersuchung über die Testaceenfauna (Protozoa, Rhizopoda) in den jüngsten Bodensee-Sedimenten. *Biol. Jb. Dononea*, **41**, 126–146.

Laminger, H. (1973c). Untersuchungen über Abundanz und Biomasse der sedimentbewohnender Testaceen (Protozoa, Rhizopoda) in einem Hochgebirgessee (Vorderer Finstertalersee, Kühtai, Tirol). *Int. Rev. Ges. Hydrobiol.*, **58**, 543–568.

Laminger, H., Schopper, M., Pipp, E., Hensler, I., and Mantl, P. (1981). Untersuchungen über Nekrozönosen und Taxozönosen der Testacea (Protozoa) im Zirbenwaldmoor (Obergurgol, Tirol/Austria). *Hydrobiologia*, **77**, 193–202.

Laminger, H., Zisette, S., Phillips, S., and Bridigam, F. (1979). Beitrag zur Kenntnis der Protozoenfauna Montanas (USA) I. *Hydrobiologia*, **65**, 257–271.

Lindberg, H. (1899). En rik torffyndighet i Jorvis-socken Savolaks. *Finska Mosskulturföreningens Årsbok 1899*, 178–213.

Martin, W. (1971). Zur Frage des ökologischen Hauptfaktors bei Testaceen-Vergemeinschaftungen. *Limnologica*, **8**, 357–363.

Meisterfeld, R. (1977). Die horizontale und vertikale Verteilung der Testaceen (Rhizopoda, Testacea) in Sphagnum (The horizontal and vertical distribution of Testacea (Rhizopoda, Testacea) in Sphagnum). *Arch. Hydrobiol.*, **79**, 319–356.

Meisterfeld, R. (1979a). Zur Systematik der Testaceen (Rhizopoda, Testacea) in Sphagnum, Eine REM-Untersuchung. *Arch. Protistenk.*, **121**, 246–269.

Meisterfeld, R. (1979b). Clusteranalytische Differenzierung der Testaceenzönosen (Rhizopoda, Testacea) in Sphagnum, (Cluster analysis of associations of testate amoeba (Rhizopoda, Testacea) in Sphagnum). *Arch. Protistenk.*, **121**, 270–307.

Olausson, E. (1957). *Das Moor Roshultsmyren: eine Geologische Botanische und Hydrologische Studie in einem Südwestschwedischen Moor mit Exzentrisch Gevölbten Mooselementen*, Publ. Inst. Miner. Paläont. Quatern. Geol., University of Lund, 39; Lund Univ. Årsskr., N.F. 53:12, Kungl. Fysiogr. Sällsk. Handl., N.F. 68:12, pp. 1–76.

Paulson, B. (1954). Some rhizopod associations in a Swedish mire. *Oikos*, **4**, 151–165.

Ruzicka, E. (1982). Die subfossilen Testaceen des Krottensees (Salzburg, Österreich). *Limnologica*, **14**, 49–88.

Schmeidl, H. (1940). Beitrag zur Frage des Grenzhorizontes im Sebastianberger Hochmoor. *Beih. Bot. Centralbl.*, **B 60**, 493–524.

Schönborn, W. (1962a). Die Ökologie der Testaceen im oligotrophen Seen, dargestellt am Beispiel des Grossen Stechlinsees. *Limnologica*, **1**, 111–182.

Schönborn, W. (1962b). Zur Ökologie der sphagnikolen, bryokolen un terrikolen Testaceen. *Limnologica*, **1**, 231–254.

Schönborn, W. (1963). Die Stratigraphie lebender Testaceen im Sphagnetum der Hochmoore. *Limnologica*, **1**, 315–321.

Schönborn, W. (1964). Lebensformtypen und Lebensraumwechsel der Testaceen. *Limnologica*, **2**, 321–335.

Schönborn, W. (1965). Untersuchungen über die zoochlorellen-Symbiose der Hochmoor-Testaceen. *Limnologica*, **3**, 173–176.

Schönborn, W. (1966a). Testaceen als Bioindikatoren im System der Seetypen:

Untersuchungen in masurischen Seen und im Suwalki-Gebiet (Poland). *Limnologica*, **4**, 1–11.

Schönborn, W. (1966b). Untersuchungen über die Testaceen Schwedisch-Lapplands: ein Beitrag zur Systematik und Ökologie der beschalten Rhizopoden. *Limnologica*, **4**, 519–559.

Schönborn, W. (1966c). Beitrag zur Ökologie und Systematik der Testaceen Spitzbergens. *Limnologica*, **4**, 463–470.

Schönborn, W. (1966d). *Beschalte Amöben (Testaceae)*, Ziemsen-Verlag Wittenberg, Lutherstadt.

Schönborn, W. (1967). Taxozönotik der beschalten Süsswasser-Rhizopoen: eine raumstrukturanalytische Untersuchung über Lebensraumerweiterung und Evolution bei der Mikrofauna. *Limnologica*, **5**, 159–207.

Schönborn, W. (1968). Vergleich der zönotischen Grössen, der Verteilungsmunster und der Anpassungsstandards der Testaceen-Taxozönosen in der Biotopreihe vom Aufwuchs bis zum Erdboden. *Limnologica*, **6**, 1–22.

Schönborn, W. (1973). Paläolimnologische Studien an Testaceen des Latnjajaure (Abisko-Gebiet, Swedish Lappland). *Hydrobiologia*, **42**, 63–75.

Schönborn, W. (1975). Studien über die Testaceen Besiedlung der Seen und Tümpel des Abisko-Gebiets (Schwedisch-Lappland). *Hydrobiologia*, **46**, 115–139.

Schönborn, W., Flössner, D., and Proft, G. (1965). Die limnologische Charakterisierung des Profundals einiger Seen mit Hilfe von Testaceen-Gemeinschaften. *Limnologica*, **3**, 371–380.

Steinecke, F. (1927). Leitformen und Leitfossilien der Zehlaubruches: die Bedeutung der fossilen Mikro-organismen für die Erkenntniss der Nekrozonosen eines Moores. *Bot. Arch. Könisb.*, **19**(5–6). 327–344.

Steinecke, F. (1929). Die Nekrozönosen des Zehlaubruches: Studien über die Entwicklung des Hochmoores an Hand der fossilen Mikro-organismen. *Schr. Phys-ökön. Ges. Königsb.*, **66**, 193–214.

Steinecke, F. (1937). Zur Geschichte der Galtgarben-Moore. *Schr. Phys-ökön. Ges. Königsb.*, **59**, 289–340.

Stepanek, M. (1953). Rhizopoda jako biologicke indicatory znecisteni vod. 1: Rhizopoda a Helizoa reky Moravice (Slezko-SR). Zvlastni otisk pirod-sborn. *Ostr. Kraje*, **14**, 470–505.

Tolonen, K. (1966). Stratigraphic and rhizopod analyses on an old raised bog, Varrassuo, in Hollola, South Finland. *Ann. Bot. Fennici*, **3**, 147–166.

Tolonen, K. (1967). Ueber die Entwicklung der Moore im finnischen Nordkarelien. *Ann. Bot. Fennici*, **4**, 219–416.

Tolonen, K. (1968). Zur Entwicklung der Binnenfinnland-Hochmoore. *Ann. Bot. Fennici*, **5**, 17–33.

Tolonen, K. (1971). On the regeneration of North-european bogs. I: Klaukkalan Isosuo in S. Finland. *Acta Agralia Fennica*, **123**, 143–166.

Tolonen, K. (1979). Peat as a renewable resource: long-term accumulation rates in North-european mires. *Proc. Int. Symp. Classification of Peat and Peatlands*, Hyytiälä, Finland, pp. 282–296. International Peat Society, Helsinki.

Tolonen, K., Huttunen, P., and Jungner, H. (1985). Regeneration of two coastal raised bogs in eastern North America. Stratigraphy, radiocarbon dates and rhizopod analysis from sea cliffs. *Ann. Acad. Sci Fennicae* Ser A. III. *Geologica-Geographica* **139**, 1–51.

Handbook of Holocene Palaeoecology and Palaeohydrology
Edited by B. E. Berglund
© 1986 John Wiley & Sons Ltd.

32

Cladocera analysis

DAVID G. FREY

*Department of Biology, Indiana University, Bloomington, Indiana,
U.S.A.*

INTRODUCTION

The microscopic animals occurring in freshwater are the Protozoa, Turbellaria, Nemata, Gastrotricha, Rotifera, Tardigrada, Cladocera, Ostracoda, Copepoda and Acari, omitting for present purposes the Anostraca, Notostraca and Conchostraca, which tend to be restricted to highly specific habitats, and the Oligochaeta and larvae of aquatic insects, particularly the Diptera. Remains of at least some representatives of each of these groups have been recovered from freshwater sediments (Frey, 1964), but certainly the Cladocera are most important in terms of the diversity and abundance of their remains. Some of the groups (e.g. Copepoda) regretfully preserve only rarely. Protozoans (such as the rhizopods, Tintinnidae, chrysophyceans and dinoflagellates) can be abundant and important in interpreting past conditions, turbellarian egg capsules and rotifer resting eggs are sometimes common but little studied to date, and Acari are common and diversified in bogs but not very frequent in lake sediments. Of all these groups of microscopic animals, with the possible exception of the Chironomidae, the Cladocera have been most useful in indicating past conditions and in helping to provide insight into the responses of aquatic ecosystems to external changes in climate and watershed processes.

The Cladocera constitute one of the major components of the offshore zooplankton (along with rotifers and copepods), besides which they occur abundantly in the littoral zone, chiefly in association with macrophytes and to a lesser extent with the sediments. The limnetic species (in the genera *Bosmina, Daphnia, Ceriodaphnia, Diaphanosoma*) primarily filter-feed on small phytoplankton, bacteria and detritus, although there are also a few predators

(*Leptodora, Bythotrephes*). The littoral species feed mainly at the substrate-
–water interface, where they most commonly obtain their food by scraping
algae and detritus from surfaces over which they crawl or from sediments in
which they burrow. Most chydorids are included in this category (Fryer, 1968).
Some filter-feeders (*Sida, Simocephalus, Scapholeberis*) also occur here, as
well as a few predators (*Polyphemus*). The macrothricids have a diversity of
habitats and food habits in the littoral (Fryer, 1974), some species being
associated with macrophytes and some with mud or other sediments. Thus, the
Cladocera can be grouped first into offshore planktonic species and inshore
littoral species, with the latter group in turn subdivided into those associated
primarily with macrophytes and those associated with sediments.

These animals, being arthropods, can grow only by periodically shedding
their exoskeletons, a process called moulting, with the in-between stages called
instars. The cast exoskeletons, or exuviae, are rapidly disarticulated into their
component pieces — headshield, shell, post-abdomen, post-abdominal claws,
antennules, antennal segments, mandibles, and portions of trunk limbs — by
biological activity. In addition, gamogenetic individuals can contribute
ephippia, copulatory hooks, and specialized headshields and post-abdomens.
It is these isolated parts of the animals rather than the intact animals themselves
that become incorporated into the sediments. Each individual generates a
variable number of exuviae, and hence of skeletal fragments, depending on
how many times it moults before death. Besides this uncertainty there is also
differential preservation of chitin and of the various exoskeletal components
from one family to another and even from one species to another within the
same family, obviously controlled by biochemical differences that affect the
ease of biodegradation. All skeletal parts of the Bosminidae and Chydoridae,
and perhaps also of the Macrothricidae, seem to preserve quantitatively except
perhaps for the first couple instars, whereas the Sididae, Holopediidae,
Daphniidae, Moinidae, Polyphemidae and Leptodoridae are variably represented by post-abdominal claws, sometimes with part of the post-abdomen still
attached, antennal segments, mandibles, caudal spines, and/or ephippia.
Many of these parts are less highly differentiated morphologically than shells
and headshields, making it considerably more difficult to determine the species
involved. Moreover, many of these fragments are very small and hence run a
greater chance of being lost in the processing of the sediments, especially if the
residues are run through a sieve.

The family Chydoridae is represented by the greatest number of species,
most of which can be identified confidently from their abundant exoskeletal
fragments in the sediments. They afford the greatest amount of information in
the sediments for interpreting palaeolimnology and also for theoretical studies
in community ecology. The Bosminidae are also well-represented by
exoskeletal fragments but with few species at any one time and considerable
uncertainty as to what they are. The Daphniidae are well-represented by

species in the living communities, but thus far have been difficult to identify from their isolated post-abdominal claws, mandibles and ephippia, although Edwards (1980) has done a nice SEM study of the molar surfaces of mandibles from 18 species of *Daphnia*. Such studies will help immensely in working with the few remains of these species preserved in lake sediments.

Typically there are several thousand fragments of Cladocera per cm^3 of fresh lake sediment, with a general range of a thousand to several hundred thousand. Densities can be much larger than this, even high enough for a special sediment type of chitin gyttja to have been proposed (Wesenberg-Lund, 1901). Densities can also be much smaller than this, down to $100/cm^3$ or even less, particularly in mineral sediments. The higher density levels are comparable to those of pollen and diatoms, thereby enabling construction of the same sort of close-interval stratigraphies. The limiting factor is not the availability of remains in the sediments but the time involved in identifying and tabulating them.

PREPARATION OF SLIDES

The data required in a cladoceran analysis are estimates of the number of exuviae per cm^3 of fresh sediment for each species present at each stratigraphic level studied. These estimates are based on microscope slide preparations, each coverslip of which represents a known fraction of the sample that was processed and hence provides a separate estimate of the population densities of the individual taxa.

Cladoceran chitin, being quite inert chemically, can be subjected to most chemical treatments used in the preparation of pollen slides. Only acetolysis must be avoided, as it can result in the differential destruction of cladoceran remains or even their complete dissolution. We now have quite a suite of procedures available, mostly borrowed from palynology. No single sequence of procedures is appropriate for all sediment types but instead must be adapted to the characteristics of the sediment being processed. The general objective, however, should be to use as few steps as are necessary to achieve satisfactory preparations.

The eight flowcharts in Figure 32.1 show the processing steps and their sequence used by various investigators over the past four decades. In the non-calcareous sediments of Linsley Pond (Deevey, 1942), deflocculation with hot KOH was adequate; but as soon as calcareous sediments were studied, such as the Madison lakes (Frey, 1960), an HCl step had to be added for removal of the carbonates. Further elaborations have involved the use of HF to dissolve mineral matter, screening at various stages in processing to remove very small or very large particles, successive decantations to remove most of the coarse mineral matter, or differential density centrifugation in a sucrose solution of density 1.2 g/cm^3 (Goulden, 1971). Ravera and Parise (1978)

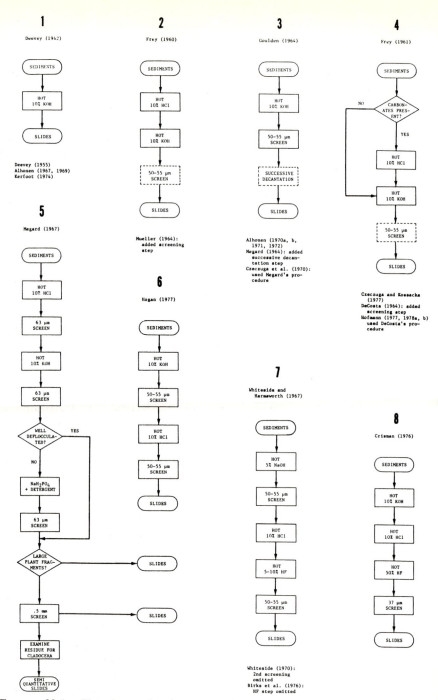

FIGURE 32.1. Flowcharts showing sequential steps in processing sediments for recovery of cladoceran remains, as used by various investigators over the past four decades. The names beneath each number indicate an early (in most instances probably the first) use of a particular processing sequence. The names beneath the flowcharts are investigators who used the sequence subsequently, sometimes with modifications

used differences in settling velocity through a vertical water column about 70 cm long to effect a separation of mineral particles from cladoceran remains in the size fraction greater than 88 μm. A recovery rate of 95% is claimed.

Whiteside and Harmsworth (1967) used a brief heating in dilute HF to eliminate mineral matter, whereas Crisman (1976) used a rather prolonged heating in full-strength HF. Our experience suggests that hot concentrated HF should be used with caution, in that it, like acetolysis, can bring about differential dissolution of cladoceran remains.

The sequence of treatment with KOH and HCl is reversed in some of the processing sequences. I believe that treatment with KOH first is preferred, as this disaggregates the sediment and thereby helps to reduce the violence of ebullition of CO_2 bubbles in the HCl treatment. This should help minimize fragmentation of the cladoceran remains during sediment processing.

Figure 32.1 shows, especially for flowchart 3, that a diversity of screen sizes has been used, mostly to achieve specific objectives. Even fine screening can cause the loss of very small exoskeletal components, such as mandibles and post-abdominal claws. As these can be very important in interpretations, such as the minute claws of *Ceriodaphnia* in Schleinsee (Frey, 1961b), all screening should be avoided if possible, or, if used, then some attempt should be made to assess the resulting losses.

Figure 32.2 shows a reasonable flowchart for preparing quantitative slides of cladoceran remains from different kinds of sediments. The various yes–no decision steps emphasize the desirability of stopping the processing as soon as possible to minimize loss of remains by dissolution or sieving and fragmentation. Additional sieving steps with coarse or fine screens may have to be inserted in any of the processing routes before slides are prepared. This is dictated by the fact that excessive organic matter cannot be eliminated chemically by acetolysis, as in pollen preparations, but only by some method of size fractionation or gravity centrifugation. Furthermore, sieving might be a better compromise than treatment with HF for removing excessive quantities of very small and very large mineral particles.

Some details concerning the various processing steps are the following:

(1) For each sample of sediment processed there should be available estimates of wet-weight per cm^3 (=bulk density), dry-weight, ash weight and organic content. If these have not been measured elsewhere in the research programme, they can be approximated easily by drying a known amount of sediment at 105°C, ashing a first time at 550°C, and then ashing a second time at approximately 1000°C. The first ashing oxidizes all the organic matter. The second ashing drives off the CO_2 of the calcium and magnesium carbonates but also drives off some water from the clay minerals (Mueller, 1964). Hence, if carbonate content is low and clay content high, the further loss in weight by ashing at the higher temperature could be misleading if ascribed entirely to carbonates. (See Chapter 21).

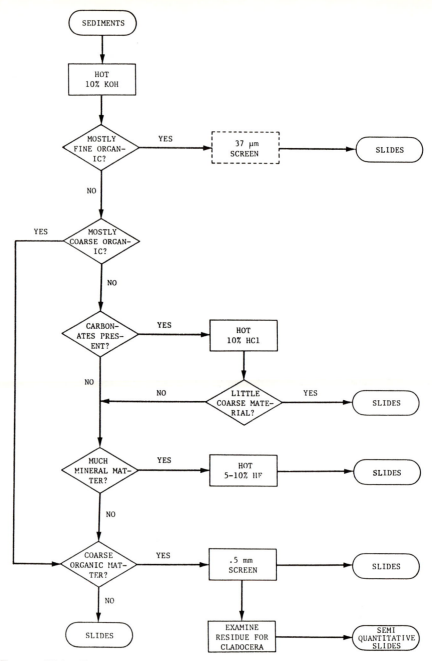

FIGURE 32.2. Recommended step procedure (algorithm) for processing various kinds of sediments to recover cladoceran and other chitinous remains. The procedure would not be applicable for calcareous and siliceous remains if the HCl and HF steps were used

(2)   The volume of sediment necessary to process varies with the density of cladoceran remains. Usually 1 cm^3 suffices, although in mineral (DeCosta, 1964) or bog sediments where the sedimentary matrix is voluminous, as much as 50 cm^3 may be necessary; and conversely in organic sediments with a high density of remains, as little as 0.2 cm^3 can be adequate. A good rule is to start with 1 cm^3 of fresh sediment, or its approximate equivalent if the sediment has dried since collecting, then to scale up or down from here as necessary.

(3)   Transfer 1 cm^3 of fresh sediment to 100 ml of 10% KOH (or NaOH) in a 250 ml beaker. Heat with gentle agitation at temperatures below boiling until the sediment is completely deflocculated. Normally this will take about 30 minutes. The process can be controlled by examining a small drop of suspended material microscopically. If all the sediment aggregates have not been dispersed, treatment should be continued. The deflocculation can be accomplished expeditiously on a stirrer hotplate, using a small flea (magnetic bar) at moderate speeds so as not to fragment the cladoceran remains. Ravera and Parise (1978) used cold 0.01N NH$_4$OH for 32 hours with agitation to deflocculate their sediments.

(4)   Spin down with a centrifuge, decant the supernatant, resuspend in distilled water to remove the KOH, spin down again, and decant. This sequence of steps is necessary after each chemical treatment to remove whatever dissolved substances have been generated as well as any excess of the chemical added.

(5)   In highly calcareous sediments the reaction of acid with the carbonates can be very vigorous, forming large and copious bubbles of CO$_2$. This is undesirable because of the risk of fragmenting the cladoceran remains as well as of losing material over the edge of the beaker. Reaction velocity can be controlled by using a lower concentration of HCl, or if critical, even by using weak organic acids, such as acetic or lactic (Palmer, 1960). Treatment with HCl must precede any treatment with HF to prevent the formation of CaF$_2$.

(6)   Whiteside and Harmsworth (1967) proposed a gentle treatment with HF to reduce the volume of mineral matter. The material to be treated is transferred to a copper beaker (Teflon or platinum can also be used), made up to 20 ml with distilled water, and then 5 ml of technical grade HF are added. Heating for 5 minutes on a hotplate was sufficient for the dissolution of most mineral matter. Crisman (1976) recommends full-strength HF, but anyone intending to use this or even weaker concentrations of HF should determine what, if any, differential losses of remains result from the treatment. DeCosta (1968a) concluded from paired treatments that it is as well to avoid HF treatment if possible.

(7)   Fine screens of bronze or stainless steel are suitable for getting rid of clays, silts and fine organic detritus (including small remains of Cladocera, unfortunately). Two sizes commonly used in North America are 250 meshes per inch (mesh size ~55 μm) and 400 meshes per inch (mesh size ~37 μm). A

cylindrical metal bucket (such as from a Clarke-Bumpus plankton sampler) 10 cm high and 5 cm in diameter with a lateral window about 4 cm high and half the circumference of the bucket covered by the desired screen is quite satisfactory. The sediment to be screened is placed in the bucket and flushed with distilled water with agitation (as by hand twirling or the use of a magnetic flea) until the water comes through the screen clear. The fine material should be examined microscopically to control the extent of loss of fine exoskeletal fragments.

(8)   The final residue of whatever procedural sequence used is transferred to a suitable bottle or vial and made up to a known volume, generally 10 or 20 cm^3, containing a couple of drops of formalin to discourage growth of bacteria and fungi. With the material being kept in uniform suspension, 0.05 ml aliquots are removed individually with a precision pipette having an aperture large enough to accommodate the largest cladoceran remains and then transferred to a microscope slide. The mounting medium that has been used most commonly is glycerine jelly, often stained with gentian violet or saffranin. Goulden (1966b) and Czeczuga et al. (1970) used glycerol. Kerfoot (1974) and Crisman (1976) used silicone oil (12,500 cs), following the procedure proposed by Davis (1966). Some workers recently have used a modified glucose solution (Karo syrup), as proposed by Taft (1978). Because glycerol and silicone oil, and to a lesser extent Karo syrup, do not harden, individual remains can be shifted into better orientation for examination of critical details by judicious pressure at the edge of the coverslip. A disadvantage of these media is the potential shifting of remains beneath the coverslip, making it difficult to relocate them from their recorded stage coordinates. The preparation slides should always be stored horizontally, rather than on edge, to minimize this condition.

Square 22 mm coverslips are very satisfactory, mounted two per slide, each coverslip representing a known fraction of the original sample. The quantity of mounting medium used should be sufficient to fill the space beneath the coverslip but not to run out excessively at the edges. The 'proper' amount can be learned only by experience. As a rule of thumb, five such double slides should be prepared for each level, although by examining the first coverslip prepared it is easy to determine if fewer slides would be sufficient or more required. But before this stage is reached the researcher should already have some idea as to the density of remains in the samples and possibly has already made adjustments in the volume of sample being processed and the final dilution volume.

(9)   Kerfoot (1974) used the palynological technique of adding a known quantity of some foreign object, in his case Eucalyptus pollen, to the sediment sample at the beginning of processing. The Eucalyptus pollen grains were counted in the final slide preparations along with the cladoceran remains. The number of the cladoceran remains counted on the slide, multiplied by the ratio of the total Eucalyptus grains added divided by the number observed, gives an

estimate of the total number of cladoceran remains in the sample, independently of the subsample volume used in preparing the slide.

(10) In addition to the quantitative slides from which the species abundance estimates are obtained, it is also desirable to make up qualitative slides of rare species, 'combination' microfossils (those containing two or more exoskeletal components still associated with one another), or any other particularly favourable remains that will help in the identification of the species present. To accomplish this, after an adequate number of quantitative slides has been prepared, some of the remaining residue is transferred to a long strip of water on a slide, which is then scanned systematically with a stereo microscope. Remains of interest are removed individually with a fine needle or a wire loop to a drop of water on a slide, accumulating a fair number in a single drop. The excess water is removed but not to complete dryness, and a drop of mounting medium added. After the remains have been arranged suitably with a needle, a coverslip is added. The mounting medium commonly used for chitinous remains is polyvinyl-lactophenol stained with lignin pink. Such preparations are particularly good for observing the fine morphological details needed to identify or characterize the species.

Needles and loops are easily made from 0.3 mm tungsten wire by electrolysis in 10–20% KOH. A 2 cm length of wire clamped to one terminal of a 6 V alternating electrical circuit is dipped into the electrolyte until the desired point has been produced. Depending on how deep the wire is inserted into the solution, the points can be short and stout or long and slender. The slender points can easily be bent into loops small enough even for handling rhizopod tests and turbellarian egg capsules. The advantage of tungsten wire over steel is that it is not affected by either acid or alkali. If the points become dull they can be resharpened by placing them in the electrolytic circuit, and if dirty they can be cleaned by placing them in a flame. The electrolytic procedure for producing points seems much better and less troublesome than using melted $NaNO_2$ in the older procedure.

EXAMINATION OF SLIDES

The slides will contain not only intact exoskeletal components but also fragments of them, the breakage having resulted mainly from predation on living animals or from the activity of various invertebrates in processing detritus and sediments containing remains. There is also a possibility of fragmentation by physical abrasion in high-energy situations, and there is an additional possibility of fragmentation during preparation of the slides. If the magnitude of the latter can be estimated, then the remaining fragmentation resulted from natural biological and physical processes.

As in pollen analysis, it is well to have a reference collection of whole animals and their exuviae against which unknowns can be compared. Exuviae are best for showing the fine morphological details used in identifying microfossils, but if they are not available, whole animals can be 'fossilized' by heating them in concentrated HCl according to the procedure of Megard (1965). At the beginning of any study it is desirable to determine from the qualitative slides what species are present using the combination microfossils and other particularly good remains picked from the sample. Literature sources that can be used to help identify the various components are Korde (1953, 1956), Frey (1958, 1959, 1960, 1961a,b, 1965, 1973, 1975, 1978, 1982a,b,c), Michael and Frey (1983, 1984), Mueller (1964), DeCosta (1964), Goulden (1964, 1966a,b), Goulden and Frey (1963), Sebestyén (1965, 1969a,b,1970), Megard (1967a,b), Kořínek (1971), Kadota (1972–76), Kubersky (1978), Smirnov (1978), Hann (1982) and Negrea (1983).

Faunistic summaries, especially those by Lilljeborg (1901), Smirnov (1971), Flössner (1972) and Negrea (1983), are useful for working with intact specimens and even the exoskeletal components, as are also the illustrated keys by Keilhack (1909), Wagler (1937), Brooks, (1959), and Scourfield and Harding (1966), and the monographs by Goulden (1968), Mordukhai-Boltovskoi (1968) and Smirnov (1976). The paper by Rossolimo (1927), in spite of its intriguing title, is of limited usefulness compared with other works mentioned, one reason being that it does not illustrate the headshields of any species. In truth everyone working on cladoceran microfossils is a pioneer, as no single comprehensive work is available that can be used as a starting point. What is needed is an atlas of the chydorids of Western Europe, this being the only part of the world where the fauna is even reasonably well-known.

So, employing whatever means are available, including the help of specialists to confirm identifications, the various exoskeletal components and their fragments encountered in scanning the quantitative slides systematically are identified and tabulated.

Table 32.1 suggests a matrix for tabulating the different kinds of cladoceran remains according to their degree of fragmentation. The objective is to arrive at estimates of the total number of exuviae of each species in the sediment, which are then equated to the number of individuals on the practical assumption that each species undergoes about the same number of moults, and that therefore each individual contributes approximately the same number of exuviae and exuvial components to the sediments. Goulden (1966a) also concluded that the variation in number of moults between species must be negligible. Where the components are fragmented, only those fragments that have some arbitrarily selected unique feature can be used in instar calculation. For example, the post-abdomen of *Eurycercus* and the shell of *Alonopsis elongata* are so highly distinctive that even minute fragments can be identified. To include all of them in the calculations would grossly distort the abundance of these species versus those (such as many species of *Alona*) that do not have exoskeletons distinctive on such a fine scale.

For each species present, separate estimates of abundance are made for each of the different skeletal components. The largest of these estimates is then taken as the most likely estimate of the abundance of the species in the sample being studied. Because of the considerable number of taxa that can be present in a single sample (often between 15 and 30, most of which will be chydorids), a rule of thumb should be to continue counting whole coverslips until at least 200 usable chydorids (=exuviae) have been tabulated. Even then all the rarer species will not be represented in the counts. For example, Frey (1961b) found in Schleinsee that the qualitative slides contained at each level 1–7 species (most commonly 3 or 4) in addition to those identified from at least 200 remains in the quantitative preparations. The decision that must be made is familiar to palynologists: at what point in an analysis is a further increment in precision of results not justified by the additional effort required to achieve it? Obviously the cutoff point will vary according to the objectives of the study.

The procedure recommended here of calculating the number of exuviae from each different skeletal component and then selecting the largest such value as the abundance estimate of that species is justified because of the complex analyses made from the data. Megard (1964), because of the extremely low numbers of remains in his sediments, as well as Whiteside and Harmsworth (1967), included all identifiable fragments in their calculations. Goulden (1966a), DeCosta (1968b) and Tsukada (1972) used both heads and shells. Deevey (1969) used heads as an index of *Bosmina* abundance, and Hofmann (1978a) used the most abundant remain of each species as an index for that species. Only the latter procedure approaches what is recommended here.

INTERPRETATION OF RESULTS

Chydorids live mainly in the littoral zone, but their exoskeletal remains are transported offshore and deposited in deepwater sediments. All samples of surficial offshore sediments yielded the same percentage composition of chydorids by species (Mueller, 1964; subsequent validation by more sophisticated analyses was made by DeCosta, 1968b, and Goulden, 1969b), which means that the relative abundance of chydorids is a good descriptor of the overall community in the lake at that point in time, integrated for differences in distribution of the various species according to habitat preferences and for seasonal variations in presence and abundance. This means also that a single offshore core of lake sediments is adequate for perceiving real changes over time in the proportional representation of the various species of chydorids. There must be some upper limit of lake size beyond which integration of relative abundance of remains for the entire lake can no longer be accomplished by water movements. We do not know what this limit is, because the largest lake Mueller studied was 204 ha. Any lakes larger than this selected

TABLE 32.1. Types of cladoceran remains recovered from sediments, their taxonomic distribution, and formulae for calculating the total number of exuviae and a fragmentation index

Skeletal component	Occurrence	Category				Total exuviae	Fragmentation index
		a	b	c	d		
Headshield[1]	Bosminidae[2]	Reasonably intact, with one or both antennules	Both antennules, but little of the head	One antennule, ditto	Others[3]	$a + b + \dfrac{c}{2}$	$\dfrac{2b + c}{2a + 2b + c}$
	Chydoridae	At least 3/4, including pores	Posterior portion with pores	Others		$a + b$	$\dfrac{b}{a + b}$
Shell	Bosminidae Chydoridae occasional Macrothricidae	Both valves reasonably whole and still attached together	One valve reasonably intact	Posterior-ventral part of one valve	Others	$a + \dfrac{b}{2} + \dfrac{c}{2}$	$\dfrac{b + c}{2a + b + c}$
Post-abdomen	Holopedium Sididae Daphniidae Moinidae some macrothricids Bosminidae Chydoridae	Complete dorsal margin with denticles	Distal portion with distal denticles	Others		$a + b$	$\dfrac{b}{a + b}$

	Both valves of ephippium still attached together	One valve intact	More than half of one valve	Others	
Ephippia	Daphniidae Moinidae occasional Macrothricidae Bosminidae Chydoridae				$a + \dfrac{b}{2} + \dfrac{c}{2}$ $\dfrac{b+c}{2a+b+c}$
Post-abdominal claws	Sididae Daphniidae Moinidae? Bosminidae Chydoridae				
Antennules	Bosminidae *Eurycercus* Macro-thricidae?	All are counted because the structures are small and seldom fragmented			
Antennal segments	Sididae				$\dfrac{a}{2}$
Caudal spines	Cercopagidae Leptodora				
Mandibles[4]	Probably general				

[1] Headshields of *Ilyocryptus* are occasionally found. Other macrothricids might turn up.

[2] In females the antennules are attached immovably to the head, whereas in males they are articulated with the head and become separated in the sediments An antennule or antennule fragment is tabulated if it contains the notch or triangular scale that covers the sensory papillae and setae about midway along the antennule. In *Bosminopsis* the two antennules are joined at their bases, yielding Y-shaped microfossils.

[3] The category 'others' indicates fragments that can be identified to species but which are not used to calculate number of instars according to the arbitrary criteria of this table.

[4] Mandibles are abundant in sediments, but as yet no studies have been undertaken (except in the genus *Daphnia*) to determine how closely they can be identified. Frequently only their distal masticatory surfaces are preserved, and because the right and left mandibles can differ significantly in the patterns of these surfaces (Fryer, 1963), identification even at the generic level may prove quite difficult. Only the mandibles of *Polyphemus*, *Bythotrephes*, *Leptodora* and *Daphnia* (Edwards, 1980) have been identified to species up to the present.

for investigation should probably be studied first for the areal homogeneity of relative abundance of remains in their surficial sediments, for without such homogeneity, many of the techniques used to analyse the data on cladoceran remains would not be applicable. On the other hand, lakes with reduced littoral zone or reduced abundance of macrophytes in the littoral zone will have only small representations of chydorids in their sediments (Boucherle and Züllig, 1983). These data on relative abundance can be analysed in various ways by computer to provide insight into:

(1) the responses of various lakes over time to external changes in climate, vegetation, erosion and weathering, internal morphometric changes resulting from accumulation of sediments or marked fluctuations in water level, and changes brought about by man (=palaeolimnology);

(2) similarity in response of groups of lakes to regional patterns of climate, vegetation and geological substrate (=regional limnology);

(3) the processes of establishment and progressive development of the chydorid taxocene and internal adjustments of the species to one another (=theoretical community ecology).

On the other hand, the concentrations of remains in the sediments, even offshore, are not uniform. Mueller (1964) found that the chydorids, as might be expected from where they live, are most abundant near shore and then decrease offshore to fairly uniform concentrations in the deep hypolimnion. Planktonic species (represented chiefly by *Bosmina* in Mueller's study), on the other hand, have a peak of abundance at depths corresponding to the upper part of the hypolimnion, with lesser concentrations toward shore and farther away from shore. Even the deep hypolimnion has a patchy distribution of remains. In only one of three lakes Mueller studied was the mean concentration of remains in the offshore samples larger than the variance of the mean (Goulden, 1969b). Thus, a single offshore core for studying lake history might yield higher concentrations at some levels and lower concentrations at others, not resulting from any real changes in production of Cladocera but only from the vagaries of a patchy distribution changing unpredictably over time. Furthermore, at falling water levels the concentration zone of *Bosmina* (and perhaps of other planktonic Cladocera) could move through the site, or at least come closer, and thereby bring about greater concentrations of both littoral and planktonic species without any real changes in production. These patterns of distribution of cladoceran remains seem contrary to the sediment focusing hypothesis of Lehman (1975), which implies that all fine constituents of the sediment should become relatively more concentrated in the deepest parts of the hypolimnion. Hence, although it is desirable to calculate annual accumulation rates of cladoceran remains, as is done for pollen and diatoms, caution must be used in interpreting the results.

Some of the techniques of analysing the data on cladoceran remains and the kinds of interpretations that have been made are given briefly below. Other more extensive reviews are given by Frey (1974, 1976), Crisman (1978), Brugam (1983), and Binford *et al.* (1983). An example of the kinds of analyses that have been made, using these various techniques, is given in Figure 32.3).

Abundance diagrams are the traditional diagrams that display changes in stratigraphic abundance of the individual taxa. Usually included also are a summary diagram of percentage planktonic versus percentage littoral, and curves for a number of chydorid taxa per level, species diversity index for chydorids, and the equitability component.

The *species diversity index* most commonly used is the Shannon–Wiener:

$$H = -\Sigma p_i \log_2 p_i$$

in which p_i is the proportion of the ith species in the total population. The magnitude of the index varies according to the number of species present and also their relative abundance. The realistic (not the mathematical) upper limit for the index is that associated with the *MacArthur Type* I *Distribution* (=broken stick model) (Goulden, 1966a, 1969b), almost as if this model had been devised with the chydorids specifically in mind. The goodness of fit of the observed distribution to that predicted by the model can be shown by graphs or by the *equitability ratio* (Lloyd and Ghelardi, 1964). Perturbations of the system, as from climate (Goulden, 1966a), agriculture (Goulden, 1966a), natural catastrophes such as episodes of volcanism (Tsukada, 1967a,b, 1972), and excessive eutrophication (Whiteside, 1970), cause lowered diversity indices as a result of the most common species becoming much more abundant than predicted and some of the less common species becoming much less common than predicted, or even being eliminated completely. Hence, progressive fluctuations in diversity and equitability over time can reflect successive disturbances and readjustments in the system, the causes of which should be revealed by other data being gathered. A connected time plot of diversity against number of species (Goulden, 1969a) can be instructive in showing how the chydorid community moves progressively closer to the enclosing envelope of maximum diversity predicted by the MacArthur model.

Species diversity of sedimentary remains has also been used to compare chydorid communities among various lakes in an attempt to discern the factors controlling diversity. For 20 lakes in Denmark and 14 in northern Indiana for which estimates of the annual production of phytoplankton were available, Whiteside and Harmsworth (1967) using the *Spearman rank correlation coefficient* found significant positive relationships between diversity and water transparency in both series of lakes, and a strong negative relationship between diversity and primary production in the Danish lakes, less strong in the Indiana lakes. They also showed that diversity was not related systematically to the area of the lake, and hence limnologists apparently do not have to be concerned

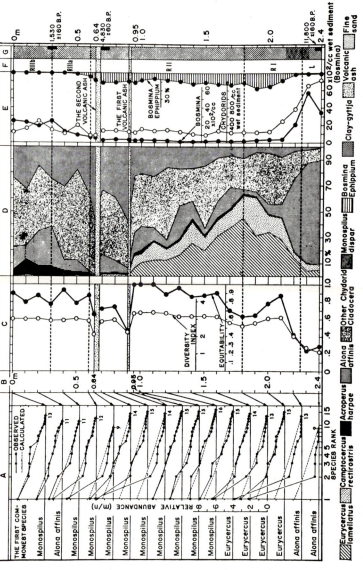

FIGURE 32.3. Diagram of sedimentary Cladocera in Lake Nojiri, Japan: (A) Actual fit of abundance data of the various species (solid line) to a predicted distribution based on the MacArthur Type I model (dashed line). (B) Depth in core. (C) Shannon–Wiener diversity index and Lloyd–Ghelardi equitability ratio. (D) Percentage composition of dominant taxa in chydorid community. (E) Concentrations of *Bosmina* and chydorid remains per cm³ of sediment. (F) Pollen zones. (G) Type of sediments and depth location of radiocarbon dates. Note the great effect on community composition of the two volcanic events (Tsukada, 1967).

with the species-area problem of terrestrial ecologists. Using *simple correlation* and *regression* analysis, Whiteside (1969) showed similar relationships for the Danish lakes, and in addition showed that there is a strong negative correlation between phytoplankton productivity and water transparency. Thus, species diversity of chydorids tends to be higher at high transparencies and low productivities.

The number of species of chydorids present depends on the structural diversity of the habitat. As most chydorids are associated with aquatic macrophytes, any factors that increase the depth distribution and diversity of species of macrophytes will be expected to increase the number of species of chydorids represented, and hence the species diversity. Hence, anything that decreases light penetration through water, such as excessive blooms of algae, increased mineral turbidity, and increased colour, can be expected to bring about a decrease in chydorid diversity and a strong departure from the Type I relative abundance. Synerholm (1979) found a negative correlation between chydorid diversity and specific conductivity for 32 Minnesota lakes, the softwater lakes having higher diversities. The explanation is the same. Softwater lakes in general have low production of phytoplankton, therefore higher transparency, and hence a greater bathymetric extent and diversity of the macrophytes.

Alhonen (1970b) and Whiteside (1970) pioneered the use of a *planktonic/ littoral ratio* to identify episodes of reduced water level, in this case the generally low water levels in subboreal time, the interpretation being that any significant lowering of water level increases the littoral habitat relative to the planktonic. Mueller (1964) had previously established a somewhat crude P/L *ratio* between the volume of the planktonic zone and the area of the lake bottom from the shoreline to the bottom of the epilimnon. For 11 lakes in Indiana and Wisconsin, the proportion of planktonic remains in the offshore sediments increased as this ratio increased. However, caution is advised in interpreting the changes in the planktonic/littoral ratio over time as resulting only from fluctuations in water level. Changes in habitat diversity at stable water levels could conceivably bring about the same changes.

At higher levels of productivity the quantity of remains in the sediments would be expected to be greater than at lower levels. This seemed true in the chain of lakes at Madison (Frey, 1960) but not in 20 Danish lakes (Harmsworth and Whiteside, 1968). However, both these studies were based on concentrations of remains in surficial sediments without any knowledge of the rate at which the sediments were accumulating, and hence there is no real basis for comparing the lakes. Here is an instance where information on the *accumulation rates* of cladoceran remains would be informative, because it is intuitive from the moulting pattern of the Cladocera and its relation to reproduction that the larger the population is at a given temperature, the more exuviae will be generated.

Sometimes the changes in occurrence and distribution of chydorids in the Holocene are so great that *faunal zones* are established, which may or may not be coincident with the pollen zones (Megard, 1964; Goulden, 1966a; DeCosta, 1968b; Harmsworth, 1968; Czeczuga *et al.*, 1970; Whiteside, 1970; Crisman, 1976). Such faunal zones should represent responses of the lake to major changes in the watershed outputs of water, dissolved substances and particulates. Most often these zones have been established by visual inspection, sometimes aided by marked discontinuities in species richness, species diversity, and equitability (DeCosta, 1968b). In an attempt to be even more objective in defining the zones, Crisman (1976) used Spearman rank correlation coefficients between successive levels to perceive the zonal boundaries. Although not used for this particular purpose, Hofmann (1978a) used a *cluster analysis* and the resulting *cluster diagram* (or dendrogram) to compare the similarity of chydorid assemblages in the various pollen zones of the Grosser Segeberger See. This technique is powerful and is destined to join the repertoire of standard procedures used in stratigraphic analysis of cladoceran remains as well as in comparing different lakes.

Another powerful tool is *discriminant analysis* for perceiving the variables that control present-day chydorid assemblages as an aid to interpreting the past. Whiteside (1970) grouped 80 Danish lakes by visual inspection into the three categories of clearwater non-polluted lakes, clearwater polluted lakes, and lakes with a perceptible brown colour. Data used in the discriminant analysis were seven commonly measured chemical and physical variables and the surficial sedimentary chydorid assemblage. In 83% of the instances the chydorid assemblage correctly predicted the type of lake, showing that community composition is by no means random, although at times it almost seems to be.

Synerholm (1979) studied the chydorid assemblages in 32 lakes along a transect extending from the prairie in eastern North Dakota through the deciduous forest zone and into the coniferous forest of northern Minnesota. She used both an *agglomerative clustering technique* (Orloci, 1967) and a *principal components analysis* (Orloci, 1966) to show the degree of relationship among the various lakes and to identify the important environmental variables. By means of a *stepwise discriminant analysis* she found that of the 34 species of chydorids present, the 15 identified by the computer as most important classified all the lakes into their correct vegetative formation, and that use of only the first four species in the computer list classified the lakes correctly by vegetation formation 81% of the time.

Hofmann (1978a) attempted an analysis of *niche overlap* in the chydorid assemblage, with moderate results. This technique has potentialities for revealing ecological similarities among the various species, but its utility needs further exploration.

Bosmina is bewildering in terms of its high diversity of morphotypes, particularly in lakes of the Baltic region, with even further morphological diversity resulting from cyclomorphosis. The origin of this great diversity in the Baltic lakes has not yet been demonstrated. Lieder (1983) claims it has arisen rapidly during the past 10,000 years or so by introgressive hybridization between one or two invading species and a long-time endemic species, whereas Hofmann (1977, 1978a,b) and Günther (1983) have found a time succession of species in lake cores from northern Germany that does not seem to agree with Lieder's prediction. In all four lakes studied, *Bosmina longirostris* was present throughout the postglacial time, along with a succession of *Eubosmina* species, from *longispina* at the beginning, through *kessleri* to *coregoni*, and possibly to *reflexa* at the present time. Hofmann regards this succession of species as indicating progressive eutrophication of the lakes.

Remains of *Bosmina* are often the most abundant remains of Cladocera in sediments. The headshields can be separated easily into *B. longirostris* and the *Eubosmina* species by the location of the lateral headpores (Goulden and Frey, 1963). Kořínek (1971) has presented a more detailed analysis of *Bosmina* headpores that may help provide some resolution of the *Eubosmina* complex. Quite typically the *Eubosmina* species occurring early in the history of a lake was replaced completely and sometimes almost instantaneously by *B. longirostris* (reviewed by Hofmann (1978a) and by Crisman and Whitehead (1978)). In three Swiss lakes with annually laminated sediments — Zürichsee, St. Moritzersee and Baldeggersee — replacement of *Bosmina (Eubosmina) longispina* by *B. longirostris* seemed to occur even in a single year (Boucherle and Züllig, 1983). This kind of replacement has usually been interpreted as the result of eutrophication, possibly mediated by the selective predation of *Chaoborus* on the larger species of *Bosmina*.

Changes in composition of total cladoceran assemblages over time should be influenced by any major changes in *predation*, as through the immigration and establishment of new predators (such as *Chaoborus*, tanypodine midges or plankton-eating fishes) or through changing conditions that might increase the efficiency of predators already present. *Chaoborus* may well have been an important agent in causing the *Bosmina* replacements in some of the instances reported, but it cannot be a universal explanation. For example, in the three north German lakes studied by Hofmann, the various species in the *Eubosmina* series increase in size toward the present, with a larger form invading the lakes and becoming abundant even when *Chaoborus* was at its peak. This is also contrary to what Kerfoot (1974) found in Frains Lake, Michigan. Here the size of mandibles in the sediments decreased toward the present, with the greatest decrease occurring in the last century at the time European man occupied and modified the watershed. Kerfoot postulates that indirectly, through increases in nutrient delivery to the lake, planktivorous fishes became more abundant and selectively preyed on the largest zooplankters present. Predation in the littoral community can also be important. Goulden (1971) found that of the various invertebrate predators in Lake Lacawac, Pennsylvania, the larvae of the tanypodine midges were most important in regulating population densities of chydorids living in the sediments.

Certainly more effort is needed to separate the effects of predation from other influences on species composition and size distribution of remains in the sediments. Nilssen (1978) has organized a helpful review on predation in relation to palaeolimnology.

There are few *indicator species* among the Cladocera, in the sense that they by themselves, independent of other species or other information, 'indicate' some special features or special events in the environment. Perhaps the closest approximation to this concept is *Chydorus sphaericus*, which often becomes superabundant in extreme eutrophication or in response to other types of severe perturbation. The only difficulty is that *C. sphaericus* is now known to be more than one species (Frey, 1980), with the common species in North America being markedly different from the common species in Europe. This is somewhat reminiscent of the *Bosmina* problem in that we are aware of the existence of an uncertain number of closely related taxa. When these species are finally differentiated, and if their remains in the sedimentary record can be identified unfailingly, then the resolution of our data and the interpretations that can be made from them will increase remarkably.

Aside from *C. sphaericus*, there is a considerable number of taxa including *Monospilus*, *Rhynchotalona*, *Chydorus piger*, *Chydorus gibbus* and others that preferentially occur on and in sediments rather than macrophytes, and hence 'indicate' sediments. Fluctuations in the ratio of these to macrophyte-associated species might be expected to occur under such conditions as pollution, which Whiteside (1970) found decreased the abundance of macrophyte-associated species but did not affect materially the abundance of the sediment-associated species. On the other hand, the proportion of these two groups of species in 'balanced' lakes might be controlled primarily by lake type or sediment type. For example, Lake Lacawac, a softwater lake, has a well-developed community of chydorids living in the sediments (Goulden, 1971), whereas Crooked Lake in northern Indiana, a hardwater lake with strongly calcareous sediments, has almost no chydorids living in the sediments (Quade, 1973).

CONCLUSIONS

It should be apparent from this discussion that the cladoceran remains in lake sediments are abundant, highly diversified, and have a very high information content. We are just beginning to develop or adapt sophisticated techniques for the interpretation of the data. The future looks bright.

Palynologists are concerned chiefly with the terrestrial part of the watershed–lake system. For them the lake sediments provide a convenient matrix for preserving in chronological sequence the pollen and spores, largely of terrestrial origin. Limnologists, on the other hand, are concerned with the lake itself and its response to changing deliveries of materials from the watershed. Community response to these changing watershed deliveries is indicated by changes in occurrence and abundance of aquatic microfossils in the sediments. Interpretation of the responses as well as the generating conditions must be based on all kinds of information available—chemical, physical and biological—with the Cladocera destined to play a prominent role.

REFERENCES

This is a 'working' list of references, by no means complete in spite of its length. It contains many of the specific studies on sedimentary Cladocera, and in addition it contains a few of the major taxonomic works that will help in identifying remains in the sediments and in understanding the relatively little that is known about the ecology of the various species. A few papers included in the references, but not in the text, are Frey (1955, 1962, 1967), Hrbáček (1969), and Zemp (1943). Many others could be mentioned, but they will be evident from the bibliographies of the individual papers. Because it is becoming apparent that many taxa in Europe and North America bearing the same names are different (Kubersky, 1977; Frey, 1980; Hann, 1982; Michael and Frey, 1983, 1984), and hence are likely to display different relative abundances within a complex community, transfer of ecological information in either direction across the Atlantic Ocean must be done with caution.

Alhonen, P. (1967). Palaeolimnological investigations of three inland lakes in south-western Finland. *Acta Bot. Fenn.*, **76**, 1–59.

Alhonen, P. (1969). The developmental history of Lake Inari, Finnish Lapland. *Annal. Acad. Scient. Fenn. A. III*, **98**, 1–18.

Alhonen, P. (1970a). The palaeolimnology of four lakes in south-western Finland. *Annal. Acad. Scient. Fenn. A. III*, **105**, 1–39.

Alhonen, P. (1970b). On the significance of the planktonic/littoral ratio in the cladoceran stratigraphy of lake sediments. *Commentat. Biol. (Helsinki)*, **35**, 1–9.

Alhonen, P. (1971). The Flandrian development of the Pond Hyrynlampi, southern Finland, with special reference to the pollen and cladoceran stratigraphy. *Acta Bot. Fenn.*, **95**, 1–19.

Alhonen, P. (1972). Gallträsket: the geological development and palaeolimnology of a small polluted lake in southern Finland. *Commentat. Biol. (Helsinki)*, **57**, 1–34.

Binford, M. W. (1982). Ecological history of Lake Valencia, Venezuela: interpretation of animal microfossils and some chemical, physical, and geological features. *Ecol. Monogr.*, **52**, 307–333.

Binford, M. W., Deevey, E. S., and Crisman, T. L. (1983). Palaeolimnology: an historical perspective on lacustrine ecosystems. *Ann. Rev. Ecol. Syst.*, **14**, 255–286.

Birks, H. H., Whiteside, M. C., Stark, D. M., and Bright, R. C. (1976). Recent palaeolimnology of three lakes in north-western Minnesota. *Quatern. Res.*, **6**, 249–272.

688 *Handbook of Holocene Palaeoecology and Palaeohydrology*

Boucherle, M. M. (1982). *An Ecological History of Elk Lake, Clearwater Co., Minnesota, Based on Cladocera Remains*, Ph.D. Thesis, Indiana University.
Boucherle, M. M., and Züllig, H. (1983). Cladoceran remains as evidence of change in trophic state in three Swiss lakes. *Hydrobiologia*, **103**, 141–146.
Bradbury, J. P., and Whiteside, M. C. (1980). Palaeolimnology of two lakes in the Klutlan Glacier Region, Yukon Territory, Canada. *Quatern. Res.*, **14**, 149–168.
Brakke, D. F. (1980). Atmospheric deposition during the last 300 years as recorded in SNSF lake sediments. III: Cladoceran community structure and stratigraphy. In: *Proc. Int. Conf. on the Ecological Impacts of Acid Precipitation*, Oslo, S.N.S.F., pp. 272–273.
Brooks, J. L. (1959). Cladocera. In: *Freshwater Biology* (2nd edn) (Ed. W. T. Edmondson), John Wiley, New York.
Brugam, R. B. (1983). Holocene palaeolimnology. In: *Late Quaternary Environments of the U.S. and the U.S.S.R. Vol. 2: Holocene Environments of the United States* (Ed. H. E. Wright), University of Minnesota Press, Minneapolis, pp. 208–221.
Crisman, T. L. (1976). *North Pond, Massachusetts: Postglacial variations in lacustrine productivity as a reflection of changing watershed–lake interactions*, Ph.D. Thesis, Indiana University, Bloomington.
Crisman, T. L. (1978). Reconstruction of past lacustrine environments based on the remains of aquatic invertebrates. In: *Biology and Quaternary Environments* (Ed. D. Walker, and J. C. Guppy), Australian Academy of Science, Canberra.
Crisman, T. L., and Whiteside, D. R. (1978). Palaeolimnological studies on small New England (U.S.A.) ponds. II: Cladoceran community responses to trophic oscillations. *Pol. Arch. Hydrobiol.*, **25** (1/2), 75–86.
Czeczuga, B., Gołebiewski, Z., and Kossacka, W. (1970). The history of Lake Wiżajny in the light of chemical investigations of the sediments and Cladocera fossils. *Schweiz. Ztschr. Hydrol.*, **32**(1), 284–299.
Czeczuga, B., and Kossacka, W. (1977). Ecological changes in Wigry Lake in the Post-glacial Period. II: Investigations of the cladoceran stratigraphy. *Pol. Arch. Hydrobiol.*, **24**(2), 259–277.
Davis, M. B. (1966). Determination of absolute pollen frequency. *Ecology*, **47**, 310–311.
DeCosta, J. J. (1964). Latitudinal distribution of chydorid Cladocera in the Mississippi Valley, based on their remains in surficial lake sediments. *Invest. Indiana Lakes & Streams*, **6**, 65–101.
DeCosta, J. J. (1968a). The history of the chydorid (Cladocera) community of a small lake in the Wind River Mountains, Wyoming, U.S.A. *Arch. Hydrobiol.*, **64**, 400–425.
DeCosta, J. J. (1968b). Species diversity of chydorid fossil communities in the Mississippi Valley. *Hydrobiologia*, **32**, 497–512.
Deevey, E. S. (1942). Studies on Connecticut lake sediments. III: The biostratonomy of Linsley Pond. *Amer. J. Sci.*, **240**, 233–264 and 313–324.
Deevey, E. S. (1955). Palaeolimnology of the Upper Swamp deposit Pyramid Valley. *Rec. Canterbury Mus.*, **6**(4), 291–344.
Edwards, C. (1980). The anatomy of *Daphnia* mandibles. *Trans. Amer. Micros. Soc.*, **99**, 2–24.
Flössner, D. (1972). Branchiopoda, Branchiura. *Die Tierwelt Deutschlands*, **60**, 1–501.
Frey, D. G. (1955). Längsee: a history of meromixis. *Mem. Ist. Ital. Idrobiol., suppl.*, **8**, 141–164.
Frey, D. G. (1958). The late-glacial cladoceran fauna of a small lake. *Arch. Hydrobiol.*, **54**, 209–275.

Frey, D. G. (1959). The taxonomic and phylogenetic significance of the head pores of the Chydoridae (Cladocera). *Int. Rev. ges Hydrobiol.*, **44**, 27–50.

Frey, D. G. (1960). The ecological significance of cladoceran remains in lake sediments. *Ecology*, **41**, 684–699.

Frey, D. G. (1961a). Differentiation of *Alonella acutirostris* (Birge, 1879) and *Alonella rostrata* (Koch, 1841) (Cladocera, Chydoridae). *Trans. Am. Micros. Soc.*, **79**, 129–140.

Frey, D. G. (1961b). Developmental history of Schleinsee. *Proc. Int. Assoc. Limnol.*, **14**, 271–278.

Frey, D. G. (1962). Cladocera from the Eemian Interglacial of Denmark. *J. Palaeontol.*, **36**, 1133–1154.

Frey, D. G. (1964). Remains of animals in Quaternary lake and bog sediments and their interpretation. *Arch. Hydrobiol., suppl. Ergebnisse der Limnologie*, **2**, 1–116.

Frey, D. G. (1965). Differentiation of *Alona costata* Sars from two related species (Cladocera, Chydoridae). *Crustaceana*, **8**, 159–173.

Frey, D. G. (1967). Phylogenetic relationships in the family Chydoridae (Cladocera). *Mar. Biol. Assoc. India, Proc. Symposium on Crustacea*, Pt I, 29–37.

Frey, D. G. (1973). Comparative morphology and biology of three species of *Eurycercus* (Cladocera, Chydoridae) with a description of *Eurycercus macracanthus* sp. nov. *Int. Rev. ges. Hydrobiol.*, **58**, 221–267.

Frey, D. G. (1974). Palacolimnology. *Mitt. Int. Verein. Limnol.*, **20**, 95–123.

Frey, D. G. (1975). Subgeneric differentiation within *Eurycercus* (Cladocera, Chydoridae) and a new species from northern Sweden. *Hydrobiologia*, **46**, 263–300.

Frey, D. G. (1976). Interpretation of Quaternary palaeoecology from Cladocera and midges, and prognosis regarding usability of other organisms. *Can. J. Zool.*, **54**, 2208–2226.

Frey, D. G. (1978). A new species of *Eurycercus* (Cladocera, Chydoridae) from the southern United States. *Tulane Stud. Zool. & Bot.*, **20**(1–2), 1–25.

Frey, D. G. (1980). On the plurality of *Chydorus sphaericus* (O. F. Müller) (Cladocera, Chydoridae), and designation of a neotype from Sjælsø, Denmark. *Hydrobiologia*, **69**, 83–123.

Frey, D. G. (1982a). Relocation of *Chydorus barroisi* and related species (Cladocera, Chydoridae) to a new genus and descriptions of two new species. *Hydrobiologia*, **86**, 231–269.

Frey, D. G. (1982b). The honeycombed species of *Chydorus* (Cladocera, Chydoridae): comparison of *C. bicornutus* and *C. bicollaris* n. sp. with some preliminary comments on *faviformis*. *Can. J. Zool.*, **60**, 1892–1916.

Frey, D. G. (1982c). The reticulated species of *Chydorus* (Cladocera, Chydoridae): two new species with suggestions of convergence. *Hydrobiologia*, **93**, 255–279.

Fryer, G. (1963). The functional morphology and feeding mechanism of the chydorid cladoceran *Eurycercus lamellatus* (O. F. Müller). *Trans. Roy. Soc. Edinburgh*, **64**(14), 335–381.

Fryer, G. (1968). Evolution and adaptive radiation in the Chydoridae (Crustacea, Cladocera): a study in comparative functional morphology and ecology. *Phil. Trans. Roy. Soc. London B.*, **254**, 221–385.

Fryer, G. (1974). Evolution and adaptive radiation in the Macrothricidae (Crustacea: Cladocera): a study in comparative functional morphology and ecology. *Phil. Trans. Roy. Soc. London B.*, **269**, 137–274.

Goulden, C. E. (1964). The history of the cladoceran fauna of Esthwaite Water (England) and its limnological significance. *Arch. Hydrobiol.*, **60**, 1–52.

Goulden, C. E. (1966a). La Aguada de Santa Ana Vieja: an interpretative study of the cladoceran microfossils. *Arch. Hydrobiol.*, **62**, 373–404.

Goulden, C. E. (1966b). The animal macrofossils. In: The history of Laguna de Petenxil, a small lake in northern Guatemala. *Mem. Connecticut Acad. Arts Sci.*, **17**, 1–126.

Goulden, C. E. (1968). The systematics and evolution of the Moinidae. *Trans. Am. Phil. Soc. NS*, **58**(6), 1–101.

Goulden, C. E. (1969a). Developmental phases of the biocoenosis. *Proc. Nat. Acad. Sci. (Washington)*, **62**, 1066–1073.

Goulden, C. E. (1969b). Interpretative studies of cladoceran microfossils in lake sediments. *Mitt. Int. Verein. Limnol.*, **17**, 43–55.

Goulden, C. E. (1971). Environmental control of the abundance and distribution of the chydorid Cladocera. *Limnol. Oceanogr.*, **16**, 320–331.

Goulden, C. E., and Frey, D. G. (1963). The occurrence and significance of lateral head pores in the genus *Bosmina* (Cladocera). *Int. Rev. ges. Hydrobiol.*, **48**, 513–522.

Günther, J. (1983). Development of Grossensee (Holstein, Germany): variations in trophic status from the analysis of subfossil microfauna. *Hydrobiologia*, **103**, 231–234.

Hann, B. J. (1982). Two new species of *Eurycercus (Bullatifrons)* from Eastern North America (Chydoridae, Cladocera). Taxonomy, ontogeny, and biology. *Int. Rev. ges. Hydrobiol.*, **67**, 585–610.

Harmsworth, R. V. (1968). The developmental history of Blelham Tarn (England) as shown by animal microfossils, with special reference to the Cladocera. *Ecol. Monogr.*, **38**, 223–241.

Harmsworth, R. V., and Whiteside, M. C. (1968). Relation of cladoceran remains in lake sediments to primary productivity of lakes. *Ecology*, **49**, 998–1000.

Hofmann, W. (1977). *Bosmina (Eubosmina)* populations of the Grosser Segeberger See (F.R.G.) during late-glacial and postglacial times. *Arch. Hydrobiol.*, **80**, 349–359.

Hofmann, W. (1978a). Analysis of animal microfossils from the Grosser Segeberger See (F.R.G.). *Arch. Hydrobiol.*, **82**, 316–346.

Hofmann, W. (1978b). *Bosmina (Eubosmina)* populations of Grosser Plöner See and Schöhsee lakes during late-glacial and postglacial times. *Pol. Arch. Hydrobiol.*, **25**(1/2), 167–176.

Hofmann, W. (1980). Tierische Mikrofossilien aus Oberflächensedimenten einiger Eifelmaare. *Mitt. Pollichia*, **68**, 177–184.

Hogan, M. E. (1977). *Studies of Cladoceran Microfossils from Short Cores of Lough Leane, Killarney*, B.Sc. Thesis, University College, Dublin.

Hrbáček, J. (1969). On the possibility of estimating predation pressure and nutrition level of populations of *Daphnia* (Crust., Clad.) from their remains in sediments. *Mitt. Int. Verein. Limnol.*, **17**, 269–274.

Kadota, S. (1972). Table 1 and Figs. 3–16, concerned with cladoceran and other animal remains recovered from Biwa core. In: *Palaeolimnology of Lake Biwa and the Japanese Pleistocene*, **I** (Ed. S. Horie), pp. 79–85.

Kadota, S. (1973). A quantitative study of microfossils in the core sample from Lake Biwa-Ko. *Jap. J. Limnol.*, **34**, 103–110. In Japanese.

Kadota, S. (1974). A quantitative study of microfossils in a 200–meter core sample from Lake Biwa. In: *Palaeolimnology of Lake Biwa and the Japanese Pleistocene*, **2** (Ed. S. Horie), pp. 236–245.

Kadota, S. (1975). A quantitative study of the microfossils in a 200-meter-long core sample from Lake Biwa. In: *Palaeolimnology of Lake Biwa and the Japanese Pleistocene*, **3** (Ed. S. Horie), pp. 354–367.

Kadota, S. (1976). A quantitative study of the microfossils in a 200-meter-long core sample from Lake Biwa. In: *Palaeolimnology of Lake Biwa and the Japanese Pleistocene*, **4** (Ed. S. Horie), pp. 297–307.

Keilhack, L. (1909). *Phyllopoda*, Die Süsswasserfauna Deutschlands, 10.

Kerfoot, W. C. (1974). Net accumulation rates and the history of cladoceran communities. *Ecology*, **55**, 51–61.

Korde, N. V. (1953). Metodika biologicheskogo analiza donnikh otlozhenii. *Metodika Izucheniya Sapropelevikh Otlozhenii*, **1**, 176–207.

Korde, N. V. (1956). Metodika biologicheskogo izucheniya donnikh otlozhenii ozer (polevaya rabota i biologicheskii analiz). *Zhizn Presnikh Vod SSSR*, **4** (1), 383–413.

Kořínek, V. (1971). Comparative study of the head pores in the genus *Bosmina* Baird (Crustacea, Cladocera). *Věst. Česk. Společ. Zool.*, **35**, 275–296.

Kubersky, E. S. (1977). Worldwide distribution and ecology of *Alonopsis* (Cladocera: Chydoridae) with a description of *Alonopsis americana* sp. nov. *Int. Rev. ges. Hydrobiol.*, **62**, 649–685.

Lehman, J. T. (1975). Interpreting lake sediment deposition rates: the sediment focusing effect. *Quatern. Res.*, **5**, 541–550.

Lieder, U. (1983). Introgression as a factor in the evolution of polytypical plankton Cladocera. *Int. Rev. ges. Hydrobiol.*, **68**, 269–284.

Lilljeborg, W. (1901). Cladocera Sueciae. *Nova Acta Soc. Sci. Upsal.*, Ser. III, **vi**, 701pp.

Lloyd, M., and Ghelardi, R. J. (1964). A table for calculating the 'equitability' component of species diversity. *J. Anim. Ecol.*, **33**, 217–225.

Megard, R. O. (1964). Biostratigraphic history of Dead Man Lake, Chuska Mountains, New Mexico. *Ecology*, **45**, 529–546.

Megard, R. O. (1965). A chemical technique for disarticulating the exoskeletons of chydorid Cladocera. *Crustaceana*, **9**, 208–210.

Megard, R. O. (1967a). Three new species of *Alona* (Cladocera, Chydoridae) from the United States. *Int. Rev. ges. Hydrobiol.*, **52**, 37–50.

Megard, R. O. (1967b). Late-Quaternary Cladocera of Lake Zeribar, western Iran. *Ecology*, **48**, 179–189.

Michael, R. G., and Frey, D. G. (1983). Assumed amphi-Atlantic distribution of *Oxyurella tenuicaudis* (Cladocera, Chydoridae) denied by a new species from North America. *Hydrobiologia*, **106**, 3–35.

Michael, R. G., and Frey, D. G. (1984). Separation of *Disparalona leei* (Chien, 1970) in North America from *D. rostrata* (Koch, 1841) in Europe (Cladocera, Chydoridae). *Hydrobiologia*, **114**, 81–108.

Mordukhai-Boltovskoi, Ph.D. (1968). On the taxonomy of the Polyphemidae. *Crustaceana*, **14**, 197–209.

Mueller, W. P. (1964). The distribution of cladoceran remains in surficial sediments from three northern Indiana lakes. *Invest. Indiana Lakes & Streams*, **6**, 1–63.

Negrea, Şt. (1983). *Cladocera*. Fauna Rep. Social. România, Crustacea, Vol. 4, Fasc. 12, 399 pp., Acad. Republ. Soc. România, Bucureşti.

Nilssen, J. P. (1978). Selective vertebrate and invertebrate predation — some palaeolimnological implications. *Pol. Arch. Hydrobiol.*, **25**(1/2), 307–320.

Orloci, L. (1966). Geometric models in ecology. I: The theory and application of some ordination methods. *J. Ecol.*, **54**, 193–215.

Orloci, L. (1967). An agglomerative method for the classification of plant communities. *J. Ecol.*, **55**, 193–206.

Palmer, A. R. (1960). Miocene copepods from the Mojave Desert, California. *J. Palaeontol.*, **34**, 447–452.

Prat, N., and Daroca, M. V. (1983). Eutrophication processes in Spanish reservoirs as revealed by biological processes in profundal sediments. *Hydrobiologia*, **103**, 153–158.

Quade, H. W. (1973). *The Abundance and Distribution of Littoral Cladocera as Related to Sediments and Plants*, Ph.D. Thesis, Indiana University.

Ravera, O., and Parise, G. (1978). Eutrophication of Lake Lugano 'read' by means of planktonic remains in the sediment. *Schweiz. Z. Hydrol.*, **40**, 40–50.

Rossolimo, L. (1927). *Atlas tierischer Überreste in Torf und Sapropel*, Volkskommissariat für Landwirtschaft R.S.F.S.R., Zentrale Torfstation, Moskau. Russian text, pp. 1–23; German text, pp. 25–48.

Scourfield, D. J., and Harding, J. P. (1966). *A Key to the British Species of Freshwater Cladocera*, Freshwater Biol. Assoc., Scient. Publ. 5, 3rd edn.

Sebestyén, O. (1965). Cladocera studies in Lake Balaton III: preliminary studies for lake history investigations. *Annal. Biol. Tihany*, **32**, 187–288. In Hungarian, with summaries in English and Russian.

Sebestyén, O. (1969a). Studies on *Pediastrum* and cladoceran remains in the sediments of Lake Balaton, with references to lake history. *Mitt. Int. Verein. Limnol.*, **17**, 292–300.

Sebestyén, O. (1969b). Cladocera studies in Lake Balaton IV: subfossil remains in the sediments of Lake Balaton I. *Anal. Biol. Tihany*, **36**, 229–256. In Hungarian, with extensive English summary and Russian abstract.

Sebestyén, O. (1970). Cladocera studies in Lake Balaton IV: subfossil remains in the sediments of Lake Balaton II. *Anal. Biol. Tihany*, **37**, 247–279. In Hungarian, with summaries in English and Russian.

Smirnov, N. N. (1971). *Chydoridae fauny mira*, Fauna S.S.S.R., Nov. Ser. 101, Rakoobraznyye, T.1, vyp. 2, 531pp. Also published as an English translation (1974) by the Israel Program for Scientific Translations, Jerusalem. Available from the U.S. Dept. of Commerce, National Technical Information Service, Springfield, Va. 22151, their no. TT-74-50007.

Smirnov, N. N. (1976). *Macrothricidae i Moinidae fauny mira*, Fauna S.S.S.R., Nov. ser. 112, Rakoobraznyye, T.1, vyp. 3, 237pp.

Smirnov, N. N. (1978). Metody i nekotoryye resul'taty istoricheskoy biotsenologii vetvistousykh rakoobraznykh. In: *Ekologiya Soobshchestv Ozera Glubokogo* (Ed. G. D. Polyakov), Izdatel'stvo Nauka, Moscow, pp. 105–173.

Synerholm, C. C. (1979). 'The chydorid Cladocera from surface lake sediments in Minnesota and North Dakota,' *Arch. Hydrobiol.*, **86**; 137–151.

Taft, C. E. (1978). 'A mounting medium for fresh-water plankton,' *Trans. Am. Micros. Soc.*, **97**, 263–264.

Tsukada, M. (1967b). (Successions of Cladocera and benthic animals in Lake Nojiri), *Jap. J. Limnol.*, **28**, 107–123. In Japanese.

Tsukada, M. (1967a). (Fossil Cladocera in Lake Nojiri and ecological order), *Quatern. Res. (Tokyo)*, 6(3), 101–110. In Japanese.

Tsukada, M. (1972). 'The history of Lake Nojiri, Japan,' *Trans. Connecticut Acad. Arts Sci.*, **44**, 339–365.

Wagler, E. (1937). *Crustacea (Krebstiere)*, Die Tierwelt Mitteleuropas, II. Bd., Lief. 2a, 224 pp.

Wesenberg-Lund, C. (1901). Studier over Søkalk, Bønnemalm og Søgytje i danske indsøer. *Medd. Dansk Geol. Foren.*, **7**, 1–180.

Whiteside, M. C. (1969). Chydorid (Cladocera) remains in surficial sediments of Danish lakes and their significance to paleolimnological interpretations. *Mitt. Int. Verein. Limnol.*, **17**, 193–201.

Whiteside, M. C. (1970). Danish chydorid Cladocera: modern ecology and core studies. *Ecol. Monogr.*, **40**, 79–118.

Whiteside, M. C., and Harmsworth, R. V. (1967). Species diversity in chydorid (Cladocera) communities. *Ecology*, **48**, 664–667.

Zemp, F. (1943). Funde von Kleintierreliken im neolithischen Wauwiler See. *Z. Hydrol.*, **9**, 50–70.

Handbook of Holocene Palaeoecology and Palaeohydrology
Edited by B. E. Berglund
© 1986 John Wiley & Sons Ltd.

33

Ostracod analysis

Heinz Löffler

Department of Limnology, University of Vienna, Wien, Austria

INTRODUCTION

The subclass Ostracoda belongs to the old group of Crustacea, occurring as early as the Cambrian. The number of recent inland water species amounts approximately to more than 2000 of a total of 35,000 taxa described, both fossil and living, from the sea and inland waters (McKenzie, 1973), and dwelling within a wide variety of habitats, including, in exceptional cases, moist terrestrial environments.

In lake sediments ostracod shells are normally preserved very well if there is a sufficient concentration of alkaline earth elements present to prevent the decomposition of the calcareous part of the carapace. Even then, however, its organic part may persist if it is strong enough, as in the Cyclocypridinae. Moreover, remains of organs and appendages such as the furca, the mandibula, the antennula, the antenna and the Zenker's organ may also be found in the sediment (Löffler, 1978b). So far, as with the persistent resting eggs, they have had little consideration.

In contrast to chydorids, bosminids and daphnids, which live in the littoral and pelagic regions of lakes, some ostracods are profundal dwellers and as such reflect the conditions of the deep benthic zones of a lake and its changes. Since remains of two other profundal groups, Turbellaria and Chironomidae, are problematic with respect to their identifications, ostracods very often offer the only access to the history of the profundal environment of a deep lake.

SAMPLING METHODS

When studying sediment from the last hundred years, samplers of the Gilson or

Kajak type may be used. If the sediment is more solid an Ekman or Auerbach sampler will be sufficient. For an analysis of the full lake history, however, core samplers such as the Kullenberg or its Livingstone modification are indispensable.

In small lakes especially, more than one core should be taken from each vertical zone. A suitable sampling strategy for each lake, however, will be essential. Landslides within the lake basin, as well as drifting of ostracod remains from the littoral to profundal zones, must be considered. The latter is insignificant in large lakes with areas of more than 20 km^2.

Since a minimum volume of 5 cm^3 of sediment is necessary to obtain a representative figure of ostracod remains, the diameter of the core sampler should have an adequate diameter, especially if one wishes to obtain a year by year history. 5–20 cm^3 (it is also desirable to indicate the dry weight of the sample from each inspected sediment layer) are then transferred to a 50 μm filter and carefully washed. For the investigation of appendages like mandibulas an even smaller mesh size (20 μm) is desirable. After washing the inorganic gyttja, frequently only the remains of organisms are left, and thus they are easy to handle. Problems occur with coarser inorganic material, but here an investigation under a binocular microscope with reflected light is helpful in sorting the ostracod shells.

Sometimes decanting procedures or treatment with alcohol through which air is bubbled may be used to concentrate ostracod shells. Organic material provides the most problems for the sorting of ostracod remains. Decanting techniques and air bubbling may also help in such cases, and sieving apparatus such as described in Löffler (1961a) may be advisable.

PRESERVATION AND IDENTIFICATION

Ostracod remains should be preserved either in alcohol or, even better, dry (Franke cells). Appendages obtained from the sediment or by dissection of the animals are best preserved in polyvinylalcohol; staining is optional. The identification of ostracod shells and other remains such as mandibulas can be performed by both simple microscopic techniques and more sophisticated techniques like scanning-electron microscope studies (e.g. Langer, 1971).

Certain features of the ostracod carapace are most conveniently studied under the binocular microscope with reflected light and a magnification of 40–50×, whereas other details of the valves can only be adequately observed in transmitted light under higher magnifications (300× or more). For detailed investigations detached valves are necessary, and the following methods may be employed to open closed carapaces (Morkhoven, 1962).

Normally the valves of the carapaces found in lake sediment are separated; if, however, this is not so the carapace may be opened by sharp needles in the middle of the ventral or dorsal margin where they cause the least damage (if at all). There is a glueing technique using Canada balsam or some other suitable medium by which the valves can be safely separated. Finally, Oertli (1956) has described a method which can be applied where sufficient material is present to allow for partial loss. This involves repeated heating and sudden cooling in water (Morkhoven, 1962).

For identification, the size, shape/outline and details of ornamentation of the valves are most important. For the study of ornamentation and the hinge, hyaline specimens are often undesirable. Therefore, procedures to obtain opaque valves have been developed. Heating by means of a 6 V electric heater on a special slide is the most convenient technique; this must be carefully carried out, since excessive heating may result in destruction of the specimen. For the study of certain features, opaque specimens may be stained with a colour of a film of silver. The dye may be a mixture of chinese ink and a weak solution of arabic gum and water or a water-base green aniline paint. Both dyes may be removed again by washing with water (Morkhoven, 1962). Silver-stained valves are obtained by painting heated specimens with a 3–5% solution of silver nitrate. After drying they are heated again, thus reducing the silver nitrate. The silver layer cannot be removed. If transmitted light investigation is desirable (lateral pores, marginal pore cannals, details of the hinge etc.) valves can be submerged in various fluids such as water, glycerine or terpineol. Sometimes sections through the peripheral or any other part of a valve must be performed by ordinary grinding methods, which are rather time-consuming (Morkhoven, 1962). Very often it is desirable to study special modifications of the carapace which provide information of ecological value (e.g. Rosenfeld, 1977; Herrig, 1975). For scanning-electron microscope studies shells have to be cleaned carefully and prepared according to the usual gold staining method (Langer, 1971).

With respect to the systematics it should be mentioned that many ostracods, like the genus *Ilyocypris*, are still very problematic and in such cases their identification is somewhat questionable.

PRESENTATION OF RESULTS

The number of ostracods per unit volume and per unit dry-weight of sediment can be given either as a percentage of the various species or as a total figure (Figures 33.1 and 33.2). It is desirable also to present the figure of different larval stages. Very often, however, this presents considerable difficulties. Likewise presentation of undamaged shells in comparison with shell fragments can be informative.

For an interpretation of the material obtained in addition to ostracod investigations, absolute dating as, for example, by the ^{14}C technique or by pollen analysis is indispensable. Of course, it is also highly desirable to gain additional information about the palaeoclimatic and trophic condition of the

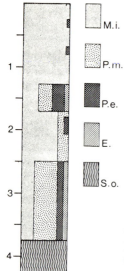

FIGURE 33.1.   Presentation of the distribution of ostracods obtained off the Gironde coast in France. The figure
represents the percentage of different ostracod associations: M.i. = marine infralittoral; P.m. = marine phytal;
P.e. = euryhaline phytal; E. = euryhaline; S.o. = without
ostracods. The same presentation could be applied for
lake cores. *Redrawn from Moyes (1974)*

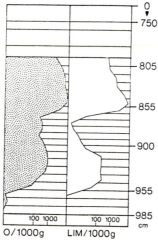

FIGURE 33.2.   Presentation of ostracods in a core form Neuwieder Becken (Kempf,
1967): O = total of ostracods per 1000 g of sediment; LIM = total of *Limnocythere*
species per 1000 g of sediment

investigation site by methods like $^{16}O/^{18}O$ ratios, pigment analysis etc. At
present, the indicative value of ostracods especially from the profundal zone is
based mainly on field data and ecological observations. Therefore, it is
essential to obtain data of different physiological parameters from laboratory
studies, such as salinity and temperature tolerance, as well as metabolic rate,
feeding habits and locomotion. So far, in comparison with other crustaceans,

data on such is limited, and even the life histories of many species are lacking (Hiller, 1972).

RESULTS OBTAINED BY OSTRACOD STUDIES IN LAKE SEDIMENTS

Although ostracods play an important role in geological investigations, particularly in oil stratigraphy, little attention has been paid to them in palaeolimnological studies, which are almost exclusively based on chydorids, chironomids and diatoms (e.g. Delorme, 1967, 1968).

Water movement, temperature, salinity and other chemical parameters obviously all influence the ostracod faunal composition and have been discussed many times in the literature (e.g. Oertli, 1971; Moyes, 1974; Swain, 1975; Hagermann, 1967; Löffler and Danielopol, 1977; Ganning, 1971).

In spite of the fact that many species are, for example, typical of certain salinities (e.g. Puri, 1964), very little use has been made of this characteristic in astatic inland waters (Figure 33.3).

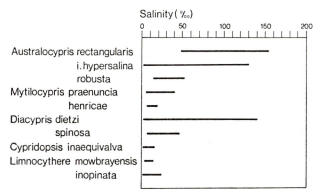

FIGURE 33.3.   Salinity range of some ostracod species (from De Dekker, 1981, and Löffler, 1961)

Long-lived species such as *Cytherissa lacustris*, which, moreover, are very unlikely to be subject to passive dispersal, may be of indicative value for the permanence of a body of water. Thus, the combination of *Cytherissa lacustris* and *Limnocythere sanctipatricii* in the no longer existing precursor of Neusiedlersee has been interpreted (Löffler, 1972) as indicating the presence of a permanent cold body of water (a maximum depth of less than 1 m is indicated by morphometric features of the landscape — Figure 33.4).

The collapse of a late Pleistocene lake (Kühnsdorfer See) is shown by a change from a profundal fauna to that of a littoral one. These results were obtained from analysis of a core taken from the shore of its remnant Klopeiner See (Löffler, 1971, 1972).

Interesting information obtained from ostracods has included the shift from

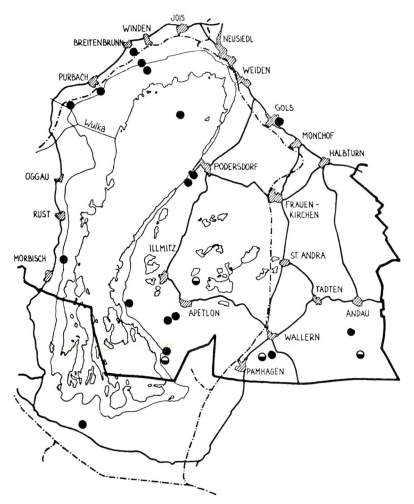

FIGURE 33.4. Subfossil distribution of *Cytherissa lacustris* (black circles) and *Limnocythere sanctipatricii* (black and white circles) in the region of Neusiedlersee. *Redrawn from Carbonel* et al. *(1984)*

a holomictic to a meromictic condition in several Austrian lakes (Figure 33.5; Löffler, 1975a, 1978a). It could be demonstrated by the profundal ostracod fauna that this shift already occurred during the preboreal period in large lakes. Moreoever, in some small lakes, sedimentation has resulted in an infilling of the monimolimnion, thus re-establishing a holomictic condition. In such a case the profundal zone has been recolonized by ostracods, very often by species which differ from those of the pre-meromictic stage (Löffler, 1977).

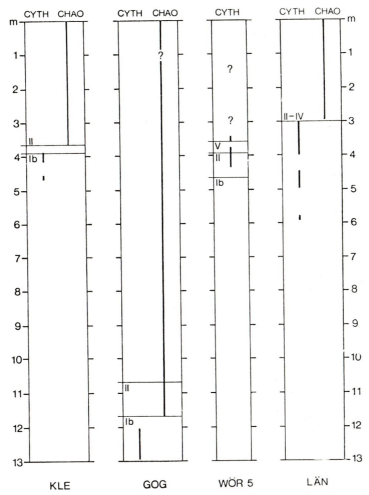

FIGURE 33.5. Distribution of *Cytherissa lacustris* (Cyth) and *Chaoborus* (Chao) in Kleinsee (Kle), Goggausee (Gog), Wörthersee (Wör 5) and Längsee (Län): Ib = Bölling; II = Alleröd; IV = Preboreal; V = Boreal (according to Firbas) (from Löffler, 1983)

In lakes which have been in existence since the Tertiary, thus offering an interesting and changing history, ostracods will be of great interest. So far this group of lakes has not been investigated. The microevolution of ostracods should be given special attention in such lakes (e.g. Benson and McDonald, 1963; Hartmann, 1975).

During eutrophication solid profundal sediments may become covered with strata of ooze, resulting from increased sedimentation of organisms. Some of the non-swimming and heavy ostracod species, like *Cytherissa lacustris*

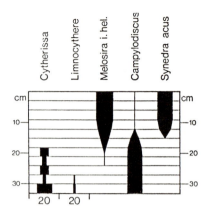

FIGURE 33.6.   Core sample from Zeller See (part of Lake Constance), width of each core scheme indicating 20 valves of *Cytherissa lacustris* and *Limnocythere sancti-particii*. Diatoms are presented semiquantitatively (*Melosira islandica helvetica*, *Campylodiscus noricus* and *Synedra acus*). *Redrawn from Löffler (1969)*

(specific weight 2.4), do not tolerate such conditions and disappear from the profundal zone. In Lake Constance *Cytherissa* has, over the last hundred years, left large areas of the benthal zone (Figure 33.6).   There is almost no record of the effect of heavy metals, pesticides, mineral oils and other more recent human influences on ostracods. In addition to relevant physiological laboratory investigations, it seems very desirable to consider quantitative aspects of the palaeoecology of ostracods; aspects of their intraspecific variations of morphology depending on the environment and population dynamics (Puri, 1964; Neale, 1969). Finally, computer simulation methods will certainly increase in importance in the palaeoecology of ostracods (Kaesler, 1969a,b).

Acknowledgement

The author is very much indebted for suggestions and criticisms made by Dan Danielopol, Mondsee.

REFERENCES

Bassiouni, M. E. A. A. (1969). Technische Hinweise zur Säuberung von Ostracoden-Schalen. *Paläont. Z.*, **43**, 3/4, 230–231.
Benson, R. H. (1966). Recent marine podocopid ostracodes. *Oceanogr. Mar. Biol. Ann. Rev.*, **4**, 213–232.
Benson, R. H., and MacDonald, H. C. (1963). Postglacial (Holocene) ostracodes from Lake Erie. *Univ. Kansas Pal. Contrib.*, 1–26.
Carbonel, P., Colin, J. P., Danielopol, D. L., Löffler, H., and Neustrueva, I. U. (1984). The Palaeoecology of limnic ostracods — a review of some major topics.
De Dekker, P. (1981). Ostracods of athalassic saline lakes. *Hydrobiologia*, **81**, 131–144.

Delorme, L. D. (1967). Field key and methods of collecting freshwater ostracodes in Canada. *Can. J. Zool.*, **45**, 1275–1281.

Delorme, L. D. (1968). Pleistocene freshwater ostracoda from Yukon, Canada. *Can. J. Zool.*, **46**, 859–876.

Ganning, B. (1971). On the ecology of *Heterocypris salinus, H. incongruens* and *Cypridopsis aculeata* (Crustacea: Ostracoda) from Baltic brackish-water rockpools. *Marine Biol.*, **8**, 271–279.

Hagerman, L. (1967). Ostracods of the Tvärminne area, Gulf of Finland. *Comm. Biologicae Soc. Sci. Fennica*, **30**, 2, 1–12.

Hartmann, G. (1975). Ostracoda. In: *Klass. Ord. Tierreichs* (Ed. Bronn), Fischer Verlag, Jena, pp. 569–789.

Hartmann, G. (Ed.) (1976). Evolution of Post-Palaeozoic Ostracoda, *Proc. 5th Int. Symp. Abh. Ver. Naturwiss. Ver. Hamburg (HF)*, 18/19, 336 pp.

Herrig, E. (1975). Über Schalen-Inkrustationen bei Ostracoden (Crustacea). *Z. Geol. Wiss.*, **3**, 671–685.

Hiller, D. (1972). Untersuchungen zur Biologie und zur Ökologie limnischer Ostracoden aus der Umgebung von Hamburg. *Arch. Hydrobiol.*, **40**, 400–497.

Kaesler, R. L. (1969a). Aspects of quantitative distributional palaeoecology. *Computer Appl. Earth Sciences*, 99–120.

Kaesler, R. L. (1969b). Ordination and character correlations of selected recent British Ostracoda. *Math. Geol.*, **1**, 97–111.

Kempf, E. K. (1967). *Ilyocypris schwarzbachi* n. sp. (Crustacea, Ostracoda) und ein vorläufiges Ostrakoden-Diagramm aus dem pleistozänen Löss von Kärlich (Neuwieder Becken). *Sonderveröff. Geol. Inst. Univ. Köln*, **13**, 65–79.

Kristan-Tollmann, E. (1977). Zur Methode der Untersuchung des Schliessmuskelfeldes von Ostracoden. *Paläont. Z.*, **51**, 3/4, 169–172.

Langer, W. (1971). Rasterelektronenmikroskopische Beobachtungen über den Feinbau von Ostracoden-Schalen. *Paläont. Z.*, **45**, 3/4, 181–186.

Löffler, H. (1961a). Vorschlag zu einem automatischen Schlämmverfahren. *Int. Rev. Hydrobiol.*, **46**, 288–291.

Löffler, H. (1961b). Beiträge zur Kenntnis der iranischen Binnengewässer II. *Int. Rev. Hydrobiol.*, **46**, 3. 309–406.

Löffler, H. (1969). Recent and subfossil distribution of *Cytherissa lacustris* (Ostracoda) in Lake Constance. *Mitt. Int. Verein. Limnol.*, **17**, 240–251.

Löffler, H. (1971). Daten zur subfossilen und lebenden Ostrakodenfauna in Wörthersee und Klopeinersee. *Carinthia Sonderh.*, **31**, 79–89.

Löffler, H. (1972). The distribution of subfossil ostracods and diatoms in pre-alpine lakes. *Int. Ass. Limnol. Trans.*, **78**, 7039–7050.

Löffler, H. (1975a). The onset of meromictic conditions in Goggausee, Carinthia. *Verh. Int. Verein. Limnol.*, **19**, 2284–2289.

Löffler, H. (1975b). The onset of meromictic conditions in alpine lakes. *Royal Soc. New Zealand*, Wellington 1975, 211–214.

Löffler, H. (1975c). The evolution of ostracod fauna in alpine and pre-alpine lakes and their value as indicators. *Bull. Am. Palaeont.*, **65**, 433–443.

Löffler, H. (1977). 'Fossil' Meromixis in Kleinsee (Carinthia) indicated by ostracodes. In: *Aspects of Ecology and Zoogeography of Recent and Fossil Ostracoda* (Eds. H. Löffler and D. L. Danielopol), Junk b.v. Publ., The Hague, pp. 321–325.

Löffler, H. (1978a). 'The palaeolimnology of some Carinthian lakes with special reference to Wörthersee. *Pol. Arch. Hydrobiol.*, **25**, 1/2, 227–232.

Löffler, H. (1978b). Limnological and palaeolimnological data on the Bale Mountain lakes (Ethiopia). *Verh. Int. Verein. Limnol.*, **20**, 1131–1138.

Löffler, H. (1983). Aspects of the history and evolution of alpine lakes in Austria. *Hydrobiologia*, **100**, 143–152.

Löffler, H., and Danielopol, D. L. (Eds.) (1977). Aspects of ecology and zoogeography of recent and fossil ostracoda, *Proc. 6th Int. Symp. Ostracoda (Saalfelden, Salzburg)*, Junk b.v. Publ., The Hague, 521 pp.

McKenzie, K. G. (1973). The biogeography of some cainozoic ostracoda. *Spec. Papers in Palaeont.*, **12**, 137–153.

Morkhoven van, F. P. C. M. (1962) *Post-Palaeozoic Ostracoda* vol. 1, Elsevier, Amsterdam, 204 pp.

Moyes, J. (1974). Un example d'étude paléoécologique et paléogéographique: la vasiere Quest-Gironde et son évolution durant l'Holocene. *Bull. Inst. Geol. Bassin Aquitaine*, **16**, 3–30.

Neale, J. W. (Ed.) (1969). *The Taxonomy, Morphology and Ecology of Recent Ostracoda*, Oliver & Boyd, Edinburgh, 553 pp.

Oertli, H. J. (1956). Ostracoden aus der oligozaenen und miozaenen Molasse der Schweiz. *Schweizer Paläont. Abh.*, 119pp.

Oertli, H. J. (Ed.) (1971). Paléoécologie Ostracodes, Pau 1970, *Bull. Centre. Rech. Pau-SNPA*, **5**, 953pp.

Puri, H. (Ed.) (1964). Ostracods as ecological and palaeoecological indicators. *Publ. Staz. Zool. Napoli*, 33 suppl., 612pp.

Rosenfeld, A. (1977). The sieve pores of *Cyprideis torosa* (Jones, 1850) from the Messimain Mavqi'im Formation in the Coastal Plain and Continental Shelf of Israel as an indicator of palaeoenvironment. *Israel J. Earth Sci.*, **26**, 89–93.

Swain, F. (Ed.) (1975). Biology and palaeobiology of ostracoda. *Bull. Am. Palaeont.*, **65**, 636pp.

Handbook of Holocene Palaeoecology and Palaeohydrology
Edited by B. E. Berglund
© 1986 John Wiley & Sons Ltd.

34

Coleoptera analysis

G. R. Coope

*Department of Geological Sciences, University of Birmingham,
Edgbaston, U.K.*

INTRODUCTION

Insects are a remarkably successful group of animals. In terms of numbers of
species they make up more than half the animals and plants known today. By
far the largest order of insects is the Coleoptera (the beetles) which contains
about 350,000 species; they are more numerous, therefore, than the flowering
plants. They are often precisely adapted to narrow environmental niches but
collectively inhabit almost all terrestrial and freshwater habitats. It is
fortunate, therefore, that this important group of animals has left an extensive
fossil record, easily accessible using simple laboratory techniques, readily
identifiable frequently to the species level and capable of providing detailed
palaeoenvironmental and palaeoclimatic information.

Fossil coleoptera may be recovered from almost any sediment that contains
macroscopic plant remains. They are often extremely abundant and the main
problem when dealing with Quaternary insect fossils is embarrassment of
riches — each animal can yield a wealth of skeletal parts and there is a fantastic
diversity exhibited by different taxonomic groups. The proper identification of
these fossils thus requires considerable experience and access to comprehensive collections of modern specimens. Of equal importance is the sympathy and
understanding of specialists who deal with the present-day animals. It is their
taxonomy, and understanding of the ecology of the species concerned, which is
the raw material on which we must base our reconstructions of past
environments.

FIELD TECHNIQUES

Because of the need to make precise comparisons between the environmental inferences drawn from both the coleoptera and plants (both pollen and macroscopic remains), it is important to select sites which permit both types of investigation and to collect samples from as nearly identical sections as is possible. Naturally the same requirements relate to samples selected for radiocarbon dating. On the whole open sections are preferable to bore-holes because moderately large samples are needed if an adequate representation of the insect fauna is to be obtained. If bore-hole samples only are available it is clearly desirable to obtain cores of the maximum diameter consistent with funds available. In fairly well understood situations core diameters of 5 cm are tolerable, but in most cases a minimum core diameter of 10 cm is required. Of course if only narrow-gauge equipment is available, adequate samples can be assessed by multiple penetrations of the sediments provided that adequate stratigraphical control can be maintained by reference to clearly recognizable marker horizons.

Best results may be obtained from sediments that contain a moderately large quantity of silt. Not only does the silt aid the process of disaggregation in the laboratory but, because of its limited compressibility, the fossils are preserved in their original shape. Sediments of high organic content are acceptable provided that the plant debris is not thoroughly felted. Deposits of this sort may have to be broken up physically with the consequent smashing of many of the fossils. In summary, the most profitable sediments to investigate for fossil insects are those which break down most readily into a slurry that can be washed conventionally through sieves.

Before commencing sampling it is important to cut back the exposed surface so that, as far as possible, material is collected that has not been wetted and dried repeatedly by the weather or exposed to fungal attack or extensive oxidation. Such surface processes can readily destroy any fossil insects that the sediment might originally have contained.

Whenever possible, sedimentary sequences are sampled continuously (i.e. without any gaps between the samples) in units of 5 cm or 10 cm, depending on the rate of sedimentation and the likely level of any faunal change. Naturally at critical horizons the sample thickness should be as small as possible, but units finer than 5 cm are not usually taken because of the necessity to collect fairly large samples and the difficulty of maintaining stratigraphical control. In the allocation of numbers to the samples it is usual practice to select a sudden lithological change within the sequence to act as sampling datum 0, and label above and below this with + and − respectively. This procedure is adopted because the upper and lower limits of sedimentary basins may be irregular.

Sample size depends largely on field circumstance, not least of which is the distance that they must be carried. In the initial exploratory phase of an investigation large samples (say 10 kg) are usually taken from selected horizons. These give some indication of the richness of the deposit. When investigating sedimentary sequences in detail a series of samples are collected that are identical in size. It is usually found easiest to collect by volume (say 5000 cm³). In all but the most poorly fossiliferous deposits, such samples provide an adequate picture of faunal changes. In well-understood situations, smaller samples may be collected and the occurrence of critical species compared with a more thoroughly investigated sequence.

The samples are packed in strong polythene bags, preferably not so stout that it is difficult to close the opening with a knot. It is vital to prevent the sediment from drying out. If necessary, double bag the samples, i.e. with one bag inside the other. Each bag should be labelled as it is collected in the field, preferably with both a tie-on label and directly on to the bag with a felt tipped pen.

LABORATORY TECHNIQUES

Storage of samples

If samples are to be stored for several months, they are best kept in a cool situation (not necessarily freezing), and preferably in the dark to prevent the growth of algae or even mosses. Fungal attack may also occur in the sample bags, destroying the fossil insects and binding the sediments in a web of hyphae. If cool storage is not available it is possible to prevent decomposition of the fossils if a little dilute formalin is introduced into each bag before it is put away.

It is most important that any sample containing clay is not permitted to dry out as the clay will shrink and harden and make extraction of insect remains almost impossible. If there are large quantities of silt or sand in the sediments, the effect of drying out is not so drastic, and well-preserved fossils can be recovered from such samples.

Extraction procedures

The description that follows is of the standard process adopted in the Quaternary Laboratory at the University of Birmingham. It is intended only as a guideline to be varied as circumstances dictate. For example, disaggregation techniques differ considerably with different lithologies. Furthermore, partially decomposed insect fossils may not be readily brought to the surface by kerosene flotation. Many skeletal elements (e.g. the heads) may have been filled with silt and are likely to remain in the heavy fraction. It is thus often profitable to examine these residues carefully (Figure 34.1).

Stage 1. The sample is broken down into a slurry. This operation is performed in a large polythene bowl fitted with a spout. If the sample will not break down by the gentle application of a jet of water, it may have to be broken down

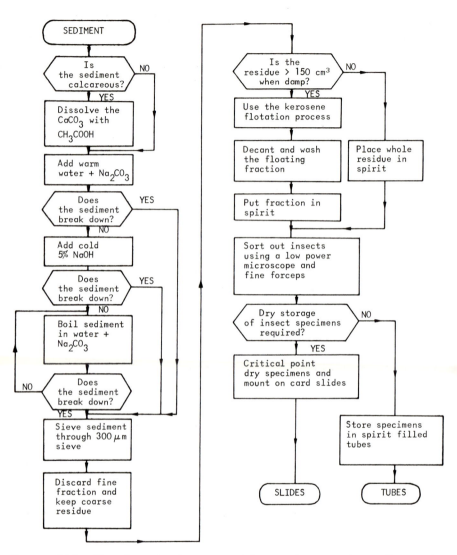

FIGURE 34.1.   Flowchart showing sequential steps in processing sediments for recovery of coleoptera remains

chemically. A solution of sodium carbonate with 5% sodium hydroxide will facilitate disaggregation. If necessary the sample may require boiling in these solutions before it will break down. Prolonged treatment with alkali can, however, make the fossil insects both thin and frail.

Stage 2. The slurry is washed through a sieve of 300 μm mesh diameter. It has been found necessary to wash small quantities at a time, otherwise the sieve becomes rapidly clogged. If the sieve does become blocked, the obstruction can often be relieved by introducing water from below. It is unprofitable to try to hurry this process by squirting water vigorously into the sieve. Gentle sideways shaking is usually adequate to clear all fine mud from the sample.

Stage 3. Shake out water from the residue held on the sieve. If there is very little material remaining on the sieve proceed directly to stage 7. If there is abundant plant and insect debris, it may be necessary to concentrate the insect fossils. To do this take the damp residue and thoroughly mix it with kerosene or any other similar grade mineral oil, in a deep polythene bowl. This is a messy process. It is best done manually wearing thin protective gloves. Thorough mixing is essential and may take several minutes. Excess oil is then poured off (it is easily recycled).

Stage 4. Add cold water to the mix obtained in stage 3, so that the surface of the water just fails to reach the lip of the spout. The cold water is best added vigorously to ensure that all plant and insect debris comes in contact with the water. If hot water is used at this stage, air bubbles caught up in the plant debris expand and float much of the plant debris.

Stage 5. Leave it to stand until the floating fraction is well-separated from the material that has sunk. This may take about half an hour.

Stage 6. Gently decant the floating fraction into the same sized sieve that was used in stage 2.

Stage 7. Thoroughly wash the decanted material, first in household detergent using hot water to remove the kerosene, and then in alcohol to remove the water. This washing is best carried out while the concentrate is still in the sieve.

Stage 8. Lastly, using alcohol, the concentrate may now be sluiced into a suitable receptacle such as a beaker or tube. All the concentrate will now sink and is easily sorted under a binocular microscope. The sorting is best carried out under alcohol; if dry-sorted the fossil insects crumple up or become badly distorted.

PRESERVATION OF FOSSIL INSECTS

Since the fossils themselves are the only objective data on which to base our interpretations, it is very important that they be preserved for future reference. We adopt two procedures.

The fossils may be stuck down on prepared cards using gum tragacanth. This ensures that there is minimal crumpling as the specimens dry, but there is almost invariably some cracking and warping. Furthermore the fossils may lose detailed structures such as tubercules that may be flattened out as the specimen shrinks down on to the card (e.g. as in *Carabus* spp.), or spurious pits may develop as a result of the collapse of the cuticle as it dries. The advantages of these carded specimens is that they are convenient to handle, need little maintenance and are essential for detailed investigation of surface sculpture. We have experimented with critical point drying in an effort to reduce the distortion on drying out, and the initial results are most promising.

Much of the problem of distortion may be avoided if the fossils are preserved permanently in alcohol. The main disadvantages of this method of keeping the fossils are that they need continuous maintenance to keep the tubes from drying and the specimens are less accessible. The addition of a very small quantity of glycerine reduces the danger of drying.

IDENTIFICATION

The taxonomy of present-day coleoptera is based on exoskeletal characteristics. For this reason the entomologist working with fossil or modern material can to a large extent use the same criteria to distinguish species. In most countries extensive keys have been published for the identification of the contemporary fauna, and these are of great value in the identification of Quaternary fossil insects. However, the keys cannot be used alone because the distinctive features selected by the authors of the keys may not be available on the fossil specimens under consideration. Such keys are best used in combination with a comprehensive modern collection, when it will be found that many characteristics may differentiate species that were not conveniently adopted in the construction of the key. Fossil identification is therefore primarily a matter of direct comparison with well-determined modern specimens.

The preservation of Quaternary coleoptera is often remarkably exact. Not only can gross morphology be used in the comparison, but also fine surface structure only available under scanning electron microscopy (details of scale structure in weevils for instance), and even the intimate construction of the male genitalia is often available in the fossils.

INTERPRETATION

All late Quaternary fossil coleoptera so far discovered can be identified with living species. In presenting the faunal lists of fossil assemblages the species can be set out, therefore, in the traditional taxonomic order adopted for the listing of living species. When dealing with localities at which a succession of horizons has been sampled, the stratigraphical range and changing numerical representation of each species may be shown on a diagram that is in some ways analogous to an absolute pollen diagram in that the numbers given refer to the actual numbers of animals counted. Abundance at each level is shown as the minimum number of individuals present in the sample, and this is estimated by taking the maximum number of any determinable skeletal element. This procedure under-represents the commoner species and, vice-versa, rather emphasizes the numbers of rarer species. If the sequence of samples from a site were originally all of equal size, such a diagram gives a measure of the changing abundance of species throughout the sedimentary sequence, and thus through time. It must be emphasized, however, that the numerical representation of species in these diagrams does not reflect regional abundance. High numbers certainly indicate that a species was common at times close to the sampling site, but low numbers do not necessarily indicate rarity; the species concerned may merely have lived further away from the place where its body was finally buried, having come more or less by accident into the deposit.

Assuming that the fossil species had the same environmental requirements as their present-day representatives, and there is now ample evidence that this is a realistic assumption, a mosaic picture can be built up of the local environmental conditions at various stages of the sedimentary sequence. In this way it is possible to build up pictures that are both detailed and complex. Numerous publications are available illustrating this procedure (Coope and Brophy, 1972; Osborne, 1972; Buckland and Kenward, 1973), and it would be unnecessarily repetitious to elaborate any further here. In general it should be pointed out that the local conditions inferred from fossil coleopteran assemblages agree well with reconstructions derived from the macroscopic plant remains that also reflect the environments available in the immediate vicinity of the investigation locality.

Whereas it is easy to devise a qualitative picture of the local environment in the way outlined above, it is not easy to give a quantitative estimate of the spatial and temporal distribution of the various habitats represented by the species in fossil assemblages. This is because of the difficulty of estimating the catchment areas of the sedimentary traps. It seems likely that sediments accumulating in open, treeless landscapes will draw on an area many hundreds of metres square, while in woodland situations the catchment areas will undoubtedly be much less than this. The difference is due primarily to the fact that many fossils represent animals incorporated passively in the sediment by being swept by wind or water from the surrounding landscape.

No doubt the most significant contribution that fossil coleoptera make to Quaternary studies is in the field of palaeoclimatology. This is because the presence of suitable climatic conditions is one of the most important factors that determine the limits of the geographical range of present-day insect species. In this respect, acceptable thermal environments and humidities are critical. As far as temperate latitudes are concerned, and Europe in particular, aridity is not a serious problem and humidity is largely a matter of local significance. In most circumstances we can assume, therefore, that there was widespread availability of adequately moist and dry habitats. In contrast the availability of suitable thermal environments has regional significance, and the distributional limits of many species parallel to a large extent the well-established temperature zonation as measured by meteorological stations.

Some explanation seems called for here about the use of climatic data from meteorological stations. The actual temperatures experienced by a beetle are those of the microenvironment, which may differ greatly from those measured at nearby meteorological stations. But to a large extent these microenvironents are controlled by macroenvironmental factors which determine the arena in which suitable microenvironments develop; in other words the macroclimate sets the broad limits to the range of a species while microclimates dictate the detailed whereabouts of suitable habitats. For this reason detailed laboratory experiments on the temperature tolerances of particular species provide little valuable information for palaeoclimatic interpretation.

A second source of difficulty stems from the fact that meteorological stations deal in averages, which are abstractions that have little impact on the life-style of the animal. But the Quaternary palaeoclimatologist has no option but to use averages. The fossil assemblages from each sample represents the accumulated fauna of several decades with no certainty that it represents modal environmental conditions, and we can have little insight into the *range* of thermal environments available during that time. For this reason, when making climatic inferences from Quaternary coleopteran assemblages I have given average July temperatures as a crude assessment of summer warmth.

As an example, Figure 34.2 shows the changes in average July temperatures estimated from coleopteran assemblages during the late-glacial and early postglacial period. These changes are set against the usual subdivisions of this period based largely on palynological criteria. It is immediately apparent that there are a number of important differences between the temperature curve derived from the coleopteran data and the traditional picture of the late-glacial climate in north-western Europe derived from pollen data. These discrepancies in timing and intensity are probably due to a variety of causes such as the rapidity of the actual climatic changes, differential mobility of colonizers, the distances that they had to cover, the availability of acceptable habitats, their thermal sensitivity, competitors and the nutritional status of the environment. All of these could have contributed to the varying response rates of the different components of the biota to changes in the thermal climate. Under conditions of rapid and intense climatic change, it is evident that the whole fauna and flora may be thrown out of harmony with the physical environment, and indeed in its ecological relationships also. Transitional assemblages of this sort may have no precise analogues at the present time.

The assessment of winter temperatures is a much more intractable problem since in northern and temperate latitudes most insects hibernate. Nevertheless, the presence of exclusively Asiatic species in deposits of the Weichselian Glaciation in western Europe strongly suggests that the climate, with its

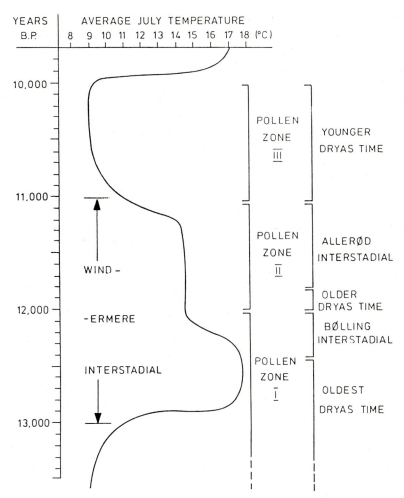

FIGURE 34.2. Fluctuations in the average July temperatures in lowland central England during the late-glacial and early postglacial periods compared with the timing of the traditional subdivisions based on palynological criteria

associated cold winters, was more continental than it is today. Exclusively Siberian species are, however, rare in Late Weichselian deposits in western Europe.

The actual procedure for the calibration of the thermal climate on the basis of Quaternary coleopteran assemblages is dealt with in Chapter 41.

PALAEOCLIMATOLOGY AND FOSSIL COLEOPTERA: SOME RESULTS

During periods of rapid climatic change the coleoptera responded more rapidly than any other component of the biota represented in the Quaternary fossil record. This response is shown by marked changes in the specific composition of the fossil assemblages in which associations of exclusively arctic/alpine animals are replaced by entirely temperate species during periods of amelioration, and vice versa at times of climatic deterioration. Thus at some time shortly before 13,000 years ago the glacial insect fauna of the British Isles was suddenly replaced by species whose ranges today barely reach as far north as southern England (Coope, 1977). This entirely thermophilous insect assemblage was associated with a largely treeless landscape with sparse tree birches as the only arborescent vegetation. Some time shortly before 12,000 years ago this temperate fauna was eliminated and in its place many species appeared that are today found to the north of the British Isles. This fauna is by no means incompatible with *Betula* forest, and indeed over much of England at this time there is evidence of extensive birch woodland. At about 11,000 years ago the arctic/alpine species became dominant in the insect fauna of the British Isles, a fauna entirely in accord with the return of harsh climates and spread of tundra-like vegetation (Coope *et al.*, 1979). The climatic amelioration at about 10,000 years ago saw the sudden elimination of all the northern species from lowland Britain and their replacement, about 9,500 years ago or even earlier, by an assemblage of species that suggests climates at least as warm as those of the present day (Ashworth, 1973; Osborne, 1974; Bishop and Coope, 1977).

During the late-glacial and early postglacial periods, therefore, there were two episodes when the total biota was evidently out of harmony with the physical environment. When making palaeoclimatic inferences for episodes such as these, it is precarious to use analogues derived from modern biotal studies in which the relationship with the physical environment has had enough time to approach equilibrium. In the interpretation of fossil data from these periods it is advisable to deal with each species and its environmental requirements on its own merits, always bearing in mind any peculiar preferences or dependencies that it might possess.

Palaeoclimatic studies of the later postglacial period are complicated by the disruptive influence of man. It is easy to demonstrate that many species of coleoptera have had their geographical ranges drastically altered during this time, but it is much more difficult to sort out the contributions made by changes in climate from those imposed by human interference. Of particular significance here is the effect of forest clearance on the distributions of those species of beetle that rely for their habitats on post-mature timber. In spite of the complications imposed by man it seems probable that some of the losses of species from the British Isles since Bronze Age times (Osborne, 1969) must be attributed to subsequent climatic deterioration.

REFERENCES

Ashworth, A. C. (1973). The climatic significance of a late Quaternary insect fauna from Rodbaston Hall, Staffordshire, England. *Ent. Scand.* **4**, 191–205.

Bishop, W. W., and Coope, G. R. (1977). Stratigraphical and faunal evidence for late glacial and early Flandrian environments in South West Scotland. In: *Studies in Scottish Late Glacial Environment* (Eds. J. M. Gray and J. J. Lowe), Pergamon Press, Oxford, pp. 61–88.

Buckland, P. C., and Kenward, H. K. (1973). Thorne Moor: a palaeoecological study of a Bronze Age site. *Nature*, **241**, 405–406.

Coope, G. R. (1977). Fossil coleopteran assemblages as sensitive indicators of climatic changes during the Devensian (last) Cold Stage. *Phil. Trans. R. Soc. Lond.*, **B. 280**, 313–340.

Coope, G. R., and Brophy, J. A. (1972). Late-glacial environmental changes indicated by a coleopteran succession from North Wales. *Boreas*, **1**, 97–142.

Coope, G. R., Dickson, J., McKutcheon, J., and Mitchell, G. F. (1979). Late glacial flora and fauna from Drumurcher, Co. Monaghan. *Proc. R. Irish Acad.*, **B79**, 63–85.

Osborne, P. J. (1969). An insect fauna of Late Bronze Age date from Wilsford, Wiltshire. *J. Anim. Ecol.*, **38**, 555–566.

Osborne, P. J. (1972). Insect faunas of Late Devensian and Flandrian age from Church Stretton, Shropshire. *Phil. Trans. R. Soc. Lond.*, **B. 263**, 327–367.

Osborne, P. J. (1974). An insect assemblage of early Flandrian Age from Lea Marston, Warwickshire, and its bearing on the contemporary climate and ecology. *Quaternary Res.*, **4**, 471–486.

Handbook of Holocene Palaeoecology and Palaeohydrology
Edited by B. E. Berglund
© 1986 John Wiley & Sons Ltd.

35

Chironomid analysis

Wolfgang Hofmann

*Max-Planck-Institut für Limnologie, Abt. Allgemeine Limnologie,
Plön, F. R. G.*

INTRODUCTION

The use of subfossil chironomids in palaeolimnological research and their ecological significance have been described by Frey (1964), Stahl (1969) and Hofmann (1971b). These authors emphasized the particular importance of these midges in relation to the bottom faunistical lake type system (Thienemann 1920, 1954; Brundin 1949, 1956). The occurrence of the profundal species is closely correlated with the oxygen condition of the hypolimnion. On the other hand, the oxygen regime of a lake generally depends on its trophic state. Thus, Thienemann referred to a eutrophic lake as *Chironomus lake* and to an oligotrophic lake as *Tanytarsus lake*, corresponding to the predominant profundal chironomid species. Brundin defined the profundal chironomid community of oligotrophic and ultra-oligotrophic lakes as being characterized by the *Tanytarsus lugens* community and by the predominance of *Heterotrissocladius subpilosus*, respectively. Chironomids are of special interest in palaeolimnology because their larval head capsules are preserved in the sediment, and the chironomid faunas of former lake stages can thus be reconstructed.

Subsequently, many workers, when dealing with subfossil chironomids from lake sediments, have used them as indicators of the earlier trophic situation of a lake in its development (Deevey, 1942; Stahl, 1959; Bryce, 1962; Goulden, 1964; Harmsworth, 1968; Hofmann, 1971a, 1978; Warwick, 1975, 1980; Carter, 1977; Brugam, 1978; Czeczuga *et al.*, 1979; Wiederholm and Eriksson, 1979; Brodin, 1982; Günther, 1983).

715

Brinkhurst (1974) recently gave a critical account of the development of lake typology based on the profundal chironomid fauna. He again stressed, referring to Brundin (1949), the point that the composition of the profundal chironomid community reflects the ecological conditions of this region (profundal) rather than the trophic state of the lake. Unlike the profundal species, the littoral chironomids have so far been neglected in palaeolimnological research, although their head capsules have been found in littoral and profundal sediments. Generally, it is difficult to interpret changes within the littoral chironomid community for the following reasons: the biotope of the littoral zone is more complicated when compared with that of the profundal zone, and most of the littoral species cannot be identified on the basis of the larval head capsules. Any ecological discussion concerning genera or species groups is therefore incomplete and inadequate. Furthermore, there is very little information on the ecology of littoral chironomids. As Saether (1975) mentioned, it is difficult to recognize indicator species of eutrophic conditions among the littoral species. Nevertheless, more attention should be paid to the chironomids of the littoral zone in this respect.

Brodin (1982) indeed has shown that littoral taxa provide valuable information for the interpretation of lake ontogeny under the condition of a profound taxonomic analysis of the subfossil material.

Much information on littoral chironomids from Swedish lakes of different trophic states and humus contents was given by Brundin (1949). A list of littoral species and their distribution in oligo-, meso- and eutrophic, and meso- and polyhumic lakes, has been compiled by Saether (1975).

LABORATORY WORK

For chironomid analysis the sediment cores should always be sampled from the deepest part of the lake. In order to separate the profundal from the littoral species, it may be necessary to examine a core taken from a shallower site for purposes of comparison. This would be possible in only a few cases owing to the time factor.

Sampling distance in the core and sample volume should provide at least 3–4 samples per pollen zone and approximately 50 specimens per sample. However, the dating of the sediments and the concentration of head capsules are not known when the samples are taken. So, a sampling distance of about 10 cm and a sample volume of 5–10 g are recommended. Owing to variations of sedimentation rates the sampling distance in lake sediments should generally be smaller in late-glacial sediments and larger in postglacial ones.

The proposed method of preparation of the samples is summarized in Figure 35.1. (Hofmann, 1971a, 1978). For deflocculation hot 10% KOH is used. Subsequently, sieving is recommended in order to reduce sample size. It is profitable to separate two fractions (>200 μm and 100–200 μm) because this

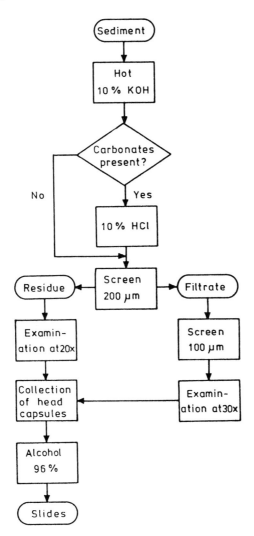

FIGURE 35.1. Summary flowchart for the laboratory preparation of samples for chironomid analysis.

separation (1) facilitates finding of the head capsules (it is rather difficult to recognize small head capsules in the midst of large and small particles), and (2) minimizes working time because the 200 μm fraction can be examined more rapidly under low magnification.

The sieved remains of each fraction are poured in small quantities into a counting chamber (about 4 × 5 cm, with grid lines on the bottom for

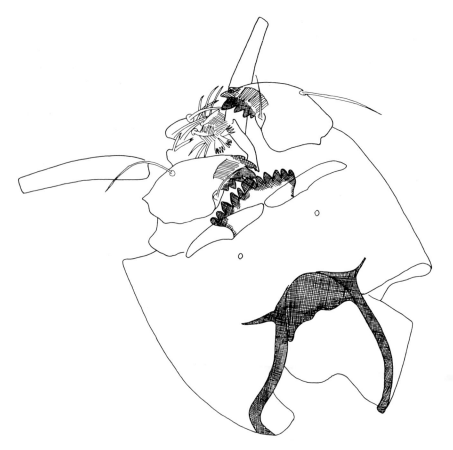

FIGURE 35.2.   Standardized sketch of an almost complete subfossil head capsule (*Tanytarsus*); in many cases only the ventro-lateral part of the head capsule with mentum is preserved

orientation) and examined under a stereo microscope with a magnification of at least 20× and 30×, respectively.

The head capsules (and male hypopyia of adults if found) must be picked out by means of fine forceps or a needle, dehydrated in 96% alcohol, and mounted with the ventral side up (Figure 35.2) in euparal. It is recommended that all head capsules be stored as permanent slides. If the concentration of head capsules is extremely high, working time can be reduced by subsampling.

In order to economize on sediment material and work, the prepared sample (after deflocculation) can be used for both chironomid and cladoceran analysis by using different size fractions from the same sample. The sieved samples can be stored in 70% ethanol or 4% formalin.

Omission of the small fraction (100–200 μm) leads to an underestimation of small head capsules (first larval stages; small species like *Corynoneura*). However, the author has found that consideration of the >200 μm fraction alone provides an adequate view of the essential way of development of the chironomid fauna. For calculation of the absolute numbers the 100–200 μm fraction has to be taken into account (Warwick, 1980).

Warwick (1980) used a more cautious and time-consuming method of preparation and gave a detailed description of his procedure. This method has also been successfully used by Brodin.

For identification of the larval head capsules numerous keys and descriptions of individual taxonomic groups exist in the chironomid literature. Three works may be recommended which cover all the taxonomic groups (Figure 35.3) of the palaearctic or holarctic region: Hofmann (1971b), Moller Pillot (1982, 1983) and Wiederholm (1983). A thorough taxonomic analysis of the subfossil material is essential for a correct interpretation of the results. Further differentiations can be obtained by, for instance: (1) looking for additional morphological characteristics in the subfossil head capsules, and (2) by comparison with (hatched) larvae from the lake under study.

In the case of subfossil male hypopygia, specialists should be asked for identification.

PRESENTATION OF RESULTS

The first step in evaluation of the results of chironomid analysis is to look for faunal changes during lake ontogeny. It has turned out that the 'traditional' plotting of abundances of individual taxa against sediment depth (time) is not an adequate way to show what was happening to the chironomid community. Calculation of relative abundances (%) of the taxa and a plot of changing percentage composition of the assemblage (Figure 35.4) (Hofmann, 1978, 1983a; Brodin, unpublished) lead to a community approach in chironomid analysis. This approach has been successfully applied and developed by further methods of community analysis (diversity index, equitability, similarity index, cluster analysis) (Warwick, 1980; Hofmann, 1983b; Brodin, unpublished). For instance, the cluster diagram (Figure 35.5) based on percentage similarities between the chironomid/ceratopogonid assemblages of samples from a core of the Lobsigensee (Switzerland) (Hofmann, 1983b) may exemplify that dynamics of community structure become distinct, in the form of stepwise changes as well as in sudden breaks in the composition of the assemblages. Such methods lead to a reduction of data and spotlight the major trends of development, hence providing an adequate background for interpretation of the results.

In addition, taxa have to be taken into consideration which, owing to their low relative abundances, have been neglected by community analysis and

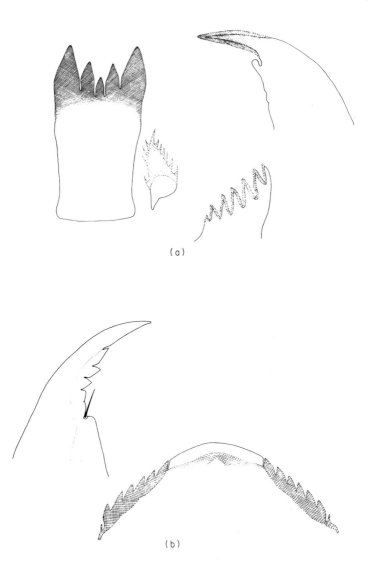

(a)

(b)

FIGURE 35.3.   Characteristics of head capsules of different taxonomic groups: (a) Tanypodinae (*Procladius*), lingua and superlingua of hypopharynx, mandible, paralabial comb. (b) Orthocladiinae (*Paracladius*), mentum, mandible. (c) Chironomini (*Chironomus*), mentum, mandible. (d) Tanytarsini (*Tanytarus*), mentum, mandible, premandible, SI bristles, antennal base. *Reproduced by permission of E. Schweizerbart'sche Verlagsbuchhandlung*

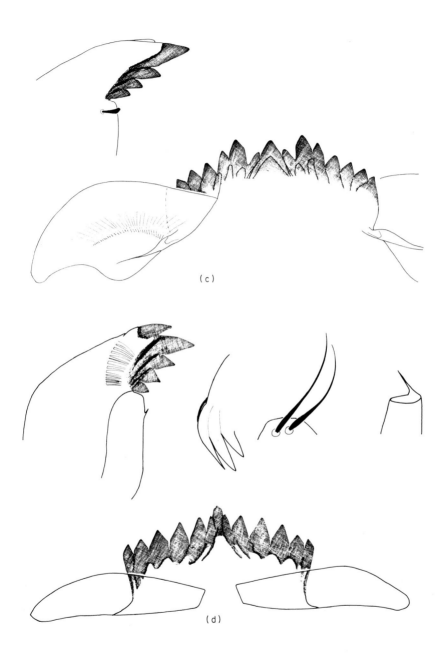

(c)

(d)

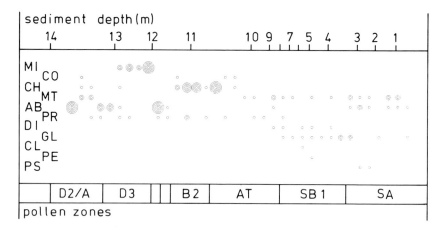

FIGURE 35.4. Late-glacial and postglacial succession of the predominant chironomid taxa (>10% of the total); each circle denotes a relative abundance of ≥10% (2 circles ≥20% ...); Poolsee, northern Germany (Hofmann, 1983a)

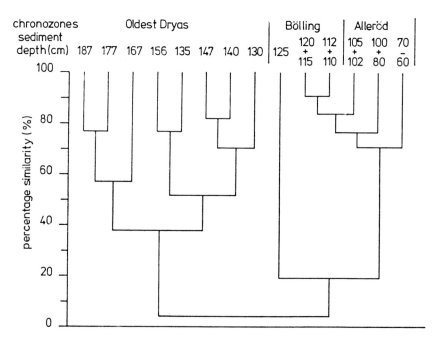

FIGURE 35.5. Cluster diagram of percentage similarity of the chironomid/ceratopogonid assemplages in late-glacial sediments from the Lobsigensee (Swiss Plateau) (Hofmann, 1983b). *Reproduced by permission of Museum d'histoire naturelle de Genève*

which are of special interest because they exhibit a particular distribution pattern or because they are indicative of certain environmental conditions.

If the total numbers of head capsules per unit of sediment have been determined, a calculation of influx rates (head capsules/cm^2 per year) is possible (Stahl 1959). The significance of those estimates depends on the accuracy of dating and on the knowledge of the pattern of sedimentation in the lake. Carter (1977) found in a short-core study in Lough Neagh similar increases of chironomid flux rates in the uppermost layers of three cores from different sites.

INTERPRETATIONS

An important point in chironomid analysis of cores from deep lakes is to find out the species composition of the former profundal chironomid taxocenosis. In this connection the species which are typical of the profundal zone of the different lake types and which have been categorized by Brundin (1949, 1956) and Saether (1975) are of special interest. Generally, a separation of the typical littoral species from profundal species is recommended. The relative abundance of the individuals of littoral species in relation to the profundal ones gives an indication of the influence of the littoral species on the particular dead assemblage.

Thienemann (1920, 1954) called the oligotrophic lakes *Tanytarus lakes* because the profundal fauna of the lakes studied by him was dominated by a species of the Tanytarsini, namely *Lauterbornia coracina*. This is not, of course, a member of the genus *Tanytarus*. Unfortunately, the term '*Tanytarus* lake' gives the impression that the genus *Tanytarsus* is implied. This point has led to some confusion, as pointed out by Hofmann (1971b). In some cases the occurrence of specimens of the genus *Tanytarsus* has been used as an indicator of oligotrophic conditions (Bryce, 1962; Goulden, 1964; Harmsworth, 1968; Brugam, 1978; Carter, 1977; Czeczuga *et al.*, 1979). This is not correct because:

(1) Almost all species of the genus *Tanytarus* are typical littoral forms. They are present in the profundal dead assemblage because their empty moulting skins have been transported from the littoral to the profundal zone. Hence, the occurrence of such specimens in the profundal assemblage of head capsules is, of course, not related to the oxygen conditions of the hypolimnion.

(2) A few *Tanytarsus* species, whose larvae cannot be identified, occur in the profundal zone of eutrophic lakes (Reiss and Fittkau, 1971).

Therefore it is recommended that members of the *Tanytarsus lugens* community (Brundin, 1956) be used as indicator species of oligotrophicity, but not the genus *Tanytarus*. In their earlier developmental stages a great number of species of this community were found in the Holstein lakes (Hofmann, 1971a, 1978).

Another point stressed by Brundin (1949) is the use of the *Tanytarsus lugens* community or the species typical of ultra-oligotrophic lakes (*Heterotrissocladius subpilosus*) as indicators of the trophic state of a particular lake. Brundin stated:

> 'On the whole, the chironomids have proved to be better indicators of the temperature and the oxygen standard than of the trophic standard.'

This has to be taken into consideration when samples from late-glacial periods are involved. Under subarctic and arctic climatic conditions the indicator species of oligotrophic and ultra-oligotrophic lakes occur in the littoral zone because they are coldstenothermic and eurybathic. In this situation their occurrence is, of course, not related to the oxygen condition of the hypolimnion. Hence, when such species are used as indicators of trophic states the climatic conditions must also be considered.

It should be pointed out again that the lake type system based on the profundal chironomid fauna applies only to deep, stratified lakes and not to shallow ones.

As for shallow, unstratified lakes, recent studies have shown that the chironomid community has been mainly influenced by changing climatic conditions which, for instance, determine water temperature as well as water level (Hofmann, 1983a,b; Brodin, unpublished).

A dramatic change in the chironomid fauna of shallow lakes generally occurred at the end of the postglacial period when the coldstenothermic species disappeared owing to the rising temperature (Walker and Paterson, 1983; Hofmann, 1983).

Besides trophic state and climatic conditions, other environmental conditions have been shown to be responsible for changes in the chironomid assemblages: Brodin, in an unpublished manuscript, discussed a case showing the effect of changing pH and humus content of a lake. The establishment of dystrophic conditions led to predominance of *Zalutschia zalutschicola* Lipina and *Tanytarsus* gr. *eminulus* Walker. However, 60–80% of the 'dystrophic' chironomid fauna became extinct owing to cultural acidification in recent times, in which two *Chironomus* species were predominating.

Walker and Paterson (1983) also studied the development of the chironomid fauna in acid environments (predominance of *Tanytarsus; Zalutschia zalutschicola* present). Henrikson *et al.* (1982) found that acidification in two Swedish lakes caused a decrease in the number of head capsules per volume of sediment and an increasing relative abundance of *Phaenopsectra* and Orthocladiinae.

Salinity is another factor determining the composition of the chironomid community. Recent core studies gave evidence of the influence of changing salinity: Paterson and Walker (1974) valued the significance of the factors salinity–competition with respect to the distribution of *Tanytarsus barbitarsis*.

Freeman (beside experiments) by means of short cores from a saline lake (Australia). Clair and Paterson (1976) reported the extinction of a freshwater chironomid community by a salt-water intrusion and its re-estabishment. A postglacial freshwater period of the Baltic Sea was clearly indicated by the occurrence of both freshwater chironomids and cladocerans (Hofmann, unpublished).

Palaeoecological studies have been more-or-less restricted to the temperate zone. Recently, Binford (1982) reported chironomids in a sediment core from the tropical Lake Valencia (Venezuela). However, the separation of faunal zones as well as the interpretation of lake ontogeny was mainly based on Ostracoda and Cladocera.

Changes in the chironomid assemblage during lake ontogeny can be considered as responses to changing environmental conditions. In this way chironomids may reflect eutrophication; that means an increasing nutrient supply and its influence on the lake ecosystem. However, as shown above, various biotic and abiotic factors may have been involved, too (Whiteside, 1983). Therefore, the interpretation of chironomid successions should be based on a thorough discussion of the major components which determined the environment of the former chironomid fauna.

REFERENCES

Binford, M. W. (1982). Ecological history of Lake Valencia, Venezuela: interpretation of animal microfossils and some chemical, physical, and geological features. *Ecol. Monogr.*, **52**, 307–333.

Brinkhurst, R. O. (1974). *The Benthos of Lakes*, Macmillan, London.

Brodin, Y. (1982). Palaeoecological studies of the recent development of the Lake Växjösjön. IV: Interpretation of the eutrophication process through the analysis of subfossil chironomids. *Arch. Hydrobiol.*, **93**, 313–326.

Brugam, R. B. (1978). Human disturbance and the historical development of Linsley Pond. *Ecology*, **59**, 19–36.

Brundin, L. (1949). Chironomiden und andere Bodentiere der Südschwedischen Urgebirgsseen. *Rep. Inst. Freshw. Res. Drottningholm*, **30**, 1–194.

Brundin, L. (1956). Die bodenfaunistischen Seetypen und ihre Anwendbarkeit auf die Südhalbkugel. *Rep. Inst. Freshw. Res.*, Drottningholm, **37**, 186–235.

Bryce, D. (1962). Chironomidae (Diptera) from freshwater sediments, with special reference to Malham Tarn (Yorks.). *Trans. Soc. Brit. Entomol.*, **15**, 41–54.

Carter, C. E. (1977). The recent history of the chironomid fauna of Lough Neagh, from the analysis of remains in sediment cores. *Freshw. Biol.*, **7**, 415–423.

Clair, T., and Paterson, C. G. (1976). Effect of a salt water intrusion on a freshwater Chironomidae community: a palaeolimnological study. *Hydrobiologia*, **48**, 131–135.

Czeczuga, B., Kossacka, W., and Niedźwiecki, E. (1979). Ecological changes in Wigry Lake in the postglacial period. III: Investigations of the Chironomidae stratigraphy. *Pol. Arch. Hydrobiol.*, **26**, 351–369.

Deevey, E. S. (1942). Studies on Connecticut lake sediments. III: The biostratonomy of Linsley Pond. *Am. J. Sci.*, **240**, 235–264.

Frey, D. G. (1964). Remains of animals in Quaternary lake and bog sediments and their interpretation. *Arch. Hydrobiol./Beih. Ergebn. Limnol.*, **2**, 1–114.

Goulden, C. E. (1964). The history of the cladoceran fauna of Esthwaite Water (England) and its limnological significance. *Arch. Hydrobiol.*, **60**, 1–52.

Günther, J. (1983). Development of Grossensee (Holstein, Germany): variations in trophic status from the analysis of subfossil microfauna. *Hydrobiologia*, **103**, 231–234.

Harmsworth, R. V. (1968). The developmental history of Blelham Tarn (England) as shown by animal microfossils, with special reference to the Cladocera. *Ecol. Monogr.*, **38**, 223–241.

Henrikson, L., Olafsson, J. B., and Oscarson, H. G. (1982). The impact of acidification on Chironomidae (Diptera) as indicated by subfossil stratification. *Hydrobiologia*, **86**, 223–229.

Hofmann, W. (1971a). Die postglaziale Entwicklung der Chironomiden- und *Chaoborus*-Fauna (Dipt.) des Schöhsees. *Arch. Hydrobiol./Suppl.*, **40**, 1–74.

Hofmann, W. (1971b). Zur Taxonomie und Palökologie subfossiler Chironomiden (Dipt.) in Seesedimenten. *Arch. Hydrobiol./Beig. Ergebn. Limnol.*, **6**, 1–50.

Hofmann, W. (1978). Analysis of animal microfossils from the Grosser Segeberger See (F.R.G.). *Arch. Hydrobiol.*, **82**, 316–346.

Hofmann, W. (1983a). Stratigraphy of Cladocera and Chironomidae in a core from a shallow north German lake. *Hydrobiologia*, **103**, 235–239.

Hofmann, W. (1983b). Stratigraphy of subfossil Chironomidae and Ceratopogonidae (Insecta: Diptera) in late glacial littoral sediments from Lobsigensee (Swiss Plateau): studies in the Late-Quaternary of Lobsigensee 4. *Rev. Paléobiol.* **2**, 205–209.

Moller Pillot, H. K. M. (1983). De larven der Nederlansee Chironomidae (Diptera), Part A3. ed. *Nederl. Faun. Meded.*, **1**.

Moller Pillot, K. M. (1984). De larven der Nederlandse Chironomidae (Diptera), Part B (Orthocladiinae). *Nederl. Faun. Meded.*, **2**.

Paterson, C. G., and Walker, K. F. (1974). Recent history of Tanytarsus barbitarsis Freeman (Diptera: Chironomidae) in the sediments of a shallow, saline lake. *Aust. J. Mar. Freshwater Res.*, **25**, 315–325.

Reiss, F., and Fittkau, E. J. (1971). Taxonomie und Ökologie europäisch verbreiteter *Tanytarsus*-Arten (Chironomidae, Diptera). *Arch. Hydrobiol./ Suppl.*, **40**, 75–200.

Saether, O. A. (1975). Nearctic chironomids as indicators of lake typology. *Verh. Internat. Verein. Limnol.*, **19**, 3127–3133.

Stahl, J. B. (1959). The developmental history of the chironomid and *Chaoborus* faunas of Myers Lake. *Invest. Indiana Lakes Streams*, **5**, 47–102.

Stahl, J. B. (1969). The uses of chironomids and other midges in interpreting lake histories. *Mitt. Internat. Verein. Limnol.*, **17**, 111–125.

Thienemann, A. (1920). Untersuchungen über die Beziehungen zwischen dem Sauerstoffgehalt des Wassers und der Zusammensetzung der Fauna in nord-deutschen Seen. *Arch. Hydrobiol.*, **12**, 1–65.

Thienemann, A. (1954). *Chironomus*. *Die Binnengewässer*, **20**, 1–834.

Walker, I. R., and Paterson, C. G. (1983). Post-glacial chironomid succession in two small, humic lakes in the New Brunswick–Nova Scotia (Canada) border area. *Freshwater Invertebr. Biol.*, **2**, 61–73.

Warwick, W. F. (1975). The impact of man on the Bay of Quinte, Lake Ontario, as shown by the subfossil chironomid succession (Chironomidae, Diptera). *Verh. Int. Ver. Limnol.*, **19**, 3127–3133.

Warwick, W. F. (1980). Palaeolimnology of the Bay of Quinte, Lake Ontario: 2800 years of cultural influence. *Can. Bull. Fish. Aquat. Sci.*, **206**, 1–117.

Whiteside, M. C. (1983). The mythical concept of eutrophication. *Hydrobiologia*, **103**, 107–111.

Wiederholm, R. (Ed.) (1983). Chironomidae of the holarctic region. Pt 1: Larvae. *Ent. scand. Suppl.* 19.

Wiederholm, T., and Erikson, L. (1979). Subfossil chironomids as evidence of eutrophication in Ekoln Bay, central Sweden. *Hydrobiologia*, **62**, 195–208.

Handbook of Holocene Palaeoecology and Palaeohydrology
Edited by B. E. Berglund
© 1986 John Wiley & Sons Ltd.

36

Mollusca analysis

VOJEN LOŽEK

Kořenského 1/1055, 150 00 Praha 5-Smíchov, Czechoslovakia

INTRODUCTION

In lake and mire environments the Mollusca are represented by two classes, Bivalvia and Gastropoda (Basommatophora, Prosobranchia). In addition, a number of land snails (Stylommatophora) are adapted to very moist conditions in the marginal zone of water bodies and in marshlands.

Molluscan shells are preserved in a wide range of sediments where there is a sufficient concentration of calcium carbonate. In acid, non-calcareous sediments they are quickly leached, so they are absent from large areas consisting of acid rocks with lime-deficient Quaternary deposits. Despite this, non-marine molluscs are one of the most common groups of macrofossils in Quaternary deposits. They dominate fossil assemblages particularly in such sediments as spring tufas or fresh water chalk (Ložek, 1964).

Where Mollusca occur they are usually found in great numbers. One of the main advantages of their study is that many larger species have characteristic shells which can be identified without difficulty in field conditions. Further, they have been studied over a long period, and much is known about their present-day ecology. Most of the old studies were in terms of faunal lists collected from sections without detailed stratification. Nevertheless, they provided a basis for a modern molluscan analysis whose results are comparable to those of quantitative palaeobotanical studies.

The analysis of land and freshwater molluscs has been placed on a firm quantitative basis particularly in Great Britain (Sparks, 1961; Sparks and West, 1972; Evans, 1972; Kerney, 1963) and in central Europe (e.g. Ložek, 1964, 1978a, 1982; Fuhrmann, 1973; Mania, 1973; Alexandrowicz, 1983). The analysis applies the technique of serial-sampling to shell-bearing deposits, and presents the results in graphical form. Individual species and groups of ecologically related species are plotted as histograms of relative or absolute abundance in a manner similar to that used in presenting pollen-analytical data.

However, the best evidence of palaeoenvironmental conditions, climatic changes and chronological position is given by land snails among which many species with restricted habitat and climatic ranges occur (Ložek, 1976). Conversely, the great majority of freshwater and marshland species will tolerate a wide range of habitats and climates. A particular species can be widely distributed from the Arctic Circle in Scandinavia southwards into the Iberian and Italian peninsulas, which over such a range is found in streams, canals, lakes, ponds and ditches. But, ignoring the climatic range for the moment, it can usually be said that a given species is characteristic of certain conditions (Sparks, 1961). With the aid of precise ecological observations published in various modern papers (e.g. Meier-Brook, 1975; Vidal *et al.*, 1966), correlations with sedimentology and with the evidence provided by terrestrial snail fauna, the freshwater and marshland molluscan assemblages can also provide reliable data on environmental and climatic changes during the Late Vistulian(=Weichselian) and Holocene, as already shown by the observations of H. Menzel (1910), N. Odhner (1910), J. Favre (1927) or D. Geyer (1927) (cf. also Waldén, in press).

FIELDWORK

Selection of sites

Fossil Mollusca may be recovered from almost any sediment that contains $CaCO_3$. Deposits in which neither shells nor features are visible, owing to a lack of lime, may therefore be tested with 10% hydrochloric acid (HCl). In a positive case, as shown by effervescence, the occurrence of shells is possible. Since shells are generally difficult to see in freshly excavated material, even when abundant, it is useful to take small spot samples from all layers which are calcareous though macroscopically seem to be shell-free. Samples are washed in the field in order to show the presence, preservation, abundance, composition, and possibly also the palaeoenvironmental significance of the fossil malacofauna. According to these preliminary results the standard sampling procedure can be modified, especially with regard to sample size. In layers poor in shell fragments the occurrence of molluscs can be ascertained only by careful examination of the surface of well-washed debris.

Mollusca are abundant particularly in sediments formed by precipitation of $CaCO_3$, such as tufas, freshwater chalk, marl, calcareous muds and fens. In inorganic clastic sediments such as lacustrine clays, sand or gravels, Mollusca are less numerous and often concentrated in individual layers, lenses etc., often only as fragments (Ložek, 1964). Open sections are preferable to boreholes because moderately large samples are needed in most cases (see Chapter 34).

It is important to select sites which permit study of other groups of fossils such as plant remains, insects or archaeological objects. The same requirements relate to samples selected for ^{14}C dating. Such conditions are provided by certain tufa series, including fossil soil horizons, various intercalations (muds, scree etc.), archaeological objects, ostracods and plant impressions. Land snails are usually the best represented, giving the possibility of correlation with terrestrial environments in the adjacent area.

Sampling techniques

The location of the sampling column is determined by (*a*) the point where the stratigraphy is most complete and representative of the deposit as a whole, and (*b*) the point of the maximum shell concentration. There are sites in which these points are situated at different places or in which local intercalations of different material occur. Then, it is useful to take additional samples beyond the main column in order to obtain a more complete faunal record.

Well-stratified late-glacial and Holocene sequences are, in most cases, appropriate for a quantitative treatment of the incorporated malacofauna. Therefore, the sampling should be undertaken with the highest possible precision in order to record in detail all changes of molluscan assemblages. From each distinctive layer the aim is to take a bulk sample of approximately the same size. Poorly stratified deposits or very thick homogeneous layers should be subdivided into units of 5, 10 or 15 cm depending on the rate of sedimentation and the likely level of any faunal change. Sample size depends largely on local circumstances, particularly on the number of shells incorporated in the deposit or in its malacologically poorest layers. Therefore a standard volume of 1000–2000 cm^3 in very rich deposits is usually sufficient, whereas in other cases larger samples (5000–10,000 cm^3) are appropriate.

The best method of sampling is to cut samples from the face of a cleaned vertical section. Samples are cut starting at the base of the section and working upwards. Fully grown shells of the larger species (*Lymnaea, Viviparus* etc.) are rarely present in sediment samples and should therefore be collected by hand during the course of excavation (Evans, 1972).

LABORATORY WORK

Extraction

The air-dried sample is placed into a bowl of water and stirred until it slakes. The majority of complete shells float to the surface and can be poured off into a hemispherical sieve with 0.5 mm mesh size. The sludge which accumulates on the bottom of the bowl is then washed, in two phases: first, through a sieve with curved long meshes (e.g. 0.5 × 5.0 cm) to separate out the coarse particles ($CaCO_3$ incrustations, stones); secondly, the humic matter and all smaller particles are removed by washing through a sieve with a 0.5 mm mesh size. When resistant crumbs are present, it is necessary to disaggregate them by using hydrogen peroxide in an appropriate concentration. The shells are also cleaned.

The washed debris is then gently dried and all shells and fragments are carefully extracted. In layers with much $CaCO_3$ precipitation (tufas), the shells often bear $CaCO_3$ incrustations which can be partly removed by briefly immersing them in acetic acid for 10–15 seconds and then in running water.

Counting

In ideal circumstances the fossil shell collection consists of individual complete shells which can be easily counted. This is particularly true of individual horizons of lacustrine deposits. In general, however, a number of shells occur in fragments which will often dominate the whole assemblage. Such fragments must be determined and, for statistical purposes, calculated as individuals. This can be best done on apical fragments or on apertures where they are diagnostic (Vertiginidae, *Carychium* etc.). Many fragments have a diagnostic surface structure, but there is still an indeterminable residue. Nevertheless, all identifiable fragments must be taken into account in order to obtain a faunal record as complete as possible. However, these percentages do not precisely reflect the original percentage representations of the species, but they do provide a basis for a graphical presentation of results.

Counting methods have been developed and successfully applied in Great Britain and central Europe. For their general description, their application and errors, reference may be made to the monographs of Evans (1972), Sparks (1961), Sparks and West (1972) and Ložek (1964).

Identification

Most of late Quaternary Mollusca can be identified with reference to living species. The routine identification of the latter is mainly based on shell characteristics and can be used for fossil material. In Europe, a number of works have been published for the identification of modern fauna and these are of great value in the identification of fossil Mollusca (e.g. Adam, 1960; Ehrmann, 1933 (+ Ergänzung by Zilch and Jaeckel, 1962); Kerney *et al.*, 1983; Licharev and Rammelmeier, 1952; Soós, 1943; Žadin, 1952).

However, these textbooks only deal with adult individuals and not with the juvenile or fragmentary shells. Moreover, some distinctive features may not be available on the fossil specimens, such as the soft parts of the body, because these are lost in the fossil state. This is also true of the periostracum, its colour and whether or not it bears hairs or spines.

Two works that deal only with Quaternary Mollusca, including keys based on characters preserved in fossil state, are the monograph *Quartärmollusken der Tschechoslowakei* (Ložek, 1964), which does not contain the descriptions of juvenile stages, and the modern textbook *Land Snails in Archaeology* (Evans, 1972), which includes many detailed descriptions and exact figures of juvenile land snails, but only exceptionally of aquatic species. Therefore, all keys may be used in combination with well-determined materials of modern Mollusca, including all the stages of development (see Chapter 34).

PRESENTATION OF RESULTS

Ecological grouping

Quantitative analyses can be presented in terms of either relative or absolute abundance. They may be shown on diagrams that are in some way analogous to pollen diagrams. Since the graphs can be constructed in different ways, it is essential that the results of analysis be recorded in tabular form. The species may be grouped into ecological units; the fundamental division into land and freshwater groups being the first step. The land species are divided into several large groups, according to their relationship to woodland and open country, and to humidity (Ložek, 1964; Evans, 1972). In central Europe, a division into four main ecological groups subdivided into ten ecological groups has been established (Ložek, 1964, 1965b; Figure 36.1).

The main group D and the groups 9, 10, and partly also 3, 5, 7 and 8, include species characteristic of freshwater and marshland environments. Analytical percentage diagrams and synthetic spectra based on the above groups have provided important information on environmental and climatic changes in central Europe and have offered the opportunity to establish a malacological subdivision of the Late Vistulian and Holocene corresponding to palaeobotanical subdivisions. This is, however, more relevant to terrestrial environments, because the freshwater and marshland faunas were rather underestimated, although in Ložek's monograph (1964) a more detailed grouping within group 10 has been made.

In contrast to this, in the area of north European glaciation the freshwater and marshland habitats are malacologically of prime importance (Ložek, 1978b). For this reason, the British workers, especially Sparks (1961), subdivided the freshwater molluscan fauna into the following four ecological groups:

(1) *slum group:* species living in habitats characterized by poor water conditions, particularly in small water bodies subject to drying: *Lymnea*

MAIN ECOLOGICAL GROUPS		ECOLOGICAL GROUPS		INDEX SPECIES
A	WOODLAND species	1	Closed forest	Acicula polita, Ruthenica, Isognomostoma, Bulgarica cana, Aegopis verticillus
		2	Predominantly woodland (scrub,mesic open ground and marginal formations)	Alinda biplicata, Helix pomatia,Cepaea hortensis, Bradybaena fruticum
		3	Moist woodland	Perforatella bidentata, Monachoides vicina, Vestia turgida and V.gulo
B	OPEN GROUND species	4	Steppe(in general) Xerothermic rocks	Chondrula, Helicellinae, Granaria, Cecilioides, Chondrina, Pyramidula
		5	Open grounds (in general)	Pupilla muscorum, Vallonia pulchella, Vertigo pygmaea
C	INDIFFERENT species	6	WOODLAND OPEN GROUNDS — Dry (xeric)	Cochlicopa lubricella Euomphalia strigella
		7	Medium (mesic) Indifferent (catholic)	Cochlicopa lubrica,Punctum, Euconulus fulvus, Vitrina pellucida
		8	Damp	Carychium tridentatum, Vertigo substriata, Nesovitrea petronella
D	AQUATIC MARSHLAND species	9	Marshlands, banks very moist habitats (in general)	Carychium minimum,Oxyloma, Vertigo antivertigo, Zonitoides nitidus
		10	Aquatic habitats (in general)	Lymnaeidae, Planorbidae, Viviparus,Bithynia,Valvata, Bivalvia

FIGURE 36.1.   Ecological groups of molluscs used as the basis for biostratigraphic and palaeogeographic evaluation. This subdivision has been established for Central Europe

truncatula (Müll.), *Aplexa hypnorum* (L.), *Anisus leucostoma* (Müll.), *Pisidium casertanum* (Poli) etc.

(2)   *moving-water group:* species found in slightly larger bodies of water, e.g. moving streams or larger ponds, where the water is stirred by currents and wind: *Valvata piscinalis* (Müll.), *Pisidium amnicum* Müll., *Bithynia*, but also *Lymnaea stagnalis* (L.), *Physa fontinalis* (L.) and large bivalves.

(3)   *ditch group:* species which are often found in ditches with clean, slowly moving water and abundant growth of aquatic plants: *Valvata cristata* Müll., *Anisus vortex* (L.), *A. vorticulus* (Trosch.), *Segmentina nitida* (Müll.), *Acroloxus lacustris* (L.).

(4)   *catholic group:* species which tolerate a wide range of habitats except the worst slums: *Lymnaea palustris* (Müll.), *L. ovata* Drap., *Gyraulus crista* (L.), *Bathyomphalus contortus* (L.), *Hippeutis complanatus* (L.),

Sphaerium corneum (L.), *Pisidium subtruncatum* Malm., *P. nitidum* Jen. and *P. milium* Held.

This subdivision reflects the fact that most aquatic molluscs tolerate a wide range of habitats and climates. It is necessary to complete this list with some continental species and to establish groups of species which are peculiar to certain well-characterized habitats (Table 36.1).

Construction of graphs

The arrangement of a diagram depends on the aspects of the pattern of change which it is intended to emphasize. As a basic graphical record an analytical histogram can be shown, showing the relative abundance of individual species in shadow graphs arranged in ecological groups ordered from left to right. These may be marked by different colours or graphical symbols (Figure 36.2). The stratigraphical column is placed on the left border and the chronological interpretation may be noted on the other side of the diagram. The synthesis is illustrated by malacospectra showing the percentage representation of individual ecological groups. The malacospectrum of species is based on numbers of species only, irrespective of the numbers of individuals; the malacospectrum of individuals is based on the numbers of individuals, i.e. a group is represented by all individuals of its species.

The evaluation of graphs of relative abundance presents some problems, well-illustrated by the arguments of Kerney (1963) who favours using absolute abundance as a basis for the histograms on the grounds that: (1) the total number of shells is often inadequate for statistically valid percentages to be calculated; (2) the species do not fall into definite ecological groups; (3) an increase in one species may cause an apparent decrease of another, whereas in fact no decrease occurs; and (4) the absolute abundance of Mollusca is often itself a reflection of climatic change. Therefore, it is useful to construct graphs of absolute abundance, at least for selected index species (Ložek, 1978a). For further information the reader should consult specialist work (e.g. Sparks, 1961; Kerney, 1963; Ložek, 1965b, 1964, 1978; Evans, 1972).

INTERPRETATION

Mollusca may be used as indicators of age, climatic conditions and local environments, the first two aspects being closely connected because the detailed Quaternary stratigraphy is based on climatic changes.

Molluscs as chronological indicators

Among freshwater and marshland molluscs only a few species are of chronological value. These are *Pisidium stewarti* Preston, which became

TABLE 36.1.   Groups of freshwater molluscs with restricted habitat range. Other freshwater molluscs tolerate a wide range of habitats except the periodic swamps and oligotrophic small streams, springs and pools. This subdivision refers to conditions mainly in Central Europe

MOVING-WATER SPECIES LIVING MAINLY IN (LARGER) STREAMS AND PARTLY IN LAKES	**SPECIES OF PERIODIC SWAMPS** *Anisus leucostoma* (Mill.)[2] *Anisus spirorbis* (L.)
Ancylus fluviatilis Müll. (also in large springs)	*Aplexa hypnorum* (L.)[2] *Valvata pulchella* (Stud.)[2]
Dreissena polymorpha (Pall.)[1] *Fagotia* spp.[1] *Lithoglyphus* spp.[1]	**SEMIAMPHIBIC SPECIES TOLER-ATING VERY POOR WATER CONDITIONS**
Margaritifera spp. *Pisidium amnicum* (Müll.)	*Lymnaea truncatula* (Müll)[2]
Pisidium supinum A. Schm. *Physella acuta* (Drap.)[1] *Pseudanodonta* spp.[1]	**SPECIES OF SMALL, MAINLY OLIGOTROPHIC WATER BODIES, TOLERATING POOR WATER CONDITIONS**
Sphaerium rivicola (Lam.) *Sphaerium solidum* Norm.	*Lymnaea peregra* (Müll.) s.str. *Pisidium personatum* Malm
Theodoxus spp.[1] *Unio crassus* Phil.	**SPECIES OF SHALLOW STAG-NANT WATER WITH ABUNDANT GROWTH OF AQUATIC PLANTS, TOLERATING PERIODIC DRYING (CONFINED MAINLY TO MARSH-LAND ENVIRONMENTS)**
Unio pictorum (L.) *Valvata naticina* Menke[1] *Viviparus viviparus* (L.)[1] *Viviparus acerosus* (Bourg.)[1]	*Anisus septemgyratus* (Rssm.) *Bithynia leachi* (Shep.)[2]
GENERA OF DWARF PROSO-BRANCHIA LIVING IN SPRINGS AND SUBTERRANEAN WATERS (SOUTHERN EUROPE)	*Gyraulus riparius* (West.) *Lymnaea glabra* (Müll.)[2] *Lymnaea occulta* (Jack.)[2] *Pisidium obtusale* (Lam.)[2]
Avenionia *Belgrandia*	*Pisidium pseudosphaerium* Schl. *Planorbis planorbis* (L.)[2]
Belgrandiella *Bythinella* *Bythiospeum*	*Segmentina nitida* (Müll.) *Sphaerium lacustre* (Müll.) *Valvata cristata* Müll.
Hauffenia *Horatia* *Iglica* *Paladilhia* *Paladilhiopsis* etc.	

[1] Thermophilous species characteristic of warm Quaternary phases.
[2] Species occurring in pleniglacial loess, i.e. tolerating the very severe conditions of cold Quaternary phases.

extinct in Europe during the late-glacial, and to a lesser extent *P. lilljeborgi* Cl., *Lymnaea glabra* (Müll.), *Valvata piscinalis alpestris* Bl., *Vertigo genesii* (Grd.), and *Columella columella* (Mart.) which occurs in the late Vistulian

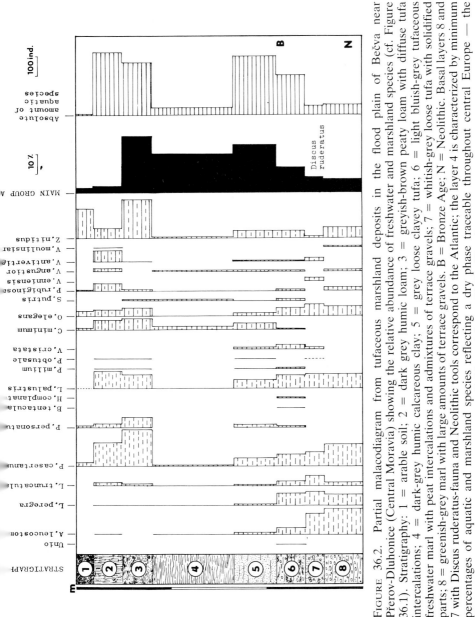

FIGURE 36.2. Partial malacodiagram from tufaceous marshland deposits in the flood plain of Bečva near Přerov-Dluhonice (Central Moravia) showing the relative abundance of freshwater and marshland species (cf. Figure 36.1). Stratigraphy: 1 = arable soil; 2 = dark grey humic loam; 3 = greyish-brown peaty loam with diffuse tufa intercalations; 4 = dark-grey humic calcareous clay; 5 = grey loose clayey tufa; 6 = light bluish-grey tufaceous freshwater marl with peat intercalations and admixtures of terrace gravels; 7 = whitish-grey loose tufa with solidified parts; 8 = greenish-grey marl with large amounts of terrace gravels. B = Bronze Age; N = Neolithic. Basal layers 8 and 7 with Discus ruderatus-fauna and Neolithic tools correspond to the Atlantic; the layer 4 is characterized by minimum percentages of aquatic and marshland species reflecting a dry phase traceable throughout central Europe — the Subboreal (sensu K.-D. Jäger)

often in the context of marshland. All these species are characteristic of the Vistulian Glacial and decline at the beginning of the Holocene in most parts of Europe. They are associated with many species living in modern times, but the composition of assemblages is different; several species, showing at present a restricted range, are abundant: *Valvata pulchella* (Stud.), *Gyraulus acronicus* (Fér.), *G. laevis* (Ald.), *Lymnaea occulta* (Jack.), whereas a number of more thermophilous species are lacking. However, some of them (e.g. *Valvata cristata* Müll., *Bithynia tentaculata* (L.)) appear during the late Vistulian in central Europe. Many aquatic species survived the Pleniglacial in the Danube lowland, as recorded in the swamp-loess deposits (Ložek, 1965a). During the late Vistulian and the earliest Holocene they gradually expanded to the north. Therefore, the molluscan chronozones are more or less metachronous in different regions.

The expansion of warmth-loving species, such as *Planorbis carinatus* Müll., *Anisus vorticulus* (Trosch.), *Viviparus contectus* (Mill.), *Acroloxus lacustris* (L.) and marshland elements, such as *Vertigo moulinsiana* (Dup.) or *Vallonia enniensis* (Grd.), starts in the earliest Holocene. These co-exist in the early Holocene with surviving glacial elements, such as *Vertigo genesii* (Grd.). The later changes are more closely linked with the development of local palaeoenvironments, i.e. hydrosere development, than with major climatic changes. Thus, the local sequences may only be applied chronostratigraphically in limited areas. In certain sites the sequence is very monotonous owing to minor habitat changes (Sparks, 1962; Marcussen, 1967).

The aquatic fauna also has its modern immigrants which expanded during the last few centuries, such as *Physella acuta* (Drap.), *Lithoglyphus naticoides* (C. Pfr), *Dreissena, Potamopyrgus* etc.

Molluscs as indicators of climate

Aquatic Mollusca are far from ideal as indicators of climate. First, it is difficult to establish the climate from freshwater faunas because this involves inferring the climate from what can be inferred about water temperatures. Secondly, many of the species have very wide climatic tolerances (Sparks and West, 1972). Ecological changes which eliminate molluscan species might easily be interpreted as climatic changes.

Nevertheless, there are several species which can be regarded as thermophilous (e.g. *Fagotia* spp., *Valvata naticina* Mke, *Viviparus acerosus* (Bourg.), *Physella acuta* (Drap.), *Dreissena, Theodoxus* spp., *Margaritifera auricularia* (Spg.), *Pseudanodonta*). Most of them are confined to larger streams in the southern half of Europe. Further north, they mostly represent the modern immigrants or interglacial index species. Additionally, incorporated land snails are usually better climatic and stratigraphic indicators than the water fauna and can help to define the conclusions more accurately.

Molluscs as indicators of local environments

Assuming that the fossil species had the same environmental requirements as their present-day representatives, a mosaic picture can be built up of the local environmental conditions at various stages of the depositional sequence. In particular, the development of different hydroseres can be traced in detail, both in the vertical and horizontal dimensions.

The shells of land snails provide very valuable information about changes in the terrestrial environment of the water habitat. In this way many problems of sedimentation processes can be elucidated. Numerous publications (e.g. Ložek, 1964, 1976; Sparks, 1961; Fuhrmann, 1973; Mania, 1973; Puisségur, 1976; Alexandrowicz, 1983) illustrate this procedure.

Correlation to other stratigraphic and palaeoenvironmental information

Local conditions inferred from fossil molluscan assemblages agree well with results of palaeopedological and sedimentological studies (Kowalkowski and Berger, 1966, 1972), and with reconstructions derived from the macroscopic plant remains and fossil coleopteran assemblages.

The advantage of malacology is that there is an enormous amount of easily recoverable material. Quaternary malacology shows the greatest promise for future palaeoecological research, inasmuch as it may solve problems that have hitherto thwarted other palaeontological studies. Until now, the regions studied in detail are few and detailed graphical representation is sparse; but the expansion of such studies will afford much valuable evidence to advance our knowledge of the late Quaternary.

REFERENCES

Adam, W. (1960). Mollusques terrestres et dulcicoles. In: *Faune de Belgique, Mollusques I*, Bruxelles.

Alexandrowicz, S. V. (1983). Malacofauna of Holocene calcareous sediments of the Cracow Upland. *Act. Geol. Polon.*, **33**, (1–4), 117–158.

Ehrmann, P. (1933). Mollusken (Weichtiere). In: *Die Tierwelt Mitteleuropas, II/1*, Quelle & Meyer, Leipzig. (Incl. Ergänzung by A. Zilch and S. Jaeckel, 1962).

Evans, J. G. (1972). *Land Snails in Archaeology*, Seminar Press, London.

Favre, J. (1927). Les Mollusques post-glaciaires et actuels du Bassin de Genève, *Mém. Soc. Phys. et Hist. Nat. Genève*, **40**, 171–434.

Fuhrmann, R. (1973). Die spätweichselglaziale und holozäne Molluskenfauna Mittel- und Westsachsens, *Freiberger Forschungshefte*, C 278-Paläontologie, 121pp.

Geyer, D. (1927). *Unsere Land- und Süsswassermollusken* (3rd ed.), K. G. Lutz, Stuttgart.

Kerney, M. P. (1963). Late-glacial deposits on the chalk of south-east England. *Phil. Trans. Royal Soc. London*, **B.730**, 203–254.

Kerney, M. P. (1977). British Quaternary non-marine Mollusca: a brief review. In: *British Quaternary Studies*, Oxford, pp. 31–42.

Kerney, M. P., Cameron, R. A. D., and Jungbluth, J. H. (1983). *Die Landschnecken Nord- und Mitteleuropas*, P. Parey, Hamburg.

Kowalkowski, A., and Berger, L. (1966). Palaeomalacological analysis in investigations on development of soils in Holocene. *Fol. Quat.*, **23**, 27 pp.

Kowalkowski, A., and Berger, L. (1972). Die Bedeutung der Conchylienfaunen für die spätpleistozäne und holozäne Sediment- und Bodenstratigraphie. *Bull. Soc. Amis d. Sc. e. d. Lettres Poznań*, **D 12/13**, 215–224.

Licharev, I. M., and Rammelmeier, E. S. (1952). Nazemnye Molljuski fauny SSSR. In: *Opredeliteli po faune SSSR* 43, AN SSSR, Moscow.

Ložek, V. (1964). Quartärmollusken der Tschechoslowakei. *Rozpr. Ústřed. Úst. Geol. Praha*, **31**, 1–374.

Ložek, V. (1965a). Das Problem der Lössbildung und die Lössmollusken. *Eisz. Gegw.*, **16**, 61–75.

Ložek, V. (1965b). *Problems of Analysis of the Quaternary Nonmarine Molluscan Fauna in Europe*, Geol. Soc. Am., Spec. Paper 84, pp. 201–218.

Ložek, V. (1976). Klimaabhängige Zyklen der Sedimentation und Bodenbildung während des Quartärs im Lichte malakozoologischer Untersuchungen. *Rozpravy ČSAV, MPV*, **86**, (8), 1–97.

Ložek, V. (1978a). Möglichkeiten und Perspektiven der Malakoanalyse in Ablagerungsfolgen des Binnenlandes. *Proc. Palaeontologická Konference '77*, Charles University, Prague, pp. 253–268.

Ložek, V. (1978b). Molluskenstratigraphie im Gebiet der skandinavischen Vereisungen. *Schriftenr. Geol. Wiss.* (Akad. Verl. Berlin), **9**, 121–136.

Ložek, V. (1982). Faunengeschichtliche Grundlinien zur spät- und nacheiszeitlichen Entwicklung der Molluskenbestände in Mitteleuropa. *Rozpravy ČSAV, MPV*, **92**, (4), 1–106.

Mania, D. (1967). Der ehemalige Ascherslebener See (Nordharzvorland) in spät-und postglazialer Zeit. *Hercynia*, **4**, 199–260.

Mania, D. (1973). Paläoökologie, Faunenentwicklung und Stratigraphie des Eiszeitalters im mittleren Elbe-Saalegebiet auf Grund von Molluskengesellschaften. *Geologie*, **21**, Beih. 78/79, 1–175.

Marcussen, I. (1967). The freshwater molluscs in the Late-glacial and early Post-glacial deposits in the bog of Barmosen, southern Sjaelland, Denmark. *Medd. Dansk Geol. Fora.*, **17**, 265–283.

Meier-Brook, C. (1975). Der ökologische Indikatorwert mitteleuropäischer Pisidium-Arten (Mollusca, Eulamellibranchia). *Eisz. Gegw.*, **26**, 190–195.

Menzel, H. (1910). Klimaänderungen und Binnenmollusken im nördlichen Deutschland seit der letzten Eiszeit. *Zt. Dtsch. Geol. Ges.*, **62**, 199–267.

Odhner, N. (1910). Die Entwicklung der Molluskenfauna in dem Kalktuffe bei Skultorp in Wästergötland. *Geol. För. Stockholm Förh.*, **32**, 1095–1138.

Puisségur, J.-J. (1976). Mollusques continentaux quaternaires de Bourgogne. *Mém. Géol. Un. Dijon*, **3**, 1–241.

Soós, L. (1943). *A Kárpát-medence Mollusca-faunája*, Magyar Tudományos Akadémia, Budapest.

Sparks, B. W. (1961). The ecological interpretation of Quaternary non-marine Mollusca. *Proc. Linn. Soc. London*, 172 Session, 1959–60, **1**, 71–80.

Sparks, B. W. (1962). Post-glacial Mollusca from Hawes Water, Lancashire, illustrating some difficulties of interpretation. *J. Conch.*, **25**, 2, 78–82.

Sparks, B. W., and West, R. G. (1972). *The Ice Age in Britain*, Methuen, London.

Vidal, H., Brunnacker, K., Brunnacker, M., Körner, H., Hartel, F., Schuch, M., and Vogel, J. C. (1966). Der Alm im Erdinger Moos. *Geol. Bavarica*, **56**, 177–200.

Waldén, H. W. (in press). A comprehensive account of the Late-Quaternary land mollusca in Scandinavia. *Malakologische Abh., Staatliches Mus. für Tierkunde in Dresden, 11.*

Žadin, V. I. (1952). Molljuski presnych i solonovatych vod SSSR. *Opredeliteli po faune SSSR* 46, AN SSSR, Moscow.

Numerical treatment of biostratigraphical data

Handbook of Holocene Palaeoecology and Palaeohydrology
Edited by B. E. Berglund
© 1986 John Wiley & Sons Ltd.

37

Numerical zonation, comparison and correlation of Quaternary pollen-stratigraphical data

H. J. B. BIRKS

Botanical Institute, University of Bergen, Norway

INTRODUCTION

Quaternary pollen-analytical and other palaeoecological data are complex, multivariate and quantitative. They consist of estimates of the abundance, defined in various ways, of many variables (e.g. pollen and spore types) in a large number of individuals (e.g. surface samples or stratigraphical levels in one or more sequences) (see Chapter 38). The variables are usually biological entities (e.g. pollen types, diatom taxa); they can also be lithological (e.g. particle sizes), chemical (e.g. elemental abundances), or physical (e.g. stable-isotope ratios). The individuals can be of two types — *non-stratigraphical* and *stratigraphical* samples.

Non-stratigraphical pollen data are estimates of pollen abundance (e.g. percentages, concentrations, or accumulation rates) in samples of the same age, usually but not invariably (e.g. Huntley and Birks, 1983) the present-day. Such contemporaneous data may be from different geographical or vegetational areas. Stratigraphical data, on the other hand, represent estimates of pollen abundance at different depths and hence ages in sequences at one or more localities. This ordering of samples with depth and age is of paramount importance. It provides the basis for presenting palynological data as stratigraphical pollen diagrams, for partitioning the data into zones or other subdivisions, for recognizing stratigraphical changes within the sequence as a whole and within sequences of individual taxa, and for interpreting the data as a temporal record of changes in past floras, populations and vegetation, and, by inference, past environments. It allows, in Deevey's (1967) words, ' a four-dimensional look at vegetation'.

The complex, multivariate nature of palynological data has encouraged the use of several multivariate analytical techniques such as principal components analysis (PCA), canonical variates analysis, and cluster analysis (see Chapter 38; Blackith and Reyment, 1971; Davis, 1973; Everitt, 1978; Gordon, 1981, for introductions to these techniques) as aids in data handling, summarization, synthesis, and interpretation, following early studies by Cole (1969), Adam (1970), Dale and Walker (1970), Webb (1971), and Cushing (unpublished). In addition new and powerful techniques such as stratigraphically constrained zonation (Gordon and Birks, 1972), sequence slotting (Gordon and Birks, 1974), sequence splitting (Walker and Wilson, 1978), and maximum-likelihood estimation of modern pollen-representation factors (Parsons and Prentice, 1981) have been developed by statisticians working in collaboration with palaeoecologists. These techniques, along with recent developments in data transformation (Gordon, 1982) and assessment of dissimilarity between pollen spectra (Prentice, 1980; Chapters 38 and 39), take account of important statistical properties of palynological data, particularly the multinomial nature of relative data, the high inherent variance of concentration and accumulation-rate estimates, and the ordering of stratigraphical data. Birks and Gordon (1985) review available numerical methods that are relevant and of proven or potential value in Quaternary palaeoecology. Several of these techniques have also been successfully used in pre-Quaternary palaeoecology (Reyment, 1980).

Quaternary palynological and other biostratigraphical data provide a unique record of biological change and, by inference, environmental history. Numerical methods can help decipher these records by, for example, distinguishing between local and regional changes and by highlighting subtle and often undetected similarities and differences in data from several sites within a study area. The advantages and limitations of numerical methods in Quaternary palaeoecology are reviewed by Birks and Birks (1980) and Birks and Gordon (1985). The most important advantage is their ability to detect in a consistent and repeatable way *patterns* of stratigraphical change within sets of complex palaeoecological data, prior to any consideration of the *processes* that may have influenced the observed patterns of change. Palaeoecology is concerned with the detection of patterns of change *and* the explanation of these patterns in terms of ecological processes. Numerical methods provide powerful means of detecting all patterns of change within a sequence, not just those changes that corroborate or conform to existing paradigms of environmental history. Ecological insights are essential in deducing and inferring the processes that produced the observed patterns. Numerical methods are no substitute for ecological knowledge; they are, however, invaluable in identifying patterns worthy of ecological attention. The careful and critical application of appropriate numerical techniques and an interaction between numerical results and ecological experience can provide new insights into the nature, timing, magnitude and temporal and spatial extent of late-Quaternary biotic and environmental changes.

The most productive applications of numerical methods to Quaternary palynology have been in four areas:

(1) The subdivision of pollen-stratigraphical sequences into local pollen zones or other stratigraphical units, and the comparison and correlation of two or more sequences for detecting similarities among profiles, delimiting regional pollen zones, and highlighting differences between sequences (e.g. Gordon and Birks, 1972, 1974; Birks and Berglund, 1979; Björck, 1979, 1981; Ritchie, 1982);

(2) The analysis of surface-pollen data and quantitative calibration of modern pollen and vegetation data as an aid in reconstructing past plant populations from pollen-stratigraphical data (e.g. Andersen, 1970; Parsons and Prentice, 1981; Ritchie, 1982; Prentice and Parsons, 1983);

(3) the analysis of modern-pollen data to detect contemporary pollen-vegetation 'corresponding patterns' (Webb, 1974; Birks *et al.*, 1975; Prentice, 1983) and the quantitative comparison of modern and fossil pollen assemblages as a basis for reconstructing, by analogy, past vegetation (e.g. Birks, 1976; Ritchie, 1977; Ritchie and Yarranton, 1978a,b; MacDonald, 1983; Lamb, 1984);

(4) the calibration of modern pollen and environmental data for reconstructing past environments, particularly regional climate (e.g. Webb and Bryson, 1972; Webb and Clark, 1977; Howe and Webb, 1983).

The appropriate numerical methods for (2) and (3) are outlined and reviewed in Chapters 38 and 39. The methodology and requisite computer programs for (4) are considered in Chapter 40. This chapter concentrates on (1), the zonation, comparison and correlation of pollen-stratigraphical sequences. Its aims are as follows:

(1) to present a general strategy for handling pollen-stratigraphical data prior to their interpretation in terms of past floras, populations, communities or environments;

(2) to outline computer-based methods for constructing pollen diagrams;

(3) to describe appropriate numerical methods for subdividing a single stratigraphical sequence into local pollen zones;

(4) to review appropriate numerical methods for comparing two or more pollen sequences, for deriving regional pollen zones, and for detecting differences between sequences;

(5) to consider additional numerical methods for analysing pollen-stratigraphical data, particularly pollen-accumulation rates;

(6) to discuss methods for comparing two or more stratigraphical data sets from the same sequence (e.g. pollen and plant macrofossils, pollen and diatoms, etc.);
(7) to list relevant computer programs that implement the recommended numerical procedures.

Although this chapter deals almost exclusively with pollen-stratigraphical data, the methods are equally applicable to other types of quantitative stratigraphical data, such as diatoms, cladocera, ostracod, mollusca, geochemical and stable-isotope data.

APPROACHES TO PRESENTATION, ZONATION, COMPARISON AND CORRELATION OF POLLEN-STRATIGRAPHICAL DATA

Pollen-stratigraphical data are conventionally presented as pollen diagrams. Construction of a pollen diagram is the first stage in handling pollen data prior to interpretation (Figure 37.1). Computer-based methods for constructing and plotting pollen diagrams are outlined later. Questions of choice of pollen sum, use of representation factors, and style and format of pollen diagrams are discussed by Faegri and Iversen (1975), Berglund and Ralska-Jasiewiczowa (Chapter 22), and Birks and Gordon (1985).

Pollen-stratigraphical data are frequently so complex that it is often necessary to partition sequences into smaller units for ease in (*a*) description, (*b*) discussion, comparison and interpretation, and (*c*) correlation in time and space of the sequences. The most convenient unit of subdivision of the vertical, temporal dimension of the pollen sequence is the *pollen zone*.

Although pollen analysts have delimited and, in some cases, defined pollen zones for over 50 years, there are very few operational definitions of what a pollen zone is. Cushing (1964) defines a pollen zone as

> 'a body of sediment distinguished from adjacent sediment bodies by differences in kind and amount of its contained fossil pollen grains and spores, which are derived from plants existing at the time of deposition of the sediment'.

Cushing (1963, 1964) discusses in detail the concept of a pollen zone, and emphasizes that a pollen zone, as defined above, is a biostratigraphical unit or biozone (Hedberg, 1976) and that a pollen zone corresponds most closely to an assemblage zone, as defined in the *International Guide to Stratigraphic Classification, Terminology and Usage* (Hedberg, 1976). Cushing (1964) recommends pollen-assemblage zones as the basic units in Quaternary pollen stratigraphy, and this view can logically be extended to other Quaternary biostratigraphical data (e.g. diatoms, cladocera and mollusca). Different types of zones are possible and, in some instances, desirable, (e.g. acme or peak,

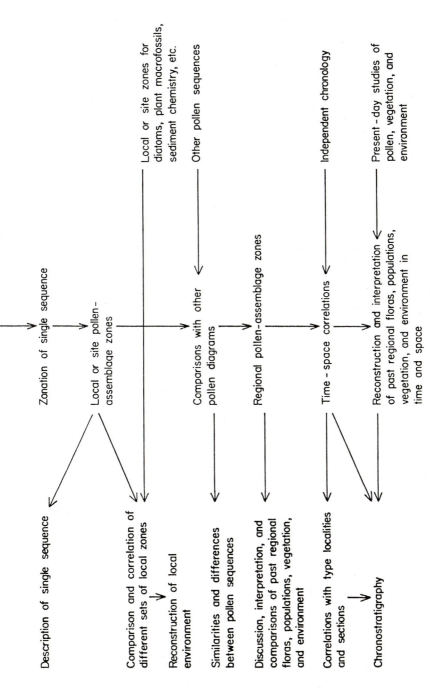

FIGURE 37.1. Stages in the handling and analysis of pollen-stratigraphical data prior to interpretation

interval, taxon-range or concurrent-range zones — see Hedberg, 1976). Peak zones can be particularly valuable in late-glacial pollen stratigraphy where several rapid but marked pollen stratigraphical changes may occur within a short interval (e.g. Watts, 1977; Craig, 1978; Prentice, 1982).

With the realization (Cushing, 1964) that pollen zones, as generally used in Quaternary palynology, are biostratigraphical assemblage zones defined solely on the observed pollen and spore content of the sediments, with no assumptions or preconceptions, explicit or implicit, about sediment lithology, past vegetation, inferred climate or assumed time equivalence, there has been much interest in developing and applying appropriate numerical methods for subdividing pollen-stratigraphical sequences (Birks and Gordon, 1985). Numerically derived zones are delimited solely on the basis of specified mathematical criteria without any reference to inferred climate, reconstructed vegetation or assumed chronology. They are repeatable and, by definition, strict pollen-assemblage zones. Numerical methods for partitioning pollen-stratigraphical sequences are considered later (see also Birks and Gordon, 1985). Numerical zonation is the second logical stage in analysing pollen-stratigraphical data (Figure 37.1).

Numerically delimited pollen-assemblage zones refer only to the sequence analysed. They are thus *local* or site pollen-assemblage zones (*sensu* Cushing, 1967; Birks, 1973; Watts, 1977) that describe and summarize the basic pollen-stratigraphical features of the sequence. Local zones are of limited use in studying broad-scale, regional changes in pollen stratigraphy and vegetational history, and in the discussion, comparison and correlation of pollen records from several sites within an area. Zonation of individual sequences is thus only a preliminary step in data analysis. The next stage is to compare pollen sequences from different sites within the area of interest in as unbiased a way as possible. If similar assemblages are found in different sequences, regional pollen-assemblage zones (*sensu* Cushing, 1967; Birks, 1973; Watts, 1977) can be delimited and defined on the basis of several sequences. Although comparisons between sequences are important in many studies, the criteria for comparing sequences and for assessing similarities and differences between sequences are rarely stated. Numerical methods for comparing sequences and for aiding in the recognition, delimitation and definition of regional pollen-assemblage zones are considered later (see also Birks and Gordon, 1985). Regional zones can be ordered stratigraphically and, if an independent chronology is available, can be correlated in time and mapped in space to produce time–space correlation diagrams (e.g. Cushing, 1967; Birks, 1973; Berglund, 1979, 1983). Such correlations can, if required, be related to standard chronostratigraphical sequences at type localities (West, 1970; cf. Birks, 1982). When considering and comparing regional vegetational changes within and between different geographical or ecological areas, regional assemblage zones are particularly appropriate as they largely reflect the 'regional pollen rain' and hence the regional vegetation of the area of interest.

In addition to comparing pollen sequences for purposes of regional zonation and correlation, it is often useful to compare profiles for ecological and interpretative purposes. Numerical comparisons of data from several sites within the same ecological region today can highlight similarities and differences in the patterns of vegetational change (Ritchie and Yarranton, 1978a, 1978b). It is also valuable to compare sequences so as to detect differences between profiles rather than solely to seek similarities, as in regional zonations. In general, similarities in pollen composition reflect regional vegetational patterns, whereas differences result from local vegetational patterns related to, for example, soils, aspect, topography, or human disturbance (e.g. Watts, 1961; Brubaker, 1975; Jacobson, 1979; Bradshaw, 1981). Numerical methods for detecting patterns of change and differences between sites are outlined later (see also Birks and Gordon, 1985).

Some workers (e.g. Watts, 1973, 1977; Walker, 1982) have questioned whether pollen zones should be delimited at all. Birks and Gordon (1985) argue that, for many purposes, there is a need to summarize the large amounts of complex, multivariate pollen-stratigraphical data. Local pollen zones, carefully defined on the basis of available data only, and their subsequent grouping into regional zones, are the most effective and convenient means currently available for summarizing succinctly the mass of information contained within pollen diagrams (Watts, 1977). If interest is centred, however, on detailed changes within *individual* pollen curves, for example when viewed as long temporal records of plant populations (Watts, 1973), alternative approaches to subdividing pollen sequences are appropriate (Walker and Wilson, 1978). These are discussed later.

For sites where more than one palaeoecological variable has been studied (e.g. pollen, diatoms, sediment chemistry), each set of stratigraphical data can be independently subdivided into local zones. Correlation diagrams of different local zonations can be constructed (see later) so as to detect any common patterns of stratigraphical change and to highlight temporally independent changes in different groups of organisms within a site. Such comparisons can aid in distinguishing between biotic changes induced by environmental change and biotic changes resulting from processes such as immigration, disease or human disturbance, and between local and regional environmental changes (see Gordon and Birks, 1972).

This strategy for handling and analysing pollen-stratigraphical data prior to interpretation (Figure 37.1) emphasizes both geographical and stratigraphical patterns. When the data are interpreted in terms of past flora, populations, vegetation and environment, these reconstructions can also be presented in terms of spatial and temporal patterns of variation. Ecological processes operative at these spatial and temporal scales can then be sought to explain the observed patterns. It is important to emphasize that subdividing pollen-stratig-raphical sequences into zones, whether local or regional, assemblage or peak, is not an end in itself but is a means to an end. Such an end is the reconstruction of past floras, populations, vegetation and environment, and of the dynamics and differentiation, in both time and space, of these reconstructions. Zonation, comparison and correlation are merely steps towards that end (Birks and Birks, 1980).

COMPUTER METHODS FOR CONSTRUCTING POLLEN DIAGRAMS

With the development of high-speed computers and increased availability of fast, axillary graph-plotters, several computer programs have been written to handle and calculate pollen-analytical data and draw pollen diagrams. Examples include Squires (1970), Squires and Holder (1970), Damblon and Schumacker (1971), Dodson (1972), Voorips (1973, 1974), King (1976), Birks and Huntley (1978), Eccles *et al.* (1979), Cushing (1979), Veldkamp *et al.* (1981), and Vuorinen and Huttunen (1981).

The aim of all these programs is to input raw pollen counts, express them as percentages of some user-specified pollen sum, concentrations or accumulation rates, print out these values, and draw a pollen diagram of the results. Programs can process large data sets (200 samples × 200 pollen and spore taxa) and plot preliminary pollen diagrams very rapidly based on different pollen sums (e.g. total pollen, total pollen excluding Cyperaceae etc.) thereby permitting the effects of different calculation bases to be explored and evaluated (cf. Wright and Patten, 1963) more readily than has hitherto been possible.

POLLDATA.MK5 (Birks and Huntley, 1978) is a versatile FORTRAN IV program for the manipulation, calculation and plotting of pollen-analytical and other palaeoecological data. It reads in basic counts, sorts samples into stratigraphical or any other specified order, and pollen and spore types into defined categories (e.g. trees, shrubs, dwarf shrubs, herbs, obligate aquatics etc.) and into any specified order, deletes any samples or pollen types if required, calculates percentages on the basis of a specified pollen sum (ΣP), draws preliminary pollen diagrams on a line-printer, and plots pollen diagrams of various sizes on a graph-plotter with either a depth or estimated age axis. Pollen and spores excluded from ΣP (e.g. obligate aquatics) are calculated as percentages of, for example, $\Sigma P + \Sigma$ Obligate aquatics (Faegri and Iversen, 1975). The program includes options to calculate and plot 95 per cent confidence intervals for percentage data (Maher, 1972), calculate pollen concentrations and pollen-accumulation rates, transform pollen counts by means of appropriate modern representation factors (e.g. Andersen, 1970), and smooth concentrations or accumulation rates by moving averages, the number of terms being specified by the user. There are several options for plotting pollen diagrams (bar histograms, joined-up silhouette or 'saw-edge' curves etc.) and for handling plant macrofossil data. The program can also process diatom, cladocera, mollusca, ostracod and other palaeoecological data. It is heavily dependent on local plotting routines exclusive to particular computer centres. It is, like other programs that plot pollen diagrams (e.g. Squires, 1970; King, 1976; Cushing, 1979), installation-dependent, large, and moderately difficult to use. Details are given later.

NUMERICAL METHODS FOR ZONATION OF
POLLEN-STRATIGRAPHICAL DATA

There are four main numerical approaches to the subdivision of pollen-stratigraphical data into local pollen zones (Table 37.1). Only one, constrained cluster analysis, takes account of the stratigraphical ordering of the samples and has as its underlying model the concept of a pollen zone as an assemblage of pollen preserved in a sediment body. The other approaches either ignore the stratigraphical order in the numerical procedure (cluster analysis and scaling or ordination methods; see Chapter 38) or emphasize boundaries between zones rather than consider the assemblage within a body of sediment (sequential correlation analysis). Although these methods may provide useful and intuitively sensible pollen zones in certain instances, they are not wholly satisfactory. The results of sequential correlation analysis can be strongly influenced by intermediate levels analysed close to suspected zone boundaries.

TABLE 37.1. Numerical approaches to zonation of pollen-stratigraphical sequences

Approach	Use of stratigraphical information	Criterion for zone delimitation	Examples
Sequential correlation analysis	Yes	Boundaries	Kershaw (1970), Yarranton and Ritchie (1972), Maher (1982)
Cluster analysis	None	Assemblages	Adam (1970, 1974), Dale and Walker (1970), Mosimann and Greenstreet (1971)
Constrained cluster analysis	Yes	Assemblages	Gordon and Birks (1972, 1974), Gordon (1973a, 1982), Birks and Gordon (1985), Cushing (unpublished)
Scaling or ordination methods	None	Assemblages	Adam (1970, 1974), Birks (1974), Birks and Berglund (1979)

Cluster analysis lacking stratigraphical constraints allows groupings of levels with similar pollen composition that may be some distance apart in the sequence. Although such groupings can be of interpretative interest, they are of limited use in zonation. The results of scaling procedures such as PCA can be presented as low-dimensional configurations of sample points (e.g. Pennington and Sackin, 1975; Gordon, 1981; Chapter 38) or as stratigraphical plots of sample scores or coordinates on each scaling axis (e.g. Adam, 1974; Birks, 1974; Birks and Berglund, 1979; Huntley, 1981; Chapter 38). This latter representation is useful if the axes effectively portray trends in the data. In procedures such as PCA the axes have a particular geometrical significance. This is not the case for other methods (e.g. non-metric multidimensional scaling). Also non-linear trends within the data may not be fully represented by any one axis (Prentice, 1980). Despite these limitations, stratigraphical plots of scaling results are useful as they are 'composite pollen diagrams' that provide a visual summary of patterns of variations within the data (see Chapter 38). Stratigraphical changes in the sign and magnitude of the sample coordinates on several axes can be used to delimit zones as groups of adjacent samples with similar coordinates (Birks, 1974). Stratigraphical plots display the relationships and transitions between zones and the variation within zones. They are particularly valuable when there are gradual and often complex transitions from one pollen assemblage to another, and are useful in assessing and interpreting results of constrained cluster analyses (e.g. Gordon, 1982).

There are many available scaling procedures (e.g. Gower, 1967; Everitt, 1975, 1980; Prentice, 1980; Chapter 38). The most useful for zonation purposes is PCA (see Blackith and Reyment, 1971; Jöreskog *et al.*, 1976; Gordon, 1981; Chapter 38). It is most conveniently implemented by a so-called R-mode algorithm. Percentage pollen data should be transformed to square-roots prior to PCA of the covariance matrix between variables. This is equivalent to a Q-mode principal coordinates analysis using chord distance (Birks, 1977, 1981; Prentice, 1980; Chapters 38 and 39) as the dissimilarity measure between samples, a coefficient with certain optimal 'signal-to-noise' properties when applied to pollen percentages. It down-weights abundant types and up-weights less common types (Prentice, 1980; Chapter 39). Aitchison (1982, 1983) proposes a log-ratio transformation for percentage data prior to PCA of the form:

$$z_{ik} = \log\ (p_{ik}/p_{i*})$$

where $\log p_{i*} = \sum\limits_{k\ =\ 1}^{t} (\log p_{ik}/t),$

p_{ik} is the percentage of pollen type k in sample i, and t is the number of pollen types. This transformation helps avoid the so-called 'closure' problem of percentage compositional data (Reyment, 1983). It has not, as yet, been applied to pollen-stratigraphical data.

Pollen concentrations or accumulation rates require a variance-stabilizing transformation prior to PCA of the covariance matrix between variables. Gordon's (1982) transformation is recommended. It is as follows:

$$x_{ik} = \tan^{-1}(y_{ik}/v)^{1/2}$$

where y_{ik} is the number of grains (cm^{-2} per year) of taxon k in sample i or the number of grains (cm^{-3}) of taxon k sample i, and v is the best available estimate of the number of exotic grains added to sample i in order to estimate y_{ik}.

Incorporation of stratigraphical constraints into a numerical classification procedure for zonation purposes was first proposed and implemented by E. J. Cushing (unpublished work cited by D. P. Adam in Mosimann and Greenstreet (1971) and used by, for example, Mehringer *et al.*, 1977) in Orloci's (1967) sum-of-squares agglomerative method. Clusters are only formed by fusing adjacent levels or groups of neighbouring levels. Several different methods of constrained classification have subsequently been developed by Gordon and Birks (1972, 1974), Gordon (1973a, 1981), and Birks and Gordon (1985) (see also Gordon, 1980a). The five main methods will now be described.

(1) *Constrained single-link cluster analysis* (CONSLINK in Gordon and Birks, 1972). This has two stages. Firstly, dissimilarities between all pairs of samples are calculated on the basis of pollen composition of the samples; secondly, stratigraphically adjacent samples are clustered into groups of samples of similar pollen composition using the single-link criterion for group membership (Gordon, 1981). Throughout only contiguous levels or groups of levels are allowed to fuse. The dissimilarity measure used by Gordon and Birks (1972) is the city-block or Manhattan metric:

$$DC_{(ij)} = \sum_{k=1}^{t} |p_{ik} - p_{jk}|$$

where p_{ik} and p_{jk} are the percentages of pollen type k in sample i and j, respectively, and there are t pollen types. Other dissimilarity measures (Everitt, 1980; Prentice, 1980; Chapters 38 and 39) can be used.

Results are usually presented as a dendrogram or 'tree diagram' (Figure 37.2). Sectioning the dendogram at a particular dissimilarity value yields the CONSLINK groupings of samples that can be regarded as zones. Although conceptually and computationally simple, it often produces results that are difficult to assess because the extent of clusters and hence of pollen zones can be difficult to delimit, particularly for large sequences (>60 samples) (Birks

ABERNETHY FOREST (H.H.Birks and R.W.Mathewes, 1978)

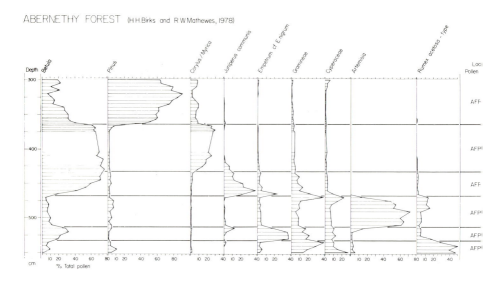

and Berglund, 1979). The method performs poorly if there are numerically small but stratigraphically consistent differences within a sequence (Gordon and Birks, 1972). The variable-barriers approach (see below) is useful in such instances. CONSLINK is effective at displaying transitional levels and highlighting samples with unusually high (or low) values of particular pollen types for which re-examination and replication are desirable. However, the final zonation is often unduly influenced by single, possibly atypical samples.

(2) *Constrained binary divisive analysis using sum-of-squares deviations as a global measure of variability within a sequence* (SPLITLSQ in Gordon and Birks, 1972). The aim is to partition a sequence into n groups by placing $n - 1$ boundaries so as to ensure that within-group variation is minimized, as assessed by total within-group sum-of-squares distance about the n centroids. This partitioning can be implemented in various ways, and in SPLITLSQ a hierarchical solution is adopted. Such a hierarchy is useful because the partitions may correspond to zones, subzones, zonules etc.

Initially the first zone boundary is placed by finding the position that gives the largest reduction in total within-group sum-of-squares, and hence the most homogeneous pair of zones. The next division is obtained by partitioning one of the existing zones into two, so as to produce the maximal decrease in sum-of-squares. The method proceeds by partitioning existing groups into smaller groups so that maximum decreases in sum-of-squares, and hence the most homogeneous partitions, result. It continues until little further reduction in variability occurs. The resulting partitions into homogeneous, stratigraphically contiguous groups of samples can be regarded as a possible zonation.

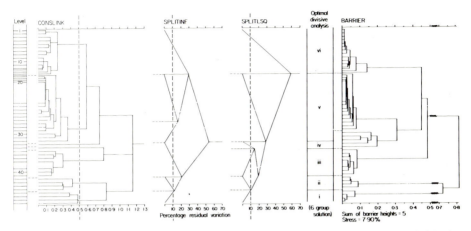

FIGURE 37.2. Pollen diagram of major taxa at Abernethy Forest, Scotland published by H. H. Birks and Mathewes (1978) and results of five numerical zonation methods (CONSLINK, SPLITINF, SPLITLSQ, Optimal divisive analysis, BARRIER) based on the nine pollen types shown. The local pollen zones delimited by H. H. Birks and Mathewes are also shown. *Redrawn and modified from Birks and Gordon (1985)*

The results can be represented as a 'block dendogram' (Figure 37.2) or in a triangular format (e.g. Figure 9 in Gordon and Birks, 1972) giving the residual variation, the residual variation as a percentage of the total variation, and the positions of the boundaries for 1,2,....*n* divisions. This and other divisive procedures have the advantage that data for a whole zone are considered in the positioning of zone boundaries (cf. CONSLINK).

(3) *Constrained binary divisive analysis using information content as a global measure of total variation within a sequence* (SPLITINF in Gordon and Birks, 1972). This is similar to (2) except that information content is used as the measure of variability.

(4) *Constrained optimal divisive analysis using a dynamic programming algorithm* (Gordon, 1980a, 1981; Birks and Gordon, 1984). If a sequence is partitioned into three zones by binary divisive procedures, corresponding to zone boundaries at positions n_1 and n_2, there is no guarantee that these nested partitions are the divisions that give the *overall* minimum total within-group variability. This is because by positioning the first boundary at n_1, the second boundary, n_2, is located in a mathematically optimal position subject to the restriction that the first boundary is fixed at n_1. Optimal partitions need not be hierarchically nested; for example, the optimal division into three groups is not necessarily obtained from the optimal partitioning into two groups by dividing

one of these two groups. Clearly examining *all* possible positions for 2,3,... boundaries is very time-consuming for large data sets. Fortunately optimal least-squares partitionings of stratigraphical data can be obtained efficiently and rapidly using a dynamic programming algorithm (Gordon, 1980a, 1981). Because optimal partitioning is often different for different numbers of zones, the results cannot be presented as a single dendrogram. They are most conveniently presented in a triangular format (Gordon, 1982), giving zone boundary positions and the residual total variation as a percentage of the initial total variability for 1,2,...,*n* divisions, or as a diagram showing positions of the zone boundaries for a specific number of divisions (Figure 37.2).

Although the method has, to date, only been used with a sum-of-squares criterion, it can be used with any criterion for which the total variation of the partitions is additive, namely the total variation is the sum of the variability within each separate group of the sequence.

(5) *Variable-barriers approach* (Gordon, 1973a; BARRIER in Gordon and Birks, 1974). In contrast to methods (2), (3) and (4) which tend to place zone boundaries in the middle of transitional intervals between zones, BARRIER attempts to distinguish groups of transitional levels from groups of levels with broadly similar pollen composition. In other words it tries to delimit two types of pollen zones; phases of relatively 'stable' pollen composition and phases of abrupt or systematic changes in pollen composition (cf. Walker, 1966; Watts, 1973).

Instead of placing boundaries between pairs of adjacent levels as in the divisive procedures, barriers of any height between 0 and 1 are positioned, subject to the constraint that their sum equals *h*, a value preset by the user. The local mean for sample *i* is influenced by all levels that can be reached from *i* by 'jumping over' barriers of total height less than 1. A measure of total variability in the pollen values from the local mean is then minimized by means of an iterative procedure, thereby ensuring that levels with dissimilar pollen composition are separated by high barriers. Samples belonging to phases of rapid transition will appear between adjacent high barriers. The method can only be implemented by an iterative function-minimization procedure that is computationally expensive. It is therefore restricted to comparatively small data sets (<50 samples). It has proved particularly useful in partitioning late-glacial sequences (e.g. Birks and Mathewes, 1978; Birks, 1981).

The results can be presented as a dendrogram with the branches corresponding to the barriers (Figure 37.2). Transitional levels are represented as late additions to clusters. In general, results are not critically dependent on *h*, the sum-of-barrier heights.

Mathematical details of the above five methods are given by Birks and Gordon (1985), and the requisite computer programs are listed later in this chapter (p. 766).

For purposes of numerical zonation and comparison, it is conventional to include only pollen and spores of taxa of non-aquatic plants. Experience with many data sets has shown that pollen types with values less than 5% are of minor *numerical* importance and exert little influence on the zonation results. Such rare types have a high relative error associated with their pollen counts (Maher, 1972; Faegri and Iversen, 1975) and are thus poorly estimated numerically unless very large pollen sums are counted. Such types may, of course, be of very considerable *ecological* importance as 'indicator species' in the interpretation of the data in terms of past flora, vegetation and environment (e.g. Iversen, 1944, 1954; Birks, 1973, 1976).

The numerical procedures outlined above have been used with a wide variety of data; see, for example, Gordon and Birks (1972, 1974), Birks (1974, 1976, 1981), Birks and Berglund (1979), Cwynar (1982), Ritchie (1982), Ovenden (1982), Hunt and Birks (1982), Pawlikowski *et al.* (1982), Bennett (1983a), and Delcourt *et al.* (1983). In all these studies the end result of this stage of analysis (Figure 37.1) is the delimitation of local pollen-assemblage zones.

Because these constrained clustering procedures make different assumptions of the data, involve different concepts and measures of dissimilarity and variability, and use different criteria for partitioning, it is essential to use several methods and to compare results. If similar results are obtained from different methods, one can be confident that the partitions are not an artifact of a particular numerical method. It is also useful to compare results of constrained clustering methods with results of scaling procedures such as PCA (see Birks and Berglund, 1979; Birks and Gordon, 1985). Zone boundaries consistently detected by several methods can then be viewed as the most effective partitions within the sequence. They provide a consistent, unambiguous and repeatable zonation based solely on the observed pollen stratigraphy and allow the delimitation of local pollen-assemblage zones for the profile of interest.

Although these methods have primarily been used to analyse percentage data, they can also be used with pollen concentrations or accumulation rates after appropriate transformation (see Gordon, 1982; Hunt and Birks, 1982). Such data are subject to large and, at present, unavoidable errors in estimation (Maher, 1981), and inevitably less confidence should be attached to the results of numerical analyses of them. Moreover, there is also very considerable within-site variation in concentrations and accumulation rates due to complex processes of pollen sedimentation and sediment focusing (Davis *et al.*, 1973; Lehman, 1975; Davis and Ford, 1982; Bennett, 1983b). Such high variability limits the value of using concentrations or accumulation rates as the primary basis of zonation.

NUMERICAL METHODS FOR COMPARING POLLEN-STRATIGRAPHICAL DATA

Comparisons of pollen sequences are essential if regional pollen-assemblage zones are to be delimited and defined. There are three main numerical approaches (Gordon and Birks, 1974).

(1) *Zone-by-zone comparison* (COMPZONE in Gordon and Birks, 1974). This compares numerically local zones defined independently for each sequence either by visual inspection or numerical zonation. It attempts to match, using information radius, a zone in one sequence with a zone in another. The method has not been widely used — see Gordon and Birks (1974) and Birks and Gordon (1985) for examples. It is appropriate for broad-scale, general comparisons and can be a useful preliminary guide in delimiting regional zones. The results are sensitive to incorrect choices of local zones for comparative purposes, and to high within-zone variability resulting from transitional samples near boundaries. Sample ordering within a zone is not considered; information about stratigraphical trends is thus lost.

(2) *Combined scaling of two or more sequences* (Gordon and Birks, 1974; Birks and Berglund, 1979). Sequences of interest are ordinated together and the basic unit of comparison is individual samples. As procedures such as PCA seek to represent in low-dimensional space similarities defined mathematically, between samples, it follows that samples of similar pollen composition, irrespective of which sequence they are derived from, will, if the ordination is an accurate and efficient representation of the data, be positioned together in the ordination. Geometrical dispostions of the sample points allow detection of groups of samples of similar pollen composition and hence of regional pollen-assemblage zones.

As sequence-comparisons frequently involve simultaneous analysis of 200 or more samples, it is computationally convenient to use an R-mode implementation of PCA because the number of variables (pollen types) is considerably less than the number of samples to be compared (Birks and Berglund, 1979). Square-root transformation of percentage data should be used prior to PCA of the covariance matrix of the variables (e.g. Birks, 1981). Alternatively log-ratio transformations of percentages may be useful.

Results are most readily presented as plots of the positions of individual samples, as represented by their component scores, in the low-dimensional space of the first few principal components (e.g. Björck, 1979, 1981; Gordon, 1981). For big data sets (>100 total samples), the number of samples is so large that these plots are difficult to assimilate and interpret. Instead mean coordinates or scores of local pollen zones at each site on the first few components can be calculated, and means and ranges for each zone plotted (see Figure 38.4; and Ritchie, 1977; Ritchie and Yarranton, 1978a, 1978b; Birks and Berglund, 1979). Zone means can be joined up in stratigraphical order. If means and ranges for zones from different sequences overlap and the ordination is an efficient representation of the data, zones of similar pollen composition, and hence possible regional pollen-assemblage zones, can be delimited. It is often useful to partition the combined data using a conventional, unconstrained sum-of-squares clustering procedure (e.g. Gordon and Henderson, 1977; Chapter 38) to aid detection of groups of samples from different sequences with similar pollen composition within the scaling results (Birks and Gordon, 1985).

This approach has been used to compare two or more late- or postglacial sequences by, for example, Gordon and Birks (1974), Ritchie (1977), Ritchie and Yarranton (1978a, 1978b), Björck (1979, 1981), Birks and Berglund (1979), Birks (1981), Gordon (1981), Lamb (1982), and Wiltshire and Moore (1983), and to compare pollen sequences from different interglacial deposits by Birks and Peglar (1979). There is considerable potential for applying it to the many problems of interglacial pollen stratigraphy and correlation.

(3) *Sequence slotting* (FITSEQ in Gordon and Birks, 1974; Gordon, 1973b; SLOTSEQ in Gordon, 1980b). In combined scalings, no stratigraphical constraint is imposed during numerical analysis. In SLOTSEQ sample ordering is an important constraint. Sequences are slotted together into a single sequence, with the property that sections of sequences with similar pollen composition must be positioned close together in the joint sequence. Gordon's (1973b) discordance statistic (Ψ) measures goodness of fit when sequences are slotted together on the basis of similarities in pollen composition as quantified by the Manhattan metric (see above). Delcoigne and Hansen (1979) present a dynamic programming algorithm for optimal sequence slotting. (See also Gordon and Reyment (1979) and Gordon (1980b).) As the method always produces a slotting, the value of Ψ should be examined to evaluate how good the fit is. The lower Ψ is the better the fit, although no formal statistical testing can yet be applied to evaluating Ψ.

Results are usefully presented as sample-by-sample matchings between sequences (Figure 37.3). Groups of samples with similar pollen composition within and between sequences can be detected and regional zones delimited; see Gordon and Birks (1974), Ritchie (1982), and Birks and Gordon (1985). Because of limitations of computer storage it can only be used to compare a few sequences simultaneously. It is fairly sensitive to different stratigraphical patterns within separate sequences. 'Blocking', where several consecutive samples from one sequence occur next to one another in the joint sequence without insertion of any samples from other sequences, can occur if there are marked differences between sequences or if there are very minor differences between consecutive samples within a sequence. In the former case Ψ is usually high, whereas in the latter Ψ is often low. The method is most robust when comparing sequences that are fairly similar but not *too* similar or dissimilar (Birks and Gordon, 1985).

The three approaches described above can aid in comparing pollen-stratigraphical sequences and in detecting and delimiting regional zones. In certain

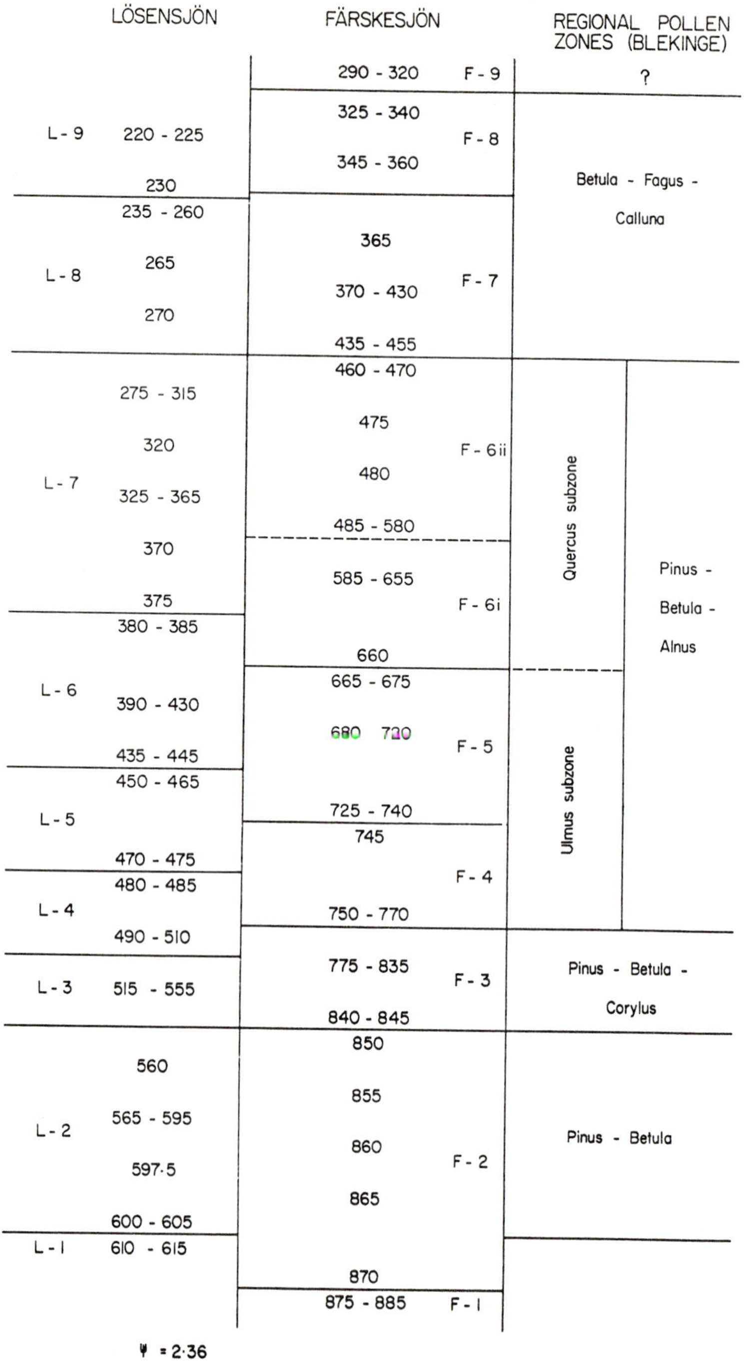

LÖSENSJÖN   FÄRSKESJÖN   REGIONAL POLLEN ZONES (BLEKINGE)

		290 - 320	F - 9	?
L - 9	220 - 225	325 - 340	F - 8	
	230	345 - 360		Betula - Fagus -
	235 - 260			Calluna
L - 8	265	365	F - 7	
	270	370 - 430		
		435 - 455		
	275 - 315	460 - 470		
	320	475	F - 6 ii	
L - 7	325 - 365	480		
	370	485 - 580		
	375	585 - 655	F - 6 i	
	380 - 385	660		
L - 6	390 - 430	665 - 675		
	435 - 445	680 720	F - 5	
	450 - 465			
L - 5		725 - 740		
	470 - 475	745		
	480 - 485		F - 4	
L - 4	490 - 510	750 - 770		
L - 3	515 - 555	775 - 835	F - 3	Pinus - Betula -
		840 - 845		Corylus
	560	850		
		855		
L - 2	565 - 595	860	F - 2	Pinus - Betula
	597·5	865		
	600 - 605			
L - 1	610 - 615	870		
		875 - 885	F - 1	

Quercus subzone   Ulmus subzone   Pinus - Betula - Alnus

Ψ = 2·36

FIGURE 37.3. Results of sequence-slotting for two Holocene sequences from Blekinge, Sweden. The numerically delimited local zones L-1 to L-9 and F-1 to F-9 and the suggested regional zones for Blekinge are also shown. *Redrawn and modified from Birks and Berglund (1979)*

cases one method is more appropriate than another; but when all three are used they are complementary and give confidence that any consistent results are not an artifact of the method used.

If an independent chronology, e.g. ^{14}C dates, is available for the sites of interest, the regional zones can be ordered stratigraphically, correlated in time, and mapped in space to produce time-space correlation diagrams (Figure 37.4). Such diagrams summarize and portray the major stratigraphical and geographical patterns of the regional zones.

Numerical methods for comparing sequences can also assist in summarizing major patterns of stratigraphical change within and between profiles from specific geographical areas. Ritchie and Yarranton (1978a, 1978b) (see also Ritchie, 1977) have used PCA of zone means to compare several pollen sequences and to summarize major patterns of stratigraphical change and, by inference, vegetational change and dynamics over long time periods (see Birks and Gordon, 1985).

Quantitative techniques can also assist in highlighting differences between sequences. Jacobson (1979) developed 'difference diagrams' for comparing pollen profiles from nearby sites on contrasting soils. Sequences are initially correlated, and groups of 2–4 stratigraphically adjacent samples are amalgamated and their pollen composition averaged. Difference diagrams between pairs of sites are then constructed, using measures of difference such as:

$$\Delta y_{ik} = y_{1ik} - y_{2ik} \qquad \text{(Jacobson, 1979)}$$

$$\Delta p_{ik} = \log(p_{1ik}/p_{2ik}) \qquad \text{(Jacobson, 1979)}$$

$$\Delta p_{ik} = p_{1ik} - p_{2ik} \qquad \text{(Bradshaw, 1981)}$$

where y_{1ik} and y_{2ik} are accumulation rates of taxon k for time interval i at sites 1 and 2 respectively, and p_{1ik} and p_{2ik} are percentages of taxon k for time i at sites 1 and 2 respectively. Some pollen types may show little difference in their values at the two sites, whereas others may have higher values at one site, suggesting local occurrence on a particular soil. Alternatively pollen sequences can be compared by superimposing curves for individual taxa from several sequences on a common age rather than depth scale (e.g Andersen, 1978). Differences in pollen composition are thus clearly displayed.

All these 'difference' methods require reliable and independent means of correlation between sequences. They provide useful means of highlighting differences between sequences and can profitably be used in studies where reconstruction of fine-scale vegetational differentiation in space and time is important.

ADDITIONAL NUMERICAL METHODS

Summarization and interpretation of concentrations and accumulation rates are aided by appropriate multivariate analyses (e.g. Mosimann and Greenstreet,

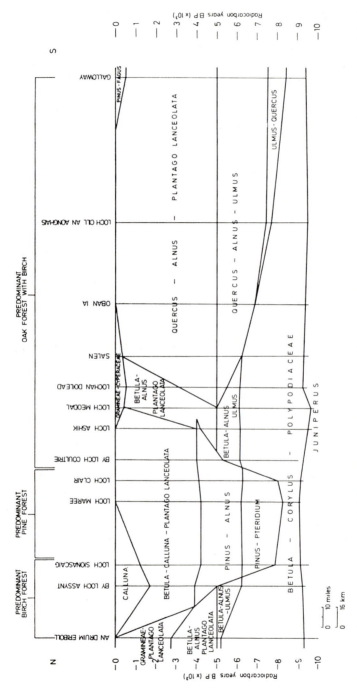

FIGURE 37.4. Time–space correlation diagram of regional pollen-assemblage zones in western Scotland. Regional zones based on the results of numerical comparisons of 13 published and unpublished sequences on a north–south transect are plotted against radiocarbon years B.P., to show the temporal and spatial extent of the various regional zones (from Birks, 1980)

1971; Gordon, 1982). Absolute data require a variance-stabilizing transformation (Gordon, 1982) prior to analysis. Quantitative relationships between samples and pollen taxa are conveniently displayed and summarized by numerical techniques that analyse samples and variables simultaneously, such as correspondence analysis (Hill, 1974) and biplots (Gabriel, 1971) (see also Jöreskog *et al.*, 1976; Gordon, 1981, 1982; Chapter 38). These methods permit rapid detection of (1) similarities between samples, (2) similarities between taxa, and (3) taxa that influence stratigraphical patterns between samples, and vice versa. They can aid interpretation of complex stratigraphical patterns directly in terms of changes in particular pollen types. Their great value lies in simultaneous analysis of samples and variables (see, for example, Melguen, 1974; Malmgren *et al.*, 1978). They are useful when emphasis is on comparing 'shapes' of pollen curves rather than their absolute magnitude (Gordon, 1982), and are particularly appropriate for detecting covariance patterns within absolute data.

Biplots have been used to analyse pollen-stratigraphical data from one (Gordon, 1982) or more (Birks and Gordon, 1985) sequences. Correspondence analysis has been used to analyse pollen-stratigraphical data (Gordon, 1982), palaeolimnological data (Elner and Happey-Wood, 1980; Peglar *et al.*, 1984), and plant-macrofossil data (H. H. Birks, 1973; Spicer and Hill, 1979; Spicer, 1981).

If reliable pollen-accumulation rates are available and the observed rates are not influenced by phenomena such as sediment focusing (Davis *et al.*, 1973; Lehman, 1975) and have not varied because of changes in pollen-recruitment processes (Pennington, 1979), accumulation rates can be viewed as long temporal records of past plant populations that have expanded, contracted, oscillated, remained stable, or become extinct through time (Watts, 1973; Bennett, 1983c). Walker and Wilson's (1978) sequence-splitting method attempts to detect stratigraphical patterns and population changes within each *individual* taxon sequence. It initially subdivides a sequence (e.g. *Pinus*) into sections on the basis of presence-or-absence data only, thereby distinguishing intervals of more or less total absence from intervals of presence. Each of the latter type is then split by a divisive algorithm into sections of distinct but homogeneous mean and standard deviation. This procedure is similar to the constrained divisive methods described earlier, except that here only a single taxon is considered. Walker and Wilson (1978) present a procedure for locating splits and testing the statistical significance of splits. Other algorithms for splitting can be used (e.g. Hawkins and Merriam, 1973). Each section is characterized by its mean, standard deviation, and mean/standard deviation ratio (Walker and Pittelkow, 1981).

When applied to all sequences within a pollen profile, the number of statistically significant splits between pairs of levels is a measure of the amount of pollen-stratigraphical and hence plant-population change occurring between these times (Green, 1982). The observed number and distribution of splits can be tested statistically (Gardiner and Haedrich, 1978) against the null hypothesis that the splits are randomly distributed, thereby distinguishing between linked and individualistic behaviour of taxa through time (Birks and J. M. Line, unpublished). Clustered distributions of splits suggest marked environmental control of changes in several populations, whereas randomly distributed splits suggest individualistic behaviour (Walker and Pittelkow, 1981; Walker, 1982; Birks and Gordon, 1985). Sequence splitting can aid in distinguishing biotic changes induced by environmental changes from biotic changes resulting from biological processes such as immigration, disease or human disturbance. It can also detect subtle stratigraphical patterns within and between individual pollen curves, expose unsuspected patterns of change between taxa, and provide insights into patterns of population change with time (Walker and Pittelkow, 1981; Green, 1982). The procedure emphasizes changes within individual sequences rather than within the fossil assemblage as a whole, thereby complementing zonation procedures. Sequence splitting allows the biologically interesting hypothesis of individualistic behaviour of taxa in time (Gleason, 1939) to be examined critically; such testing is more difficult if stratigraphical data are partitioned into pollen zones (cf. Birks, 1976; Walker and Pittelkow, 1981). It retains information that is lost in zonation.

Birks and Gordon (1985) review the assumptions and limitations of the approach. The main assumption is that observed pollen-accumulation rates are a consistent and reliable record of population sizes within the pollen-catchment area of the site. At present there is no satisfactory quantitative means (Lehman, 1975; Green, 1983) of distinguishing between changes in pollen-accumulation rates caused by variations in pollen-recruitment processes and sediment focusing from changes reflecting population expansion or decline, other than by analysing several cores from the same basin (Davis and Ford, 1982; Bennett, 1983b) and from different basins in the same area (Pennington, 1979).

Applications of sequence splitting include: Walker and Wilson (1978), Walker and Pittelkow (1981), Green (1982, 1983), and Birks and Gordon (1985). Although it has only been used to analyse pollen-accumulation rates, it can be usefully applied, for example, to absolute diatom and cladocera data as a means of exploring local population dynamics over long time periods.

COMPARISON OF STRATIGRAPHICAL DATA SETS FROM THE SAME SEQUENCE

Zonation methods can be used to partition stratigraphical sequences of other palaeoecological variables, such as plant macrofossils, diatoms, cladocera, ostracods and sediment chemistry. If, for example, pollen and plant-macrofossil analyses are available from the same sequence, it is valuable to discover if

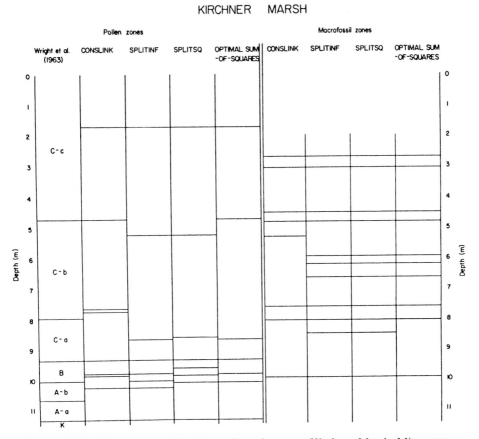

FIGURE 37.5.   Comparison of local zonation schemes at Kirchner Marsh, Minnesota, U.S.A., based on pollen stratigraphy (left) and macrofossil stratigraphy (right). The original pollen zones delimited visually by Wright *et al.* (1963) are also shown.
Redrawn and updated from Gordon and Birks (1972)

major stratigraphical patterns in the pollen data correspond to major stratigraphical patterns in the macro fossil data. The easiest way of answering this question is to partition independently the pollen data into local pollen zones and the macrofossil data into local macrofossil zones, and to compare directly the stratigraphical patterns of two zonations by means of correlation diagrams (Figure 37.5; Gordon and Birks, 1972; Birks, 1976; Ovenden, 1982). The hypothesis of stratigraphically independent boundaries in two or more sets of variables can be tested statistically using a modified version of Gardiner and Haedrich's (1978) method for analysing transect data.

In some instances pollen and macrofossil zones may correspond, in other instances they may not. As pollen largely reflects regional, upland vegetation

Handbook of Holocene Palaeoecology and Palaeohydrology

and as macrofossils tend to reflect local vegetation, correspondence in stratigraphical patterns between the two suggest major vegetational changes both regionally and locally. Changes in macrofossils that are independent of pollen changes imply local changes only, whereas changes in pollen stratigraphy that are independent of changes in the macrofossils may reflect forest succession, tree immigration, or other processes such as disease, soil maturation or human disturbance. Comparisons of independent zonations for the same sequence can also provide insights into the timing and response of stratigraphical change in different ecological groups.

Alternatively, different data sets can be ordinated independently using PCA (e.g. Pennington and Sackin, 1975) or correspondence analysis (e.g. Elner and Happey-Wood, 1980) and stratigraphical plots of sample scores for different sets of variables compared visually or numerically. These comparisons can detect common patterns of stratigraphical changes between sets of variables and may highlight any unique patterns of change within particular sets of variables. Such plots can also provide useful quantitative assessment of the underlying directions of variation within the data that may relate, for example, to climatic change, human impact, or local limnological changes (e.g. Binford, 1982).

Numerical techniques are particularly valuable in interdisciplinary palaeolimnological studies where large numbers of variables are often studied, by providing a means of detecting stratigraphical similarities and differences between variables. Numerical zones are invaluable for comparing two or more independent sets of stratigraphical data because these zones are not influenced by any preconceptions about relationships between variables (Birks, 1976; Pawlikowski et al., 1982).

COMPUTER PROGRAMS

Only a few of the multivariate statistical techniques discussed above (e.g. PCA, correspondence analysis, biplots) are available in program packages such as GENSTAT or CLUSTAN. GENSTAT is versatile but difficult to use, and neither package allows for appropriate transformation options for absolute pollen data or outputs the results in useful stratigraphical formats.

Programs written specifically to analyse biostratigraphical data are more useful. They are easier to use, they implement all appropriate transformations and mathematical procedures relevant for zonation and comparison of pollen-stratigraphical data, and they produce output in a stratigraphical format.

All the programs listed below are written in FORTRAN IV by the person whose initials are given first. Many of the programs have been modified, updated, or include subroutines written by other people whose initials follow the main author's initials. The initials refer to D. P. Adam, H. J. B. Birks, J.

C. Davis, A. D. Gordon, B. Huntley, J. E. Klovan, J. M. Line and R. A. Reyment.

BARRIER	variable-barriers constrained zonation. ADG; HJBB, BH, JML.
BENZECRI	correspondence analysis of pollen-stratigraphical data. JEK and RAR; HJBB, JML.
BIPLOT	biplots of pollen-stratigraphical data. ADG; JML.
BOUND2	statistical testing of locations of stratigraphical boundaries for different sets of variables (Gardiner and Haedrich, 1978). JML.
COMPZONE	numerical comparisons of pollen zones using information radius and concordance. ADG.
PCARMODE	principal components analysis implemented by an R-mode algorithm of pollen-stratigraphical data and with a wide variety of input and output options. HJBB; DPA, JCD, BH, JML, RAR. (See also Chapter 38.)
POLLDATA	standard program for processing pollen-stratigraphical data, for calculating percentages, concentrations or accumulation rates, and for plotting pollen diagrams (Birks and Huntley, 1978). HJBB and BH; JML.
SLOTSEQ	slotting of two pollen sequences. A simple version is listed in Birks (1979). A more elaborate version of the program with various types of extrinsic constraints is listed in Gordon (1980b). ADG; HJBB.
SPLIT1	splitting of stratigraphical sequences of pollen-accumulation rates using the algorithms of Walker and Wilson (1978) and Hawkins and Merriam (1973). Part of the program for the Walker and Wilson (1978) method is based on the programs CLUBIN and CLUVAR written by Y. Pittelkow and listed in Walker and Wilson (1978). JML.
ZONATION	constrained zonation incorporating CONSLINK, SPLITINF and SPLITLSQ of Gordon and Birks (1972). An early version is listed in Birks (1979) and a later version implementing optimal divisive analysis using a dynamic programming algorithm is given in Birks and Gordon (1985). The latest version, with a wide variety of output facilities and transformation for absolute data, is also available. ADG; HJBB, BH, JML.

Acknowledgements

I am deeply indebted to my friends Emma and Henry Lamb for providing a suitable environment in which to write this chapter. I am also indebted to Hilary Birks, Björn Berglund, Francis Gilbert, Allan Gordon and Sylvia Peglar for their multifarious support and help with the writing and preparation of the manuscript. I dedicate this chapter to Björn Berglund, in recognition of his many contributions to international scientific collaboration and his concern for people and in appreciation of his support and friendship.

REFERENCES

Adam, D. P. (1970). *Some Palynological Applications of Multivariate Statistics*, Ph.D. Thesis, University of Arizona.

Adam, D. P. (1974). Palynological applications of principal component and cluster analyses. *J. Res. of the United States Geological Survey*, **2**, 727–741.

Aitchison, J. (1982). The statistical analysis of compositional data. *J. Royal Statistical Society B*, **44**, 139–177.

Aitchison, J. (1983). Principal component analysis of compositional data. *Biometrika*, **70**, 57–65.

Andersen, S. Th. (1970). The relative pollen productivity and representation of North European trees, and correction factors for tree pollen spectra. *Danmarks Geologiske Undersøgelse*, II, **96**, 99pp.

Andersen, S. Th. (1978). Local and regional vegetational development in eastern Denmark in the Holocene. *Danmarks Geologiske Undersøgelse Årbog 1976*, 5–27.

Bennett, K. D. (1983a). Devensian late-glacial and Flandrian vegetational history at Hockham Mere, Norfolk, England. I. Pollen percentages and concentrations. *New Phytologist*, **95**, 457–487.

Bennett, K. D. (1983b). Devensian late-glacial and Flandrian vegetational history at Hockham Mere, Norfolk, England. II. Pollen accumulation rates. *New Phytologist*, **95**, 489–504.

Bennett, K. D. (1983c). Postglacial expansion of forest trees in Norfolk, UK. *Nature*, **303**, 164–167.

Berglund, B. E. (1979). The deglaciation of southern Sweden 13,500–10,000 B.P. *Boreas*, **8**, 89–118.

Berglund, B. E. (1983). Palaeohydrological studies in lakes and mires — a palaeoecological research strategy. In: *Background to Palaeohydrology* (Ed. K. J. Gregory), John Wiley, Chichester, pp. 237–254.

Binford, M. W. (1982). Ecological history of Lake Valencia, Venezuala: interpretation of animal microfossils and some chemical, physical, and geological factors. *Ecological Monographs*, **52**, 307–333.

Birks, H. H. (1973). Modern macrofossil assemblages in lake sediments in Minnesota. In: *Quaternary Plant Ecology* (Eds. H. J. B. Birks and R. G. West), Blackwell, Oxford, pp. 173–190.

Birks, H. H., and Mathewes, R. W. (1978). Studies in the vegetational history of Scotland. V. Late Devensian and early Flandrian pollen and macrofossil stratigraphy at Abernethy Forest, Inverness-shire. *New Phytologist*, **80**, 455–484.

Birks, H. J. B. (1973). *Past and Present Vegetation of the Isle of Skye — a Palaeoecological Study*, Cambridge University Press, London.

Birks, H. J. B. (1974). Numerical zonations of Flandrian pollen data. *New Phytologist*, **73**, 351–358.

Birks, H. J. B. (1976). Late-Wisconsinan vegetational history at Wolf Creek, central Minnesota. *Ecological Monographs*, **46**, 395–429.

Birks, H. J. B. (1977). Modern pollen rain and vegetation of the St. Elias Mountains, Yukon Territory. *Can. J. of Botany*, **55**, 2367–2382.

Birks, H. J. B. (1979). Numerical methods for the zonation and correlation of biostratigraphical data. In: *Palaeohydrological Changes in the Temperate Zone in the last 15,000 years. Subproject B: Lake and Mire Environments*, vol. 1, (Ed. B. E. Berglund), Department of Quaternary Geology, University of Lund, pp. 99–123.

Birks, H. J. B. (1980). Quaternary vegetational history of western Scotland, *Vth International Palynological Conference*, Cambridge, Excursion Guide C8.

Birks, H. J. B. (1981). Late Wisconsin vegetational and climatic history at Kylen Lake, north-eastern Minnesota. *Quaternary Research*. **16**, 322–355.

Birks, H. J. B. (1982). Holocene (Flandrian) chronostratigraphy of the British Isles: a review. *Striae*, **16**, 99–105.

Birks, H. J. B., and Berglund, B. E. (1979). Holocene pollen stratigraphy of southern Sweden: a reappraisal using numerical methods. *Boreas*, **8**, 257–279.

Birks, H. J. B., and Birks, H. H. (1980). *Quaternary Palaeoecology*, Edward Arnold, London.

Birks, H. J. B., and Gordon, A. D. (1985). *Numerical Methods in Quaternary Pollen Analysis*, Academic Press, London (in press).

Birks, H. J. B., and Huntley, B. (1978). Program POLLDATA.MK5: documentation relating to FORTRAN IV program of 26 June 1978. Sub-Department of Quaternary Research, University of Cambridge.

Birks, H. J. B., and Peglar, S. M. (1979). Interglacial pollen spectra from Sel Ayre, Shetland. *New Phytologist*, **83**, 559–575.

Birks, H. J. B., Webb, T., and Berti, A. A. (1975). Numerical analysis of pollen samples from central Canada: a comparison of methods. *Review of Palaeobotany and Palynology*, **20**, 133–169.

Björck, S. (1979). *Late Weichselian Stratigraphy of Blekinge, S.E. Sweden, and Water Level Changes in the Baltic Ice Lake*, LUNDQUA Thesis, Dept. of Quaternary Geology, Lund University, pp. 1–248.

Björck, S. (1981). A stratigraphic study of Late Weichselian deglaciation, shore displacement and vegetation history in south-eastern Sweden. *Fossils and Strata*, **14**, 1–93.

Blackith, R. E., and Reyment, R. A. (1971). *Multivariate Morphometrics*, Academic Press, London.

Bradshaw, R. H. W. (1981). Quantitative reconstruction of local woodland vegetation using pollen analysis from a small basin in Norfolk, England. *J. of Ecology*, **69**, 941–955.

Brubaker, L. B. (1975). Postglacial forest patterns associated with till and outwash in north-central Upper Michigan. *Quaternary Research*, **5**, 499–527.

Cole, H. S. (1969). *Objective Reconstruction of the Paleoclimatic Record through the Application of Eigenvectors of Present-day Pollen Spectra and Climate to the Late-Quaternary Pollen Stratigraphy*, Ph.D. Thesis, University of Wisconsin.

Craig, A. J. (1978). Pollen percentage and influx analyses in south-east Ireland: a contribution to the ecological history of the late-glacial period. *J. of Ecology*, **66**, 297–324.

Cushing, E. J. (1963). *Late-Wisconsin Pollen Stratigraphy in East Central Minnesota*, Ph.D. Thesis, University of Minnesota.

Cushing, E. J. (1964). Application of the Code of Stratigraphic Nomenclature to Pollen Stratigraphy. Unpubl. manuscript.

Cushing, E. J. (1967). Late-Wisconsin pollen stratigraphy and the glacial sequence in

Minnesota. In: *Quaternary Paleoecology* (Eds. E. J. Cushing and H. E. Wright), Yale University Press, New Haven, pp. 59–88.

Cwynar, L. C. (1982). A late-Quaternary vegetation history from Hanging Lake, northern Yukon. *Ecological Monographs*, **52**, 1–24.

Dale, M. B., and Walker, D. (1970). Information analysis of pollen diagrams, I. *Pollen et Spores*, **12**, 21–37.

Damblon, F., and Schumacker, R. (1971). New prospects for the study of palynological data: the use of computers. *Pollen et Spores*, **13**, 609–614.

Davis, J. C. (1973). *Statistics and Data Analysis in Geology*, John Wiley, London.

Davis, M. B., Brubaker, L. B., and Webb, T. (1973). Calibration of absolute pollen influx. In: *Quaternary Plant Ecology* (Eds. H. J. B. Birks and R. G. West), Blackwell, Oxford, pp. 9–25.

Davis, M. B., and Ford, M. S. (1982). Sediment focusing in Mirror Lake, New Hampshire. *Limnology and Oceanography*, **27**, 137–150.

Deevey, E. S. (1967). Introduction. In: *Pleistocene Extinctions* (Eds. P. S. Martin and H. E. Wright), Yale University Press, New Haven, pp. 63–72.

Delcoigne, A., and Hansen, P. (1975). Sequence comparison by dynamic programming. *Biometrika*, **62**, 661–664.

Delcourt, H. R., Delcourt, P. A., and Spiker, E. C. (1983). A 12,000-year record of forest history from Cahaba Pond, St. Clair County, Alabama. *Ecology*, **64**, 874–887.

Dodson, J. R. (1972). Computer programs for the pollen analyst. *Pollen et Spores*, **14**, 455–465.

Eccles, M., Hickey, M., and Nichols, H. (1979). *Computer Techniques for the Presentation of Palynological and Paleoenvironmental Data*, Institute of Arctic and Alpine Research Occasional Paper, 16, pp. 1–139.

Elner, J. K., and Happey-Wood, C. M. (1980). The history of two linked but contrasting lakes in North Wales from a study of pollen, diatoms and chemistry in sediment cores. *J. of Ecology*, **68**, 95–121.

Everitt, B. S. (1978). *Graphical Techniques for Multivariate Data*, Heinemann, London.

Everitt, B. S. (1980). *Cluster Analysis* (2nd edn), Heinemann, London.

Faegri, K., and Iversen, J. (1975). *Textbook of Pollen Analysis* (3rd edn), Blackwell, Oxford.

Gabriel, K. (1971). The biplot graphic display of matrices with application to principal components analysis. *Biometrika*, **58**, 453–467.

Gardiner, F. P., and Haedrich, R. L. (1978). Zonation in the deep benthic megafauna. *Oecologia*, **31**, 311–317.

Gleason, H. A. (1939). The individualistic concept of the plant association. *American Midland Naturalist*, **21**, 92–110.

Gordon, A. D. (1973a). Classification in the presence of constraints. *Biometrics*, **29**, 821–827.

Gordon, A. D. (1973b). A sequence-comparison statistic and algorithm. *Biometrika*, **60**, 197–200.

Gordon, A. D. (1980a). Methods of constrained classification. In: *Analyse de Données et Informatique* (Ed. R. Tomassone), I.N.R.I.A., Le Chesnay, pp. 161–171.

Gordon, A. D. (1980b). SLOTSEQ: a FORTRAN IV program for comparing two sequences of observations. *Computers and Geosciences*, **6**, 7–20.

Gordon, A. D. (1981). *Classification: Methods for the Exploratory Analysis of Multivariate Data*, Chapman and Hall, London.

Gordon, A. D. (1982). Numerical methods in Quaternary palaeoecology. V. Simultaneous graphical representations of the levels and taxa in a pollen diagram. *Review of Palaeobotany and Palynology*, **37**, 155–183.

Gordon, A. D., and Birks, H. J. B. (1972). Numerical methods in Quaternary palaeoecology. I. Zonation of pollen diagrams. *New Phytologist*, **71**, 961–979.

Gordon, A. D., and Birks, H. J. B. (1974). Numerical methods in Quaternary palaeoecology. II. Comparison of pollen diagrams. *New Phytologist*, **73**, 221–249.

Gordon, A. D., and Henderson, J. T. (1977). An algorithm for Euclidean sum of squares classification. *Biometrics*, **33**, 355–362.

Gordon, A. D., and Reyment, R. A. (1979). Slotting of borehole sequences. *J. of Mathematical Geology*, **11**, 309–327.

Gower, J. C. (1967). Multivariate analysis and multidimensional geometry. *The Statistician*, **17**, 13–28.

Green, D. G. (1982). Fire and stability in the postglacial forests of south-west Nova Scotia. *J. of Biogeography*, **9**, 29–40.

Green, D. G. (1983). The ecological interpretation of fine resolution pollen records. *New Phytologist*, **94**, 459–477.

Hawkins, D. M., and Merriam, D. F. (1973). Optimal zonation of digitized sequential data. *J. of Mathematical Geology*, **5**, 389–395.

Hedberg, H. D. (Ed.) (1976). *International Stratigraphic Guide: a Guide to Stratigraphic Classification, Terminology and Procedure*, John Wiley, London.

Hill, M. O. (1974). Correspondence analysis: a neglected multivariate method. *Applied Statistics*, **23**, 340–354.

Howe, S., and Webb, T. (1983). Calibrating pollen data in climatic terms: improving the methods. *Quaternary Science Reviews*, **2**, 17–51.

Hunt, T. C., and Birks, H. J. B. (1982). Devensian late-glacial vegetational history at Sea Mere, Norfolk. *J. of Biogeography*, **9**, 517–538.

Huntley, B. (1981). The past and present vegetation of the Caenlochan National Nature Reserve, Scotland. II. Palaeoecological investigations. *New Phytologist*, **87**, 189–222.

Huntley, B., and Birks, H. J. B. (1983). *An Atlas of Past and Present Pollen Maps for Europe: 0–13,000 Years Ago*, Cambridge University Press, Cambridge.

Iversen, J. (1944). *Viscum, Hedera* and *Ilex* as climatic indicators: a contribution to the study of post-glacial temperature climate. *Geologiske Föreningens i Stockholm Förhandlingar*, **66**, 463–483.

Iversen, J. (1954). The late-glacial flora of Denmark and its relation to climate and soil. *Danmarks Geologiske Undersøgelse*, II, **80**, 87–119.

Jacobson, G. L. (1979). The palaeoecology of white pine (*Pinus strobus*) in Minnesota. *J. of Ecology*, **67**, 697–726.

Jöreskog, K. G., Klovan, J. E., and Reyment, R. A. (1976). *Geological Factor Analysis*, Elsevier, Amsterdam.

Kershaw, A. P. (1970). A pollen diagram from Lake Euramoo, north-east Queensland, Australia. *New Phytologist*, **69**, 785–805.

King, L. (1976). Pollenanalyse und Computer: Erfahrungen mit PALYNO, Programme zur Berechnung und Darstellung Pollenanalytischer Daten. *Pollen et Spores*, **18**, 93–104.

Lamb, H. F. (1982). *Late Quaternary Vegetation History of the Forest-Tundra Ecotone in North Central Labrador*, Ph.D. Thesis, University of Cambridge.

Lamb, H. F. (1984). Modern pollen spectra from Labrador and their use in reconstructing Holocene vegetational history. *J. of Ecology*, **72**, 37–59.

Lehman, J. T. (1975). Reconstructing the rate of accumulation of lake sediment: the effect of sediment focusing. *Quaternary Research*, **5**, 541–550.

MacDonald, G. M. (1983). Holocene vegetation history of the Upper Natla River Area, Northwest Territories, Canada. *Arctic and Alpine Research*, **15**, 169–180.

Maher, L. J. (1972). Nomograms for computing 0.95 confidence limits of pollen data. *Review of Palaeobotany and Palynology*, **13**, 85–93.

Maher, L. J. (1981). Statistics for microfossil concentration measurements employing samples spiked with marker grains. *Review of Palaeobotany and Palynology*, **32**, 153–191.

Maher, L. J. (1982). The palynology of Devils Lake, Sauk County, Wisconsin. In: *Quaternary History of the Driftless Area* (Eds. J. C. Knox, L. Clayton and D. M. Mickleson), Wisconsin Geological and Natural History Survey Field Trip Guide Book 5, Madison, pp. 119–135.

Malmgren, B., Oviatt, C., Gerber, R., and Jeffries, H. P. (1978). Correspondence analysis: applications to biological oceanographic data. *Estuarine and Coastal Marine Science*, **6**, 429–437.

Mehringer, P. J., Arno, S. F., and Petersen, K. L. (1977). Postglacial history of Lost Trail Pass Bog, Bitterroot Mountains, Montana. *Arctic and Alpine Research*, **9**, 345–368.

Melguen, M. (1974). Facies analysis by correspondence analysis: numerous advantages of this new statistical technique. *Marine Geology*, **17**, 165–182.

Mosimann, J. E., and Greenstreet, R. L. (1971). Representation-insensitive methods for paleoecological pollen studies. In: *Statistical Ecology. 1: Spatial Patterns and Statistical Distributions* (Eds. G. P. Patil, E. C. Pielou and W. E. Waters), Pennsylvania State University Press, University Park, pp. 23–58.

Orloci, L. (1967). An agglomerative method for classification of plant communities. *J. of Ecology*, **55**, 193–206.

Ovenden, L. (1982). Vegetation history of a polygonal peatland, northern Yukon. *Boreas*, **11**, 209–224.

Parsons, R. W., and Prentice, I. C. (1981). Statistical approaches to R-values and the pollen-vegetation relationship. *Review of Palaeobotany and Palynology*, **32**, 127–152.

Pawlikowski, M., Ralska-Jasiewiczowa, M., Schönborn, W., Stupnicka, E., and Szeroczyńska, K. (1982). Woryty near Gietrzwald, Olsztyn Lake District, NE Poland — vegetational history and lake development during the last 12,000 years. *Acta Palaeobotanica*, **22**, 85–116.

Peglar, S. M., Fritz, S. C., Alapieti, T., Saarnisto, M., and Birks, H. J. B. (1984). Composition and formation of laminated sediments in Diss Mere, Norfolk, England. *Boreas* **13**, 13–28.

Pennington, W. (1979). The origin of pollen in lake sediments: an enclosed lake compared with one receiving inflow streams. *New Phytologist*, **83**, 189–213.

Pennington, W., and Sackin, M. J. (1975). An application of principal components analysis to the zonation of two Late-Devensian profiles. *New Phytologist*, **75**, 419–453.

Prentice, H. C. (1982). Late Weichselian and early Flandrian vegetational history of the Varanger peninsula, north-east Norway. *Boreas*, **11**, 187–208.

Prentice, I. C. (1980). Multidimensional scaling as a research tool in Quaternary palynology: a review of theory and methods. *Review of Palaeobotany and Palynology*, **31**, 71–104.

Prentice, I. C. (1983). Pollen mapping of regional vegetation patterns in south and central Sweden. *J. of Biogeography*, **10**, 441–454.

Prentice, I. C., and Parsons, R. W. (1983). Maximum likelihood linear calibration of pollen spectra in terms of forest composition. *Biometrics*, **39**, 1051–1057.

Reyment, R. A. (1980). *Morphometric Methods in Biostratigraphy*, Academic Press, London.

Reyment, R. A. (1983). Moors and Christians: an example of multivariate analysis applied to human blood-groups. *Annals of Human Biology*, **10**, 505–522.

Ritchie, J. C. (1977). The modern and late Quaternary vegetation of the Campbell-Dolomite Uplands, near Inuvik, N.W.T., Canada. *Ecological Monographs*, **47**, 401–423.

Ritchie, J. C. (1982). The modern and late-Quaternary vegetation of the Doll Creek area, North Yukon, Canada. *New Phytologist*, **90**, 563–603.

Ritchie, J. C., and Yarranton, G. A. (1978a). The late-Quaternary history of the boreal forest of central Canada, based on standard pollen stratigraphy and principal components analysis. *J. of Ecology*, **66**, 199–212.

Ritchie, J. C., and Yarranton, G. A. (1978b). Patterns of change in the late-Quaternary vegetation of the western interior of Canada. *Can. J. of Botany*, **56**, 2177–2183.

Spicer, R. A. (1981). *The Sorting and Deposition of Allochthonous Plant Material in a Modern Environment at Silwood Lake, Silwood Park, Berkshire, England*, United States Geological Survey Professional Paper 1143, 77pp.

Spicer, R. A., and Hill, C. R. (1979). Principal components and correspondence analysis of quantitative data from a Jurassic plant bed. *Review of Palaeobotany and Palynology*, **28**, 273–299.

Squires, R. H. (1970). *A Computer Program for the Presentation of Pollen Data*, University of Durham Geography Department Occasional Paper 11, 53pp.

Squires, R. H., and Holder, A. P. (1970). The use of computers in the presentation of pollen data. *New Phytologist*, **69**, 875–883.

Veldkamp, A. C., Hagen, T., and van der Woude, J. D. (1981). Laser plotting of pollen diagrams. *Review of Palaeobotany and Palynology*, **32**, 441–443.

Voorips, A. (1973). An ALGOL-60 program for computation and graphical representation of pollen analytical data. *Acta Botanica Neerlandica*, **22**, 645–654.

Voorips, A. (1974). An ALGOL-60 program for pollen analytical data; the CDC version. *Acta Botanica Neerlandica*, **23**, 701–704.

Vuorinen, J., and Huttunen, P. (1981). The desktop computer in processing biostratigraphic data. *Pollen et Spores*, **23**, 165–172.

Walker, D. (1966). The late Quaternary history of the Cumberland Lowland. *Phil. Trans. Royal Soc. London B*, **251**, 1–210.

Walker, D. (1982). Vegetation's fourth dimension. *New Phytologist*, **90**, 419–429.

Walker, D., and Pittelkow, Y. (1981). Some applications of the independent treatment of taxa in pollen analysis. *J. of Biogeography*, **8**, 37–51.

Walker, D., and Wilson, S. R. (1978). A statistical alternative to the zoning of pollen diagrams. *J. of Biogeography*, **5**, 1–21.

Watts, W. A. (1961). Post-Atlantic forests in Ireland. *Proc. Linnean Society of London*, **172**, 33–38.

Watts, W. A. (1973). Rates of change and stability in vegetation in the perspective of long periods of time. In: *Quaternary Plant Ecology* (Eds. H. J. B. Birks and R. G. West), Blackwell, Oxford, pp. 195–206.

Watts, W. A. (1977). The late Devensian vegetation of Ireland. *Phil. Trans. Royal Soc. London B*. **280**, 273–293.

Webb, T. (1971). *The Late- and Post-Glacial Sequence of Climatic Events in Wisconsin and East-Central Minnesota: Quantitative Estimates Derived from Fossil Pollen Spectra by Multivariate Statistical Analysis*, Ph.D. Thesis, University of Wisconsin.

Webb, T. (1974). Corresponding patterns of pollen and vegetation in Lower Michigan: a comparison of quantitative data. *Ecology*, **55**, 17–28.

Webb, T., and Bryson, R. A. (1972). Late- and post-glacial climatic change in the northern Midwest, USA: quantitative estimates derived from fossil pollen spectra by multivariate statistical analysis. *Quaternary Research*, **2**, 70–115.

Webb, T., and Clark, D. R. (1977). Calibrating micropaleontological data in climatic terms: a critical review. *Annals of the New York Academy of Sciences*, **288**, 93–118.

West, R. G. (1970). Pollen zones in the Pleistocene of Great Britain and their correlation. *New Phytologist*, **69**, 1179–1183.

Wiltshire, P. E. J., and Moore, P. D. (1983). Palaeovegetation and palaeohydrology in upland Britain. In: *Background to Palaeohydrology* (Ed. K. J. Gregory), John Wiley, Chichester, pp. 433–451.

Wright, H. E., and Patten, H. L. (1963). The pollen sum. *Pollen et Spores*, **5**, 445–450.

Wright, H. E., Winter, T. C., and Patten, H. L. (1963). Two pollen diagrams from south-eastern Minnesota: problems in the regional lateglacial and postglacial vegetational history. *Geological Society of America Bulletin*, **74**, 1371–1396.

Yarranton, G. A., and Ritchie, J. C. (1972). Sequential correlation as an aid in placing pollen zone boundaries. *Pollen et Spores*, **14**, 213–223.

Handbook of Holocene Palaeoecology and Palaeohydrology
Edited by B. E. Berglund
© 1986 John Wiley & Sons Ltd.

38

Multivariate methods for data analysis

Iain Colin Prentice

Institute of Ecological Botany, University of Uppsala, Uppsala, Sweden

INTRODUCTION

Quaternary palaeoecology requires the analysis and interpretation of multi-variate data, in the form of biotic assemblages sampled at many points in space and time. Multivariate data arise in many branches of the biological, environmental and human sciences. A matrix of multivariate data typically consists of tens to hundreds of 'objects', each characterized by a moderate to large number of 'variables'. In palaeoecology applications, each object is typically a biotic assemblage, sampled at a particular level in a particular core. The 'objects' in a given study may be all the samples from one or more cores, or they may be a set of samples (e.g. surface samples) from different locations; the variables are usually taxa. The data matrix may therefore be written out like this:

		Sample no.				
		1	2	3	...	n
Taxon no.	1	x_{11}	x_{12}	x_{13}	...	x_{1n}
	2	x_{21}	x_{22}	x_{23}	...	x_{2n}
	3	x_{31}	x_{32}	x_{33}	...	x_{3n}
	$\vdots$	$\vdots$	$\vdots$	$\vdots$	$\vdots$	$\vdots$
	t	x_{t1}	x_{t2}	x_{t3}	...	x_{tn}

Here there are n samples and t taxa, and the matrix entry x_{ik} denotes the amount of taxon i in sample k. These amounts are most often proportions or percentages of the total for each sample; they may also be estimates of absolute amounts (concentration or accumulation rates), or presence–absence records ($x_{ik} = 1$ for presence, 0 for absence). This report will concentrate on the case where the data are percentages, denoted by p_{ik}, the percentage of taxon i in sample k.

The need for special data-analysis techniques for multivariate data matrices arises because of problems in display and communication. Data in which there are just two variables are easy to represent graphically: a scatter diagram showing values of one variable plotted against values of the other is a complete summary, which clearly shows any clusters or trends that may exist. If the data contain three variables but are in percentage form, then a complete summary can still be obtained by plotting the samples on isometric (triangular) graph paper — as is commonly done for sand, silt and clay percentages. Occasionally biotic assemblage data are as simple as this. Hicks (1977), for example, used an isometric graph to display the pollen percentages of *Pinus, Betula* and *Picea* in surface samples from north Finland, where these three taxa account for most of the pollen assemblage. No such simple method is available when there are more than three taxa to be considered, and the problem of display becomes rapidly acute as the number of taxa increases. The need to mentally integrate information on (say) twenty or more major taxa in a pollen or diatom diagram also creates a barrier to communication between specialists in different disciplines.

Multivariate data analysis methods assist in display and communication, by providing straightforward graphical summaries that can be published alongside the primary data. The simplest use of principal components analysis (PCA), for example, is to generate two-dimensional scatter diagrams which can be interpreted in much the same way as an ordinary scatter plot or isometric graph of two or three variables. This use of PCA illustrates the way in which multivariate methods exploit the correlations between variables in any one data set in order to 'squash' what would otherwise be a multidimensional configuration of points into a small number of dimensions.

The methods described here are applicable to many kinds of multivariate data. Birks (Chapter 37) has already described and given computer programs for the zonation and sequence-slotting methods specially devised for stratigraphic data by Gordon and Birks (1972, 1974). This chapter deals with general-purpose *cluster analysis* and *ordination* methods, which are also widely used in palaeoecology. Most applications have been to pollen data, following pioneer work by Adam (1970), Dale and Walker (1970), Mosimann and Greenstreet (1971), Gordon and Birks (1974), Webb (1974) and others (reviewed by Prentice, 1980). Multivariate methods are equally applicable to data on plant macrofossils (e.g. H. H. Birks, 1973), cladocera (Whiteside,

1970; Beales, 1976; Hofmann, 1978; Synerholm, 1979; Norton *et al.*, 1981), diatoms (Davis and Norton, 1978; Jatkar *et al.*, 1979; Brugam, 1980; Bruno and Lowe, 1980; Elner and Happey-Wood, 1980; Pirttiala, 1980; Norton *et al.*, 1981), ostracods (Kaesler, 1966; Maddocks, 1966) and other fossil groups and to abiotic data including sediment chemical and physical characteristics (e.g. Pennington and Sackin, 1975; Elner and Happey-Wood, 1980; Davis, 1970; Brown, 1983).

<h3 style="text-align:center">PRINCIPLES</h3>

General

Cluster analysis methods classify samples into more-or-less homogeneous groups (clusters). Many methods create nested series of clusters which can be represented as a hierarchy or dendrogram (tree diagram) like Figure 38.1. The main usefulness of ordinary cluster analysis methods in palaeoecology is in grouping modern assemblages. For example Birks *et al.* (1975) showed using cluster analysis that the surface pollen samples from central Canada published by Lichti-Federovich and Ritchie (1968) fell into groups which corresponded closely with the vegetational regions from which the samples were taken. Several of the studies cited above involved clustering modern assemblages of lake-dwelling organisms and characterizing the resulting groups in terms of features of the lake environment. Cluster analysis of purely stratigraphic data is less useful, because clusters that do not consist of contiguous samples are a little difficult to represent in a stratigraphic diagram.

Zonation (in the sense of forming local pollen assemblage biozones) is constrained cluster analysis: the constraint is that all clusters must consist of contiguous samples. Numerical zonation techniques were introduced by Gordon and Birks (1972) (see Chapter 37). Caseldine and Gordon (1978) used slight modifications of these techniques for 'zoning' surface samples along transects. Cluster analysis can also incorporate two-dimensional constraints, e.g. surface samples from a network of sites can be grouped into conterminous regions. Geographers call this process 'regionalization'. Howe (1978) developed one such regionalization procedure for samples not located on a regular grid. Constrained cluster analysis is not considered further in this chapter.

Ordination methods represent samples as points in a two- or few-dimensional space, similar samples being located together and dissimilar samples apart. The term 'multidimensional scaling', in its broad sense, is equivalent to ordination. Ordination methods are widely used in palaeoecology; the most widely used is principal components analysis (PCA).

Ordination is equally applicable to surface sample data, stratigraphic data or combinations of both. Birks (1974) suggested the use of PCA to complement zonation. PCA of a single profile gives information that zonation alone does not, because it illustrates variation within as well as between zones and shows the interrelations among the zones. Birks and Berglund (1979) and Gordon (1981) illustrated the use of PCA for comparing two or more profiles. Again PCA gave information that zonation plus sequence slotting did not, and provided an effective overview. Gordon and Birks (1974) used another ordination method, non-metric multidimensional scaling, in a similar way. PCA has become a standard tool in the analysis of surface pollen data, where the lack of a time-scale makes representation especially difficult without recourse to multivariate methods. Ritchie and Yarranton (1978) and others have used PCA to compare modern and fossil assemblages. PCA can also be applied to networks of contemporaneous samples at, say 1000-year or 2000-year intervals (e.g. Birks and Saarnisto, 1975; Huntley and Birks, 1983) in order to show changes in geographic patterns through time. An improvement on this method would be to perform PCA on a 'space–time' data matrix (e.g. pollen spectra at 1000-year intervals throughout a region). Mapped patterns of principal components would then refer to a common set of components across time-planes.

The workings of cluster analysis and ordination methods are outlined in the rest of this section. Prominence is given to Ward's method of cluster analysis, and to PCA, since these are the most commonly used multivariate techniques. A short note on *canonical analysis* is appended at the end of the chapter. Canonical analysis can be used to characterize the differences between predetermined groups of samples (i.e. as an ordination method applicable when there are *a priori* groups), or to investigate the distinguishability of samples classified according to some external criterion. For further information on multivariate techniques, see Everitt (1978, 1980), Webster (1977), Gordon (1981) and Gower and Digby (1981).

Geometric models and dissimilarity coefficients

Many multivariate methods, including Ward's method and PCA, are most easily explained in terms of a geometric model. Each sample is considered as a point in a multidimensional space, whose axes are the variables (taxa). The samples are located on these axes according to the abundances of each taxon.

In this model, the distances between points are *Euclidean distances*, given by Pythagoras' formula (Chapter 39). Euclidean distance is one possible *dissimilarity coefficient* (DC) between samples. The model can be changed by transformations or standardizations of the data. For example, all abundances may be replaced by their square-roots. If so, the distances between points become *chord distances*. Alternatively, the abundances of each taxon may be scaled to unit standard deviation. Then the distances between points become *standardized Euclidean distances*. Formulae for all three DCs are given in Chapter 39.

There are also DCs that do not relate to a simple geometric model; for example, the Manhattan metric used in CONSLINK and SLOTSEQ (Chapter 37), the chi-square coefficient used by Prentice (1978), and the information radius used by Pirttiala (1980). However, dissimilarity coefficients can be shown to belong to 'families' with common properties (Prentice, 1980; Overpeck *et al.*, 1985). Simple (unweighted) DCs include Euclidean distance; signal-to noise ratio DCs include chord distance; equal-weight DCs include standardized Euclidean distance. (The Manhattan metric is another unweighted DC.) These families of DCs differ in the weight given to less abundant taxa. Unweighted DCs give rare taxa minimal weight; signal-to-noise ratio DCs give to them somewhat more weight; equal-weight DCs give all taxa equal weight, so that the rarest must usually be omitted.

Ward's method and PCA in their 'raw' forms are based on the simple geometric model in which the distances between points are Euclidean distances. It is also possible to transform or standardize the data before analysis — e.g. square-root transformation, which converts these distances to chord distances, or standardizing taxa to unit standard deviation, which converts them to standardized Euclidean distances. (This third option is very commonly used in PCA, often without comment.) *The user must make the choice.* If in doubt, try different options. Sometimes the results may be robust, i.e. essentially the same whatever option is used. Other times the results may differ — in which case the differences may be interesting! The final case-study in the 'Examples' section illustrates the properties of different options in PCA.

Absolute pollen data obtained by counting with exotic markers (chapter 22) have a peculiar error structure, which could lead to analyses becoming dominated by 'noise'. Gordon (1982) suggested that such data should be transformed before any multivariate treatment. See Gordon's paper for details.

Cluster analysis

The kinds of cluster analysis that are useful with biotic assemblage data are those that aim simply *to divide the samples into relatively homogeneous groups.* The single-link method is not particularly useful because it looks for *disjoint*, not necessarily homogeneous clusters. A better choice is Ward's sum of squares method (Ward, 1963; Wishart, 1969), as described, for example, by Gordon (1981). Ward's method tries to form groups whose total within-group sum of squares is low. The within-group sum of squares for a group g containing $n(g)$ samples is defined as:

$$S(g) = \sum_{k=1}^{n(g)} \sum_{i=1}^{t} (P_{ik} - q_i)^2$$

where q_i is the mean percentage of taxon i in the group. $S(g)$ is the sum of squared Euclidean distances from the samples in a group to the centre of gravity of the group; the more compact (homogeneous) the group, the lower $S(g)$ is. Finding the strictly optimal groups (i.e. the groupings that minimize $\Sigma_g S(g)$) is a difficult mathematical problem, but two 'quick methods' give adequate results in practice. *Divisive algorithms* split the samples into two, then each group into two again, and so on. (The SPLITLSQ zonation program, for example, does this in such a way that each division into two causes the greatest possible reduction in the total within-group sum of squares.) But even divisive algorithms are impractical unless (as in zonation) there is some constraint limiting the number of possible groupings. Ward's method instead uses an agglomerative algorithm, which builds up a dendrogram 'from the bottom' (starting at the tips of the branches) instead of from the top, so at the outset each sample is in a cluster to itself, and the total within-group sum of squares is zero. Clustering proceeds by repeated fusions of samples with other samples, or (later on) samples with already-formed clusters or clusters with other clusters. Each fusion that occurs is the one that causes the *least possible increase* in the total within-group sum of squares. In this way a complete dendrogram is built up, the vertical axis representing the increase in sum of squares caused by each fusion. Any horizontal 'slice' through the dendrogram then gives clusters that are reasonable homogeneous according to the sum of squares criterion.

$S(g)$ is a simple (unweighted) heterogeneity measure, based on Euclidean distance. One can also optionally transform or standardize data before applying Ward's method. Taking square-roots of all data values will make $S(g)$ into a signal-to-noise ratio type measure, based on chord distance; standardizing each taxon to unit standard deviation will make $S(g)$ into an equal-weight measure based on standardized Euclidean distance. There are also heterogeneity measures, not related to Euclidean distance. SPLITINF uses total within-group information content, another measure with good signal-to-noise ratio properties. The unconstrained agglomerative counterpart of SPLITINF is sometimes called 'information analysis', although this term is ambiguous. Finally there are methods that work directly on DC values, and will use any DC that is supplied. Single-link cluster analysis is one such method. Otherwise the best-known is the group average (UPGMA) method, whose properties are intermediate between single-link and Ward's method. There is still controversy over what kind of cluster analysis is theoretically best, but the literature shows that scientists in a wide variety of fields have voted for Ward's method.

Ward's method is widely used in many branches of science. Programs for Ward's method — and indeed all methods that seek homogeneity within groups — should carry warnings along the following lines:

(1) Hierarchical structure will be *forced* on the data. (There is no reason why the true *optimal* sum-of-squares groups should be hierarchically nested.)

The dendrogram will certainly indicate 'good' divisions, but not necessarily unique or optimal ones.

(2)   Cluster analysis can do an effective job of dividing up a continuum of variation into moderately homogeneous groups. There may be no real discontinuities between clusters. Indeed, the precise boundaries between clusters can be quite arbitrary.

(3)   It is hazardous to compare clusterings of different data sets (e.g. surface pollen and vegetation from the same sites) because if variation between samples is more-or-less continuous then there is no reason why the cluster boundaries should agree. A better procedure would be to cluster the samples on the basis of vegetation data, then see if the groups can be characterized and differentiated in terms of pollen data. A neat way to do this is provided by canonical analysis.

Principal components analysis

Returning to the geometric model: PCA projects the sample points perpendicularly on to a plane chosen so that the *sum of squared distances from the points to the plane is minimal.* The first principal component is the line of best fit through the points; the second is at right-angles to the first, and together with the first defines the plane of best fit; and so on. The coordinates of the samples on each principal component are called principal component *scores.*

If the abundances of taxa varied at random with respect to one another there would be little point in doing PCA. In practice there tend to be strong (positive or negative) correlations among the taxa. As a result, the plane of best fit may fit rather well — the sample scores on the first two or three principal components may give a reasonably accurate picture of the distances among samples. The goodness of fit of a PCA representation based on the first k principal components is measured by the ratio of the sum of the k *eigenvalues* associated with those components to the sum of all the eigenvalues. Here the eigenvalue is simply the variance of the scores on a given principal component, and the eigenvalues sum up to the sum of the variances of all the taxa. This ratio is therefore called the *proportion of variance accounted* for by the first k principal components.

In general there can be as many principal components (with positive eigenvalues) as there are variables. With untransformed percentage data, the maximum number is one less than the number of variables (Gower, 1967): just as only two dimensions are needed to represent percentages of three taxa, so only t − 1 are needed for t taxa. This point is usually academic, however, because two or three dimensions often give a representation that is good enough for practical purposes.

A sample's score on any given principal component works out as the sum of a series of quantities, each of which is the abundance of one taxon (minus that taxon's mean value) multiplied by a *loading* for that taxon. PCA programs also output these loadings, which serve to indicate the relative contribution of each taxon to each principal component. PCA is usually *computed* via an *R-mode algorithm*, which first computes taxon loadings from the $t \times t$ matrix of variances and covariances between taxa. These loadings are then used to generate sample scores. At this stage one can also 'add' new samples, which will be given scores in the same way as the original set. This facility is useful if one has a standard set of surface samples with which to compare some fossil assemblages, or a standard set of profiles with which one wants to compare a range of surface assemblages (e.g. Ritchie and Yarranton, 1978).

Data may be transformed or standardized before PCA in order to change the weighting of rare taxa — for example, by taking square-roots of all data values, or standardizing each of the taxa to unit standard deviation. In ordinary PCA, the distances between points approximate to the Euclidean distances between the samples. With square-root transformation (Birks, 1977, 1981) the distances approximate to chord distances. With standardization, they approximate to standardized Euclidean distances. The variance–covariance matrix after standardization of taxa is simply the correlation matrix, so the option of standardization is effected in most R-mode PCA programs by computing the correlation matrix instead of the variance–covariance matrix.

The *scaling of axes* in PCA needs some attention. The scores of samples on a given principal component can be scaled in two ways, so that their sum of squares equals either 1 or λ, the eigenvalue. If 1, the samples are spread out equally on all principal components. This is not necessarily what is wanted. If λ, the successive principal components diminish in length — as they certainly should, if the distances between points are to accurately reflect the dissimilarities between samples. In ordination plots, the sums of squares of sample scores should normally equal λ for each component. Scores that are scaled in the other way should be multiplied by $\sqrt{\lambda}$.

Taxon loadings can also optionally be scaled to sum of squares 1 (e.g. Tables 38.1, 38.2) or λ (Figures 38.3 and 38.5). If the loadings will only be tabulated, then the scaling may not be critical, but the scaling *is* important if sample scores and taxon loadings are to be represented on the same plot. This type of plot is called a *biplot* (e.g. Gower and Digby, 1981; Gordon, 1981, 1982; Ter Braak, 1983). The variables in a biplot are represented by vectors radiating from the centroid. There are two types of PCA biplot. Covariance biplots (Ter Braak, 1983) have scores scaled to sum of squares 1, loadings to sum of squares λ, and preserve information about the variances and covariances of variables at the expense of accurate display of distances between samples. Distance biplots have scores scaled to sum of squares λ, loadings to sum of squares 1, and preserve distances between samples at the expense of accurate information on the variables. See the references cited above for further information on biplots.

Other ordination methods

Principal coordinates analysis is a PCA-like analysis which allows the use of most DCs, including many that do not relate to a geometric model. Principal coordinates analysis is computed by a *Q-mode algorithm* which computes sample scores directly from the $n \times n$ matrix of DC values. It is therefore limited in the number of samples it can process, and does not provide taxon loadings.

Non-metric multidimensional scaling (NMDS) also allows the use of any DC and works directly from a DC matrix. It permits non-linear relationships between input DCs and output distances. Prentice (1980) suggested that NMDS might be superior to PCA for ordinating certain types of palaeoecological data — notably pollen spectra encompassing a wide range of variation. However, like principal coordinates analysis, NMDS can process only a limited number of samples, and does not provide any equivalent of taxon loadings. It is also expensive in computer time.

Correspondence analysis (CA — e.g. Greenacre, 1981; Gordon, 1982) and *detrended correspondence analysis* (DCA — Hill and Gauch, 1980; Gauch, 1982) are more promising and are suitable for the analysis of any kind of frequency, abundance or presence–absence data. These techniques have recently become widely used in plant community ecology. CA is related to PCA. Roughly, CA is a generalized PCA in which samples have implicit weights (all equal if the data are in percentage form) and distance between sample-points approximate *chi-squared distances* (a signal-to-noise ratio DC) between samples. Taxa also have implicit weights — their total abundances — and distances between taxon-points approximate chi-squared distances between taxa. The method simultaneously provides two sets of points —sample points and species points — which can be superimposed. The two sets of points can be scaled in different ways — symmetrically (e.g. Gordon, 1982), or asymmetrically so that each sample-point lies at the centre of gravity of the species-points weighted according to their abundance in the sample. This second option is helpful for interpretation, and is adopted in the program DECORANA, widely used for CA and DCA.

Hill (1973) popularized CA in ecology, under the name 'reciprocal averaging' (e.g. H. H. Birks, 1973; Elner and Happey-Wood, 1980). DCA is a modification of CA. It was designed to eliminate the problem that both PCA and CA tend to reproduce long gradients — that is, a series of samples with many successive species turnovers — in curved form ('arches' and 'horse shoes'). DCA has proved to be extremely effective in recovering underlying structure in vegetation data, and DECORANA can process much larger numbers of samples and taxa than can most ordination programs. DCA lacks the conceptual simplicity of PCA, but may become the preferred method for analysing highly heterogeneous sets of samples in ecology and palaeoecology.

EXAMPLES

Surface pollen samples from Scotland

The first example is taken from O'Sullivan and Riley's (1974) work on surface pollen samples — mostly moss polsters — from an area of primary *Pinus sylvestris* forest in Scotland. A dendrogram (Figure 38.1) obtained by Ward's method was used to divide the samples initially into five groups, labelled A–E.

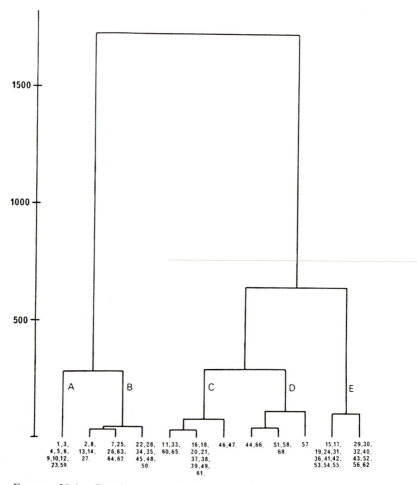

FIGURE 38.1.   Dendrogram (Ward's sum of squares method, unweighted) of surface pollen samples from a primary pine forest area in Scotland. The dendrogram is truncated so that the tips of the branches — corresponding to classifications into more than twelve groups — are not shown (from O'Sullivan and Riley, 1974). *Reproduced by permission of Pollen et Spores*

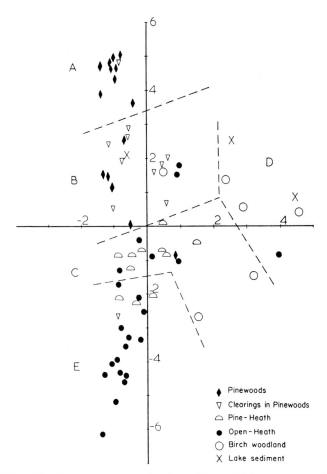

FIGURE 38.2. The five-group clustering from Figure 38.1, and vegetation types of origin, superimposed on an unweighted PCA of surface pollen samples from Scotland (from O'Sullivan and Riley, 1974). *Reproduced by permission of Pollen et Spores*

These groups are indicated on Figure 38.2, which is a plot of sample scores on the first and second principal components. No transformation or standardization was done in either analysis, so both are based on unweighted Euclidean distances between samples. The first two principal components accounted for 76% and 19% of total variance — 95% in all, an unusually high figure resulting from the relatively small number of species dominating the pollen spectra.

Figure 38.2 shows (*a*) that the groups identified by Ward's method correspond to distinct regions in the PCA representation, and (*b*) that the different vegetation subtypes produce characteristic pollen spectra, the samples from each subtype occupying a particular region of the plot and corresponding almost exactly to a single cluster analysis group. The three lake sediment samples reflect the predominant vegetation of their catchments. O'Sullivan and Riley rightly stressed the *complementarity* of the cluster analysis and ordination. The cluster analysis is effective in classifying the samples according to vegetation type, but the PCA gives further information about the pattern of variation within and between groups. Component 1 has loadings of high magnitude for *Pinus* and *Calluna* only; Figure 38.2 shows that groups A, B, C and E lie on this axis, which is essentially a gradient from *Pinus* to *Calluna* dominance in the pollen spectra. Samples 'misclassified' by Ward's method are seen to be transitional in pollen composition. This simple example also illustrates how cluster analysis imposes discrete structure on continuous variation.

Holocene pollen stratigraphy in southern Sweden

The second example is from Birks and Berglund's (1979) reappraisal of the Holocene pollen stratigraphy of southern Sweden. Birks and Berglund based their study on two standard profiles from the Blekinge region and one from the Skåne region. Their strategy of analysis was as follows:

(1) zonation and PCA of each profile by itself;
(2) establishment of local pollen assemblage zones for each profile, on the basis of (1);
(3) comparison of the two Blekinge profiles on a level-by-level basis, using PCA and sequence-slotting. Both methods indicated an excellent correlation between the two sites;
(4) establishment of regional pollen assemblage zones for Blekinge on the basis of (2) and (3);
(5) comparison of one Blekinge profile with the Skåne profile by PCA. This comparison showed that biostratigraphic correlation between the two regions is impossible. Birks and Berglund concluded that separate sets of regional pollen assemblage zones should be defined, and correlated between regions by radiocarbon dating.

All PCAs were based on the principal 14 taxa, and were 'standardized'. Figure 38.3 shows stratigraphic plots of principal component scores for the three profiles Bjärsjöholmssjön (Skåne), Färskesjön and Lösensjön (Blekinge). Beneath each graph is given the percentage of variance accounted for by the individual component, and a diagram indicating the signs and relative magnitudes of the loadings of taxa on each component. Figure 38.4 illustrates the use of PCA to compare two profiles; the correlation is clear. Note that for ease of representation Birks and Berglund showed just the mean and range of scores for each (numerically delimited) local pollen assemblage zone, and joined up the mean-points in stratigraphic sequence within each profile. Figure 38.5 gives loadings and other statistics.

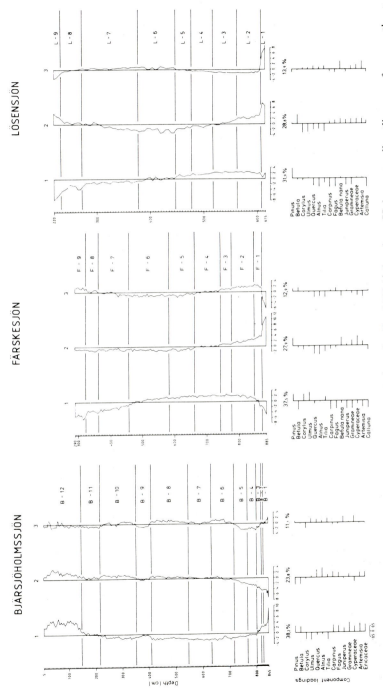

FIGURE 38.3.   Stratigraphic plots of principal components (standardized PCA) for three Holocene pollen diagrams from southern Sweden. Percentages of variance and the relative magnitudes of loadings are indicated below the graph for each component (from Birks and Berglund, 1979)

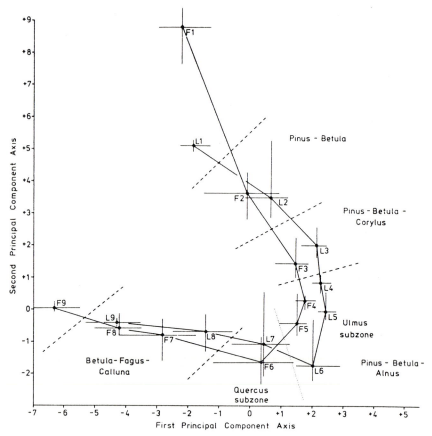

FIGURE 38.4.   Comparison of two pollen diagrams from Blekinge, southern
Sweden, using standardized PCA (from Birks and Berglund, 1979)

Regional pollen mapping in Sweden

Figure 38.6 reverts to surface pollen samples, again from Sweden (Prentice, 1983), and shows how maps of principal components can be used to summarize geographic trends. These maps were based on PCA, with square-root transformation, of 40 surface mud samples from moderate-sized lakes. Component 1 has large positive loadings (see Table 38.1) for *Pinus* and *Picea* and large negative loadings for *Alnus, Quercus* and *Fagus*, reflecting the major gradient from boreal coniferous forest through mixed coniferous–deciduous forest to temperate deciduous forest in the extreme south. Component 2, dominated by *Betula* (positive loading) and Gramineae, cereal type, *Rumex* and Filicales (negative loadings), reflects primarily human impact through forest clearance and farming, followed in some southern areas by reversion to successional woodland. Such maps summarize geographic trends in the

	1	2	3	4	5
Pinus	0.7470	0.5813	0.1823	0.1781	0.0614
Betula	−0.7356	0.2262	0.3734	−0.3276	−0.0665
Corylus	0.4242	−0.6100	−0.1660	0.0055	−0.2657
Ulmus	0.5932	−0.3851	−0.2017	0.2759	0.2467
Quercus	−0.3684	−0.7849	−0.2928	−0.1770	−0.0383
Alnus	−0.2325	−0.8653	−0.2945	−0.0245	0.0005
Tilia	0.2232	−0.7392	−0.3928	0.0297	0.2368
Carpinus	−0.8290	−0.1898	0.1319	−0.0604	−0.1522
Fagus	−0.7652	−0.0666	0.2412	0.4434	0.2153
Betula nana	−0.1357	0.5891	−0.7473	0.1223	−0.0466
Juniperus	−0.7240	0.1327	−0.0552	0.4953	0.2326
Gramineae	−0.7058	0.2708	−0.4241	−0.1743	−0.2213
Cyperaceae	−0.1916	0.6033	−0.7257	0.1470	−0.1404
Artemisia	−0.2331	0.2982	−0.2347	−0.5429	0.6479
Calluna	−0.8515	−0.2797	0.1181	0.2051	0.0314
Variance (eigenvalue)	5.000	3.830	1.980	1.097	0.812
Percent of total variance	33.34	25.54	13.20	7.32	5.42
Cumulative percent of total variance	33.34	58.87	72.07	79.39	84.81

Total variance (trace) = 15.0

FIGURE 38.5. Loadings and percentages of variance explained by successive principal components of two pollen diagrams from Blekinge, southern Sweden. The first two components are plotted in Figure 38.4. (from Birks and Berglund, 1979)

regional pollen spectrum, help to establish the scale of the relationship between pollen and vegetation, indicate areas of rapid change (which would benefit from denser sampling), and pinpoint anomalous sites.

Surface pollen samples from northern Fennoscandia

The final example (Figure 38.7, Table 38.2) is based on surface pollen samples from Finland and Finnmark, northern Norway (Prentice, 1978) and illustrates general properties of three variants of PCA — (i) unweighted, (ii) with square-root transformation, (iii) standardized — corresponding to the DCs (i) Euclidean distance, (ii) chord distance, (iii) standardized Euclidean distance. Thus the distances between points in the unweighted PCA plot are approximations to the Euclidean distances between the samples they represent; the same relationship applies for chord distances in the 'square-root' PCA plot and standardized Euclidean distances in the standardized PCA plot. All three analyses were based only on the 25 taxa that exceed 1% of total land pollen and pteridophyte spores in one or more samples.

Unweighted PCA gives large loadings on the first two principal components to a small number of taxa only. Essentially p.c.1 is *Pinus* versus *Betula* and

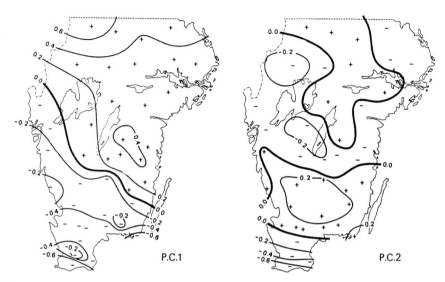

FIGURE 38.6. Maps of principal component scores (square-root option) for 40 surface pollen samples from lakes in southern Sweden (from Prentice 1983). *Reproduced by permission of Blackwell Scientific Publications Ltd*

p.c.2 is *Pinus* and *Betula* versus *Picea, Alnus* and Gramineae. The various vegetational regions are not as effectively separated as they are in the other two analyses. In this case the major pollen types — to which unweighted analysis gives most importance — are not the most interesting, and the unweighted analysis not as informative or useful as the others. Standardized PCA separates the vegetation regions more effectively, and orders them in a north–south sequence. Many more pollen types participate in the configuration. The tundra samples are extremely spread out, because of their differences in a number of minor taxa. The analysis based on chord distance is intermediate in character. The regions are still well-separated, but the 'good' tundra samples are now placed together. Most of the pollen types participate, but the rarer taxa have less weight than in the standardized analysis.

The above three analyses emphasize different aspects of the data. Note that the relative magnitudes of the proportions of variance explained in two dimensions — (i) > (ii) > (iii) — illustrate a general feature of such comparisons, resulting from the fact that more variables contribute substantially to type (ii) than to type (i) analyses, and more to type (iii) than to type (ii) analyses. One should *not* deduce from this that unweighted PCA is the best! This example also illustrates the fact that chord distance often provides an effective compromise between the extreme properties of the other two dissimilarity measures.

TABLE 38.1. Principal component loadings and percentages of variance explained for surface pollen samples from lake sediments in southern Sweden. Loadings of magnitude 0.2 or more are shown in italic (see Figure 38.6)

	p.c.1	p.c.2
Pinus	*0.62*	0.10
Betula	−0.17	*0.50*
Picea	*0.43*	−0.15
Alnus	*−0.23*	0.08
Corylus	−0.18	0.13
Fraxinus	−0.05	0.00
Quercus	*−0.32*	0.16
Tilia	−0.09	0.15
Ulmus	−0.08	−0.01
Fagus	*−0.36*	−0.16
Carpinus	−0.13	0.00
Juniperus	−0.01	−0.19
Salix	−0.04	−0.08
Ericales	−0.10	0.10
Artemisia	−0.05	−0.11
Cereal type	−0.05	*−0.32*
Cyperaceae	−0.06	−0.18
Gramineae	−0.14	*−0.44*
Rumex	−0.06	*−0.26*
Filicales	−0.07	*−0.38*
Variance explained (%)	50.3	16.5
Cumulative (%)	50.3	66.8

PROGRAMS

CLUSTAN is the most widely available cluster analysis package. For Ward's method, use the CLUSTAN procedures FILE, CORREL and HIERARCHY. FILE allows standardization, if needed. Specify 'squared Euclidean distance' in CORREL, and 'Ward's method' in HIERARCHY. SYN-TAX (Podani, 1980) is another package which can be used to perform various cluster analyses, including Ward's method.

PCA is available (usually as a variety of 'factor analysis') in standard statistical packages, but the options may be inconveniently restrictive. For example, the SPSS program for PCA automatically uses the correlation matrix, and both the SPSS and BMDP programs scale component scores to unit sum of squares. SAS is recommended, among the widely available packages. There are also many published computer programs for PCA. Another, unpublished, program (PCARMODE, written mostly by H. J. B. Birks and J. M. Line) is available on request. This program allows a wide range of options including

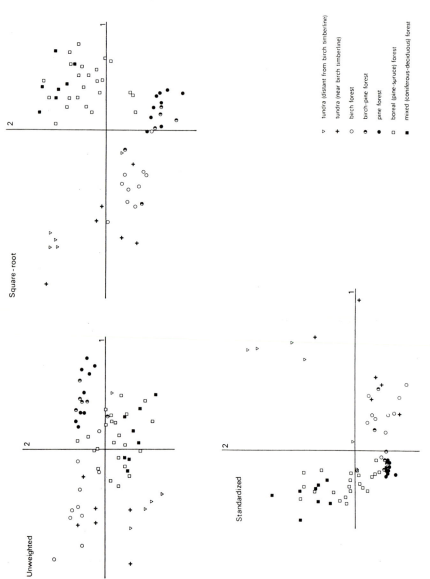

Square-root

Unweighted

Standardized

▽   tundra (distant from birch timberline)

+   tundra (near birch timberline)

○   birch forest

◑   birch-pine forest

●   pine forest

□   boreal (pine-spruce) forest

■   mixed (coniferous-deciduous) forest

FIGURE 38.7.   Comparison of PCA options using 67 surface pollen samples from lakes in Finland and northern Norway

TABLE 38.2. Principal component loadings and percentages of variance explained for the Finland/Finnmark surface pollen sample data of Prentice (1978). Loadings of magnitude 0.2 or more are shown in italic (see Figure 38.7)

	Unweighted		Square-root		Standardized	
	p.c.1	p.c.2	p.c.1	p.c.2	p.c.1	p.c.2
Betula	-0.48	*0.74*	*-0.26*	-0.14	0.18	*-0.20*
Pinus	*0.85*	*0.43*	*0.43*	*-0.44*	*-0.26*	*-0.20*
Picea	0.10	*-0.42*	*0.45*	*0.41*	*-0.25*	0.14
Alnus	0.02	-0.15	*0.26*	*0.40*	*-0.21*	*0.26*
Corylus/Myrica	0.00	-0.02	0.06	0.13	-0.12	*0.22*
Juniperus	-0.01	-0.03	-0.04	0.07	0.03	0.10
Quercus	0.00	-0.02	0.08	0.11	-0.18	*0.23*
Salix	-0.02	-0.03	-0.14	0.14	*0.28*	*0.29*
Angelica type	0.00	-0.01	-0.05	0.06	0.15	*0.27*
Calluna	0.00	0.00	0.02	0.03	-0.05	0.02
cereal type	0.00	-0.03	0.08	0.19	-0.15	*0.28*
Cyperaceae	-0.12	-0.01	*-0.38*	0.00	*0.29*	-0.01
Empetrum	-0.02	0.01	-0.16	-0.07	*0.29*	0.13
Ericales undiff.	-0.03	0.04	-0.15	-0.12	*0.21*	-0.17
Filipendula	0.00	0.00	-0.01	0.05	0.03	0.15
Gramineae	-0.08	*-0.22*	-0.18	*0.43*	0.18	*0.38*
Oxyria/Rumex	-0.00	-0.02	0.01	0.17	-0.06	*0.32*
Ranunculus acris type	-0.01	-0.03	-0.08	0.08	0.15	*0.23*
Rubus chamaemorus	-0.01	0.01	-0.07	-0.01	0.14	-0.11
Solidago type	-0.04	-0.10	*-0.22*	0.16	0.19	*0.20*
Diphasium	-0.02	0.01	*-0.21*	-0.08	*0.27*	-0.03
fern spores undiff.	-0.07	-0.13	*-0.21*	*0.29*	0.18	*0.25*
Gymnocarpium dryopteris type	-0.05	-0.03	*-0.23*	0.04	*0.28*	0.05
Lycopodium annotinum	-0.01	0.00	-0.08	-0.03	*0.24*	-0.08
Selaginella	-0.01	-0.01	-0.12	0.01	*0.24*	-0.03
Variance explained (%)	65.9	18.1	40.7	19.4	25.9	15.2
Cumulative (%)	65.9	84.0	40.7	60.1	25.9	41.1

computation of percentages, square-root transformation, Gordon's transformation for absolute pollen data, facility to add points, etc., and accepts raw counts in the format specified in Chapter 37.

Programs for PCA biplots and correspondence analysis may be obtained form Dr. A. D. Gordon, Department of Statistics, University of St. Andrews, St. Andrews KY16 9SS, U.K. The program DECORANA, and other ecological data analysis programs in the Cornell series, may be obtained from Dr. H. D. Gauch, Jr., Department of Ecology and Systematics, Cornell University, Ithaca, NY 14853, U.S.A.

A NOTE ON CANONICAL ANALYSIS

Canonical analysis, or multiple discriminant analysis, is a multivariate statistical technique which presupposes an existing classification of samples into groups. For example, surface pollen spectra may have been categorized according to vegetational regions, as in the applications by Adam (1970) and Birks *et al.* (1975).

Like PCA, canonical analysis can be explained *via* a geometric model (Campbell and Atchley, 1981). The first principal component of a set of samples is (among other things) the axis of maximum variance. The scores obtained by projecting the sample points on to the first principal component have a greater variance than scores along any other possible axis. The first canonical variate is analogously the axis of maximum discrimination among groups, i.e. the axis along which the projections of samples from different groups are optimally separated in terms of the ratio of between- to within-group sum of squares. If there are just two groups then there is only one such axis, called the discriminant function. In general there are no more than $r - 1$ canonical variates, where r is the number of groups. But the canonical variates — unlike principal components — are not perpendicular to one another in the original space. When the canonical variate scores are plotted on rectangular axes the samples in each group fall roughly within circular or spherical envelopes; in other words the canonical variates are uncorrelated within groups. Interpretation is not difficult. Scatter diagrams of canonical variate scores indicate the interrelationships and the amount of overlap between groups, and each canonical variate has an eigenvalue (indicating its contribution to the separation of groups) and a set of loadings which indicate the diagnostic variables for distinguishing between groups. The loadings also allow new samples to be assigned scores; so, for example, fossil pollen spectra can now be positioned on canonical variates derived from groups of surface pollen spectra (Birks, 1976). See Webster (1977) for a fuller introduction.

Canonical analysis is likely to find applications in two classes of problems in palaeoecology:

(1) representation of the differences between, and the diagnostic taxa for,

groups of assemblages categorized according to some external classification, e.g. surface pollen spectra grouped by vegetational units;
(2) characterization of *clusters* in terms of an external set of variables (e.g.
clusters of modern diatom assemblages in terms of ion concentrations).

Programs for canonical analysis are available in standard statistical packages
including SPSS, BMDP and SAS.

Acknowledgements

This chapter was first written (Prentice, 1982) during tenure of a Hartley
fellowship in the Department of Geography, University of Southampton. It
was revised with support from the Swedish Natural Science Research Council
(NFR) within the framework of the project 'Simulation modelling of natural
forest dynamics'. John Birks drew my attention to some relevant literature;
Allan Gordon reviewed the manuscript.

REFERENCES

Adam, D. P. (1970). *Some Palynological Applications of Multivariate Statistics*, Ph.D.
Thesis, University of Arizona.
Beales, P. W. (1976). *Palaeolimnological Studies of a Shropshire Mere*. Ph.D. Thesis,
University of Cambridge.
Birks, H. H. (1973). Modern macrofossil assemblages in lake sediments in Minnesota.
In: *Quaternary Plant Ecology*. (Eds. H. J. B. Birks and R. G. West), Blackwell,
Oxford, pp. 173–190.
Birks, H. J. B. (1974). Numerical zonations of Flandrian pollen data. *New Phytologist*,
73, 351–358.
Birks, H. J. B. (1976). Late-Wisconsinan vegetational history at Wolf Creek, central
Minnesota. *Ecological Monographs*, **46**, 365–429.
Birks, H. J. B. (1977). Modern pollen rain and vegetation of the St. Elias Mountains,
Yukon Territory. *Can. J. of Botany*, **55**, 2367–2382.
Birks, H. J. B. (1981). Late Wisconsin vegetational and climatic history at Kylen Lake,
north-eastern Minnesota. *Quaternary Research*, **16**, 322–355.
Birks, H. J. B., and Berglund, B. E. (1979). Holocene pollen stratigraphy of southern
Sweden: a reappraisal using numerical methods. *Boreas*, **8**, 257–279.
Birks, H. J. B., and Saarnisto, M. (1975). Isopollen maps and principal components
analysis of Finnish pollen data for 4000, 6000 and 8000 years ago, *Boreas*, **4**, 77–96.
Birks, H. J. B., Webb, T., and Berti, A. A. (1975). Numerical analysis of surface pollen
samples from central Canada: a comparison of methods. *Review of Palaeobotany
and Palynology*, **20**, 133–169.
Brown, A. G. (1983). The use of traditional and multivariate techniques in the
identification and interpretation of floodplain sediment grain size variations. In:
Abstracts of papers Severn 1983. (Eds. K. E. Barber and K. J. Gregory), Dept. of
Geography, Univ of Southampton, pp. 55–57.
Brugam, R. B. (1980). Postglacial diatom stratigraphy of Kirchner Marsh, Minnesota.
Quaternary Research, **13**, 133–146.
Bruno, M. G., and Lowe, R. L. (1980). Differences in the distribution of some bog
diatoms: a cluster analysis. *American Midland Naturalist*, **104**, 70–79.
Caseldine, C. J., and Gordon, A. D. (1978). Numerical analysis of surface pollen
spectra from Bankhead Moss, Fife. *New Phytologist*, **80**, 435–453.

Campbell, N. A., and Atchley, W. R. (1981). The geometry of canonical variate analysis. *Systematic Zoology*, **30**, 268–280.

Dale, M. B., and Walker, D. (1970). Information analysis of pollen diagrams, I. *Pollen et Spores*, **12**, 21–37.

Davis, J. C. (1970). Information contained in sediment-size analyses. *Mathematical Geology*, **2**, 105–112.

Davis, R. B., and Norton, S. A. (1978). Palaeolimnologic studies of human impact on lakes in the United States with emphasis on recent research in New England. *Polskie Archivum Hydrobiologii*, **25**, 99–115.

Elner, J. K., and Happey-Wood, C. M. (1980). The history of two linked but contrasting lakes in North Wales from a study of pollen, diatoms and chemistry in sediment cores. *J. of Ecology*, **68**, 95–121.

Everitt, B. S. (1978). *Graphical Techniques for Multivariate Data*, Heinemann, London.

Everitt, B. S. (1980). *Cluster Analysis*, Heinemann, London.

Gauch, H. G. (1982). *Multivariate Analysis in Community Ecology*, Cambridge University Press, Cambridge.

Gordon, A. D. (1981). *Classification*, Chapman and Hall, London.

Gordon, A. D. (1982). Numerical methods in Quaternary Palaeoecology. V: Simultaneous graphical representation of the levels and taxa in a pollen diagram. *Review of Palaeobotany and Palynology*, **37**, 155–183.

Gordon, A. D., and Birks, H. J. B. (1972). Numerical methods in Quaternary palaeoecology. I: Zonation of pollen diagrams. *New Phytologist*, **71**, 961–979.

Gordon, A. D., and Birks, H. J. B. (1974). Numerical methods in Quaternary palaeoecology. II: Comparison of pollen diagrams. *New Phytologist*, **73**, 221–249.

Gower, J. C. (1967). Multivariate analysis and multidimensional geometry. *The Statistician*, **17**, 13–28.

Gower, J. C., and Digby, P. G. N. (1981). Expressing complex relationships in two dimensions. In: *Interpreting Multivariate Data* (Ed. V. Barnett), John Wiley, London, pp. 83–118.

Greenacre, M. J. (1981). Practical correspondence analysis. In: *Interpreting Multivariate Data* (Ed. V. Barnett), John Wiley, London, pp. 119–146.

Hicks, S. (1977). Modern pollen rain in Finnish Lapland investigated by analysis of surface moss samples. *New Phytologist*, **78**, 715–734.

Hill, M. O. (1973). Reciprocal averaging: an eigenvector method of ordination. *J. of Ecology*, **61**, 237–249.

Hill, M. O., and Gauch, H. G. (1980). Detrended correspondence analysis: an improved ordination technique. *Vegetatio*, **42**, 47–58.

Hofmann, W. (1978). Analysis of animal microfossils from the Segeberger See (F.R.G.). *Archiv für Hydrobiologie*, **82**, 316–346.

Howe, S. E. (1978). *Estimating Regions and Clustering Spatial Data: Analysis and Implementation of Methods using the Voronoi Diagram*, Ph.D. Thesis, Brown University.

Huntley, B., and Birks, H. J. B. (1983). *An Atlas of Past and Present Pollen Maps for Europe 0–13,000 Years Ago*, Cambridge University Press, Cambridge.

Jaktar, S. A., Rushforth, S. R., and Brotherson, J. D. (1979). Diatom floristics and succession in a peat bog near Lily lake, Summit County, Utah. *Great Basin Naturalist*, **39**, 15–43.

Kaesler, R. L. (1966). Quantitative re-evaluation of ecology and distribution of recent foraminifera and ostracoda of Todos Santos Bay, Baja California, Mexico. *University of Kansas Palaeontological Contributions*, **10**, 1–50.

Lichti-Federovich, S., and Ritchie, J. C. (1968). Recent pollen assemblages from the Western Interior of Canada. *Review of Palaeobotany and Palynology*, **7**, 297–344.

Maddocks, R. F. (1966). Distribution patterns of living and subfossil podocopid ostracodes in the Nosy Be' area, northern Madagascar. *University of Kansas Palaeontological Contributions*, **12**, 1–72.

Mosimann, J. E., and Greenstreet, R. L. (1971). Representation-insensitive methods for paleoecological pollen studies. In: *Statistical Ecology* (Eds. G. P. Patil, E. C. Pielou and W. E. Waters), volume 1, pp. 23–58.

Norton, S. A., Davis, R. B., and Brakke, D. F. (1981). *Responses of Northern New England Lakes to Atmospheric Inputs of Acids and Heavy Metals*, Land and Water Resources Center, University of Maine at Orono.

O'Sullivan, P. E., and Riley, D. H. (1974). Multivariate numerical analysis of surface pollen spectra from a native Scots pine Forest. *Pollen et Spores*, **16**, 239–264.

Overpeck, J. T., Webb, T., and Prentice, I. C. (1984). Quantitative interpretation of fossil pollen spectra: dissimilarity coefficients and the method of modern analogs, *Quaternary Research*, **23**, 87–108.

Pennington, W., and Sackin, M. J. (1975). An application of principal components analysis to the zonation of two Lake-Devensian profiles. I: Numerical analysis. *New Phytologist*, **75**, 419–440.

Pirttiala, K. (1980). *Numerical Methods Applied to Diatom Data*, Nordic meeting of diatomologists 1980, Lammi Biological Station, Finland (preprint).

Podani, J. (1980). SYN-TAX: Számitógépes programcsomag ökológiai, cönológiai és taxonómiai osztalyozások végrehajtására (SYN-TAX: computer program package for cluster analysis in ecology, phytosociology and taxonomy). *Abstracta Botanica*, **6**, 1–158.

Prentice, I. C. (1978). Modern pollen spectra from lake sediments in Finland and Finnmark, north Norway. *Boreas*, **7**, 131–153.

Prentice, I. C. (1980). Multidimensional scaling as a research tool in Quaternary pollen analysis: a review of theory and methods. *Review of Palaeobotany and Palynology*, **31**, 71–104.

Prentice, I. C. (1982). Multivariate methods for the presentation and analysis of data. In: *Palaeohydrological Changes in the Temperate Zone in the Last 15,000 Years. Subproject B, Lake and Mire Environments* (Ed. B. E. Berglund), volume 3, Department of Quaternary Geology, University of Lund, pp. 25–51.

Prentice, I. C. (1983). Pollen mapping of regional vegetation patterns in south and central Sweden. *J. of Biogeography*, **10**, 441–454.

Ritchie, J. C. and Yarranton, G. A. (1978). The Late-Quaternary history of the boreal forest of central Canada, based on standard pollen stratigraphy and principal components analysis. *J. of Ecology*, **66**, 199–212.

Synerholm, C. C. (1979). The chydorid cladocera from surface lake sediments in Minnesota and North Dakota. *Archiv für Hydrobiologie*, **86**, 137–151.

Ter Braak, C. J. F. (1983). Principal components biplots and alpha and beta diversity. *Ecology*, **64**, 454–462.

Ward, J. H. (1963). Hierarchical grouping to optimize an objective function. *J. Am. Statistical Association*, **58**, 236–244.

Webb, T. (1974). Corresponding patterns of pollen and vegetation in Lower Michigan: a comparison of quantitative data. *Ecology*, **55**, 17–28.

Webster, R. (1977). *Quantitative and Numerical Methods in Soil Classification and Survey*, Oxford University Press, Oxford.

Whiteside, M. C. (1970). Danish chydorid cladocera: modern ecology and core studies. *Ecological Monographs*, **40**, 79–118.

Wishart, D. (1969). An algorithm for hierarchical classifications. *Biometrics*, **25**, 165–170.

Handbook of Holocene Palaeoecology and Palaeohydrology
Edited by B. E. Berglund
© 1986 John Wiley & Sons Ltd.

39

Forest-composition calibration of pollen data

IAIN COLIN PRENTICE

Institute of Ecological Botany, University of Uppsala, Uppsala, Sweden

INTRODUCTION

Pollen assemblages in late-quaternary sediments provide a quantitative historical record of vegetation composition. The record is systematically biased by differences in the pollen productivity and dispersal characteristics of taxa. This chapter describes techniques designed to measure and correct such biases.

Not all sources of bias can be eliminated — for example, genetic and environmental heterogeneity presumably cause some variation in the pollen productivity of each taxon. But the existence of such variation does not remove the need to correct for differences *between* taxa. Taxa may depart from proportional representation by as much as an order of magnitude. The importance of estimating and correcting for these differences has been recognized since the earliest days of pollen analysis, although the use of explicit mathematical techniques is a more recent development. The R-value method as originally applied by Davis (1963) met serious difficulties caused by statistical estimation problems (Parsons *et al.* 1983) and by failure to take account of the long-distance component of pollen spectra (Faegri, 1966; Livingstone, 1968; Webb *et al.*, 1981; Parsons and Prentice, 1981). Applications of more recent techniques have shown approximate consistency between the relative pollen representation of the same tree taxa in comparable sampling sites located in different areas (e.g. Andersen, 1970; Bradshaw and Webb, 1985; Prentice and Webb, 1985).

Some plants, including many herbs, seldom appear in the pollen record and constitute 'blind spots' in pollen analysis; but most forest trees produce abundant, well-dispersed pollen. Tree pollen spectra from forested areas can therefore be calibrated quantitatively against forest composition. The resulting calibrations can be used to estimate past forest composition from fossil pollen spectra. Techniques for calibration are summarized here; see Andersen (1970), Webb *et al.* (1981), Prentice and Parsons (1983), Prentice (1984) and Prentice and Webb (1985) for discussion of the theory behind them. This chapter also deals with the associated topics of surface sampling methodology and the identification of modern analogs for fossil pollen spectra.

SURFACE POLLEN AND VEGETATION DATA FOR CALIBRATION

Principles of survey design

Jacobson and Bradshaw (1981), Prentice (1985) and Bradshaw and Webb (1985) have discussed the relationship between basin size and pollen source area in forested regions. ('Basin size' means effectively the distance between the sampling point and the forest edge.) According to theory developed by Tauber (1965) and Prentice (1985), the proportion $F_i(X)$ of the 'canopy component' input of pollen of a taxon i originating within X metres of the centre of a basin, in a uniformly forested region, is approximately:

$$F_i(X) = 1 - e^{-75(v_{g(i)}/u)(X^{1/8} - r^{1/8})} \tag{1}$$

where $v_{g(i)}$ is the deposition velocity of the pollen type, u is the wind speed, and r is the radius of the basin. Heavy pollen grains fall fast; lighter pollen grains therefore have larger source areas. Furthermore the source areas of all pollen types become disproportionately larger as basin size increases. This effect is strongest at small radii, so there is a qualitative difference between the source area of a site with zero radius and a site that is tens or hundreds of metres across.

The same theory predicts that the relative representation of different pollen types can be affected not only by differences in pollen productivity ('production bias') but also by differences in deposition velocity, causing differential losses between the forest edge and the sampling location ('dispersal bias'). In a uniform forest:

$$y_i = P_i x_i e^{-75(v_{g(i)}/u)r^{1/8}} \tag{2}$$

where y_i is the predicted pollen accumulation rate of taxon i, P_i is its pollen productivity, and x_i is its abundance in the forest. The exponential term in equation (2) represents dispersal bias. When $r = 0$, there is only production bias. As r increases, dispersal bias increasingly favours light pollen grains. The problem of calibration is essentially the problem of how to estimate and correct for the joint effects of production and dispersal bias in sites of a particular size, given modern pollen and forest composition data from the real (non-uniform!) landscape.

The surface samples used in practical calibration studies should have the same pollen-collecting properties as the fossil samples to which the calibrations will be applied. The forest composition data should refer to areas roughly equivalent to those 'sensed' by the pollen spectra. For practical purposes, surface sample surveys carried out in connection with calibration studies fall into two categories (Wright, 1967). In *local-scale* surveys, pollen spectra in moss polsters from the forest floor ($r\sim0$) are compared with forest composition within 20–30 m (Andersen, 1967, 1970; Bradshaw, 1981a,b; Heide and Bradshaw, 1982). Local forest composition is estimated at the same time as the samples are collected, and all the samples may be from a relatively small area. These surveys yield calibrations applicable to pollen profiles from small woodland hollows or mor humus on the forest floor (e.g. Bradshaw 1981b, Baker *et al.*, 1978). In *regional-scale* surveys, pollen spectra in moderate-sized lakes or bogs ($r \sim 10^2$ to 10^3 m) are compared with average forest composition within a much larger radius, typically 20–30 km, to yield calibrations applicable to pollen profiles from similar basins (Kabailiene, 1969; Parsons *et al.*, 1980; Webb *et al.*, 1981; Delcourt *et al.*, 1983; Delcourt *et al.*, 1984; Bradshaw and Webb, 1985; Prentice and Webb, 1985; Prentice *et al.*, 1985). Such data are commonly obtained from regional or national forest surveys.

Pollen traps (e.g. Tauber, 1974) provide an alternative way to sample atmospheric pollen fallout, in or outside the forest, but pollen trapping results need to be averaged over 5–10 years to smooth out annual variations in flowering; pollen trapping studies are unlikely to supplant sediment-based calibration studies.

Field and laboratory methods

Moss polsters, and surface samples from bogs, present no particular sampling problems. Moss polsters should ideally include only green parts, to avoid problems of preservation. Any reasonably large species could be used, because species and growth form do not affect pollen collecting or preserving properties (Bradshaw, 1981a). Pollen is extracted from moss by stirring in distilled water and then sieving through a screen of mesh greater than 150 μm. The subsequent procedure is as usual for organic sediments. It is normally worth doing only relative counts, not absolute counts (but see Andersen, 1970).

Surface mud from lakes may be obtained by any of the standard coring techniques, but it is often more convenient to use lightweight, free-fall samplers designed to take short cores (typically 20–40 cm) of flocculent surface material (e.g. Hakala, 1971; Berggren, 1972). Perspex core tubes allow one to check whether the sediment surface is undisturbed. For many purposes only the top 2–3 cm need to be retained; but if more precise time control is important then the whole short core should be kept for dating — for example, by ^{210}Pb, ^{137}Cs (Chapter 4) or using a marker horizon. The best time control is available in laminated sediments, which are ideally sampled by freezing methods (Chapter 17). Pollen can be extracted from surface mud by standard methods, with exotic marker added if required for absolute counting and determination of pollen concentrations and accumulation rates (Chapter 22).

Forest composition is commonly characterized as total or percentage basal area for each taxon. Basal area can be estimated from diameter. Alternatively, forest inventories often publish results in terms of growing-stock volume. (Average basal areas and volumes are normally highly correlated between stands.) Crown area is another possible measure, again normally well-correlated with basal area. Forest inventories use ground sampling, aerial photography or a combination of both. Procedures vary according to terrain, the economic value of the forest surveyed, and so on. Sampling methods and data storage conventions also vary between countries. See references already cited for examples of the compilation of data from forest inventories for regional-scale calibration; also Delcourt *et al.* (1981, 1984) for information on the North American continuous forest inventory.

Little calibration work has been attempted in non-forest situations, partly because of the difficulty in quantifying the abundance of shrubs, dwarf-shrubs and herbs. The problem essentially is how to measure 'landscape occupancy'. The 1–4 m^2 scale of most ecological descriptive work on non-arboreal vegetation is a finer scale than that recorded by pollen sites of any type.

Preliminary data analysis

Surface sample and forest data can be displayed as transect diagrams, isopoll and isophyte maps and maps of principal components of pollen and vegetation data (e.g. Webb, 1974; Webb and McAndrews, 1976; Bernabo and Webb, 1977; Prentice, 1978, 1983; Huntley and Birks, 1983; Delcourt *et al.*, 1984). Such techniques can establish general correspondences, and help to detect aberrant samples or regions.

Problems caused by human impact

Human impact on the modern landscape causes two distinct types of problem for calibration work. First, the vegetation may have been so altered that it is unlike that of any previous time. Many areas of Europe are so deforested that regional-scale calibration work is impossible (Andersen, 1980). Local-scale work is possible within remnant wooded areas, but local-scale calibration functions are not simply transferable to the regional scale. Calibration is possible in regions that are only partly deforested; but if forest composition is changing rapidly due to selective felling, planting or clearance, then some care is necessary in synchronizing pollen and vegetation measurements (Donner,

1972; Prentice, 1978). Second, human activity may have altered the representation characteristics of the trees. This alteration can happen in several ways: increase of effective basin size due to clearance (Jacobson and Bradshaw, 1981); interference with flowering by coppicing or other management (Bradshaw, 1981a); gross distortion of age structure or density through management (Parsons *et al.*, 1980; Bradshaw, 1981a). Such complications limit the precision of calibration functions estimated in areas with a history of intensive land-use. Comparisons of results from different areas, however, should help to identify any serious anomalies.

Short cores, especially from laminated sediments (Parsons *et al.*, 1980), should help in tracing palynological changes through recent historic time; these changes could be compared with documented changes in forest composition. Studies of *pre-settlement* pollen spectra in the midwestern U.S.A. (Webb, 1973; Brubaker, 1975) have been reassuring in that isopoll and principal component patterns similar to those of today have emerged.

THE ANALOG METHOD

General

In the analog method one tries to match a given fossil pollen spectrum with modern pollen spectra from a range of vegetation types. The analog method is not strictly a calibration method, but one can use it to give a rough reconstruction of past vegetation — on the assumption that if two pollen spectra are similar, they must have been produced by similar vegetation. The analog method is commonly used to assign fossil pollen spectra to broad categeories of vegetation (formations). Some pollen spectra, however, have no modern analogs (Overpeck *et al.*, 1985).

In order to identify modern analogs, one should compare each fossil pollen spectrum with each pollen spectrum in a large modern database. It is instructive to discover what areas on the modern landscape provide analogs for the various levels in a profile; and to observe changes in the residual dissimilarity between each fossil pollen spectrum and its closest modern analog (Overpeck *et al.*, 1985).

Measurement of resemblance between pollen spectra

The degree of resemblance between two pollen spectra can be measured using any one of a variety of numerical coefficients. Davis *et al.* (1975), Ogden (1977) and others have used correlation coefficients to compare modern and fossil pollen spectra. Prentice (1980) and Overpeck *et al.* (1985) recommended instead the use of dissimilarity coefficients.

Different dissimilarity coefficients give different amounts of weight to the less abundant pollen types (see also Chapter 38). *Unweighted* coefficients include Euclidean distance:

$$d^2(j,k) = \sum_i (p_{ij} - p_{ik})^2 \tag{3}$$

where $d^2(j,k)$ is the squared Euclidean distance between pollen spectra j and k, and p_{ij} and p_{ik} are the pollen proportions or percentages of taxon i in j and k. Unweighted coefficients are most strongly dependent on the percentages of the most abundant taxa. *Equal-weight* coefficients include standardized Euclidean distance:

$$d^2(j,k) = \sum_i \left[(p_{ij} - p_{ik})/s_i \right]^2 \tag{4}$$

where s_i is the standard deviation for taxon i within a specified data set — conventionally the modern pollen data set (Overpeck *et al.*, 1985). Equal-weight coefficients are sensitive to the rare taxa, and to the choice of pollen sum. *Signal-to-noise ratio* coefficients adopt an intermediate weighting strategy. This category includes chord distance:

$$d^2(j,k) = \sum_i (\sqrt{p_{ij}} - \sqrt{p_{ik}})^2 \tag{5}$$

Overpeck *et al.* (1985) found that signal-to-noise ratio coefficents gave consistently the most reliable discrimination among vegetational units.

Squared chord distance ranges from zero to 2. It is zero if and only if the two pollen spectra j and k are identical, and 2 if and only if they have no taxa in common. Studies on transects of modern pollen spectra across eastern North America (Overpeck *et al.*, 1985) showed that pollen spectra with squared chord distance less than 0.15 are likely to be from the same formation. Pollen spectra with squared chord distance less than 0.12 are sure to be from the same formation, and likely to be from the same forest type. (These numerical values apply when pollen *proportions* are used in equation (5).) A computer program is needed to search a modern pollen database for closest analogs to fossil pollen spectra. One such program, using a variety of dissimilarity coefficients including Euclidean distance, standardized Euclidean distance and chord distance, has been written by R. Arigo and J. T. Overpeck and is available on request.

Further notes on the analog method

Multivariate data anlaysis methods can be used to *supplement* the use of dissimilarity coefficients. For example, principal components analysis can be used to display the relationships among modern pollen spectra, or canonical analysis to display the relationships among sets of pollen spectra from different vegetation units. Fossil pollen spectra may then be positioned on the principal components or canonical variates (e.g. Birks, 1976). Note, however, that these techniques inevitably introduce some distortion, so that it is possible for pollen spectra that are dissimilar to be placed close together.

The analog method can suggest possible abundances of taxa that are too poorly represented to allow quantitative calibration. Examples include *Populus*, whose pollen does not preserve well in most sediments, and the insect-pollinated *Liriodendron*, both of which reach high percentages in some present-day forests. However, such inferences are dubious unless backed up by independent evidence, e.g. from macrofossils.

ANDERSEN'S METHOD

The quantitative calibration method described by Andersen (1967, 1970) can be applied when the surface samples are moss polsters from woodland areas surrounded by a deforested landscape. The resulting calibrations are relevant to fossil pollen data from small hollows and mor humus profiles under forest canopies.

In Andersen's method, present 'non-local' pollen input is assumed constant over the area studied. Bradshaw (1981a) defined 'tree taxa not recorded in the woods sampled, and pollen from herbs most commonly recorded in non-woodland habitats' as non-local. Quantities proportional to pollen accumulation rates (influx indices) are then estimated as ratios of local tree pollen counts to non-local pollen counts (cf. the use of exotic markers, Chapter 22).

Linear regression is used to estimate the slopes (α_i) in the equations

$$y'_{ik} = \alpha_i x_{ik} + w'_i \tag{6}$$

where y'_{ik} is the influx index at sample k for taxon i, and x_{ik} its basal area (or other measure) in the surrounding 20–30 m. A separate regression is carried out for each taxon. Outliers, if any, may be detected in scatter plots and deleted before regression.

For Andersen's method to succeed, scatter plots of y'_{ik} against x_{ik} must show a linear relationship with positive slope (α_i) and *small*, i.e. negligible, intercept (w'_i) (Figure 39.1). If so, the relative magnitudes of the α_i should correspond to the relative magnitudes of the pollen productivities P_i in equation (2). Fossil pollen spectra may be corrected by dividing each taxon's pollen count by the appropriate slope factor α_i, and adjusting the result to add up to 100% in each sample. The results of this procedure are estimates of past percentage basal area within 20–30 m (Figures 39.2 and 39.3).

EXTENDED R-VALUE METHOD

General

Andersen's method is applicable to a strictly limited number of calibration

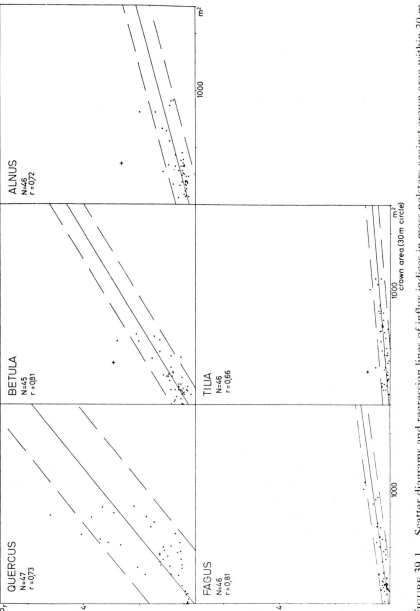

FIGURE 39.1.   Scatter diagrams and regression lines of influx indices in moss polsters against crown area within 30 m for the major tree taxa in Draved Forest, Denmark (from Andersen, 1970)

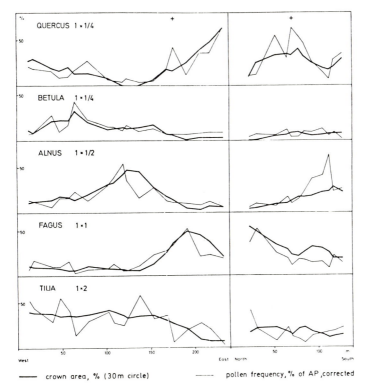

FIGURE 39.2.   Transect diagrams from Draved Forest, Denmark, comparing *corrected* pollen percentages from moss polsters (thin lines) with true crown areas within 30 m (thick lines) (from Andrsen, 1970)

problems. Often there is no way to estimate pollen accumulation rates in local-scale surveys. Absolute pollen data can be obtained from lakes and bogs but are often too variable to allow quantitative calibration (Davis *et al.*, 1973; Webb *et al.*, 1978, 1981). A generally applicable calibration method must work with percentage data directly.

Three main methods have been developed for direct calibration of pollen percentage data: the R-value method of Davis (1963), the regression method of Webb *et al.* (1981), and the extended R-value method (ERV method). The ERV method as described here is equivalent to the 'background method' of Parsons and Prentice (1981) and Model 1 of Prentice and Parsons (1983) and Prentice and Webb (1985). It combines features from the other two methods.

The ERV method estimates parameters α_i and z_i for each taxon i in the equations:

$$p_{ik} = \alpha_i v_{ik} f_k + z_i \qquad (7)$$

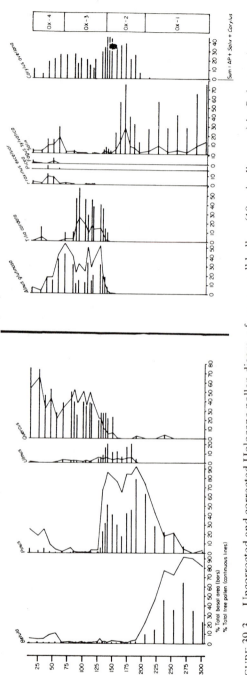

FIGURE 39.3. Uncorrected and corrected Holocene pollen diagram from a small hollow (10 m in diameter) in Oxborough Wood, southern England. The continuous lines are pollen percentages. The bars represent estimates of percent basal area within 20 m, obtained by applying slope coefficients as follows: *Betula* 16.0, *Pinus* 7.7, *Alnus* 1.5, *Fraxinus* 1.1, *Quercus* 1.0, *Salix* 0.8, *Fagus* 0.5 (all from Bradshaw, 1981a) and *Ulmus* 1.2, *Tilia* 0.3 (Andersen, 1970) (from Bradshaw, 1981b). *Reproduced by permission of Blackwell Scientific Publications Ltd*

where p_{ik} is the pollen proportion or percentage of taxon i in site k, v_{ik} is the corresponding proportion or percentage basal area (within an appropriate distance), and f_k is a site factor which ensures that the right-hand-sides of (7) sum to 1 (100%). It can be shown that:

$$f_k = \left(1/\sum_j \alpha_j v_{jk}\right)\left(1 - \sum_j z_j\right) \tag{8}$$

ERV programs obtain simultaneous maximum-likelihood estimates of all the α_i and z_i, given a range of surface pollen samples and associated forest composition data.

Comparison with other methods

The R-value method of Davis (1963) and Parsons and Prentice (1981) estimates parameters R_i in:

$$p_{ik} = R_i v_{ik} g_k \tag{9}$$

where g_k is a site factor. The estimates, however they are obtained, can be highly unstable if v_{ik} is small and there is a significant background component.

Unlike the R-value method, ERV estimates a 'background' component (z_i) which takes care of pollen input from outside the area used to estimate basal area percentages. ERV is equivalent to the R-value method if all z_i are set to zero.

The regression method of Webb *et al.* (1981) estimates parameters a_i and b_i in:

$$p_{ik} = a_i v_{ik} + b_i \tag{10}$$

by separate linear regression for each taxon. In practice, scatter plots of pollen percentages against vegetation percentages may show linear relationships (Figure 39.4). Reduced-major-axis regression (= geometric mean regression — e.g. Imbrie, 1956) has been used in several calibration studies (Birks, 1980; Webb *et al.*, 1981; Delcourt *et al.*, 1983, 1984). Regression based on equation (10) provides a first approximation to ERV. However, equation (7) predicts non-linear scatters for taxa with *large or small* α_i and v_{ik} values ranging to 0.3 (30%) or more. In such cases (e.g. *Pinus, Acer* in the northern Midwest of the USA) the regression method gives systematically biased estimates, and poor fit to the data, in comparison with ERV (Prentice and Webb, 1985).

How to use the ERV method

ERV requires more computer time than ordinary regressions, and needs a special computer program. Prentice and Webb (1985) suggested the following procedure:

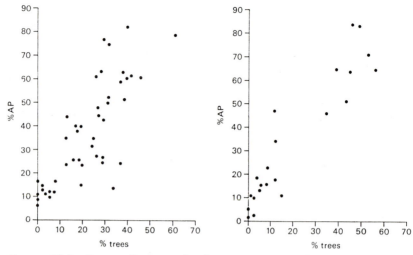

FIGURE 39.4.   Scatter diagrams of pollen percentages against tree percentages may show roughly linear relationships. Left: *Quercus* in moss polsters against crown area percentages within 30 m (Andersen, 1970). Right: *Quercus* in moderate-sized basins against average basal area within 30 km (Webb *et al.*, 1981)

(1)   Plot simple scatter diagrams of pollen percentages against vegetation percentages (like Figure 39.4) for each taxon. Use these to locate outliers, errors and any other obvious problems.

(2)   Optionally carry out linear regressions, to obtain first estimates of the parameters.

(3)   To estimate final calibration parameters, apply ERV to selected data sets. To see how well the ERV model fits the data, plot scatter diagrams of pollen percentages against *site-adjusted* vegetation percentages, $v_{ik}f_k$. These plots should show linear relationships (Figure 39.5), even when simple scatter diagrams like Figure 39.4 do not.

Some experimentation is likely to be needed in order to get best results from ERV. The fit should be good, and the z_i values small. Results may be improved by manipulating the sampling radius used for basal area estimates (Prentice *et al.*, 1985) or by changing the region(s) used for calibration. Note that if the data come from too small a region, the range of vegetation percentages may be too narrow for ERV to give good estimates. On the other hand, if data from too large a region are included in a single ERV analysis, the requirement for a constant background pollen percentage (z_i) may be seriously violated, causing excessive scatter. Data sets from very large regions should therefore be subdivided. Ideally, the results from different subregions should agree as to the relative magnitudes of the α_i for different taxa, although the z_i might vary between regions.

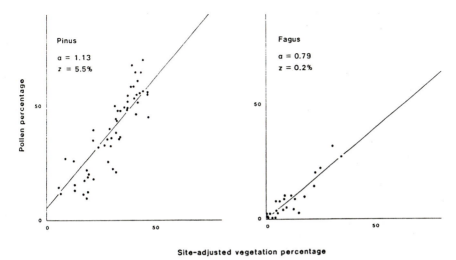

FIGURE 39.5. Scatter diagrams showing goodness of fit to the extended R-value model. Left: *Pinus* in moderate-sized basins in southern and central Sweden, against 'site-adjusted' volume percentages within 50 km. Right: *Fagus* from the same analysis. The lines have slopes α_i and intercepts z_i (from Prentice *et al.*, 1985)

Correction of fossil pollen spectra

Moderate-sized basins receive significant input of lighter pollen types from more than 30 km away (Prentice, 1985) and significant z_i for such taxa arise in regional-scale calibration studies (e.g. Bradshaw and Webb, 1985). Bradshaw and Webb also found that background pollen inputs varied in predictable ways with changes in basin size and sampling radius. Generally, background pollen cannot be ignored; *but nor can it be assumed constant* except over short periods in limited regions.

ERV estimates slopes (α_i) and background pollen percentages (z_i). How does one obtain from these values the best estimates of 'correction factors' for fossil pollen spectra, given that background pollen input (from any distance) can change in space and time? There are several possible methods:

(1) Choose the sampling radius large enough that all the z_i become small. Then use the α_i estimates as correction factors, in the same way that one would use the R-values of Davis (1963), or the slopes estimated by Andersen's method.

(2) Choose the sampling radius in the same way, but then obtain new estimates of the α_i *with all z_i set to zero*. These new estimates are, of course, true R-values. The use of the R-value model is acceptable when ERV has shown that the sampling radius is appropriate, i.e. that background pollen input is not likely to be a problem.

(3)   Choose the sampling radius by experiment (or informed guesswork) to optimize goodness-of-fit to the ERV model. The optimal sampling radius may well be different for each taxon. For example, Prentice *et al.* (1985) found that good fits can often be obtained when $F_i(X)$ in equation (1) is on the order of 0.75, i.e. when roughly three-quarters of the pollen input of taxon *i* has originated within X, where X is now the sampling radius. Then adjust each estimate of α_i for long-distance transport (divide each α_i by $F_i(X)$) and use the resulting, long-distance-adjusted estimates as R-values.

Experience may in the future indicate which of these methods gives the most consistent results.

A computer program

At the time of writing a new, portable computer program for ERV is under development at Brown University. The program will carry out all three steps of the suggested practical procedure for application of ERV, including pre-and post-analysis scatter plots, and will allow parameters to be fixed as specified (e.g. z_i values to zero for the estimation of true R-values). Copies of the program and documentation will be available to interested users.

FURTHER NOTES

The procedure described in the preceding sections are designed (*a*) for analysis of the various pollen representation characteristics of taxa, and (*b*) to correct pollen spectra for production and dispersal bias. Corrected pollen percentages should be unbiased, but still represent a 'pollen-sample's view' of the surroundings. This view is a distance-weighted average of the vegetation, with a different distance-weighting for each taxon. The pollen sample 'sees' mostly nearby trees of taxa with heavy pollen grains (*Abies*, *Acer*) but also sees a wider area of forest when it comes to taxa with light pollen grains, such as *Pinus*. If far-distant pine forests contribute pollen to a pollen spectrum, they will also contribute to the corrected pollen spectrum; there is no way that a single pollen sampling site can give information about the spatial distribution of vegetation around the site. If corrected pollen spectra from a network of sites are used to generate reconstructed isophyte maps, these maps will contain different levels of smoothing for different taxa. The taxa with the heaviest pollen grains will generally show the sharpest changes from site to site.

   Jacobson (1979) showed how pollen data from *paired* basins can be combined in order to obtain some degree of separation between the local and more distant components of pollen input for each taxon. Underlying Jacobson's method is the principle that local pollen input varies more quickly in space than more distant input. To a first approximation, two moderate-sized basins that are (say) 10 km apart might be assumed to have different inputs from 5–10 km, but a common source vegetation for 'background' pollen input. This reasoning would lead to a set of simultaneous equations which could be solved for two components at different spatial scales. The same approach could be extended to handle networks of sites. Similar proposals have been made by Livingstone (1969), Kabailiene (1969) and Janssen (1973) (see also Jacobson and Bradshaw, 1981). The full use of the spatial information in networks of pollen sites is a possibility that would repay further theoretical study and practical exploration. One immediate objective might be to construct deconvoluted palaeoisophyte maps containing equal levels of spatial smoothing for all taxa. Another might be to generate more detailed palaeovegetation maps, based on several sites in the same region.

The final note concerns important differences between vegetational and climatic calibration of pollen data. The methods used are indeed different, with good reason. The calibration of pollen spectra in terms of species abundance relies on the fact that only pine trees can produce *Pinus* pollen—likewise for the other taxa — and the assumption that the amount of pollen produced and dispersed by a given taxon depends on the abundance of that taxon and not on the abundance of any other taxon. The underlying models used in this type of calibration rely on linear relationships that apply separately to each taxon; interactions between taxa are confined to the purely numerical effects of the percentage constraint.

In climatic calibration, no direct relationship can be assumed between the pollen percentage of any taxon and any climatic variable. In the underlying conceptual model a number of unique, dome-shaped or more complex response surfaces represent the abundance of each taxon as a function of several climatic variables. Such response surfaces are non-linear by definition. The current methodology for climatic calibration (Chapter 40) involves choosing geographic regions, representing restricted regions of 'climate space', within which the abundances of pollen types are either approximately linear in the climate variable, or can be transformed to linearity: a 'piecewise linear' approach. It may be reasonable to seek pollen-tree calibration 'slopes' or α_i values that are widely applicable (e.g. across a continent) to sites of comparable size; but it is unreasonable to expect climatic calibrations obtained by linear models to be applicable to pollen spectra very different from those used in the derivation of the calibrations. One way to use the methods described in the next chapter therefore involves deriving separate sets of equations for different regions, and using the analog method (as described in this chapter) to help decide which equations should be applied to a given fossil pollen spectrum.

Acknowledgements

I wish to thank Richard Bradshaw, Hazel Delcourt, Paul Delcourt, Allan Gordon and Tom Webb—for sharing their ideas, and asking difficult questions.

An earlier version was written during tenure of a Hartley fellowship in the Department of Geography, University of Southampton. Research results and manuscript revision were supported by the Swedish Natural Science Research Council (NFR) under the projects 'Calibration of recent pollen spectra through analysis of the pollen/vegetation relationship' and 'Simulation modelling of natural forest dynamics'.

REFERENCES

Andersen, S. T. (1967). Tree-pollen rain in a mixed deciduous forest in south Jutland (Denmark). *Review of Palaeobotany and Palynology*, **3**, 267–275.

Andersen, S. T. (1970). The relative pollen productivity and pollen representation of north European trees, and correction factors for tree pollen spectra. *Danmarks Geologiske Undersøgelse*, II, **96**, 1–99.

Andersen, S. T. (1980). The relative pollen productivity of the common forest trees in the early Holocene. *Danmarks Geologiske Undersøgelse, Årbog 1979*, 5–19.

Baker, C. A., Moxey, P. A., and Oxford, P. M. (1978). Woodland continuity and change in Epping Forest. *Field Studies*, **4**, 645–699.

Berggren, H. (1972). Sedimentprovtagning med rörhämtare. *Vatten*, **4**, 374–377.

Bernabo, J. C., and Webb, T. (1977). Changing patterns in the Holocene pollen record of northeastern North America: a mapped summary. *Quaternary Research*, **8**, 64–96.

Birks, H. J. B. (1976). Late Wisconsinan vegetational history at Wolf Creek, central Minnesota. *Ecological Monographs*, **46**, 395–429.

Birks, H. J. B. (1980). *Quaternary Vegetational History of West Scotland*, Vth International Palynological Conference, Cambridge, Excursion Guide C8.

Bradshaw, R. H. W. (1981a). Modern pollen representation factors for woods in south-east England. *J. of Ecology*, **69**, 45–70.

Bradshaw, R. H. W. (1981b). Quantitative reconstruction of local woodland history using pollen analysis from a small basin in Norfolk, England. *J. of Ecology*, **69**, 941–955.

Bradshaw, R. H. W., and Webb, T. (1985). Relationships between contemporary pollen and vegetation data from Wisconsin and Michigan, U.S.A. *Ecology*, in press.

Brubaker, L. B. (1975). Postglacial forest patterns associated with till and outwash in northcentral Upper Michigan. *Quaternary Research*, **5**, 499–527.

Davis, M. B. (1963). On the theory of pollen analysis. *A. J. of Science*, **261**, 897–912.

Davis, M. B., Brubaker, L. B., and Webb, T. (1973). Calibration of absolute pollen influx. In: *Quaternary Plant Ecology* (Eds. H. J. B. Birks and R. G. West), Blackwell, Oxford, pp. 395–434.

Davis, R. B., Bradstreet, T. E., Stuckenrath, R., and Borns, H. W. (1975). Vegetation and associated environments during the past 14,000 years near Moulton Pond, Maine. *Quaternary Research*, **5**, 435–463.

Delcourt, H. R., West, D. C., and Delcourt, P. A. (1981). Forests of the southeastern United States: quantitative maps for aboveground woody biomass, carbon and dominance of major tree taxa. *Ecology*, **62**, 879–887.

Delcourt, P. A., Delcourt, H. R., and Davidson, J. L. (1983). Mapping and Calibration of modern pollen-vegetation relationships in the southeastern United States. *Review of Palaeobotany and Palynology*, **39**, 1–45.

Delcourt, P. A., Delcourt, H. R., and Webb, T. (1984). Atlas of mapped distributions of dominance and modern pollen percentages for important tree taxa of eastern North America. *American Association of Stratigraphic Palynologists, Contribution Series*, **14**.

Donner, J. J. (1972). Pollen frequencies in the Flandrian sediments of Lake Vakojärvi, south Finland, *Commentationes Biologicae*, **53**, 1–19.

Faegri, K. (1966). Some problems of representativity in pollen analysis. *Palaeobotanist*, **15**, 135–140.

Hakala, I. (1971). A new model of the Kajak bottom sampler and other improvements in the zoobenthos sampling technique. *Annales Zoologici Fennici*, **8**, 422–426.

Heide, K. M., and Bradshaw, R. H. W. (1982). The pollen-tree relationship within forests of Wisconsin and Upper Michigan, U.S.A. *Review of Palaeobotany and Palynology*, **36**, 1–23.

Huntley, B., and Birks, H. J. B. (1983). *An Atlas of Past and Present Pollen Maps for Europe: 0–13,000 Years Ago*, Cambridge University Press, Cambridge.

Imbrie, J. (1956). Biometrical methods in the study of invertebrate fossils. *Bulletin of the American Museum of Natural History*, **108**, 214–252.

Jacobson, G. L. (1979). The palaeoecology of white pine (*Pinus strobus*) in Minnesota. *J. of Ecology*, **67**, 697–726.

Jacobson, G. L., and Bradshaw, R. H. W. (1981). The selection of sites for paleovegetational studies. *Quaternary Research*, **16**, 80–96.

Janssen, C. R. (1973). Local and regional pollen deposition. In: *Quaternary Plant Ecology* (Eds. H. J. B. Birks and R. G. West), Blackwell, Oxford, pp. 31–42.

Kabailiene, M. B. (1969). Formirovanie pyltsevykh spektrov i metody vosstanovleniya paleorastitel'nosti (On formation of pollen spectra and restoration of vegetation), *Ministry of Geology of the U.S.S.R., Institute of Geology (Vilnius) Transactions*, **11**, 1–147.

Livingstone, D. A. (1968). Some interstadial and postglacial pollen diagrams from eastern Canada. *Ecological Monographs*, **38**, 87–125.

Livingstone, D. A. (1969). Communities of the Past. In: *Essays in Plant Geography and Ecology* (Ed. K. N. H. Greenidge), Nova Scotia Museum, Halifax, Nova Scotia, pp. 83–104.

Ogden, J. G. (1977). Pollen analysis: state of the art. *Géographie Physique et Quaternaire*, **31**, 151–159.

Overpeck, J. T., Prentice, I. C., and Webb, T. (1985). Quantitative interpretation of fossil pollen spectra: dissimilarity coefficients and the method of modern analogs, *Quaternary Research*, **23**, 87–108.

Parsons, R. W., and Prentice, I. C. (1981). Statistical approaches to R-values and the pollen-vegetation relationship. *Review of Palaeobotany and Palynology*, **32**, 127–152.

Parsons, R. W., Gordon, A. D., and Prentice, I. C. (1983). Statistical uncertainty in forest composition estimates obtained from fossil pollen spectra *via* the R-value model. *Review of Palaeobotany and Palynology*, **40**, 177–189.

Parsons, R. W., Prentice, I. C., and Saarnisto, M. (1980). Statistical studies on pollen representation in Finnish lake sediments in relation to forest inventory data. *Annales Botanici Fennici*, **17**, 379–393.

Prentice, I. C. (1978). Modern pollen spectra from lake sediments in Finland and Finnmark, north Norway. *Boreas*, **7**, 131–153.

Prentice, I. C. (1980). Multidimensional scaling as a research tool in Quaternary pollen analysis. *Review of Palaeobotany and Palynology*, **31**, 71–104.

Prentice, I. C. (1983). Pollen mapping of regional vegetation patterns in south and central Sweden. *J. of Biogeography*, **10**, 441–454.

Prentice, I. C. (1985). Pollen representation, source area and basin size: towards a unified theory of pollen analysis. *Quaternary Research*, **23**, 76–86

Prentice, I. C., Berglund, B. E., and Olsson, T. (1985). Quantitative forest-composition sensing characteristics of pollen samples from Swedish lakes. MS.

Prentice, I. C., and Parsons, R. W. (1983) Maximum likelihood linear calibration of pollen spectra in terms of forest composition. *Biometrics*, **39**, 1051–1059.

Prentice, I. C., and Webb, T. (1984) Pollen percentages, tree abundances and the Fagerlind effect. MS.

Tauber, H. (1965). Differential pollen dispersion and the interpretation of pollen diagrams. *Danmarks Geologiske Undersøgelse*, II, **89**, 1–69.

Tauber, H. (1974). A static non-overload pollen collector. *New Phytologist*, **73**, 359–369.

Webb, T. (1973). A comparison of modern and presettlement pollen from southern Michigan (U.S.A.). *Review of Paleobotany and Palynology*, **16**, 137–156.

Webb, T. (1974). Corresponding patterns of pollen and vegetation in Lower Michigan: a comparison of quantitative data. *Ecology*, **55**, 17–28.

Webb, T., Howe, S. E., Bradshaw, R. H. W., and Heide, K. M. (1981). Estimating plant abundances from pollen percentages: the use of regression analysis. *Review of Palaeobotany and Palynology*, **34**, 269–300.

Webb, T., Laseski, R. A., and Bernabo, J. C. (1978). Sensing vegetational patterns with pollen data: choosing the data. *Ecology*, **59**, 1151–1163.

Webb, T., and McAndrews, J. H. (1976). Corresponding patterns of contemporary pollen and vegetation in central North America. *Geological Society of America Memoirs*, **145**, 267–299.

Wright, H. E. (1967). The use of surface samples in Quaternary pollen analysis. *Review of Palaeobotany and Palynology*, **2**, 321–330.

Handbook of Holocene Palaeoecology and Palaeohydrology
Edited by B. E. Berglund
© 1986 John Wiley & Sons Ltd.

40

Climatic calibration of pollen data: an example and annotated computing instructions

R. Arigo

Department of Geological Sciences,
Brown University, Providence, R.I., U.S.A.

S. E. Howe

National Bureau of Standards,
Washington, D.C., U.S.A.

T. Webb, III

Department of Geological Sciences, Brown University,
Providence, R.I., U.S.A.

INTRODUCTION

In certain regions, pollen data from radiocarbon-dated sediments can provide estimates of past temperature, rainfall and moisture balance. By recording variations in the vegetation over the past 15,000 years, pollen data provide information about many of the climatic changes that forced alterations in the vegetation. Because pollen data are quantitative, multiple regression procedures can be used to calibrate the geographic distribution of modern pollen data in terms of temperature or rainfall (Webb and Bryson, 1972; Webb, 1980). The resultant regression equations can then be applied to fossil samples, and time series and maps of past temperature or precipitation patterns can be calculated. This chapter describes how these regression equations are produced. Full descriptions of the methods for collecting and counting pollen data appear in Birks and Birks (1980) and elsewhere in this book (Chapters 1, 5, 6, 7 and 22).

Development of calibration functions that transform pollen percentages into estimates of climatic variables requires a sequence of computer programs. These make a series of calculations that help the analyst select the samples, pollen types and climatic variables to include in the data set before a multiple regression equation is computed and checked for assumption violations (Webb and Clark, 1977; Howe and Webb, 1977, 1983). Our chapter describes the sequence of steps and associated computer programs that were in use at Brown University in May 1981. Bartlein and Webb (1984a) have since updated this sequence of programs by introducing programs from BMDP (Biomedical Computer Programs P-Series — Dixon and Brown, 1979). Other procedures exist for the climatic interpretation of pollen data, and some of these do not require use of regression analysis (Iversen, 1944; Grichuk, 1969). Certain of these other procedures, as well as regression analysis, have been reviewed and discussed in Webb and Clark (1977), Davis (1978), Webb (1980), Birks and Birks (1980), Birks (1981), Prentice (1983), and Howe and Webb (1983).

The pollen and climate data used for calibration work at Brown University are stored in an SPSS (Statistical Package for Social Scientists — Nie *et al.*, 1975) file with 260 pollen types and 100 climatic variables observed at 3300 sites in eastern North America (Webb and McAndrews, 1976). Within SPSS, programs are available for executing the first seven steps for obtaining a regression equation (Figure 40.1). These steps begin with data selection and end with various tests of the regression equation to show that it fulfils various statistical assumptions (Howe and Webb, 1983). FORTRAN programs designed by S. Howe and R. Arigo can then be used for further testing of the equations.

In this chapter, we illustrate the several steps in the calibration procedure by way of an example in which pollen and climate data from lower Michigan (Webb, 1974) are used to develop an equation for estimating the mean July daily temperature. The results of using this and other equations are described in the final section of this chapter.

In the description of the calibration procedure, we have assumed that the data are aready stored in an SPSS file with associated labels for the pollen types and climatic variables. With the data in an SPSS file, SPSS commands make it relatively easy to select the samples and variables for use in a given analysis. Those researchers who cannot use or do not elect to use SPSS will still find our documentation useful in spelling out the many decisions needed to gain a calibration equation. These analysts can either write their own programs or use available programs to do the operations that we do in SPSS.

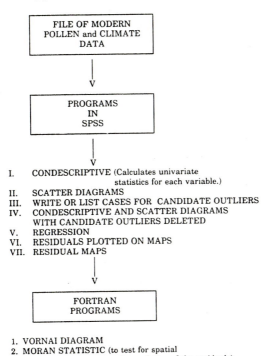

FIGURE 40.1. Sequence of computer programs used to develop a regression equation for calibrating pollen data in climatic terms

SELECTION OF SAMPLING SITES AND POLLEN TYPES

Initial selection of the set of modern data for analysis

Knowledge of the fossil pollen data to be calibrated is a prerequisite to selection of the region of modern samples for calculation of a calibration function or a set of calibration functions. When the fossil data are from a single site or several nearby sites, the chosen set of modern samples should, if possible, contain a range of pollen percentages broad enough to match the range of the pollen percentages of the major types within the fossil data (Webb and Bryson, 1972). Such a match is a necessary, although not a sufficient, condition for the modern samples to be analogs for the fossil samples (Neter and Wasserman, 1974). The geographical extent of the modern samples should initially be chosen large enough to yield the desired range in the pollen percentages for each of the major pollen types. Upon later

TABLE 40.1. Standard statistics for pollen types from lower Michigan: an ordered listing of the output from CONDESCRIPTIVE

Name	Mean	Standard deviation	Maximum
Quercus	24.726	12.906	51.685
Pinus	23.504	18.688	67.682
Betula	6.435	4.578	19.851
Gramineae	6.247	5.505	29.487
Ulmus	5.494	4.082	29.101
Fagus	3.608	3.238	14.748
Acer	2.960	1.890	8.798
Tsuga	2.828	3.577	16.172
Fraxinus	2.826	2.123	10.582
Polygonaceae	2.270	2.081	10.092
Alnus	1.873	1.363	5.338
Salix	1.544	1.490	8.072
Carya	1.529	1.665	6.969
Populus	1.486	1.577	5.714
Unidentified	1.445	1.284	8.273
Ostrya/Carpinus	1.407	0.862	3.704
Cyperaceae	1.306	2.873	22.192
Chenopodeaceae	1.287	0.742	3.374
Compositae	0.763	0.584	2.667
Juniperus/Thuja	0.751	0.682	2.913
Misc. Herbs	0.629	0.898	4.103
Artemisia	0.622	0.498	1.972
Platanus	0.617	0.649	2.878
Juglans	0.546	0.518	2.158
Plantago	0.535	0.573	2.158
Picea	0.431	0.570	3.185
Myrica	0.387	0.609	2.711
Unknown	0.374	0.579	3.333
Tilia	0.272	0.493	3.175
Rosaceae	0.255	0.380	1.695
Other Trees	0.200	0.317	1.302
Abies	0.188	0.361	1.911
Rhamnus/Vitis	0.184	0.326	1.778
Corylus	0.150	0.232	1.058
Morus	0.125	0.238	0.990
Liquidambar	0.040	0.107	0.360
Celtis	0.032	0.113	0.719
Aquafoliaceae	0.032	0.112	0.704
Thalictrum	0.029	0.080	0.342
Zea	0.019	0.066	0.282
Ericaceae	0.017	0.064	0.277
Larix	0.012	0.057	0.345
Taxus	0.009	0.073	0.587
Ephedra	0.005	0.040	0.324
Sarcobatus	0.004	0.032	0.256

analysis with the scatter diagrams (Step II), the size of the initial region may need to be modified, but selection of modern samples that match the fossil data is a good place to start the analysis.

Dissimilarity coefficients that directly match fossil pollen spectra to modern spectra provide another method for choosing the set of modern samples to use (Overpeck *et al.*, 1985; Chapter 39). These coefficients are particularly helpful when fossil data from broad networks of samples are being calibrated and calibration functions from different regional sets of modern data are available. Bartlein and Webb (1985b) describe how a continent-wide set of modern pollen samples can be divided into regional subsets for calculation of calibration functions.

By setting the limiting values for the latitude and longitude of the region (within the SPSS program, use SELECT IF LATITUDE GE 40.0, etc,; see Step I, CONDESCRIPTIVE below), an analyst can choose the geographic region containing paired observations of modern pollen and climatic data. The analyst may also choose the type of pollen sites (e.g. lake, bog etc.) to be included in the study and should, if possible, use modern data from a sediment type similar to the sediments in which the fossil data accumulated.

Calculation of pollen percentages

SPSS commands are used to calculate pollen percentages. For our example of the pollen data from lower Michigan, the percentages were calculated from a sum of total pollen minus spores and obligate aquatic types. In order to minimize the effect of human disturbance on the percentages of various pollen types, *Ambrosia* (ragweed) pollen was deleted from the pollen sum. Its current high value in most samples from Michigan results from human land-clearance and agriculture and makes these modern data non-analagous to most fossil data. Deletion of ragweed pollen helps to minimize this problem (Webb, 1973; van Zant *et al.*, 1978). After selection of pollen samples and calculation of the pollen percentages, the data are then ready to be processed by the SPSS routine CONDESCRIPTIVE. (Note: in the text the names of SPSS routines, commands and variable names appear in capital letters.)

COMPUTING INSTRUCTIONS

Step I: CONDESCRIPTIVE

Purpose

This program calculates the maximum value for each pollen type and such basic univariate statistics as the mean and variance (Figure 40.2). The pollen types meeting certain minimum criteria are then selected for further analysis.

Sample Output

VARIABLE PICEA   PICEA • SPRUCE

MEAN	0.431	STD ERROR	0.071	STD DEV	0.570
VARIANCE	0.325	KURTOSIS	8.122	SKEWNESS	2.422
RANGE	3.185	MINIMUM	0.0	MAXIMUM	3.185
SUM	27.607				

VALID OBSERVATIONS -   64   MISSING OBSERVATIONS -   0

- -

VARIABLE ABIES   ABIES • FIR

MEAN	0.188	STD ERROR	0.045	STD DEV	0.361
VARIANCE	0.130	KURTOSIS	8.167	SKEWNESS	2.627
RANGE	1.911	MINIMUM	0.0	MAXIMUM	1.911
SUM	12.047				

VALID OBSERVATIONS -   64   MISSING OBSERVATIONS -   0

- -

VARIABLE LARIX   LARIX • LARCH

MEAN	0.012	STD ERROR	0.007	STD DEV	0.057
VARIANCE	0.003	KURTOSIS	24.317	SKEWNESS	4.922
RANGE	0.345	MINIMUM	0.0	MAXIMUM	0.345
SUM	0.771				

VALID OBSERVATIONS -   64   MISSING OBSERVATIONS -   0

- -

VARIABLE JUNIPER   JUNIPERUS,THUJA • JUNIPER, ARBOR VITAE

MEAN	0.751	STD ERROR	0.085	STD DEV	0.682
VARIANCE	0.466	KURTOSIS	2.018	SKEWNESS	1.366
RANGE	2.913	MINIMUM	0.0	MAXIMUM	2.913
SUM	48.083				

VALID OBSERVATIONS -   64   MISSING OBSERVATIONS -   0

- -

FIGURE 40.2.   Sample of output from the SPSS program CONDESCRIPT giving values for standard univariate statistics. STD = standard; DEV = deviation

Our current practice is to select only those types whose mean is greater than 1.0% or whose mean is less than 1.0% but whose maximum value is greater than 5.0%. In general the pollen types that do not meet these criteria have too weak a numerical relationship with the particular climatic variable to be of use in the regression equation (Howe and Webb, 1983).

Computer commands

Command	**Argument**	**Program and task: comments**
RUN NAME	CONDESCRIPTIVE FOR MICHIGAN TEMPERATURE RUN	Label of run (should be well-selected for easy reference).
GET FILE	MODERN3D	Modern3d is the SPSS file of modern data used in this example. It contained 256 variables for each of 1312 cases or sites. (The modern file of surface data is fre-

quently updated, and the current version contains many more variables and cases.)

ALLOCATE   TRANSPACE=16000

This command allocates sufficient space for memory. (Less space may be adequate.)

COMPUTE   OLDSEQ=SEQNUM

SEQNUM is an intrinsic SPSS variable which numbers the cases within a given SPSS system file. When a 'SELECT IF' card is encountered, SPSS renumbers the remaining cases and changes the values of variable SEQNUM accordingly. Because the original values of SEQNUM are used to identify modern site locations, another variable 'OLDSEQ' must be created to preserve those values of SEQNUM.

SELECT IF   (LATITUDE GE 41.5 AND LE 46.0 AND LONGITUD GE-86.5 AND LE-83.0 AND SEQNUM LE 600 AND NE 483)

Selects data from the input file according to constraints listed. In this example, all cases located in the region 41.5 to 46.0°N and 83.0 and 86.5°W (except cases with sequence number 483 or with sequence numbers above 600) are selected from MODERN3D. This particular 'SELECT IF' chooses 64 cases at sites in the Lower Peninsula (LP) of Michigan by selecting first all cases at sites in a region containing the LP, then deleting those in the Upper Peninsula (UP) and

		one site (seq. num. 483 = Murry Lake) at the same location as Frains Lake. (In tests for spatial autocorrelation, only one of several coincident sites can be used.)
RECODE	PICEA TO ARTEMISI, IVA TO CYPERACE, MISCHERB TO UNKNOWN (−1,−2,−3=0)	In the list of pollen types (as they are labelled in the SPSS file) PICEA ... ARTEMISI, IVA ... CYPERACE, MISCHERB ... UNKNOWN, all values of −1,−2 and −3 will be made into 0's. The negative values are used in SPSS to indicate missing values, but none exist for these pollen types among the data from lower Michigan.
COMPUTE	TOTAL=0.0	
DO REPEAT	VI=PICEA TO ARTEMISI, IVA TO CYPERACE, MISCHERB TO UNKNOWN	
COMPUTE	TOTAL=TOTAL+V1	
END REPEAT		In the above four steps, the sum of all pollen counts (excluding spores and aquatic pollen) is accumulated in variable TOTAL.
COMPUTE	TOTAL=(TOTAL/100.0)	TOTAL is divided by 100 in order to yield percentages, not proportions.
DO REPEAT	V1=PICEA TO ARTEMISI, IVA TO CYPERACE, MISCHERB TO UNKNOWN	
COMPUTE	V1=V1/TOTAL	This command converts values of pollen counts of

		all types in the pollen sum to percentages of the TOTAL.
END REPEAT		
WRITE CLASS	(/4X,F4.0,4(2X,F6.3), 4X,3(2X,A4)) OLDSEQ, LONDEQ, LONMIN, LATDEG, LATMIN,STATE, NAME1,NAME2	This step is optional and gives a listing of the whole data set.
CONDE- SCRIPTIVE	PICEA TO ARTEM- ISI, IVA TO CYPERACE, MIS- CHERB TO UN- KNOWN	
STATISTICS	ALL	These two commands give the following statistics for each pollen type in the requested list: mean, variance, range, sum, standard error, kurtosis, minimum value, standard deviation, skewness, and maximum value.
FINISH		End card for SPSS deck.

Results

For the pollen data from lower Michigan, the following types satisfied the initial criteria for being included in the further analysis needed to calculate a calibration equation: *Acer* (maple), *Alnus* (alder), *Betula* (birch), *Carya* (hickory), Chenopodiaceae/Amaranthaceae (pigweed/amaranth families), Cyperaceae (sedge family), *Fagus* (beech), *Fraxinus* (ash), Gramineae (grass family), *Ostrya/Carpinus* (hornbeam), *Pinus* (pine), Polygonaceae (buckwheat family), *Populus* (aspen), *Quercus* (oak), *Salix* (willow), *Tsuga* (hemlock), and *Ulmus* (elm). (In SPSS, these pollen types have the following labels: ACER, ALNUS, BETULA, CARYA, CHENOPOD, CYPERACE, FAGUS, FRAXINUS, GRAMINEA, OSTRYACA, PINUS, POLYGO-NA, POPULUS, QUERCUS, SALIX, TSUGA and ULMUS.) The category of Unidentifiable pollen also satisfied the criteria but was excluded because it contains a mixture of ecologically dissimilar pollen types and therefore is an

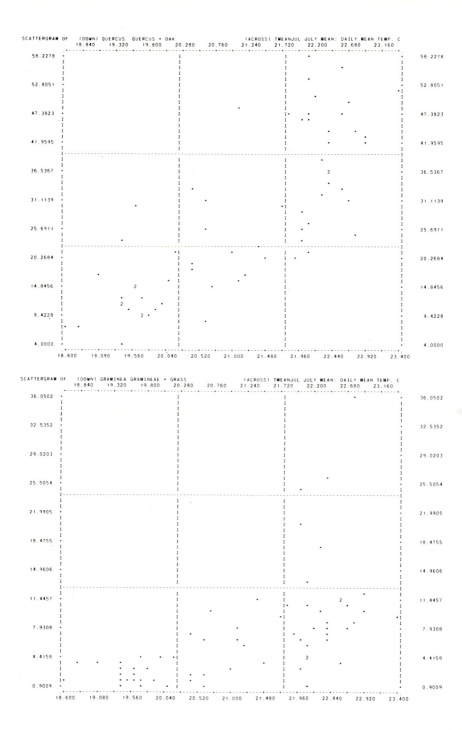

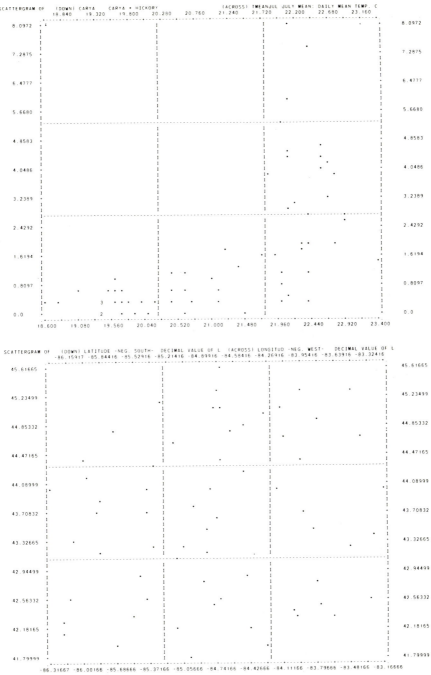

FIGURE 40.3. Scatter diagrams of *Quercus*, *Carya* and Gramineae versus mean July temperature and of latitude versus longitude for lower Michigan. The final plot provides a map of the sites

undesirable type to retain as a candidate for the regression equation. Had the categories, Other Trees, Miscellaneous Herbs, or Unknown pollen, satisfied the selection criteria, they would also have been excluded at this stage for the same reason. Including these four types in the CONDESCRIPTIVE step is useful in revealing any sample with high values for one of these categories. Such samples might best be excluded from further analysis. No such examples existed in the data set from lower Michigan.

For the types that were retained, the next question was to find out what type of quantitative relationship they had with July mean temperature (TMEANJUL) and whether any sites contained outlier values (i.e. anomalous values). For some of the types with means less than 1.0%, the samples with values above 5.0% may be outliers. If so, removal of these samples will mean that these types must be deleted from further analysis. Scatter diagrams help to answer these questions and aid the decision of which types and which sites to retain for the regression analysis.

Step II: PLOTTING SCATTER DIAGRAMS WITH THE SCATTERGRAM PROGRAM

Purpose

One purpose of plotting scatter diagrams (Figure 40.3) is to identify candidate outliers. A case is a candidate outlier if, for at least one pollen type, the paired pollen percentage and TMEANJUL value are 'very' different from those for most other cases. More information about candidate outliers is obtained in Step III (WRITE or LIST CASES) by listing the name, location, and type of the site along with its pollen percentages and climatic value.

A second purpose of printing the scatter diagrams is to identify the type of relationship between the climatic variable (TMEANJUL) and each pollen type. This relationship (after exclusion of the candidate outliers) may be:

(1)   linear (see *Quercus* in Figure 40.3);
(2)   curvilinear, which can be made linear by proper transformation of the pollen type (see *Carya* and Gramineae in Figure 40.3); or
(3)   neither of the above.

Pollen types with linear or curvilinear relationships are most likely to appear as terms in the regression equation, whereas pollen types of category (3) probably will not appear. Knowledge of these relationships can be helpful in the interpretation of the regression equation.

Computer commands

Command	**Argument**	**Program and task: comments**
RUN NAME	SCATTER DIA-GRAMS FOR...	

GET FILE	MODERN3D
ALLOCATE	TRANSPACE=
SELECT IF	(LATITUDE GE
	41.5...
	... ... 483)
RECODE	PICEA to ...
	(−1,−2,−3=0)
COMPUTE	TOTAL=0.0
DO REPEAT	V1=ACER,ALNUS,
	BETULA,CARYA,
	CYPERACE,
	FAGUS,
	FRAXINUS,
	GRAMINEA,
	OSTRYACA,
	PINUS,
	POLYGONA,
	POPULUS,
	QUERCUS,
	ULMUS,
	SALIX,TSUGA
COMPUTE	TOTAL=TOTAL+V1
END REPEAT	

These commands calculate percentages based on the X types selected for the pollen sum in the previous CONDESCRIPTIVE step. The types that have not been deleted are arranged in alphabetical order for easy access.

COMPUTE	TOTAL=TOTAL/100.0
DO REPEAT	V1=ACER,ALNUS,
	BETULA,CARYA,
	CYPERACE,
	FAGUS,
	FRAXINUS,
	GRAMINEA,
	OSTRYACA,
	PINUS,
	POLYGONA,
	POPULUS,
	QUERCUS,
	ULMUS,
	SALIX,TSUGA

COMPUTE	V1=V1/TOTAL	
END REPEAT		
SCATTER-GRAM	LATITUDE WITH LONGITUD	This command maps the sites associated with the cases retained by the SELECT IF statement.
SCATTER-GRAM	V1=ACER,ALNUS, BETULA,CARYA, CYPERACE, FAGUS, FRAXINUS, GRAMINEA, OSTRYACA, PINUS, POLYGONA, POPULUS, QUERCUS, ULMUS, SALIX,TSUGA	This command plots scatter diagrams for each type listed in the scattergram (on Y-axis) against TMEANJUL (mean July temperature) on X-axis. (Note that our practice has been to plot the dependent regression variable along the X-axis and therefore differs from conventional statistical practice.) V1 gives the label for the scatter diagrams.
STATISTICS	ALL	
FINISH		

Step III: WRITE CASES FOR CANDIDATE OUTLIERS

Preparation

The purpose of using the WRITE CASES program is to obtain more information about candidate outliers. Candidate outliers are identified on the scatter diagrams. In the scatter diagram (Figure 40.3) of GRAMINEA (grass pollen) versus TMEANJUL (mean July temperature), for example, the case plotted near the top right corner could be an outlier. This case is identified in a WRITE CASES run by selecting both TMEANJUL and GRAMINEA to be greater than 20. For every pollen type with a candidate outlier, a sequence of commands similar to the following example for the GRAMINEA (8 letter name for Gramineae in the SPSS file) variable can be executed:

*SELECT IF	(TMEANJUL GT20 AND GRAMINEA GT 20)	
WRITE CLASS	(/4X,F4.0,4X, 2A4,4X, F4.2, /12X,8(F6.3,2X) /12X,8(F6.3)	The first line of the WRITE CASES command gives the printing format for the variables listed on the following lines.

LONMIN,
LATDEG,
LATMIN,
STATE,NAME1,
NAME2,
TMEANJUL,
GRAMINEA

Computer commands

Command	Argument	Program and task: comments
RUN NAME	WRITE CANDIDATE OUTLIERS	
GET FILE	MODERN3D	
ALLOCATE	TRANSPACE=1600	
COMPUTE	OLDSEQ=SEQNUM	This COMPUTE card is needed to ensure that the numbering or sequencing of the input files is not altered internally by SPSS during a SELECT IF command. (See note about this command in CONDESCRIPTIVE stage.)
SELECT IF	(LATITUDE ... ... NE 483)	
RECODE	PICEA ... (-1,-2,-3=0)	
COMPUTE	TOTAL...	
END REPEAT		
COMPUTE	TOTAL	
DO REPEAT	V1=... ...TSUGA	
COMPUTE	V1=V1/TOTAL	
END REPEAT		
WRITE CASES	(/4X,F4.0,4X, 2A4,4X,F4.2, /12X,8(F6.3,2X) /12X,8(F6.3)	The first line of the WRITE CASE command gives the printing format for the variables listed on the following lines.

```
                    MODSEQ,NAME1,
                    NAME2,TMEANJUL
                    ACER,
                    ALNUS,
                    BETULA,
                    CARYA,
                    CYPERACE,
                    FAGUS,
                    FRAXINUS,
                    GRANMINEA,
                    OSTRYACA,
                    PINUS,
                    POLYGONA,
                    POPULUS,
                    QUERCUS
                    SALIX,TSUGA,
                    ULMUS
FINISH
```

Step IV: CONDESCRIPTIVE AND SCATTER DIAGRAMS WITH OUTLIERS DELETED

Purpose

The purpose of recomputing the CONDESCRIPTIVE statistics and scatter diagrams is to check the pollen types that remain in the data set in the light of the new pollen sum and the deletion of the outliers. If no types are deleted from the pollen sum as a result of Steps II and III, this step may not be necessary. In the example from lower Michigan, five sites and four pollen 'types (Chenopodiaceae/Amaranthaceae, Polygonaceae, Gramineae, and Cyperaceae) were deleted. After study of the scatter diagrams, the first three types were judged to be like *Ambrosia* pollen in primarily reflecting human disturbance. Cyperaceae pollen was judged to be of wetland origin and not useful in climatic calibration in Michigan. The pollen sum was reduced to 13 pollen types.

Pollen types identified as having a curvilinear relationship with the climatic variable are transformed as part of this step, and the transformed variable is plotted against TMEANJUL to verify that the relationship is linear. (The transformed variables are also used in Step V, the REGRESSION ANALYSIS.) In the data from lower Michigan, square-root transformations were chosen for *Fraxinus* and *Carya*, and a cubic root for *Carya* was also included.

Preparation

After study of the scatter diagrams and the listed cases, the outliers are chosen and deleted from the analysis. If a site is to be deleted, adjust the first SELECT IF command to delete it. If the whole taxon is to be deleted, remove the variable name from the two DO REPEAT loops that compute the sums.

The cases that appear to be outliers on the scatter diagrams may be deleted from the subsequent analysis if as samples they are anomalous or contain errors. Reasons for deleting sites with unusual values include:

(1) anomalous basin features (e.g. dammed river, cattle trough, when all other samples are from lakes);
(2) anomalous edaphic features (e.g. the only sample in an area of sandy outwash);
(3) anomalous elevations;
(4) errors in the data (correction is preferable).

Computer commands

Command	**Argument**	**Program and task: comments**
RUN NAME	...	
GET FILE	...	
ALLOCATE	...	
SELECT IF	(LATITUDE LONGITUD... SEQNUM... NE 454 AND 459 AND 461 AND 462 AND 467 AND 483)	From the analysis of the previous runs, we are deleting sites 454, 459, 461, 462 and 467 in addition to site 483, which was deleted in the previous runs also.
RECODE	... (−1,−2,−3=0)	
COMPUTE	OLDSEQ...	
COMPUTE	TOTAL= ...	
DO REPEAT	V1=ACER,ALNUS, BETULA,CARYA, FAGUS, FRAXINUS, OSTRYACA, PINUS,POPULUS, QUERCUS,SALIX, TSUGA,ULMUS	Notice CHENOPOD, CYPERACE, GRA-MINEA, POLYGONA, were totally deleted from the pollen sum.
COMPUTE	TOTAL ...	
END REPEAT		

COMPUTE	TOTAL=	
DO REPEAT	ACER,ALNUS,	
	BETULA,CARYA,	
	FAGUS,	
	FRAXINUS,	
	OSTRYACA,	
	PINUS,POPULUS,	
	QUERCUS,SALIX,	
	TSUGA,ULMUS	
COMPUTE	V1=V1/TOTAL	
END REPEAT		Notice again, CHENO-POD, CYPERACE, GRAMINEA, POLYGO-NA were deleted from the pollen sum.
CONDE-SCRIPTIVE	ACER,ALNUS, BETULA,CARYA, FAGUS, FRAXINUS, OSTRYACA, PINUS,POPULUS, QUERCUS,SALIX, TSUGA,ULMUS	
STATISTICS	ALL	Compute the statistics for the remaining pollen types from the remaining sites.
SCATTER-GRAM	ACER,ALNUS, BETULA,CARYA, FAGUS, FRAXINUS, OSTRYACA, PINUS,POPULUS, QUERCUS,SALIX, TSUGA,ULMUS WITH TMEANJUL	
STATISTICS	ALL	Make scatter diagrams for TMEANJUL vs. each remaining pollen type.
FINISH		

Step V: REGRESSION

Purpose

The purpose of this program is to calculate a series of multiple regression equations, from which one is chosen as a candidate equation for calibrating pollen data in the pollen diagrams that are being studied. The regression program in SPSS uses a forward-selection stepwise multiple regression procedure. Backwards elimination and other procedures for multiple regression might also be used, if programs for these procedures are available (Bartlein and Webb, 1985a). Once a regression equation is chosen, the program is rerun to gain a scatter plot for checking the residuals of this equation.

Preparation

Included among the possible independent variables in the regression analysis are all types in the pollen sum as well as any of those transformed variables that exhibit an approximately linear relationship wih TMEANJUL. For example, CARYA, the square-root of CARYA, or the cube-root of CARYA, may be in the list of candidate variables. (In Bartlein and Webb (1985a) only one entry for each pollen type is used, whichever one is judged to yield the best linear relationship.) A sample of the output from the regression program is given in Figure 40.4.

Computer commands

Command	*Argument*	*Program and task: comments*
RUN NAME	REGRESSION RUN FOR MICHIGAN TEMPERATURE REGRESSION	
GET FILE	MODERN3D	
ALLOCATE	TRANSPACE=1600	
SELECT IF	(LATITUDE ... LONGITUD ... SEQNUM ... NE ... AND ...)	
RECODE	PICEA TO ARTEM-ISI ... $(-1,-2,-3=0)$	

Sample Output

```
* * * * * * * * * * * * * * * * * * * * * * * * * M U L T I P L E   R E G R E S S I O N * * * * * * * * * * * * *   VARIABLE LIST   1
                                                                                                                REGRESSION LIST 1
DEPENDENT VARIABLE..     TMEANJUL    JULY MEAN: DAILY MEAN TEMP. C

VARIABLE(S) ENTERED ON STEP NUMBER  4..     TSUGA        TSUGA * HEMLOCK
```

			ANALYSIS OF VARIANCE	DF	SUM OF SQUARES	MEAN SQUARE	F
MULTIPLE R	0.92890		REGRESSION	4.	87.14790	21.78697	84.93692
R SQUARE	0.86286		RESIDUAL	54.	13.85142	0.25651	
ADJUSTED R SQUARE	0.85270						
STANDARD ERROR	0.50647						

------------------ VARIABLES IN THE EQUATION ------------------

VARIABLE	B	BETA	STD ERROR B	F
QUERCUS	0.4281467D-01	0.61079	0.00560	58.523
FAGUS	0.9228251D-01	0.29098	0.01753	27.728
ULMUS	0.7773806D-01	0.21926	0.02187	12.631
TSUGA	-0.6566986D-01	-0.20600	0.02345	7.839
(CONSTANT)	19.02248			

------------- VARIABLES NOT IN THE EQUATION -------------

VARIABLE	BETA IN	PARTIAL	TOLERANCE	F
ACER	-0.02509	-0.05689	0.70510	0.172
ALNUS	-0.05454	-0.12750	0.74952	0.876
BETULA	-0.07017	-0.10746	0.32162	0.619
CARYA	-0.11610	-0.18755	0.35788	1.932
FRAXINUS	0.04792	0.08727	0.45480	0.407
OSTRYACA	0.03640	0.08471	0.74277	0.383
PINUS	0.04985	0.03598	0.07146	0.069
POPULUS	0.06823	0.18073	0.96223	1.790
SALIX	0.01989	0.04238	0.62230	0.095
SCARYA	-0.09100	-0.13357	0.29544	0.963
CCARYA	-0.07125	-0.11345	0.34768	0.691
SFRAX	0.07296	0.13055	0.43915	0.919
LATITUDE	-0.59066	-0.49243	0.09532	16.966
LONGITUD	-0.06539	-0.15720	0.79258	1.343

FIGURE 40.4. Sample of output after the fourth step of the forward selection stepwise multiple regression program in SPSS. MULTIPLE R refers to the multiple correlation coefficient, R SQUARE is the percent variance explained by regression equation, ADJUSTED R SQUARE is R SQUARE adjusted for the number of degrees of freedom (DF), B refers to the regression coefficients, and BETA refers to the standardized regression coefficients. The standard error of the regression coefficients is STD ERROR B; F is the F-statistic; BETA IN is the standardized regression coefficient, were it entered in the next step; PARTIAL is the partial correlation coefficient with TMEANJUL; TOLERANCE is used in the stepwise scheme (see Nie *et al.*, 1975). The information needed for an analysis of variance of the regression is also listed

COMPUTE	OLDSEQ=SEQNUM	
COMPUTE	TOTAL=0.0	
DO REPEAT	V1= ...	
	FRAXINUS ...	
	POPULUS ...	
COMPUTE	TOTAL= ...	
END REPEAT		Same 13 pollen types as previous run, and also same sites.
COMPUTE	TOTAL= ...	
DO REPEAT	V1= ...	
	FRAXINUS ...	
	QUERCUS	
COMPUTE	V1= ...	
END REPEAT		Same pollen types as above.
COMPUTE	SCARYA=SQRT (CARYA)	This command calculates the square-root of CARYA to try to eliminate the curvature when *Carya* percentages are plotted against mean July temperature.
COMPUTE	CCARYA=CARYA **(1./3.)	Calculates the cube-root of CARYA.
COMPUTE	SFRAX=SQRT (FRAXINUS)	Calculates the square-root of FRAXINUS to try to eliminate curvature when *Fraxinus* percentages are plotted against TMEANJUL.
SCATTER-GRAM	SCARYA, CCARYA,SFRAX WITH TMEANJUL	Makes scatter diagrams of SCARYA versus TMEANJUL, CCARYA versus TMEANJUL, SFRAX versus TMEANJUL to illustrate whether the curvature remains.
STATISTICS	ALL	This command gives all the statistics to go along with the scattergrams.
REGRESSION	VARIABLES= TMEANJUL, ACER,ALNUS,	This command for stepwise forward multiple regression first identifies the variables

BETULA,CARYA, FAGUS, FRAXINUS, OSTRYACA, PINUS,POPULUS, QUERCUS,SALIX, TSUGA,ULMUS, SCARYA,C-CARYA, SFRAX, LATITUDE, LONGITUD/REGRESSION=TMEANJUL WITH ACER, ALNUS,BETULA, CARYA,FRAGUS, FRAXINUS, OSTRYACA, PINUS,POPULUS, QUERCUS,SALIX, ULMUS,SCARYA, CCARYA,SFRAX (1) LATITUDE,LONGITUD (0) RESID=0

and then designates TMEANJUL as dependent variable and all variables listed after WITH as independent variables. The code 1 indicates that the variables preceding the (1) are to be entered sequentially. At each step the variable that enters the regression has the highest F-to-enter value. LATITUDE and LONGITUD have code 0, so their F-to-enter values are given, but the variables never enter the regression. These values indicate how much spatial dependency remains in the data set during the regression.

STATISTICS   6
FINISH

Use of regression output

One regression equation is selected. This choice is helped by inspection of (*a*) the percentage variance explained that is adjusted for the number of degrees of freedom (ADJUSTED R SQUARE) and (*b*) the standard error of estimate (STANDARD ERROR) (see Figure 40.4). For example, the equation is selected for the regression step (often when 3–6 pollen types are in the equation) at which changes in these two values stabilize to 2–3% or to 0.1 or 0.2°C. The size of these limiting numbers depends on the size of the study area and the strength of the pollen/climate relationship. In the example, regression equation 4 was chosen with variables QUERCUS, FAGUS, ULMUS and TSUGA because addition of another variable, POPULUS, in regression equation 5 made little change in the ADJUSTED R SQUARE or the STANDARD ERROR.

Step VI: RESIDUAL CHECK

Purpose

Once the regression equation is chosen, the regression program should be rerun to print out a diagram of the standardized residuals (*y*-axis) versus the standardized magnitude of the dependent variable (mean July temperature

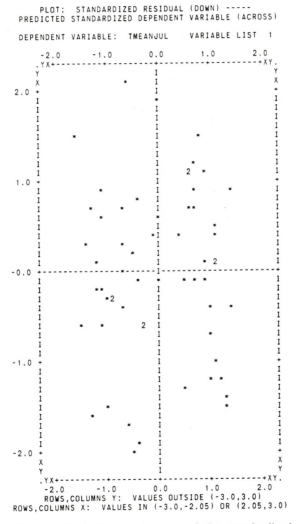

FIGURE 40.5. Scatter diagram of the standardized residuals versus the standardized dependent variable (mean July temperature)

along the *x*-axis). The residuals are observed minus estimated temperatures and are standardized by dividing each by the standard deviation of all residuals, and the dependent variable is standardized by subtracting its mean and dividing by its standard deviation. This scatter plot (Figure 40.5) shows whether or not the variance of the residuals is homogeneous and therefore similar for all values of the dependent variable (Neter and Wasserman, 1974). If the variance is inhomogeneous because the largest positive residuals only occur for low values of the dependent variable, then the regression equation is biased and may need recalculation (Howe and Webb, 1983). The plot of residuals for the lower Michigan data showed the variance to be homogenous.

Computer commands

Command	**Argument**
RUN NAME	REGRESSION RUN FOR MICHIGAN TEMPERATURE REGRESSION
GET FILE	MODERN3D
ALLOCATE	TRANSPACE=16000
SELECT IF	(LATITUDE GE 41.5 AND LE 46.0 AND LONGITUD GE −86.5 AND LE −83.0 AND SEQNUM LE 600 AND NE 454 AND 459 AND 461 AND 462 AND 467 AND 483)
RECODE	PICEA TO ARTEMISI,IVA TO CYPERACE,MISCHERB TO UNKNOWN (−1,−2,−3=0)
COMPUTE	OLDSEQ=SEQNUM
COMPUTE	TOTAL=0.0
DO REPEAT	V1=ACER,ALNUS, BETULA,CARYA, FAGUS, FRAXINUS, OSTRYACA,

```
                          PINUS,POPULUS,
                          QUERCUS,SALIX,
                          TSUGA, ULMUS
COMPUTE                   TOTAL=TOTAL+V1
END REPEAT
COMPUTE                   TOTAL=TOTAL/100.0
DO REPEAT                 V1=ACER,ALNUS,BETULA,
                          CARYA,FAGUS,
                          FRAXINUS,
                          OSTRYACA,
                          PINUS, POPULUS,
                          QUERCUS,SALIX,
                          TSUGA, ULMUS
COMPUTE                   V1=V1/TOTAL
END REPEAT

REGRESSION                VARIABLES=TMEANJUL,
                          FAGUS,
                          QUERCUS,
                          TSUGA,ULMUS/
                          REGRESSION=TMEANJUL
                          WITH FAGUS,
                          QUERCUS,
                          TSUGA,
                          ULMUS (2)
                          RESID=0
STATISTICS                6
```

Step VII: RESIDUAL PLOT AND RESIDUAL MAPS

Purpose

This program computes the residuals associated with the regression equation chosen in the previous step. It produces a sequence of maps of these residuals showing the spatial distribution of residuals in intervals, such as $(-\inf., -2]$, $(-2, -1]$, etc. It then splits the residuals into subsets and maps each of these subsets (Figure 40.6). If the residuals in these subsets are distributed somewhat evenly, then the regression equation chosen is a good one, because it exhibits little spatial autocorrelation in its residuals. Calculation of the Moran statistic (Cliff and Ord, 1981) provides an explicit statistical test for spatial autocorrelation among the residuals. We have written a FORTRAN program that calculates the Moran statistic.

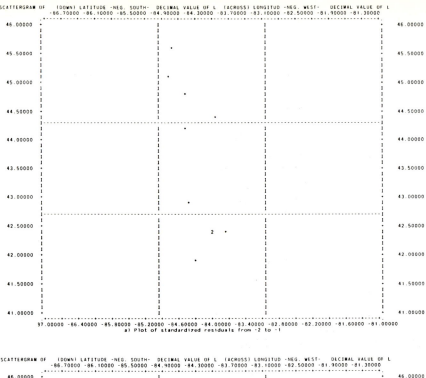

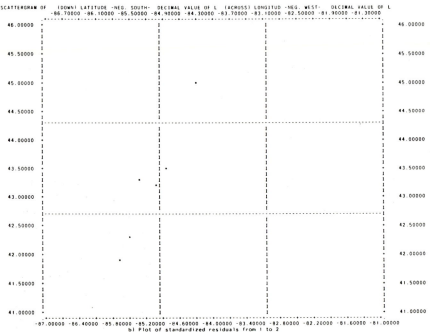

FIGURE 40.6. Map of standardized residuals (a) from −2 to −1 and (b) from 1 to 2 that are plotted on scatter diagrams of latitude versus longitude

Computer commands

Command	Argument	Program and task: comments
RUN NAME	RESIDUAL ANALYSIS FOR MICHIGAN ...	
GET FILE	MODERN3D	
ALLOCATE	TRANSPACE=1600	
SELECT IF	(LATITUDE ... LONGITUD ... SEQNUM ... NE 454 ... AND 467 ...)	
RECODE	PICEA ... (−1,−2,−3=0)	
COMPUTE	OLDSEQ= ...	
COMPUTE	TOTAL=...	
DO REPEAT	V1=ACER ... FRAXINUS ... POPULUS ...	
COMPUTE	V1= ...	
END REPEAT		
COMPUTE	EST=19.02248 +0.0428*QUERCUS +0.0922*FAGUS +0.0777*ULMUS −0.0656*TSUGA	This command computes the temperature estimate for each case, and the next command calculates the standardized residuals.
COMPUTE	SRES=(TMEANJUL -EST)/0.50647	
SCATTER-GRAM	SRES WITH EST	The estimated mean July daily temperature is EST=..., the regression equation selected from the regression runs. SRES gives the standardized residual TMEANJUL-EST. The scatter diagram of SRES versus EST shows how the residuals are distributed.

SCATTER-GRAM	LATITUDE (41.0,46.0) WITH LONGITUD (−87.0,−81.0)	Map of the sites.
*SELECT IF	(SRES LT −2.0)	
SCATTER-GRAM	LATITUDE (41.0,46.0) WITH LONGITUD (−87.0,−81.0)	Map of the sites associated with residuals less than −2.
*SELECT IF	(SRES GT −2.0 AND LT −1.0)	
SCATTER-GRAM	LATITUDE ...	Map of the sites associated with residuals between −2 and −1.
*SELECT IF	(SRES GT −1.0 AND LT 0.0)	
SCATTER-GRAM	LATITUDE (41.0 ...	
*SELECT IF	(SRES GT 0.0 AND LT 1.0)	
SCATTER-GRAM	LATITUDE ...	
*SELECT IF	(SRES GT 1.0 AND LT 2.0)	
SCATTER-GRAM	LATITUDE ...	
*SELECT IF	(SRES GT 2.0)	
SCATTER-GRAM FINISH	LATITUDE ...	

DISCUSSION

Use of the above sequence of programs will yield a multiple regression equation that can be used to calibrate fossil pollen data in terms of climatic estimates. An important task within these steps is the close examination of the data used in calculating the regression equation. The final regression equation is most likely to be useful (1) if the pollen types or transformed pollen types are linearly related to the dependent climatic variable, (2) if outliers that may disproportionately influence the regression results are identified and deleted, and (3) if the large residuals are carefully checked to indicate whether they cause a poor or biased performance of the regression equation (Howe and Webb, 1983). This careful checking of the modern data will help produce a robust unbiased regression equation. Further tests for

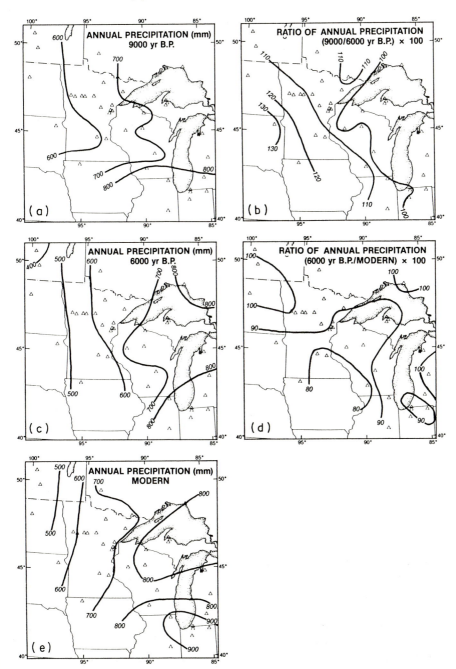

FIGURE 40.7.   Maps of annual precipitation (mm.) for (a) 9000 B.P., (b) the ratio of its values between 9000 and 6000 B.P., (c) 6000 B.P., (d) the ratio between 6000 B.P. and today, and (e) today (from Bartlein *et al.*, 1984; *reprinted by permission of Academic Press Inc.*)

finding the best regression equation are still being developed (Bartlein and Webb, 1985a), and we are continuing to improve our sequence of programs to incorporate these tests (Figure 40.1).

The temperature equation calculated in our example was derived for illustrative purposes. It is similar to the equation described in Howe and Webb (1983), and can be used on pollen data that is younger than 6000 B.P. in central lower Michigan. Howe and Webb (1983) used programs from both SPSS and BMDP in order to derive their equation (Bartlein and Webb, 1984a) and chose an equation with six terms, including three (*Fagus*, *Quercus* and *Ulmus*) that appeared in the equation in Figure 40.4. Their equation had an adjusted percentage-variance explanation of 86%, which matches the performance of the equation in our example (Figure 40.4). Differences in the programs used can therefore result in slight differences in the final equation that is selected for use. Were either equation important for paleoclimatic reconstruction, then further testing might be needed to choose between them. The equations were only developed to illustrate the calibration procedures, and other equations calculated from much larger calibration regions than lower Michigan are now in use for producing isotherm maps in the Midwest (Bartlein *et al.*, 1984) and eastern North America (Bartlein and Webb, 1985b).

The contoured maps (Figures 40.7 and 40.8) of estimated precipitation and temperature for 6000 and 9000 B.P. in central North America (Bartlein *et al.*, 1984) illustrate a key product of using regression equations for palaeoclimatic estimation. These maps provided a climatic interpretation for certain patterns in the Holocene pollen record for this region (Wright, 1968; Webb *et al.*, 1983). The maps show that the eastward movement of the prairie/forest border from 9000 to 6000 B.P. was associated with estimated precipitation decreases of 10–30% (Figure 40.7(*b*)) and the subsequent westward return of the prairie border was associated with precipitation increases of 20% (Figure 40.7(*d*)). Changes in temperature were relatively minor in the past 9000 years, and the difference maps (Figure 40.8(*b*) and (*d*)) show patterns of temperature change in which temperatures increased in some areas and decreased in others. Bartlein *et al.* (1984) described the possible changes in atmospheric circulation that were associated with these estimated changes in precipitation and temperature.

The work of Bartlein *et al.* (1984) also demonstrated the steps in data analysis that must precede the decision to transform fossil data into temperature and precipitation estimates. These steps allow for checking the ecological assumptions required by calibration work. (See Davis (1978), Birks (1981), Webb (1980), Howe and Webb (1983) and Prentice (1983) for a description and discussion of these assumptions.) Bartlein *et al.* (1984) first presented isopoll maps from central North America (Webb *et al.*, 1983) to illustrate broad-scale patterns of change in the fossil pollen record that are indicative of climate as an important causative factor (Webb, 1980). The

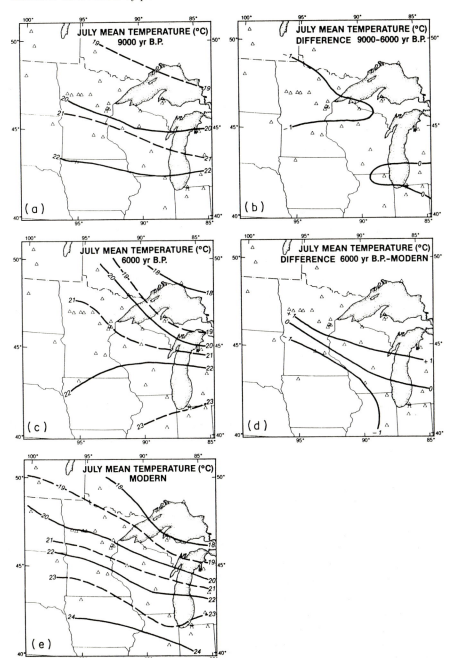

FIGURE 40.8. Maps of July mean temperature (°C) for (a) 9000 B.P., (b) the difference between 9000 and 6000 B.P., (c) 6000 B.P., (d) the difference between 6000 B.P. and today, and (e) today (from Bartlein *et al.*, 1984; *reprinted by permission of Academic Press Inc.*)

patterns further suggested that both temperature and precipitation had changed, and the authors cited evidence from other data (e.g. lake-levels) in the Midwest to support this fact. They next showed that the modern data contained patterns related to temperature and precipitation and that the variation in these modern data matched the range of variation in the fossil data from the past 9000 years. Only after this evaluation of the central North American pollen data did Bartlein *et al.* (1984) use statistical methods to calibrate the fossil data in climatic terms. They thus illustrated that the critical examination of the modern and fossil data being calibrated is a key first step in a quantitative climatic interpretation of pollen data. Once this first step is completed and the modern and fossil data are shown to be both sensitive to climate and adequate for calibration work, the statistical analysis described in this chapter should yield regression equations useful for the climatic calibration of pollen data.

Acknowledgements

Grants (ATM-79-1623 and ATM-81-11870) from the NSF Program of Climatic Dynamics and Contract (DE-ACO2-79EV 10097) from the Carbon Dioxide and Climatic Research Program of the U.S. Department of Energy supported the development of this sequence of programs. We thank R. McKendall, M. Ryall, and R. M. Mellor for their technical assistance.

REFERENCES

Bartlein, P. J., and Webb, T. (1985a). Annotated computer programs for climatic calibration of pollen data: a user's guide. American Association of Stratigraphic Palynologists, Contribution Series in press.

Bartlein, P. J., and Webb, T. (1985b). Mean July temperature at 6000 yr. B.P. in eastern North America: regression equations for estimates from fossil pollen data. In: *Syllogeus*, **55**, 301–342.

Bartlein, P. J., Webb, T., and Fleri, E. (1984). Holocene climatic change in the northern Midwest: pollen-derived estimates. In: *Quaternary Research* in press.

Birks, H. J. B. (1981). The use of pollen analysis in the reconstruction of past climates: a review. In: *Climate and History* (Eds. T. M. L. Wigley, M. J. Ingram and G. Farmer), Cambridge University Press, Cambridge, pp. 111–138.

Birks, H. J. B., and Birks, H. H. (1980). *Quaternary Palaeoecology*, Edward Arnold, London.

Cliff, A. D., and Ord, J. K. (1981). *Spatial Processes, Models and Applications*, Pion, London.

Davis, M. B. (1978). Climatic interpretation of pollen in Quaternary sediments. In: *Biology and Quaternary Environments* (Eds. D. Walker and J. C. Guppy), Australian Academy of Science, Canberra, pp. 35–51.

Dixon, W. J., and Brown, M. B. (Eds.). (1979). *BMDP-79 Biomedical Computer Programs P- Series* University of California Press, Berkeley.

Grichuk, V. P. (1969). An attempt to reconstruct certain elements of the climate of the northern hemisphere in the Atlantic Period of the Holocene. In: *Golotsen* (Ed. M. I. Neustadt), Izd-vo Nauka, Moscow, pp. 41–57.

Howe, S. and Webb, T. (1977) Testing the statistical assumptions of palaeoclimatic calibration functions. In: *Preprint Volume Fifth Conference on Probability and Statistics in Atmospheric Science*, American Meteorological Society, Boston, pp. 152–157.

Howe, S. E., and Webb, T. (1983). Calibrating pollen data in climatic terms: improving the methods. In: *Quaternary Science Reviews*, **2**, 17–51.

Iversen, J. (1944). *Viscum, Hedera*, and *Ilex* as climatic indicators. In: *Geol. Foren. Forhandl. Stock.*, **66**, 463–483.

Nie, N. H., Hull, C. H., Jenkins, J. G., Steinbrenner, K., and Bent, D. H. (1975). *SPSS Statistical Package for the Social Sciences* (2nd ed.), McGraw-Hill, New York.

Neter, J., and Wasserman, W. (1974). *Applied Linear Statistical Models*, R. D. Irwin Inc., Homewood, IL.

Overpeck, J. T., Webb, T., and Prentice, I. C. (1985). Quantitative interpretation of fossil pollen spectra: dissimilarity coefficients and the method of modern analogs. In: *Quaternary Research*, **23**, 87–108.

Prentice, I. C. (1983). Postglacial climatic change: vegetation dynamics and the pollen record. In: *Progress in Physical Geography*, **7**, 273–286.

van Zant, K. L., Webb, T., Peterson, G. M., and Baker, R. G. (1978). Increased *Cannabis/Humulus* pollen, an indicator of European agriculture in Iowa. *Palynology*, **3**, 227–233.

Webb, T. (1973). A comparison of modern and presettlement pollen from southern Michigan, U.S.A. *Review of Palaeobotany and Palynology*, **16**, 137–156.

Webb, T. (1974). Corresponding distributions of modern pollen and vegetation in lower Michigan. *Ecology*, **55**, 17–28.

Webb, T. (1980). The reconstruction of climatic sequences from botanical data. *J. of Interdisciplinary History*, **10**, 749–772.

Webb, T., and Bryson, R. A. (1972). Late- and postglacial climatic change in the northern Midwest, U.S.A.: quantitative estimates derived from fossil pollen spectra by multivariate statistical analysis. *Quaternary Research*, **2**, 70–115.

Webb, T., and Clark, D. R. (1977). Calibrating micropalaeontological data in climatic terms: a critical review. *Annals of the New York Academy of Science*, **288**, 93–118.

Webb, T., Cushing, E. J., and Wright, H. E. (1983). Holocene changes in the vegetation of the Midwest. In: *Late Quaternary Environments of the United States. 2: The Holocene* (Ed. H. E. Wright), University of Minnesota Press, Minneapolis, pp. 142–165.

Webb, T., and McAndrews, J. H. (1976). Corresponding patterns of contemporary pollen and vegetation in central North America. *Geological Society of America Memoirs*, **145**, 267–299.

Wright, H. E. (1968). History of the prairie peninsula. In: *The Quaternary of Illinois* (Ed. R. E. Bergstrom), Special Report 14, College of Agriculture, University of Illinois, Urbana, pp. 78–88.

Handbook of Holocene Palaeoecology and Palaeohydrology
Edited by B. E. Berglund
© 1986 John Wiley & Sons Ltd.

41

Climatic calibration of coleopteran data

T. C. ATKINSON, K. R. BRIFFA

*Climatic Research Unit and School of Environmental Sciences,
University of East Anglia, Norwich, U.K.*

and

G. R. COOPE, M. J. JOACHIM AND D. W. PERRY

*Department of Geological Sciences,
University of Birmingham, Edgbaston, U.K.*

INTRODUCTION

The study of Quaternary coleoptera has established this group of organisms as one of the most suitable with which to attempt to reconstruct past climates. They are extremely diverse, identifiable to species level in many fossil occurrences, and have undergone very few extinctions and almost no morphological evolution for the past several hundred millenia. The present-day geographical distributions of many species appear to be controlled on a global scale by the thermal climate. From their styles of distribution, present-day species can be grouped according to their preferences for warm, temperate or cool summers and oceanic or continental climates. In fossil assemblages, species of coleoptera with similar modern climatic distributions tend to occur together, while assemblages made up of species with incompatible distributions are remarkably rare. Thus, the suite of species in a fossil assemblage is a good guide to the general character of the thermal climate at the time of its burial. These considerations are reviewed in detail by Coope (1977; Chapter 34).

Numerous attempts have been made to quantify the palaeoclimatic interpretations of Quaternary coleoptera. At all times we have avoided the use of 'indicator species' and concentrated on analysis of whole assemblages using presence–absence of a species rather than changes in its relative abundance. Initially climatic inferences were made by determining the area of overlap of modern geographical ranges of the species in the fossil assemblage (Coope, 1959; Bryant *et al.*, 1983). Attempts to avoid problems associated with changes of latitude (e.g. angle of insolation) involved the plotting of the present-day occurrences of species in the floral zones of the Scandinavian mountains and the correlation of zonal range with climatic parameters. Comparative plotting of species from fossil assemblages then provided the palaeoclimatic interpretation (Lindroth, 1948; Coope, 1959; Coope, 1968; Coope and Sands, 1966). Unfortunately many assemblages included species that are absent from the modern fauna of Scandinavia. Grouping the species in assemblages according to different biogeographical categories has also been used in attempts to quantify the interpretation of changes in thermal climate (Morgan, 1973). The present technique, here called the Mutual Climatic Range Method, is the most recent of these attempts and offers, for the first time, the possibility of testing the general accuracy of climatic reconstructions.

MUTUAL CLIMATIC RANGE METHOD (M.C.R.M.)

The basic principle of the method is to establish the range of climates occupied at the present-day by each beetle species represented in a fossil assemblage. The climate indicated by the whole assemblage can then be taken to lie within the area of overlap of the climatic ranges of all the species present. This approach is conceptually simple and easy to apply but has hitherto been almost entirely neglected because of the difficulty of obtaining reliable information on the present climatic ranges of individual species. A similar method has been applied to pollen data by Iversen (1944) and Grichuk (1969), but doubts have been expressed because of the possibility that some pollen types may have been transported to a site beyond the geographic and hence climatic range of the parent plant (Webb and McAndrews, 1976). The simple fact that present-day coleopteran assemblages when treated by the M.C.R.M. provide figures that are compatible with actual measurements of the thermal climate suggests that this objection is not a severe problem here.

Methodology

Suppose a fossil assemblage contains two species of beetle whose distributions are known with respect to the mean temperature of the warmest month (TMAX, an index of summer warmth) and the temperature range between warmest and coldest months (TRANGE, an index of continentality). The climatic distribution of each beetle can be represented as an enveloped area on a graph of

OVERLAP OF CLIMATIC ENVELOPES

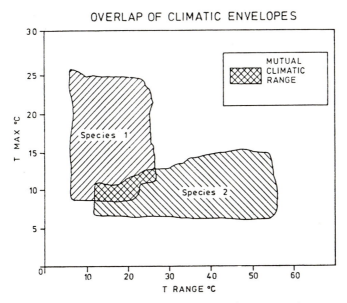

FIGURE 41.1.   Mutual climatic range for two species

TMAX against TRANGE (Figure 41.1). The range of climates in which the two species could coexist is given by the overlap between the two envelopes. This is the *mutual climatic range*. An example of the mutual climatic range of six species is shown in Figure 41.2. Here the contours map the percentage overlap of the six climatic ranges on axes calibrated in intervals of 1°C. The best estimate of palaeotemperature represented by this fossil assemblage is that it lay somewhere within the mutual climatic range as defined by the 100% contour. It is convenient to quote the median values of the mutual overlap of TMAX and TRANGE, together with the limits given by the extremes of overlap on each axis. It is also possible to estimate the temperature of the coldest month (TMIN) by constructing diagonal isolines and again taking the median value and extremes covered by the mutual overlap. The values reconstructed for the assemblage in Figure 41.2 are:

TMAX $= 10.5 \pm 2.5°C$
TRANGE $= 28.5 \pm 10.5°C$
TMIN $= -18.5 \pm 8.5°C$

While the basic principles are extremely simple, the task of assembling adequate data and actually determining the mutual overlap of species is very laborious. It divides into four stages as follows.

Stage 1: Compilation of distribution map for each species
We have compiled maps of the present-day geographic distributions of over

MUTUAL CLIMATIC RANGE RECONSTRUCTION PACKAGE   Climatic reconstruction for GLANLLYNNAU 138-143

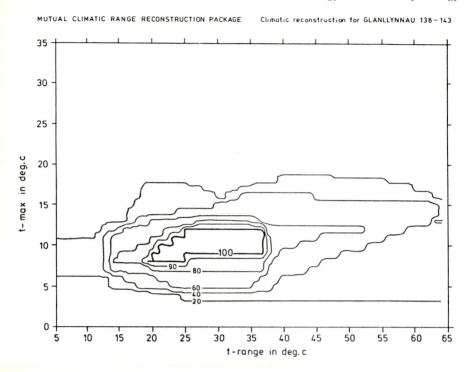

FIGURE 41.2.   Contour plot showing the mutual climatic range for 6 species found in the lowest layers of sediment filling a kettle-hole in North Wales, 14,500 B.P. (Coope and Brophy, 1972)

300 species from six families (Carabidae, Dytiscidae, Hydrophilidae, Haliplidae, Gyrinidae, Scarabaediae) within the palaearctic region. The sources used were treatises and catalogues of coleoptera, vice-county records, some museum collections and original collectors' reports. Account was taken of the varying reliability and precision of these sources by using different shading symbols on the maps. New maps are being compiled in order to expand the database, and the original ones are checked and updated as information becomes available.

Stage 2: Determination of climatic range for each species

The geographical range of each species is converted into a climatic range with the aid of a base map showing the location of meteorological stations for which data is available. By comparing the two maps the stations can be listed according to whether they lie within the species' geographical range or outside it. The computer then processes this information and produces a plot of the two groups of stations on a graph of TMAX against TRANGE. An envelope can then be drawn which delimits the climatic range of the species. Account is taken of the altitudinal distribution of the beetle relative to the altitudes of the meteorological stations.

This procedure automatically groups together widely separated geographical locations where a species occurs under similar climates, and reveals the climatic homogeneity underlying many species' disjunct geographical distributions.

Stage 3: Computer storage and retrieval of species climatic envelopes

To facilitate the rapid calculation of palaeotemperatures by the computer the climatic range envelope for each species is coded and stored in numeric form. This is achieved by superimposing a grid of squares (36 × 60 in 1 deg. C units) on the TMAX/TRANGE plot and designating each element by a 'zero' or 'one' according to whether it lies outside or within the envelope.

Stage 4: Climatic reconstruction from an assemblage of named species

Given a list of species the computer retrieves and superimposes their numeric envelopes to produce a TMAX/TRANGE graph of percentage overlap. Visual inspection of this allows the area of maximum overlap to be determined and the values of TMAX, TRANGE and TMIN to be identified.

Calibration and checking of accuracy

The M.C.R.M. lends itself to rigorous checking by reconstructing present-day climates from modern faunas collected within restricted areas, and comparing the results with the mean temperatures recorded at nearby meterological stations. The result obtained from a small northern British fauna is given in Figure 41.3. The actual and reconstructed temperatures are:

	TMAX	TRANGE	TMIN (in deg C)
Actual	14.6	10.9	3.7
Reconstructed	15.5 ± 0.5	14 ± 1	2 ± 1

This, and other work to date, suggests that the accuracy of the method is acceptable with the database in its present form. With further additions and refinement the accuracy will be improved, although there will probably always be a slight tendency for the median of reconstructed TMAX to overestimate the true value in cooler climates, while underestimating TMIN. This factor must be borne in mind when interpreting exactly where within the mutual range of a fossil fauna TMAX and TMIN values are most likely to have been in reality.

WORKED EXAMPLE

To illustrate the value of the M.C.R.M. applied to assemblages of Quaternary coleoptera, we have selected the closing phase of the last Glaciation and early

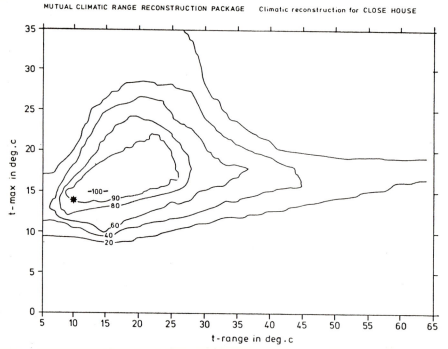

FIGURE 41.3.   Contour plot showing the M.C.R.M. estimated thermal climate for the modern fauna recorded from Close House (Research Station of the University of Newcastle, U.K.). Actual climate indicated by *

Holocene, an episode that has previously been subject to intense geological and palaeontological research effort.

Figure 41.4 shows a composite picture of reconstructions of TMAX and TMIN from a number of sites in Britain plotted against radiocarbon years. Though these results must be viewed as preliminary, since many more species have yet to be processed, they show several clear episodes of change in the thermal environment. The results clearly indicate a marked climatic warming at about 13,000 years B.P., temperate conditions to 12,250 B.P., followed by progressive decline in temperatures into cold conditions between 11,000 B.P. and 10,000 B.P. The marked change in winter temperature at about 13,000 B.P. strongly indicates a change from a continental climatic regime to one of considerable oceanicity. The cold episode between 11,000 and 10,000 B.P. likewise shows a return of more continental conditions, though seemingly not so extreme as during the period prior to 13,000 B.P. The opening of the Holocene at about 10,000 B.P. is dramatically sudden and conditions as warm as present-day were established by 9500 B.P.

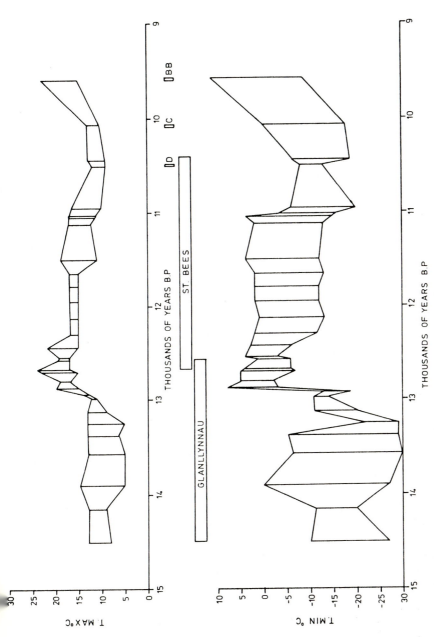

FIGURE 41.4. M.C.R.M. reconstruction of the late-glacial thermal environment as indicated by the coleoptera from five British sites: Glanllynnau (Coope and Brophy, 1972); St. Bees (Coope and Joachim, 1980); Drumurcher (D) (Coope et al., 1979); Croydon (C) (Peake and Osborne, 1971); Brighouse Bay (BB) (Bishop and Coope, 1977)

CONCLUSION

We have presented here a numerically based technique for the calibration in terms of thermal climate of the changes in species composition of Quaternary fossil insect assemblages. It must be emphasized that the procedure is still in its infancy and is continuously being updated and modified; but already there are clear enough results to show that there is here a valuable source of objective palaeoclimatic information.

REFERENCES

Bishop, W. W., and Coope, G. R. (1977). Stratigraphical and faunal evidence for Glacial and Early Flandrian environments in South-West Scotland. In: *Studies in the Scottish Lateglacial Environment*, (Eds. J. M. Gray and J. J. Lowe), Pergamon, Oxford, pp. 61–68.

Bryant, I. D., Holyoak, D. T., and Moseley, K. A. (1983). Late Pleistocene deposits at Brimpton, Berkshire, England. *Proc. Geol. Ass.*, **94**, 321–343.

Coope. G. R. (1959). A Late Pleistocene insect fauna from Chelford, Cheshire. *Proc. R. Soc. B*, **151**, 70–86.

Coope. G. R. (1968). An insect fauna from Mid-Weichselian deposits at Brandon, Warwickshire, *Phil. Trans. R. Soc. B*, **254**, 425–456.

Coope, G. R. (1977). Fossil coleopteran assemblages as sensitive indicators of climatic changes during the Devensian (Last) cold stage. *Phil. Trans. R. Soc. B*, **280**, 313–348.

Coope, G. R., and Brophy, J. A. (1972). Late Glacial environmental changes indicated by a coleopteran succession from North Wales. *Boreas*, **1**, 97–142.

Coope, G. R., Dickson, J. H., McKutcheon, J. A., and Mitchell, G. F. (1979). The Lateglacial and Early Postglacial deposit at Drumurcher, Co. Monaghan. *Proc. R. I. Acad.*, **79**, 63–85.

Coope, G. R., and Joachim, M. J. (1980). Late Glacial environmental changes interpreted from fossil coleoptera from St. Bees, Cumbria, N.W. England. In: *Studies of the Lateglacial of N.W. Europe* (Eds. J. J. Lowe, J. M. Gray and J. E. Robinson), Pergamon, Oxford, pp. 55–68.

Coope, G. R., and Sands C. H. S. (1966). Insect faunas of the last glaciation from the Tame valley, Warwickshire. *Proc. R. Soc. B*, **165**, 389–412.

Grichuk, V. P. (1969). An attempt to reconstruct certain elements of the climate of the northern hemisphere in the Atlantic Period of the Holocene. In: *Golotsen* (Ed. M. I. Neishadt), Izd-vo Nauka, Moscow.

Iversen, J. (1944). *Viscum, Hedera* and *Ilex* as climatic indicators. *Geol. Foren. Forhandl. Stock.*, **66**, 463–483.

Lindroth, C. H., (1948). Interglacial insect remains from Sweden. *Arsbok Sveriges Geologisken Undersokn*, **C42**, 1–28.

Morgan, A. (1973). Late Pleisticene environmental changes indicated by fossil insect faunas of the English Midlands. *Boreas*, **2**, 173–212.

Peake, D. S., and Osborne, P. J. (1971). The Wandle Gravels in the vicinity of Croydon. Their age and their insect faunas. *Proceedings and Transactions of the Croydon Natural History and Scientific Society*, **14**, 145–176.

Webb, T., and McAndrews, J. A. (1976). Corresponding patterns of contemporary pollen and vegetation in central North America. *Geol. Soc. Amer. Memoir*, **145**, 267–299.

Index

Subjects indexed below are restricted to major references. Locations and authors are not included. In general, names of biological taxa and geological time periods are also excluded.

859